680
P343
1997

Paleontological Events

Stratigraphic, Ecological, and Evolutionary Implications

Edited by Carlton E. Brett
and Gordon C. Baird

Columbia University Press
New York

Columbia University Press
Publishers Since 1893
New York Chichester, West Sussex
Copyright © 1997 Columbia University Press
All rights reserved
Library of Congress Cataloging-in-Publication Data
Paleontological events : stratigraphic, ecological, and evolutionary implications / edited by Carlton E. Brett and Gordon C. Baird.
p. cm.
Includes bibliographical references and index.
ISBN 0-231-08250-9
1. Paleontology, Stratigraphic. 2. Taphonomy. I. Brett, Carlton E. (Carlton Elliot) II. Baird, Gordon C.
QE711.2.P32 1996
560'.17—dc20 96-1588
CIP

∞
Casebound editions of Columbia University Press books are printed on permanent and durable acid-free paper.
Printed in the United States of America
c 10 9 8 7 6 5 4 3 2 1
p 10 9 8 7 6 5 4 3 2 1

Contents

Preface

In the nineteenth and first half of the twentieth centuries stratigraphers focused primarily on a form of biological event stratigraphy to establish a global correlation system. Enthusiasm in this effort burgeoned owing to early successes in the matching of system-series and stage-scale units between basins and continents in the nineteenth century (see Rudwick 1985). Many early workers thought that a predictable succession of zones would be rigorously demonstrated between basins and continents, assuming both global or near-global distribution for zonal taxa and development of single intervals of peak fossil abundance within each zone. One could, in theory, work out a maximum number of successive peak zones or "acme zones" (hemerae) in a region that would correlate into similar successions within stratigraphically complete sections in other parts of the world. This approach, which was popular with many nineteenth-century geologists, came to be known as the "layer cake" paradigm.

Although parts of this logic were sound, the lack of knowledge of Waltherian facies relationships, the paleoecology of ancient organisms, or the dynamic factors controlling fossil distribution led to numerous correlational mistakes. Refined correlational work by numerous workers in the twentieth century showed that "acme zones" could be reversed and that, for a variety of environmental and taphonomic factors, both zonal and nonzonal taxa are absent from large regions. Biostratigraphic work continued unabated, mainly as a result of energy exploration, but much mid-twentieth-century stratigraphic research was centered on the application of various sedimentologic and facies models to the study of sedimentary basins.

By the 1970s various authors (Dennison and Head 1975; Heckel 1977; Vail et al. 1977b) recognized the connection between sedimentary cyclicity and inferred global eustasy at varying temporal scales, thus triggering a resurgence in stratigraphic research that continues to the present. The past three decades have witnessed a renewed appreciation for the episodic or catastrophic nature of the geological record (see Ager 1973, 1974, 1993). Similarly, paleontologists focused their interests on paleontological events, such as mass extinction and evolutionary radiation events, which also triggered a resurgence of interest in correlation. In particular, some paleontologists (Baird 1978, 1981; Kauffman 1986, 1988; Walliser 1986; Brett et al. 1986; Kidwell 1989, 1991a, 1991b; Einsele and Seilacher et al. 1982, and papers therein; Aigner 1985) recognized a variety of bed-scale paleontological events that are widespread within basins and essentially synchronous. These events, which are the result of taphonomic, ecological, and migrational factors (see Kauffman 1986, 1988), provide a correlational framework with a much higher resolution than that offered by traditional zonation. Moreover, these event beds are probably closely linked to eustatic changes and should provide a complimentary link to correlations derived from the sequence stratigraphy paradigm (see Kidwell 1989, 1991a, 1991b; Sageman et al., this volume).

Superficially, many sedimentary deposits, particularly in offshore marine strata, appear undifferentiated and monotonous. Careful field study reveals, however, that these intervals are actually composed of a succession of distinctive, and often persistent, layers of fossils, which occur as complex shell beds (Kidwell 1991a); horizons with unique taphonomic signatures (Brett et al. 1986; Brett and Seilacher 1991); intervals with distinctive taxa, or epiboles; and levels marking major faunal extinctions, originations, and incursions (Kauffman 1986, 1988). Such beds form key elements of the newly established discipline of high-resolution event stratigraphy (see Kauffman 1988). Moreover, paleontological event beds occur nonrandomly in depositional sequences (as defined by Vail et al. 1991) and thus may aid in recognition of condensed stratigraphic intervals (Banerjee and Kidwell 1991; Kidwell 1991a, 1991b). This approach is redefined to allow for high-resolution chronostratigraphy in basins using a host of new parameters (discussed later by the other volume contributors).

Paleontological events are linked to a variety of shorter-term phenomena, from storms to climate-forcing cycles, that produce mappable bed-scale features of variable spatial/regional extent. Nonetheless, the biological and taphonomic significance of many such horizons remains poorly understood. All these phenomena represent forms of paleontological event horizons, in that they are distinctive, geologically short-lived (hours to about 100 kyr), and commonly widespread biological events. However, superficially similar fossil event horizons may represent disparate processes and temporal scales. The various types of biotic and taphonomic events need to be distinguished carefully before questions of their genesis can be addressed (Fürsich 1978). To that end we present a preliminary classification of paleontological event horizons, based largely on our experience with the Lower and Middle Paleozoic rocks of eastern North America but integrated with other studies of bioevents (see Walliser 1986; Kauffman and Walliser 1990).

We distinguish three broad categories of paleontological events, each of which is subdivisible. These are (1) *fossil beds* or *Lagerstätten*, (2) *epiboles*, and (3) *major bioevents*. At this time we are primarily concerned with recognizing patterns of occurrence in the geologic record. Recent developments in the field of high-resolution event stratigraphy (Kauffman 1988; Sageman et al. 1991) have promoted a much more thorough examination of the geologic record of particular fossil-bearing strata. The record is highly episodic and composed of a relatively finite number of discrete and often traceable units that often may contain distinctive paleontological and taphonomic features. Authors of this volume consider the influence of short-, intermediate-, and long-term geological events in shaping fossil beds and shell-rich horizons.

The first nine papers involve distinctly short-term, but widespread, paleontological event layers. These generally include beds enriched in fossils that record the effects of major storms or hurricanes (Ager 1974; Aigner 1985). Storms have a dual influence on seafloor communities. In areas above storm wave base, organism skeletons tend to be aggregated and fine-grained sediments tend to be winnowed, producing shell pavements that may armor the bottom and aggregate over a series of depositional and winnowing events to form widespread, complex, shell-rich layers. In deeper water areas below storm wave base affected by gradient currents, the influence of storms is one of sediment blanketing. This blanketing provides an important taphonomic agent responsible for the preservation of so-called obrution (rapid burial) layers. Some obrution layers, particularly in shallow-water facies, may be local and not persistent. However, work by several authors points to the possibility that certain storm-generated sediment layers may be traceable over tens to even hundreds of kilometers. Erosion and deposition also have ecological effects that may be recorded in the fossil record. The development of winnowed pavements provides substrates for encrusting hard-bottom biota (Kidwell 1991c). In contrast, the sudden blanketing by siliciclastic mud may alter an otherwise diverse firm substrate community to an assemblage dominated by burrowing infaunal organisms. In many environments storms also provide major disturbances, which may disrupt or kill organisms or may overturn colonial organisms into new orientations followed by regrowth.

Certain shell-rich horizons are highly condensed and are the mark of longer-term sediment starvation, commonly associated with rapid sea-level-rise events (see Fürsich 1971, 1978; Kidwell 1991b). In addition to accumulations of skeletal material, such beds may contain concentrations of early diagenetic clasts, such as phosphatic or even pyritic fossil molds. Surfaces of such condensed beds are often of critical importance to biostratigraphy, for these provide rich samples of fossils, including index fossils. This is particularly pronounced where index fossils consist of chemically resistant skeletal materials, as in the case of conodonts.

Epibole is an old biostratigraphic term that implies a thin stratigraphic interval characterized by an extraordinary abundance of fossil taxa that are generally uncommon or absent in a particular sedimentary basin. The second section of the book, with eleven papers, focuses on epibole phenomena. Such horizons may be simply an artifact of unique taphonomic conditions. Some fragile skeletons are normally disarticulated or fragmented and are thus unrecognizable, at least to a species level. However, extraordinary burial events and/or early diagenetic mineralization may lead to preservation of these otherwise cryptic fossils. A second type of epibole appears to record unique ecological conditions that favored faunal dominance by an unusual taxon or morphotype. Still more intriguing are epiboles that reflect abrupt, brief, and widespread incursion of a species (or group of species) that was previously absent. These intriguing horizons are important stratigraphically, corresponding in part with traditional biozones. Such incursion epiboles are poorly understood; however, they may represent episodes of interconnection of formerly separate provinces that permitted influx of exotic larvae. The immigrant taxa appear to have been unable to establish themselves for extended periods in the new area.

The broadest category of paleontological events consists of global bioevents. These are widespread events of accelerated originations and/or extinctions. Such global bioevents, especially mass extinctions, have been the subject of several recent symposium volumes (see, for example, Nitecki 1984; Bayer and Seilacher 1985; Elliott 1986; Donovan 1989). Consequently, they are not the primary focus of this volume. However, selected papers (Jeppsson, McGhee, this volume) have been included to illustrate the overlap with incursion epiboles and to consider some of the processes that generate longer-term paleontological events.

As a personal matter, the present authors began fieldwork in the Paleozoic strata of the northern Appalachian foreland basin during the early and mid-1970s well aware of regionally diachronous magnafacies and of the daunting thicknesses of the "monotonous" marine mudrock, siltstone, and sandstone. Using such tools as mud-floored diastem horizons, colonization surfaces rich in large corals, beds containing rare and exotic organisms, and smothered bottom assemblages of large regional extent, we were able to assemble a correlation framework that was detailed and largely isochronous. Our experience brought us to the early realization that numerous variably extensive event beds could be mapped across magnafacies in the Silurian and Devonian sections in upstate New York and that magnafacies were only a secondary aspect of sedimentary deposits superimposed upon a chronological succession of physical and biological events. When strata are examined at a fine scale, magnafacies come to appear as generalizations of regional conditions.

The supposedly "random" succession beds in many marine sections constitute, in fact, a hierarchical succession and system of regionally mappable event beds. Vagaries of exposures, as well as local amalgamation of beds in these sections, act to conceal this order. This apparent randomness led earlier stratigraphers to concentrate on magnafacies, suitable only for process sedimentology or paleoecology studies. Our work and that of the other authors in this volume should help put this misperception to rest. Chaos is organized!

ACKNOWLEDGMENTS

We express our deepest appreciation to the authors of these articles, who not only have put their time and energy into producing interesting papers, but also have patiently borne with us in the prolonged process of compiling and editing the volume. We also appreciate the efforts of outside reviewers, whose names are acknowledged in each of the contributions. The final project benefited especially from the technical reviews of David Lehmann, Mary Nardi, and Ron Harris, who spent many hours reviewing the papers and ferreting out errors. Richard Alexander, David Bottjer, Art Boucot, Bob Linsley, and Curt Teichert reviewed original proposals and encouraged us to undertake this project. We appreciate their support. Our own efforts have been supported by grants from the Petroleum Research Fund, American Chemical Society, and Natural Science Foundation grants EAR 8313103 and EAR 524841. Finally, we extend our sincere thanks to Ed Lugenbeel of Columbia University Press for his extreme patience and encouragement throughout the project.

REFERENCES

Ager, D. V. 1973. *The Nature of the Stratigraphic Record.* New York: Wiley.

———. 1974. Storm deposits in the Jurassic of the Moroccan High Atlas. *Palaeogeography, Palaeoclimatology, Palaeoecology* 15:83–93.

———. 1993. *The New Catastrophism.* New York: Cambridge University Press.

Aigner, T. 1985. Storm depositional systems: Dynamic stratigraphy in modern and ancient shallow-marine sequences. *Lecture Notes in the Earth Sciences,* vol. 3. Berlin: Springer-Verlag.

Banerjee, I. and S. M. Kidwell, 1991. Significance of molluscan shell beds in sequence stratigraphy: An example from the Lower Cretaceous Mannville Group of Canada. *Sedimentology* 38:913–34.

Baird, G. C. 1978. Pebbly phosphorites in shale: A key to recognition of a widespread submarine discontinuity. *Journal of Sedimentary Petrology* 48:105–22.

———. 1981. Submarine erosion on a gentle Paleoslope: A study of two discontinuities in the New York Devonian. *Lethaia* 14:105–22.

Bayer, U. and A. Seilacher, eds. 1985. *Sedimentary and Evolutionary Cycles. Lecture Notes in Earth Sciences,* vol. 1, Berlin: Springer-Verlag.

Brett, C. E. and A. Seilacher, 1991. Fossil Lagerstätten: A taphonomic consequence of event sedimentation. In G. Einsele, W. Ricken, and A. Seilacher, eds., *Cycles and Events in Stratigraphy.* Berlin: Springer-Verlag.

Brett, C. E., S. E. Speyer, and G. C. Baird, 1986. Storm-generated sedimentary units: Tempestite proximality and event stratification in the Middle Devonian Hamilton Group of New York. In C. E. Brett, ed., *Dynamic Stratigraphy and Depositional Environments of the Hamilton Group (Middle Devonian) in New York State,* part 1. *N.Y. State Museum Bulletin* 457:129–55.

Conway Morris, S. 1986. Community structure of the Middle Cambrian phyllopod bed (Burgess Shale), *Palaeontology* 29:423–58.

Dennison, J. M. and J. W. Head, 1975. Sea-level variations interpreted from the Appalachian Basin Silurian and Devonian, *American Journal of Science* 275:1089–1120.

Donovan, S. K., ed. 1989. *Mass Extinctions.* New York: Columbia University Press.

Einsele, G. and A. Seilacher, eds. 1982. *Cyclic and Event Stratification.* Berlin: Springer-Verlag.

Elliot, D. K., ed. 1986. *Dynamics of Extinction.* New York: Wiley.

Fürsich, F. T. 1971. Hartgründe und Kondensation in Dogger von Calvados. *Neues Jahrbuch fur Geologie und Paleontologic Abhandlungen* 138:313–42.
———. 1978. The influence of faunal condensation and mixing on the preservation of fossil benthic communities. *Lethaia* 11:243–50.
Goodwin, P. W. and E. J. Anderson, 1986. Punctuated aggradational cycles: A general hypothesis of episodic stratigraphic accumulation. *Journal of Geology* 93:515–33.
Heckel, P. H. 1973. Nature, origin, and significance of the Tully Limestone: An anomalous unit in the Catskill delta, Devonian of New York. *Geological Society of America Special Paper* No. 138.
———. 1977. Origin of black phosphatic shale facies in Pennsylvanian cyclotherns of the midcontinent, North America. *American Association of Petroleum Geologists Bulletin* GI:1045–1068.
Jeppsson, L. 1987. Lithological and conodont distributional evidence for episodes of anomalous oceanic conditions during the Silurian. In R. J. Aldridge, ed., *Palaeobiology of Conodonts.* Chichester, England: Ellis Horwood Ltd., 129–145.
Johnson, M. E. 1988. Why are rocky shores so uncommon? *Journal of Geology* 96:469–86.
Kauffman, E. G. 1986. High-resolution event stratigraphy: Regional and global Cretaceous bio-events. In O. Walliser, ed., *Global Bioevents, Lecture Notes in Earth Sciences,* vol. 8, pp. 279–335. Berlin: Springer-Verlag.
———. 1988. Concepts and methods of high-resolution stratigraphy. *Annual Review of Earth and Planetary Sciences* 16:605–54.
Kauffman, E. and O. H. Walliser, 1990. *Extinction Events in Earth History, Lecture Notes in Earth Sciences,* vol. 30. Berlin: Springer-Verlag.
Kidwell, S. M. 1989. Stratigraphic condensation of marine transgressive records: Origin of major shell deposits in the Miocene of Maryland. *Journal of Geology* 97:1–24.
———. 1991a. The stratigraphy of shell concentrations. In P. Allison and D. E. G. Briggs, eds. *Taphonomy: Releasing the Data Locked in the Fossil Record.* New York: Plenum.
———. 1991b. Condensed deposits in siliciclastic sequences: Expected and observed features. In G. Einsele, W. Ricken, and A. Seilacher, eds., *Cycles and Events in Stratigraphy,* pp. 682–95. Berlin: Springer-Verlag.
———. 1991c, Taphonomic feedback (live/dead interactions) in the genesis of bioclastic beds: Keys to reconstructing sedimentary dynamics. In G. Einsele., W. Ricken, and A. Seilacher, eds., *Cycles and Events in Stratigraphy,* pp. 268–82. Berlin: Springer-Verlag.
Nitecki, M., ed. 1984. *Extinctions.* Chicago: University of Chicago Press.
d'Orbigny, A., 1849–1852. *Cours Elementaire de Paléontologie et de Géologie Stratigraphique,* pp. 299, 382, 383–841. Paris: Masson,
Rudwick, M. J. S. 1985. *The Great Devonian Controversy.* Chicago: University of Chicago Press.
Sageman, B. B., P. B. Wignall, and E. G. Kauffman. 1991. Biofacies model for oxygen-deficient facies in epicontinental seas: Tool for paleonenvironmental analysis. In G. Einsele, W. Ricken, and A. Seilacher, eds., *Cycles and Events in Stratigraphy,* pp. 542–64. Berlin: Springer-Verlag.
Vail, P. R., R. M. Mitchum, Jr., and S. Thompson. 1977a. Seismic stratigraphy and global changes of sea level. Part 3. Relative changes of sea level from coastal onlap. In C. E. Payton, ed., *Seismic Stratigraphy.* American Association of Petroleum Geologists Memoir No. 26, pp. 63–81.
———. 1977b. Seismic stratigraphy and global changes of sea level. Part 4. Global cycles of relative changes of sea levels. In C. E. Payton, ed., *Seismic Stratigraphy. American Association of Petroleum Geologists Memoir* No. 26, pp. 83–97.
Vail, P. R., F. Audemard, S. A. Bowman, P. N. Eisner, and C. Perez-Cruz. 1991. The stratigraphic signatures of tectonics, eustasy and sedimentology: An overview. In G.

Einsele, W. Ricken, and A. Seilacher, eds., *Cycles and Events in Stratigraphy*, pp. 617–81. Berlin: Springer-Verlag.

Walliser, O. H. 1986. *Global Bio-events, Lecture Notes in the Earth Sciences*, vol. 8. Berlin: Springer-Verlag.

———. 1990. How to define "global bio-events." In E. G. Kauffman and O. H. Walliser, eds., *Extinction Events in Earth History, Lecture Notes in the Earth Sciences*, vol. 30, pp. 1–3. Berlin: Springer-Verlag.

Paleontological Events

Stratigraphic, Ecological, and Evolutionary Implications

Part One

Fossil Beds: Lagerstätten and Their Genesis

1

Fossil Lagerstätten: Stratigraphic Record of Paleontological and Taphonomic Events

Carlton E. Brett, Gordon C. Baird, and Stephen E. Speyer

ABSTRACT

Fossil Lagerstätten, beds yielding unusual paleontological information, range in time significance from single-event beds, representing deposition at scale of hours to days real time, to time-rich multievent units encompassing several zones. This requires that events producing fossil concentrations be described in a hierarchical sense. This paper reviews existing categories of marine Lagerstätten and presents some new concepts within a hierarchical classification of paleontological events. Single-event Lagerstätten include mass mortality events that usually involve rapid burial (obrution), although some mass mortality events are related to other causes. Obrution assemblages can be further classified by the nature of the burial event, spatial extent of the Lagerstätten, and postburial, diagenetic events. We recognize four categories of obrution deposits: Type 1 deposits are patchy lenses produced by current surges and storm wave- impingement in shallow, high-energy shelf settings; type 2 concentrations are proximal storm tempestites that often display sedimentary grading and pararipple development; type 3 deposits are distal mud tempestites deposited at or below the limit of storm wave base; type 4 deposits are extremely distal tempestites or turbidites deposited in quiet, dysoxic to anoxic settings. These are thin, organic-rich mud blankets sometimes yielding important preservational Lagerstätten.

Conservation Lagerstätten require favorable early diagenesis in addition to obrution. Burial of organisms in sediment low in organic content allows for establishment of strong diffusion gradients around carcasses and subsequent development of interior pyrite or carbonate cements and/or the formation of enclosing concretions. Burial of organisms within organic-rich muds sometimes leads to formation of thin, carbonate- and/or sulfide-rich mineral films around carcasses and even preservation of soft parts. Single-event concentrations are here classified as allochthonous and parautochthonous. Allochthonous concentrations reflect significant transport of shells due to traction of fallout. Parautochthonous assemblages are composed both of *in situ* and neighborhood-derived bioclasts that reflect some mixing due to burrowing and/or minor storm influences.

Longer-term (multievent) shell accumulations are discussed using the existing classification scheme of Kidwell (1991a); we distinguish shell beds of increasing implied temporal magnitude that correspond, respectively, to *composite, hiatal,* and *lag concentrations* of her classification. Composite concentrations record multiple storm obrution events as well as episodes of winnowing. Assemblages are complex and record "time averaging" and blending of the paleocommunity record. Exposure of shell pavements through winnowing sets the stage for new colonization by sessile taxa; this taphonomic feedback further increases the complexity and diversity of assemblages in composite shell beds.

Hiatal shell concentrations are complex, time-rich accumulations recording periods of sediment starvation and sometimes sediment bypass conditions. Sedimentary condensation, punctuated by minor, local erosion is the process by which shells become concentrated. Lag concentrations are even more time-rich; these are produced through extensive erosion of older beds, typically resulting in a mix of pristine, highly degraded and prefossilized shells within the lag assemblages. Beds of this type often contain fossils from two or more biostratigraphic zones.

Distribution of Lagerstätten types is distinctly nonrandom within sedimentary cycles. Early transgressive to peak highstand systems tracts favor preservation of obrution (rapidly buried) assemblages owing to increased sediment accommodation space and lower energy conditions. Late highstand to lowstand systems tracts occasionally yield types 1 and 2 obrution deposits, but burial events are usually not preserved because of reduced sediment accommodation space and higher energy conditions. Composite shell beds are common in these facies both because of frequent alternation of obrution and winnowing events and because of major activity of benthic organisms. Hiatal and lag concentrations, respectively, mark widespread condensed beds and regional unconformities. These usually link to sediment bypass facies, downlap surfaces, and flooding surfaces, as well as sequence bounding unconformities recognized by sequence stratigraphers.

Although Lagerstätten are commonly thought of as representing unusual, often local events, many link to widespread biotic events of unknown cause. Some are presently known to constitute *taphonomic epiboles* (see Brett and Baird, paper 10, this volume) while others may link to sea level and paleoclimatic events that have evolutionary significance. Hence Lagerstätten may be linked to *ecological* and *incursion epiboles*. Moreover, many minor and major "extinction" and "origination" events may be artifacts of submarine erosion and sedimentary condensation within the rock record. The recognition of hiatal and lag concentrations is critical for identifying discontinuity of section and loss of biostratigraphic succession where none had been previously recognized. Conversely, we suspect that hiatal and lag concentrations may record unusual conditions, which, indeed, may have precipitated evolutionary restructuring and turnover within the marine biosphere. This latter possibility demands rigorous future analysis of shell concentrations both in the rock record and in modern oceans.

Fossil beds, or Lagerstätten, are horizons or thin intervals displaying enrichment in fossil remains (see Seilacher et al. 1985; Kidwell and Jablonski 1983; Kidwell 1991a, 1991b; Kidwell and Bosence 1991). These beds encompass varying temporal scales, ranging from geologically rapid event deposits to long-term condensed beds that record minimal net deposition over thousands of years. Skeletal concentrations or shell beds have been classified by Kidwell (1991a) as *event, composite, hiatal,* and *lag* concentrations. The term *event* has come to take on a wide array of meanings, ranging from nearly instantaneous episodes of erosion/sedimentation to longer but still geologically brief intervals up to hundreds of thousands of years of unusual conditions (e.g., mass extinction "events" or bioevents; see Walliser 1990). To avoid possible confusion among these different meanings we modify Kidwell's term *event concentrations* to *single-event concentrations* but otherwise adopt and amplify her classification scheme for fossil Lagerstätten (Table 1.1).

Single-event concentrations include mass mortality or census assemblages (Kidwell and Bosence 1991) that are commonly obrution deposits (Seilacher et al. 1985; Brett and Seilacher 1991), as well as simple shell layers (Kidwell 1982; Parsons et al. 1988). Composite concentrations are represented by slightly more complex shell layers that record multiple events (major simple shell beds of Kidwell 1982), as well as complexly amalgamated shell-rich intervals (major complex beds of Kidwell 1982). Hiatal or condensed beds typically display environmental or biostratigraphic scales of fossil condensation (Kidwell and Bosence 1991). Finally, lag concentrations result from erosional truncation of older strata and reworking of prefossilized bioclasts.

From the standpoint of stratigraphy, any of these four types of fossil concen-

TABLE 1.1 *Categories of Fossil Lagerstätten*

1. Mass mortality horizons
   - A. Without rapid burial (non-obrutionary)
   - B. With rapid burial (obrution deposits) types 1–4
2. Skeletal concentrations
   - A. Allochthonous fossil beds
   - B. Parautochthonous
     1. Single event
     2. Composite
     3. Hiatal (condensed) concentrations
     4. Lag concentrations

trations may be recognizable as paleontologic event strata (*sensu lato*). That is, they may form regionally traceable marker horizons or thin intervals that are uniquely characterized by particular fossil taxa and/or modes of preservation. However, the degree of time-stratigraphic resolution afforded obviously varies greatly, from hours or days for single-event intervals to thousands of years of skeletal accumulation for sediment-starved intervals. In the present paper we review and further classify short- to intermediate-scale paleontological events (10^{-1} to 10^4 years) recorded in skeletal accumulation, beginning with single events and concluding with longer-term hiatal and lag deposits. For each category we cite specific examples and discuss their utility for stratigraphic and paleoecological study. Single-event concentrations or beds provide the most precise time resolution in any given outcrop and may yield "snapshots" of paleocommunities (generally lacking soft tissue and color). Unfortunately, from the standpoint of event stratigraphy these beds are commonly very thin and not persistent even at the scale of a single outcrop (Seilacher et al. 1985; Kidwell and Aigner 1985).

However, this is not invariably the case. In a number of instances it has been possible to correlate evident single-event obrution layers by direct tracing in single cliff exposures for up to a few kilometers (Kauffman 1988). Moreover, with careful attention to details of unique taphonomic or taxonomic features and stacking patterns relative to other markers, it is also possible to correlate individual beds between outcrops across distances of tens of kilometers. This procedure may be aided by distinctive diagenetic overprints, such as carbonate or pyrite concretions (e.g., Waage 1964; Brett et al. 1986; Parsons et al. 1988; Brett and Taylor, paper 8, this volume).

Four general types of single-event fossil concentrations can be recognized: (1) mortality horizons without burial, (2) obrution deposits that represent mass mortality associated with burial, (3) allochthonous coquinas, and (4) parautochthonous coquinas. Longer-term concentrations of skeletons may also be of considerable value in paleontological "event" stratigraphy. These beds, including *composite, hiatal,* and *lag* concentrations of Kidwell (1991a, 1991b), represent prolonged scales of time-averaging ranging from thousands to perhaps a few million years (Kidwell and Bosence 1991). Thus they do not afford as much stratigraphic resolution as do single-event beds. Yet they are typically more widespread and thus are especially valuable in establishing broad-based correlations of sequences over large areas. These longer-term paleontological "event" beds will be considered briefly at the end of this paper.

Mass Mortality Horizons

Mass mortality and rapid, permanent burial are both necessary conditions for the formation of conservation Lagerstätten, but neither is sufficient in itself (fig. 1.1). Whether or not they are killed as a result of mass mortalities, organism carcasses readily disintegrate on the sediment surface in the absence of burial. Conversely, rapid burial pulses will preserve intact skeletal remains, but if burial is not accompanied by a mortality of organisms these remains will consist only of scattered, partly articulated remains of recently dead organisms and/or such items as intact molt elements of arthropods (Brett and Speyer 1989). Such nonlethal burial layers might also feature escape traces. However, these assemblages, although distinct from slowly buried material, do not display concentrations of exceptionally preserved fossils. If mass mortality occurs a considerable time before deposition of sediment, a " spike," or abundance peak

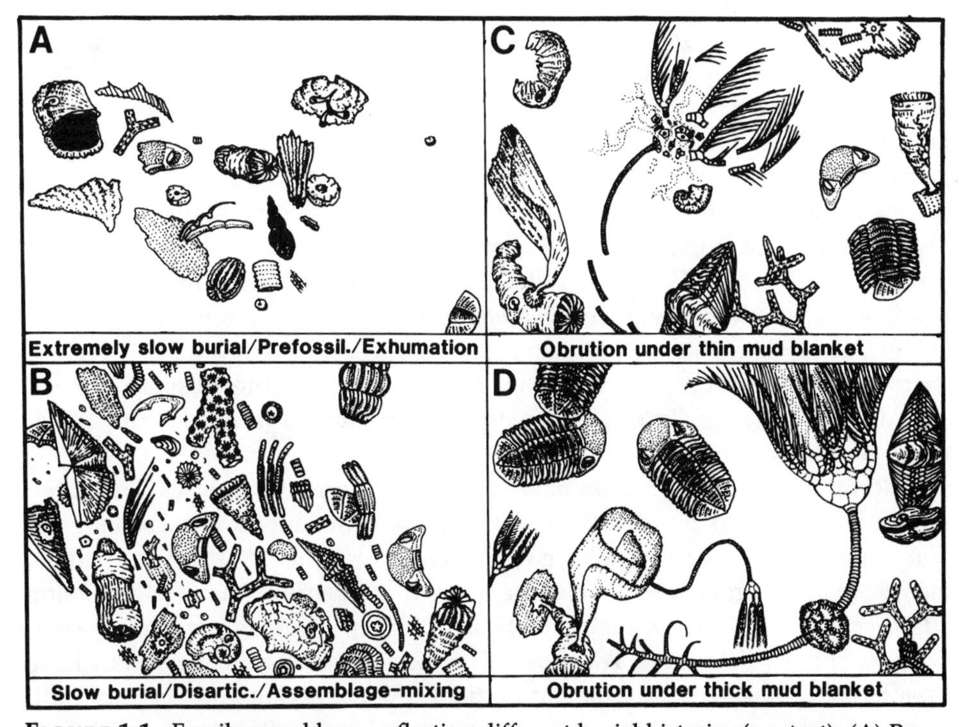

FIGURE 1.1. Fossil assemblages reflecting different burial histories (see text). (A) Prolonged exposure of skeletal material showing extensive degradation of shell as well as exhumation of prefossilized remains (black phosphatic steinkerns) on the seafloor. This is a common condition of hiatal and lag assemblages. (B) Mixed (time-averaged) assemblage typical of many shell beds. Variable shell degradation plus commingling of normally discrete communities indicates that these shell beds formed over a long time interval. Note contrasting biocorroded corals and brachiopods in association with partly intact multielement crinoid arm fragment. (C) Obrution assemblage covered by thin sediment blanket. Note intact multielement organisms, particularly molt ensemble of trilobite that was definitely buried at the site of arthropod ecdysis. Postburial disturbance of crinoid by "grave robbing" scavengers is evidence for obrution under shallow sediment cover. Offset crinoid columnals suggest that some skeletal decay preceded the burial event. (D) Undisturbed obrution assemblages showing not only multielement skeleton preservation, but also connected organisms (attached bryozoans, corals, and brachiopods) in life position.

of disarticulated/fragmented skeletal material, may occur. Only when a mass mortality event immediately precedes or accompanies a sedimentation event will an obrution Lagerstätten be preserved (*obrutus,* "burial"; see Seilacher et al. 1985; Brett 1990; Brett and Seilacher 1991).

Mass Mortalities Without Rapid Burial

A variety of environmental perturbations may produce mass mortalities of either benthic or pelagic organisms or both (see Brongersma-Sanders 1957; Brett and Seilacher 1991, for review of mortality agents). Normally, if these events are not closely associated in time with episodes of burial, these mass mortalities will not be recorded as event beds (see later). However, under rare circumstances the mortality may produce an abrupt pulse of skeletal remains into benthic environments. If this input pulse is not obliterated by scavenging, burrowing, or physicochemical breakdown of skeletons, it may appear as a discrete thin layer.

One illustration of such mortalities is the presence of single layers of largely disarticulated and commonly fragmented shells; see, for example, layers of small bivalves in the Cretaceous of the Western Interior (Kauffman 1986) or bedding planes covered with *Leiorhynchus* brachiopod valves from the Devonian of New York (Thompson and Newton 1987). These beds also typically represent short-term colonization events and may form "epiboles" if they display unusual abundance of a normally rare taxon through an interval (see Brett and Baird, paper 10, this volume).

Colonization surfaces may occur as single bedding planes in black, laminated shales. Some authors previously ascribed such accumulations to the sedimentation of epiplanktonically rafted organisms (e.g., Seilacher 1982a). However, in many cases the organisms involved appear to be benthic fauna (e.g., *Bositra* ["*Posidonia*"]) bivalves (Wignall and Hallam 1991)]. Thus alternative explanations have been proposed. The facies in which these assemblages occur have been termed *exaerobic* by Savrda and Bottjer (1987), implying low oxygen conditions in bottom waters and anoxia within the substrate. But their model does not fully explain the occurrence of shells on discrete bedding planes. An alternative model invokes temporary oxygenation events on normally low-oxygen or anoxic seafloors to allow brief colonization of the substrate by opportunistic, low-oxygen-tolerant epifauna (poikilaerobic or fluctuating aerobic/anaerobic zones of Oschmann 1991; see also Wignall and Hallam 1991). Sageman et al. (1991) distinguish three categories of fossil assemblages in low-oxygen, organic-rich facies: resident or background assemblages, event assemblages, and allochthonous deposits. The latter, composed of transported fossils, may yield bedding plane assemblages but are rare except in areas with substantial paleoslopes.

Thus most bedding plane assemblages and thin bioturbated layers within otherwise sparsely fossiliferous, organic-rich shales can be attributed to brief oxygenation events. For example, Kauffman (1986) and Sageman et al. (1991) describe persistent layers of inoceramids and layers with mixed epibenthic mollusks and trace fossils in the Cretaceous of the Western Interior. Savrda and Bottjer (1991) discuss redox-related events, recorded by trace fossils in black shales. Provided that such layers can be correlated laterally, they may provide precise bioevent markers in shale successions.

Such bedding planes typically bear evidence for abrupt termination of colonization events. In the case of single bedding plane assemblages, typically only a single or a few generations of organisms may be present, as indicated by a single size class. Bedding planes covered with diminutive (juvenile?) brachiopods noted from the Devonian Hamilton Group (Brett et al. 1991) are comparable to assemblages of diminutive inoceramids, or *Bositra,* from the Jurassic Posidonia black shales (Savrda and Bottjer 1991), or to tiny paper pectens from the Cretaceous Hartland Shale (Sageman et al. 1991). In all cases, these assemblages appear to represent opportunistic colonization of the seafloor followed by mass mortalities. The termination event may represent deoxygenation of the bottom water, event burial, or both. However, it may be possible to discriminate between the two. Savrda and Bottjer (1991) note that *Bositra* ("*Posidonia*") from the Jurassic Posidonia shales of Germany have the valves splayed open ("butterflied positions"; Allmon 1985). Hence their mass mortality was followed by a period of decay prior to burial. However, in some cases the mass mortality appears to have coincided with burial. Examples, including bedding plane assemblages with articulated crinoids and even pelagic organisms such as fish or ichthyosaurs, cannot simply be attributed to deoxygenation or toxicity events (Brett and Seilacher 1991). They are a form of obrution deposit (see later) in which mass mortality, even in the water column, is associated with a sedimentation event.

Mass mortalities of pelagic organisms without burial may also rarely be recorded as event horizons. An excellent example is provided by the "Fish Scale Marker bed" of the Cretaceous Mowry Shale, a thin bed found near the Albian/Cenomanian Stage boundary at all outcrops from northern Alberta to central Colorado. Kauffman (1986) attributes this bed to a regional oxygen overturn event that virtually exterminated fish. This event was followed by an interval of unknown duration in which all remains of the fish except keratinous scales were destroyed.

Mass Mortalities with Rapid Burial: Obrution Deposits

General Aspects

In certain instances unique conditions of rapid burial and early diagenesis lead to the deposition of thin horizons of fossils that display exceptional preservation (obrution deposits *sensu* Seilacher et al. 1985; Brett and Seilacher 1991). Obrution layers are among the most precisely resolvable horizons in event stratigraphy. The fact that they often display unique faunal or taphonomic fingerprints makes them more readily recognized and distinguished than bentonites, at least in the field.

For example, many horizons in the Upper Ordovician and Middle Devonian of the eastern North America region contain an unusual abundance of complete flat or enrolled trilobites, sometimes in species-segregated clusters (see Brandt 1985; Speyer and Brett 1985; Speyer 1987; Hickerson, paper 9, this volume). These horizons can be correlated for tens of kilometers at least along facies strike and reflect regional storm burial events.

Burial is a critical prerequisite to any fossil preservation, and obrution deposits reflect exceptionally rapid burial events in which the bodies of organisms were abruptly and permanently entombed in the sediment. Keys to the recognition of obrution layers in the field include complete articulated multielement skeletons, such as those of echinoderms and arthropods (figs. 1.1C, 1.1D), as well as burial of other taxa *in situ*. For example, Meyer et al. (1989) have devised a taphonomic index based on the degree of articulation of Mississippian crinoids that permits discrimination of instances of "instantaneous" burial from cases of longer residence times as skeletal remains of the seafloor. Rare preservation of weakly skeletonized or even soft body parts may point to extraordinary geochemical conditions associated with the burial (Seilacher et al. 1985; Allison 1990; Allison and Briggs 1991). Other clues include the presence of articulated spar-filled shells and/or fossils preserved in life orientation (Alexander 1986; Brett and Baird 1986).

Obrution deposits include the classic conservation Lagerstätten (see Whittington and Conway Morris, eds., 1985, and papers therein; Conway Morris 1990; Baird et al. 1986; Allison and Briggs 1991). They provide dramatic "snapshots" of ancient organisms and their paleobiology and may form important marker horizons in event stratigraphy. Such deposits are also extremely useful in reconstructing paleoenvironmental processes. Factors critical to the formation of obrution layers include the following: (1) the presence of numerous intact carcasses of organisms on the substrate at the times of rapid burial; (2) the rapid accumulation of a relatively thick (1-mm to 1-cm) sediment blanket (figs. 1.1 and 1.2); (3) the absence of later physical or biological reworking; and (4) a favorable early diagenetic environment (Brett and Speyer 1989; Allison 1986, 1988; Allison and Briggs 1991). Obrution deposits require the coincidence of these factors and therefore occur preferentially in certain portions of facies gradients and selectively in certain facies tracts within sedimentary cycles.

Causes of mass mortality in marine organisms have been reviewed by Brett and Seilacher (1991) and include burial itself, increased turbidity, changes in water chemistry, temperature, and other factors. But only those factors actually associated with rapid sedimentation events are of importance in forming Lagerstätten. Others, such as disease, commonly will not be recorded as mortality horizons (see, for example, Greenstein 1989). Agents of obrution-related mor-

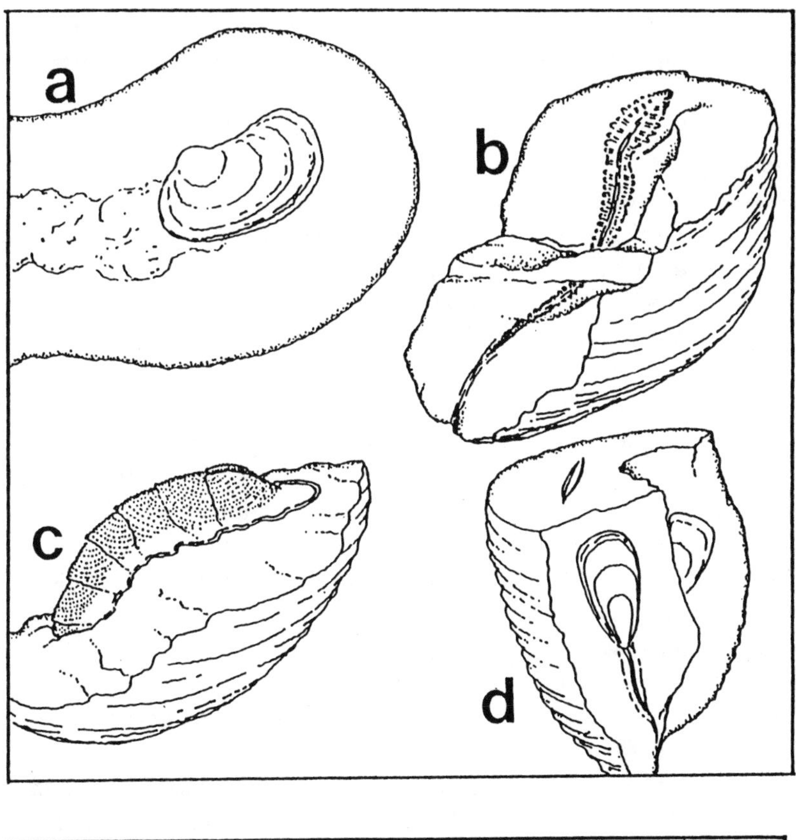

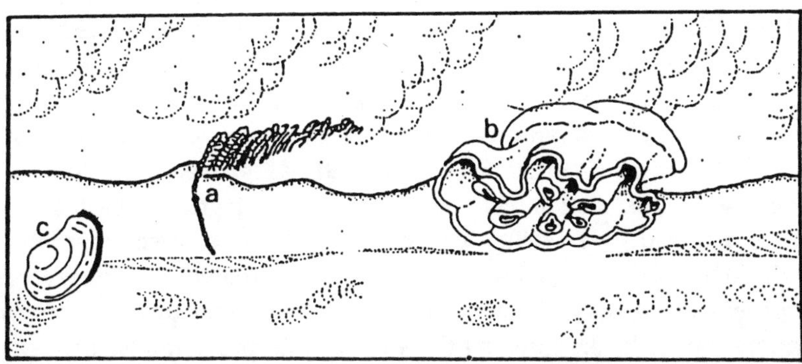

FIGURE 1.2. Extremely rapid obrution events, in part, responsible for unusual soft-bodied preservation at Middle Pennsylvanian (Westphalian) Mazon Creek localities in Illinois. Fossils occur in sideritic concretions within estuarine deltaic deposits of the Francis Creek Shale Member. (Upper figure) Unsuccessful escape attempts by four different animals recorded in Mazon Creek concretions: (a) bivalve with associated fugichnial burrow; (b) polychaete worm buried oblique to bedding; (c) chiton oblique to bedding; (d) *Lingula* upright with preserved pedicle. (Lower figure) Francis Creek organisms in the process of being buried. (a) Fern leaf part being buried in upended orientation; (b) jellyfish being smothered; (c) bivalve generating escape burrow (from Baird et al. 1986).

tality will preferentially affect certain organism guilds relative to others. Sedentary benthic organisms, such as attached brachiopods or various echinoderms, are most sensitive to all factors and will commonly be represented in obrution layers (for Ordovician, Devonian, and Mississippian examples, respectively, see Guensburg 1984; Koch and Strimple 1968; Horowitz and Waters 1972). In contrast, death of organisms by smothering is much less likely for vagrant forms, such as ophiuroids, which may escape up to 10 cm of sediment deposited in less than an hour (see Schäfer 1972). Kranz (1974) observed that many bivalves can escape burial, thus biasing the record of smothered bottom assemblages. Burial usually cannot affect pelagic organisms. Nonetheless, many obrution deposits do contain well-preserved remains of pelagic forms.

Many fossils in obrution layers show other indications that the organisms were already dead and had undergone slight decay prior to burial. For these cases we must invoke other causes of mortality, such as rapid chemical or turbidity changes that occurred prior to sediment deposition.

The mortality factor, at the outset, will limit the types of obrution deposits that can form. Obrution-related mortality that is preservable should be most common in offshore environments (see figs. 1.3A and 1.3B). Areas slightly below average storm wave base will experience frequent episodes of sediment smothering that may take place close to the oxygen minimum zone (see fig. 1.3B). In addition, storm disturbances may produce brief intervals of toxicity in the water column.

The remains of organisms killed by mass mortalities must be buried rapidly if they are to be recorded. Modern experimental studies (Allison 1986, 1988; Meyer 1971; Liddell 1975; Plotnick 1986; Kidwell and Baumiller 1990) indicate that multielement skeletons of arthropods and echinoderms, for example, begin to disarticulate within a few hours to days after death. Even under anaerobic conditions ligaments and connective tissues are completely destroyed within days to weeks (Allison 1986). In such cases, weak currents and gas refloating scatter the skeletal elements even in "stagnant" basinal settings (Zangerl and Richardson 1963). Consequently, the rate of sedimentation will control the degree of articulation. Completely articulated multielement skeletons represent burial within several hours up to several days in some cases. Furthermore, burial layers must be thick enough to prevent later disruption by bioturbation and physical disturbance (see fig. 1.1C for effect of insufficient burial).

Potential burial agents include seismites, turbidites, and ash falls or flows (Seilacher 1982a; Clifton 1988). However, we assume that in most marine shelf settings, storms will be the major source of event burial. Furthermore, we can model the distribution of obrution deposits in relation to tempestite proximality (see Brett and Seilacher 1991; fig. 1.3).

Storms produce a predictable spectrum of effects on shallowly sloping seafloors (Aigner 1985; Jennette and Pryor 1993). Areas subjected by extensive storm wave impingement experience erosional winnowing, but organisms may commonly be killed and covered by rapidly shifting coarse sediments (fig.

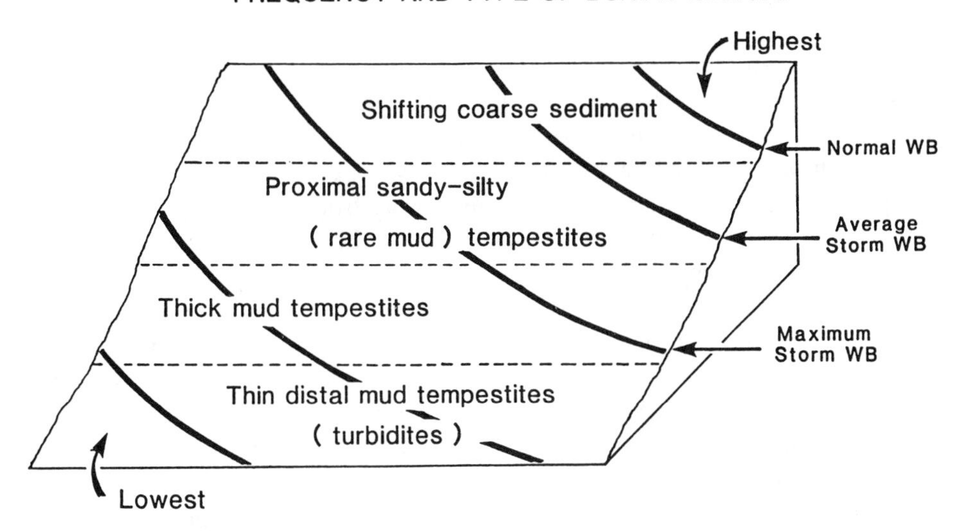

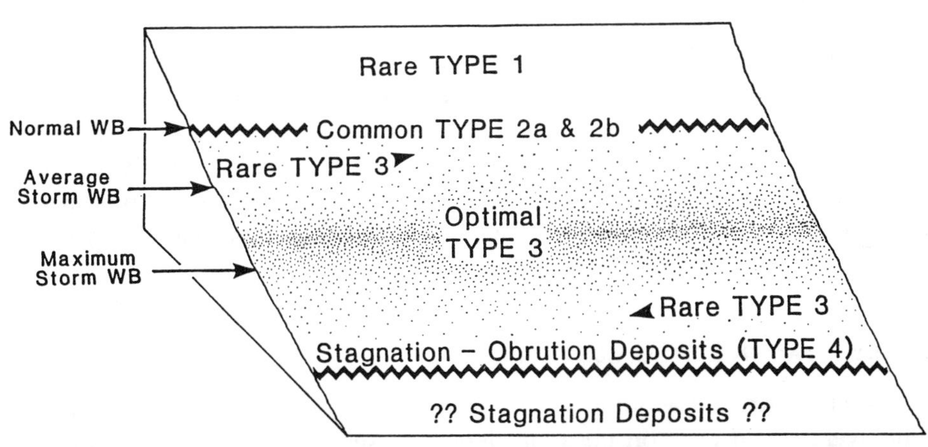

FIGURE 1.3. Frequency and type of burial events with respect to depth and wave base dynamics. (A) Frequency and type of burial events in relation to energy regime. Because limits of fair weather wave base (FWWB) and maximum storm wave base (MSWB) vary regionally and temporally, they are shown as crossing depth lines on the diagram. Although obrution is most frequent in shallow, high-energy settings within FWWB, these rapidly buried assemblages have low preservation potential. Conversely, obrution is infrequent in low-energy, distal settings below MSWB-limit, but preservation potential is much higher (see text). Overlap of different types of obrution deposits (bounded by dashed lines) into zones of varying energy (solid-line boundaries) reflect reality that diagram generalizations are based on probabilities of occurrence for different types of obrution deposits. (B) Occurrence of different categories of obrution deposits with regard to energy and depth regime. Type 1 obrution deposits are patchy and nonextensive, reflecting rapid shifting of sediment and skeletal material in high-energy settings. Type 2 deposits, usually expressed as storm sand layers, are proximal tempestites. Type 3 deposits, which are the most common type of obrution deposits, are mud tempestites, deposited in lower-energy settings. Type 4 deposits represent organisms buried in organic-rich mud by distal tempestites or turbidites under quiet conditions. See discussion in text. (Modified from Speyer and Brett 1988.)

1.3A). As a given storm proceeds, muds and silts winnowed from the upslope areas may be rapidly deposited from gradient currents below storm wave base owing to barometric effects of coastal setup (Aigner 1985; Clifton 1988). Such setup is compensated for by the basinward-flowing gradient currents that often transport fine-grained suspended sediment away from shore (fig. 1.3A).

Assuming that most obrution deposits record the effects of storms, particularly the blanketing by sediment fallout following the disturbance, their frequency and type can be modeled as a function of tempestite proximality and proximity to sediment input. The frequency of burial events by shifting, coarse sediments is predicted to be highest in relatively shallow storm wave–affected areas with high sediment input (fig. 1.3A). Thus overall, obrution deposits should form most commonly in nearshore environments. However, other factors are critical to long-term preservation of obrution sediments. Shallow-water, high-energy environments are conducive to rapid burial but also experience high rates of reworking of event deposits. This may lead to the destruction of all but rare remnants of former obrution layers. In general, the reworking potential will decrease downslope or offshore in marine environments, approximately converse to the burial frequency. The net effect of these two processes is that optimal conditions for preservation of burial events will occur at moderate depths, near the lower end of the storm wave base (fig. 1.3B). Such areas will receive frequent blankets of fine-grained burial sediments from upslope areas, but these will not be reworked as commonly by later events (Meyer et al 1981; Aigner 1985). Anoxic or dysoxic conditions in the upper sediment and the attendant lack of deep bioturbation will also prevent biogenic reworking of the deposits.

Early Diagenetic Overprints

The geochemistry of the burial environment strongly influences the early diagenetic overprint, if any, on obrution deposits. Aerobic decay rapidly destroys soft tissues of organisms (Allison 1986, 1988). But burial of organic remains in anaerobic sediments inhibits decay slightly and may trigger a number of early diagenetic processes in the upper few centimeters of the sediment that enhance preservation potential (Allison 1988; Allison and Briggs 1991; Canfield and Raiswell 1991a, 1991b). The type of mineralization will depend upon the sediment's organic content and the rate of sediment accumulation subsequent to the burial event. Sudden emplacement of organic matter into low organic, nonsulfidic sediment may lead to spectacular mineralization, such as early diagenetic nodules and pyritized fossils (Hudson 1982; Brett and Baird 1986; Baird et al. 1986; Canfield and Raiswell 1991a; Bjerreskov 1991). This takes place because the entombed organisms provide local sites of sulfate reduction or methanogenic activity and may serve as "ion pumps" relative to the surrounding sediment. Again, when obrution is followed by prolonged periods of nondeposition, a number of effects may occur. For example, in calcareous sediments early diagenesis may result in the formation of carbonate concretions or nodules that encapsulate the fossils, ensuring their preservation in a noncompacted state (Baird et al. 1986; fig. 1.4). Where postevent accumulation is very low, phosphatic steinkerns and nodules may form, in microzones between the aerobic bottom water and anoxic sediment, particularly in organic or phosphate-rich deposits (Lucas and Prévôt 1991). Phosphatic nodule zones are associated with periods of sediment starvation and so occur nonrandomly associated with transgressive phases of cycles. Alternations of reworked and episodically buried skeletal material should be particularly typical of condensed intervals associated with sea-level rise (see fig. 1.5; inset E).

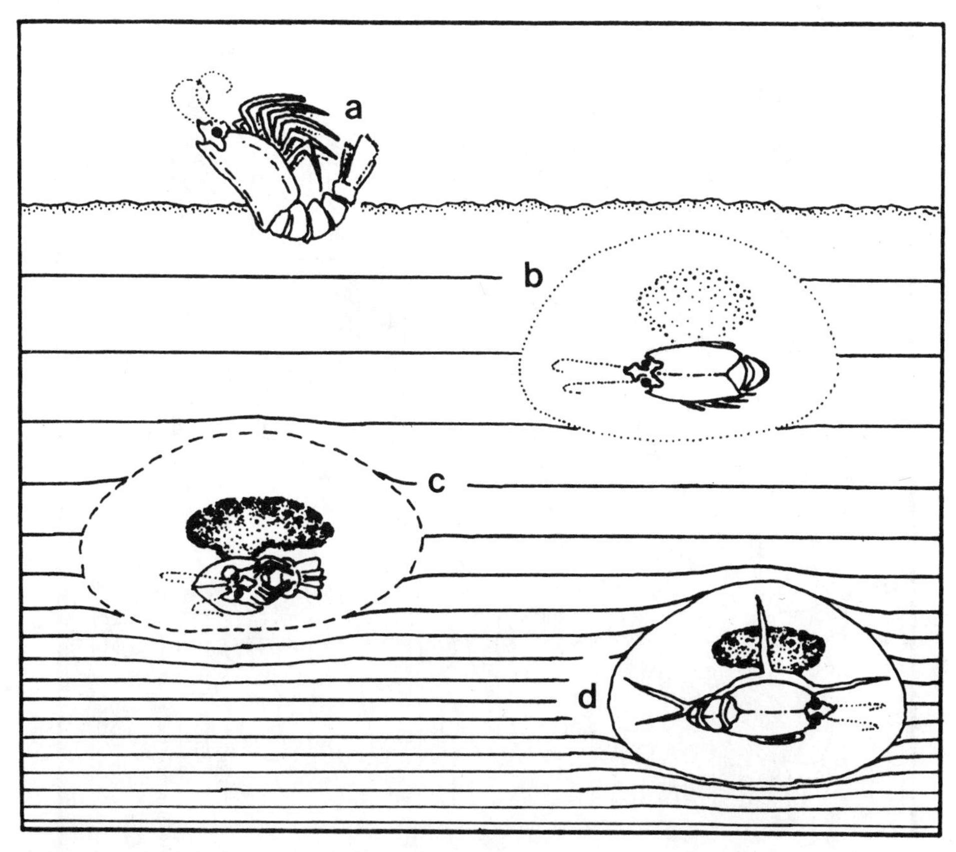

FIGURE 1.4. Preservation of soft-bodied and lightly sclerotized organisms by very early diagenetic mineralization. Example shown is formation of a sideritic concretion around rapidly entombed shrimp in the Middle Pennsylvanian (Westphalian) Francis Creek Shale, northeast Illinois. Pyrite (dark stipples above shrimp) and siderite formation, respectively, reflect sulfidic and methanogenic activity in this environment. Decay of the organism may have triggered or accelerated the bacterially mediated reactions with formation of a strong local chemical diffusion gradient within sediment low in organic matter (see text). (a) Burial of organisms; (b–c) formation of protoconcretion and pyrite (or pyrite precursors) around fossil as well as initial dewatering of sediment around nodule; (d) formation of septarial cracks within concretion. (From Baird et al. 1986.)

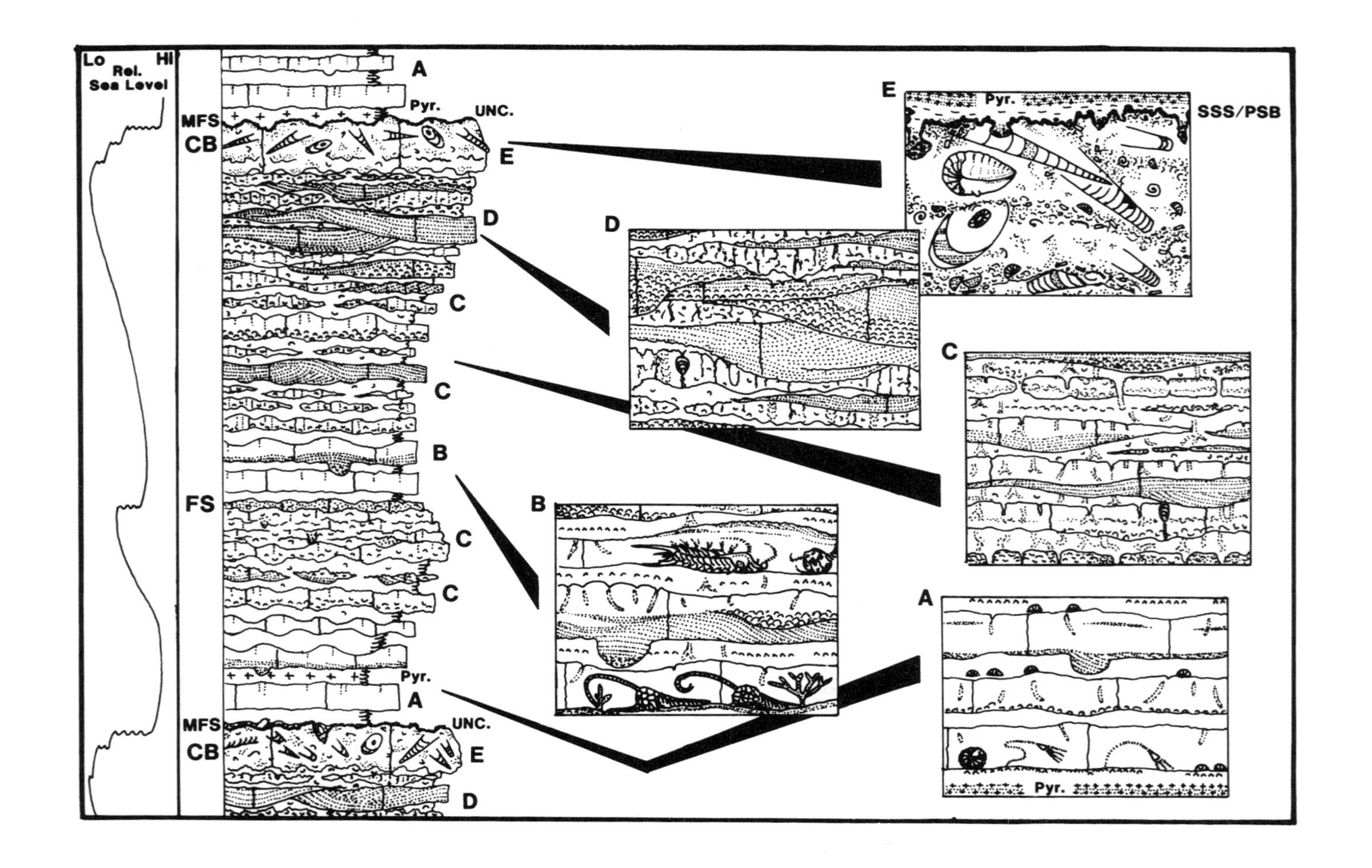
Lo
Hi
Rel.
Sea Level
MFS
CB
FS
MFS
CB
A
Pyr.
UNC.
E
D
C
C
B
C
C
Pyr.
A
UNC.
E
D
E
Pyr.
SSS/PSB
D
C
B
A
Pyr.

Finally, entombment of organism remains in organic-rich, sulfidic sediments usually results in minor mineralization resulting from weak concentration gradients between the organism bodies and surrounding organic-rich sediments (Fisher 1986; Canfield and Raiswell 1991a, 1991b). However, rapid burial may lead to the preservation of organic matter by carbonization or by coating of skeletons with phosphate, pyrite, or other minerals (Allison and Briggs 1991). Allison (1988) has emphasized that soft-tissue decay can only be halted by such early mineralization. Burial of organic remains in anoxic sediments may trigger a number of geochemical processes within the upper few centimeters of the sediment, leading to extraordinary Lagerstätten deposits (see, for example, Seilacher et al. 1985). However, this does not guarantee unusual preservation unless burial is rapid (see later).

Categories of Obrution Deposits

The various aspects of tempestite proximality and early diagenesis may be combined to derive a general classification of obrution deposits (figs. 1.3B, 1.5, 1.6). The figures place obrution deposits into a taphofacies framework (see Speyer and Brett 1988). Type 1 obrution deposits consist of localized patchy remains of organisms typically killed by mechanical means, such as wave or current surges, and buried rapidly by shifting coarse-grained sediments. Pockets of complete crinoids occurring in amalgamated grainstones illustrate this point particularly well. Type 2 deposits consist of remains of organisms caught

Figure 1.5. Generalized fifth-order transgressive/regressive sedimentary cycle in the Late Middle Ordovician Trenton Limestone Group of central New York State. Insets A–E show representative taphonomy associated with different phases of the cycle. Storm-generated obrution layers reflect changing balance between control factors, including energy levels, sediment supply, and accommodation space for sediment accumulation during cycle assimilation. Inset A records early highstand conditions with low energy conditions, abundant accommodation space, and deposition of type 3 and sometimes type 4 obrution layers; these are typically tabular to lenticular micritic limestones with rapidly buried, articulated, multielement organisms; B records obrution under more proximal, late-highstand to early-lowstand conditions with adequate accommodation space to preserve buried organisms. A mixture of mud (calcilutite) tempestites and graded storm layers accumulates in this regime, reflecting conditions of deposition for type 3, and occasionally, type 2 obrution beds; C records conditions of late-highstand to early-lowstand transition. Graded tempestite and skeletal storm sand sheets and pararipple bedforms predominate. Accommodation space is limited, sediment bypass predominates, and only a few type 1 and 2 obrution assemblages are preserved. D records high-energy lowstand grainstone facies with minimal accommodation space. Only a few spatially limited (patchy) type 1 and rare type 2 assemblages will be preserved here. E records transgressive conditions with development of sediment-starved condensed nodular carbonate facies, rich in cephalopod conchs. Although this is a moderate- to low-energy deposit, obrution assemblage preservation potential is low owing to low sediment supply and extensive sediment churning. Nonetheless, spotty obrution assemblages can be found in this facies as indicated by the enrolled trilobite (see discussion in text).

up and buried in proximal storm deposits. These may occur as well-preserved fossils on the bases of storm sand layers. For example, Hickerson (paper 9, this volume) describes trilobite clusters preserved in tubular tempestites along a discontinuity surface. Remains may also be buried on the tops of debris layers. These horizons are commonly associated with hardgrounds and therefore may be *in situ* (see Brett and Liddell 1978 for Ordovician hardground example). The third category (type 3), which comprises the majority of well-known obrution deposits, consists of mud tempestites and is well illustrated by beds of clustered trilobites; examples include trilobite beds from the Cambrian of Utah (Gunther and Gunther 1981), the Ordovician of Oklahoma (Laudon 1939), and the Devonian in New York State (Speyer and Brett 1985). Type 3 beds grade into the previous type of deposits. Here nearly optimal conditions exist for preserving articulated, in situ skeletons. Conditions conducive to this type of deposit occur slightly below storm wave base in areas of abundant outfall of fine-grained sediments carried in suspension by gradient currents. Also, the low frequency

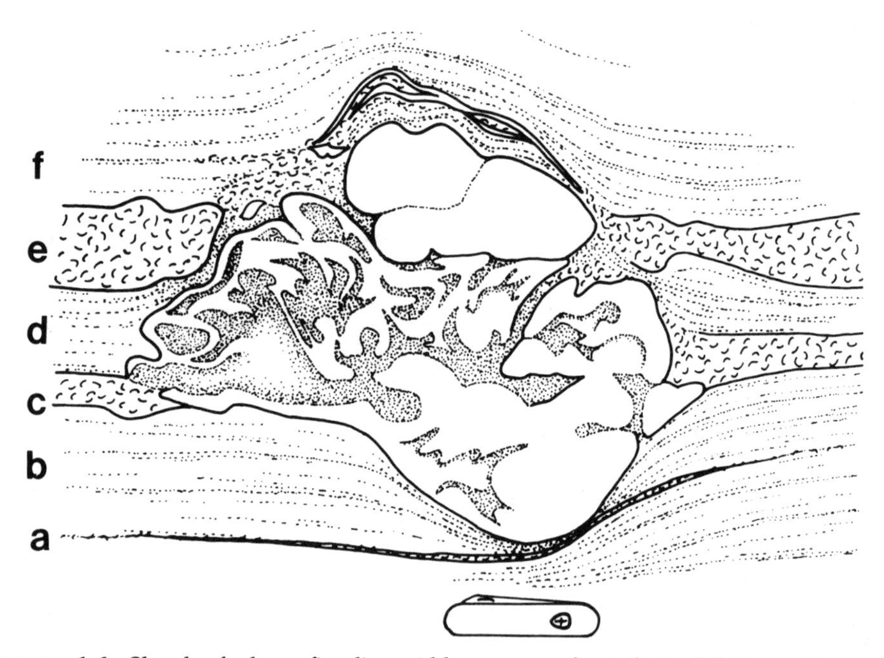

FIGURE 1.6. Sketch of a large fistuliporoid bryozoan colony that exhibits continuous, but perturbed, growth through several storm-related abrution layers. Bryozoan growth was apparently initiated on the lowest shell bed (a). After an interval of mud deposition (b) the colony expanded greatly. Colony expansion was coincident, in part, with development of the second shell bed (c). Lenses of mud within the colony (dark stippling) indicate two or more depositional events not recorded in the shell bed (c). Discontinuity within the colony (arrows) records a major mud burial event (d) that permanently smothered shell bed c. However, the colony reestablished itself during the time of development of highest shell bed (e) before both were buried by mud layer (f). Example shown is from the Middle Devonian Windom Shale at Moravia, N.Y. Jacknife gives scale. (Modified from Parsons et al. 1988.)

of storm wave disturbance of the sea bottom in these areas ensures that most such deposits will not be disinterred later. Note that, although some early diagenetic minerals may form under these settings, no preservation of soft parts occurs here or in the previous two types. However, in a fourth type of obrution deposit associated with distal tempestites or microturbidites, burial in organic-rich sediments may preserve soft parts as carbonized films or pyritic coatings. The famous Beecher trilobite beds (Frankfort shale, Ordovician, New York State; see Cisne 1973; Briggs et al. 1991) and the Devonian Hunsrück shale (see Seilacher and Hemleben 1966; Seilacher et al. 1985) provide examples of this type. These represent stagnation/obrution deposits, and, in some instances organisms within them may be allochthonous remains that are buried in anoxic muds. This category includes some of the most important conservation Lagerstätten (Seilacher et al. 1985).

From the standpoint of event stratigraphy, two key aspects of obrution deposits are critical. First, because various types of rapid-event deposits are related to sedimentary environment and particularly to tempestite proximality, they will occur nonrandomly in sedimentary cycles (see figs. 1.5 and 1.6). Optimal conditions for preserving obrution deposits may occur in a relatively narrow belt below normal wave base downward to slightly below the maximum wave base associated with moderate rates of background sedimentation (see Brett and Baird, paper 8, this volume). Thus obrution deposits may be most typical of transgressive and highstand facies of transgressive/regressive cycles, rather than strictly at maximum transgression or maximum regression. They would be associated most particularly with early transgressive deposits that overlie and sometimes underlie surfaces of maximum starvation or marine flooding surfaces near the base of highstand deposits (see Van Wagoner et al. 1988; Vail et al. 1991). In these settings firm or hard substrates plus low background sedimentation rates provide excellent conditions for the development of diverse epifaunal communities. Moreover, relatively low energy conditions and an aggradational setting below average storm wave base should aid in the preservation of burial horizons. During rapid transgressions, deepening should enhance preservation potential of obrution deposits because of increased accommodation space due to elevation of normal or storm wave base above the seafloor. The infrequency of disturbances and low sedimentation rate may also set the stage for early cementation of obrution layers that enhances their preservability (Walker and Diehl 1985).

Skeletal Concentrations

A second broad category of paleontological event beds consists of skeletal concentrations (concentration Lagerstätten, often referred to as *shell beds* or coquinas). Shell beds obviously record varied temporal scales. For excellent reviews see Kidwell and Aigner (1985), Kidwell (1991b), and Kidwell and Bosence (1991). In the simplest cases, a thin shell pavement was produced and buried by sediments. Such fossil layers, often displaying spatfall cohorts, reflect primarily single events. These shell beds may be further characterized as *allochthonous* (transported assemblages) or *parautochthonous* (near-neighborhood) concentrations.

Shell-rich beds, potentially recognizable as "events," may accumulate over considerably longer spans of time (tens to thousands of years) in response to relative sediment starvation, winnowing (sediment bypass), or erosion lag concentration (Kidwell 1991a). Although these do not provide as much time resolution as single events, they may be much more widespread and thus of considerable importance in event stratigraphy. In the following sections we review single-event allochthonous and parautochthonous skeletal beds from the standpoint of event stratigraphy.

Allochthonous Fossil Beds

In certain environments allochthonous skeletal layers may be common or even the norm. An example of allochthonous single-event beds is represented by shell-rich turbidites or debris flow beds. Given appropriate paleoslopes, particularly large events may carry shells as sediment particles for great distances. Pilkey and Curran (1986) discuss examples of shell-rich turbidites that have been correlated for more than 500 km. Such beds obviously form excellent single-event markers for stratigraphy in normally fossil-poor facies. Supratidal and intertidal flat facies may also contain thin and possibly extensive layers of exotic shells cast up by particularly severe storms. As a specific case, thin layers of *Tentaculites*, leperditian ostracodes, and bivalves found in the peritidal Manlius Limestone (Lower Devonian) of New York may represent examples of such allochthonous storm lags (Brett, personal observation); these beds are traceable, at least within outcrops, but have not been used for regional correlations.

Another category of allochthonous fossil accumulation that may give rise to laterally extensive horizons consists of pelagic "rain-out" events. For example, black shale successions commonly display bedding planes crowded with fragments of carbonized wood or terrestrially derived animals (Brower and Nye 1991; Kauffman 1981). These could be associated with the aftermath of storm flooding events that carried large amounts of plant material seaward.

Layers with particular concentrations of pelagic organisms may record spans of low sedimentation in which skeletal remains accumulated on the seafloor by settling out from suspension. Certain cephalopod-rich beds, in which shells display varied modes of preservation, could reflect simple gradual accumulations of this sort (Wendt and Aigner 1982, 1985). However, in many cases articulated preservation of pelagic organisms (e.g., complete crinoids attached to logs in the Jurassic Posidonia shale) suggests a mass mortality event (Seilacher 1982a).

In the case of microfossils, thin enrichment horizons may record short-lived productivity events or plankton blooms (see Kauffman 1986). Again, under appropriate depositional conditions short-term blooms of even years of duration might be recorded as thin laminae. For example, varvelike laminae in the upper meter of dark, organic-rich sediments in the Black Sea record periodic, possibly seasonal blooms of coccoliths (Lyons 1992); individual enrichment laminae can be precisely correlated for more than 300 km in sediment cores. Similarly, Kauffman (1986) records abrupt contacts between coccolithic or foraminiferal limestones with dark shales that may reflect abrupt productivity events.

Possible Paleozoic analogs would include bedding planes of fragmentary graptolites or styliolinids within black shales (Williams and Rickards 1984; Brett et al. 1991). In many such cases a brief interval of sediment starvation may also be indicated. However, these layers may be traceable, at least in local areas, and could provide important correlation markers in otherwise featureless shales.

Parautochthonous Single-Event Fossil Beds

Studies of modern subtidal shelf environments indicate that most skeletal remains accumulate within the habitat of the producing organisms and in most cases in close proximity to their living site (see Kidwell and Bosence 1991, for extensive review and compilation). Even where storm waves suspend and redeposit skeletal remains, the lateral transport is commonly on the order of meters.

Single-event shell layers may be the result of simple in situ accumulation due to normal attritional mortality. Such accumulations will display varying amounts of time-averaging. Depending upon the duration, shells may be concentrated or mixed into older or younger sediment activity by burrowing (see Cutler and Flessa 1990; Flessa et al. 1993). Such background accumulations may yield a variable package of sediment, up to meters thick, in which skeletons are more or less randomly scattered or display only minor local burrow-fill concentrations or small-scale winnowed aggregations. In such cases no distinct event beds are produced. However, if normal background accumulation is abruptly terminated by a pulse of sediment or an abrupt change in environmental conditions (e.g., water column becoming anoxic), then a sharp upper terminus to the accumulation will be produced. This bedding plane is an event surface that may be usable as a stratigraphic marker. Such a bedding plane may also coincide with an obrution deposit, at least in some localities, if the terminating agent was an episode of rapid deposition.

Simple parautochthonous shell beds may be produced by winnowing events that concentrate skeletal debris produced over a variable period of background accumulation. In such cases a single-event bed can be produced. Agents responsible for winnowing events include both storm waves that touch bottom and gradient currents. Many parautochthonous coquinites record storm events and constitute classic examples of shelly tempestites (see Aigner 1985; Kreisa 1981; Kreisa and Bambach 1982; Jennette and Pryor 1993; Cherns 1988).

Typical features of parautochthonous shell layers produced by winnowing include sharp, scoured bases combined with gutter or pot-cast features, sometimes with distinct hypichnial traces (Aigner 1985). Internally, the beds, which range from millimeters to several centimeters in thickness, may display grading that involves skeletal remains and fine-grained sediments. Evidence for hydraulic concentration and local transport of shells may also be present in the form of shells in unstable oblique or edgewise orientations, preferred long-axis azimuths, and nested stacking of concavo-convex valves. If shells have been subject to in situ winnowing with little transport, they may display mud sheltering (i.e., the conservation of fine-grained sediment in protected cavities of shells). Conversely, if winnowing is followed by an interval of gradual resettlement and infiltration of fine sediments into the winnowed shell lag, shelter voids may occur beneath shells. Here shells provide "umbrellas" to prevent infiltration of sediment beneath them (Kreisa 1981). Reworking of previously buried skeletons may also be evident from rotated geopetal fillings, sediment infillings that do not resemble the matrix of the beds, and various aspects of early diagenesis or prefossilization (*sensu* Seilacher 1973). Internal laminations and bedforms in these event beds may also reflect wave, current, or combined flow associated with many phases of the events. Planar and small- to large-scale

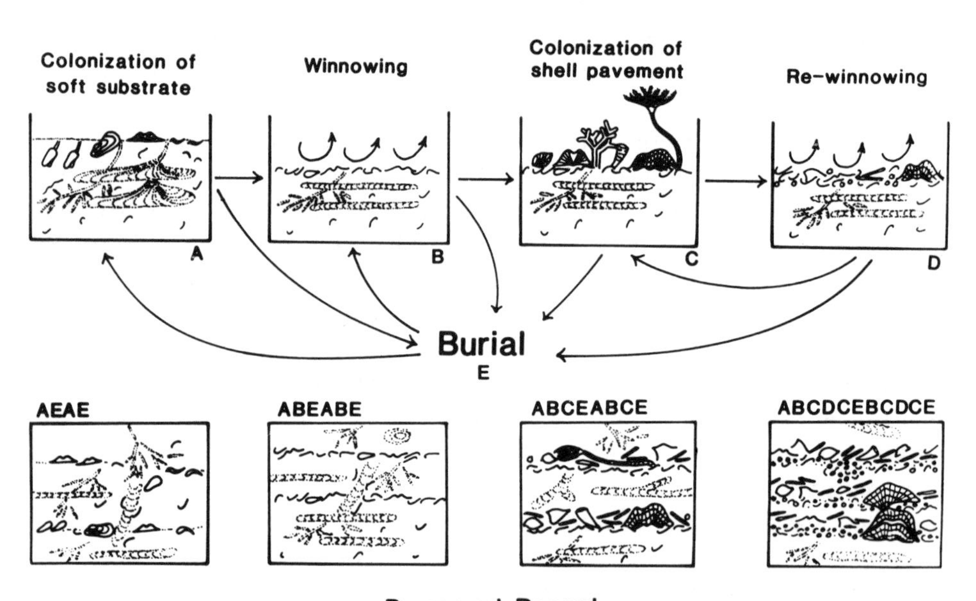

FIGURE 1.7. Schematic flow chart showing interpreted sequence of events responsible for generating a variety of preserved shell bed types. The trend, left to right, shows a spectrum from simple (colonization) burial motif to increasingly dynamic settings where winnowing and taphonomic feedback are dominant processes. (From Miller et al. 1988.)

hummocky cross-laminations are common where shelly coquinas are overlain by silt or sandy caps (Aigner 1985; Kreisa 1981).

Evidence of disruption of growth in survivors of turbulence or sedimentation pulses may also point to the existence of events that would otherwise go unrecorded. For example, Cuffey (paper 5, this volume) points to cases of overturning or partial burial followed by regrowth in Ordovician bryozoans. Parsons et al. (1988) and Miller et al. (1988) provide examples of burial disruption and regrowth in Devonian fistuliporoid bryozoans (fig. 1.7).

These varied lines of evidence permit unambiguous recognition of single-event skeletal concentrations resulting from turbulence and sedimentational events. Unfortunately, many or most such beds, at least in proximal areas, display discontinuous lenticular and complex cross-cutting patterns. Hence they are of limited applicability as marker beds for regional event stratigraphy.

However, in some cases particularly outsized storms may produce an isolated, widely traceable bed of winnowed/graded coquinite in an otherwise rather homogeneous background. Such beds may provide important paleontological event horizons even if they are lenticular, because they stand out from surrounding strata (e.g., beds of brachiopod coquina in the Silurian Williamson Shale; Eckert and Brett 1989).

Composite Shell Beds/Obrution Deposits

Although thin shell layers are commonly interpreted as simple storm-generated coquinites, the complexity and ecological disparity of certain shell-rich layers suggest that they represent a complex series of burial and reworking events instead (fig. 1.8; see Walker and Alberstadt 1975, Walker and Parker 1976). Storm concentration of skeletal debris into pavements provides substrates for colonization of new epifaunal organisms (Miller et al. 1988; Kidwell 1991b). This process of taphonomic feedback (Kidwell and Jablonski 1983) is undoubtedly important in generating many diverse skeletal accumulations. Repetitive mud blanketing of shell pavements and rewinnowing of the muds by storms produce complex shell beds and generate a variety of different shell layers (Miller et al. 1988; Kidwell 1991b; figs. 1.6, 1.8). Shell beds thus range from simple pavements a few millimeters thick to more complex shell beds up to tens of centimeters thick. Some of these are traceable laterally for tens or hundreds of kilometers. Hence the actual shell layer represents a long-duration feature, but the mud drape covering it may represent a geologically instantaneous single pulse of siliciclastic sediments (see fig. 1.8). For example, Masaurel (1987) and Webby and Percival (1983) noted well-preserved brachiopods on the tops of skeletal hash beds, indicating abrupt burial of a final community after long-term accumulation of debris. Similarly, Meyer (1991) discussed edrioasteroid assemblages attached to strophomenid brachiopod pavements from the Ordovician of Ohio. Such contacts may provide outstanding event markers, traceable for tens or even hundreds of kilometers (Parsons et al. 1988).

In more complex intervals a thick winnowed shell layer forms a "reference

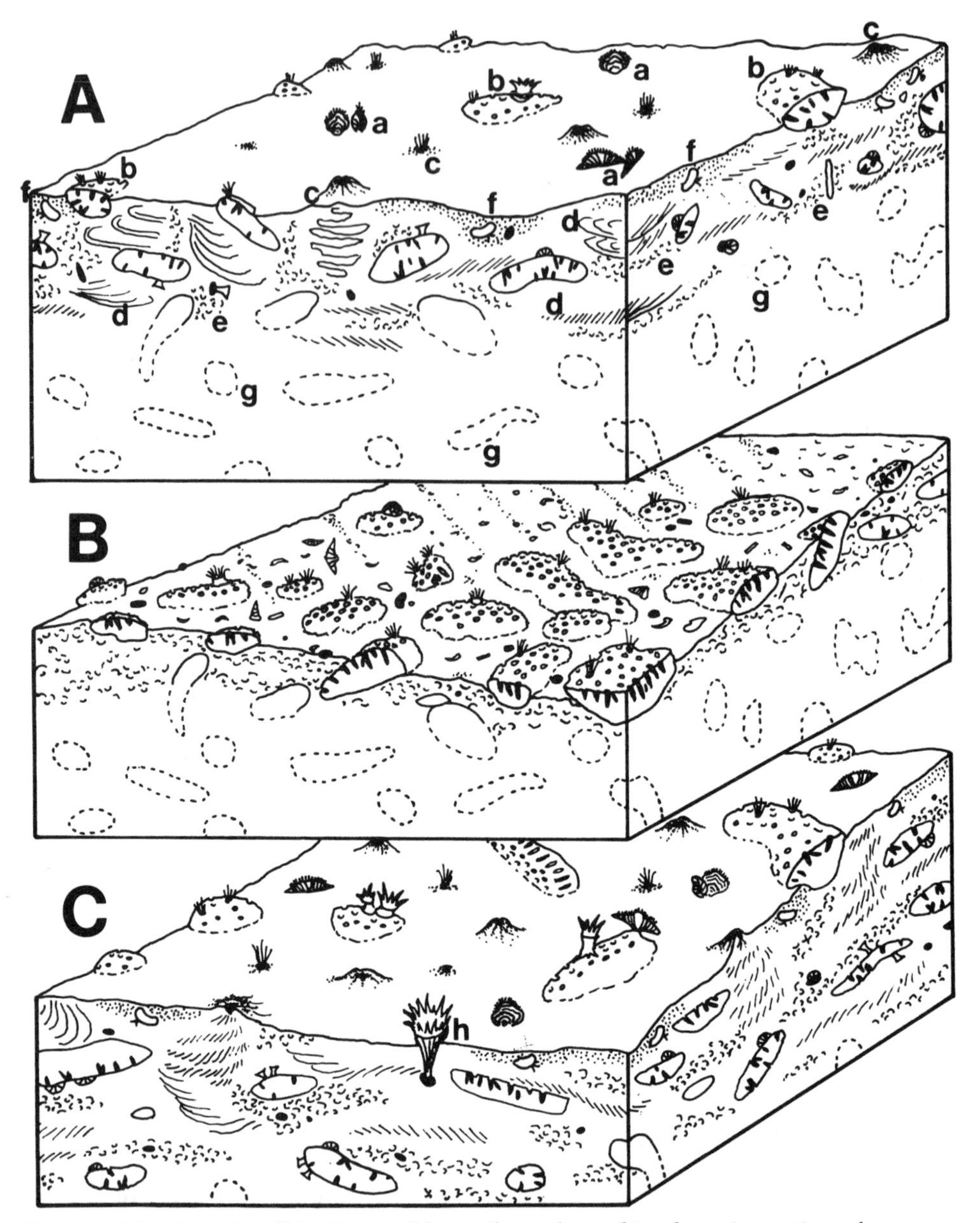

FIGURE 1.8. Genesis of hiatal assemblages through combined erosion action of currents and burrowing activity. (A) is a shale-on-shale diastem in the Middle Devonian (Givetian) Ludlowville Formation that is marked by diffuse erosional boundary, shell debris ranging from pristine (a,e) to prefossilized, and abundant reworked concretions (b) that are encrusted (h) and bored. (B) The sloped erosion surface was subjected to intense activity of mud burrowers (c,d,f) and episodic periods of scour by currents that exposed the concretions and removed the water-rich burrowed sediment. (C) The process continued with renewed burial of concretions and subsequent scour. In essence, burrowers softened the substrate and facilitated the erosion process. The fossil assemblage in a bed of this type has aspects of two assemblage categories discussed herein; it reflects both sedimentary condensation and exhumation of older shell material. Condensation is reflected in the prolonged concentration of shells over a long period within the zone of sediment reworked by infauna; in this sense, the bed has a hiatal character. Genuine downward scour and removal of units beneath the bed, however, prove that it is also erosional in character and that much of the shell concentration is lag debris. This is exemplified by occurrence of abundant phosphatic fossil steinkerns exhumed from underlying beds and sometimes encrusted, e.g., by a rugose coral (h). (from Baird 1981).

horizon" as in the Jeram model of Seilacher (1985). Subsequent storms may winnow muds, allowing shell debris to settle downward onto the earlier horizon and gradually building up a layer of shells of considerable thickness and complexity. The fact that skeletons of infaunal organisms (such as burrowing bivalves) are commingled with epifaunal remains proves that more than one community may be represented in given shell beds; that is, they are ecologically condensed (*sensu* Kidwell and Bosence 1991). Such shell beds are commonly overlain by barren mudstone layers that represent deposition of sediment that was sufficiently rapid and thick to form a permanent blanket over the last generation of shell-bed-producing organisms. Organisms of this last generation may be preserved intact and articulated on the upper surface of the shell layer (see discussion of Miller et al. [1988] and example of Jurassic Gmeund echinoderm locality discussed by Seilacher et al. [1985]).

Beds of fossils often display features of diagenetic enhancement or " overprinting," such as the development of nodules and/or tabular layers of cemented concretionary carbonate (fig. 1.6). In carbonate-dominated sediments, hardgrounds are commonly formed at the tops of shell-rich beds (Fürsich 1979; Brett and Brookfield 1984). Similarly, concretionary underbeds form beneath many coquinites (Aigner 1985). Early cementation may be favored by the open framework of skeletal gravels as well as by the concentration of carbonates that serve as nucleation sites for cements (Aigner 1985; Kidwell 1991a; Wilson and Palmer 1992). Burial and anaerobic decay of organic matter may lead to buildup of sulfides and bicarbonates in pore waters, favoring precipitation of carbonate and, in some cases, pyrite within fossil beds (Waage 1964; Baird et al. 1986; Canfield and Raiswell 1991a, 1991b).

Hiatal (Condensed) Concentrations

Certain enrichment horizons record the combined long-term effects of sedimentary reworking and condensation. These are referred to as hiatal, or condensed, concentrations (Fürsich 1978; Kidwell 1991a). Condensed beds of this type are recognizable on the basis of several criteria. They may cross-cut facies and typically are very widespread (identifiable over hundreds or thousands of square kilometers). In addition, they may be associated with physical evidence for diastems or disconformities and omission surfaces, such as firmgrounds or hardgrounds (Kidwell 1991a). Condensed shell beds may contain *diaclasts,* such as reworked concretions, phosphatic pebbles, glauconite granules, and/or reworked sedimentary pyrite (Fürsich 1971, 1978; Seilacher et al. 1985; Landing and Brett 1987; Baird and Brett 1991). They may be enriched in rare but geochemically resistant fossils, such as crinoid ossicles, conodonts, fish teeth, and bones. Hiatal concentrations often pass laterally into lag accumulations as conditions of alternating sedimentary condensation and erosion become solely erosional (see "Lag Concentrations").

Taphonomically, condensed beds display a wide array of different grades (*sensu* Brandt 1989) of preservation ranging from complete and even articulated skeletons to heavily corroded fragments (Fürsich 1971, 1978; Holland 1988; Kidwell 1991a). Skeletal remains in condensed beds may be heavily encrusted, bored, and pitted (e.g., Holland 1988; Brett and Bordeaux 1990). Fossils also may display evidence of early diagenesis and reworking ("prefossilization" of Seilacher 1973); for example, phosphatic steinkerns of mollusks may be common (Snyder and Bretsky 1971; Baird 1978; see fig. 1.1A).

Taphonomically condensed cephalopod-rich limestones (cephalopodenkalk) form widespread marker beds in some Paleozoic and Mesozoic pelagic facies (Wendt and Aigner 1982, 1985; Wendt et al. 1984; Wendt 1988.) The origin of these beds is enigmatic but probably involves episodes of burial, diagenesis, and "prefossilization" followed by exhumation by winnowing and reburial.

Many hiatal concentrations display features indicative of long-term time-averaging, including the intimate commingling of fossils normally found in distinct biofacies; for example, phosphatic steinkerns of mollusks may be common (Snyder and Bretsky 1971; Baird 1978). Such mixtures indicate "environmental condensation" (i.e., faunas from successive habitats accumulated in situ [Kidwell and Bosence 1991]). Normally, this further implies the passage of considerable amounts of time, on the order of thousands to tens of thousands of years, based on dated Pleistocene examples of environmentally condensed assemblages in shallow-shelf to peritidal areas (Einsele et al. 1974; Cadée 1984; Meldahl 1987).

In rare cases, hiatal beds may display truly *biostratigraphic condensation* (i.e., commingling of zonally distinct fossils). For example, Fürsich (1971) cites an example of a thin Jurassic limestone bed containing well-preserved remains of four distinct ammonoid zones. Moreover, the 0.8-m-thick Lower Mississippian Chappel Limestone in Texas contains three conodont zones as well as abundant reworked Devonian elements (Hass 1959; Merrill 1980). Obviously, these shells have become mixed by multiple erosional exhumation and reburial events. Their good preservation is strong evidence that many of these fossils were never exposed for any substantial time period on the seafloor.

Obviously, the temporal resolution afforded by hiatal or condensed beds is one or more orders of magnitude less precise than that for single-event deposits. However, this drawback is offset by the fact that many condensed beds provide distinct markers that can be correlated for considerable distances across facies. At least for many of the environmentally condensed assemblages there is strong evidence that the beds record isochronous intervals of sediment starvation and thus provide stratigraphic markers with much greater time resolution than that afforded by biostratigraphy (Kidwell 1991a). For example, many condensed beds that we have recognized in the Paleozoic of the Appalachian Basin display consistent lateral gradients of faunal change that mirror those of synjacent facies (e.g., Baird 1981; Lafferty et al. 1994). Despite some evidence for mixing of different biofacies, there is generally a very strong dominance of fossils from a single community type at any given locality. Moreover, the fauna of the condensed bed typically shows fairly strong affinities to those of overlying less-condensed mudrock facies. This observation suggests that, despite evidence for long-term sediment starvation, fossils representing the last generations of organisms in a given habitat are strongly overrepresented. Most older bioclasts are represented as highly comminuted and corroded fragments that are easily factored out in making formal gradient studies. In some instances the upper laminae or top surface of a condensed bed is an obrution layer.

Because of the "swamping" of the faunal signature by late-appearing biotas, ecologically condensed beds may be usable as natural *transects* for gradient analysis. That is, these beds display gradational lateral faunal changes of time-averaged assemblages that probably parallel original biofacies gradients (Baird 1981; Lafferty et al. 1994).

Condensed shell beds also occur predictably within stratigraphic sequences (see Kidwell 1991a, 1991c; Banerjee and Kidwell 1991). Thus they provide important clues to recognizing key stratigraphic surfaces (figs. 1.5, 1.6). Kidwell (1991a, 1991c) distinguishes several different processes that produce hiatal accumulations, including dynamic processes of sediment starvation, and sediment bypass through winnowing. She then relates these processes and the hiatal beds to cycles of relative sea-level fluctuation and thus to depositional sequences (*sensu* Van Wagoner et al. 1988; Vail et al. 1991).

Condensed shell accumulations produced by winnowing through transgressional ravinement, and sediment starvation typically occur at bases of transgressive systems tracts or flooding surfaces (e.g., base of cycle shell beds of Banerjee and Kidwell 1991; fig. 1.5: inset E, A; fig. 1.6: insets A, B). These beds may be slightly diachronous. Very thin conodont bone and phosphate-nodule-rich beds occur in mid-cycle positions commonly associated with surfaces of maximum sediment starvation (mid-cycle coquinas; also termed *backlap* and *downlap* shell beds by Kidwell 1991c). Finally, complex discontinuous, sometimes amalgamated, shell gravels may occur near regressive maxima; these are produced by winnowing and bypass effects in high-energy settings (end-cycle coquinas or toplap beds [Banerjee and Kidwell 1991; fig. 1.5: see insets D–E; fig. 1.6: inset F]). In Paleozoic sequences of the Appalachian Basin we have recognized a third type of condensed shell-rich bed that we term *mid-highstand,* or "*precursor, beds*" (Brett and Baird 1990). These enigmatic shell beds occur in offshore facies associated with abrupt pulses of relative shallowing (fig. 1.6: see inset D). Commonly, the shell gravels and associated reworked concretions abruptly overlie black or dark gray mudrocks and are overlain by gray mudstones at the *bases* of shallowing-upward successions. At this time, the processes producing sediment starvation and erosion during episodes of sea-level fall are enigmatic. Yet such mid-highstand hiatal beds are recognizable in most third- and fourth-order sequences and form very useful regional markers.

The thin shell-bone beds found at mid-cycle (base of highstand) positions, on the precursor beds (mid-highstand), and minor shell gravels that mantle marine flooding surfaces of parasequences have proven most useful as stratigraphic markers of brief duration. By careful correlation of these beds, it is not only possible to delineate approximately coeval sedimentary packages but also to determine gradients of general biofacies for paleoecological analysis. In this manner condensed beds may be used in gradient analysis in much the same way that Cisne and Rabe (1978) used ash layers to identify gradient transects in the middle Ordovician of the Mohawk Valley area in New York. The potential of gradient transects is only beginning to be realized (Lafferty et al. 1994).

Lag Concentrations

As defined by Kidwell (1991a), *lag concentrations* are thin skeletal accumulations that represent reworking of fossils associated with substantial erosional truncation of older strata. Lag concentrations differ from hiatal concentrations in being more explicitly erosional in origin rather than the product of sedimentary condensation. Lag concentrations do, however, grade laterally into hiatal concentrations. Thus the difference between lag and hiatal concentrations is often subtle and is revealed through the textural and zonal aspects of component bioclasts, as well as the stratigraphic context.

Lag beds commonly display a very strong degree of biasing toward highly resistant bioclasts, and they often contain diaclasts. They may also display biostratigraphic condensation produced by erosional concentration of bioclasts of differing age. Lag deposits include conodont and bone beds, exemplified by the Devonian (early Frasnian) North Evans Limestone or "conodont bed" of western New York (Baird and Brett 1986), and bone beds that mantle discontinuities in the Triassic Muschelkalk discussed by Aepler and Reif (1971) and Reif (1982). In many cases, the concentration of pyritized and/or phosphatized skeletal remains suggests long-term winnowing and erosion under submarine conditions. In the case of reworked pyritic concentrations, erosion is believed to have taken place even under anoxic or minimally oxygenated water conditions (see Baird and Brett 1986, 1991).

Beds of reworked fossils take on a larger meaning in the context of sequence stratigraphy. Typically, shell-rich or bone-rich lag beds are associated with surfaces of sediment starvation, which correspond to flooding surfaces, in minor cycles, or maximum flooding surfaces (downlap and backlap surfaces) of major cycles (see Banerjee and Kidwell 1991; Kidwell 1991a; fig. 1.5: inset E; fig. 1.8: inset B). Such horizons record times of most rapid rise in relative sea level, which produces maximum entrapment of siliciclastic sediments in near-shore bays and estuaries and excludes sediments from offshore areas. They are an outcrop expression of downlap surfaces recognized by seismic stratigraphers and therefore have considerable stratigraphic significance, albeit poor time resolution (Kidwell 1991c).

Discussion

Paleontological events may be viewed in a hierarchical perspective. Examples of obrution deposits and storm layers discussed in this paper and others in this volume record mass mortalities of a local to regional scope that take place virtually instantaneously. As noted, storm events may preserve local faunal gradients (Miller, paper 3, this volume) or have ecological effects on individual organisms or colonies (paper 5, this volume). Hence these events provide resolution of details at an ecological time scale of days to years. In some cases obrutionary "snapshots" may record details of communities with startling clarity. However, the biased distribution of such short-term events in stratigraphic sections points to a higher-order control on their preservation (see discussions in Brett and Taylor or LoDuca and Brett, paper 8 or paper 14, this volume). Appropriate combinations of environmental parameters for Lagerstätten preservation recur predictably in sedimentary cycles. Evidence suggests that obrution events occur preferentially in certain portions of cycles especially associated with the first sedimentary deposits overlying marine flooding surfaces (see Brett and Taylor, paper 8, this volume). This predictability aids not only in discovery of new Lagerstätten but also in interpretation of sedimentary dynamics.

The link between ecological and/or biogeographic events occurring at relatively short temporal and local geographic scales to larger-scale ecological evolutionary events must be more fully explored. At present there has been a considerable amount of research on large-scale mass extinctions and intermediate-level global bioevents but much less emphasis on smaller-scale or local extinctions, incursions, and originations. Consequently, it may be premature to attempt any synthesis that relates these levels. Nevertheless, we can make some preliminary conjectures about their relationship.

First, groups of single-event beds that preserve extraordinary biotas chiefly because of obrution and associated early diagenetic effects may constitute a type of epibole (i.e., an interval typified by an abundance of a normally rare, absent, or at least unrecognized taxon). These represent *taphonomic epiboles* and are in large measure an artifact of preservation. However, inasmuch as the occurrence of these beds is also related to environmental controls they may also partially reflect ecological conditions (see the discussion by Brett and Baird, paper 10, this volume).

Second, single, very large physical events may in rare cases coincide with larger-scale biotic events. Extremely large ashfalls could produce regional extinction of an entire biota. For example, Sloan (1992) argued that the Middle Ordovician Deike k-bentonite represents a single ashfall that exterminated benthic fauna throughout the upper Mississippi Valley. According to Sloan, recovery required a period of millions of years ! If this is true, then a single event could affect an evolutionary scale change. Another notorious putative example is the possibility that mass extinction at the Cretaceous Tertiary boundary is associated with a single bolide impact event recorded by tsunamis and by a globally distributed layer enriched in iridium, shocked quartz grains, and ash (Alvarez et al. 1984; Bourgeois et al. 1988). Setting aside these types of highly speculative connections, it seems unlikely that most convulsive geologic events have more than local, if short-lived impacts on organism populations or community structure and no influence on longer-term processes. For example, certain Devonian trilobite beds containing clusters of complete phacopid trilobites have been mapped over about 1,000 km^2 (Speyer and Brett 1985). Assuming, as do Speyer and Brett, that these reflect single burial events, and given concentrations of up to hundreds of individuals in single clusters in a few square decimeters, these represent truly catastrophic mortalities of up to several billion individuals. Yet all species involved reappear abundantly in beds a few millimeters to centimeters above the obrution horizons.

Third, the abrupt alteration of substrate associated with sedimentologic events can produce sweeping changes in benthic communities. As noted by Miller et al. (1988) and Kidwell (1991a, 1991b), these events may blanket seafloors with a mantle of fine-grained sediment, or alternatively, turbulence may produce winnowed skeletal lags. The first situation may alter seafloor benthos by favoring soft-substrate vagrant infaunal and semi-infaunal species. In contrast, winnowed skeletal pavements provide opportunities for positive taphonomic feedback, favoring sessile epifaunal taxa. Thus single events may lead to a form of community succession in subtidal seafloors (Johnson 1972; Kidwell and Jablonski 1983). However, in some cases single events may trigger colonization by extraordinary forms, such as specialized burrowers or epifaunal species that are normally rare. Such a case is well illustrated by the Cretaceous x-bentonite of the Western Interior. By blanketing the seafloor this ash layer sealed off seepage of toxic waters and permitted colonization by *Ostrea beloiti* biostromes, which form an epibole horizon (Hattin 1965; Sageman and Johnson 1985). Thus a link may exist between simple winnowing or depositional events and the establishment of short-term epiboles.

Fourth, composite and longer-term hiatal skeletal accumulations may represent extraordinary environmental conditions, especially in terms of very low sediment accumulation rates, low turbidity, firming or lithification of sediment, and so on. These conditions may permit the colonization or proliferation of unusual taxa (see Kauffman 1986). The environmental fluctuations (e.g., relative sea-level rise or fall) that produce composite and hiatal concentrations may also control other ecological attributes (e.g., the upward growth of bioherms). This association permits a link between longer-term sedimentological paleontological event beds and the phenomenon of ecological epiboles (see Brett and Baird, paper 10, this volume).

Finally, the recognition of time-rich condensed intervals is critical in interpreting bioevents because they may introduce spurious punctuations in community structure into stratigraphic sections. Failure to recognize unconformities and condensed intervals may lead to interpretation of abrupt bioevent boundaries of various scales (Schindel 1980, 1982; Sadler, 1981). For example, the abrupt changes in community composition may occur above a major flooding surface because many or all the transitional biofacies are absent because of nondeposition or erosion. In the absence of sedimentation, few relics of these transitional faunas will remain except for rare phosphatic teeth, conodonts, etc. (see Davies et al. 1989). The abrupt "jump" is an artifact of sediment starvation or erosion. Such spuriously abrupt boundaries are even more of a problem in the study of larger bioevents, such as the boundaries of ecological/evolutionary units. This is particularly so because real biological changes may commonly coincide with geological processes, such as major sea-level fluctuations that lead to development of discontinuities in sedimentation and/or erosion.

Thus careful study of skeletal concentrations may be a critical prerequisite to study of bioevents in local sections. Short-term storm concentrations must be distinguished from ecologically condensed or even long-term biostratigraphically condensed beds. Taphonomic attributes of contained fossils may permit an estimate of scales of time-averaging in skeletal accumulations (Kidwell 1982; Brandt-Velbel 1984). If condensed intervals are recognized in a section, then they may provide an obvious explanation for seemingly punctuational patterns in community structure or morphology within species. Hence study of concentration Lagerstätten may elucidate limitations on the resolution of bioevents. In any case, the possibility of major discontinuities must be worked out before bioevents at any scale can be viewed as abrupt.

On the other hand, recognition of condensed and hiatal shell beds may prove very useful in delineating boundary surfaces, essential to sequence stratigraphy. One of the most important aspects of the sequence approach is an attempt to distinguish "time-rich" process beds or surfaces from rapidly and continuously deposited "time-poor" successions. In this regard, detailed study of skeletal beds has a very key role to play.

Summary

Fossil Lagerstätten, as discussed earlier, are expressed in many different ways in the rock record and differ greatly in terms of their sedimentological and paleontological significance. Fossil assemblages can be single events recording mass mortality usually, but not always, involving rapid burial, or multievent concentrations. However, they can also be longer-term multievent concentrations involving the physical mixing of different communities, the stacking of storm beds, successional in situ communities and/or taphonomic feedback. Finally, skeletal concentrations can be time-rich assemblages associated with sediment starvation, winnowing, and erosion. Hiatal and lag concentrations may erroneously appear to evolutionary workers as times marked by extinction or intervals of rapid community restructuring when, in fact, most time is bound up in such units. However, such concentrations can also be linked through sequence stratigraphic work to global paleoclimate and sea-level events that presumably may have coincided with important evolutionary events.

We observe that Lagerstätten distribution is distinctly nonrandom in transgressive regressive cycles; types 1 and 2 storm obrution beds are most likely to occur in the regressive or lowstand portions of cycles, and types 3 and 4 beds are to be formed in the early to late highstand phase. More significantly, preservation of obrution Lagerstätten is also nonrandom in cycles. The greatest probability of their preservation is in transgressive systems tract intervals and in highstand systems tract successions, owing to abundance of accommodation space and high preservation potential. Conversely, time-averaged, multievent shell beds and longer-term winnowed hiatal and lag beds should predominate in the regression to lowstand phase of cycles, owing to reduced accommodation space for accumulation, sediment bypass effect, and erosion.

Paleontological event horizons reflect a wide range of taphonomic, ecological, and evolutionary processes that together yield discrete, sometimes widespread marker beds. These have been recognized to some extent and used by stratigraphers for decades. As we have noted, very early pioneering workers noted many key bioevents in defining the concepts of zonal acme and stage that we are now beginning to rediscover. Until recently, knowledge of these beds has largely escaped the attention of paleoecologists and evolutionary paleobiologists. Herein, we underscore the empirical reality of these beds, which have at least a twofold significance. First, paleontological event horizons provide an outstanding tool and, together with physical and chemical events, allow for the refinement of regional correlations of stratigraphy. Even without fully understanding the genesis of these event horizons, they can be used to help construct a high-resolution-event stratigraphy (see Kauffman 1988). Second, the recognition of widespread events demands a thorough investigation of their underlying taphonomic and biotic causes. What is the real meaning of mass mortality horizons, epiboles, and condensed beds, not to mention mass extinction horizons? Obviously, we are only just beginning to find out answers to these questions.

ACKNOWLEDGMENTS

We wish to express sincere thanks to Brad Sageman and David Lehmann for critical reviews of the manuscript. Mary Nardi assisted with technical editing. Heidi Kimble and Margaret Teichmann processed the manuscript. Wendy Taylor aided in figure preparation. Our research has been generously supported by grants from the donors to the Petroleum Research Fund, American Chemical Society, and by NSF grants EAR 8313103 and 8816856.

References

Aepler, R. and W. E. Reif. 1971. Origin of bone-beds. In *Abstracts VIII, International Sedimentology Congress*, vol. 1.

Aigner, T. 1985. *Storm Depositional Systems: Dynamic Stratigraphy in Modern and Ancient Shallow-Marine Sequences. Lecture Notes in the Earth Sciences*, vol. 3. Berlin: Springer-Verlag.

Alexander, R. R. 1986. Life orientation and post-mortem reorientation of Chesterian brachiopod shells by paleocurrents. *Palaios* 1:303–11.

Allison, P. A. 1986. Soft-bodied animals in the fossil record: The role of decay in fragmentation during transport. *Geology* 14:139–54.

———. 1988. The decay and mineralization of proteinaceous macrofossils. *Paleobiology* 14:139–54.

———. 1990. Variation in rate of decay and disarticulation in echinodermata: implications for taphonomic experiments. *Palaios* 5:432–40.

———, and D. E. G. Briggs. 1991. A taphonomy of nonmineralized tissues. In P. A. Allison and D. E. G. Briggs, eds., *Taphonomy: Releasing the Data Locked in the Fossil Record. Topics in Geobiology*, vol. 9. New York: Plenum Press.

Allmon, R. A. 1985. "Butterflied" bivalves as paleoenvironmental indicators. *Geological Society of America Abstracts with Programs* 17:512.

Alvarez, L. W., W. Alvarez, F. Asaro, and H. V. Michel. 1980. Extraterrestrial cause for the Cretaceous–Tertiary extinction. *Science* 208:1095–1108.

Alvarez, W., E. G. Kauffman, F. Surlyk, L. W. Alvarez, F. Asaro, and H. V. Michel. 1984. Impact theory of mass extinction and the invertebrate fossil record. *Science* 223:1135–41.

Baird, G. C. 1978. Pebbly phosphorites in shale: A key to recognition of a widespread submarine discontinuity. *Journal of Sedimentary Petrology* 48:105–22.

———. 1981. Submarine erosion on a gentle paleoslope: A study of two discontinuities in the New York Devonian. *Lethaia* 14:105–22.

Baird, G. C. and C. E. Brett. 1986. Erosion on the anaerobic sea floor: Significance of reworked pyrite deposits from the Devonian of New York State. *Palaeogeography, Palaeoclimatology, Palaeoecology* 57:157–93.

———. 1991. Submarine erosion on the anoxic seafloor: Stratinomic palaeoenvironmental and temporal significance of reworked pyrite-bone deposits. In R. V. Tyson and T. H. Pearson, eds., *Modern and Ancient Continental Shelf Anoxia. Geological Society Special Publication* 58:233–58.

Baird, G. C., S. D. Sroka, C. W. Shabica, and G. J. Kuecher. 1986. Taphonomy of Middle Pennsylvanian Mazon Creek fossil localities, northeast Illinois: Significance of exceptional fossil preservation in syngenetic concretions. *Palaios* 1:271–85.

Banerjee, I. and S. Kidwell. 1991. Significance of molluscan shell beds in sequence stratigraphy: Example from the Lower Cretaceous Mannville Group of Canada. *Sedimentology* 38:913–34.

Bjerreskov, M. 1991. Pyrite in Silurian graptolites from Bornholm, Denmark. *Lethaia* 24:351–61.

Bourgeois, J., T. A. Hansen, P. L. Wiberg,, and E. G. Kauffman. 1988. A tsunami deposit at the Cretaceous-Tertiary boundary in Texas. *Science* 241:561–70.

Brandt-Velbel, D. 1984. On defining limits to paleoecological interpretation in the fossil record. *Géobios Mémoire Spéciale* 8:415–18.

Brandt, D. S. 1985. Ichnologic, taphonomic, and sedimentologic clues to the deposition of Cincinnatian Shales (Upper Ordovician) Ohio, U.S.A. In *Society of Economic Paleontologists and Mineralogists Special Publication*, pp. 299–307.

———. 1989. Taphonomic grades as a classification for fossilferous assemblages and assemblages and implications for paleoecology. *Palaios* 4:303–309.
Brett, C. E. 1990. Obrution deposits. In D. E. G. Briggs and P. R. Crowther, eds., *Palaeobiology: A Synthesis*, pp. 239–43. Oxford, England: Blackwell Scientific Publications.
Brett, C. E. and G. C. Baird. This volume. Epiboles, outages and ecological evolutionary bioevents: Taphonomic, ecological and biogeographical factors. In C. E. Brett and G. C. Baird, eds., *Paleontological Events: Stratigraphic, Ecologic, and Evolutionary Implications*. New York: Columbia University Press.
———. 1986. Comparative taphonomy: A key to paleoenvironmental interpretation based on fossil preservation. *Palaios* 1:207–27.
———. 1990. Submarine erosion and condensation in a foreland basin: Examples from the Devonian of Erie County, New York. *New York State Geological Association 62nd Annual Meeting Field Trip Guidebook*, pp. A1–A56.
Brett, C. E. and Y. L. Bordeaux. 1990. Taphonomy of brachiopods from a Middle Devonian shell bed: Implications for the genesis of skeletal accumulations. *Proceedings of the Second International Brachiopod Congress*. Dunedin, New Zealand: Balkema Press.
Brett, C. E. and M. E. Brookfield. 1984. Morphology, faunas, and genesis of Ordovician hardgrounds from Southern Ontario, Canada. *Paleogeography, Palaeoclimatology, Palaeoecology* 46:233–90.
Brett, C. E., V. B. Dick,, and G. C. Baird. 1991. Comparative taphonomy and paleoecology of Middle Devonian dark gray and black shale facies from western New York. In E. Landing and C. E. Brett, eds., Dynamic stratigraphy and depositional environments of the Hamilton Group (Middle Devonian) in New York State. *New York State Museum Bulletin* 469:5–36.
Brett, C. E. and W. D. Liddell, 1978. Preservation and paleoecology of a Middle Ordovician hardground community. *Paleobiology* 4:329–48.
Brett, C. E. and A. Seilacher. 1991. Fossil Lagerstätten: A taphonomic consequence of event sedimentation. In: G. Einsele, W. Ricken, and A. Seilacher, eds., *Cycles and Events in Stratigraphy*. New York: Springer-Verlag.
Brett, C. E. and S. E. Speyer. 1989. Facies distribution of obrution deposits: Toward a predictive model for fossil burial events. *28th International Geological Congress Abstracts*, Washington, D.C., pp. 1–197.
Brett, C. E., S. E. Speyer, and G. C. Baird. 1986. Storm-generated sedimentary units: Tempestite proximality and event stratification in the Middle Devonian Hamilton Group of New York. In C. E. Brett, ed., Dynamic stratigraphy and depositional environments of the Hamilton Group (Middle Devonian) in New York State, part I. *New York State Museum Bulletin* 457:129–55.
Brett, C. E. and W. L. Taylor. This volume. The Homocrinus beds: Silurian crinoid Lagerstätten of western New York and southern Ontario. In C. E. Brett and G. C. Baird, eds., *Paleontologic Events: Stratigraphic, Ecologic, and Evolutionary Implications*. New York: Columbia University Press.
Briggs, D. E. G., S. H. Bottrell, and R. Raiswell. 1991. Pyritization of soft-bodied fossils: Beecher's trilobite bed Upper Ordovician, New York State. *Geology* 19:1221–24.
Brongersma-Sanders, M. 1957. Mass mortality in the sea. In J. W. Hedgepeth, ed., Treatise on Marine Ecology and Paleoecology, *Geological Society of America, Memoir No. 67*, pp. 941–1010.
Brower, J. C. and O. B. Nye, Jr. 1991. Quantitative analysis of paleocommunities in the lower part of the Hamilton Group near Cazenovia, New York. In E. Landing and C. E. Brett, eds., Dynamic stratigraphy and depositional environments of the Hamilton Group (Middle Devonian) in New York State, part II. *New York State Museum Bulletin* 469:37–75.

Brower, J. D., and J. Veinus. 1978. Middle Ordovician crinoids from the Twin Cities area of Minnesota, *Bulletins of American Paleontology* 74(304):372–506.
Busch, R. G. and H. B. Rollins. 1984. Correlation of Carboniferous strata using a hierarchy of transgressive-regressive units, *Geology* 12:471–74.
Cadée, G. C. 1984. Macrobenthos and macrobenthic remains on the Oyster Ground, North Sea, *Netherlands Journal of Sea Research* 18:160–78.
Canfield, D. E. and R. Raiswell. 1991a. Pyrite formation and fossil preservation. In P. A. Allison and D. E. G. Briggs, eds., *Taphonomy: Releasing the Data Locked in the Fossil Record*. Topics in Geobiology, vol. 9. New York: Plenum Press.
———. 1991b. Carbonate precipitation and dissolution: Its relevance to fossil preservation. In P. A. Allison and D. E. G. Briggs, eds., *Taphonomy: Releasing the Data Locked in the Fossil Record. Topics in Geobiology*, vol. 9. New York: Plenum Press.
Cherns, L. 1988. Faunal and facies dynamics in the upper Silurian of the Anglo-Welsh Basin. *Palaeontology* 31:451–502.
Cisne, J. L. 1973. Beecher's trilobite bed revisited: Ecology of an Ordovician deep water fauna. *Postilla* 160:25.
Cisne, J. L. and B. D. Rabe. 1978. Coenocorrelation: Gradient analysis of fossil communities and its applications to stratigraphy. *Lethaia* 11:341–64.
Clifton, H. E., ed. 1988. Sedimentologic consequences of convulsive geologic events. *Geological Society of American Special Paper* 229:157.
Conway Morris, S. 1990. Late Precambrian and Cambrian soft-bodied faunas. *Annual Reviews and Planetary Sciences* 18:101–22
Cuffey, R. J. This volume. *Prasopora*—bearing event for the Coburn Limestone (Bryozoan; Ordovician; Pennsylvania). In C. E. Brett and G. C. Baird, eds., *Paleontological Events: Stratigraphic, Ecologic, and Evolutionary Implications*. New York: Columbia University Press.
Cutler, A. H. and K. W. Flessa. 1990. Fossils out of sequence: computer simulations and strategies for dealing with stratigraphic disorder. *Palaios* 5:227–35.
Davies, D. J., E. N. Powell, and R. J. Stanton, Jr. 1989. Relative rates of shell dissolution and net sediment accumulation-a commentary: Can shell beds form by the gradual accumulation of biogenic debris on the seafloor? *Lethaia* 22:207–212
Eckert, B. Y. and C. E. Brett. 1989. Bathymetry and palaeoecology of Silurian benthic assemblages, Late Llandoverian, New York State. *Palaeogeography, Palaeoclimatology, Palaeoecology* 74:297–26.
Einsele, G., D. Herm, and H. U. Schwartz. 1974. Holocene eustatic sea level fluctuation at Mauritania, *Meteor Forschungsarbeit Reihec* 18:43–62.
Fisher, I. St. J. 1986. Pyrite replacement of mollusc shells from the Lower Oxford Clay (Jurassic) of England, *Sedimentology* 33:575–85.
Flessa, K. W., A. H. Cutler, and K. H. Meldahl. 1993. Time and taphonomy: Quantitative estimates of time-averaging and stratigraphic disorder in a shallow marine habitat. *Paleobiology* 19:266–86.
Fürsich, F. T. 1971. Hartgründe und Kondensation im Dogger von Calvados. *Neues Jahrbuch für Geologie und Paläeontologie, Abhandlungen* 138:313–42.
———. 1978. The influence of faunal condensation and mixing on the preservation of fossil benthic communities, *Lethaia* 11:243–50.
———. 1979. Genesis, environments and ecology of Jurassic hardgrounds. *Neues Jahrbuch für Geologie und Paläeontologie, Abhandlungen* 158:1–63.
Greenstein, B. J. 1989. Mass mortality of the West-Indian echinoid *Diadema antillarum* (echinodermata: echinoidea): A natural experiment in taphonomy. *Palaios* 4:487–92.
Guensburg, T. E. 1984. Echinodermata of the Middle Ordovician Lebanon Limestone, central Tennessee. *Bulletins of American Paleontology* 86(319):1–100.

Gunther, L. F. and L. I. Gunther. 1981. Some Middle Cambrian fossils of Utah. *Brigham Young University Geological Studies* 28:1–87.
Hass, W. H. 1959. Conodonts from the Chappel Limestone of Texas. *U.S. Geological Survey Professional Paper* 29 A-J:365–99.
Hattin, D. E. 1965. Stratigraphy of the Graneros Shale (Upper Cretaceous) in central Kansas. *Kansas Geological Survey Bulletin* 178:83.
Hickerson, W. J. This volume. Middle Devonian trilobite cluster from Rock Island County, Illinois. In C. E. Brett and G. C. Baird, eds., *Paleontological Events: Stratigraphic, Ecological and Evolutionary Implications.* New York: Columbia University Press.
Holland, S. M. 1988. Taphonomic effects of seafloor exposure on an Ordovician brachiopod assemblage. *Palaios* 3:588–97.
Horowitz, A. S. and J. A. Waters. 1972. A Mississippian echinoderm site in Alabama. *Journal of Paleontology* 46:660–65.
Hudson, J. D. 1982. Pyrite in ammonite-bearing shales from the Jurassic of England and Germany. *Sedimentology* 25:339–69.
Jennette, D. C. and W. A. Pryor. 1993. Cyclic alternations of proximal and distal storm facies on a prograding ramp: Examples from the Kope and Fairview Formations (Upper Ordovician), Ohio and Kentucky. *Journal of Sedimentary Petrology* 63:183–203.
Johnson, R. G. 1972. Conceptual models of benthic marine communities. In T. J. M. Schopf, ed., *Models in Paleobiology.* San Francisco: Freeman Cooper.
Kauffman, E. G. 1981. Ecological reappraisal of the German Posidonienschiefer (Toarcian) and the stagnant basin model. In J. Gray, A. J. Boucot, and W. B. N. Berry, eds., *Communities of the Past.* Stroudsburg, Pa.: Hutchinson Ross.
———. 1986. High-resolution event stratigraphy: Regional and global Cretaceous bioevents. In O. Walliser, ed., *Global Bioevents,* vol. 8. New York: Springer-Verlag.
———. 1988. Concepts and methods of high-resolution event stratigraphy. *Annual Review of Earth Planetary Sciences* 16:605–54.
Kauffman, E. G. and O. H. Walliser. 1990. *Extinction Events in Earth History.* Lecture Notes in Earth Sciences, vol. 30. New York: Springer-Verlag.
Kidwell, S. M. 1982. Time scales of fossil accumulation: Patterns from Miocene benthic assemblages. *Proceedings of the Third North American Paleontological Convention, Montreal* 1:295–300.
Kidwell, S. M. 1991a. The stratigraphy of shell concentrations. In P. Allison and D. E. G. Briggs, eds., *Taphonomy: Releasing the Data Locked in the Fossil Record.* New York: Plenum Press.
———. 1991b. Taphonomic feedback (live/dead interactions) in the genesis of bioclastic beds: Keys to reconstructing sedimentary dynamics. In G. Einsele, W. Ricken, and A. Seilacher, eds., *Cycles and Events in Stratigraphy.* New York: Springer-Verlag.
———. 1991c. Condensed deposits in siliciclastic sequences: Expected and observed features. In G. Einsele, W. Ricken, and A. Seilacher, eds., *Cycles and Events in Stratigraphy.* New York: Springer-Verlag.
Kidwell, S. M. and T. Aigner. 1985. Sedimentary dynamics of complex shell beds: Implications for ecologic and evolutionary patterns. In U. Bayer and A. Seilacher, eds., *Sedimentary and Evolutionary Cycles. Lecture Notes in Earth Sciences,* vol. 1, pp. 382–95. New York: Springer-Verlag.
Kidwell, S. M. and D. Bosence. 1991. Taphonomy and time-averaging of marine shelly faunas. In P. A. Allison and D. E. G. Briggs, eds., *Taphonomy: Releasing the Data Locked in the Fossil Record.* New York: Plenum Press.
Kidwell, S. M. and T. Baumiller. 1990. Experimental disintegration of regular echinoids: Roles of temperature, oxygen and decay thresholds. *Paleobiology* 16:247–71.
Kidwell, S. M., F. T. Fürsich,, and T. Aigner, 1986. Conceptual framework for the analysis of fossil concentrations. *Palaios* 1:228–38.

Kidwell, S. M. and D. Jablonski. 1983. Taphonomic feedback: Ecological consequences of shell accumulation. In M. J. S. Tevesz and P. L. McCall, eds., *Biotic Interactions in Recent and Fossil Benthic Communities*. New York: Plenum Press.

Koch, D. L. and H. L. Strimple. 1968. A new Upper Devonian cystoid attached to a discontinuity surface. *Iowa Geological Survey, Report of Investigations* 5:1–49.

Kranz, P. M. 1974. The anastrophic burial of bivalves and its paleoecological significance. *Journal of Geology* 82:237–65.

Kreisa, R. D. 1981. Storm-generated sedimentary structures in subtidal marine facies with examples from the Middle and Upper Ordovician of southwestern Virginia. *Journal of Sedimentary Petrology* 51:823–48.

Kreisa, R. D. and R. K. Bambach. 1982. The role of storm processes in generating shell beds in Paleozoic shelf environments. In G. Einsele and A. Seilacher, eds., *Cyclic and Event Stratification*. New York: Springer-Verlag.

Lafferty, A., A. I. Miller, and C. E. Brett, 1994. Spatial variability in faunal composition along two Middle Devonian paleoenvironmental gradients. *Palaios* 9:224–36.

Landing, E. and C. E. Brett. 1987. Trace fossils and regional significance of a Middle Devonian (Givetian) disconformity in southwestern Ontario. *Journal of Paleontology* 61:205–30.

Laudon, L. R. 1939. Unusual occurrences of *Isotelus gigas* Dekay in the Bromide Formation (Ordovician) of southern Oklahoma. *Journal of Paleontology* 13:211–13.

Liddell, W. D. 1975. Recent crinoid biostratinomy. *Geological Society of America, Abstracts with Programs* 1:1169.

LoDuca, S. T. and C. E. Brett. This volume. Conservation Lagerstätten in the Late Silurian of eastern North America and the *Medusaegraptus* epibole. In C. E. Brett and G. C. Baird, eds., *Paleontological Events: Stratigraphic, Ecologic and Evolutionary Implications*. New York: Columbia University Press.

Loutit, T. S., J. Hardenbol, P. R. Vail, and P. Baum. 1988. Condensed sections: The key to age dating and correlation of continental margin sequences. In C. Wilgus et al., eds., *Sea-Level Changes: An Integrated Approach. Society of Economic Paleontologists and Mineralogists Special Publication* No. 42, pp. 183–213.

Lucas, J. and L. E. Prévôt. 1991. Phosphates and fossil preservation. In P. A. Allison and D. E. G. Briggs, eds., *Taphonomy: Releasing the Data Locked in the Fossil Record*. New York: Plenum Press.

Lyons, T. 1992. Contrasting sediment types from oxic and anoxic sites of the modern Black Sea: Geochemical and sedimentological criteria. Ph.D. dissertation, Yale University.

Masaurel, H. 1987. Macrofossils and their paleoecology in deltaic sequences of the Lower Carboniferous Yoredale Series, Yorkshire, England, *Geologie Mijnboukundige* 66:221–37.

Meldahl, K. H. 1987. Sedimentologic and taphonomic implications of biogenic stratification. *Palaios* 2:350–58.

Merrill, G. K. 1980. Road log—Day two. In *Geology of the Llano Region, Central Texas, Guidebook to the Annual Field Trip of the West Texas Geological Society*, pp. 60–199.

Meyer, D. L. 1971. Post-mortem disarticulation of recent crinoids and ophiuroids under natural conditions. *Geological Society of America, Abstracts with Programs* 3:645.

———. 1991. Population paleoecology and comparative taphonomy of two edrioasteroid pavements: Upper Ordovician of Kentucky and Ohio. *Historical Biology* 4:155–78.

Meyer, D. L., A. I. Ausich, and R. I. Terry. 1989. Comparative taphonomy of echinoderms in carbonate facies: Fort Payne Formation (Lower Mississippian) of Kentucky and Tennessee. *Palaios* 4:533–52.

Meyer, D. L., R. C. Tobin, W. A. Pryor, W. B. Harrison, and R. G. Osgood. 1981. Stratigraphy, sedimentology, and paleoecology of the Cincinnatian Series (Upper Ordovi-

cian) in the vicinity of Cincinnati, Ohio. In T. G. Roberts, ed., *Geological Society of America, Cincinnati. 1981 Field Trip Guidebooks*, vol. 1, American Geological Institute, Virginia, pp. 31–71.
Miller, A. I., this volume. Counting fossils in a Cincinnatian storm bed: Spatial resolution in the fossil record. In C. E. Brett and G. C. Baird, eds., *Paleontological Events: Stratigraphic, Ecologic, and Evolutionary Implications.* New York: Columbia University Press.
Miller, K. B., C. E. Brett, and K. M. Parsons. 1988. The paleoecological significance of storm-generated disturbance within a Middle Devonian epeiric sea. *Palaios* 3:35–52.
Mortimore, R. N. and R. Pomerol. 1991. Stratigraphy and eustatic implications of trace fossil events in the Upper Cretaceous chalk of northwestern Europe. *Palaios* 6:216–32.
Oschmann, W. 1991. Poikiloaerobic-Aerobic: A new facies zonation for modern and ancient neritic facies. In G. Einsele, W. Ricken, and A. Seilacher, eds., *Cycles and Events in Stratigraphy.* New York: Springer-Verlag.
Parsons, K. M., C. E. Brett, and K. B. Miller. 1988. Taphonomy and depositional dynamics of Devonian shell-rich mudstones. *Palaeogeography, Palaeoclimatology, Palaeoecology* 63:108–139
Pemberton, S. G. This volume. The ichnological signature of storm deposits: The use of trace fossils in event stratigraphy. In C. E. Brett and G. C. Baird, eds., *Paleontologic Events: Stratigraphic, Ecologic, and Evolutionary Implications.* New York: Columbia University Press.
Pilkey, O. H. and H. A. Curran. 1986. Molluscan shell transport: You ain't seen nothin' yet. *Palaios* 1:197.
Plotnick, R. E. 1986. Taphonomy of a modern shrimp: Implications for the arthropod fossil record. *Palaios* 1:286–93.
Reif, W. E. 1982. Muschelkalk/Keuper bone-beds (Middle Triassic, s.w. Germany) storm condensation in a regressive cycle. In G. Einsele and A. Seilacher, eds., *Cyclic and Event Stratification.* New York: Springer-Verlag.
Sadler, P. J. 1981. Sediment accumulation rates and the completeness of stratigraphic section. *Journal of Geology* 89:569–84.
Sageman, B. B. and C. C. Johnson. 1985. Stratigraphy and paleobiology of the Lincoln limestone member, Greenhorn limestone, Rock Canyon Anticline, Colorado. In L. M. Pratt, E. G. Kauffman, and F. B. Zelt, eds. Fine-grained deposits and biofacies of the Cretaceous Western Interior Seaway: Evidence of cyclic sedimentary processes. *Society for Economic Paleontology and Mineralogy Field Trip Guidebook* 4:100–109.
Sageman, B. B., B. P. Kauffman, and E. G. Wignall. 1991. Biofacies model for oxygen-deficient facies in epicontinental seas: Tool for paleoenvironmental analysis. In G. Einsele, W. Ricken, and A. Seilacher, eds., *Cycles and Events in Stratigraphy.* New York: Springer-Verlag.
Savrda, C. E. and D. J. Bottjer. 1987. The exaerobic zone, a new oxygen-deficient marine biofacies. *Nature* 327:54–56.
———. 1991. Oxygen-related biofacies in marine strata: An overview. In R. V. Tyson and T. H. Pearson, eds., Modern and ancient continental shelf anoxia. *Geological Society Special Publication* 58:291–310.
Schäfer, W. 1972. *Ecology and Palaeoecology of Marine Environments.* Chicago: University of Chicago Press, 568 p.
Schindel, D. E. 1980. Micro-stratigraphic sampling and the limits of paleontological resolution. *Paleobiology* 6:408–26.
———. 1982. Resolution analysis: A new approach to the gaps in the fossil record. *Paleobiology* 8:340–53.
Seilacher, A. 1973. Biostratinomy the sedimentology of biologically standardized parti-

cles. In R. N. Ginsburg, ed., *Evolving Concepts in Sedimentology*. Baltimore: The Johns Hopkins University Press.

———. 1982a. Posidonia shale (Toarcian, s. Germany)—stagnant basin revalidated. In E. Montanaro-Gallitelli, ed., *Paleontology: Essentials of Historical Geology*, Modena.

———. 1982b. General remarks about event deposits. In G. Einsele and A. Seilacher, eds., *Cycles and Events in Stratigraphy*. New York: Springer-Verlag.

———. 1985. The Jeram model: Event condensation in a modern intertidal environment. In U. Bayer and A. Seilacher, eds., *Sedimentary and Evolutionary Cycles*. New York: Springer-Verlag.

Seilacher, A. and Hemleben, C. 1966. Spurenfauna und Bildungstiefe der Hunsrückschiefer Unterdeum), *Notizblatt Landesamt Bodenforschung* 94:40–53.

Seilacher, A., W. E. Reif, and F. Westphal. 1985. Sedimentological, ecological and temporal patterns of fossil Lagerstätten. In H. B. Whittington and S. Conway Morris, eds., Extraordinary biotas: Their ecological and evolutionary significance, *Philosophical Transactions, Royal Society of London B* 311:5–23.

Sloan, R. E. 1992. The Deike K-bentonite and the repopulation of the Trenton Sea, *Geological Society of America, Abstracts with Programs* 24(7):A197.

Snyder, J. and P. W. Bretsky. 1971. Life habits of diminutive bivalve molluscs in the Maquoketa Formation (Upper Ordovician). *American Journal of Science* 271:227–51.

Speyer, S. E. 1987. comparative taphonomy and paleoecology of trilobite Lagerstätten. *Alcheringa* 11:205–32.

Speyer, S. E. and C. E. Brett. 1985. Clustered trilobite assemblages in the Middle Devonian Hamilton Group. *Lethaia* 18:85–103.

———. 1988. Taphofacies models for epeiric sea environments: Middle Paleozoic examples. *Palaeogeography, Palaeoclimatology, Palaeoecology* 63:225–62.

Thompson, J. B. and C. R. Newton. 1987. Ecological reinterpretation of the dysaerobic Leiorhynchus fauna: Upper Devonian Geneseo black shale, central New York. *Palaios* 2:274–81.

Vail, P. R., F. Audemard, S. A. Bowman, P. N. Eisner, and C. Perez-Cruz. 1991. The stratigraphic signatures of tectonics, eustacy and sedimentology-an overview. In G. Einsele, W. Ricken, and A. Seilacher, eds., *Cycles and Events in Stratigraphy*. New York: Springer-Verlag.

Van Wagoner, J. C., H. W. Posamentier, R. M. Mitchum, P. R. Vail, J. F. Sarg, T. S. Loutit, and J. Hardenbol. 1988. An overview of the fundamentals of sequence stratigraphy and key definitions, *Society of Economic Palaeontologists and Mineralogists Special Publication* No. 42, pp. 39–45.

Waage, K. M. 1964. Origin of repeated fossiliferous concretion layers in the Fox Hills Formation. *Kansas Geological Survey Bulletin* 169:541–63.

Walker, K. R. and L. P. Alberstadt. 1975. Ecological succession as an aspect of structure in fossil communities. *Paleobiology* 1:238–57.

Walker, K. R. and W. W. Diehl. 1985. The role of cementation in the preservation of Lower Paleozoic assemblages. *Philosophical Transactions of the Royal Society of London* B311:143–53.

Walker, K. R. and W. C. Parker. 1976. Population structure of a pioneer and a later stage species in an Ordovician ecological succession. *Paleobiology* 2:191–201.

Walliser, O. H., ed. 1986a. *Global Bio-events*. Lecture Notes in Earth Sciences. New York: Springer-Verlag.

———. 1986b. Towards a more critical approach to bio-events. In O. H. Walliser, ed., *Global Bio-events: A Critical Approach. Lecture Notes in Earth Sciences*. New York: Springer-Verlag.

———. 1990. How to define "global bioevents." In E. G. Kauffman, and O. H. Walliser, eds., *Extinction Events in Earth History, Lecture Notes in Earth Science*, vol. 30, pp. 1–4. New York: Springer-Verlag.

Webby, B. D. and I. G. Percival. 1983. Ordovician trimerellacean brachiopod shell beds. *Lethaia* 16:215–32.

Wendt, J. 1988. Condensed carbonate sedimentation in the Late Devonian of the eastern Anti-Atlas (Morocco). *Ecologae Geologic Helvetica* 81:155–73.

Wendt, J. and T. Aigner. 1982. Condensed Griotte facies and cephalopod accumulations in the Upper Devonian of the eastern Anti-Atlas, Morocco. In G. Einsele and A. Seilacher, eds., *Cyclic and Event Stratification.* New York: Springer-Verlag.

Wendt, J. and T. Aigner. 1985. Facies patterns and depositional environments of Paleozoic cephalopod limestones. *Sedimentary Geology* 44:263–300.

Wendt, J., T. Aigner, and J. Neugebauer. 1984. Cephalopod limestone deposition on a shallow pelagic ridge: The Tifilalt Platform (Upper Devonian), eastern Anti-Atlas, Morocco). *Sedimentology* 31:601–25.

Whittington, H. B. and S. Conway Morris, eds. 1985. Extraordinary fossil biotas: Their ecological and evolutionary significance. *Philosophical Transactions of the Royal Society of London* B311:192.

Wignall, P. B. and A. Hallam. 1991. Biofacies, stratigraphic distribution and depositional models of British onshore Jurassic black shales. In R. V. Tyson and T. H. Pearson, eds., *Modern and Ancient Continental Shelf Anoxia, Geological Society Special Publication* No. 58, pp. 291–310.

Williams, S. H. and R. G. Rickards. 1984. Paleoecology of graptolitic black shales. In D. L. Bruton, ed., *Aspects of the Ordovician System, Paleontological Contributions of the University of Oslo* No. 295, pp. 154–66.

Wilson, M. A. and T. J. Palmer. 1992. Hardgrounds and hardground faunas, *University of Wales, Aberystwyth, Institute of Earth Studies Publications* 9:1–131.

Wolosz, T. H. This volume. Thicketing events: A key to understanding the ecology of the Edgecliff reefs (Middle Devonian Onondaga Formation of New York and Ontario, Canada. In C. E. Brett and G. C. Baird, eds., *Paleontological Events: Stratigraphic, Ecologic and Evolutionary Implications.* New York: Columbia University Press.

Zangerl, R. and E. S. Richardson, Jr. 1963. The Paleoecological history of two Pennsylvanian Black Shales, *Fieldiana Geological Memoir* No. 4, p. 352.

2

Taphonomic Constraints on Event Horizons: Short-Term Time Averaging of Anadara brasiliana *Valves, St. Catherines Island, Georgia*

Harold B. Rollins and Ronald R. West

EVENT horizons, the products of geologically rapid processes, have justifiably received special attention from stratigraphers and paleoecologists. However, they may record many different temporal scales, ranging from a few hours to hundreds of thousands of years. They may result from diverse physical, chemical, and/or paleobiological phenomena, including storms, changes in ocean currents or stratification, tectonic uplift, volcanism, punctuational evolution, faunal or floral immigration or emigration, population "bursts," and catastrophic burial. As noted by Seilacher (1982) and Miller et al. (1988), event processes may have effects that extend all the way from individual organisms to biofacies. Resulting deposits may dominate the sedimentary rock record (Kauffman 1986,1988; Dott 1983). Recognition of short-term effects not only is important for stratigraphic correlation, but strongly influences our perceptions of benthic community patterns, community evolution, basin analysis, evolutionary models, and so on (Miller et al. 1988; Seilacher 1982: Davies et al. 1989; Ruffell 1988).

Event horizons can be double-edged swords for the paleoecologist. Rapid burial is valued for harboring ecological census data and primary information on intra- and interspecific interactions. On the other hand, such short-term processes as storms commonly impart profoundly disruptive taphonomic signatures upon death assemblages, and the frequency of storms may dramatically affect the retrieval of ecological information from the marine benthic environment (Brett and Baird 1986; Miller et al. 1988; Parsons et al. 1986). Generally, careful observation of the taphonomic attributes of a fossil assemblage permits a reasonably accurate assessment of contained ecological information, although recent paleoecological literature contains many examples of subtle, even cryptic, nuances of taphonomic overprinting (Kidwell and Jablonski 1983; Kidwell 1986; Cummins et al. 1986; West and Rollins 1989; Powell et al. 1989).

The goal of this paper is to illustrate how a taphonomically imprinted death assemblage, the product of only days of formation, can appear to be ecologically intact. In a sense the study waves a cautionary flag at unquestioned acceptance of even some of the more convincing conventional indicators of an in situ assemblage.

St. Catherines Island, Georgia

Recent shell deposits of the ark *Anadara brasiliana* were examined along Middle Beach, St. Catherines Island, Georgia. St. Catherines Island is a barrier island nestled in the apex of the Georgia Bight about 30 miles southeast of Savannah (fig. 2.1). The island experiences a mean tidal range of 2.4 m (Howard and Frey 1980; Frey and Howard 1988). This part of the Georgia coast is protected from intense wave activity by an extensive shallow shelf, and in a bathymetric sense it resembles many epeiric foreland basins of the geologic record. St. Catherines Island is bordered on the west by an extensive salt marsh network and on the east by long, narrow quartz sand beaches punctuated by two major oceanic inlets, Seaside and McQueens (fig. 2.1). Middle Beach, situated between the two inlets, has extensively developed overwash storm deposits and surf-beveled relict marsh muds (Deery and Howard 1977; Pemberton and Frey 1985; Sherrod et al. 1989; West et al. 1990). This beach receives the brunt of storm energy reaching the island, but wave energy is normally slight, about 0.25 m amplitude during nonstorm conditions (Kuroda and Marland 1973). Storm activity is greater during winter months, which have predominantly northeasterly wind and littoral drift patterns. Southeasterly drift is common during the summer. Relict marsh mud deposits along Middle Beach preclude easy characterization of energy relationships along the littoral shoreface. The spectrum of energy distribution ranges from protected runnels behind relict mud exposures to higher-energy surge channels funneled between promontories of relict mud. As such, the exposed stretches of sand along Middle Beach do not fit into existing zonation schemes, such as that of McLachlan (1990). On the other hand, as noted by Frey and Howard (1988:638), "omission surfaces and palimpsest substrates may be key records of incipient transgressive pulses within paralic sequences of this type."

Wrack Line Transects of Anadara brasiliana

A large population of the ark *Anadara brasiliana* inhabits the shallow subtidal sands along Middle Beach. After every tidal exchange empty valves of this

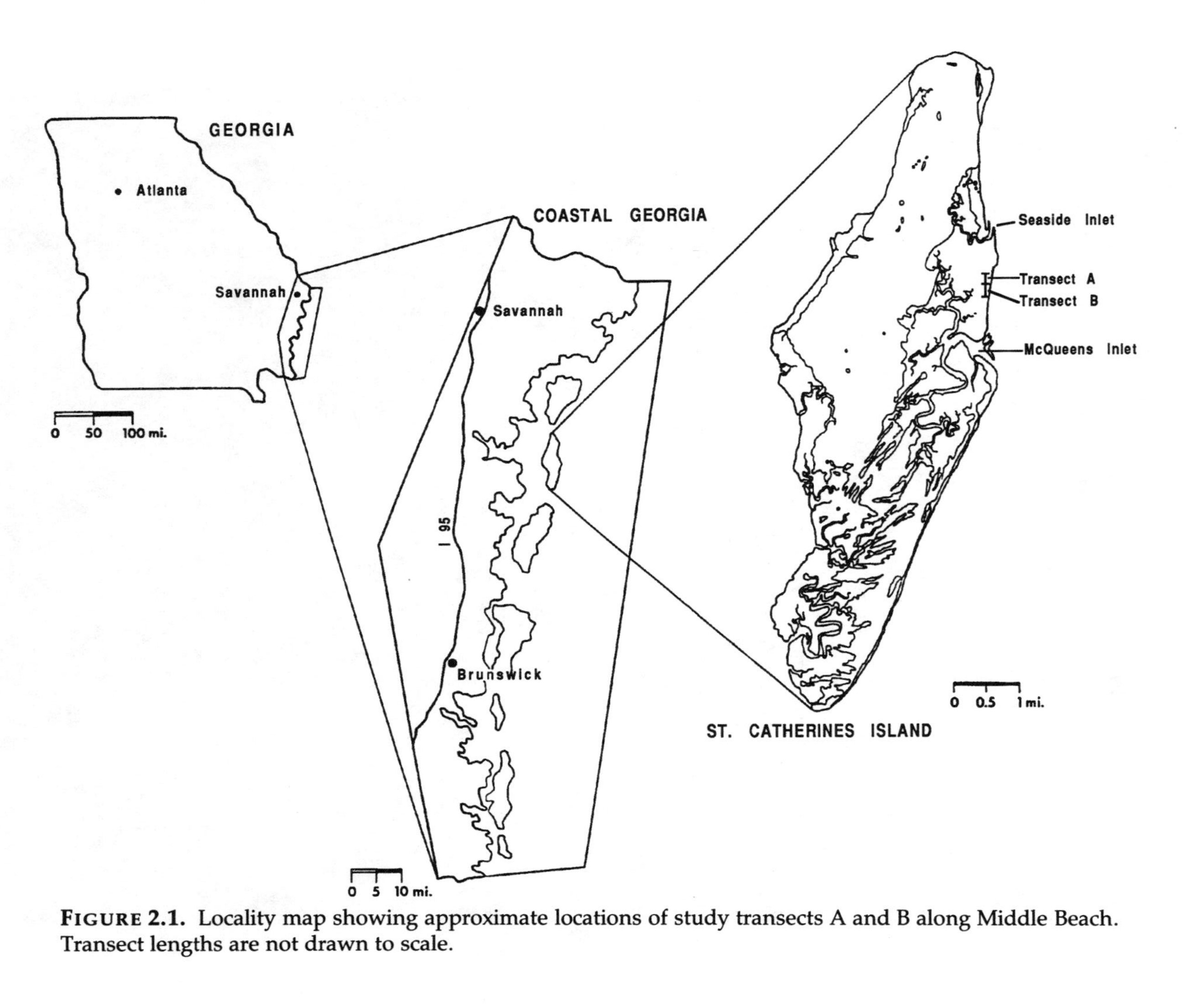

FIGURE 2.1. Locality map showing approximate locations of study transects A and B along Middle Beach. Transect lengths are not drawn to scale.

species are transported from the site of the living population and deposited along a wrack line on the upper shoreface of the beach. This ongoing process makes *A. brasiliana* the most commonly encountered large bivalve along extensive stretches of Middle Beach. As emphasized by Frey (1987), wrack line shell accumulations afford useful paleoecological baselines for assessment of taphonomic phenomena in modern coastal environments. The wrack line is a readily identifiable indicator of the previous high-tide position. Our attention was directed toward the abundant *A. brasiliana* specimens deposited along Middle Beach because most had been transported as articulated valves and had achieved a hydrodynamically stable attitude along the wrack line with the relatively thick umbos pointing downward (fig. 2.2), mimicking the living orientation. Ligamental tissue of *A. brasiliana* is quite resistant, and generally the ligaments become brittle enough to break and permit valve separation only after weeks of exposure on the upper shoreface.

In August 1988 we made wrack line counts along two transects on Middle Beach (fig. 2.1). Transect A extended along a high wrack line accumulation on a berm adjacent to the beach ridge dune field. A survey of total megafaunal and megafloral components was completed over a two-day interval for each of 58-m^2 quadrants along the wrack line (table 2.1). Relative proportions of articulated and disarticulated *A. brasiliana* valves were noted. This high wrack line survey reflected either the last storm or spring tide accumulation along this stretch of beach. Transect B extended for about 400 m along a wrack line accumulation

Figure 2.2. Photograph of *Anadara brasiliana* wrack line assemblage along transect A.

TABLE 2.1 *Species List—Transect A*

Bivalvia

Anadara brasiliana
Anadara ovalis
Barnea truncata*
Crassostrea virginica*
Cyrtopleura costata*
Dinocardium robustum
Dosinia discus
Geukensia demissa*
Hyalina sp.*
Mercenaria mercenaria*
Mulinia lateralis
Petricola pholadiformis*
Tellina sp.
Tagelus plebeius*

Gastropoda

Busycon carica
Busycon canaliculata
Ilyanassa obsoleta*
Littorina irrorata*
Melampus bidentatus*
Polinices duplicatus
Sinum perspectivum

Arthropoda

Callinectes sapidus*
Libinia dubia
Limulus polyphemus
Menippe mercenaria

Cnidaria

Leptogorgia virgulata

Echinodermata

Mellita quinquiesperforata

Annelida

Diopatra cuprea*

Vertebrata

Bone fragments
Feathers

Plant

Spartina alterniflora*
Wood fragments*
Pine cones*

Traces

Ocypode quadrata burrows*

*Species incompatible with subtidal Anadara brasiliana association.

that reflected the previous high tide. *A. brasiliana* specimens were collected along transect B during a single low-tide interval. Laboratory analysis of *A. brasiliana* included measurement of maximum width of each articulated specimen, and observations of the number and positions of boreholes and epibionts. In addition, chipping and scissoring of valves were noted, if present.

Results

Two hundred forty-four articulated specimens of *A. brasiliana* were counted along transect A (fig. 2.3). Of these, 130 (53%) were bored. Five percent had incomplete boreholes, suggesting unsuccessful gastropod predation. Only one articulated specimen recorded multiple boring attempts. Disarticulated, or single, valves comprised only 8% of total *A. brasiliana* specimens. Density of articulated specimens varied greatly from quadrant to quadrant along transect A, ranging from none to 29 per m^2 (fig. 2.3). Fifty-five percent of the 104 articulated *A. brasiliana* specimens collected from transect B were bored (figs.

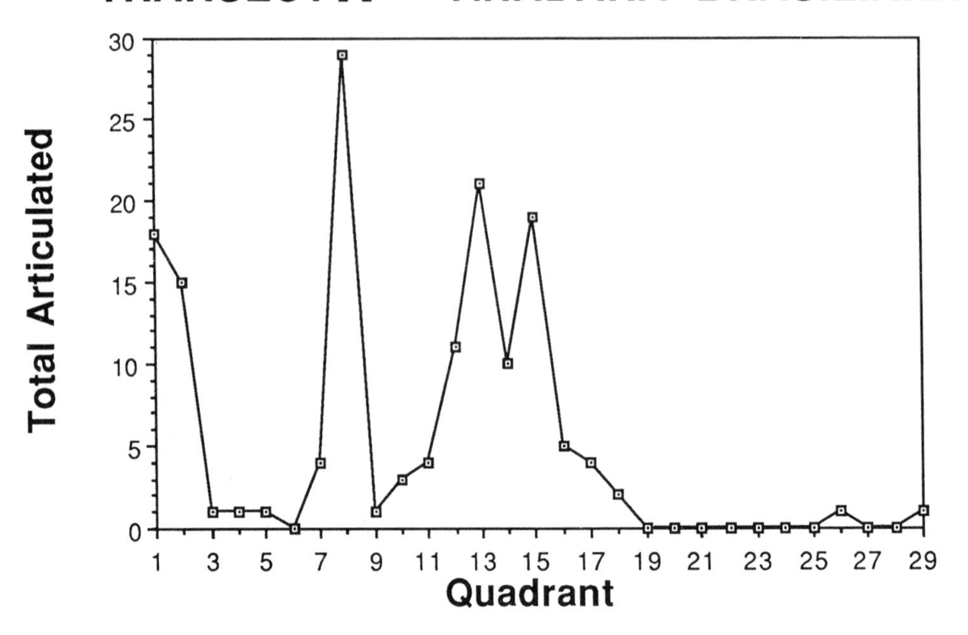

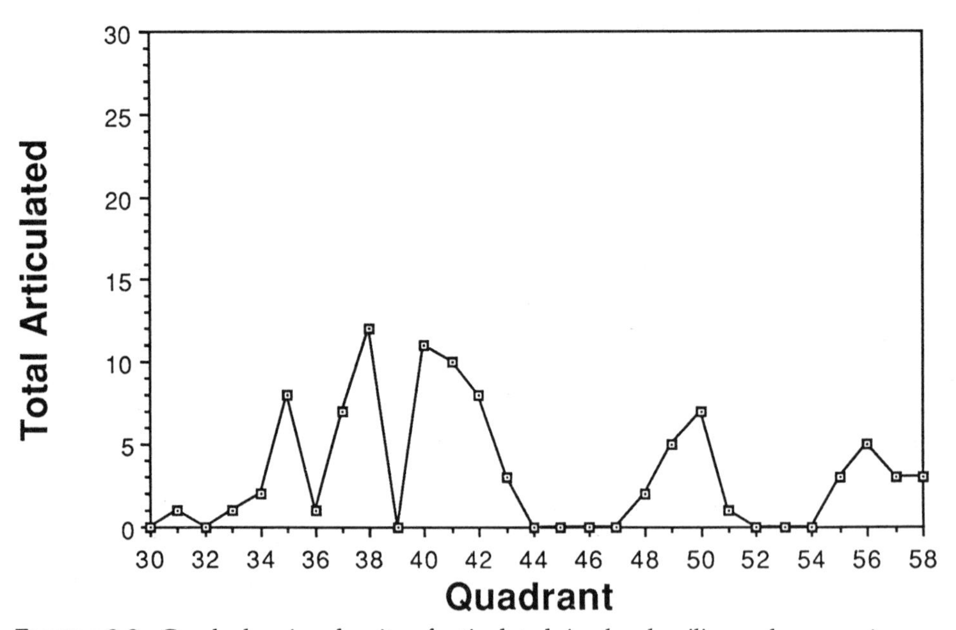

FIGURE 2.3. Graph showing density of articulated *Anadara brasiliana* valves per size class along transect A.

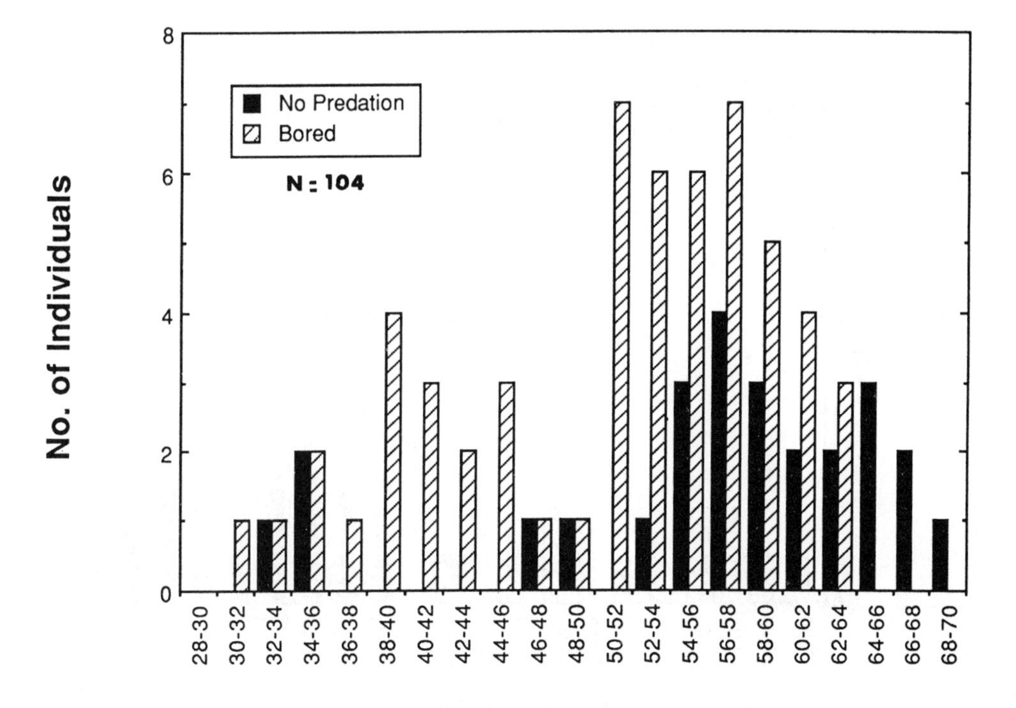

FIGURE 2.4. Histogram of relative frequencies of bored and nonpredated *Anadara brasiliana* valves per size class, transect B.

2.4 and 2.5). Sixty-two percent of these were bored on the left valves (fig. 2.6). Fifty-eight percent of the articulated specimens from transect B possessed encrusting epibionts, commonly sandworms, on the valve interiors.

In both transects, the drillhole positions in *A. brasiliana* were highly stereotypical, nearly all in the umbonal region of the shell. All boreholes had tapered sides, characteristic of naticids. The preferred umbo drillhole positions suggest a typical naticid attack of infaunal bivalve prey from the underside, although considerable differences apparently exist in styles of naticid predation (Ansell 1960; Paine 1963; Thomas 1976; Carriker 1981; Savazzi and Reyment 1989; Kabat 1990). Moreover, we cannot rule out possible naticid predation of *A. brasiliana* through a commissural gape without drilling (Frey et al. 1986). The common occurrence of *Polinices duplicatus* shells in the surveyed quadrants provided circumstantial evidence of naticid predation. No muricid shells were observed, although another naticid, *Sinum perspectivum*, occurred rarely. Muricids do occur locally in coastal Georgia environments. Walker (1981), for example, noted *Urosalpinx cinerea, Eupleura caudata, Fasciolaria hunteria,* and *Thais haemastoma*

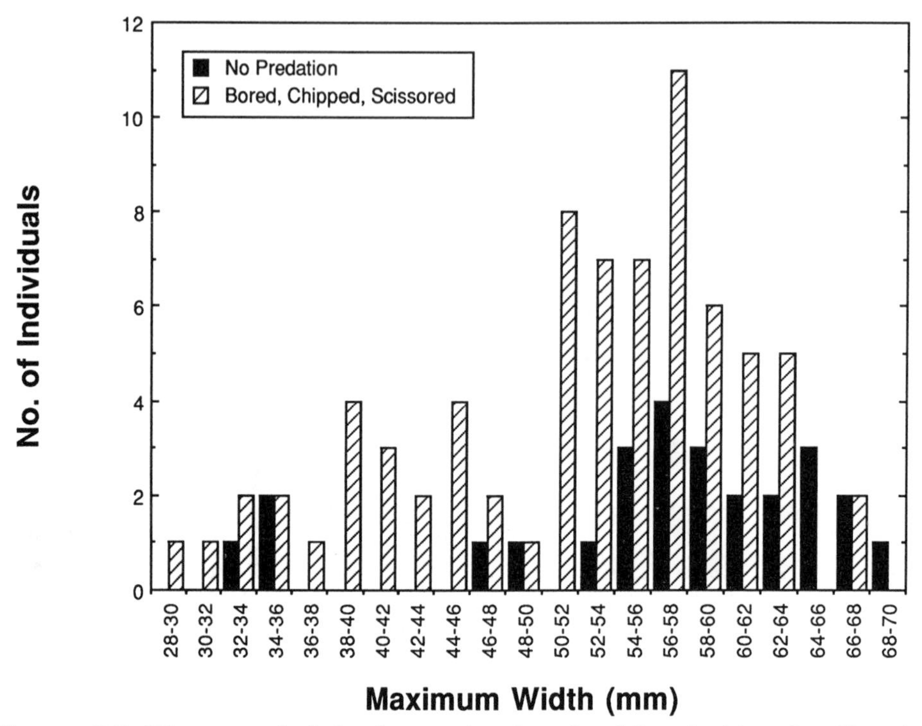

FIGURE 2.5. Histogram of relative frequencies of predated (bored, chipped and/or scissored) and nonpredated *Anadara brasiliana* valves per size class, transect B.

floridiana in Wassaw Sound. The observed stereotypic drilling is also more consistent with manipulative naticid predation than with more randomly cited muricid drillholes (Ansell 1960; Negus 1975; Alba and Reyment 1988). The small number of incomplete boreholes also supports naticid predation (Aitken and Risk 1988; Carriker and Van Zandt 1972).

The largest *A. brasiliana* specimens in the wrack line counts tended to be unbored (fig. 2.4), suggesting a possible size refuge of this species regarding naticid predation (Kitchell et al. 1981).

Discussion and Paleoecological Implications

Because there is little evidence of spatial averaging in the Middle Beach assemblage, if preserved it would likely be viewed as a life assemblage and subjected to paleoecological analyses to the point of ecological reconstruction. We were able to compare this taphonomically modified assemblage with a relatively unmodified *A. brasiliana* subtidal association because Dorjes (1972) published the results of subtidal faunal sampling from box-coring transects adjacent to Sapelo Island, immediately south of St. Catherines Island. We combined the results of the upper offshore portions of the northern transects of the Dorjes survey (see composite faunal list, table 2.2), as these are most comparable to the ark habitat opposite Middle Beach, St. Catherines Island. A comparison of tables 2.1 and 2.2 emphasizes the vast difference in species composition between the Middle Beach wrack line assemblage and the assemblage accumulating at the likely "source" of the ark shells. Approximately one-half of the 63 species listed in table 2.2 contain no preservable hard parts; thus the number of preservable species are comparable in the two assemblages. However, only seven (25%) of the preservable wrack line species occur in the subtidal *A. brasiliana* association. This demonstrates the ecological infidelity of the Middle Beach wrack line assemblage and highlights the necessity for very thorough assessment of taphonomic overprinting. It is important to be aware of how this affects the integrity of the contained ecological information. Of major concern, of course, is the obvious distortion provided by the habitat heterogeneity com-

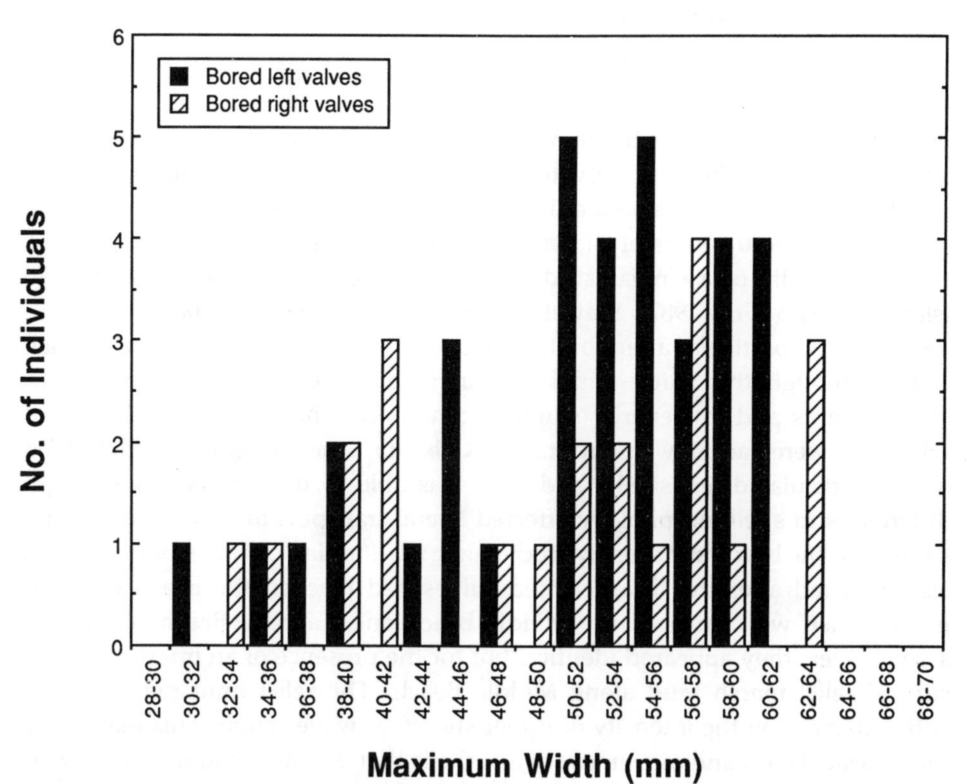

FIGURE 2.6. Histogram showing ratios of bored left valves and bored right valves of *Anadara brasiliana* per size class, transect B.

TABLE 2.2 *Northern Transect-Combined Dorjes, 1972*

Bivalvia	Crustacea
Abra aequalis	*Callianassa biformis*
Anadara ovalis	*Corophium tuberculatum*
*Anadara transversa**	*Edotea montosa*
Ensis directus	*Idunella* sp.
Mulinia lateralis	*Monoculodes* sp. I
Noetia ponderosa	*Ogyrides alphaerostris*
Solen viridus	*Oxyurostylis smithi*
Spisula solidissima	*Pinnixa chaetoperana*
Tellina alternata	
Tellina iris	Cnidaria
Tellina texana	*Edwardsia* sp.
	Haliactis sp.
Gastropoda	*Renilla reniformis*
Busycon carica	
Epitonium angulatum	Polychaeta
Epitonium rupicola	Twenty-nine species
Nassarius trivittatus	
Natica pusilla	Echinodermata
Olivella multica	
Olivella nivea	One species
Polinices duplicatus	
Retusa canaliculata	
Sinum perspectivum	
Terebra dislocata	

*Includes *Anadara brasiliana*.

bined within the Middle Beach wrack line assemblage. If interpreted as an autochonous shell bed, preservation of the Midde Beach assemblage would result in inaccurate assessment of interspecific compatibility.

Wrack line occurrences of *A. brasiliana* along Middle Beach can be compared with the results of a similar study conducted along Cabretta Beach, Sapelo Island, Georgia (Frey 1987). Frey studied the distribution of *A. brasiliana* and *A. ovalis*, paying particular attention to taphonomic input in the formation of shell beds. Although this study and Frey's involved generally comparable beach environments and low-energy summer conditions, the death assemblages of ark shells were radically different. The Cabretta Beach assemblage contained only disarticulated ark shells, and Frey was able to demonstrate how slight differences in shell morphology affected lateral transport of valves from living habitats to the beach surface. Hydrodynamic partitioning of valve morphologies also occurred after the disarticulated valves had reached the beach environment. When we compared the Middle Beach and Cabretta Beach wrack line assemblages, they appeared identical but for the presence of an intertidal exposure of relict marsh mud along Middle Beach. The relict mud may act as a baffle, decreasing the intensity of upper shoreface wave action. This may delay the disarticulation and movement of ark shells that become exhumed from their shallow subtidal habitat and become entrained behind the mud barrier. These two beach environments probably provide much information about the significance of even subtle hydrodynamic differences in the formation of death assemblages.

If preserved in the fossil record, the Middle Beach *A. brasiliana* assemblage would present a serious challenge to accurate paleoenvironmental reconstruction. Generally, the clues to spatial averaging of fossil assemblages are more obvious than those indicating time-averaging. Conventional indicators include evidence of transport of fossil shells, such as abrasion, breakage, sorting, and so on. These, if coupled with evidence of the co-occurrence of incompatible taxa, inconsistent matrix type, monospecificity, and so on, provide the paleoecologist with an effective arsenal that can be aimed at a fossil assemblage prior to engagement in paleocommunity reconstruction, trophic assessment, and so on. Short-term events that generate death assemblages (e.g., storms) are commonly accompanied by heightened energy conditions; thus evidence of spatial averaging is strong. The *A. brasiliana* assemblage on Middle Beach would preserve few of these warning signs. The enclosing matrix would be a medium-grained quartz sand, similar to that of the "living" population just offshore. Most of the preserved bivalves would be articulated and be situated in a life position, posterior margin down. The small number of single valves would not be at all unusual in a life context. Abrasion and breakage would be minimal. Density distributions would be consistent with that of a living population, showing expected patchiness (see fig. 2.3). Thus clues of intense taphonomic overprinting would be few but evidence might be available from careful assessment of the taxonomic content of the assemblage. Although species diversity would be high (table 2.1), the contained taxa would present a wide spectrum of habitats, ranging from back barrier salt marsh to subtidal marine. The assemblage would also contain a complex temporal overprinting and taphonomic feedback history, involving exhumation of preserved older marsh taxa by modern marine processes and even subsequent attempted recolonization of marine intertidal habitats by modern marsh taxa (Frey and Basan 1981; Pemberton and Frey 1985; West et al. 1990; Rollins et al. 1990; West and Rollins 1991). However, even the more ecologically incompatible taxa might also occur admixed with the cojacent shallow subtidal *A. brasiliana* population. Accurate assessment would likely depend on the recognition and evaluation of contextual sedimentary structures, both physical and biogenic, such as beach-type cross-bedding, ghost crab burrows, amphipod burrows, and so on, or upon evaluation of the fossil assemblage within the context of the stratigraphic sequence (e.g., subjacent beach deposits and superjacent marsh deposits).

Admittedly, as noted by Frey (1987:161) "wrack-line shell accumulations stand little chance for direct preservation in the fossil record," but "general processes and parameters revealed by such studies remain useful in taphonomic interpretations of ancient paralic assemblages." The *A. brasiliana* accumulation described earlier would probably be highly modified prior to final burial. Although Frey and Dorjes (1988:570) discussed potential processes of preservation of beach shell accumulations, none of these applies directly to the Middle

Beach occurrence, which is complicated by the presence of relict muds. It is difficult to construct a scenario that would lead to preservation of the *A. brasiliana* accumulation without additional remobilization and dispersion of the different biotic elements. We do not, however, want to preclude such preservation, especially in view of the many unanswered questions regarding the assembly and preservation of shell beds (e.g., the Pliocene Pinecrest beds of Florida discussed by Geary and Allmon [1990], possibly explained by the complex lateral variation of shore dynamics, as modeled by Seilacher [1985] and discussed by Frey and Dorjes [1988]). Indeed, subtidally assembled shell beds, which have higher preservational potential, also probably formed under a similar open system and therefore possess similar limits to ecological reconstruction. The examples presented by Miller et al. (1988) for the Middle Devonian may well be the rule in the formation of shallow subtidal shell beds. If the Middle Beach *A. brasiliana* assemblage can form in the taphonomically active zone (TAZ of Powell et al. 1989), it can occur also in the so-called zone of accumulation (Nittrouer and Sternberg 1981; Powell et al. 1989). If, in fact, we subscribe to the increasingly accepted view that "most death assemblages are probably accumulations of microtempestites even though they may not appear to be" and the "bulk of these death assemblages, exclusive of some deep infauna, accumulates from the remains of pulses that are reworked deeper into the TAZ where taphonomic disintegration rates become slow relative to long-term rates of net sedimentation" (Powell et al. 1989:570), then the dynamics of accumulation of the Middle Beach *A. brasiliana* assemblage might be mirrored in the shallow subtidal parautochthonous regime. If so, under certain conditions of pulse-type leakage of shell debris deeper into the TAZ, many aspects of the Middle Beach *A. brasiliana* assemblage might be preserved. If we suspect that many shell beds, even those that appear to be ecologically intact, actually were assembled under conditions of spatial and temporal averaging of open systems, we must address the necessary consequences for ecological reconstruction (even if we ignore many aspects of taphonomic input such as selective preservation resulting from relative proportion of hard- and soft-bodied organisms, rate of shell dissolution, etc.). The recent review by Powell et al. (1989), and references contained therein, demonstrates how pervasively taphonomy and temporal/spatial averaging affect the transferral of ecological paradigms or "rules" to the death assemblage and fossil record (see also Rollins et al. 1990).

The mode of accumulation of the Middle Beach wrack line assemblage also engenders rather subtle ecological distortion, illustrated by consideration of evidence of predation. Any inference of rate of predation must consider that the assemblage accumulated in an open system. Stochastic events (storms, tides, etc.) "harvested" the parent subtidal population of live and dead individuals. Live, unbored individuals could reburrow. Moreover, harvesting most likely did not "grab" random proportional subsamples of bored and unbored *A. brasiliana* shells and thus contrasts sharply with a more nearly closed system that might be time-averaged but not subjected to episodic postmortem transport. Therefore, as noted by Powell et al. (1989), predator ratios in paleoecology are quite different from their ecologically measured counterparts. The former represent "lifetime-integrated values" (Powell et al. 1989:577), and, even ignoring taphonomic intervention, the effect of many generations of death and accumulation results in a time-averaged net measure of predator interaction. Presence of gastropod boreholes in molluscan valves has been touted as one of the clearest evidences of biotic interaction contained in a death assemblage (Kabat and Kohn 1986), providing an example of what Paine (1983:87) called "hallmarks of ecological processes entombed in the fossil record." But if shell beds are commonly products of spatial and temporal averaging of death assemblages (i.e., open systems), predation ratios *(sensu lato)* have little validity and comparisons through time based on comparison of such ratios must remain suspect. On the other hand, evidence of predation and/or predation attempts (shell repairs following unsuccessful predation attempts, incomplete boreholes, etc.) contained on single individuals affords some of the best examples of biotic interaction preserved in fossil assemblages. In a sense, the latter represent small-scale closed systems with concomitantly enhanced ecological fidelity.

The Middle Beach *A. brasiliana* assemblage, we feel, provides the following insight: (1) Taphonomic overprinting of death assemblages can be incredibly subtle, requiring careful and complete taphonomic assessment prior to reconstruction of ecological interactions. (2) Even an allochthonous product of episodic hydrodynamic events, such as the Middle Beach wrack line assemblage, may appear only slightly impacted taphonomically. (3) Attempts to standardize evaluations of taphonomic intensities of fossil assemblages (e.g., Powell et al. 1989; Brett and Baird 1986) should be encouraged. (4) All available tools for the assessment of taphonomic impact should be used prior to ecological interpretation.

Acknowledgments

We are grateful for grants from the St. Catherines Island Research Program of the American Museum of Natural History supported by the Edward John Noble Foundation. We appreciate the help of Mr. Royce Hayes, superintendent of St. Catherines Island, who provided us with necessary equipment, transportation, and valuable advice. Comments by the late Dr. R. W. Frey and an anonymous reviewer greatly improved the manuscript. We are also grateful to Ms. Judith Rollins for field and laboratory assistance, Ms. Janet Baker for compiling faunal data,and Ms. Wendy Brindle and Ms. Cindy Venn for help with illustrations.

References

Aitken, A. E. and M. J. Risk. 1988. Biotic interactions revealed by macroborings in arctic bivalve molluscs. *Lethaia* 21:339–50.

Alba, S. G. and R. A. Reyment. 1988. Differentiation between the traces of predation of muricids and naticids in Spanish Pliocene *Chlamys*. *Estudios Geologicos* 44:317–28.

Ansell, A. D. 1960. Observations on predation on *Venus striatula* (da Costa) by *Natica alderi* (Forbes). *Proceedings, Malacological Society of London* 34:157–65.
Brett, C. E. and G. C. Baird. 1986. Comparative taphonomy: A key to paleoenvironmental interpretation based on fossil preservation. *Palaios* 1:207–27.
Carriker, M. R. 1981. Shell penetration and feeding by naticacean and muriacean predatory gastropods: A synthesis. *Malacologia* 20:403–22.
Carriker, M. R. and D. Van Zandt. 1972. Predatory behavior of a shell boring gastropod. In H. E. Winn and B. L. Olla, eds., *Behaviour of Marine Animals: Current Perspectives in Research*, pp. 157–244. New York: Plenum Press.
Cummins, H., E. N. Powell, R. J. Stanton Jr., and G. Staff. 1986. The size-frequency distribution in palaeoecology: The effects of taphonomic processes during formation of death assemblages in Texas bays. *Palaeontology* 29:495–518.
Davies, D. J., E. N. Powell, and R. J. Stanton Jr. 1989. Taphonomic signature as a function of environmental process: Shells and shell beds in a hurricane-influenced inlet on the Texas coast. *Palaeogeography, Palaeoclimatology, Palaeoecology* 72:317–56.
Davies, D. J., G. M. Staff, W. R. Callender, and E. N. Powell. 1990. Description of a quantitative approach to taphonomy and taphofacies analysis: All dead things are not created equal. In W. Miller III, ed., *Paleocommunity Temporal Dynamics: The Long-Term Development of Multispecies Assemblies, Paleontological Society Special Publication* No. 5, pp. 328–50, University of Tennessee, Knoxville.
Deery, J. R. and J. D. Howard. 1977. Origin and character of washover fans on the Georgia coast, U.S.A. *Gulf Coast Association Geological Societies, Transactions* 27:259–71.
Dorjes, J. 1972. Georgia coastal region, Sapelo Island, U.S.A.: Sedimentology and biology. VII. Distribution and zonation of macrobenthic animals. *Senckenbergiana Maritima* 4:183–216.
Dott, R. H., Jr. 1983. 1982 SEPM presidential address: Episodic sedimentation—how normal is average? How rare is rare? Does it matter? *Journal of Sedimentary Petrology* 53:5–23.
Frey, R. W. 1987. Distribution of ark shells (Bivalvia: *Anadara)*, Cabretta Island Beach, Georgia. *Southeastern Geology* 27:155–63.
Frey, R. W. and P. B. Basan. 1981. Taphonomy of relict Holocene salt marsh deposits, Cabretta Island, Georgia. *Senckenbergiana Maritima* 13:111–55.
Frey, R. W. and J. Dorjes. 1988. Fair- and foul-weather shell accumulations on a Georgia beach. *Palaios* 3:561–76.
Frey, R. W., J. D. Howard, and J. Hong. 1986. Naticid gastropods may kill solenid bivalves without boring: Ichnologic and taphonomic consequences. *Palaios* 1:610–12.
Frey, R. W. and J. D. Howard. 1988. Beaches and beach-related facies, Holocene barrier islands of Georgia. *Geological Magazine* 125:621–40.
Geary, C. H. and W. D. Allmon. 1990. Biological and physical contributions to the accumulation of strombid gastropods in a Pliocene shell bed. *Palaios* 5:259–72.
Hoffman, A. and J. Martinell. 1984. Prey selection by naticid gastropods in the Pliocene of Emporda (Northeast Spain). *Neues Jahrbuch für Gedogie und Paläontologie, Muratshefte* 7:393–99.
Howard, J. D. and R. W. Frey. 1980. Holocene depositional environments of the Georgia coast and continental shelf. In J. D. Howard, C. B. DePratter, and R. W. Frey, eds., *Excursions in Southeastern Geology: The Archaeology-Geology of the Georgia Coast, Georgia Geological Survey, Guidebook* 20, pp. 66–134.
Kabat, A. R. 1990. Predatory ecology of naticid gastropods with a review of shell boring predation. *Malacologia* 32:155–93.
Kabat, A. R. and A. J. Kohn. 1986. Predation on early Pleistocene naticid gastropods. *Palaeogeography, Paleoclimatology, Palaeoecology* 53:255–69.
Kauffman, E. G. 1986. High-resolution event stratigraphy: Regional and global bio-

events. In O. H. Walliser, ed., *Global Bioevents. Lecture Notes Earth History*, pp. 279–335. Berlin: Springer-Verlag.

———. 1988. Concepts and methods of high-resolution event stratigraphy. *Annual Reviews Earth and Planetary Science* 16:605–54.

Kidwell, S. M. 1986. Taphonomic feedback in Miocene assemblages: Testing the role of dead hardparts in benthic communities. *Palaios* 1:239–55.

Kidwell, S. M. and D. Jablonski. 1983. Taphonomic feedback: Ecological consequences of shell accumulation. In M. J. S. Tevesz and P. L. McCall, eds., *Biotic Interactions in Recent and Fossil Benthic Communities*, pp. 195–248. New York: Plenum Press.

Kitchell, J. A., C. H. Boggs, J. F. Kitchell, and J. A. Rice. 1981. Prey selection by naticid gastropods: Experimental tests and application to the fossil record. *Paleobiology* 7:533–52.

Kuroda, R. and F. C. Marland, 1973. Physical and chemical properties of the coastal waters of Georgia, Georgia Institute Technology, Environmental Research Center; University of Georgia Marine Institute.

McLachlan, A. 1990. Dissipative beaches and macrofauna communities on exposed intertidal sands. *Journal of Coastal Research* 6:57–71.

Miller, K. B., C. E. Brett, and K. M. Parsons. 1988. The paleoecologic significance of storm-generated disturbance within a Middle Devonian muddy epeiric sea. *Palaios* 3:35–52.

Negus, M. 1975. An analysis of boreholes drilled by *Natica catena* (da Costa) in the valves of *Donax vittatus* (da Costa). *Proceedings Malacological Society of London* 41:353–56.

Nittrouer, C. A. and R. W. Sternberg. 1981. The formation of sedimentary strata in an allochthonous shelf environment: The Washington continental shelf. *Marine Geology* 12:201.

Paine, R. T. 1963. Trophic relationships of eight sympatric predatory gastropods. *Ecology* 44:63–73.

———. 1983. On paleoecology: An attempt to impose order on chaos. *Paleobiology* 9:86–90.

Parsons, K. M., C. E. Brett, and K. B. Miller. 1986. Comparative taphonomy and sedimentary dynamics in Paleozoic marine shales, *Fourth North American Paleontological Convention, Abstracts with Programs*, pp. A34–A35.

Pemberton, S. G. and R. W. Frey. 1985. The *Glossifungites* ichnofacies: Modern examples from the Georgia coast, U.S.A. In H. A. Curran, ed., *Biogenic Structures: Their Use in Interpreting Depositional Environments. Society of Economic Paleontologists and Mineralogists, Special Paper* No. 35, pp.237–59.

Powell, E. N., G. M. Staff, D. J. Davies, and W. R. Callender. 1989. Macrobenthic death assemblages in modern marine environments: Formation, interpretation, and application. *CRC Critical Reviews in Aquatic Sciences* 1:555–89.

Rollins, H. B., R. R. West, and R. M. Busch. 1990. Hierarchical genetic stratigraphy and marine paleoecology. In W. Miller III, ed., *Paleocommunity Temporal Dynamics: The Long-Term Development of Multispecies Assemblies, Paleontological Society Special Publication* No. 5, pp. 273–08, University of Tennessee, Knoxville.

Ruffell, A. H. 1988. Palaeoecology and event stratigraphy of the Wealden-Lower Greensand transition in the Isle of Wight. *Proceedings Geological Association* 99:133–40.

Savazzi, E. and R. A. Reyment. 1989. Subaerial hunting behaviour in *Natica gualteriana* (naticid gastropod). *Palaeogeography, Palaeoclimatology, Palaeoecology* 74:355–64.

Seilacher, A. 1982. General remarks about event deposits. In Einsele, G. and A. Seilacher, eds., *Cyclic and Event Stratification*, pp. 161–74. New York: Springer-Verlag.

———. 1985. The Jeram model: Event condensation in a modern intertidal environment. In U. Bayer and A. Seilacher, eds., *Sedimentary and Environmental Cycles*, pp. 336–441. New York: Springer-Verlag.

Sherrod, B. L., H. B. Rollins, and S. K. Kennedy. 1989. Subrecent intertidal diatoms from St. Catherines Island, Georgia: Taphonomic implications. *Journal of Coastal Research* 5:665–77.

Staff, G. M. and E. N. Powell. 1990. Taphonomic signature and the imprint of taphonomic history: Discriminating between taphofacies of the inner continental shelf and a microtidal inlet. In W. Miller III, ed., *Paleocommunity Temporal Dynamics: The Long-Term Development of Multispecies Assemblies*, pp. 370–90. *Paleontological Society Special Publication* No. 5, University of Tennessee, Knoxville.

Thomas, R. D. K. 1976. Gastropod predation on sympatric Neogene species of *Glycymeris* (Bivalvia) from the eastern United States. *Journal of Paleontology* 50:488–99.

Walker, R. 1981. Distribution of oyster drills, *Urosalpinx cinerea* (Say), in Wassaw Sound, Georgia. *Georgia Journal of Science* 39:127–39.

West, R. R. and H. B. Rollins. 1989. The "anatomy" of a condensed sequence: An example from coastal Georgia. Symposium 16, *28th International Geological Congress, Abstracts* 3:3–350.

———. 1991. Temporal resolution of bedding surface events: Examples from the Georgia Bight. *Geological Society of America, Abstracts with Programs* 23(1):147.

West, R. R., H. B. Rollins, and R. M. Busch. 1990. Taphonomy and an intertidal palimpsest surface: Implications for the fossil record. In W. Miller III, ed., *Paleocommunity Temporal Dynamics: The Long-Term Development of Multispecies Assemblies*, pp. 351–69. Paleontological Society Special Publication No. 5, University of Tennessee, Knoxville.

3

Counting Fossils in a Cincinnatian Storm Bed: Spatial Resolution in the Fossil Record

Arnold I. Miller

ABSTRACT

Despite a growing body of actualistic evidence suggesting that the fossil record exhibits excellent spatial fidelity on confined lateral scales, small-scale (i.e., tens of meters or less) spatial patterns of faunal distribution are typically unexplored in ancient strata. Lateral spacing of samples in paleoecological investigations is commonly on the order of tens of kilometers, suggesting that biologically significant spatial variability at finer scales has been ignored.

To explore the question of spatial resolving power in ancient fossil assemblages, I investigated spatial patterns in faunal composition along a storm-deposited horizon of the Upper Ordovician Fairview Formation in the Cincinnati, Ohio, area. At one locality, a 150-m sampling transect, trending roughly from north to south, was established and bulk faunal samples were collected at 10-m lateral intervals on the sampling horizon; for comparative purposes, five additional samples were collected from a second locality. Multivariate analyses of faunal data revealed significant spatial variation in abundances of the brachiopod genera *Platystrophia* and *Rafinesquina* along the primary transect: Samples from the central third of the transect were dominated by *Platystrophia,* whereas samples from the southern third were dominated by *Rafinesquina* and those from the northern third exhibited more equitable abundances of the two genera. The limited sample set from the second locality did not exhibit a unique faunal signature relative to the primary locality. Although a precise biological explanation for spatial patterns observed on the sampling transect has yet to be determined, it is unlikely that they were consequences of postmortem transport.

> The primary locality was selected for logistical ease of sampling; there was no advance sense that the patterns recognized would, indeed, be exhibited. Thus, results of this preliminary investigation suggest that spatial variability in faunal composition on the scale explored here may be quite common in the fossil record and is worth investigating in more detail.

To evaluate the fidelity of the marine fossil record, myriad actualistic studies have explored the correspondence between present-day environmental patterns on the seafloor and spatial distributions of accumulating shelly material (e.g., Johnson 1965; Lawrence 1968; Warme 1969, 1971; Macdonald 1976; Peterson 1976; Stanton 1976; Warme et al. 1976; Bosence 1979; Cummins et al. 1986a, 1986b; Miller 1988; Meldahl and Flessa 1990; Miller et al. 1992). Although the scopes and specific questions addressed by such investigations vary considerably, the conclusions reached have generally been encouraging. In most cases, spatial transitions exhibited by subfossil assemblages faithfully reflect measured environmental changes (see detailed review in Kidwell and Bosence 1991), suggesting that ancient fossil assemblages commonly exhibit excellent spatial fidelity to environmental patterns, albeit time-averaged, at even limited scales (10 m or less in some instances).

Despite these optimistic assessments, spatial patterns on similar scales have generally not been explored in the ancient. In fact, given the results of actualistic studies, an entire class of potentially meaningful faunal variability in the fossil record may have been overlooked. Typically, study areas for actualistic investigations are contained in bays and lagoons (see Kidwell and Bosence 1991), with an average spacing of tens of meters between samples. In contrast, localities in ancient paleoecological studies are commonly tens of kilometers apart, with no assessment of outcrop-scale variability in faunal composition along individual bedding planes. In most instances, entire study areas of actualistic investigations could fit easily between any two locations from which lateral samples are collected in the ancient. Wide lateral spacing is the rule even in studies of systematic spatial transitions in fossil assemblages, generally referred to as "gradient analyses," which have emerged as valuable investigative tools in paleoecology and evolutionary paleobiology (e.g., Cisne and Rabe 1978; Cisne and Chandlee 1982; Cisne et al. 1982a, 1982b; Springer and Bambach 1985).

Lateral postmortem movement of skeletal material is obviously limited to some degree; large-scale faunal gradients would not otherwise be detectable in the fossil record. Even in the case of storm deposition, transport may not be pronounced in many instances (Aigner 1979, 1980, 1985; Kreisa 1981a; Kreisa and Bambach 1982; Aigner and Reineck 1982; Miller et al. 1992; but see Westrop 1986, for an alternative view). Despite these findings, it has been assumed implicitly, with a few notable exceptions (e.g., Feldman 1985, 1988), that on the scale of a single outcrop, meaningful or useful spatial patterns are not preserved.

The purpose of this study was to explore outcrop-scale faunal variability in a storm-deposited horizon of the Upper Ordovician Fairview Formation in the Cincinnati area. Most samples were confined to a 150-m transect established at a single outcrop, although a smaller suite was collected for comparative purposes from the same horizon at another locality.

Below, I provide a brief description of the stratigraphic and methodological frameworks for the investigation. I then present an analysis and discussion of compositional variability among collected faunal samples. Results of this preliminary analysis demonstrate that there is notable, outcrop-scale spatial variability along the primary transect, although the specific explanation for this pattern remains elusive.

Stratigraphy and Localities

The classic Upper Ordovician fossil assemblages of the Cincinnati Arch were deposited in habitats profoundly influenced by storms. Indeed, many fossils are preserved in basal packstone and grainstone lags of storm-deposited strata (Kreisa 1981b; Tobin and Pryor 1981; Tobin 1982; Jennette 1986). In turn, these storm deposits are packaged in a series of "large-scale" regressive sequences (three to five; Tobin 1982; Holland 1990), which Tobin (1982) and Tobin and Pryor (1985) call "shoaling-upward cycles." Stratigraphically, the lowermost of these regressions initiates in the Edenian Kope Formation and continues upward through the Maysvillian Fairview Formation, culminating in the Maysvillian Bellevue Formation (Tobin and Pryor 1981; Tobin 1982). These units are thought to have been deposited on a northward-dipping ramp, and the boundaries between them are time transgressive. Thus, during the regression, deposition of the shale-dominated Kope facies was taking place in a down-ramp position, whereas deposition of the limestone-dominated Bellevue facies was occurring up-ramp; deposition of the mixed limestones and shales of the Fairview was ongoing in a mid-ramp position. As the regression proceeded, the entire package migrated northward, generating the vertical Kope through Bellevue sequence recognized at localities throughout the area.

Superimposed on this large-scale regression is a series of smaller-scale cycles, on the order of 1 m thick, that have thus far been recognized throughout the Kope and Lower Fairview Formations and are readily visible at outcrops throughout the Cincinnati area (Tobin and Pryor 1981, 1985; Tobin 1982; Jennette 1986). Tobin (1982) and Tobin and Pryor (1985) refer to these smaller-scale cycles as "megacycles." Although Tobin and Pryor (1981:53) suggested that such cycles are not traceable laterally, subsequent analyses by Tobin (1982) clearly showed otherwise. Within the Kope, the caps of each cycle are recognizable as laterally persistent limestones (grainstone/packstone); lower portions of each cycle are dominated by shale and wackestones. Within the Fairview, the cycles are more limestone-rich but still contain some mud; individual beds become less muddy and more laterally persistent near the cycle tops, as in the Kope (fig. 3.1). By applying proximality concepts developed by Aigner (1979, 1980, 1985) to these storm-deposited units, Jennette (1986; Jennette and Pryor 1993) demonstrated that these cycles were, in fact, small-scale regressive sequences superposed on the larger-scale, ongoing regression. Furthermore, based on characteristics unique to particular capping limestones, Jennette correlated these cycles throughout the Cincinnati area (fig. 3.2).

Because individual cycle caps can be traced readily throughout the region, they provide opportunities to sample virtually the same horizon at numerous localities in the Cincinnati metropolitan area. In this preliminary study, a capping packstone in the lower Fairview Formation was selected for analysis at two localities (figs. 3.1 and 3.2). Most faunal sampling of this bed was concentrated along a transect at a locality in northern Kentucky (corner of Kentucky Route 16 and Mason Road; locality 1 in fig. 3.3). A smaller-sample suite was collected from the same horizon, approximately 10 km to the north-northwest (down-dip on the paleoramp) in Cincinnati (Bald Knob; locality 2 in fig. 3.3).

The specific horizon selected for analysis was the cap to Cycle B, the second of four cycles recognized in the lower Fairview by Jennette. At outcrop, this bed appears as a laterally persistent packstone, rich in bryozoans and brachiopods; its thickness averages approximately 30 cm but varies somewhat along the outcrop face. It is overlain directly by an intermittent, fossil-poor, laminated

FIGURE 3.1. Photograph of upper Kope and lower Fairview Formations at locality 1 (fig. 3.3), illustrating stratigraphic cyclicity of both units (see text for description). The sampling interval for the present study is marked at the northern (left) end of the outcrop with two asterisks.

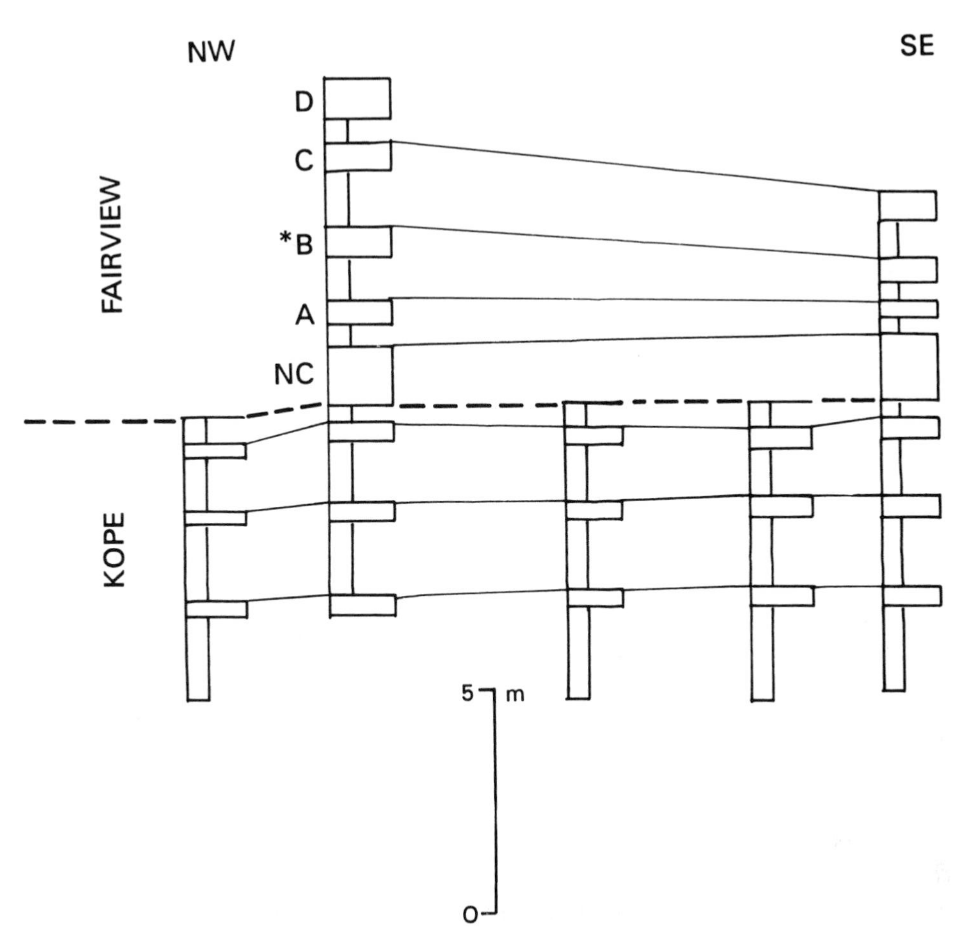

FIGURE 3.2. Stratigraphic cross section of the upper Kope and lower Fairview Formations for a series of localities arrayed along depositional dip on the Late Ordovician paleoramp; these five localities are joined by a dotted line in fig. 3.3. Intervals that protrude on the cross section are limestone rich; recessed intervals are shale rich. Correlation of limestone ledges that comprise cycle caps was accomplished by Jennette (1986), based on characteristics unique to particular caps, as well as their stratigraphic positions with respect to the Kope/Fairview boundary. Jennette designated the base of the Fairview, which is a thick limestone interval, as " noncyclic" (NC on figure). Above this zone he recognized four cycles (A-D) in the lower Fairview. The sampling interval for the present study is at the top of cycle B (after Jennette 1986).

siltstone, which is characteristic of storm deposition (Kreisa 1981a). However, on closer inspection it is evident that the sampling horizon does not preserve just a single storm. Rather, slabs of the bed reveal a complex internal structure, with as many as five distinct concentrations of skeletal material and occasional laminated silts between them, indicating an amalgamation of several storm events. This suggests a proximal environment of storm deposition (Aigner 1985; Jennette 1986); its implications for the present study are discussed later.

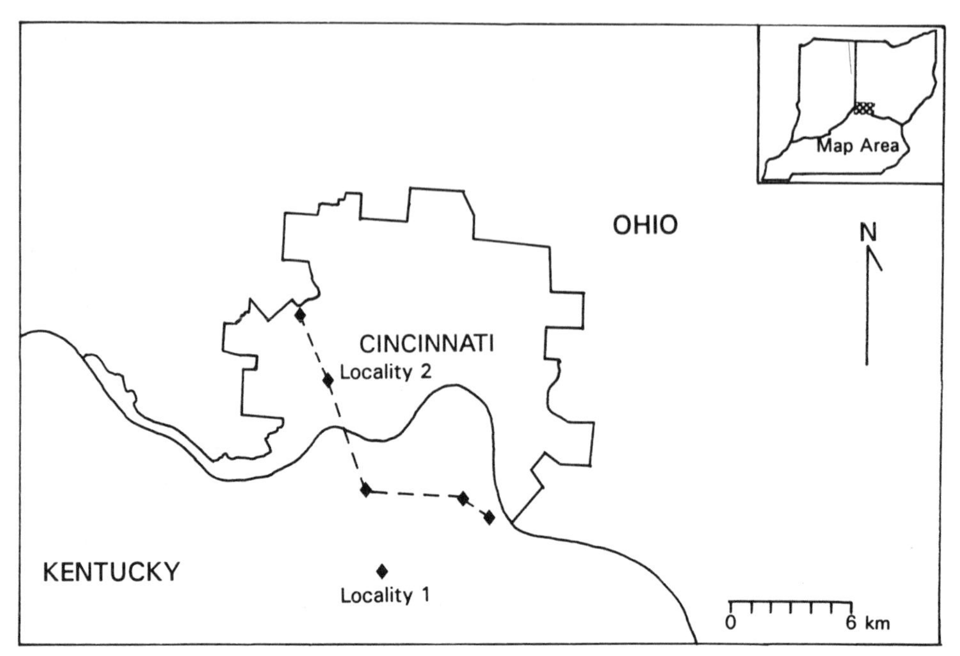

FIGURE 3.3. Locality map for the study area; two localities were sampled in the present study. The five localities illustrated in fig. 3.2 are joined by a dotted line (after Jennette 1986).

Methods

At locality 1, a 150-m transect, trending N22°W, was established along the outcrop face on the east side of Kentucky Route 16 (fig. 3.1); a 40-m transect that trended N25°E was also set up on the east slope of Bald Knob at locality 2. Both transects were sampled at 10-m lateral intervals. Thus, 16 samples were collected at locality 1 and five were collected at locality 2. Individual samples consisted of bulk pieces of the entire bed, ranging in volume from approximately 2 to 10 liter, depending upon ease of extraction.

In the laboratory, faunal contents of samples were censused as completely as possible. Among solitary organisms that remain fairly intact, only articulate brachiopods and bivalve molluscs were present; sample preparation involved mechanically breaking the majority of the collected sample and etching pieces with a weak solution of acetic acid to help clean individual specimens. Subsequently, these specimens were identified and their abundances determined by tabulating the number of valves possessing an umbo. Censusing of bryozoans was accomplished with point counting. Within each sample, the zone of heaviest bryozoan concentration was located, and the sample was slabbed parallel to bedding at this subhorizon. Given the orientation of the majority of bryozoan material, this maximized the probability of exposing longitudinal and tangential sections through colony branches, which are desirable for identification. Slabs were polished and etched, permitting the preparation of acetate peels, which were point-counted on a microfiche reader using a 50-point, 3- by 7-cm grid overlay. Identifications of bryozoans were facilitated with illustrations provided in references, including Anstey and Perry (1972) and Karklins (1984). Faunal censuses were conducted at the taxonomic level of genus when possible (some trepostome bryozoan material could not be further identified). While identification to species level would have been desirable, species designations are uncertain for some of the genera recognized in the sampling horizon. Moreover, among these samples, there was no apparent indication that genera were represented by more than one species, thereby minimizing the importance of species-level designations for the purposes of this limited study.

Solitary organisms that routinely disaggregate (e.g., trilobites and crinoids) were meager to absent among the samples, as were nonbryozoan colonial elements (e.g., graptolites). Thus, they were not included in the analysis.

The resulting data matrix from laboratory analysis was evaluated numerically to compare samples on the basis of faunal content. Although a variety of multivariate techniques was utilized for this purpose, only the results of a three-axis polar ordination are presented to avoid redundancy. The polar ordination algorithm utilized here is known as POLAR II, contained in a FORTRAN computer program written by J. Sharry and J. J. Sepkoski, Jr., in 1976, modified by Sepkoski in 1980, and revised for microcomputer by A. Miller in 1988. The method differs somewhat from more "traditional" polar ordination algorithms (e.g., Bray and Curtis 1957; Beals 1965); a detailed technical description is available upon request.

To understand better the sample distributional pattern exhibited by polar ordination, correlations of individual variables (genera) with each ordination axis were determined by utilizing the Pearson product-moment correlation coefficient to compare variable abundances among samples directly with sample coordinate values. In general, a large positive coefficient value for a genus on a particular axis suggests that the genus is relatively abundant in samples that have large coordinate values on that axis. A large negative value indicates that the genus is relatively abundant in samples with small coordinate values (see below).

Results

The 21 samples contained a total of 997 brachiopod and bivalve specimens belonging to five genera (table 3.1); the vast majority were brachiopods. Of the 1,050 points counted in the bryozoan census (50 points for each sample), bryozoans were present at 455; 359 could be identified further and belonged to five genera (table 3.1).

Polar ordination of the samples, based on faunal compositions, is presented in figures 3.4 and 3.5. (In all figures and text discussion, the sample number

TABLE 3.1 *Genera Identifed in Samples Collected at Localities 1 and 2*

Brachiopods	Bivalves	Bryozoans
Rafinesquina	*Caritodens*	*Parvohallopora*
Platystrophia	*Modiolopsis*	*Heterotrypa*
Plectorthis		*cf. Dekayia*
		Constellaria
		Escharopora

indicates locality [before the hyphen], and distance [after the hyphen], in meters, from the southernmost end of the applicable transect.) On these ordination scatter plots, proximity of samples reflects relative similarity in composition. With respect to the 16 samples from locality 1, an interesting pattern is immediately apparent: the transect is divisible roughly into thirds on the basis of transitions in faunal composition. In fig. 3.4, which is a two-dimensional rendition of sample coordinate values on axes 1 and 2, samples 1–0 through 1–40 are grouped in the lower right corner of the plot, 1–100 to 1–150 in the center, and 1–50 to 1–90 in the in the upper left. This pattern is further demonstrated in fig. 3.5, which provides a perspective view of samples from locality 1 on all three ordination axes.

Correlations of genera with axes 1 and 2, in cases where they were especially diagnostic, are sketched directly on fig. 3.4 and indicate the pattern of faunal distribution responsible for the ordination of samples. The brachiopod *Rafinesquina* exhibits a large positive correlation with axis 1, suggesting that samples with large coordinate values on this axis, toward the right side of fig. 3.4, are relatively rich in this genus. The brachiopod *Platystrophia* and the bryozoan *Parvohallopora* exhibit large negative correlations with this axis, suggesting that they are abundant in samples with small coordinate values, toward the left. On axis 2, correlation values suggest that samples near the top are rich in *Platystrophia* and cf. *Dekayia*. Thus at locality 1, samples 1–0 to 1–40 should, on average, be fairly rich in *Rafinesquina*, whereas samples 1–50 to 1–90 should be relatively abundant in *Platystrophia*, cf. *Dekayia*, and *Parvohallopora*). In general,

FIGURE 3.4. Bivariate scatter plot of samples from localities 1 and 2, based on coordinate values for axes 1 and 2 derived from polar ordination of these samples. (Sample numbers are explained in text.) For this analysis, samples were percent-transformed: Solitary specimen counts and bryozoan point counts were percent-transformed separately and given equal weight. Similarities among samples were calculated with the quantified Dice coefficient. The cophenetic correlation coefficient for three axes = 0. 92. Genera exhibiting particularly diagnostic correlations with axes 1 and 2 are listed directly on the plot, with arrows indicating the "direction" of this correlation. Key to symbols for sample points: + = 1–0 to 1–40, x = 1–50 to 1–90; ◆ = 1–100 to 1–150; ▼ = 2–0 to 2–20, 2–40; and * = overlap of samples 1–90 and 2–30.

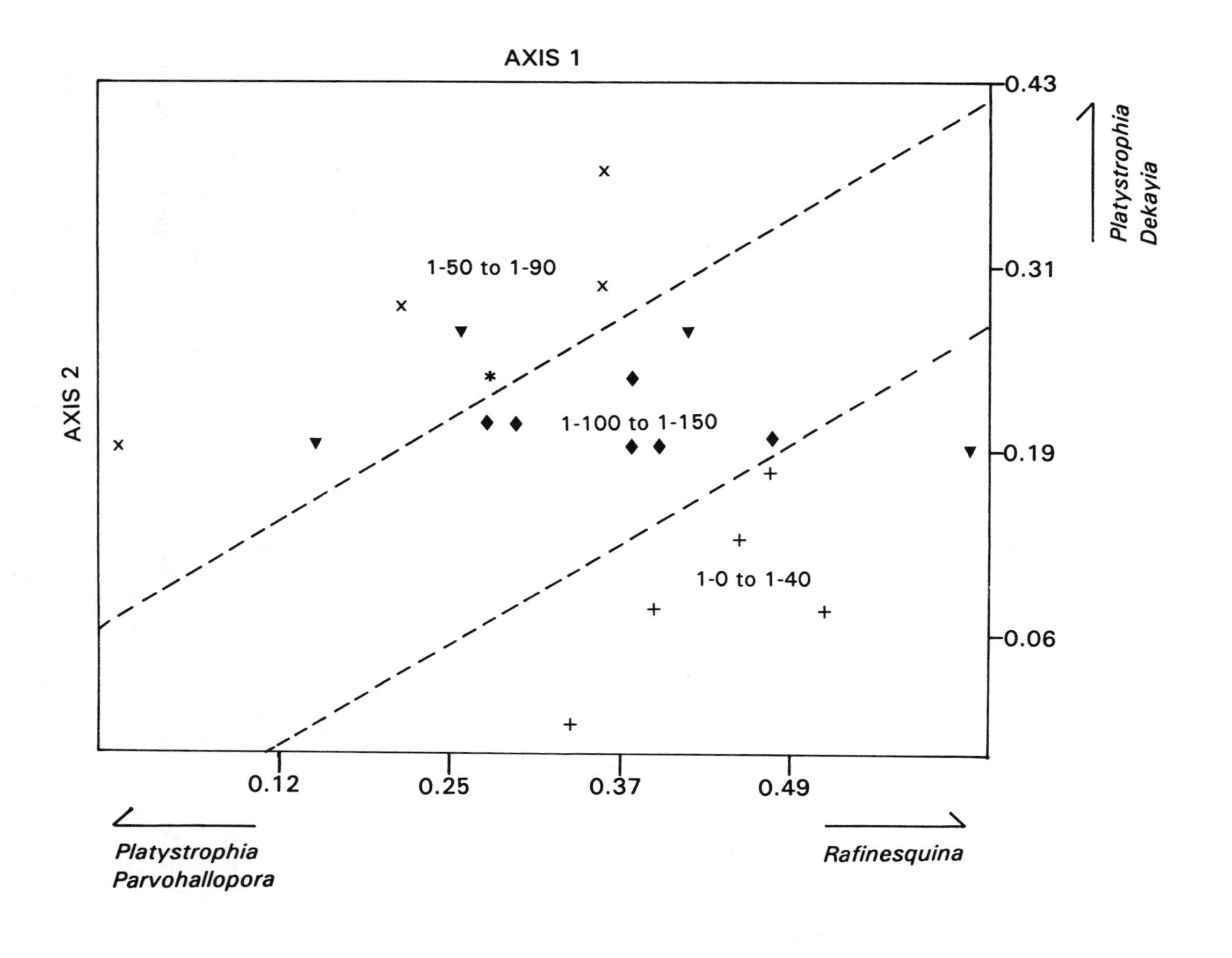

AXIS 1
AXIS 2
0.43
0.31
0.19
0.06
0.12
0.25
0.37
0.49
1-50 to 1-90
1-100 to 1-150
1-0 to 1-40
Platystrophia
Dekayia
Platystrophia
Parvohallopora
Rafinesquina

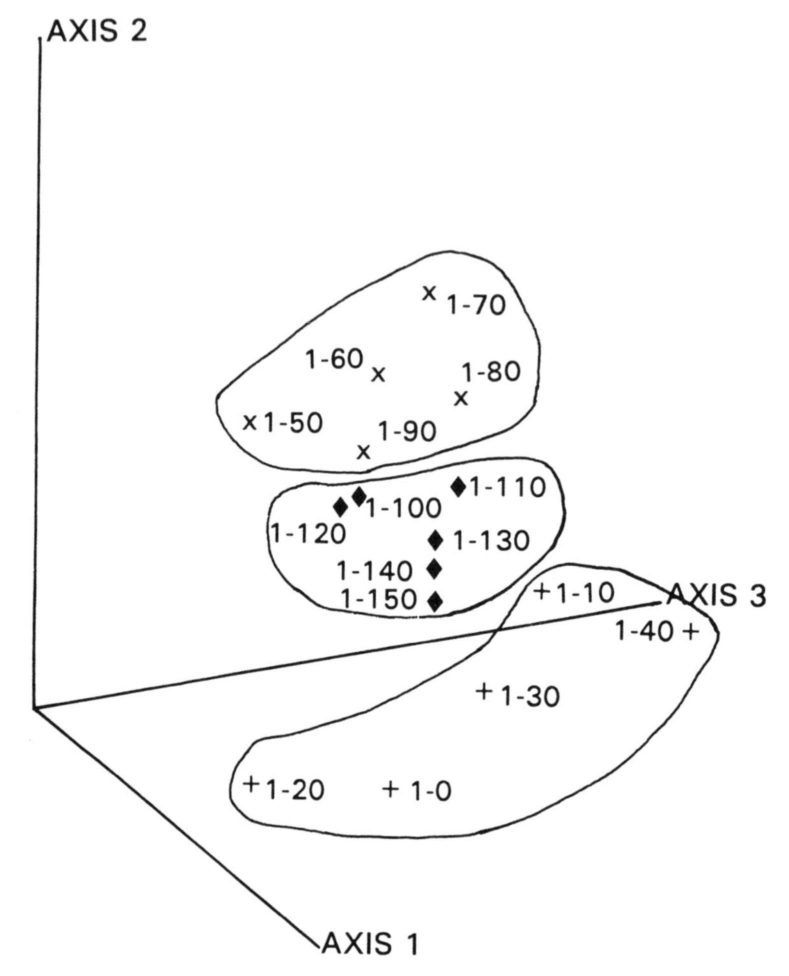

FIGURE 3.5. Perspective scatter plot of samples from locality 1, based on coordinate values for axes 1, 2, and 3, derived from polar ordination. (See caption to fig. 3.4 for details of ordination.)

the compositions of samples 1–100 to 1–150 should be intermediate to those of the other two groups.

These patterns are partly confirmed by the data in table 3.2. Of the three sample groups at locality 1, the mean abundance of *Rafinesquina* is greatest in 1–0 to 1–40 and smallest in 1–50 to 1–90; the opposite is true, as had been

TABLE 3.2 *Mean Abundances of Four Genera in Three Sample Groups at Locality 1 (for brachiopods, number of specimens; for bryozoans, number of points)*

Group	*Rafinesquina*	*Platystrophia*	*Parvohallopora*	*cf. Dekayia*
1–0 to 1–40	25.0	11.8	9.9	1.0
1–50 to 1–90	16.6	28.2	12.5	3.2
1–100 to 1–150	24.5	23.0	9.7	0.2

suggested by ordination, in the case of *Platystrophia*. The two bryozoan genera are more problematic: cf. *Dekayia* and *Parvohallopora* appear to be only slightly more abundant, on average, in samples 1–50 to 1–90 than elsewhere. Moreover, cf. *Dekayia* is a relatively uncommon faunal element at locality 1, except in a single sample (1–70; 8 points).

Thus, on the basis of abundance patterns for *Platystrophia* and *Rafinesquina*, faunal composition along the sampling horizon at locality 1 does not appear simply to vary randomly from sample to sample. Rather, there is compositional zonation on the transect: Among samples in the southern zone (1–0 to 1–40), abundances of *Rafinesquina* substantially exceed those of *Platystrophia*, whereas the opposite is the case among samples in the central zone (1–50 to 90); in the northern zone, abundances of these two genera are more equitable. This tendency for *Platystrophia* to be abundant where *Rafinesquina* is not was confirmed statistically by comparing their abundances among samples with the nonparametric Spearman's rank correlation coefficient. For these 16 samples, the coefficient had a value of -0.922, which is highly significant at $\alpha = .01$.

The two-dimensional scatter plot of ordination axes 1 and 2 (fig. 3.4) shows that the five samples from locality 2 did not group separately from those at locality 1. In fact, the locality 2 sample points are scattered among the others. Thus the ordination demonstrates that samples at locality 2 were not compositionally unique from those at locality 1.

Discussion

There are several potential explanations for the spatial variability recognized at locality 1:

1. *Patterns are generated by postmortem transport.* Because of substantial differences in size and shape (in this analysis, *Rafinesquina* was large and flat, whereas *Platystrophia* was smaller and somewhat inflated), it is arguable that variation in composition along the transect was a consequence of postmortem hydraulic sorting associated with storm deposition. However, the recognized distributional pattern suggests that this is not likely. The transition from *Rafinesquina* dominance to *Platystrophia* dominance is neither continuous nor monotonic along the transect, as might be expected in the event of storm sorting (Westrop 1986). Compositionally, the central zone of the transect is not transitional between the northern and southern zones. Rather, the northern zone is intermediate in composition between the southern and central zones.
2. *Transitions in brachiopod abundance are related to spatial patterns among bryozoans.* The species of *Platystrophia* recognized in this study is morphologically comparable to those found by Richards (1972) to have been attached to bryozoans and large shell fragments. Thus, it is intuitively appealing to suggest that increased numbers of *Platystrophia* in the central zone of the transect are associated with increased abundances of bryozoans, which

could have served as attachment substrates. However, the data do not support this scenario; although the polar ordination suggests some relationship between *Platystrophia* abundances and those of *Parvohallopora* and *Dekayia,* the actual abundance data do not (table 3.2). No clear-cut spatial pattern is indicated with respect to bryozoan abundances along the transect, although this may reflect a shortcoming of the sampling scheme (see below).

3. *The faunal transitions may be mirroring environmental transitions on the Ordovician seafloor.* Environmental zonation on seafloors, at the spatial scale of the faunal zonation at locality 1 (ca. 50 m), is quite common. Moreover, it has been demonstrated in the Recent that spatial patterns in subfossil accumulations may faithfully reflect small-scale environmental transitions, even in the wakes of storms (Miller 1988; Miller et al. 1992). Thus, spatial faunal patterns on the transect may be associated with small-scale changes in the physical environment. Although there were no obvious transitions in the character of the sampling horizon along the transect, a petrographic analysis of the bed has not been performed and might yield valuable insight into small-scale environmental changes.
4. *The faunal transitions may simply reflect random faunal variability on the Ordovician seafloor, although this record is obviously time-averaged.* It is generally accepted that random patchiness among benthic marine organisms is ephemeral and may not be preserved because of time-averaging (Walker and Bambach 1971; Peterson 1976; Staff et al. 1986). However, if the patches were sufficiently large, they might *not* have been ephemeral on the time scale of skeletal accumulation that comprises the sampling horizon. In a sense, this scenario represents a fallback position if all other possible interpretations fail. It is not directly testable, because there is no way of assuredly documenting random variability in an environmentally "homogeneous" setting on an ancient seafloor. If lithologic or other data were to suggest environmental homogeneity, it could always be argued that evidence of some subtle environmental transition has been obscured by a plethora of processes involved in the formation of the unit in question.

Further complicating any interpretation is the matter of storm amalgamation, described earlier. Given that the bed may encompass several storm events, it can logically be questioned whether the same overall pattern would emerge if, rather than sampling the entire bed, discrete, "microstratigraphic" skeletal concentrations within each sample were examined individually and treated as independent samples. This relates to the question of environmental "lifespan" on the Ordovician seafloor. Although there may be no direct way to measure the degree to which such environments were ephemeral, my experience working in the modern carbonate environments of St. Croix, U.S. Virgin Islands, suggests that benthic environmental zones on the scale of 50–100 m in diameter have sufficient temporal "staying power" to outlast several storm events. The implication is that microstratigraphic evaluation of the Fairview sampling horizon would yield a spatial picture comparable to that exhibited with the relatively coarse stratigraphic sampling of the present study. A test of this hypothesis is ongoing; the transect has been resampled for the purpose of microstratigraphic analysis. This further evaluation will also permit a more rigorous assessment of patterns among bryozoans. As noted earlier, in the present study, the zone of greatest apparent bryozoan concentration was singled out in each sample for preparation of acetate peels. However, because of amalgamation, there were typically several concentrations in a bulk sample, and the specific choice of a single horizon was somewhat arbitrary. Possible associations between *Platystrophia* and certain bryozoan genera may have been overlooked if the bryozoan horizon selected was microstratigraphically removed from concentrations of *Platystrophia* within the sample. This potential problem can be overcome by separately evaluating *every* bryozoan concentration, as is the case in the new analysis.

It is interesting that there were no net differences in faunal composition between localities 1 and 2, given that they were separated by 10 km, and locality 2 is thought to have been located further down the paleoramp (see earlier description). Because the extent of the transect at locality 2 was limited, it is difficult to assess the degree of within-locality spatial variability relative to that at locality 1.

Concluding Remarks

The localities and horizon evaluated in this initial study were chosen strictly because of logistical ease of sampling; there was no advance sense that they would yield the observed pattern. Thus it is tempting to suggest that lateral variability at this spatial scale is common in the fossil record, although considerably more work is required to substantiate such a claim. Despite the uncertainty regarding explanation of the spatial patterns at locality 1, the mere presence of measurable variability suggests that further exploration in the ancient at this limited spatial scale is worth the effort, when feasible logistically. The traditional limitation of lateral sampling to widely spaced outcrops, with no lateral sampling along individual horizons at outcrop scale, ignores a potential source of meaningful data for a variety of paleontological investigations. For example, paleobiologists continue to be concerned with determining the relative contributions of temporal changes in within-community (alpha) and between-community (beta) diversity to Phanerozoic transitions in faunal marine diversity recognized on a global scale (see Sepkoski 1988). A calibration of such changes that is more precise than has heretofore been possible will necessarily require direct assessment of small-scale spatial variability in faunal content, because it is only at this limited scale, where communities and intergradations among them are potentially discernible, that alpha diversity and beta diversity can be distinguished definitively. The present study suggests that the possibility of making such distinctions at even fine scales is not simply a pipe dream.

Finally, this initial study suggests a general caution: If two widely spaced samples from the same horizon are highly comparable in composition, this does not necessarily imply that the unsampled area between them is just as comparable.

ACKNOWLEDGMENTS

I thank Sharon Diekmeyer, Mike Armstrong, and Rich Terry for assistance in field work and laboratory preparation of samples. Acknowledgment is made to the donors of the Petroleum Research Fund, administered by the American Chemical Society, for support of this research.

REFERENCES

Aigner, T. 1979. Schill-Tempestite im Oberen Muschelkalk (Trias, SW-Deutschland). *Neues Jahrbuch Fuer Geologie und Palaeontologie, Abhandlungen* 157:326–43.

———. 1980. Storm deposits as a tool in facies analysis. I. Calcareous Tempestites, *International Association of Sedimentologists*, 1st European Meeting, Bochum, *Abstract*, pp. 44–46.

———. 1985. *Storm Depositional Systems*. Berlin: Springer-Verlag.

Aigner, T. and H. Reineck. 1982. Proximality trends in modern storm sands from the Helgoland Bight (North Sea) and their implications for basin analysis. *Senckenbergiana maritima* 14:183–215.

Anstey, R. L. and T. G. Perry. 1972. Eden Shale Bryozoans: A Numerical Study (Ordovician, Ohio Valley). *Publications of the Museum, Michigan State University*, 1:1–80.

Beals, E. W. 1965. Ordination of some corticolous cryptogamic communities in south-central Wisconsin. *Oikos* 16:1–8.

Bray, J. R. and J. T. Curtis. 1957. An ordination of the upland forest communities of southern Wisconsin. *Ecological Monographs* 27:325–49.

Bosence, D. W. J. 1979. Live and dead corals from coralline algal gravels, Co. Galway. *Palaeontology* 22:449–78.

Cisne, J. L. and G. O. Chandlee. 1982. Taconic Foreland Basin graptolites: Age zonation, depth zonation, and use in ecostratigraphic correlation. *Lethaia* 15:343–63.

Cisne, J. L. and B. D. Rabe. 1978, Coenocorrelation: Gradient analysis of fossil communities and its applications in stratigraphy. *Lethaia* 11:341–64.

Cisne, J. L., G. O. Chandlee, B. D. Rabe, and J. A. Cohen. 1982a. Clinal variation, episodic evolution, and possible parapatric speciation: The trilobite *Flexicalymene senaria* along an Ordovician depth gradient. *Lethaia* 15:325–41.

Cisne, J. L., D. E. Karig, B. D. Rabe, and B. J. Hay. 1982b. Topography and tectonics of the Taconic outer trench slope as revealed through gradient analysis of fossil assemblages. *Lethaia* 15:229–46.

Cummins, H., E. N. Powell, H. J. Newton, R. J. Stanton, Jr., and G. Staff. 1986a. Assessing transport by the covariance of species with comments on contagious and random distributions. *Lethaia* 19:1–22.

———. 1986b. The rate of taphonomic loss in modern benthic habitats: How much of the potentially preservable community is preserved? *Palaeogeography, Palaeoclimatology, Palaeoecology* 52:291–320.

Feldman, H. 1985. Benthic faunal patchiness on soft substrates in modern and Paleozoic communities. *Geological Society of America Abstracts with Programs* 17:580.

———. 1988. Scales of patchiness in benthic marine faunas with an example from the

Waldron Shale (Silurian, Indiana). *Geological Society of America Abstracts with Programs* 20:255.
Holland, S. M. 1990. Distinguishing eustasy and tectonics in foreland basin sequences: The Upper Ordovician of the Cincinnati Arch and Appalachian Basin. Ph.D. dissertation, University of Chicago.
Jennette, D. 1986. Storm-Dominated Cyclic Sedimentation on an Intracratonic Ramp: Kope-Fairview Transition (Upper Ordovician), Cincinnati, Ohio Region. Unpublished M.S. Thesis at the University of Cincinnati.
Jennette, D. C. and W. A. Pryor. 1993. Cyclic alternation of proximal and distal storm facies: Kope and Fairview formations (Upper Ordovician), Ohio and Kentucky. *Journal of Sedimentary Petrology* 63:183–203.
Johnson, R. G. 1965, Pelecypod death assemblages in Tomales Bay, California. *Journal of Paleontology* 39:80–85.
Karklins, O. L. 1984, Trepostome and cystoporate bryozoans from the Lexington Limestone and the Clays Ferry Formation (Middle and Upper Ordovician) of Kentucky. United States Geological Survey Professional Paper, 1066–I.
Kidwell, S. M. and D. W. J. Bosence. 1991. Taphonomy and time-averaging of marine shelly faunas. In P. A. Allison and D. E. G. Briggs, eds., *Taphonomy: Releasing the Data Locked in the Fossil Record,* pp. 115–209. New York: Plenum Press.
Kreisa, R. D. 1981a, Storm-generated sedimentary structures in subtidal marine facies with examples from the Middle and Upper Ordovician of southwestern Virginia. *Journal of Sedimentary Petrology* 51:823–48.
———. 1981b. The origin of stratification in a Paleozoic epicontinental sea: The Cincinnatian Series. *Geological Society of America Abstracts with Programs* 13:491.
Kreisa, R. D. and R. K. Bambach. 1982. The role of storm processes in generating shell beds in Paleozoic shelf environments. In G. Einsele and A. Seilacher, eds., *Cyclic and Event Stratification,* pp. 200–207. Berlin: Springer-Verlag.
Lawrence, D. R. 1968. Taphonomy and information losses in fossil communities. *Geological Society of America Bulletin* 79:1315–30.
MacDonald, K. B. 1976. Paleocommunities: Towards some confidence limits. In R. W. Scott and R. R. West, eds., *Structure and Classification of Paleocommunities,* pp. 87–106. Stroudsburg, PA: Dowden, Hutchinson and Ross.
Meldahl, K. H. and K. W. Flessa. 1990. Taphonomic pathways and comparative biofacies and taphofacies in a Recent intertidal/shallow shelf environment. *Lethaia* 23:43–60.
Miller, A. I. 1988. Spatial resolution in subfossil molluscan remains: Implications for paleobiological analyses. *Paleobiology* 14:91–103.
Miller, A. I., G. Llewellyn, K. M. Parsons, H. Cummins, M. R. Boardman, B. J. Greenstein, and D. K. Jacobs. 1992. Effect of Hurricane Hugo on molluscan skeletal distributions, Salt River Bay, St. Croix, U.S. Virgin Islands. *Geology* 20:23–26.
Peterson, C. H. 1976. Relative abundances of living and dead molluscs in two California lagoons. *Lethaia* 9:137–48.
Richards, R. P. 1972. Autecology of Richmondian brachiopods (Late Ordovician of Indiana and Ohio). *Journal of Paleontology* 46:386–405.
Sepkoski, J. J., Jr. 1988. Alpha, beta, or gamma: Where does all the diversity go? *Paleobiology* 14:221–34.
Springer, D. A. and R. K. Bambach. 1985. Gradient versus cluster analysis of fossil assemblages: A comparison from the Ordovician of southwestern Virginia. *Lethaia* 18:181–98.
Staff, G. M., R. J. Stanton, Jr., E. N. Powell, and H. Cummins. 1986. Time-averaging, taphonomy, and their impacts on paleocommunity reconstruction. *Geological Society of American Bulletin* 97:428–43.

Stanton, R. J., Jr. 1976. Relationship of fossil communities to original communities of living organisms. In R. W. Scott and R. R. West, eds., *Structure and Classification of Paleocommunities*, pp. 107–142. Stroudsburg, Pa.: Dowden, Hutchinson and Ross.

Tobin, R. C. 1982. A Model for Cyclic Deposition in the Cincinnatian Series of Southwestern Ohio, Northern Kentucky, and Southeastern Indiana. Unpublished Doctoral Dissertation at the University of Cincinnati.

Tobin, R. C. and W. A. Pryor. 1981, Sedimentological interpretation of an Upper Ordovician carbonate-shale vertical sequence in northern Kentucky. In D. L. Meyer, R. C. Tobin, W. A. Pryor, W. B. Harrison, and R. G. Osgood. *Stratigraphy, Sedimentology, and Paleoecology of the Cincinnatian Series (Upper Ordovician) in the Vicinity of Cincinnati, Ohio.* Geological Society of America 1981 Field Trip Guidebook, Field Trip No. 12, pp. 49–57.

Tobin, R. C. and W. A. Pryor. 1985. Cincinnatian series-model for cyclic and episodic deposition of carbonates and shales on a storm-dominated ramp. *American Association of Petroleum Geologists Bulletin* 69:311–12.

Walker, K. R. and R. K. Bambach. 1971. The significance of fossils from fine-grained sediments: Time-averaged communities. *Geological Society of America Abstracts with Programs* 3:783–84.

Warme, J. E. 1969. Live and dead molluscs in a coastal lagoon, *Journal of Paleontology* 43:141–50.

———. 1971. Paleoecological aspects of a modern coastal lagoon, University of California Publications in Geological Sciences 87.

Warme, J. E., A. A. Ekdale, S. F. Ekdale, and C. H. Peterson. 1976. Raw material of the fossil record. In R. W. Scott and R. R. West, eds., *Structure and Classification of Paleocommunities*, pp. 143–70. Stroudsburg, Pa.: Dowden, Hutchinson and Ross.

Westrop, S. R. 1986. Taphonomic versus ecologic controls on taxonomic relative abundance patterns in tempestites. *Lethaia* 19:123–32.

4

The Ichnological Signature of Storm Deposits: The Use of Trace Fossils in Event Stratigraphy

S. George Pemberton and James A. MacEachern

ABSTRACT

Trace fossils are proving to be powerful tools for the recognition and interpretation of event beds. Tempestites contain a mixed trace fossil assemblage that reflects fluctuations in energy levels. The two different ichnocoenoses reflect varying behavioral responses of the organisms colonizing two successive, individually distinct habitats. The resident or fair-weather ichnocoenose can be considered representative of a stable benthic community, within which individual populations are at or near their carrying capacity. Periodic generation of the storm ichnocoenose, on the other hand, represents the flourishing of a community of opportunistic organisms in an unstable, high-stress, physically controlled environment. The general succession, typical of tempestites, consists of (1) a fair-weather resident trace fossil suite; (2) a sharp basal contact, with or without a basal lag; (3) parallel to subparallel laminations (reflecting hummocky or swaley cross-stratification); (4) common escape structures; (5) the dwelling burrows of opportunistic organisms that colonize the unexploited storm unit; (6) gradational burrowed tops, representative of bioturbation resulting from subsequent burrowing by organisms from higher colonization levels; and (7) a fair-weather resident trace fossil suite indicative of a return to quiescent conditions following storm abatement.

In his recent book on the Burgess Shale Stephen Jay Gould wrote: "Taxonomy (the science of classification) is often undervalued as a glorified form of filing—with each species in its folder, like a stamp in its prescribed place in an album; but taxonomy is a fundamental and dynamic science, dedicated to exploring the causes of relationships and similarities among organism. Classifications are theories about the basis of natural order, not dull catalogues compiled to avoid chaos" (1989:98). The same could be said about stratigraphy. It was once considered a somewhat routine and mundane discipline, consisting mainly of the dry cataloguing of lithostratigraphic units. However, it has recently undergone a dramatic renaissance, and during the last decade, stratigraphers have altered the way we perceive and therefore interpret the rock record. Once almost exclusively the domain of biostratigraphers, the stratigraphic column has been subjected to new ideas and methods. A synergistic approach has resulted in the development and refinement of new stratigraphic paradigms such as seismic stratigraphy, allostratigraphy, tephrastratigraphy, magnetostratigraphy, sequence stratigraphy, and event stratigraphy, to name just a few.

The stratigraphic utility of trace fossils can take on many guises, and their significance varies, depending on what stratigraphic paradigm one is employing. In the past, trace fossils were considered to be almost useless in stratigraphy because (1) most have long temporal ranges; (2) they are somewhat facies dependent; (3) a particular structure may be produced by the work of two or several different organisms living together, or in succession, within the structure; (4) the same individual or species of organism may produce different structures corresponding to different behavior patterns; (5) the same individual may produce different structures corresponding to identical behavior but in different substrates (e.g., in sand, in clay, or at sand and clay interfaces); and (6) identical structures may be produced by the activity of systematically different trace-making organisms, where behavior is similar (Bromley and Fürsich 1980). These factors combined to make their biostratigraphic value negligible. Traditionally, there were only three ways in which trace fossils were thought to be utilized in chronostratigraphy: (1) as the means to trace the evolution of behavior; (2) as morphologically defined entities (with no assumptions concerning their genesis); and (3) as substitutes for the trace-making organisms (Magwood and Pemberton 1990). On the other hand, trace fossils are proving to be one of the most important groups of fossils in delineating stratigraphically important boundaries related to sequence stratigraphy (MacEachern et al. 1991, 1992; Savrda 1991a, 1991b) and allostratigraphy (Pemberton et al. 1992a). Likewise, event stratigraphy is another area where trace fossils have proven to be a powerful tool for the recognition and interpretation of event beds (Seilacher 1981, 1982a, 1982b; Wheatcroft 1990; Pemberton et al. 1992b; Frey and Goldring 1992).

Event layers are generally regarded as strata that are attributed to rapid, episodic deposition and include such diverse entities as volcanic ash beds, or tephra deposits (i.e., Pedersen and Surlyk 1983), beds resulting from seismic shocks, or seismites (Seilacher 1969), and episodic sedimentation events such as turbidites (Seilacher 1962); tempestites, or storm deposits (Aigner 1985); phytodetritus pulses (Rice et al. 1986); and inundites, or flood deposits (Leithold 1989). Although trace fossils are known to be significant features of most of these deposits, this paper deals mainly with the ichnological signature of storm deposits.

Episodic Depositional Events

During the last two decades geologists have begun to recognize the extent and significance of relatively rapid influxes of sediment (episodic depositional events) in the rock record. In fact, Dott (1983) has successfully argued that most of the sedimentary record, rather than reflecting day-to-day, steady-state conditions, typically represents episodic or discontinuous depositional events. Initial research was confined almost exclusively to the deep-water, outer continental shelf and slope environments, where gravity-induced turbidity currents constitute one of the more important operating mechanisms of deposition. More recently, however, sedimentologists have focused on episodic depositional events in shallow, coastal areas. Such events, generally the result of major storms or hurricanes, are well documented in recent environments (e.g., Hayes 1967; Morton and Paine 1985; Morton 1988) and are being recognized with more frequency in the rock record (e.g., Dott and Bourgeois 1982; Walker 1984a; Duke 1985; and Duke et al. 1991).

In a seminal paper, Seilacher (1982a) succinctly pointed out that episodic sedimentation events (turbidites, tempestites, inundites, and phytodetritus pulses) have a number of common characteristics: (1) they reflect the onset, culmination, and waning of water turbulence during the event, by distinctive erosional and depositional structures; (2) they redistribute the organic and inorganic sediment material along a vertical (bottom to top) and horizontal (shallow to deep) gradient; and (3) they change the ecological situation for benthic organisms by altering the consistency and/or the food content of the bottom for a biologically relevant period of time after the event. The two most prolific types of event beds are turbidites and tempestites.

Turbidites

Perhaps the best-known event bed is the turbidite, which results from a single, short-lived event, a turbidity current. First recognized by Kuenen and Migliorini (1950) turbidity currents are a form of density current that flows downslope on the ocean floor driven by gravity acting on the density difference between the suspended sediment and the surrounding seawater. Walker (1984b) stressed that turbidity currents can be considered commonplace in modern oceans and that their resultant deposits are extensive and volumetrically important.

Walker (1984b) summarized a set of descriptors that characterize the features of classical turbidites:

1. Turbidite successions consist of sandstones and shales that are monotonously interbedded through many tens or hundreds of meters. (The beds typically have flat tops and bottoms with no scouring or channeling greater than a few centimeters.)
2. Sandstone beds have sharp, abrupt bases and typically grade upward into finer sand, silt, and clay.

3. Sandstone soles have abundant markings that reflect either physical (tool marks, scour marks) or biogenic (trails or burrows) activity.
4. Within the sandstone beds combinations of parallel lamination, ripple cross-lamination, climbing ripple cross-lamination, convolute lamination, and graded bedding result in a recurring sequence that reflects stages in waning current flow. This idealized sequence was first described by Bouma in 1962 and is known as the Bouma sequence.

The Bouma sequence represents one of the most elegant facies models, as it acts as a basis for hydrodynamic interpretations. The sequence can be divisible into a number of discrete zones or divisions using textural and structural features: (1) division A is distinguished by a graded division that lacks internal lamination and is either massive or normally graded; (2) division B is typified by parallel laminations with parting lineations; (3) division C is distinguished by current ripple cross lamination; (4) division D is distinguished by the interlamination of clean silt and mud; and (5) division E consists only partly of mud introduced by the current (Bouma 1962).

The ichnology of turbidites is very well known, and a number of excellent recent papers describe pre- and postturbidite ichnofossil assemblages (Wetzel 1991; Miller 1991a, 1991b; and Leszczynski and Seilacher 1991).

TEMPESTITES

Major problems exist in the recognition and interpretation of storm events because shallow coastal zones comprise a multitude of depositional environments commonly differing from one another in only subtle ways. Successions of rock types and sedimentary structures that are of value in delineating deep-water turbidity currents have, at present, somewhat limited significance in shallow-water deposits. One particular type of bedding (hummocky cross-stratification), however, seems to be formed exclusively by storm waves (Walker 1984a). Although well studied, the genesis of hummocky cross-stratification and the associated swaley cross-stratification (cf. Leckie and Walker 1982) remains the topic of considerable debate. For details concerning specific issues in this controversy see Walker (1984b), Leckie and Krystinik (1989), Duke (1990), and Duke et al. (1991).

Duke et al. (1991) indicated that the main points of contention involve (1) the class of flow that generates hummocky cross-stratification; (2) the bed configuration(s) that produce(s) hummocky cross-stratification; and (3) the nature of the storm-induced currents. Further complicating matters is the fact that although Walker (1984a) suggested that hummocky cross-stratification can form only below fair-weather wave base, Harms (1979) estimated that it can form in water depths ranging from 5 to 30 m, and Hunter and Clifton (1982) indicated a depth as shallow as 2 m. Recently, Cotter and Graham (1991) recognized hummocky cross-stratification in what they interpreted as late Devonian fluvial deposits from Ireland. Traditional paleontological control is rarely useful in such determinations because most sandy clastic units are commonly devoid of body fossils. These same units, however, are commonly replete with trace fossils (Pemberton and Frey 1984; Vossler and Pemberton 1988; Dam 1990; Frey 1990; Frey and Goldring 1992).

Storm units represent individual sandstone beds that were deposited rapidly (from a few hours to several days) from single waning flow events, which generally possess a strong oscillatory flow component (Dott 1983). They can superficially resemble turbidites but are generally distinguished by the presence of wave-generated and/or combined-flow sedimentary structures and shallow marine faunas in the intercalated shale deposits (Johnston and Baldwin 1986). Conceptual models of storm-derived units have been constructed (e.g., Dott and Bourgeois 1982) and interpreted. These idealized successions generally show the following characteristics:

1. *Storm erosion:* a basal erosion surface that may be undulatory, with sole marks, and intraclasts of pebbles, shells, or mudstone, locally sideritized.
2. *Main storm deposition:* main hummocky or swaley cross-stratification interval, locally with associated horizontal or parallel lamination.
3. *Waning storm deposition:* combined-flow and/or wave-rippled sand layer, indicating a progressive return to lower flow regime oscillatory conditions.
4. *Poststorm—fair-weather mud deposition:* reflecting either the final suspension fall-out of storm-derived sediment (i.e. , poststorm mud) or the return to normal background sedimentation (i.e. , fair-weather mud) (Johnston and Baldwin 1986).

Biological Consequences of Event Beds

Population Strategies

In recent ecological studies of benthic organisms, equilibrium (or K-selected) species have been distinguished from opportunistic (or r-selected) species (table 4.1). In general, opportunistic species can respond rapidly to an open or unexploited niche and are characterized by (1) a lack of equilibrium population size, (2) a density-independent mortality, (3) the ability to increase abundance rapidly, (4) a relatively poor competitive ability, (5) high dispersal ability, and (6) a high proportion of resources devoted to reproduction (Grassle and Grassle 1974). Opportunistic organisms display an r-strategy in population dynamics, emphasizing rapid growth rate (r), whereas equilibrium species adopt a K-strategy, based on the carrying capacity of the environment (K) (Boesch and Rosenberg 1981). Short generation span is the most important mechanism for increasing population size in an r-strategy; therefore life spans of opportunistic species are shorter, and sexual maturation is reached earlier (Rees et al. 1977). Broad environmental tolerances and generalized feeding habits facilitate rapid colonization of open niches (Pianka 1970; Wolff 1973).

Opportunistic organisms are tolerant of conditions that are physiologically

TABLE 4.1 *Characteristics of r-Selected and K-Selected Populations (Modified After Pemberton* et al., *1992b).*

r-Selected (opportunistic)	K-selected (equilibrium)
1. Responds rapidly to open niche	1. Slow to colonise
2. Lacks equilibrium population size	2. Reach population size equilibrium
3. Density-independent mortality	3. Density-dependent mortality
4. Relatively poor competition ability	4. Excellent competitors
5. High dispersal ability	
6. Efficient reproductive system	

stressful. As such, their lifestyles and feeding strategies can be extremely variable (Grassle and Grassle 1974). In most cases, opportunistic colonists are either suspension-feeding or surface deposit-feeding polychaetes (table 4.2). However, Cadée (1984) stressed that opportunistic organisms have a great capacity to vary their feeding habits, depending on food availability. Thus if suspension feeding is the most viable strategy, opportunistic organisms will utilize it. Many organisms that inhabit brackish water environments can be considered omnivores or trophic generalists. Wolff (1973) found in the Dutch estuaries that 35% of the species were omnivores, whereas this percentage was only 6% to 16% in adjacent freshwater or fully marine areas. This occurrence of wide niches is related to an opportunistic life history. Goerke (1971) found that in estuaries the polychaete *Nereis diversicolor* can employ several feeding techniques and may act as a deposit feeder, a suspension feeder, a scavenger, and a predator, depending on prevailing conditions.

Rhoads et al. (1978) found that opportunistic colonists usually live in dense clusters. Such gregarious colonization can inhibit competitors from settling. Whitlatch and Zajac (1985) stated that among opportunistic species, individuals usually settle near conspecifics, indicating that gregarious settling behavior is based on the presence of others rather than on the availability of a preferred substrate. They went on to conclude that since these organisms usually brood their larvae, they can settle immediately upon release from the adult, allowing rapid crowding in a given space. The initial colonists thus preempt space, inhibiting the settlement of other species.

RECOLONIZATION FOLLOWING DISRUPTION

The mechanics of initial larvae settlement are presently undergoing intense scrutiny. In the past, the most favorable hypothesis involved active habitat selection by the organism. In this hypothesis the larvae actively select the habitat by employing a variety of strategies, such as swimming, substrate selection, and inhibiting metamorphosis until a suitable area was found (Butman 1987). Currently, however, benthic ecologists are realizing the importance of hydrodynamics in larval settlement, and a new hypothesis, termed *passive*

TABLE 4.2 *Modern Examples of Recolonisation Following Defaunation (modified after Vossler and Pemberton, 1988).*

Author	Setting	Stress	First colonist	Second colonist	Others
Grassle & Grassle 1974	Estuary nearshore	Oil spill	1st month: *Capitella capitata*	8 months: *Polydora ligni*	*Microphthalmus aberrans, Syllides verilli, Streblospio benedicti*
Dauer & Simon 1976	Intertidal	Red tide	1st month: *Polydora ligna*	2–3 months: *Eteone* heteropod *Nereis succinea*	year 2: *Capitata ambiseta, Minuspio cirrifera, Trovisia* sp.
Boesch et al. 1976	Embayment	Storm	2 months: *Melita nitida, Streblospio benedicti, Scoloplo fragilis, Glyeinde solitaria*		
Pearson & Rosenberg 1978	Microtidal fiord	Pollution	3 years: *Capitella capitata*	*Scolepsis fulignosa*	
Grassle 1977	Deep sea (1760m)	Experimental defaunation	2 months: *Priapulis atlantsi*	28 months: *Capitella* sp.	
McCall 1977	Embayment	Experimental defaunation	10 days: *Streblospio benedicti, Capitella capitata, Ampelisca abdita*	50 days: *Nucula proxima*	
Rees et al. 1977	Shallow marine	Storms			Mobile predators with short life cycles
Rhoads et al. 1978	Estuary	Dredging	10 days: *Streblospio benedicti*	29–50 days: *Capitella capitata Ampelisca abdita*	50 days: *Nucula annulata* 86 days: *Tellina agilis* 175 days: *Nephthys incisa*
Sanders et al. 1980	Marine estuary	Oil spill	1st 11 months: *Capitella*	2d year *Mediomastus*	
Desbryers et al. 1980 (*in* Thistle, 1981)	Offshore (2160m)	Experimental defaunation	6 months: *Prionspio* sp. *Ophryaboche puerilis*		
Bonsdorff 1983	Shallow, brackish	Dredging	*Nereis diversicolor Corophium volutator Macoma balthica*		
Bonvicini Paglial et al. 1985	Shallow marine	Dredging	6 months: *Corbula gibba, Lumbrineresis gracilis Nephthys sphaerocirrata Prionospio malmgreni Scolelepsis fuliginosa*		
Whitlatch & Zajac 1985	Estuary	Experimental defaunation	20 days: *Streblospio benedicti, Hobsonia florida, Polydora ligni Capitella capitata*	30 days: *Microdeutopus gryllotalpa, Corophium insidiosum, Nemetostella vectensis*	
Berry 1989	Shallow marine	Storms	2 months: *Phragmatopema lapidosa, californica*		

deposition, is gaining wide acceptance. Butman (1987) summarized the two main points of view regarding passive deposition as the following: either (1) larvae are deposited over broad areas but differentially survive only in hospitable adult habitats or (2) species-specific larval fall velocities correspond with particular sediment fall velocities, such that hydrodynamically similar particles and larvae are deposited in the same environment. Recent work on larvae in sediment flumes (Jumars and Nowell 1984) has confirmed that larvae respond to prevailing hydrodynamic conditions. Episodic depositional events, like storms and turbidites, therefore can have a powerful effect on the redistribution of the larvae of benthic organisms (Rees et al. 1977; Hagerman and Rieger 1981; Dobbs and Vozarik 1983). In fact, Berry (1989) stressed that adaptations shown by opportunistic organisms are analogous to those exhibited by several species of terrestrial plants in fire-disturbed communities. Intense wave events and fires are analogous disturbances, in that both have similar temporal patterns and ecological effects, and both exert similar selection pressures on species in each community (Berry 1989).

At present, most benthic biologists believe that larval settlement involves a complex interaction between active and passive processes. Competent planktonic larvae initially reach the seafloor at sites where passively sinking particulates, having fall velocities similar to the larvae, initially settle (Hannan 1984). Other biological or physical processes may then redistribute them. In this way, larvae are passively deposited and accumulate at the large spatial scales (tens of meters or kilometers) that apply to sediment transport and deposition. Active habitat selection occurs over much smaller scales (centimeters or meters) within these broad depositional areas (Butman 1987).

In most event deposits, therefore, initial larval settlement may be a function of the hydrodynamics of the event bringing both larvae and sediment into the area (table 4.3). After initial settlement, exploitation of the open niche becomes more a function of the reproductive characteristics of the individual species. In most cases, because of their efficient reproduction cycles, opportunistic species quickly dominate the initial stages of recolonization. Table 4.4 summarizes the life history characteristics of seven of the most common colonizers. Zajac (1986) reported that adults of opportunistic species become sexually active after a physical disturbance or after seasonal depopulation; this allows for a rapid increase in the size of the larval pool. Gray (1974) noted that opportunistic organisms can have both planktonic and benthic larvae, thereby giving them flexible reproductive capabilities.

Adult recruitment from surrounding areas into naturally depopulated zones can also be appreciable (Santos and Simon 1980). This can be facilitated by the

TABLE 4.3 *Major Parameters Affecting the Rate and Pattern of Colonization Following a Disturbance (Adapted from Sousa 1984)*

1. The morphological and reproductive traits of species that are present on the site when the disturbance occurs. Such traits determine, in part, the likelihood that these species will survive the event and rapidly reoccupy the site.
2. The reproductive biology of species that were not present on the site when it is disturbed but that have occupied it previously or live within dispersal distance of it.
3. Characteristics of the disturbed patch, including
   a. The intensity and severity of the disturbance that created it.
   b. Its size and shape.
   c. Its location and degree of isolation from sources of colonists.
   d. The heterogeneity of its internal environment.
   e. The time of year it was created (reflecting seasonal controls on recolonisation).

TABLE 4.4 *Life-History Characteristics of the Seven Most Common Macrofaunal Taxa Found After Defaunation of an Estuary (Modified After Whitlatch and Zajac 1985).*

Taxon	Feeding/mortality type	Reproductive features
Polychaetes		
Streblospio benedicti	Bipalpate; surface deposit feeder; tybe-dwelling	Larviparous; *planktonic phase: 1–14 days. generation time: 30–50 days*
Hobsonia florida	Multitentaculate; surface deposit feeder; tube-dwelling	Tube brooding; planktonic phase (?); generation time: 25–35 days
Polydora ligni	Bipalpate; surface deposit feeder; tube-dwelling	Tube brooding; plankton phase: 2–10 days; generation time: 30–40 days
Capitella capitata	Subsurface deposit feeder	Tube brooding; plankton phase: several hours; generation time: 30–40 days
Amphipod crustaceans		
Corophium insidiosum	Surface deposit feeder; tube-dwelling	Brooding; generation time: 30–90 days
Microdeutopus gryllotalpa	Surface deposit/suspension feeder; tube-dwelling	Brooding generation time: 30–50 days
Anthozoan		
Nematostella vectensis	Infaunal; zooplanktivore (?)	?

*Approximate times based upon summer water temperatures.

transport and subsequent relocation of adults by storm currents. Rees et al. (1977) noted that after storm abatement there was a large mass stranding and redistribution of many species, with both adult organisms and their larval forms being displaced to open ocean areas, thus facilitating rapid recolonization. Dobbs and Vozarik (1983) concluded that disturbance by storms may be a mechanism for wide postlarval dispersal of ostensibly obligate infauna.

Studies of recolonization rates of stable and unstable (e.g., fluctuating ecological parameters, such as salinity, sedimentation rate, and temperature) modern environments show that organisms in stable environments are more adversely affected by physiological stress. Species present in unpredictable environments (e.g., estuaries and river-dominated deltas), usually have broad environmental tolerances and can recover from disturbances quickly (Jernelöv and Rosenberg 1976). For instance, the benthic population of relatively stable deep-sea environments can take more than 2 years to recover completely, while the benthic population of relatively unstable estuarine environments may recover in less than 11 months (Dauer and Simon 1976; Grassle 1977). Marginal marine organisms are subject to relatively high physiological stress on a continual basis; therefore the resident population is likely to exhibit some degree of r-selected strategy (Ekdale et al. 1984; Ekdale 1985; Beynon et al. 1988). For example, Rhoads et al. (1985) found small, shaft-dwelling organisms comprising the normal, resident community in nearshore areas with a rapid rate of sedimentation. Recolonization, if it progresses without disturbance, will follow a trend to make complex and deep-burrowing behaviors.

Paleontological Evidence for Opportunistic Behavior

The characteristics of an opportunistic body fossil assemblage were summarized by Pemberton and Frey (1984) as (1) of limited aerial distribution (Waage 1968); (2) occurring in a continuous, thin isochronous horizon (Waage 1968); (3) abundant in several otherwise distinct faunal assemblages (Levinton 1970); (4) having great abundance in a facies with which it is not generally associated (Levinton 1970); and (5) numerical domination of one species within the fossil assemblage (Levinton 1970).

In the Maastrichtian Fox Hills Sandstone of South Dakota, Waage (1968) reported possible opportunistic body fossil occurrences. Assemblage zones, attributed to mass mortality, were found bounded by sparsely populated to barren zones. Waage (1968) attributed the sudden abundance of fauna to a slight decrease in stress that allowed colonization of the newly opened niche.

Rollins et al. (1979) documented a faunal succession within a community that contained an opportunistic brachiopod-gastropod assemblage; ecological stress was applied by shallow marine conditions associated with early transgression. This ecological sere was dominated by epifaunal suspension feeders. Later, an equilibrium population appeared to have developed, but with the transgressive nature of the unit, there is some doubt that interspecific competition (necessary for a true ecological succession) was responsible for the biotic change.

Diminutive bivalves with paedomorphic traits have been interpreted as opportunistic species in an area of low oxygen concentrations (Snyder and Bretsky 1971). This relationship between paedomorphosis and ecology was discussed by Gould (1977). Gould's theory that accelerated sexual maturity is associated with r-selecting paedomorphs, whereas the hypothesis that retarded somatic development characterizes K-selecting paedomorphs is supported by recent models on the ecological causation of heterochrony (McKinney 1986). The example cited by Snyder and Bretsky (1971) is suggestive of accelerated sexual development, diagnosed by reduced size of descendants relative to their ancestors (McNamara 1986). Reduced oxygen conditions are also invoked by Brezinski (1986) to explain an interpreted opportunistic assemblage of trilobites in the Upper Ordovician of Missouri.

Ichnology of Storm Deposits

The activities of soft-bodied infauna, which have low preservational potential as body fossils, represent a significant component of any biological succession. It has already been shown that most opportunistic species that initially recolonize defaunated environments are tube-dwelling polychaetes (table 4.2). Thus the ichnofossil record (fig. 4.1) may be the best place to find evidence of opportunistic colonists in an ecological succession (Vossler and Pemberton 1988). Ekdale (1985) summarized a number of possible examples resulting from

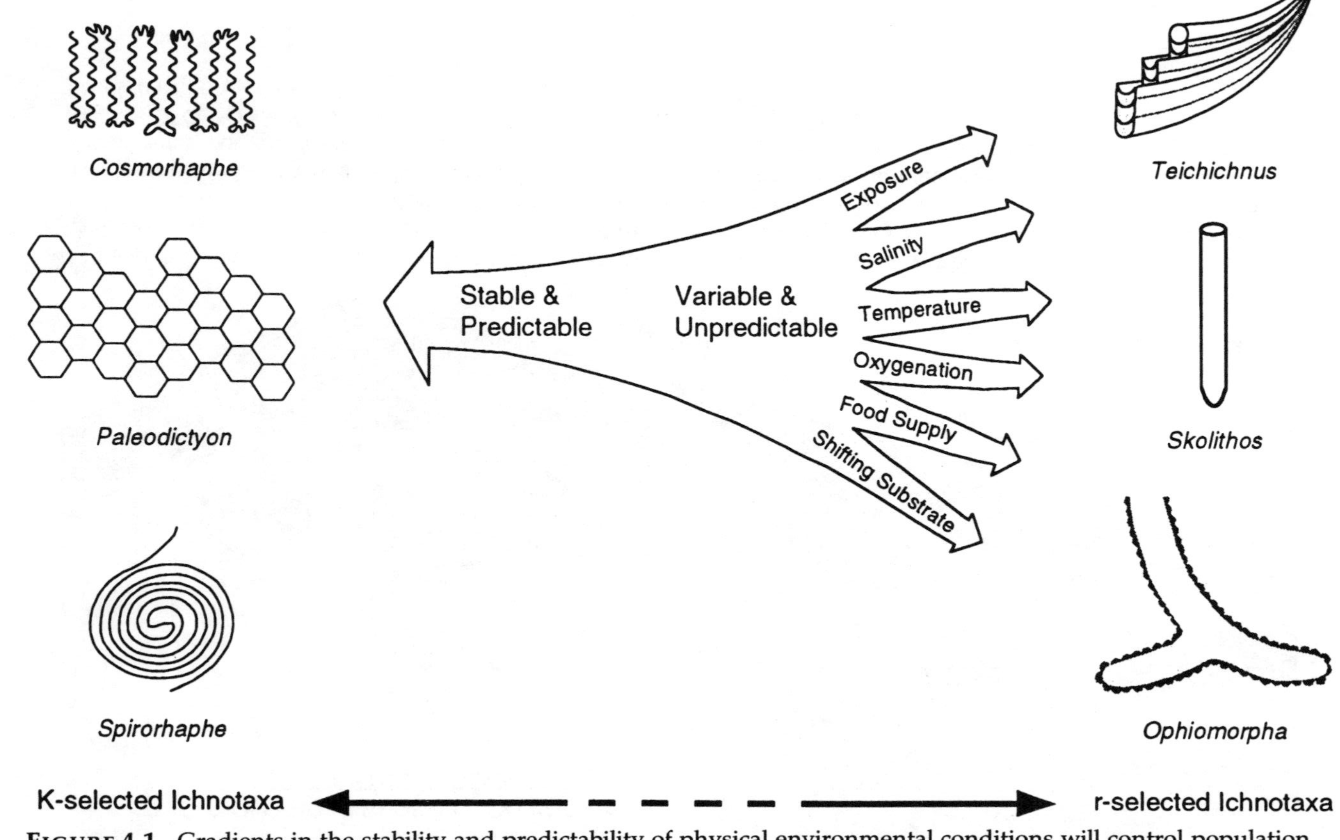

FIGURE 4.1. Gradients in the stability and predictability of physical environmental conditions will control population strategies among burrowing organisms. Equilibrium (K-selected) trace fossils flourish in high-diversity assemblages under very stable and predictable conditions. Opportunistic (r-selected) trace fossils rise to prominence in low-diversity assemblages under extremely variable and unpredictable conditions. (Modified from Ekdale 1985.)

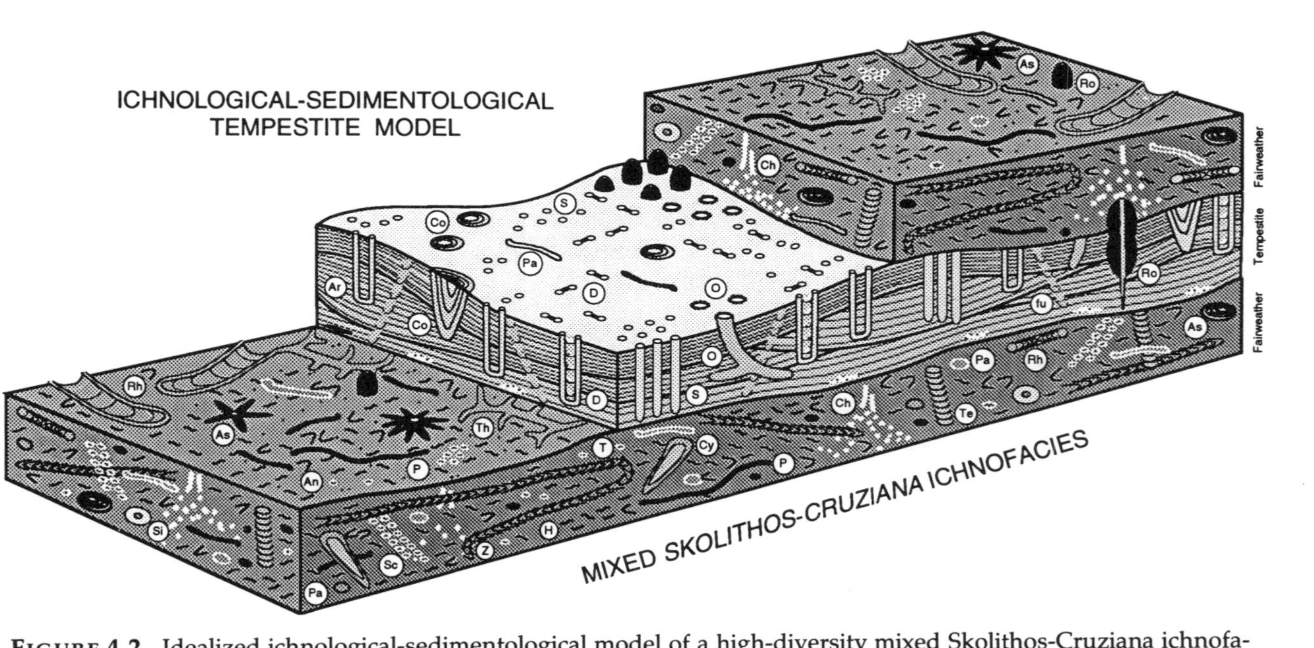

FIGURE 4.2. Idealized ichnological-sedimentological model of a high-diversity mixed Skolithos-Cruziana ichnofacies, based on observations of Cretaceous strata of the Western Interior Seaway of North America. Fair-weather deposits are characterized by thorough bioturbation with a high diversity of deposit feeding and grazing structures. The tempestite is burrowed with a lower-diversity suite of mainly suspension feeding and dwelling structures. An: *Anconichnus*, Ar: *Arenicolites*, As: *Asterosoma*, Ch: *Chondrites*, Co: *Conichnus*, D: *Diplocraterion*, fu: fugichnia, H: *Helminthopsis*, O: *Ophiomorpha*, P: *Planolites*, Pa: *Palaeophycus*, Rh: *Rhizocorallium*, Ro: *Rosselia*, Sc: *Schaubcylindrichnus*, Si: *Siphonichnus*, Sk: *Skolithos*, T: *Terebellina*, Te: *Teichichnus*, Th: *Thalassinoides*, Z: *Zoophycos*.

(1) turbidite deposition (i.e., Seilacher 1962); (2) salinity variations (i.e., Beynon et al. 1989); (3) oxygen deficiencies (Bromley and Ekdale 1984); and (4) storm deposition (i.e., Pemberton and Frey 1984; Vossler and Pemberton 1988, 1989). In all these cases, some sort of physical and/or chemical stress was applied to the benthic population.

To illustrate the characteristic trace fossil patterns discerned in storm-influenced settings, reference is made to the well-preserved suite of trace fossils from the Upper Cretaceous (Turonian) Cardium Formation of Alberta, Canada (Pemberton and Frey 1984; Vossler and Pemberton 1988). Trace fossil assemblages in this unit are complex, because they contain elements of both the *Skolithos* and *Cruziana* ichnofacies (fig. 4.2). Fine-grained siltstones and mudstones are characterized by a diverse suite of trace fossils, including *Chondrites, Cochlichnus, Gyrochorte, Palaeophycus, Phoebichnus, Planolites, Rhizocorallium, Rosselia, Taenidium, Thalassinoides,* and *Zoophycos.* This association is indicative of the *Cruziana* ichnofacies, in which the diversity is high and individual densities of most ichnogenera are typically low. In general, these trace fossils suggest activities of mobile carnivores and deposit feeders exploiting relatively nutrient-rich, fine-grained sediments, deposited in low-energy, offshore environments.

Interspersed throughout the sequence are sandstones that contain a different trace fossil association consisting of *Skolithos, Ophiomorpha,* and escape structures. This assemblage is dominated by dwelling burrows and is indicative of the *Skolithos* ichnofacies. Individual densities of traces are high, but overall ichnotaxonomic diversity is low.

Therefore parts of the Cardium Formation collectively contain a mixed trace fossil assemblage within a single depositional cycle. This alternation of two typically depth-related ichnocoenoses is commonly attributed to repeated rises and falls of sea level, but a more likely interpretation involves the alternation of sand and silt sedimentation caused by hydrodynamic energy fluctuations. The two different ichnocoenoses reflect varying behavioral response of the animals colonizing two successive, individually distinct habitats. The resident ichnocoenose (i.e., elements of the *Cruziana* ichnofacies) can be considered representative of a stable benthic community, within which individual populations are at or near their carrying capacity. Such associations, which typically are regarded to be resource limited (Levinton 1970), occur in habitats having low physiochemical stress. Periodic generation of elements of the *Skolithos* ichnofacies, on the other hand, represents the flourishing of a community of opportunistic organisms in an unstable, high-stress, physically controlled environment. These trends have been recognized in 15 Cretaceous siliciclastic wave-dominated shoreface intervals in the Western Interior Seaway of North America (table 4.5). This basic pattern has also been recognized in numerous other deposits, ranging in age from Ordovician to Oligocene (table 4.6).

Such laminated to burrowed successions (fig. 4.3) are typical of many interpreted shoreface environments in strata from the Cretaceous of the Western Interior. Pemberton et al. (1992c) described storm-dominated deposits from the

Table 4.5 *Storm (Opportunistic) and Fair-weather Trace Fossil Suites from Cretaceous Strata of the Western Interior Seaway of North America. Al: Altichnus, An: Anconichnus, Anc: Ancorichnus, Ar: Arenicolites, As: Asterosoma, Au: Aulichnites, Be: Bergaueria, Ch: Chondrites, Co: Conichnus, Coc: Cochlichnus, Cosm: Cosmorhaphe, Cy: Cylindrichnus, D: Diplocraterion, fu: fugichnia, Fus: Fustiglyphus, Gy: Gyrochorte, H: Helminthopsis, Lo: Lockeia, M. simplicatus: Macaronichnus simplicatus, Ma: Mammilichnus, O. borneenis: Ophiomorpha borneensis, O. irregulaire: Ophiomorpha irregulaire, O. nodosa: Ophiomorpha nodosa, P: Planolites, Pa: Palaeophycus, Pho: Phoebichnus, Rh: Rhizocorallium, Ro: Rosselia, Sc: Scolicia, Sch: Schaubcylindrichnus, Si: Siphonichnus, Sk: Skolithos, T: Terebellina, Ta: Taenidium, Te: Teichichnus, Th: Thalassinoides, Z: Zoophycos.*

			Trace fossil suite	
Age	Unit	Location	Storm	Fairweather
U. Aptian L. Albian	Bluesky Fm	WCSB W-C Alberta	*O. nodosa, Ar, Sk, D, Co, fu*	Distal: *H, An, Ch, P, Th, Te.* Proximal: *Ch, T, P, As, Ro, Te, Rh, Cy, Pa, D, M. simplicatus.*
Albian	Cadotte Mbr	WCSB W-C Alberta	*O. nodosa, Ar, Sk, D, Co, fu*	Proximal: *T, P, Ro, Te, Pa, M. Simplicatus, O. irregulaire.*
U. Albian	Viking Fm	WCSB Alberta	*O. nodosu, Ar, Sk, D, Co, fu.*	Distal: *An, Z, Ch, P, H, T, Te, Th.* Proximal: *H, Ch, T, Ro, As, P, Te, Sk, Rh, Pa, Ar, Cy, Si, Sch, o. irregulaire.*
U. Albian	Newcastle Sst	Montana	*O. nodosa, Co, fu.*	Proximal: *Ch, T, P, Te, O. irregulaire.*
Turonian	Cardium Fm	WCSB Alberta	*O. nodosa, D, Sk, fu.*	Distal: *H, An, Z, Ch, P, T, Ro, Te, Gy, Cy, Th.* Proximal: *H, An, Z, Ch, P, T, As, Ro, Te, Rh, Cy, Pa, Sch, Gy, Sk, D, Ar, Si, O. irregulaire.*
Campanian	Spring Canyon Mbr	Book Cliffs Utah	*O. nodosa, D, Ar, Sk, Pa, fu.*	Distal: *H, An, P, T, Te, Th.* Proximal: *H, An, Ch, P, T, Th, As, Ro, Rh, Pa, Sk, Ar, D, O. irregulaire.*
Campanian	Aberdeen Mbr	Book Cliffs Utah	*O. nodosa, Ar, Sk, fu.*	Distal: *H, Z, Ch, P, As, T, Te, Th.* Proximal: *P, Ta, Lo, Pa, Sch, Te, Anc, Gy, Fus, O. irregulaire.*
Campanian	Kenilworth Mbr	Book Cliffs Utah	*O. nodosa, Sk, Pa, Be, fu.*	Proximal: *H, Ch, P, As, Ro, Te, Ta, Cy, Pa, Sch, Gy, Sc, Be, Al, Ma, Cosm, Th, Aul, O. irregulaire.*
Campanian	Sunnyside Mbr	Book Clifs Utah	*O. nodosa, Sk, fu.*	Proximal: *H, Ch, P, Pa,Sch, Te, O. irregulaire.*
Campanian	Eagle Sst	Montana	*O. nodosa, Co, fu.*	Proximal: *Pa, Cy, Sch, Te, Anc, O. irregulaire.*
Campanian	Virgelle Mbr	Writing-On-Stone	*O. nodosa, Sk, Be, fu.*	Proximal: *P, T, Ro, Be, Cy, Pa*
Campanian	Shannon Sst	Wyoming	*O. nodosa, Sk, Pa, Be, Co, D, fu.*	Proximal: *H, Cosm, P, Ro, T, Te, Gy, Pa, Sch, Anc, Cy, M. simplicatus.*
Campanian	Sussex Sst	Wyoming	*O. nodosa, Sk, Pa,Co, fu.*	Proximal: *H, P, T, Sc, Lo, Cy, Pa, Sch, Gy.*
Campanian	Blood reserve	Jenson Reservoir Alberta	*O. nodosa, Sk, Co, Be, fu.*	Proximal: *H, Ch, P, Lo, Be, Ro, O. irregulaire.*
Campanian	Appaloosa Set	Drumheller Alberta	*O. borneensis, Sk, Co, fu.*	Proximal: *H, P, Te, Rh, Pa, As, Ro, O. irregulaire, M. simplicatus.*

TABLE 4.6 *Comparative Ichnology of Storm* Versus *Fairweather Deposits (Modified from Pemberton and Frey 1984 and Vossler and Pemberton 1988).*

Age	Formation	Ichnofossil (fair-weather)	Ichnofossils (storm)	Reference
Mid-Upper Ordovician	Martinsburg Formation	None given	*Planolites*, fugichnia	Kreisa 1981
Lower Silurian	Ross Brook Formation	Mottling, *Chondrites, Helminthopsis*	*Skolithos, Palaeophycus*	Hurst & Pickerill 1986
Lower Silurian	Hughley Formation	*Scolicia, Walcottia, Rusophycus, Cruziana*	*Diplocraterion, Palaeophycus, Chondrites, Skolithos.*	Benton & Gray 1981
Middle Silurian	Thorold Formation	*Arthrophycus, Chondrites, Cruziana, Daedalus, Diplichnites, Dolopichnus, Arenicolites, Lobichnus, Lingulichnus, Monomorphichnus, Palaeophycus, Planolites, Polycylindrichnus, Rusophycus, Teichichnus*	*Skolithos, Diplocraterion*	Penberton & Risk, 1982
Lower Carboniferous	Courceyan & Arundian Formations	None given	*Rhizocorallium, Zoophycos, Planolites*, fugichnia.	Wu 1982
Lower Jurassic	Neill Klinter Formation	*Rhizocorallium irregulare, Gyrochorte comosa, Parahaentzschelinia surlyki, Curvolithus multiplex, Taenidium serpentinum, Gyrophyllites kwassicensis, Planolites beverleyensis, Helminthopsis magna, Nereites* sp., *Phycosiphon* sp., *Scolicia* sp.,*Phoebichnus trochoides*	*Arenicolites, Ophiomorpha*	Dam 1990
Lower Cretaceous	Choshi Group	None given	*Planolites, montanus, Nankaites*, steeply inclined *Palaeophycus*	Katsura et al. 1984
Upper Cretaceous	Frontier Formation	*Asterosoma, Teichichnus*	*Ophiomorpha.*	Winn et al. 1983
Upper Cretaceous	Shannon Sandstone	*Asterosoma, Skolithos*, gastropod trails, *Teichichnus* 'donut burrows' (possibly *Schaubcylindrichnus*), *Thalassinoides*	*Skolithos.*	Tillman & Martinsen 1985
Upper Cretaceous	Cardium Formation	*Chondrites, Planolites, Cochlichnus, Phoebichnus, Meunsteria, Rhizocorallium, Zoophycos, Planolites, Thalassinoides, Cylindrichnus, Rosselia, Gyrochorte*	*Skolithos, Diplocraterion, Palaeophycus.*	Pemberton & Frey, 1984a; Vossler & Pemberton 1988, 1989
Upper Cretaceous	Star Point Formation	*Scolicia, Palaeophycus herberti, Palaeophycus tubularis, Cylindrichnus concentricus, Thalassinoides suevicus, Teichichnus rectus, Planolites beverleyensis, Planolites montanus, Ancorichnus capronus, Ophiomorpha annulata*	*Ophiomorpha nodosa, Skolithos linearis.*	Howard & Frey 1984
Upper Cretaceous	Cape Sebastian Sandstone	*Ophiomorpha, Planolites, Scolicia, Sabellarites Thalassinoides, Macaronichnus segregatis*	*Ophiomorpha*	Hunter & Clifton 1982
Upper Cretaceous	Cape Sebastian Sandstone	*Ophiomorpha, Planolites, Scolicia, Sabellarites Thalassinoides*	fugichnia, *Scolicia, Ophiomorpha*	Bourgeois, 1980
Upper Cretaceous	Blackhawk Formation	*Ophiomorpha annulata, Rosselia socialis, Terebellina* sp., *Planolites montanus, Planolites beverleyensis, Cylindrichnus concen-*	*Ophiomorpha nodosa, Cylindrichnus concentricus*	Frey 1990

TABLE 4.6 *(Continued)*

Age	Formation	Ichnofossil (fair-weather)	Ichnofossils (storm)	Reference
		tricus, Palaeophycus heberti, Chondrites sp.		
Mid-Tertiary	Arno	*Imbrichnus, Scolicia, Planolites*	*Ophiomorpha.*	Ward & Lewis, 1975
Oligocene	Magazine Pt. Formation	*Thalassinoides*	*Ophiomorpha.*	Lewis 1980

Spring Canyon Member of the Blackhawk Formation in Utah. Figure 4.4 depicts the development of laminated to burrowed (lam-scram) beds typical of the lower shoreface showing storm (lam) and fairweather (scram) trace fossil assemblages and sedimentary structures.

In core, it is difficult to recognize hummocky and swaley cross-stratification because of scale problems. However, characteristic trace fossil patterns can be used to assist in the interpretation of storm deposits. A core (Pan Am B-1 Ricinus 10–01–35–8W5) from the Cardium Formation displays a diverse and well-preserved trace fossil suite, including representatives of at least 15 ichnogenera: *Anconichnus, Asterosoma, Chondrites, Diplocraterion, Helminthopsis, Palaeo-*

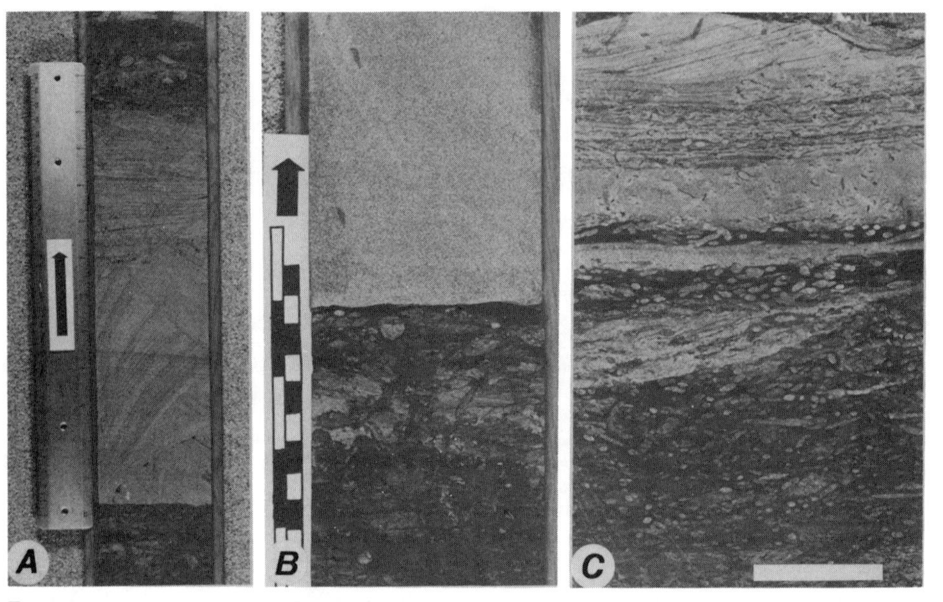

FIGURE 4.3. Laminated to burrowed units in the Price River C core, Spring Canyon Member, Blackhawk Formation, Utah. (A) Laminated to burrowed sequence showing a sharp base, hummocky cross-stratified sandstone, and a burrowed gradational top; scale is 60 cm long. (B) Tempestite with a sharp, storm-scoured base in the proximal lower shoreface; the resident background assemblage gives way to an opportunistic suite that characterizes the storm bed. Scale in inches and centimeters. (C) Entrained assemblage, representing doomed pioneers of *Anconichnus* (arrows), in very fine-grained proximal offshore, storm-generated sandstone beds. Scale bar equals 2 cm.

phycus, Planolites, Rhizocorallium, Rosselia, Skolithos, Subphyllochorda, Teichichnus, Terebellina, Thalassinoides, and *Zoophycos* (fig. 4.5). Near the top of the core are a number of thin, discrete sandstone beds that have been interpreted as storm deposits. These beds are characterized by (1) sharp bases, with or without basal lags; (2) low angle parallel to subparallel laminations; (3) escape structures; (4) the dwelling burrows of opportunistic organisms; and (5) gradational burrowed tops (fig. 4.6).

The trace fossils in the sandstone beds are interpreted to represent the activities of opportunistic organisms that rapidly colonize the surface of the storm sand. Characteristic forms include *Skolithos* and *Ophiomorpha.* These sandstones are interbedded with bioturbated fine-grained deposits dominated by *Anconichnus, Asterosoma, Chondrites, Helminthopsis, Planolites, Rhizocorallium, Rosselia, Subphyllochorda, Teichichnus, Terebellina, Thalassinoides,* and *Zoophycos.* The suite represents the resident suite and is indicative of stable, fair-weather deposition. Proximal-distal trends can be recognized by documenting changes in the resident trace fossil suite (fig. 4.7).

Recently, Bromley and Asgaard (1991) introduced a new ichnofacies, the *Arenicolites* ichnofacies, to account for short-term, opportunistic occurrences of trace fossils in incongruous settings, such as storm deposits. Such a designation is confusing because it cannot be separated from the present *Skolithos* ichnofacies concept. The present array of ichnofacies is adequate to explain such occurrences.

Föllmi and Grimm (1990) have postulated the "Doomed Pioneer" concept to explain anomalous trace fossils exclusively associated with gravity-flow deposits within otherwise unbioturbated sediments (table 4.7). Such a scenario would indicate that event deposition in oxygen-depleted sedimentary environments may be accompanied by the appearance of allochthonous infauna. The concept (fig. 4.8) involves the following: (1) the entrainment of living decapod crustaceans (or presumably other hardy infauna) took place in some type of turbulent sedimentation event; (2) these imported burrowers reworked substantial quantities of laminated, commonly organic-rich sediment, in an environment from which they were previously excluded; and (3) the persistence of oxygen-depleted environmental conditions limited the survival time of these transported infaunal dwellers and rendered them "doomed pioneers" (Grimm and Föllmi 1990).

From the standpoint of event deposition this concept may well be important in understanding the occurrence of other anomalous trace fossil associations. Likewise, as pointed out by Föllmi and Grimm (1990), the occurrence of *Thalassinoides* in anoxic hemipelagic host sediments has previously been incorrectly interpreted to indicate bottom-water ventilation and reoxygenation on a broad regional and/or temporal scale. The recognition of doomed pioneer trace fossil associations as ephemeral ecological phenomena that are prominent in some laminated sequences should permit a more precise understanding of paleo-oxygen levels and basin history (Grimm and Föllmi 1990).

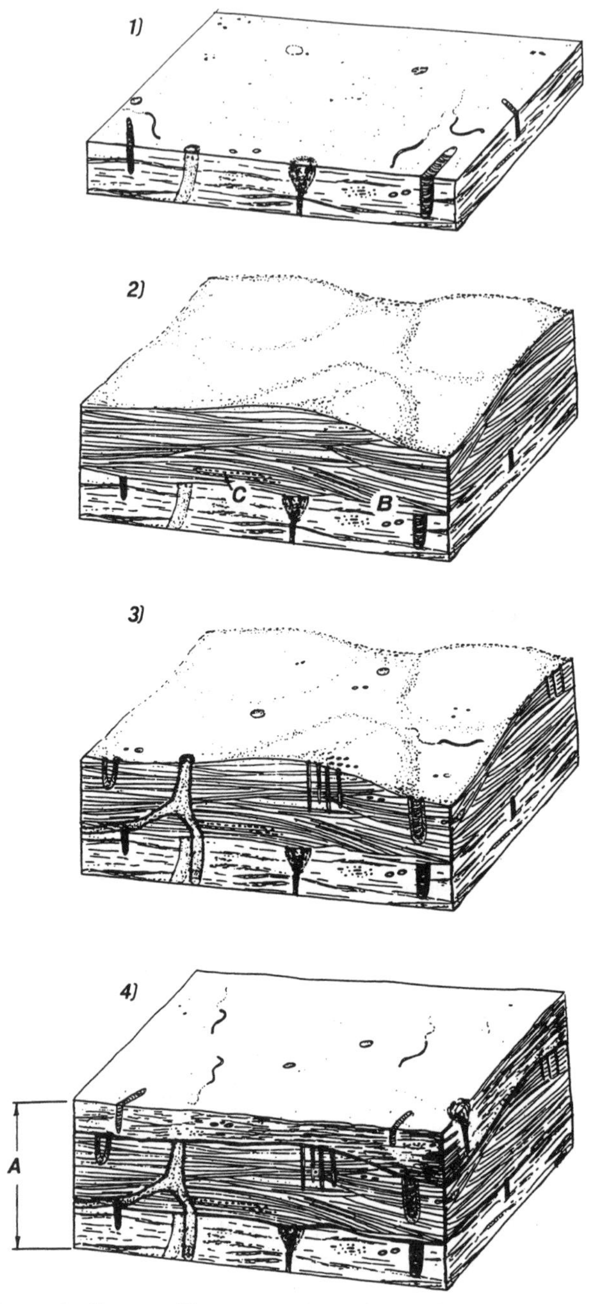

FIGURE 4.4. Schematic diagram illustrating the genesis of tempestites in typical shoreface settings from the Cretaceous of the Western Interior. (1) Fair-weather sedimentation-background resident assemblage consisting of *Anconichnus, Asterosoma, Chondrites, Helminthopsis, Palaeophycus, Planolites, Rosselia, Teichichnus, Terebellina,* and *Thalassinoides.* (2) Storm deposition of fine-grained sand with a sharp base (B) and possibly with an entrained fauna (C) of both larvae and adults. (3) Cessation of storm deposition and colonization by opportunistic fauna represented here by *Arenicolites, Diplocraterion, Ophiomorpha, Palaeophycus,* and *Skolithos.* (4) Resumption of fair-weather conditions and sedimen-

Preservational Potential

The increasing awareness of event bed deposition in the ancient record, or "uniformitarianist catastrophism" (Kumar and Sanders 1976), has correspondingly led to much consideration of the preservability of such beds (e.g., Dott 1983, 1988; Wheatcroft 1990). Observations of tempestites in various depositional settings demonstrate a wide variety of preservation styles. The fundamental parameters revolve around the net sedimentation rate, the biogenic mixing rate, and the magnitude of physical reworking, although most proposed quantitative models operate under the assumption that the latter factor is absent. Wheatcroft (1990) demonstrated that neither sedimentation rate nor biogenic mixing rate are easily defined, both of them masking important, complex affiliated processes and interactions. Few of these have been addressed in detail, despite the effect they have on event bed preservation. In contrast to previous models, which have concentrated mainly on layer thicknesses of event beds and mixing zones, Wheatcroft (1990) focused on time scales. The result has been a model relating transit time (i.e., the time necessary to bury the event bed beyond the reach of the burrowing infauna) and dissipation time (i.e., the time required to destroy the event bed biogenically). Clearly, in circumstances where the event bed is thicker than the zone of burrowing, the fraction that exceeds the burrowing zone will be preserved regardless of the duration of biogenic reworking. Compelling though this model is, the inherent variability of the physical and biological responses to the depositional event greatly limits the accuracy of this quantification and thus limits its utility to the interpretation of the ancient record.

Transit Time

Previous workers have calculated the transit time by dividing the thickness of the zone of biogenic reworking by the sedimentation rate. Wheatcroft (1990) pointed out that the thickness of the event bed itself was important in the calculation of transit time and proposed that the combined thickness of the biogenic reworking zone and event layer, less one-half of the thickness of the event layer, all divided by the sedimentation rate was a more accurate estimation of the transit time.

Several interactions exist in the determination of transit time, which are probably impossible to quantify in the ancient record. The first is the thickness of the biogenic mixing zone. Clearly, this is related in part to the size of the

tation results in the elimination of the opportunistic organisms as the resident background burrowers are reestablished. The resident fauna burrows down from the top, resulting in the generation of a gradational upper contact (A). (Modified from Pemberton et al. 1992c.)

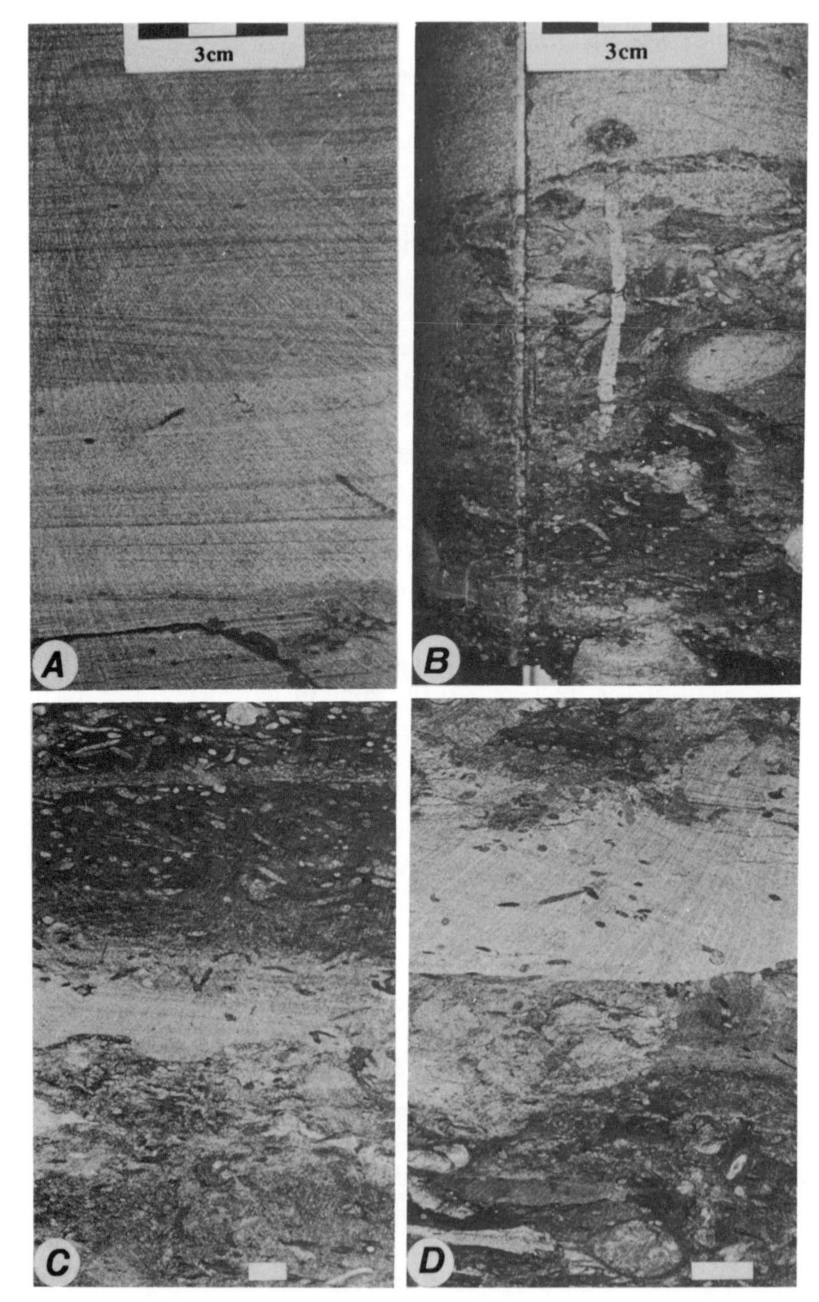

FIGURE 4.5. Ichnological features of storm-derived beds in the Cardium Formation, Well 10–1–35–8W5: (A) Parallel to low-angle laminations (hummocky cross-stratification), 2569. 2 m. (B) *Skolithos* part of the opportunistic storm assemblage that cross-cuts the resident assemblage, 2570. 2 m. (C) Laminated to burrowed sequence, fair-weather assemblage dominated by *Planolites, Chondrites,* and *Helminthopsis,* storm unit has sharp base and burrowed gradational top, 2570. 3 m.; scale bar equals 1 cm. (D) Storm sandstone with sharp base, burrowed top, and opportunistic trace fossil assemblage; scale bar equals 1 cm.

animals constituting the benthic community. Although burrow size can be related to organism size in some instances, in many it cannot, and the common overprinting of the trace suite by later deep burrowers serves to obscure much of these data. Further, the degree of bed penetration is not simply a function of the animal size; it is also closely related to the behavior engaged in during its generation. Many grazing and foraging structures are often restricted to the sediment-water interface and do not deeply penetrate the bed. Mobile deposit feeders and carnivores are often more deeply penetrating. In contrast, vertical dwelling structures of suspension feeders or passive carnivores may be very deeply penetrating (e.g., *Ophiomorpha*), but once constructed, they have little more effect on the bed. Unless the dwelling structures become very closely spaced and overprint one another, it is unlikely that such suites will totally obscure an event bed. The nature of organism behavior also has important implications for dissipation time (see later).

In general, the zone of biogenic reworking is thinnest in the lower offshore to abyssal depths, where diminutive grazers and foragers dominate the benthic community. The zone thickens through the upper offshore, where more robust and mobile deposit feeders dominate. The fair-weather *Skolithos* ichnofacies may extend the zone of biogenic reworking much deeper in lower shoreface and shallower settings but is also characterized by a decrease in effective homogenization of the bed at those depths.

Sedimentation rate is highly problematic to resolve quantitatively, even using modern settings as analogs. The most obvious problem revolves around the use of the average or fairweather (ambient) sedimentation rates. The mere presence of an event bed in the interval in question demonstrates that sedimentation rates are inherently unsteady and unpredictable in the setting. An offshore zone may experience months of fairweather deposition at a rate that is totally overshadowed by the accumulation of a single tempestite deposited over the course of days. Since these events are unpredictable, with respect to both their frequency and the magnitude of deposition, they currently defy reliable quantification and inclusion into calculated sedimentation rates of most settings.

Dissipation Time

Dissipation time corresponds to the time required to destroy an event bed biogenically. Wheatcroft (1990) demonstrated that there is a fundamental difference between destruction of an event bed characterized by a discrete mineralogy and chemistry (e.g., ash beds or iridium anomalies) and destruction of an event bed distinguishable only by its lithology and fabric (e.g., HCS beds and turbidites). The former anomalies may survive considerable biogenic reworking, becoming mixed over a considerable thickness of interval and still remaining detectable. In contrast, the latter is recognized on the basis of far more easily obliterated criteria (e.g., sedimentary structures) and is therefore

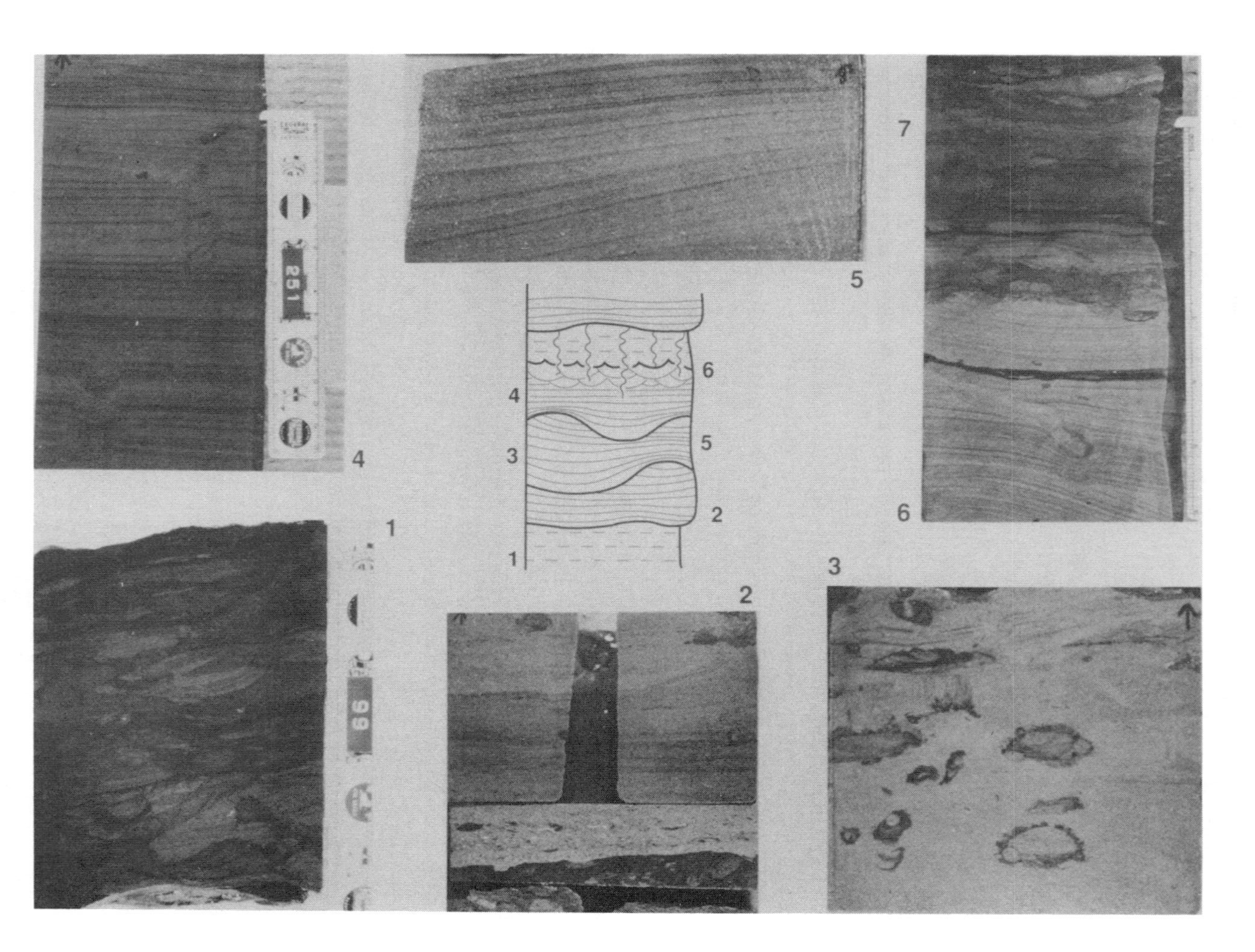
4
5
7
6
1
2
3
1
2
3
4
5
6

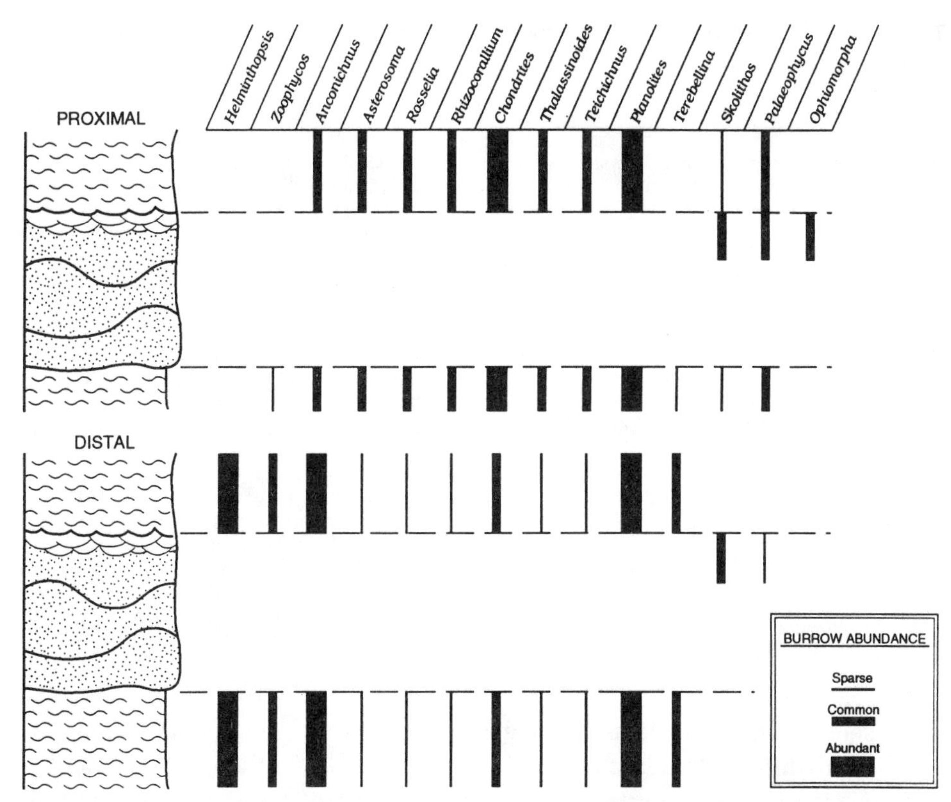

FIGURE 4.7. Proximal-distal trends in the 10–1–35–8W5 core are associated with changes in the resident (or fair-weather) assemblage; going up core this assemblage changes from a distal *Cruziana* assemblage to a proximal *Cruziana* assemblage. (Modified from Pemberton et al. 1992b.)

FIGURE 4.6. Summary of the ichnological features associated with an idealized storm succession, as seen in core from the Cretaceous Cardium Formation of Alberta. (1) The succession starts with a diverse, fair-weather resident ichnocoenose comprising elements of the Cruziana ichnofacies. Dominant forms might include *Zoophycos, Asterosoma, Helminthopsis, Anconichnus, Rhizocorallium,* etc. (2) This assemblage is interrupted by the storm event. It is initiated with a sharp base, commonly overlain by a basal lag consisting of rip-up clasts and shell debris. (3) The main body of the storm deposit consists of parallel to subparallel lamination, representing hummocky cross-stratification. (4) The storm-generated stratification may contain escape structures, recording the attempt of organisms to burrow through the event bed. These organisms may comprise elements of the resident assemblage, or fauna that were entrained in the sediment of the storm deposit itself. (5) The top of the storm bed is commonly biogenically disturbed by dwelling burrows of opportunistic organisms that rapidly colonize the bed and flourish for a small period of time; common representatives include *Skolithos* and *Ophiomorpha.* (6) Following storm abatement, fair-weather depositional patterns resume and the equilibrium resident assemblage is reestablished, producing a gradational burrowed top to the event bed. (7) The fair-weather assemblage continues to flourish, and depending upon the thickness of the storm layer and the frequency of the storm events, the entire event bed could become biogenically homogenized. (Modified from Pemberton et al. 1992b.)

TABLE 4.7 *Doomed Pioneers Concept: Critical Observations and Interpretations (After Grimm and Föllmi 1990)*

Observations	Interpretations
1. Bioturbation only within event strata and in directly subjacent sediments	1. Conclude a genetic relationship between event deposition and bioturbation; infer causal relationship between inputs of allochthonous sediment and importation of burrowing infauna
2. *Thalassinoides* and *Gyrolithes*	2. Crustacean dwelling burrows; foreign to anoxic basinal environment, derived from neritic setting
3. Selective bioturbation of fine-grained (=organic rich) intervals; avoidance of tuffaceous deposits.	3. Probably indicates successful colonization and exploitation of available nutrients in basinal paleoenvironment
4. Bioturbated intervals are sharply and comfortably overlain by unbioturbated hemipelagic sediment	4. Abrupt return to or persistence of anoxic conditions sharply limits temporal duration of burrowing episode
5. Shelly benthos and tiered infauna are absent	5. Abrupt burrowing episode was not the onset of substantial ecological reorganization on the basin floor, absence of ecological complexity may reflect limited survivability of infaunal immigrants and/or insufficient opportunity for subsequent larval recruitment

more sensitive to the duration of biogenic activity. Dissipation time is comparatively more difficult to determine than transit time.

Previous discussion regarding the characteristics of the benthic community touched on two parameters strongly affecting dissipation time. The most obvious parameter is the size of the animals constituting the infauna. Robust organisms can displace or mix sediment a greater distance per unit time than can

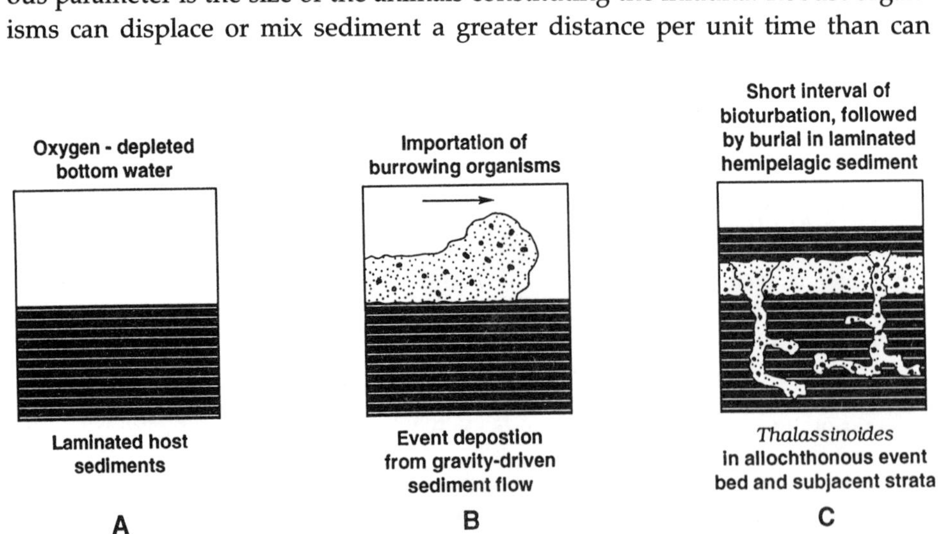

FIGURE 4.8. Schematic diagram illustrating the doomed pioneers model to explain the exclusive association of allochthonous event deposits with bioturbation; this example relates to the introduction of burrowers into oxygen-depleted environments. (Modified from Grimm and Föllmi 1990.)

diminutive ones. In a very general sense, organism sizes tend to be larger in shallow water than in deep water, although each benthic community typically contains a wide range of infauna sizes, attributable to a large number of factors, ranging from differences in genera or species to relative proportions of juveniles and adults (Dörjes and Hertweck 1975).

More important with regard to dissipation time is the behavior engaged in by the organism. Mobile deposit feeders and carnivores tend to produce much greater disruption of the bed, in both a vertical and a lateral sense, than are generated by grazers and foragers. The grazing-foraging behaviors may result in intense reworking of the sediment, but this is generally restricted close to the sediment-water interface (Ekdale 1980). Continuous, slow accumulation of fair-weather deposition may promote thorough homogenization of the sediment laterally, but the tracemakers usually show little propensity to penetrate deeply into the substrate (Howard 1975; Dott 1983, 1988). An exception is the *Zoophycos* tracemaker, which may display substantial vertical penetration into the substrate (Kotake 1989). However, the biogenic structure rarely shows intense reworking of the event bed, which typically has little in the way of deposited food (Frey and Goldring 1992). Typically, the *Zoophycos* trace crosses through the event bed with minimal disturbance, until a resource-rich interval, generally constituting a buried fair-weather or ambient deposit, is encountered. Thorough reworking of the buried fair-weather sediment may then occur (Vossler and Pemberton 1988).

Dwelling structures and suspension feeding structures may show appreciable vertical penetration into an event bed, but typically they display minimal volumetric modification of the bed's fabric. The structures may deepen or extend areally over the lifetime of the tracemaker, but such structures clearly do not result in the degree of homogenization imparted by mobile infauna (Pemberton and Frey 1984; Frey 1990). As such, the actual thickness of the biogenic mixing zone is less important than the nature of the behavior employed by organisms within the mixing zone.

One of the most underappreciated factors regarding the dissipation time is associated with the dynamic interplay between the benthic community and the depositional event itself. In many cases the event bed follows a disturbance that largely displaces or destroys the resident benthic community. Biogenic modification may not simply begin immediately after the event. Instead the site may remain unpopulated for a considerable time. Once recolonization is initiated it may take even more time for the community to achieve the density it possessed prior to the disturbance (Sousa 1984). Shallower-water settings, which are characterized by persistent and generally higher degrees of stress, may show repopulation within 11 months of a disturbance, whereas more highly sensitive deep-sea settings may take more than twice that time to recover (Dauer and Simon 1976; Grassle 1977).

Complete destruction of the benthic community may occur in two main ways. The most obvious is related to the severity of the disturbance producing the event bed. This is particularly evident in shallower-water settings, where erosional amalgamation of beds is more pronounced, and locally, cannibalistic (Aigner and Reineck 1982; MacEachern et al. 1991). Under these conditions, the entire benthic community is typically washed out of the substrate and transported elsewhere. The severity of such disturbances decreases basinward, such that distal tempestites, and, in particular, turbidites may show well-preserved fair-weather-resident trace fossil suites on their basal surfaces (e.g., Seilacher 1962; Crimes 1977; Crimes et al. 1981; Pemberton and Frey 1984; Frey and Goldring 1992).

The second major factor resulting in destruction of the benthic community is associated with the thickness of the event bed laid down. Once the organisms become buried, they attempt to burrow up through the bed to reach the new sediment water interface. If the event bed is too thick, none of the original resident community will survive to repopulate the post-storm substrate, minimizing the time of "effective" bioturbation (Kranz 1972). On the other hand, thinner event beds may be repopulated more quickly, since most of the benthic community may escape burial. Some biogenic modification of the event bed will occur, purely as a function of the escape structures.

Tempestites show a general basinward thinning . Therefore it may seem initially that the thinner beds of the shelf and lower shoreface may be rapidly colonized. In actuality, because organisms of these basinal benthic communities are usually diminutive , few animals may be able to reach the top of even a relatively thin event bed. Moreover, few tracemakers of the shelf and more basinal settings possess carapaces; therefore they are unlikely to survive entrainment in a tempestite or turbidite flow, unlike some organisms of shallower-water settings.

Another important factor surrounding the dissipation time is associated with the desirability of the event bed as a site for colonization and as a repository of food resources. The event bed generally varies markedly from the normal ambient (fair-weather) sediment, to which the resident community is suited. Even those organisms that burrow out of the event bed may, upon reaching the sediment-water interface, find a substrate devoid of suitable food resources, and subsequently die. Both those that do survive and opportunistic organisms whose larvae managed to find their way to the unexploited substrate are trophic generalists. They typically do not burrow through the bed; instead they establish domiciles and engage in suspension feeding or carry out surface deposit feeding and scavenging strategies (Jumars 1993). Penetrating domiciles show minimal lateral disturbance of the bed, and exceedingly dense populations are needed to obliterate a bed entirely. Once suitable amounts of ambient sediment accumulate and the resident faunal community reestablishes itself, the infauna generally show little inclination to penetrate deeply into the event bed; rather, they engage in behaviors suited to the ambient conditions (Rees et al. 1977). This avoidance pattern appears most pronounced in basinal settings, where the contrast between behaviors suited to the event bed and those suited to the ambient substrate are greatest. This is most common where turbidites are deposited in abyssal settings. In upper offshore conditions, where robust and diverse tracemakers of more proximal *Cruziana* ichnofacies predominate, the differences between event bed and ambient deposit behaviors are less pronounced. In lower shoreface and middle shoreface settings, no preference for fair-weather burrowing over event bed burrowing may be noted (e.g., Pemberton et al. 1992c), and for all intents and purposes, both suites are identical.

The final factor regarding dissipation time is the absolute time available for burrowing. As in the discussion of sedimentation rates, there is obviously considerable variability in burrowing time in any setting characterized by episodic deposition. In the case of tempestites, one may suspect minimal burrowing times to occur in shallower-water settings, with progressively more time available basinward. This primarily reflects the greater probability of storm interaction with the seafloor in shallower water, regardless of storm magnitude. In the case of turbidites, however, variability in burrowing time is more closely related to basin paleogeography and, if present, the configurations of any submarine fan systems.

In general, the preceding factors impose fundamental controls on the effectiveness of biogenic modification of event beds, and consequently, on their preservation potential. Unfortunately, these factors appear to resist quantification and therefore severely limit the effectiveness of existing mathematical models to explain event bed preservation in the rock record.

Physical reworking of the substrate is typically avoided in most of the models proposed, but it clearly has a profound effect on the preservation of both the event bed and the fair-weather deposits. In the case of tempestites, enhanced physical reworking is associated with shallower-water settings, with higher-magnitude storms and with higher-frequency storms. The latter factor principally reflects minimal fair-weather accumulation on the tempestite and minimal colonization of the substrate, both of which serve to enhance the ability of successive storm events to rework earlier tempestites. Under proximal conditions, successive tempestites may become highly cannibalistic (Aigner and Reineck 1982), resulting in only minor preservation of the beds.

Despite the difficulties in mathematically modeling preservation potential of tempestites, a number of proximal-distal trends are evident on the basis of empirical observations. In general, shallow-water settings favor the preservation of event beds, since higher numbers of storms enhance erosional amalgamation and minimize the time available for burrowing, factors that are also responsible for the minimal preservation of fair-weather deposits. These appear to overcome the factors favoring biogenic modification of event beds, such as larger animal sizes, greater bed penetration, suitability of the event bed to colonization by the resident benthic community, and higher rates of recolonization following a disturbance. Shallow-water deposits are dominated by vertical to subvertical domiciles, minimizing event bed modification. These settings typically correspond to proximal lower-shoreface and middle-shoreface environments.

Distal settings also favor preservation of event beds, but for markedly different reasons. Fair-weather deposits also have a high preservation potential, unlike shallow-water settings. Factors favoring preservation of tempestites are minimal erosional amalgamation, minimal bed penetration by infauna, diminutive size of infaunal organisms, general avoidance of event beds as a viable substrate, sensitivity of the benthic community to environmental disturbances, and slow rates of recolonization following a disturbance. These serve to overshadow conditions favoring event bed modification, such as long duration of fair-weather conditions, slow rates of ambient sediment accumulation, and reduced potential of event bed accumulation. Such settings are typical of shelf and lower offshore settings in the Cretaceous of the Western Interior Seaway.

The lowest preservation potential for tempestites appears to occur somewhere between the proximal and distal extremes. Under such conditions, erosional amalgamation is not as effective, and displacement of the resident suite is not as common as at the extremes. Furthermore, the reduced sensitivity of the benthic community to disturbances and generally higher rates of recolonization, all favor more rapid biogenic modification of the event bed. The variability in organism type and behaviors employed ensures that the tempestite will be suitable to some of the benthic community as a viable medium for burrowing. The higher numbers of mobile deposit feeders and carnivores favors rapid and thorough modification of bed fabrics within the biogenic mixing zone, and the introduction of suspension feeders also promotes the development of deeply penetrating vertical structures. Such zones appear to correspond to proximal *Cruziana* ichnofacies suites of the upper offshore and distal portions of the lower shoreface.

The tempestite preservation potential (fig. 4.9) seems to be consistent with observations of many modern and ancient examples (Kumar and Sanders 1976; Pemberton and Frey 1984; MacEachern and Pemberton 1992). Dott (1988) used the deposit of Hurricane Carla to outline this variability of preservation potential. McGowan (reported in Dott 1988) was unable to identify the prominent tempestite resulting from Hurricane Carla, reported on by Hayes (1967), a mere 15 years after the event. Bioturbation had apparently obliterated much of the record. However, Nummedal (reported in Dott 1988) managed to recognize the event bed further offshore (beyond 18–20 m of water depth), where biogenic modification of the tempestite had been less intense. Dott (1983) indicated that this alternating interplay of deposition, burrowing, and scouring can produce a very subtle record of event deposition (fig. 4.10). Concealed bed-junction preservation of the trace fossils, represented by truncated burrow tops and exotic burrow fills may represent the only clue to the episodic history of such amalgamated sequences. Figure 4.11 schematically shows the progressive colo-

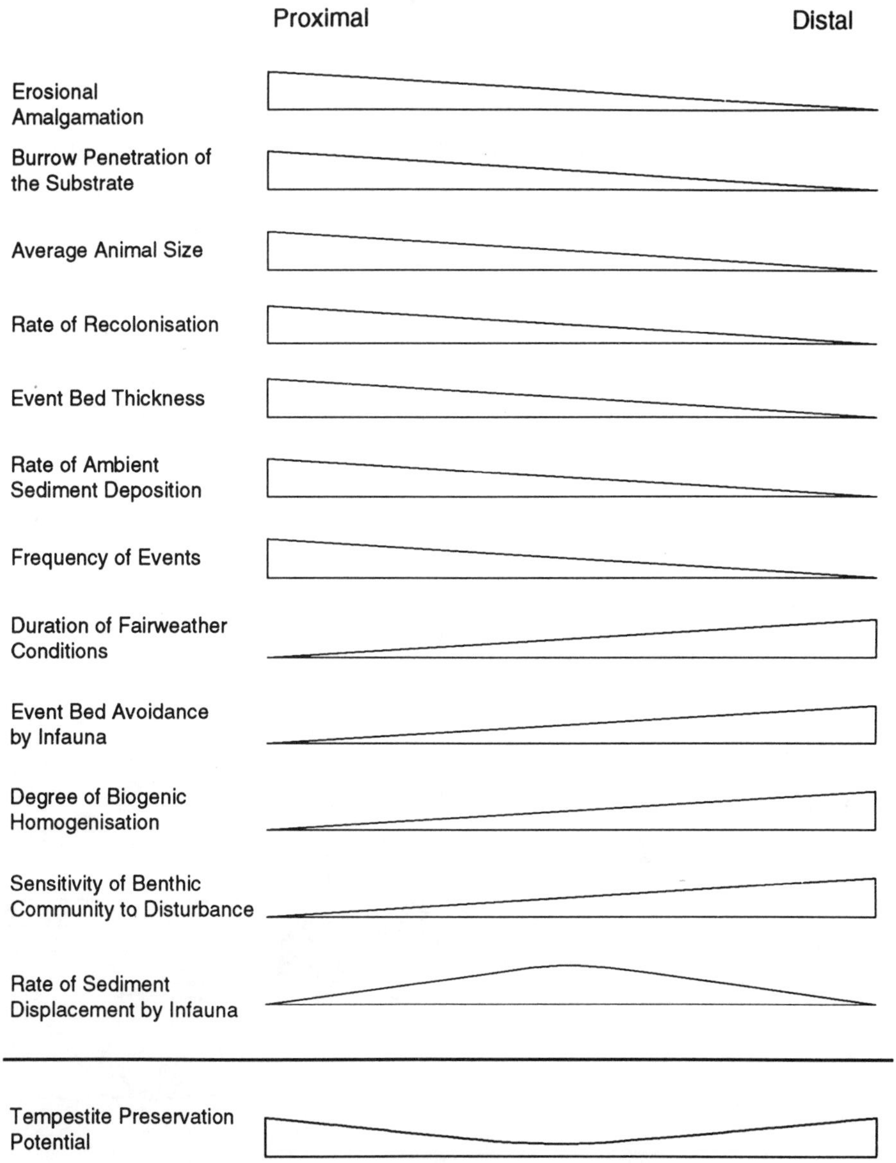

FIGURE 4.9. Schematic representation of the main factors affecting tempestite preservation and their relative importance with respect to proximal and distal depositional positions. Note that in general, tempestite preservation potential appears lower in intermediate depositional settings. Higher preservability occurs in proximal and distal settings, although for markedly different reasons.

nization of a tempestite and the variable effects of latter erosional amalgamation on the preserved record.

The low preservation potential of hurricane-induced tempestites may also reflect the infrequent nature of such disturbances. In these settings, the tempestite is exposed to long periods of biogenic colonization and modification before it is buried below the reach of infauna. In contrast, many high-latitude settings are characterized by cyclic variations in storm activity. Such settings are characterized by a winter storm season, where disturbances are both frequent and of high magnitude, and a summer fair-weather season, where storms are less frequent and of lower intensity (Owens 1977). Under such conditions tempestites accumulate rapidly during winter seasons with little or no time for

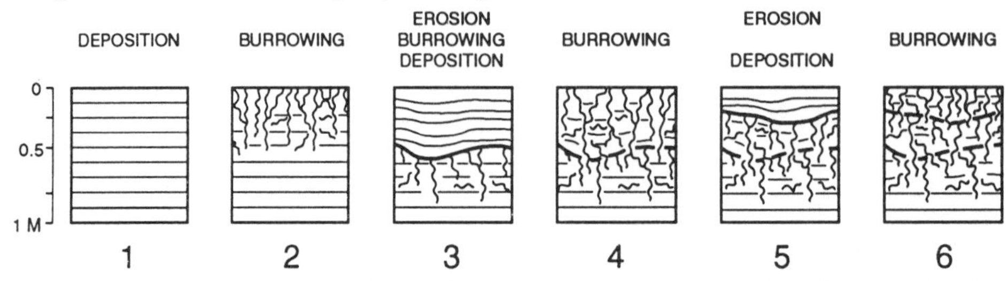

FIGURE 4.10. Amalgamation of strata by intense organism burrowing; the record of deposition punctuated by two events is obscured by the overprint of bioturbation. (Modified from Dott 1983.)

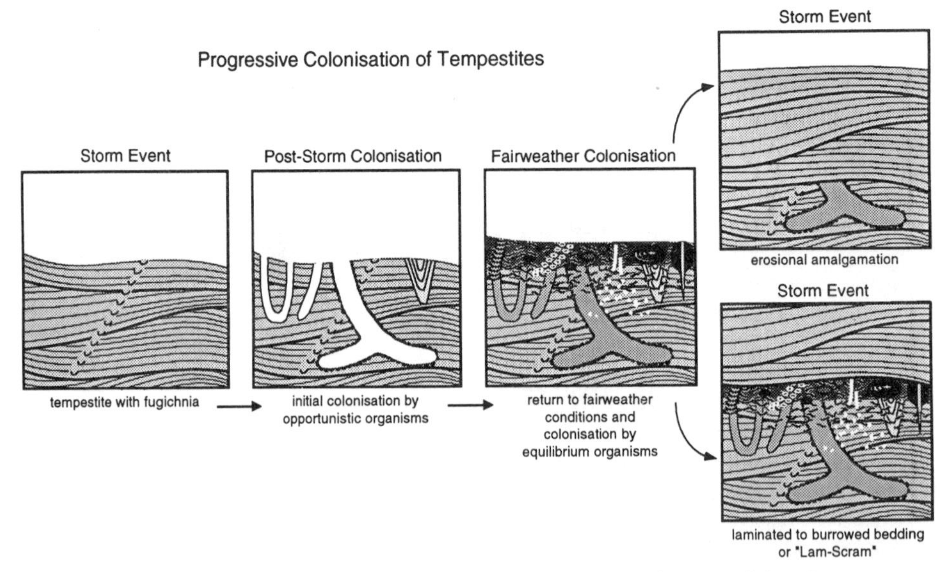

FIGURE 4.11. Progressive colonization of Tempestites. Initial stage of development shows emplacement of the tempestite with internal truncation surfaces. Fugichnia record the escape of organisms entrained in the flow or buried by the bed. The second stage of development reflects initial colonization of the unexploited storm bed under poststorm conditions. This suite is dominated by the suspension feeding and dwelling structures of opportunistic organisms. The third stage of development reflects fair-weather conditions and reestablishment of the resident equilibrium community. Subsequent storm events may result in (1) complete or virtually complete removal of evidence of poststorm colonization (some deeply penetrating structures and escape structures may survive) or (2) minimal erosional amalgamation of storm beds, preservation of the opportunistic storm suite, and locally, partial preservation of the fair-weather suite, producing "laminated-to-burrowed" bedding.

biogenic colonization. Summer fair-weather seasons favor biogenic colonization and modification of the tempestites; however, this is largely restricted to the top of the uppermost event bed. Any tempestites that may accumulate in response to the infrequent and lower-intensity storms typical of the fair-weather season may show significantly greater degrees of biogenic mottling.

Summary

Storms represent an important mode of deposition in most basins and are being recognized with great frequency in the rock record. In one of the best books written on stratigraphy, Ager (1981) concluded, "Nothing is world-wide, but everything is episodic. In other words, the history of any one part of the earth, like the life of a soldier, consists of long periods of boredom and short periods of terror" (Ager 1981, pp. 106–107). Tempestites contain a characteristic trace fossil suite that consists of a stable fair-weather assemblage and an unstable storm assemblage. The fair-weather (or resident) assemblage is dominated by traces of equilibrium (or K-selected) species, while the storm (or pioneer) assemblage is dominated by traces of opportunistic (or r-selected) species.

Tempestites show the following physical and ichnological characteristics: (1) a sharp base, with or without a basal lag; (2) parallel to subparallel laminations (hummocky or swaley cross-stratification); (3) common escape structures; (4) dwelling burrows of opportunistic organisms; and (5) gradational burrowed tops. Proximal-distal trends are generally discerned by changes in the character of the fair-weather suite. The integration of ichnological characteristics with the physical sedimentary features is particularly important in recognizing tempestites

ACKNOWLEDGMENTS

Funding was provided by Esso Resources Canada Ltd., PanCanadian Petroleum Ltd., Husky Oil Operations Ltd., and Canadian Hunter Exploration Ltd. Research support was also provided through a Natural Science and Engineering Research Council of Canada Operating Grant (No. A0816) to S. G. Pemberton, a Sir Izaack Walton Killam Memorial Scholarship, and a Province of Alberta Graduate Fellowship, both awarded to J. A. MacEachern. We are grateful to these agencies for their generous support of our research. This paper was completed during the senior authors stay as a visiting scientist with Exxon Production Research and he thanks John Van Wagoner, Kirt Campion, and Bob Todd for their generous hospitality and support.

REFERENCES

Ager, D. V. 1981. *The Nature of the Stratigraphic Record,* 2d ed. London: Macmillan.
Aigner, T. 1985. *Storm Depositional Systems.* Berlin: Springer-Verlag.
Aigner, T. and H. E. Reineck. 1982. Proximality trends in modern storm sands from the

Helgoland Bight (North Sea) and their implications for basin analysis. *Senckenbergiana Maritima* 14:183–215.

Benton, M. J. and D. I. Gray. 1981. Lower Silurian distal shelf storm-induced turbidites in the Welsh Borders: Sediments, tool-marks and trace fossils. *Journal of the Geological Society of London* 138:675–94.

Berry, J. P. 1989. Reproductive response of a marine annelid to winter storms: An analog to fire adaptation in plants? *Marine Ecology Progress Series* 54:99–107.

Beynon, B. M. , S. G. Pemberton, D. A. Bell, and C. A. Logan. 1988. Environmental implications of ichnofossils from the Lower Cretaceous Grand Rapids Formation, Cold Lake Oil Sands Deposit. In D. J. James and D. A. Leckie, eds., *Sequences, Stratigraphy, Sedimentology: Surface and Subsurface. Canadian Society of Petroleum Geologists Memoir* No. 15, pp. 275–90.

Boesch, D. F., R. J. Diaz, and R. W. Virnstein. 1976. Effects of tropical storm Agnes on soft-bottom macrobenthos communities of the James and York Estuaries and the Lower Chesapeake Bay. *Chesapeake Science* 17:246–59.

Boesch, D. F. and R. Rosenberg. 1981. Response to stress in marine benthic communities. In G. W. Barrett and R. Rosenberg, eds., *Stress Effects on Natural Ecosystems*, pp. 179–99. New York: Wiley.

Bonsdorff, E. 1983. Recovery potential of macrozoobenthos from dredging in shallow brackish waters. *Oceanological Acta*, Special Volume, Fluctuation and Succession in Marine Ecosystems, December 1983:27–32.

Bonvicini Pagliai, A. M. Cognetti Varriale, R. Crema, M. Curini Galletti, and R. Vandini Zunarelli. 1985. Environmental impact of extensive dredging in a coastal marine area. *Marine Pollution Bulletin* 16:483–88.

Bouma, A. H. 1962. *Sedimentology of Some Flysch Deposits*. Amsterdam: Elsevier.

Bourgeois, J. 1980. A transgressive shelf sequence exhibiting hummocky cross-stratification: The Cape Sebastian Sandstone (Upper Cretaceous), southwestern Oregon. *Journal of Sedimentary Petrology* 50:681–707.

Brezinski, D. K. 1986. An opportunistic Upper Ordovician trilobite assemblage from Missouri. *Lethaia* 19:315–26.

Bromley, R. G. and U. Asgaard. 1991. Ichnofacies: A mixture of taphofacies and biofacies. *Lethaia* 24:153–63.

Bromley, R. G. and A. A. Ekdale. 1984. *Chondrites:* A trace fossil indicator of anoxia in sediments. *Science* 224:872–74.

Bromley, R. G. and F. T. Fürsich. 1980. Comments on the proposed amendments to the international Code of zoological nomenclature regarding ichnotaxa. *Bulletin of Zoological Nomenclature* 37:6–10.

Butman, C. A. 1987. Larval settlement of soft-sediment invertebrates: The spatial scales of pattern explained by active habitat selection and the emerging role of hydrodynamical processes. *Annual Review of Oceanography and Marine Biology* 25:113–65.

Cadée, G. C. 1984. Opportunistic feeding, a serious pitfall in trophic structure analysis of (paleo)faunas. *Lethaia* 17:289–92.

Crimes, T. P. 1977. Trace fossils in an Eocene deep-sea sand fan, northern Spain. In T. P. Crimes and J. C. Harper, eds., *Trace fossils 2. Geological Journal*, Special Issue 9:71–90.

Crimes, T. P., R. Goldring, P. Homewood, J. van Stuijvenberg, and W. Winkler. 1981. Trace fossil assemblages of deep-sea fan deposits, Grunigel and Schlieren flysch (Cretaceous-Eocene, Switzerland). *Eclogae Geologicae Helvetiae* 74:953–95.

Cotter, E. and J. R. Graham. 1991. Coastal plain sedimentation in the late Devonian of southern Ireland; hummocky cross-stratification in fluvial deposits. *Sedimentary Geology* 72:201–24.

Dam, G. 1990. Paleoenvironmental significance of trace fossils from the shallow marine

Lower Jurassic Neill Klinter Formation, East Greenland. *Palaeogeography, Palaeoclimatology, Palaeoecology* 79:221–48.

Dauer, D. M. and J. L. Simon. 1976. Repopulation of the polychaete fauna on an intertidal habitat following natural defaunation: Species equilibrium. *Oecologia* 22:99–117.

Dobbs, F. C. and J. M. Vozarik. 1983. Immediate effects of a storm on coastal ichnofaunas. *Marine Ecology Progress Series*, vol. 11, pp. 273–79.

Dörjes, J. and G. Hertweck. 1975. Recent biocoenoses and ichnocoenoses in shallow-water marine environments. In R. W. Frey, ed., *The Study of Trace Fossils*, pp. 459–91. New York: Springer-Verlag.

Dott, H. R. Jr. 1983. 1982 SEPM Presidential address: Episodic sedimentation: How normal is average? How rare is rare? Does it matter? *Journal of Sedimentary Petrology* 53:5–23.

———. 1988. An episodic view of shallow marine clastic sedimentation. In P. L. de Boer, A. van Gelder, and S. D. Nio, eds., *Tide-Influenced Sedimentary Environments and Facies*, pp. 3–12. Dordrecht: D. Reidel Publishing Company.

Dott, R. H. Jr. and J. Bourgeois. 1982. Hummocky stratification: Significance of its variable bedding sequences. *Geological Society of America Bulletin* 93:663–80.

Duke, W. L. 1985. Hummocky cross-stratification, tropical hurricanes, and intense winter storms. *Sedimentology* 32:167–94.

———. 1990. Geostrophic circulation or shallow marine turbidity currents? The dilemma of paleoflow patterns in storm-induced prograding shoreline systems. *Journal of Sedimentary Petrology* 60:870–83.

Duke, W. L., R. W. C. Arnott, and R. J. Cheel. 1991. Shelf sandstones and hummocky cross-stratification: New insights on a storm debate. *Geology* 19:625–28 .

Ekdale, A. A. 1980. Graphoglyptid burrows in modern deep-sea sediment. *Science* 207:304–306.

———. 1985. Paleoecology of marine endobenthos. *Palaeogeography, Palaeoclimatology, Paleoecology* 50:63–81.

Ekdale, A. A., R. G. Bromley, and S. G. Pemberton. 1984. *Ichnology. Society of Economic Paleontologists and Mineralogists Short Course 15.*

Föllmi, K. B. and K. A. Grimm. 1990. Doomed pioneers: Gravity-flow deposition and bioturbation in marine oxygen-deficient environments. *Geology* 18:1069–72.

Frey, R. W. 1990. Trace fossils and hummocky cross-stratification, Upper Cretaceous of Utah. *Palaios* 5:203–18.

Frey, R. W. and R. Goldring. 1992. Marine event beds and recolonization surfaces as revealed by trace fossil analysis. *Geological Magazine* 129:325–35.

Goerke, H. 1971. Die Ernahrungsweise der Nereis-Arten (Polychaeta, Nereidae) der deutschen Kusten. *Veroffentlichungen Institut Meersforschung Bremerhaven* 13:1–50.

Gould, S. J. 1977. *Ontogeny and Phylogeny*. Cambridge: Harvard University Press.

———. 1989. *Wonderful Life*. New York: Norton .

Grassle, J. F. 1977. Slow recolonization of deep-sea sediment. *Nature* 265:618–19.

Grassle, J. F. and J. P. Grassle. 1974. Opportunistic life histories and genetic systems in marine benthic polychaetes. *Journal of Marine Research* 32:253–84.

Gray, J. S. 1974. Animal-sediment relationships. *Oceanographic Marine Biology Annual Review* 12:223–61.

Grimm, K. A. and K. B. Föllmi. 1990. Doomed pioneers: Event deposition and bioturbation in anaerobic environments. *American Association of Petroleum Geologists Bulletin* 74:666.

Hagerman, G. M. and R. M. Rieger. 1981. Dispersal of benthic meiofauna by wave and current action in Bogue Sound, North Carolina, U.S.A. *P.S.Z.N. Marine Ecology* 2:245–70.

Hannan, C. A. 1984. Planktonic larvae act like passive particles in turbulent near-bottom flows. *Limnology and Oceanography* 29:1108–15.

Harms, J. C. 1979. Primary sedimentary structures. *Annual Review of Earth and Planetary Science* 7:227–48.

Hayes, M. O. 1967. Hurricanes as geological agents: Case studies of Hurricanes Carla, 1961 and Cindy, 1963. *Bureau of Economic Geology, The University of Texas, Report of Investigations* 61:1–56.

Howard, J. D. 1975. The sedimentological significance of trace fossils. In R. W. Frey, ed., *The Study of Trace Fossils*, pp. 131–46. New York: Springer-Verlag.

Howard, J. D. and R. W. Frey. 1984. Characteristic trace fossils in nearshore to offshore sequences, Upper Cretaceous of east-central Utah. *Canadian Journal of Earth Sciences* 21:200–19.

Hunter, R. E. and H. E. Clifton. 1982. Cyclic deposits and hummocky cross-stratification of probable storm origin in Upper Cretaceous rocks of the Cape Sebastian area, southwestern Oregon. *Journal of Sedimentary Petrology* 52:127–43.

Hurst, J. M. and R. K. Pickerill. 1986. The relationship between sedimentary facies and faunal associations in the Llandovery siliciclastic Ross Brook Formation. Arisaig, Nova Scotia. *Canadian Journal of Earth Sciences* 23:705–26.

Jernelöv, A. and R. Rosenberg. 1976. Stress tolerance of ecosystems. *Environmental Conservation* 3:43–46.

Johnston, H. D. and C. T. Baldwin. 1986. Shallow siliciclastic seas. In H. G. Reading, ed., *Sedimentary Facies and Environments*, 2d ed., pp. 229–82. Oxford: Blackwell Scientific Publications.

Jumars, P. A. 1993. *Concepts in Biological Oceanography*. New York: Oxford University Press.

Jumars, P. A. and A. R. M. Nowell. 1984. Fluid and sedimentary dynamic effects on marine benthic community structure. *American Zoologist* 24:45–55.

Katsura, Y., F. Masuda, and I. Obata. 1984. Storm dominated shelf sea from the Lower Cretaceous Choshi Group, Japan. *Annual Report of the Institute of Geosciences, University of Tsukuba* 10:92–95.

Kotake, N. 1989. Paleoecology of the *Zoophycos* producers. *Lethaia* 22:327–41.

Kranz, P. M. 1972. The anastrophic burial of bivalves and its paleoecological significance. *Journal of Geology* 82:237–65.

Kreisa, R. D. 1981. Storm-generated sedimentary structures in subtidal marine facies with examples from the middle and upper Ordovician of southwestern Virginia. *Journal of Sedimentary Petrology* 51:823–48.

Kuenen, P. H. and C. Migliorini. 1950. Turbidity currents as a cause of graded bedding. *Journal of Geology* 58:91–127.

Kumar, N. and J. E. Sanders. 1976. Characteristics of shoreface storm deposits: Modern and ancient examples. *Journal of Sedimentary Petrology* 46:145–62.

Leckie, D. A. and L. A. Krystinik. 1989. Is there evidence for geostrophic currents preserved in the sedimentary record of inner to middle-shelf deposits? *Journal of Sedimentary Petrology* 59:862–70.

Leckie, D. A. and R. G. Walker. 1982. Storm-and tide-dominated shorelines in Cretaceous Moosebar-Lower Gates interval-outcrop equivalents of deep basin gas trap in Western Canada. *Bulletin of American Association of Petroleum Geologists* 66:138–57.

Leithold, E. L. 1989. Depositional processes on an ancient and modern muddy shelf, northern California. *Sedimentology* 36:179–202.

Leszczynski, S. and A. Seilacher. 1991. Ichnocoenoses of a turbidite sole. *Ichnos* 1:293–303.

Levinton, J. S. 1970. The paleoecological significance of opportunistic species. *Lethaia* 3:69–78.

Lewis, D. W. 1980. Storm generated graded beds and debris flow deposits with *Ophiomorpha* in a shallow offshore Oligocene sequence at Welsa, South Island, New Zealand. *New Zealand Journal of Geology and Geophysics* 23:353–69.
MacEachern, J. A., S. G. Pemberton, and I. Raychaudhuri. 1991. The substrate-controlled *Glossifungites* ichnofacies and its application to the recognition of sequence stratigraphic surfaces: Subsurface examples from the Cretaceous of the Western Canada Sedimentary Basin, Alberta, Canada. In D. A. Leckie, H. W. Posamentier, and R. W. Lovell, eds., *NUNA Conference on High Resolution Sequence Stratigraphy, Geological Association of Canada, Program, Proceedings and Guidebook,* pp. 32–36. Calgary, Canada: Geological Association of Canada.
MacEachern, J. A., I. Raychaudhuri, and S. G. Pemberton. 1992. Stratigraphic applications of the *Glossifungites* ichnofacies: Delineating discontinuities in the rock record. In S. G. Pemberton, ed., *Applications of Ichnology to Petroleum Exploration, Society of Economic Paleontologists and Mineralogists Core Workshop Notes* 17:169–98.
Magwood, J. P. A. and S. G. Pemberton. 1990. Stratigraphic significance of *Cruziana:* New data concerning the Cambrian-Ordovician ichnostratigraphic paradigm. *Geology* 18:729–32.
McCall, P. L. 1977. Community patterns and adaptive strategies of the infaunal benthos of Long Island Sound. *Journal of Marine Research* 35:221–26.
McKinney, M. L. 1986. Ecological causation of heterochrony: A test and implications for evolutionary theory. *Paleobiology* 12:282–89.
McNamara, K. J. 1986. A guide to the nomenclature of heterochrony. *Journal of Paleontology* 60:4–13.
Miller, W. 1991a. Intrastratal trace fossil zonation, Cretaceous flysch of northern California. *Ichnos* 1:161–71.
———. 1991b. Paleoecology of graphoglyptids. *Ichnos* 1:305–12.
Morton, R. A. 1988. Nearshore responses to great storms. *Geological Society of America Special Paper* 229:7–22.
Morton, R. A. and J. G. Paine. 1985. Beach and vegetation line changes at Galveston Island Texas; Erosion, deposition, and recovery from Hurricane Alicia. *The University of Texas at Austin Bureau of Economic Geology, Geological Circular* 85–5:1–39.
Owens, E. H. 1977. Temporal variations in beach and nearshore dynamics. *Journal of Sedimentary Petrology* 47:168–90.
Pearson, T. H. and R. Rosenberg. 1978. Macrobenthic succession in relation to organic enrichment and pollution of the marine environment. *Oceanography and Marine Biology Annual Review* 16:229–311.
Pedersen, G. K. and F. Surlyk. 1983. The Fur Formation, a late Paleocene ash-bearing diatomite from northern Denmark. *Bulletin of the Geological Society of Denmark* 32:43–65.
Pemberton, S. G. and R. W. Frey. 1984. Ichnology of storm-influenced shallow marine sequence: Cardium Formation (Upper Cretaceous) at Seebe, Alberta. In D. F. Stott and D. J. Glass, eds., *The Mesozoic of Middle North America. Canadian Society of Petroleum Geologists Memoir* No. 9, pp. 281–304.
Pemberton, S. G., J. A. MacEachern, and R. W. Frey. 1992a. Trace fossil facies models: Environmental and allostratigraphic significance. In R. G. Walker and N. P. Jones, eds., *Facies Models: Response to Sea Level Change,* pp. 47–72. Calgary, Canada: Geological Association of Canada.
Pemberton, S. G., J. A. MacEachern, and M. J. Ranger. 1992b. Ichnology and event stratigraphy: The use of trace fossils in recognizing tempestites. In S. G. Pemberton, ed., *Applications of Ichnology to Petroleum·Exploration,* pp. 85–117. *Society of Economic Paleontologists and Mineralogists, Core Workshop* 17.
Pemberton, S. G. and M. J. Risk. 1982. Middle Silurian trace fossils in the Hamilton, On-

tario region: Their identification, abundance and significance. *Northeastern Geology* 45:98–104.
Pemberton, S. G., J. C. Van Wagoner, and G. D. Wach. 1992c. Ichnofacies of a wave-dominated shoreline. In S. G. Pemberton, ed., *Application of Ichnology to Petroleum Exploration*, pp. 339–82. *Society of Economic Paleontologists and Mineralogists, Core Workshop* 17.
Pianka, E. R. 1970. On r and k selection. *American Naturalist* 104:592–97.
Rees, E. I. S., A. Nicholaidou, and P. Laskeridou. 1977. The effects of storms on the dynamics of shallow water benthic associations. In B. F. Keegan, P. O. Ceidigh, and P. J, Boaden, eds., *Biology of Benthic Organisms*, pp. 465–74. Oxford: Pergamon Press.
Rhoads, D. C., D. F. Boesch, T. Zhican, X. Fengshan, H. Liqiang, and K. J. Nilsen. 1985. Macrobenthos and sedimentary facies on the Changjiang delta platform and adjacent continental shelf. *Continental Shelf Research* 4:189–213.
Rhoads, D. C., P. L. McCall, and J. Y. Yingst. 1978. Disturbance and production on the estuarine seafloor. *American Scientist* 66:577–86.
Rice, A. L., D. S. M. Billett, J. Fry, A. W. G. Joh, R. S. Lampitt, R. F. C. Mantoura, and R. J. Morris. 1986. Seasonal deposition of phytodetritus to the deep-sea floor. *Royal Society of Edinburgh Proceedings* 88B:205–79.
Rollins, H. B., M. Carothers, and J. Donahue. 1979. Transgression, regression, and fossil community succession. *Lethaia* 12:89–104.
Sanders, H. L., J. F. Grassle, G. R. Hampson, L. S. Morse, S. Garner-Price, and C. C. Jones. 1980. Anatomy of an oil spill: Long-term effects from the grounding of barge Florida off West Falmouth, Massachusetts. *Journal of Marine Research* 38:265–82.
Santos, S. L. and J. L. Simon. 1980. Marine soft-bottom community establishment following annual defaunation: Larval or adult recruitment. *Marine Ecology—Progress Series* 2:235–41.
Savrda, C. E. 1991a. Ichnology in sequence stratigraphic studies: An example from the Lower Paleocene of Alabama. *Palaios* 6:39–53.
———. 1991b. *Teredolites*, wood substrates and sea-level dynamics. *Geology* 19:905–908.
Seilacher, A. 1962. Paleontological studies in turbidite sedimentation and erosion. *Journal of Geology* 70:227–34.
———. 1969. Fault-graded beds interpreted as seismites. *Sedimentology* 13:155–59.
———. 1981. Towards an evolutionary stratigraphy. *Acta Geologica Hispanica* 16:39–44.
———. 1982a. General remarks about event beds. In G. Einsele and A. Seilacher, eds., *Cyclic and Event Stratification*, pp. 161–74. Berlin: Springer-Verlag.
———. 1982b. Distinctive features of sandy tempestites. In G. Einsele and A. Seilacher, eds., *Cyclic and Event Stratification*, pp. 333–49. Berlin: Springer-Verlag.
Snyder, J. and R. W. Bretsky. 1971. Life habits of diminutive bivalve mollusks in the Maquoketa Formation (Upper Ordovician). *American Journal of Science* 271:227–51.
Sousa, W. P. 1984. The role of disturbance in natural communities. *Annual Review Ecology Systematics* 15:353–91.
Thistle, D. 1981. Natural physical disturbances and communities of marine soft bottoms. *Marine Ecology—Progress Series* 6:223–28.
Tillman, R. W. and R. S. Martinsen. 1985. Shannon Sandstone, Hartzog Draw Field core study. In R. W. Tillman, D. P. J. Swift, and R. G. Walker, eds., *Shelf Sands and Sandstone Reservoirs. Society of Economic Paleontologists and Mineralogists Short Course*, 13:577–644.
Vossler, S. M. and S. G. Pemberton. 1988. *Skolithos* in the Upper Cretaceous Cardium Formation: An ichnofossil example of opportunistic ecology. *Lethaia* 21:351–62.
———. 1989. Ichnology and paleoecology of offshore siliciclastic deposits in the Cardium Formation (Turonian, Alberta, Canada). *Palaeogeography, Palaeoclimatology, Palaeoecology* 74:217–29.

Waage, K. M. 1968. The type Fox Hills Formation, Cretaceous (Maestrichtian), South Dakota. Part 1. Stratigraphy and Paleoenvironments. *Peabody Museum of Natural History, Yale University, Bulletin* 27.

Walker, R. G. 1984a. Shelf and shallow marine sands. In R. G. Walker, ed., *Facies Models*, 2d ed., Reprint Series 1, pp. 141–70. Toronto: Geoscience Canada.

———. 1984b. Turbidites and associated coarse clastic deposits. In R. G. Walker, ed., *Facies Models*, 2d ed., Reprint Series 1, pp. 171–88. Toronto: Geoscience Canada.

Ward, D. M. and D. W. Lewis. 1975. Paleoenvironmental implications of storm-scoured, ichnofossiliferous Mid-Tertiary limestone, Waihao District, South Canterbury, New Zealand. *New Zealand Journal of Geology and Geophysics* 18:881–908.

Wetzel, W. 1991. Ecologic interpretation of deep-sea trace fossil communities. *Palaeogeography, Palaeoclimatology, Palaeoecology* 85:47–69.

Wheatcroft, R. A. 1990. Preservation potential of sedimentary event layers. *Geology* 18:843–45.

Whitlatch, R. B. and R. N. Zajac. 1985. Biotic interactions among estuarine infaunal opportunistic species. *Marine Ecology—Progress Series* 21:299–311.

Winn, R. D. Jr., S. P. Stonecipher, and M. G. Bishop. 1983. Depositional environments and diagenesis of offshore sand ridges, Frontier Formation, Spearhead Ranch Field, Wyoming. *The Mountain Geologist* 20:41–58.

Wolff, W. J. 1973. The estuary as a habitat: An analysis of data on the soft-bottom macro-fauna of the estuarine area of the rivers Rhine, Meuse, and Scheldt. *Zoologische Verhandlungen Leiden* 126:1–242.

Wu, X. 1982. Storm-generated depositional types and associated trace fossils in Lower Cretaceous shallow marine carbonates of Three Cliffs Bay and Ogmore-by-Sea, South Wales. *Palaeogeography, Palaeoclimatology, Palaeoecology* 39:187–202.

Zajac, R. N. 1986. The effects of intra-specific density and food supply on growth and reproduction in an infaunal polychaete, *Polydora ligni* Webster. *Journal of Marine Research* 44:339–59.

5

*Prasopora-*Bearing Event Beds in the Coburn Limestone (Bryozoa; Ordovician; Pennsylvania)

Roger J. Cuffey

Abstract

The Coburn Limestone (late Middle Ordovician, central Pennsylvania) is characterized by thin fossil-coquinite rudstones, which form the basal part of repetitive carbonate tempestite (storm) sequences that comprise the formation. Randomly oriented hemispherical colonies of the trepostome bryozoan *Prasopora simulatrix* occur in many of these basal coquinites; other hemispherical *Prasopora* are found in upright growth position in the finer-grained micritic tops of the tempestites and in the overlying intertempestite shales. Taphonomic occurrence of the *Prasopora* colonies suggests that storm waves stirred up the sea bottom and redeposited skeletal material, including the hemispherical colonies, in jumbled fashion. Storm sedimentation was followed by much longer intervals of quiet water when *Prasopora* larvae colonized the fresh substrate, grew into large colonies, and were eventually buried by fine sediment. Later storms repeated this depositional pattern and created the stacked tempestite sequences characteristic of the Coburn, in which *Prasopora* participate as both sedimentary particles and biological colonizers.

As defined in the preface to this volume, paleontological event horizons or beds are relatively thin but regionally persistent stratigraphic surfaces or intervals characterized by unusual paleontological features. Some of these thin units were deposited in hours, others over millennia, but both extremes carry the connotations of geochronologically instantaneous or rapid happenings, and some difference from ordinary processes or the ambient sedimentation.

The highest Middle Ordovician carbonate formation in central Pennsylvania, the Coburn Limestone, includes a number of thin beds containing the unusual hemispherical bryozoan *Prasopora*. Each bed has only local extent, but the interval carrying them all extends regionally. The richest beds are virtually *Prasopora* conglomerates, but in others the colonies can be widely scattered.

Close examination of the distribution, orientations, and matrix of the Coburn *Prasopora* indicates that each bed is part of a package of storm sediments, representing a single storm event (hours to days in duration). Thus the Coburn Limestone represents a period (probably many millennia) during which storms periodically swept the shallow carbonate platform. Each storm event bed thus captures a single moment—a snapshot—in the life of these ancient organisms; the focus of this paper is on those individual event beds and their paleoecological and taphonomic interpretation.

Coburn Limestone

In central Pennsylvania, the Lower and Middle Ordovician consists of thick carbonates, overlain by Upper Ordovician detrital clastics. The uppermost limestones, the Coburn and underlying Salona formations, are exposed along strike (fig. 5.1) in the Appalachian Valley and Ridge Province. The Coburn yields abundant fossil invertebrates, including *Prasopora*, other bryozoans (Arens and Cuffey 1989a), brachiopods, gastropods, trilobites, and crinoid ossicles. The underlying Salona is mostly unfossiliferous but does yield an occasional *Prasopora* locally (Arens and Cuffey 1989a, 1989b; Thompson 1963).

The Coburn Limestone is late (but not latest) Trentonian in age, specifically Denmarkian, and probably diachronously Shorehamian southeastward (fig. 5.2), as discussed by Arens and Cuffey (1989a:118). The Coburn thus belongs in the Middle Ordovician of classical North American usage. Relabeling it and much of the underlying Middle Ordovician as "Upper" Ordovician (Keith 1989:1–2) should be vigorously rejected because such relabeling will introduce massive confusion into this system's biostratigraphic literature, much of which records the rich shelly American fauna like the Coburn *Prasopora* species. (A better solution would be to split the overly long, less fossiliferous British Caradocian into two series, separated at the same horizon as the American Middle/Upper Ordovician contact.)

The Coburn Limestone occurs in rhythmic or repetitive beds 10–30 cm thick (E to E in fig. 5.3; also fig. 5.4). Beds pinch and swell laterally over tens of meters and are commonly lenticular. Individual beds cannot be reliably traced laterally between outcrops, as Thompson (1963) also noted. Each Coburn sequence begins with a sharp erosional base, commonly with intraclasts of underlying micritic material, but without sole marks. Overlying this base is a fining-upward sequence of coquinite or bioclastic calcirudite (rudstone), calcarenite and calcisiltite (packstone to wackestone; plane-parallel-laminated below, cross-

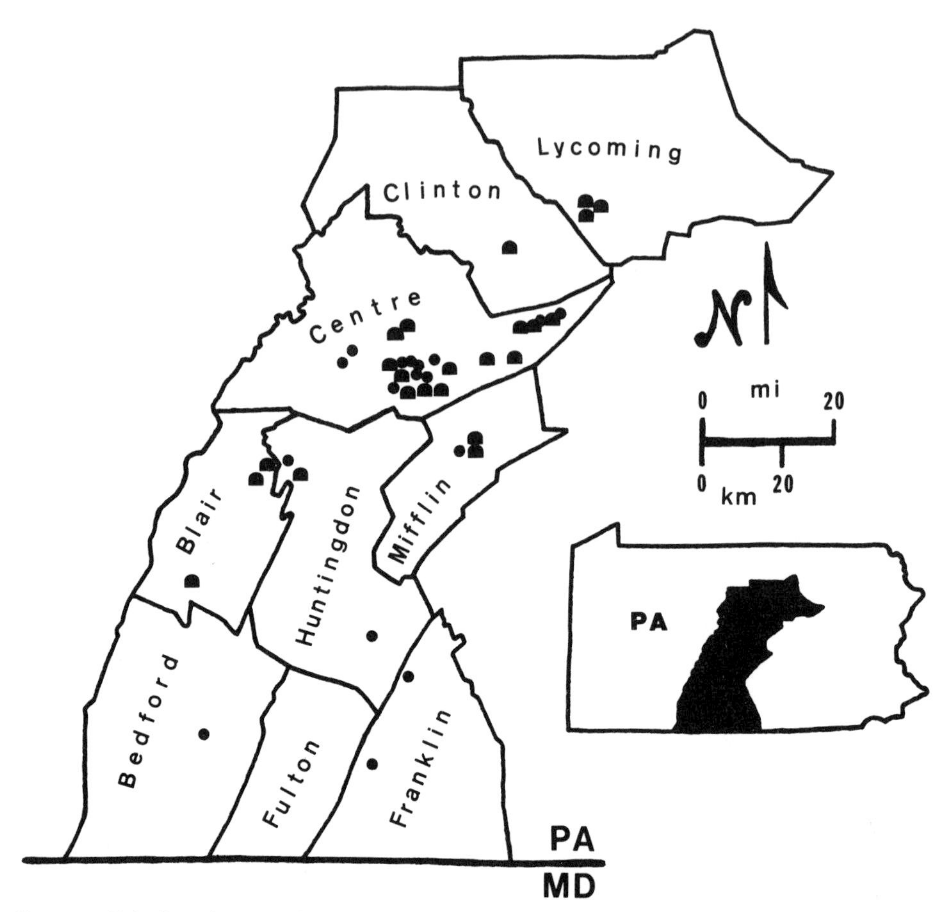

FIGURE 5.1. Localities and counties yielding bryozoans from the Coburn and Salona limestones; hemispheres indicate *Prasopora simulatrix* sites. (Modified from Arens and Cuffey [1989a:113]. Reproduced with permission of Pennsylvania Academy of Science.)

laminated, possibly hummocky above), and frequently a capping calcilutite (micstone).

Calcareous shale or mudstone interbeds commonly occur between the repetitive limestone sequences; these interbeds become thicker, more numerous, and more siliciclastic up section. These repetitive sequences are interpreted as storm-wave-produced tempestites, because of their similarity to the products of ancient and modern storm sedimentation as described by Aigner (1985) and discussed by Arens and Cuffey (1989c; also note Gardiner-Kuserk 1989).

Thin (3–20 cm) bentonites are locally present within the Coburn Limestone (Thompson 1963; Smith et al. 1986; Arens and Cuffey 1989c:222). Many of these bentonites cannot be correlated between outcrops, thus suggesting that the once uniform blankets of precursor ash were reworked, possibly by wave action, at some localities but not at others. This further suggests that the storm wave base hovered near the Coburn sea bottom, with local submarine topography determining whether or not bentonites were preserved at a given location.

The Coburn Limestone represents the transition between an earlier very-shallow-water carbonate platform and subsequent deeper-water detrital clastics of the Taconic foreland basin. Despite its setting on a foundering continental margin, the Coburn appears still to have been deposited in relatively shallow

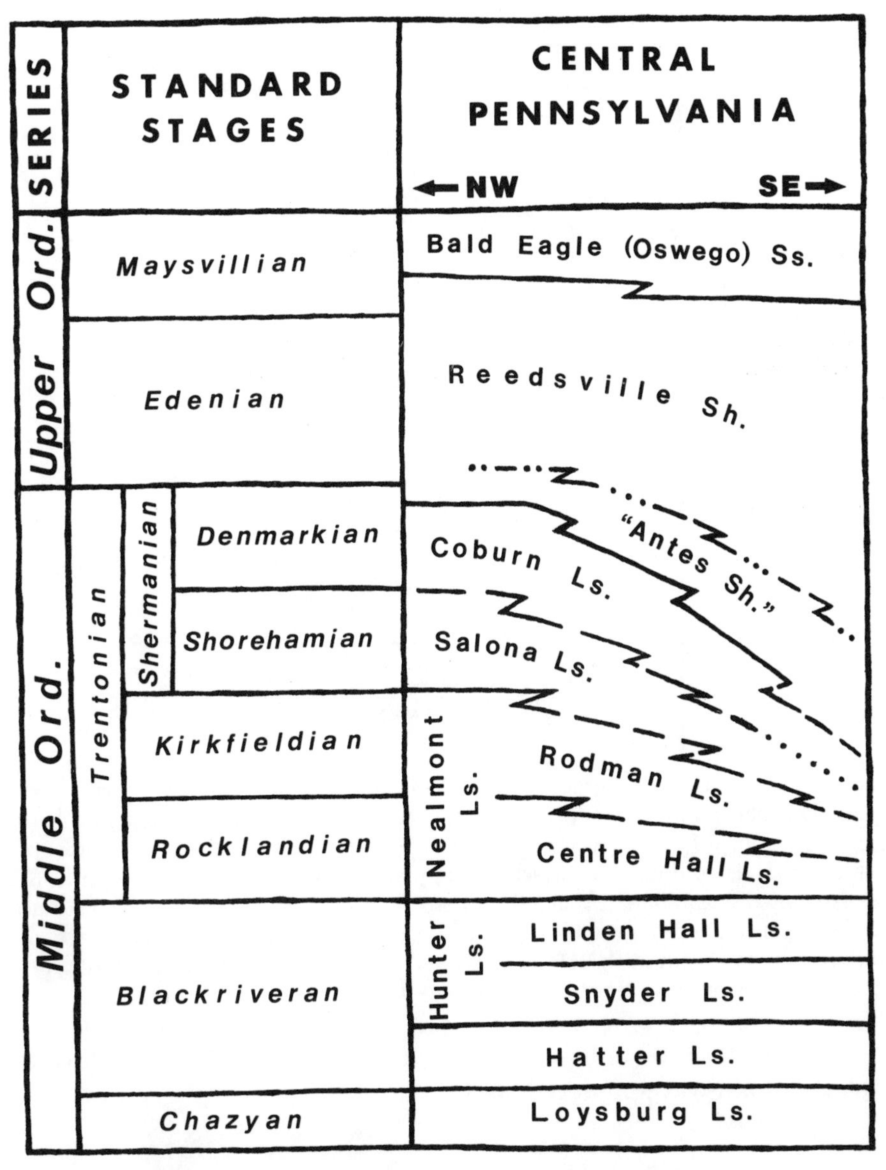

FIGURE 5.2. Correlation of the Coburn Limestone and related formations with the standard Ordovician time scale. (From Arens and Cuffey [1989a:119]. Reproduced with permission of Pennsylvania Academy of Science.)

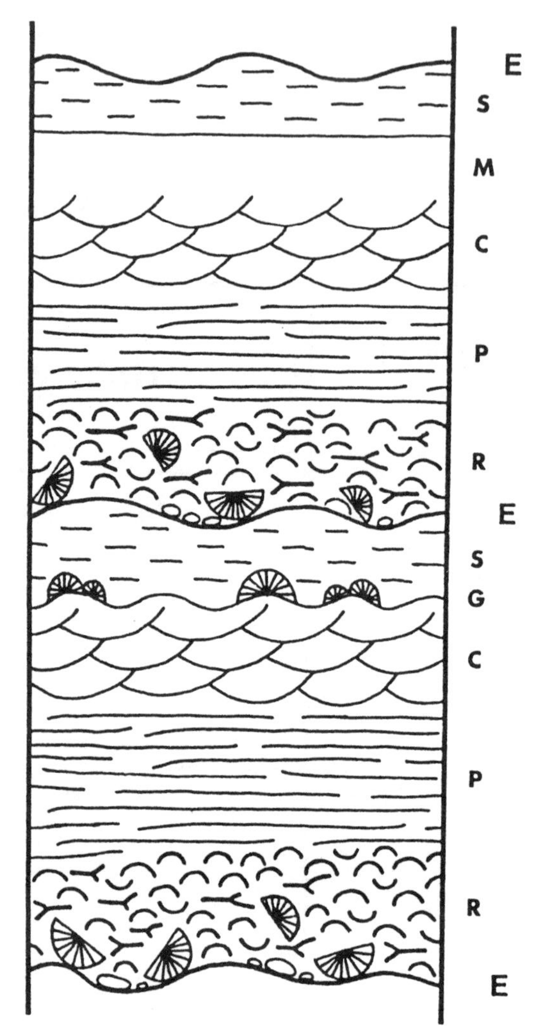

FIGURE 5.3. Idealized stratigraphic sequence illustrating two *Prasopora*-bearing event beds in the Coburn Limestone: (E) erosional base of bed; (R) bioclastic coquinite or rudstone with jumbled *Prasopora* colonies, other bryozoan branch fragments, convex-up brachiopod shells, and micritic intraclasts; (P) plane-parallel-laminated calcarenite; (C) cross-laminated calcarenite; (M) massive or structureless micstone; (G) rippled surface with *Prasopora* colonies in growth position; (S) calcareous shale or mudstone; each event bed (E to E) is 10–30 cm thick. (Modified from Arens and Cuffey [1989c:220]. Reproduced with permission of Northeastern Science Foundation.)

water (above storm wave base), because it is characterized by repetitive tempestites or storm-wave-produced beds.

HEMISPHERICAL *PRASOPORA*

Understanding the critical role of *Prasopora* in the Coburn event beds requires a brief summary of these fossils' morphology and distribution. Establishing the internal morphology of the hemispherical *Prasopora simulatrix* colonies is essential for determining the orientations of the many specimens observed in the field. Establishing the high variability of this species is essential for ensuring that all the colonies observed were conspecific and hence ecologically coherent (i.e., will have responded to environmental stimuli in the same way, so that all their distributional data can be integrated into a consistent picture). Furthermore, establishing this species' taxonomic history is essential for integrating the implications of earlier studies into a fuller understanding of the Coburn *Prasopora* event beds.

In the Coburn and underlying Salona limestones, the most noticeable bryozoans are hemispherical or domal colonies 1–4 cm in diameter (figs. 5.5, 5.6A,B), virtually all of which represent *Prasopora simulatrix* Ulrich. This species is conspicuous within later Middle Ordovician bryozoan faunas because of its

FIGURE 5.4. Three stacked repetitive tempestite sequences in the Coburn Limestone (PENMIF-20 locality); note erosional bases, coarse-textured coquinite or rudstone, fining-upward limestone sequence, and overlying intersequence shales. (From Arens and Cuffey [1989c:221]. Reproduced with permission of Northeastern Science Foundation.)

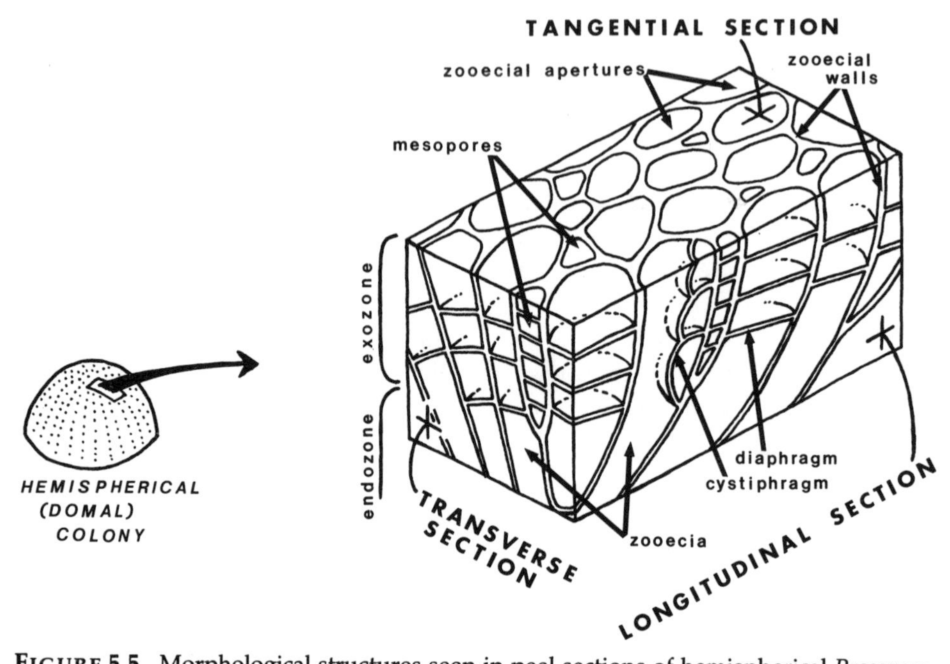

FIGURE 5.5. Morphological structures seen in peel sections of hemispherical *Prasopora* colonies. (Modified from Arens and Cuffey [1989a:119]. Reproduced with permission of Pennsylvania Academy of Science.)

unusually large size, shape, abundance, and ubiquity. Hemispherical *Prasopora* colonies now weather out from eroding shale banks in the Midwest and etch into positive relief on exposed limestone surfaces in the Northeast. These colonies have attracted much attention, both scientific and amateur (as "fossil gumdrops").

Prasopora simulatrix is characterized by bubblelike cystiphragms, numerous diaphragms, thin zooecial walls, round zooecial apertures, and closely tabulated mesopores (fig. 5.6C–G; Ross 1967:407, 412–414, text-fig. 2a, pl. 46, figs. 1, 2, 4, pl. 47, fig. 1, 3–5, pl. 49, fig. 5, 7, 9, pl. 50, fig. 1, 2; Bork and Perry 1968:1058–1061, pl. 137, fig. 8–10, pl. 138, fig. 5–7; Ross 1970:352–353; Arens 1988; Arens and Cuffey 1989a:114–115, 119, Fig. 2A–2J).

Prasopora simulatrix is classified in the family Monticuliporidae within the leptaulate tubulobryozoan order Trepostomida, because of the presence of cystiphragms and diaphragms in the elongate tubular zooecia (Arens and Cuffey 1989a:114; Bassler 1953; Cuffey 1973). Monticuliporids superficially resemble small tabulate corals, but thin sections reveal typical bryozoan internal structures.

Examination of the many hemispherical *Prasopora* colonies (in this study, more than 200 from the Coburn Limestone) reveals that *Prasopora simulatrix* is an extremely variable species. Tangential sections exhibit variations in zooecial diameter and wall thickness. In longitudinal sections, zooecial walls may be either straight or undulating ("wavy"). Zooecial tubes may have few to many diaphragms, which may be evenly or irregularly spaced, and range from horizontal to inclined to chevron shaped. Hemiphragms may be present. Cystiphragms may be present, and they may line one or both sides of the zooecial cavity. They may occur alone or with diaphragms that connect them to the opposite wall or to other cystiphragms, or even themselves touch cystiphragms from the other side of the zooecial cavity. Finally, overall colony form can also vary, as illustrated by Sardeson (1935:175) and Kobluk and Nemcsok (1982:685).

Despite this large range of morphological variation, the *Prasopora* specimens examined clearly represent a single species. Evidence leading to this conclusion includes series of colonies from the same stratigraphic horizon, thus representing essentially contemporaneous and potentially interbreeding local populations, which can be arranged to show complete morphological intergradation between the observed extremes. Additional evidence is the observed wide range of variation within certain single colonies; because all zooids within one colony result from asexual budding from the same initial larva, all the various

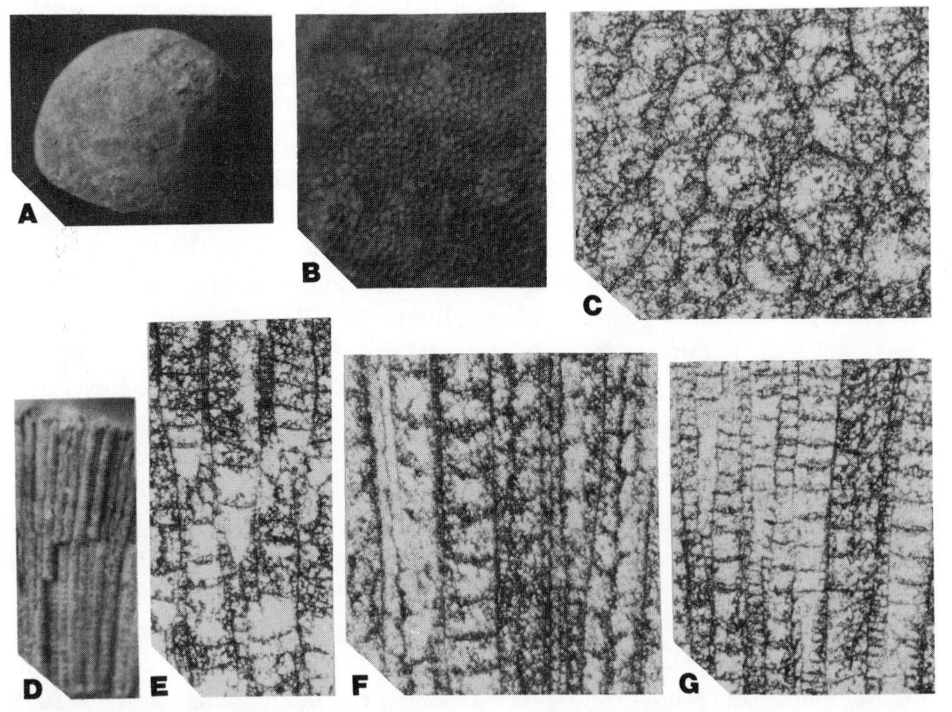

FIGURE 5.6. Morphology of *Prasopora simulatrix:* (A) Hemispherical colony, ×0.8; (B) external surface, ×4; (C) tangential peel section, ×24; (D) broken weathered section, ×5; (E–G) longitudinal peel sections of different colonies showing intraspecific variability, all ×24. Specimens in the Paleobryozoological Research Collection (PBRC) at Pennsylvania State University ([A,B] PENCEN-51–A-hds; [C] PENLYC-5–Ct5–29; [D] PENCEN-52–hds; [E] PENLYC-5–Ct5–19; [F] PENLYC-5–Ct5–15; [G] PENCEN-114–fl-1Ca). (From Arens and Cuffey [1989a:115]. Reproduced with permission of Pennsylvania Academy of Science.)

zooecia in that colony obviously represent the same species. Thus *Prasopora simulatrix* appears to be a highly variable species, similar to other well-studied long-tubed bryozoan species (Cuffey 1967; Horowitz 1968; Warner and Cuffey 1973; Cuffey and Sorrentino 1985; Madsen 1986).

The substantial variability observed within the Coburn-Salona population embraces several variants that have been given separate species names but are more appropriately synonymized within *Prasopora simulatrix,* as recommended also by Ross (1967) and Bork and Perry (1968). These include *P. conoidea, P. contigua, P. similis,* and *P. sinclairi* (Ulrich 1886, 1893; Bassler 1919; Allen and Lester 1954; Fritz 1957; Boardman 1971). Published illustrations of other named *Prasopora* species that have been listed as *Prasopora simulatrix* synonyms do not match any specimens within the Coburn-Salona population, which suggests that those may represent different real species.

A tiny minority—less than 1%—of the Coburn-Salona hemispherical bryozoans do actually belong to other trepostome species: *Batostoma* cf. *B. sheldonensis, Cyphotrypa pachymuralis, Diplotrypa westoni,* and *Mesotrypa orientalis* (Arens and Cuffey 1989a:114–116, fig. 2K–2N, 3A–3D). These species are indistinguishable from the predominant *Prasopora* in external appearance and distribution, and can fit into the paleoecological and sedimentologic roles inferred for *Prasopora simulatrix.*

In addition to the taxonomic implications of its great variability, *Prasopora simulatrix* has been known to science for an unusually long time. This has introduced further complications that must be addressed.

Prasopora colonies are common and conspicuous fossils in the type Trenton strata of upstate New York (see fig. 5.9F in following section), a region heavily involved in the initial industrialization of the United States (projects such as the Erie Canal come to mind immediately). It was thus inevitable that these unusual fossils would come to the early attention of workers there, and James Hall's (1847) recognition of them counts as one of the first Paleozoic bryozoan records from the North American continent. Further indication of this bryozoan's importance in that region is early workers' use of the name *Prasopora* beds for the type Trenton strata (Titus 1989:81).

Hall (1847:64–67; pl. 23, fig. 1[*sic*]–1f; pl. 24, fig. 1a–1d) originally included what we know today as *Prasopora simulatrix* in his broader species concept *Chaetetes lycoperdon* (fig. 5.7), which also contained several other branching trepostome species (Hall 1847:276; pl. 75, fig. 2a–f) not now part of the genus *Prasopora,* and which now can be restricted to a chaetetid sponge species (Brett 1991, personal communication). Hall (1847) attributed the trivial name *lycoperdon* to Say; Ulrich, years later (1890:318), credited Vanuxem with a similar name for these fossils, *Favosites lycopodites. Lycoperdon* is a puffball fungus, and the analogy with the bryozoan's colony shape is apt.

However, Ulrich (1886) introduced another species concept, *Prasopora simulatrix,* for identical-looking hemispherical bryozoans from Trenton equivalents in the Upper Mississippi Valley, later explaining that *lycoperdon* and *lycopodites*

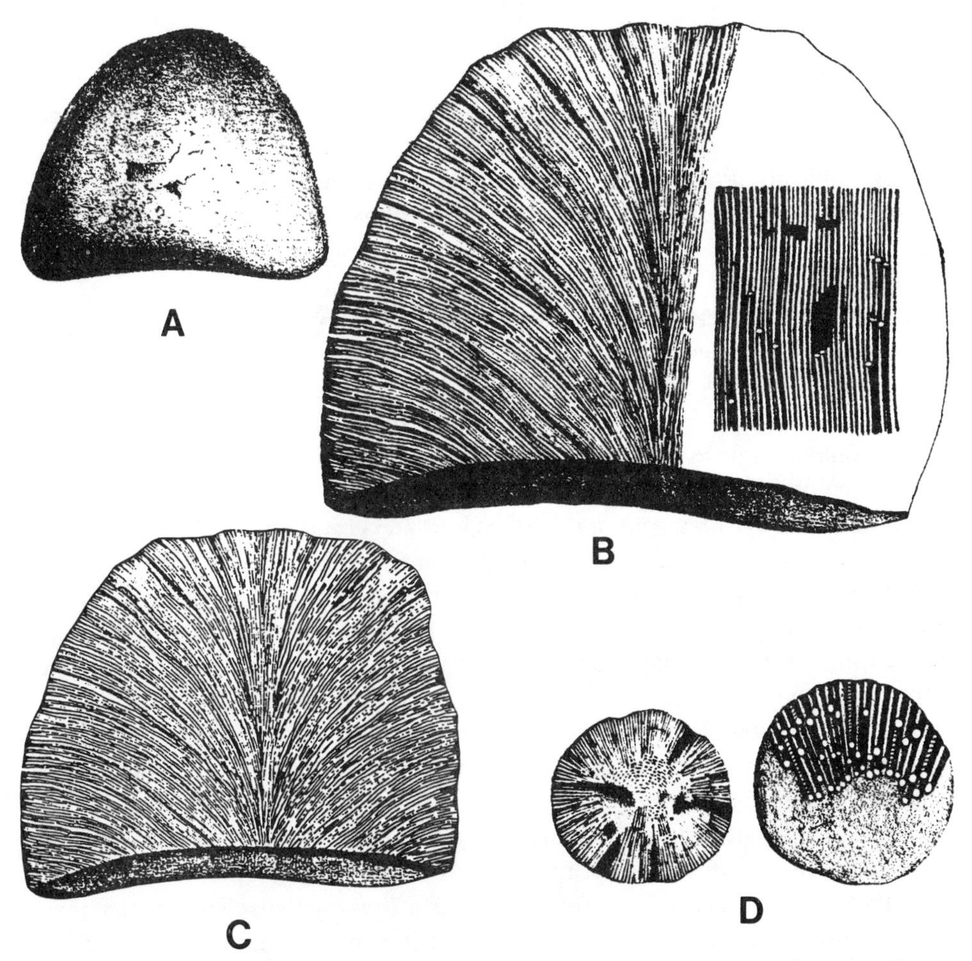

FIGURE 5.7. Hall's original figures illustrating hemispherical *Prasopora* from the Trenton Limestone of upstate New York (approximately ×1, or if "enlarged" about ×5): (A) exterior lateral view of colony; (B) upwardly or distally flaring zooecial tubes giving fibrous texture to vertical (= longitudinal) broken surface, and inset showing tubes enlarged; (C) same but with right half completed by the present author to assist field recognition of orientations; (D) horizontal (= transverse) broken surface of colony showing radiating fibrous texture due to elongate zooecial tubes (enlarged on right). (From Hall [1847, pl. 23; fig. 1d; 1[*sic*] and 1a; 1 again; 1b and 1c, respectively]. Reproduced with permission of New York State Museum and Science Service.)

were unrecognizable (Ulrich 1893:246–48). He soon elucidated his species with then-new thin-section techniques (Ulrich 1893:245–248; pl. 16, figs. 1–10), and paleobryozoologists since have utilized his concept extensively, either by itself or as a geographic eastern variety "*orientalis*" (for complete synonomy see Brown, 1965:983; Ross 1967:412–13; Bork and Perry 1968:1058–59; Arens 1988:78–79).

Marintsch (1981) suggested that *Prasopora falesi* (James 1884, from the Lexington Limestone in Kentucky) and *Prasopora simulatrix* (Ulrich 1886) are actually subjective synonyms. If so, the unfamiliar earlier name would take priority over the widely used *simulatrix*, thereby complicating the species-level literature for this paleoecologically important genus. Karklins (1984), however, maintained the two as separate species on morphological grounds.

Geographically, *Prasopora simulatrix* ranges across the eastern half of the mid-Ordovician North American craton, from southeastern Ontario and upstate New York, westward to Minnesota and Iowa, and southward to Georgia and Tennessee.

In New York particularly, the fossil assemblages containing this species have been analyzed for paleocommunities; *Prasopora simulatrix* occurs in several such communities but does not seem to have played a dominant or defining role in any (Titus 1986:820). However, in eastern Tennessee, *Prasopora* colonies are conspicuous enough to lend their name to one paleocommunity, the *Zygospira-Prasopora* community (Walker and Diehl 1986:68–69).

Stratigraphically, *Prasopora simulatrix* is known certainly from the basal Trentonian into the lower Edenian, and apparently from as old as the middle Blackriveran. A few early reports extend the species's range back into the lower Blackriveran and even upper Chazyan, but these are questionable because of uncertainties in the regional correlations involved. This species's entire range falls within the European Caradocian, but not coinciding with either of its boundaries, thereby illustrating the imprecision of utilizing the longer European stage terminology in this region in contrast to the more finely subdivided American nomenclature.

Very late in Trentonian time, *Prasopora simulatrix* gave rise to several descendant species (Ross 1967:411; Sparling 1964; Rudolph 1975; Mehrtens and Barnett 1979, 1986). All the many Coburn *Prasopora* are *Prasopora simulatrix* rather than any of those evolutionary descendants, and hence the youngest part of the Coburn must be at least mid-Upper Trentonian (middle Denmarkian) or older.

To summarize, the numerous hemispherical bryozoan colonies found in the Coburn tempestite sequences belong almost entirely to the morphologically variable, widely distributed, single species *Prasopora simulatrix*. It is therefore both reasonable and necessary to interpret their ecological and sedimentologic roles in coherent, unified fashion.

Distribution and Orientation of Coburn Prasopora

Within each *Prasopora*-bearing Coburn event bed, *Prasopora simulatrix* colonies are found in two distinct stratigraphic positions, with differing characteristic colony orientations, and fitting consistently into the same portion of the repetitive tempestite sequences. These observations are summarized graphically in fig. 5.3.

First, hemispherical *Prasopora* occur in the fossil coquinite at the bases of the repetitive storm sequences (R in fig. 5.3). Within the basal coquinite, *Prasopora*

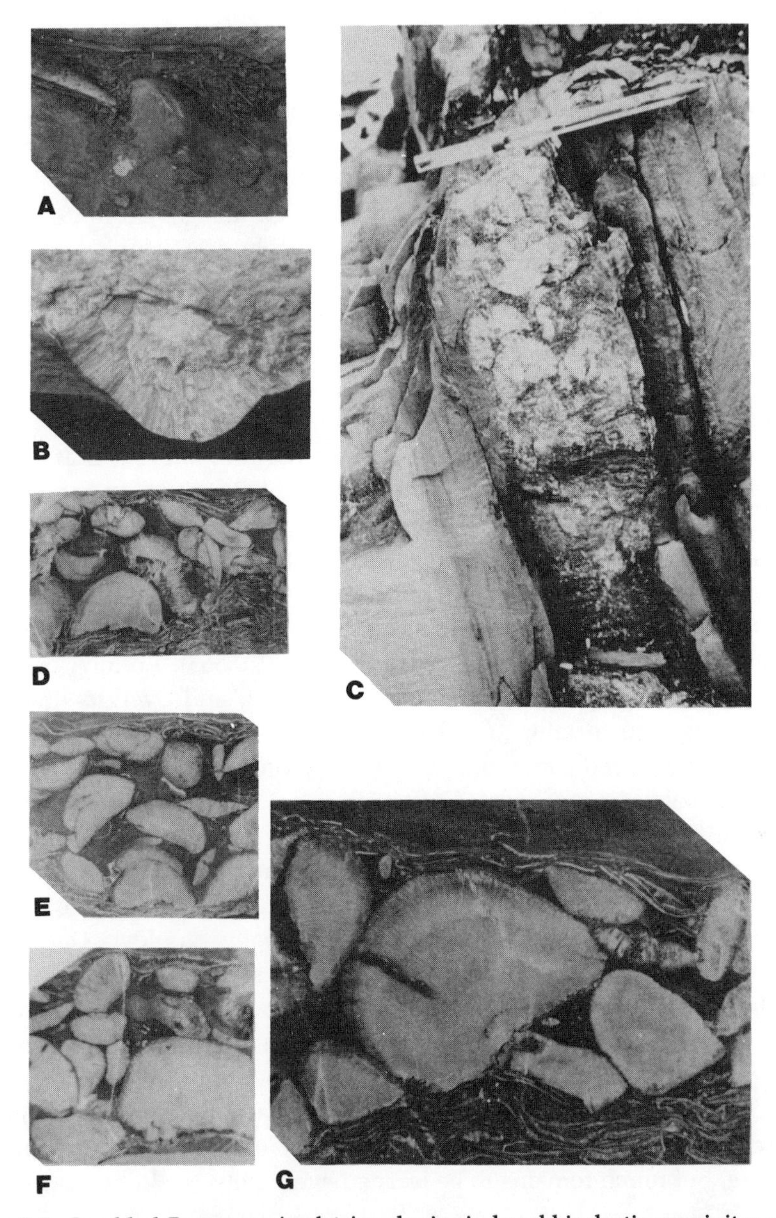

FIGURE 5.8. Jumbled *Prasopora simulatrix* colonies in basal bioclastic coquinites or rudstones of Coburn event beds; stratigraphic up is toward top of page in A, B, D–G, and to right in C: (A) tumbled colony within bed, ×0.4; (B) overturned colony embedded in bottom of bed, ×1; (C) outcrop face exposing lenticular *Prasopora* conglomerate, ×0.4; D–G, polished surfaces prepared from same bed as C, ×0.6, ×0.6, ×0.6, ×1.3, respectively.
(A, PBRC PENLYC-5–Ct5; B, PENLYC-5–Ct80–hds; C-G, PENCLI-8–0′-hds.) (In part modified from Arens and Cuffey [1989a:115; 1989c:221]. Reproduced with permission of Pennsylvania Academy of Science [B] and Northeastern Science Foundation [C].)

colonies are randomly and variably oriented, from upright to slightly tilted, tumbled onto their sides, and even overturned (fig. 5.8A–B, 5.8D–G). Locally, these colonies are so numerous that they form packed *Prasopora* conglomerate beds (fig. 5.8C–G). The *Prasopora simulatrix* are mostly unbroken, unabraded, and generally well preserved, suggesting relatively short-distance transport and rapid burial (i.e., a disturbed-neighborhood assemblage; Scott 1970; also see Carter, Miller, and Smosna 1988:34). The hemispherical colonies are accompanied by convex-up brachiopod and gastropod shells, crinoid and trilobite debris, and broken branch fragments of other bryozoans. These last lie parallel to bedding, but with no preferred azimuthal orientation; this supports deposition under turbulent flow, such as the circular motion of storm waves, rather than the unidirectional flow characteristic of storm or turbidity currents.

The second within-bed position for hemispherical *Prasopora* is at the tops of the repetitive storm sequences (G and S in fig. 5.3), either in the uppermost finest-grained limestone or in the overlying calcareous shale. Here the *Prasopora* colonies are uniformly upright, in growth position (fig. 5.9A–F). If the uppermost limestone is finely calcarenitic, its upper surface may be rippled, and then the *Prasopora* prefer the ripple crests (in contrast to New York specimens that more often occur in the troughs instead; Ross 1970:353). Locally, colonies are clustered together (fig. 5.9D–F), even overlapping and overgrowing one another but nowhere forming true bioherms. These specimens suggest normal growth in undisturbed conditions for intervals longer than the life span of such colonies (probably several years; Cuffey 1967).

Far west from Pennsylvania, in Iowa, *Prasopora simulatrix* colonies occur in similar circumstances, in approximately correlative carbonates (Kolata and Sloan 1987:99; Karklins 1987:174, 176), which could well be storm beds (R. L. Anstey 1991, personal communication). Additionally, the Iowa Decorah Shale yields *Prasopora*, but most weathered free; future studies should examine their orientations in situ, compare their sizes and shapes with the Coburn colonies, and consider what characteristics storm beds would show if composed entirely of terrigenous muds. Studies of bryozoan taphonomy and colony-fragment sedimentology have begun only recently (Anstey and Rabbio 1989; Cuffey et al. 1981; Errett and Cuffey 1989; Ettensohn et al. 1986), so that generalizations are premature, but much remains to be learned eventually.

Conclusion: Storm Events and Prasopora Growth

During the later Middle Ordovician, eastern North America was covered by extensive, shallow, equatorial, carbonate-depositing, epicontinental seas. To the east, compression associated with the Taconic Orogeny produced a subsiding foreland basin that began to fill with detrital clastic sediments. The presence of storm-generated sedimentary features indicates that central Pennsylvania

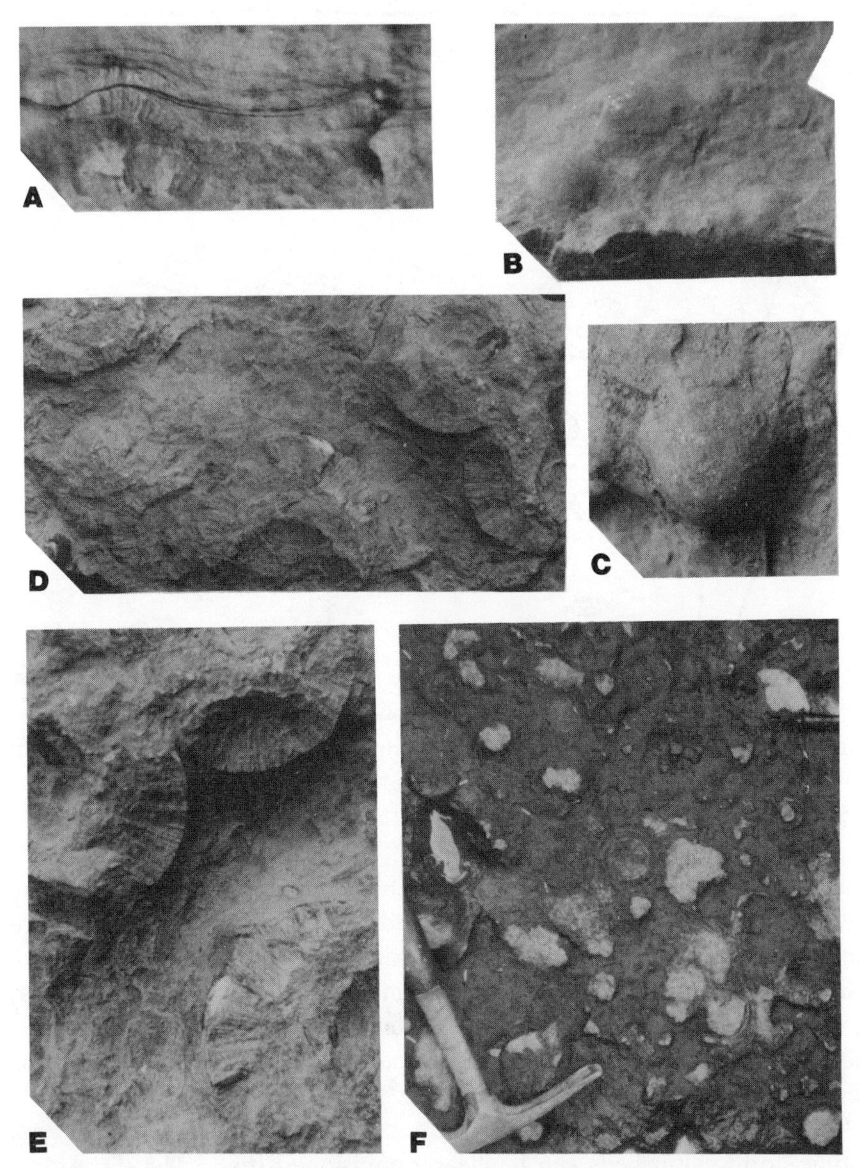

FIGURE 5.9. Upright *Prasopora simulatrix* colonies in growth position in tops of Coburn event beds: (A) upright colonies in outcrop vertical section, stratigraphic up toward top of page, ×1.3; (B,C) intact hemispherical colonies on upper surface of bed (B, oblique view, ×0.7; (C) vertical view, ×2); (D–F) horizontally broken in-place colonies embedded in upper surface of bed, all vertical views, ×0.7, ×1, ×0.2, respectively. ([A] PBRC PENMIF-20–J; [B,C] PENLYC-5–Ct5–hds; [D,E] PENCEN-50–hds; [F] bed of Cincinnati Creek, Trenton, New York). (In part modified from Arens and Cuffey [1989a:115]. Reproduced with permission of Pennsylvania Academy of Science [A,B].)

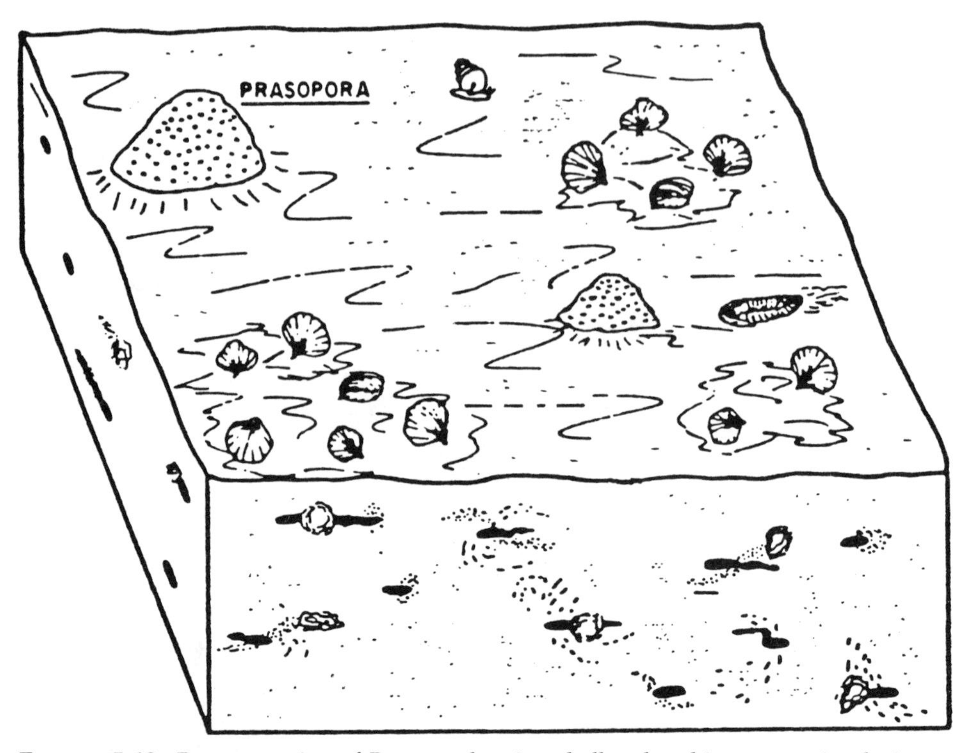

Figure 5.10. Reconstruction of *Prasopora*-bearing shallow benthic community during life. (Modified from Walker and Diehl [1986:69]. Reproduced with permission of SEPM—Society for Sedimentary Geology and those authors.)

remained a shallow carbonate platform while comparable areas in present-day New York and Tennessee fragmented and foundered into the foreland basin along syndepositional normal faults (Cisne et al. 1982; Mehrtens 1984, 1989; Shanmugam and Walker 1980; Shanmugam and Lash 1982). Central Pennsylvania remained relatively shallow after carbonate sedimentation was finally overwhelmed by the initial Taconic detritals (the lower Reedsville or "Antes" shales; Arens and Cuffey 1989c:219–22; Conrad 1985).

With this paleoenvironment in mind, the origin of the Coburn *Prasopora*-bearing storm event beds can be clearly envisioned (fig. 5.11A–D; Arens and Cuffey 1989b). Hemispherical *Prasopora simulatrix* colonies grew, with other shelly invertebrates, on the open sea bottom (figs. 5.10, 5.11A). A severe storm eventually passed through the area, its waves extensively disrupting the bottom for hours or a few days (fig. 5.11B). *Prasopora* colonies, shells, and sediment were swept up by the scouring waves, briefly suspended and transported, and then redeposited in size- and density-graded fashion as the water quieted after the storm. The heavy *Prasopora* and shells settled out almost immediately, the colonies landing in jumbled or random orientations, along with the coarsest skeletal debris, to form the fossil coquinite. The living zooids would have been smothered by the entombing sediment, so that no overgrowths could develop to change the overturned colonies into more spherical masses (in contrast to those noted elsewhere; Dade and Cuffey 1984; Gyllenhaal and Kidwell 1989). Progressively finer carbonate sediments subsequently settled out (fig. 5.11C), the sandy ones shaped partly by weakening waves and currents as storm-generated flow ebbed, the muddy ones settling last to form a massive or structureless micritic cap to most storm sequences.

The seafloor then would have been barren lime mud or very fine-grained lime sand. A few small or broken *Prasopora* colonies by fortunate chance may have landed right side up on top of the redepositing carbonate sediment and were able to continue growth with little or no interruption of zooecial development. One colony examined was in growth position but had zooecial tubes disrupted by a 1–mm lime-mud layer, above which the tubes continued with only slight offset. More *Prasopora* colonies would soon colonize the available substrate, recruited from meroplanktic larvae settling to the bottom; this recruitment probably accounts for most of the sequence-cap and interstitial *Prasopora* colonies. Such larvae may have already been in the area and survived the storm

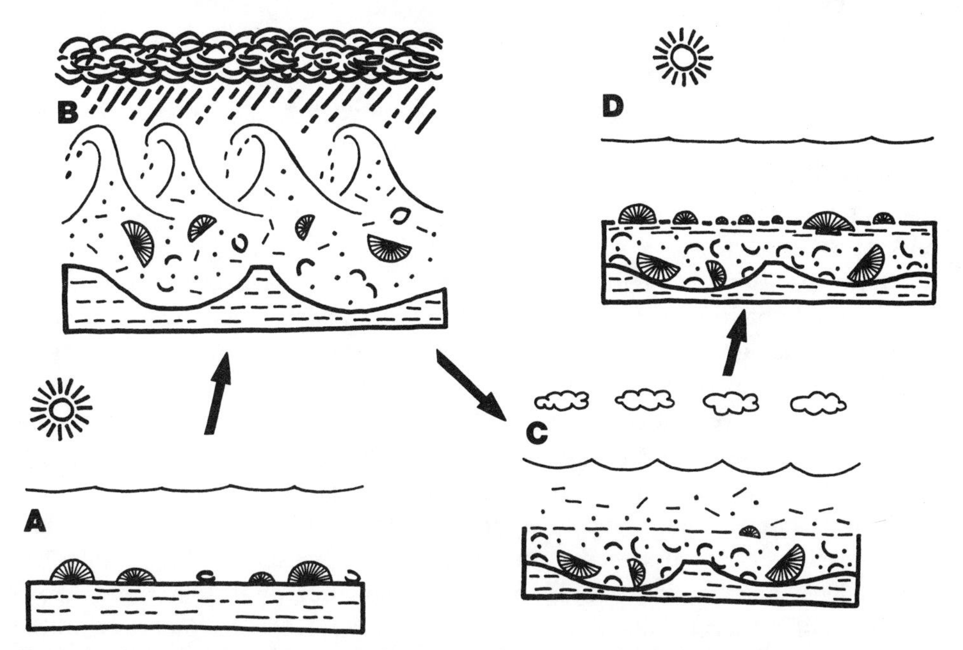

FIGURE 5.11. Origin of *Prasopora*-bearing Coburn event beds: (A) shallow open bottom with flourishing hemispherical *Prasopora simulatrix* colonies; (B) storm waves scouring bottom and suspending *Prasopora*, shells, and sediments; (C) skeletal materials immediately settling out in jumbled fashion, with carbonate sediments overall settling into a size-graded fining-upward bed, as water movements weaken or ebb following the storm; (D) shallow open bottom with recovered and recruited hemispherical *Prasopora* colonies growing amidst slowly accumulating ambient detrital and carbonate mud long after the storm, and constituting step A for the next storm sequence.

by having been up in the water column, or they may have come into the area later on currents from undisturbed *Prasopora* communities outside the storm's track. All the *Prasopora* colonies would then have grown upward (fig. 5.11D) so as to avoid being smothered by the normal, very slow, continuing accumulation of the ambient detrital mud in the several- to many-years-long quiet interval before the next major storm to affect the particular locality. (Meanwhile, many other storms might move across other parts of the overall central Pennsylvania platform, but far enough away not to affect this site.) Eventually, however, another hurricane would threaten this locality, so as to repeat the entire sequence and thereby produce another *Prasopora*-bearing storm-event bed to add to the accumulating Coburn succession.

Finally, storm events and bryozoan growth occur within time frames that are at least somewhat understood, and therefore make possible the estimation of Coburn depositional rates in a way that is not available for most formations. Both storm and growth processes involve variations and uncertainties; such estimates thus must be regarded as highly speculative and order-of-magnitude only, but they do suggest a possible method for future refinement.

Each Coburn tempestite is 10–30 cm thick (fig. 5.3). The Coburn's maximum thickness is roughly 100 m (Thompson 1963). At any given locality, the formation thus records on the order of 1,000–300 major storms. Because of the erosive scouring potential of such meteorologic events, part of the previous accumulation would be destroyed and hence the higher figure seems to be the more realistic (even a minimal) estimate. Each storm lasts only a few days, so that the total time represented by all the storms (even though producing the bulk of the thickness of each storm deposit) would be negligible. In contrast, the time represented by in-place growth of the *Prasopora* colonies after the storms would have been much greater. If each colony's life span was around 10 years (by analogy with other trepostome species; Cuffey 1967), clusters of overlapping colonies covering the top of a tempestite could represent several decades of calm conditions between hurricanes. Storm recurrence intervals for specific sites in the modern tropics vary greatly but are comparable, 25–100–200 years between strikes (although several hurricanes develop each year, their tracks are widely enough separated that any given locality suffers direct hits only occasionally). If 10 decades, or 100 years, is a reasonable approximation, the 1,000 Coburn storms would have stretched out over something like a minimum of at least 100,000 years. Longer recurrence intervals, more reworking of the deposits, or greater water depths (so that only the most severe and hence rarest storms would have had any effect on the bottom) could push the total duration of the Coburn much higher, perhaps approaching the order of a million years or so. This all is clearly speculative, but it does suggest the potential benefits that can result from combining paleobiological and sedimentologic thinking.

ACKNOWLEDGMENTS

N. C. Arens documented field occurrences, internal variability, and current synonyms for *Prasopora simulatrix*, recognized Coburn tempestite sequences, and critically read the initial version of this manuscript. E. Cotter, M. Miller, C. Morgeneier, S. Shaw, F. Swartz, and P. Zell contributed many additional *Prasopora* to the Coburn collections; C. Brungardt, E. Messersmith, and K. Phillips furnished field assistance; R. Anstey, D. Gold, and A. Traverse commented on earlier stages of this study. The Pennsylvania Academy of Science (S. Majumdar), Northeastern Science Foundation (S. Buttner), New York State Museum and Science Service (E. Landing), and SEPM—Society for Sedimentary Geology (R. Dixon, K. Walker) graciously gave permission to reproduce figures for which they hold the copyright.

REFERENCES

Aigner, T. 1985. *Storm Depositional Systems: Dynamic Stratigraphy in Modern and Ancient Shallow-Marine Sequences.* New York: Springer-Verlag.

Allen, A. T. and J. G. Lester. 1954. Contributions to the paleontology of northwest Georgia. *Georgia Geological Survey Bulletin* 62:1–166.

Anstey, R. L. and S. F. Rabbio. 1989. Regional bryozoan biostratigraphy and taphonomy of the Edenian stratotype (Kope Formation, Cincinnati area): Graphic correlation and gradient analysis. *Palaios* 4:574–84.

Arens, N. C. 1988. Salona-Coburn bryozoans: Systematics, paleoecology and sedimentologic interpretation of a Middle Ordovician fauna from central Pennsylvania. Master's thesis, Pennsylvania State University.

Arens, N. C. and R. J. Cuffey. 1989a. Bryozoan fauna of the Coburn and Salona Limestones (Middle Ordovician) of central Pennsylvania. *Journal of the Pennsylvania Academy of Science* 63:113–21; abstract also 1990, *Geological Society of America Abstracts with Programs* 22(2):2.

———. 1989b. Life and death in the late Middle Ordovician: *Prasopora simulatrix* (Bryozoa) in central Pennsylvania. *Geological Society of America Abstracts with Programs* 21(2):2.

———. 1989c. Shallow and stormy: Late Middle Ordovician paleoenvironments in central Pennsylvania: *Northeastern Geology* 11:218–224; abstract also 1989, *Geological Society of America Abstracts with Programs* 21(2):2.

Bassler, R. S. 1919. *Cambrian and Ordovician.* Baltimore: Maryland Geological Survey and Johns Hopkins University Press.

———. 1953. *Bryozoa: Treatise on Invertebrate Paleontology* G:1–253.

Boardman, R. S. 1971. Mode of growth and functional morphology of autozooids in some Recent and Paleozoic tubular Bryozoa. *Smithsonian Contributions to Paleobiology* 8:1–51.

Bork, K. B. and T. G. Perry. 1968. Bryozoa (Ectoprocta) of Champlainian age (Middle Ordovician) from northwestern Illinois and adjacent parts of Iowa and Wisconsin; Part III, *Homotrypa, Orbignyella, Prasopora, Monticulipora,* and *Cyphotrypa. Journal of Paleontology* 42:1042–65.

Brown, G. D. 1965. Trepostomatous Bryozoa from the Logana and Jessamine Limestones (Middle Ordovician) of the Kentucky bluegrass region. *Journal of Paleontology* 39:974–1006.

Carter, B., P. Miller, and R. Smosna. 1988. Environmental aspects of Middle Ordovician limestones in the central Appalachians. *Sedimentary Geology* 58:23–36.

Cisne, J., D. Karig, R. Rabe, and B. Jay. 1982. Topography and tectonics of the Taconic outer trench slope as revealed through gradient analysis of fossil assemblages. *Lethaia* 15:229–46.

Conrad, J. A. 1985. Shelf sedimentation above storm wave base in the Upper Ordovician Reedsville Formation in central Pennsylvania: *Geological Society of America Abstracts with Programs* 17:13.

Cuffey, R. J. 1967. Bryozoan *Tabulipora carbonaria* in Wreford Megacyclothem (Lower Permian) of Kansas. *University of Kansas Paleontological Contributions Article* 43:1–96.

———. 1973. An improved classification, based upon numerical-taxonomic analyses, for the higher taxa of entoproct and ectoproct bryozoans. In G. P. Larwood, ed., *Living and Fossil Bryozoa*, pp. 549–64. London: Academic Press.

Cuffey, R. J. and A. V. Sorrentino. 1985. Globular *Ceriopora* species (Cyclostomida, Bryozoa) from the Virginia-Carolinas Pliocene, and the status of *Atelesopora*. In C. Nielsen and G. P. Larwood, eds., *Bryozoa: Ordovician to Recent*, pp. 79–86. Fredensborg, Denmark: Olsen and Olsen.

Cuffey, R. J., C. J. Stadum, and J. D. Cooper. 1981. Mid-Miocene bryozoan coquinas on the Aliso Viejo Ranch, Orange County, southern California. In G. P. Larwood and C. Nielsen, eds., *Recent and Fossil Bryozoa*, pp. 65–72. Fredensborg, Denmark: Olsen and Olsen.

Dade, W. B. and R. J. Cuffey. 1984. Holocene multilaminar bryozoan masses—The "rolling stones" or "ectoproctaliths" as potential fossils in barrier-related environments of coastal Virginia. *Geological Society of America Abstracts with Programs* 16:132.

Errett, D. H. and R. J. Cuffey. 1989. Bryozoan mode of occurrence relative to individual beds in the type Cincinnatian (Upper Ordovician; Ohio, Indiana, Kentucky). *Geological Society of America Abstracts with Program* 21(4):11.

Ettensohn, F. R., B. C. Amig, J. C. Pashin, S. F. Greb, M. Q. Harris, J. C. Black, D. J. Cantrell, C. A. Smith, T. M. McMahan, A. G. Axon, and G. J. McHargue. 1986. Paleoecology and paleoenvironments of the bryozoan-rich Sulphur Well Member, Lexington Limestone (Middle Ordovician), central Kentucky. *Southeastern Geology* 26:199–219.

Fritz, M. A. 1957. Bryozoa (mainly Trepostomata) from the Ottawa Formation (Middle Ordovician) of the Ottawa–St. Lawrence Lowland. *Geological Survey of Canada Bulletin* 12:1–41.

Gardiner-Kuserk, M. A. 1989. Cyclic sedimentation patterns in the Middle and Upper Ordovician Trenton Group of central Pennsylvania. *American Association of Petroleum Geologists Studies in Geology* 29:55–76.

Gyllenhaal, E. D. and S. M. Kidwell. 1989. Growth histories of subspherical bryozoan colonies: Ordovician and Pliocene evidence for paleocurrent regimes and commensalism. *Geological Society of America Abstracts with Programs* 21(6):A112–13.

Hall, J. 1847. Descriptions of the organic remains of the lower division of the New York System. *New York Natural History Survey, Paleontology of New York* 1:1–362.

Horowitz, A. S. 1968. The ectoproct (bryozoan) genus *Actinotrypa* Ulrich. *Journal of Paleontology* 42:356–73.

James, U. P. 1884. Descriptions of four new species of fossils from the Cincinnati Group. *Journal of the Cincinnati Society of Natural History* 7:21–24.

Karklins, O. L. 1984. Trepostome and cystoporate bryozoans from the Lexington Limestone and the Clays Ferry Formation (Middle and Upper Ordovician) of Kentucky. *U.S. Geological Survey Professional Paper* 1066–I:1–105.

———. 1987. Bryozoa from Rocklandian (Middle Ordovician) rocks of the Upper Mississippi Valley region. *Minnesota Geological Survey Report of Investigations* 35:173–76.

Keith, B. D. 1989. Regional facies of Upper Ordovician series of eastern North America. *American Association of Petroleum Geologists Studies in Geology* 29:1–16.

Kobluk, D. R. and S. Nemcsok. 1982. The macroboring ichnofossil *Trypanites* in colonies of the Middle Ordovician bryozoan *Prasopora:* Population behaviour and reaction to environmental influences. *Canadian Journal of Earth Sciences* 19:679–88.

Kolata, D. R. and R. E. Sloan. 1987. The Middle and Late Ordovician strata and fossils of Iowa. *Minnesota Geological Survey Guidebook Series* 15:97–121.

Madsen, L. 1986. Variation in some trepostome bryozoans—Taxonomic implications. *International Bryozoology Association Conference Abstracts* 7:36.

Marintsch, E. J. 1981. Taxonomic reevaluation of *Prasopora simulatrix* Ulrich (Bryozoa: Trepostomata). *Journal of Paleontology* 55:957–61.

Mehrtens, C. J. 1984. Foreland basin sedimentation in the Trenton Group of central New York. *New York State Geological Association Guidebook* 56:59–98.

———. 1989. Bioclastic turbidites in the Trenton Limestone: Significance and criteria for recognition. *American Association of Petroleum Geologists Studies in Geology* 29:87–112.

Mehrtens, C. J. and S. G. Barnett. 1979. Evolutionary change in the bryozoan genus *Prasopora* as a tool for correlating within the Trenton Group (Middle Ordovician). *Geological Society of America Abstracts with Programs* 11:44.

———. 1986. Evolutionary change in *Prasopora* as a tool for refined correlation within the Middle Ordovician Trenton Group. *International Bryozoology Association Conference Abstracts* 7: supplemental p.

Ross, J. P. 1967. Evolution of ectoproct genus *Prasopora* in Trentonian time (Middle Ordovician) in northern and central United States. *Journal of Paleontology* 41:403–16.

———. 1970. Distribution, paleoecology, and correlation of Champlainian Ectoprocta (Bryozoa), New York State, Part III. *Journal of Paleontology* 44:346–82.

Rudolph, R. 1975. Evolution, plasticity, and function of the bryozoan genus *Prasopora* (Order Trepostomata) from the Trenton Group (Middle Ordovician) in central New York. *Geological Society of America Abstracts with Programs* 7:113.

Sardeson, F. W. 1935. Behavior of the bryozoan *Prasopora simulatrix. Pan-American Geologist* 63:173–88.

Scott, R. W. 1970. Paleoecology and paleontology of the Lower Cretaceous Kiowa Formation, Kansas. *University of Kansas Paleontological Contributions Article* 52:1–94.

Shanmugam, G. and G. C. Lash. 1982. Analogous tectonic evolution of the Ordovician foredeeps, southern and central Appalachians. *Geology* 10:562–66.

Shanmugam, G. and K. R. Walker. 1980. Sedimentation, subsidence, and evolution of a foredeep basin in the Middle Ordovician, southern Appalachians. *American Journal of Science* 280:479–96.

Smith, R. C., J. H. Way, and S. W. Berkheiser. 1986. Summary of Union Furnace bentonite and possible bentonitic horizons. *Annual Field Conference of Pennsylvania Geologists Guidebook* 51:115.

Sparling, D. R. 1964. *Prasopora* in a core from the Northville area, Michigan. *Journal of Paleontology* 38:1072–81.

Thompson, R. R. 1963. Lithostratigraphy of the Middle Ordovician Salona and Coburn Formations in central Pennsylvania. *Pennsylvania Topographic and Geological Survey (4th ser.) Bulletin* G-38:1–154.

Titus, R. 1986. Fossil communities of the upper Trenton Group of New York state. *Journal of Paleontology* 60:805–824.

———. 1989. Facies of the Trenton Group of New York. *American Association of Petroleum Geologists Studies in Geology* 29:77–86.

Ulrich, E. O. 1886. Report on Lower Silurian bryozoans with preliminary descriptions of some new species. *Minnesota Geology and Natural History Survey Annual Report* 14 (for 1885):57–103.

———. 1890. Paleontology of Illinois, Part II, section VI, Paleozoic Bryozoa: *Geological Survey of Illinois* 8:283–688.

———. 1893. On Lower Silurian Bryozoa of Minnesota. *Geological and Natural History Survey of Minnesota—Paleontology* 3:96–332.

Walker, K. R. and W. W. Diehl. 1986. The effect of synsedimentary substrate modification on the composition of paleocommunities: Paleoecologic succession revisited. *Palaios* 1:65–74.

Warner, D. J. and R. J. Cuffey. 1973. Fistuliporacean bryozoans of the Wreford Megacyclothem (Lower Permian) of Kansas. *University of Kansas Paleontological Contributions Paper* 65:1–24.

6

Paleoecology and Comparative Taphonomy of an Isotelus *(Trilobita) Fossil Lagerstätten from the Waynesville Formation (Upper Ordovician, Cincinnatian Series) of Southwestern Ohio*

Gregory A. Schumacher and Douglas L. Shrake

ABSTRACT

Two distinct *Isotelus*-bearing trilobite shales occur 1 and 7 m below the top of the Upper Ordovician Waynesville Formation of southwestern Ohio. The lower *Isotelus*-bearing shale was studied in outcrop through the excavation and detailed mapping of 1 m^2 of this shale exposed in the emergency spillway for the Caesar Creek Reservoir in Warren County, Ohio. The upper shale was examined in outcrop but was not excavated. In the subsurface both *Isotelus*-bearing shales were studied in three continuous cores located in Warren County and adjacent Clinton County, Ohio.

Correlation between field exposures and continuous cores using stratigraphic position and shale-percentage logs indicates that both *Isotelus*-bearing shales are laterally correlative for at least 40 km along the outcrop belt of the Waynesville Formation.

The fauna inhabiting the lower *Isotelus*-bearing shale consists of annelid worm tubes, bivalves, brachiopods, bryozoans, cephalopods, chitinozoans, conodonts, crinoids, eurypterids, ostracodes, scolecodonts, trace fossils, and trilobites. Three types of fossil assemblages inhabited the sediments and water column during the accumulation of this bed. Soft muds were inhabited by a mixture of sessile, semi-infaunal, suspension-feeding organisms and vagrant, deposit-feeding organisms. Firm or hard substrates were colonized by sessile, epifaunal suspension-feeding organisms. The water column was inhabited by suspension-feeding and predatory organisms.

Within the lower *Isotelus*-bearing shale, four discrete stratigraphic intervals containing autochthonous fossil assemblages with sessile, epifaunal, suspension-feeding organisms preserved in life position are recognized. Two additional stratigraphic intervals contain encrusted, disarticulated skeletal elements and occur slightly above shallow, horizontal feeding traces. These six intervals represent periods of low sedimentation. The fossil assemblages were smothered by blankets of storm-resuspended mud or mud transported by storm-generated density currents. These sedimentation episodes incorporated and preserved the articulated carcasses of multielement organisms and preserved the fossils of the autochthonous fossil assemblages in life position. Seven tempestites are recognized in the lower *Isotelus*-bearing shale.

"Trilobite shales" containing abundant, well-preserved trilobites occur sporadically throughout the Upper Ordovician Cincinnatian Series exposed in southwestern Ohio (Austin 1927; Wolford 1930; Caster et al. 1955; Brandt Velbel 1985; Frey 1987a, 1987b). In 1919 August Foerste published the description of *Isotelus brachycephalus* (Foerste) from a number of specimens, including one that is still one of the largest trilobites ever found. This specimen, measuring 25.3 cm in width and 36.8 cm in length, was collected from a "trilobite shale" located in the upper part of the Waynesville Formation during the construction of Huffman Dam of the Miami Conservancy in Greene County, Ohio (fig. 6.1, site 1).

Foerste (1919), Austin (1927), Wolford (1930), and Robert C. Frey (written communication 1978) reported additional, well-preserved specimens of *I. brachycephalus* and/or *I. maximus* (Locke), which approached the size of Foerste's specimen, from a "trilobite shale" located in the upper 1 m of the Waynesville Formation and another 7 m below the top of the Waynesville Formation in Greene, Clinton, and Warren counties, Ohio (fig. 6.1).

In 1986 amateur paleontologist and professional collector Thomas T. Johnson, in cooperation with the U.S. Army Corps of Engineers and the U.S. National Museum, excavated a large portion of the lower *Isotelus*-bearing shale from the Waynesville Formation. His excavation was in the emergency spillway for Caesar Creek Reservoir located in Warren County, Ohio (fig. 6.2). We were granted permission by Mr. Johnson, the U.S. Army Corps of Engineers, the U.S. National Museum, and the Ohio Department of Natural Resources, Division of Geological Survey to conduct a systematic excavation of 1 m^2 of this unique bed.

The objectives of our study are (1) to describe the composition, paleoecology, and comparative taphonomy of the fauna preserved in the lower *Isotelus*-bearing shale; (2) to address the origin of the shale; (3) to determine if the interval is laterally persistent as implied by Austin (1927); (4) to compare the lithologic and taphonomic features of this bed to other trilobite-bearing shales in the Cincinnatian Series and other Paleozoic marine stratigraphic sequences; and (5)

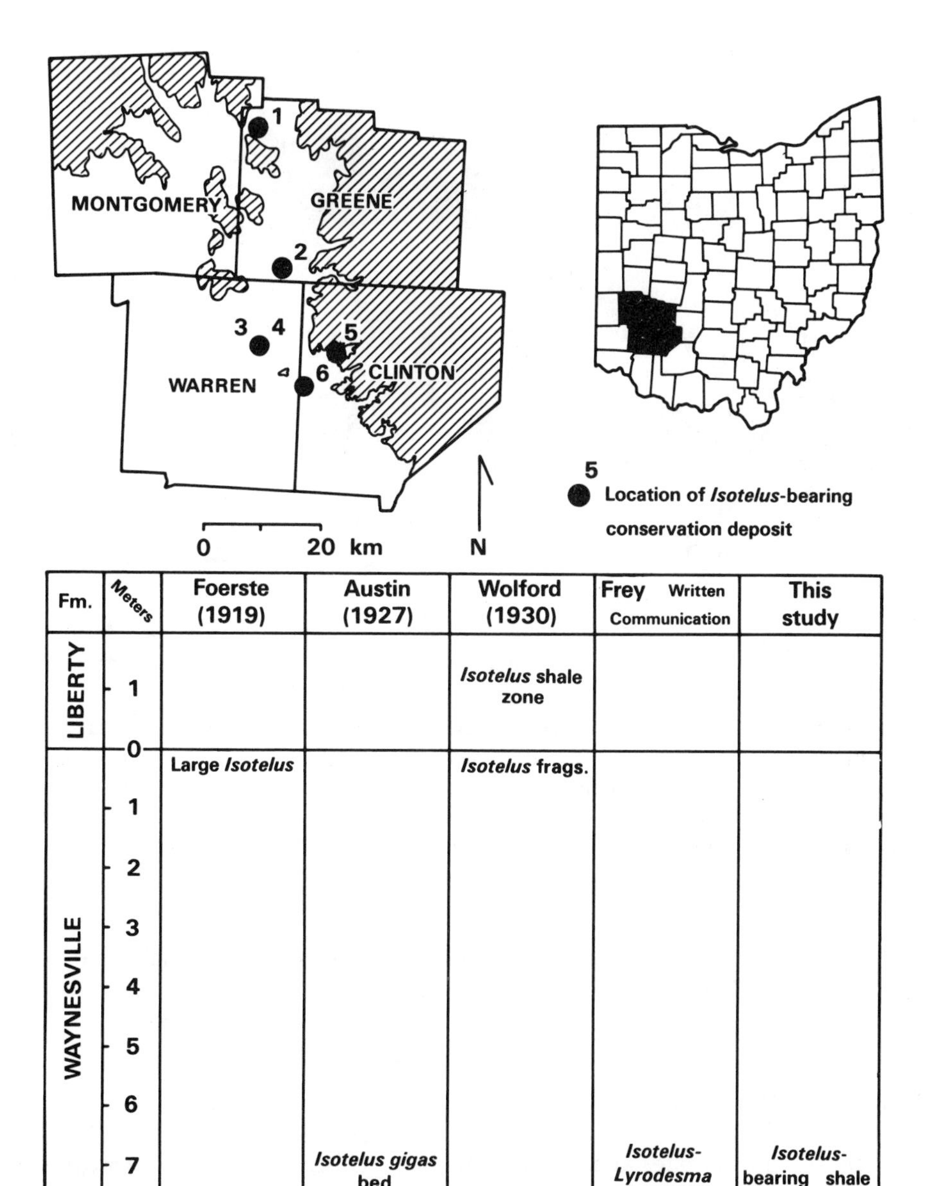

FIGURE 6.1. Historical summary of the location and stratigraphic positions of *Isotelus*-bearing conservation deposits from the upper 7 m of the Waynesville Formation of southwestern Ohio. Site 1 is the Huffman Dam of the Miami Conservancy locality discussed by Foerste (1919). Sites 2, 3, and 4 are the Roxanna Road, Caesar Creek emergency spillway, and Flat Fork localities, respectively, of this study (also see appendix 1). Sites 4, 5, and 6 are the Flat Fork, Cowan Creek, and Clarksville, Ohio, localities, respectively, of Austin (1927) and Wolford (1930). Cross-hatched area represents surface rocks of Silurian age.

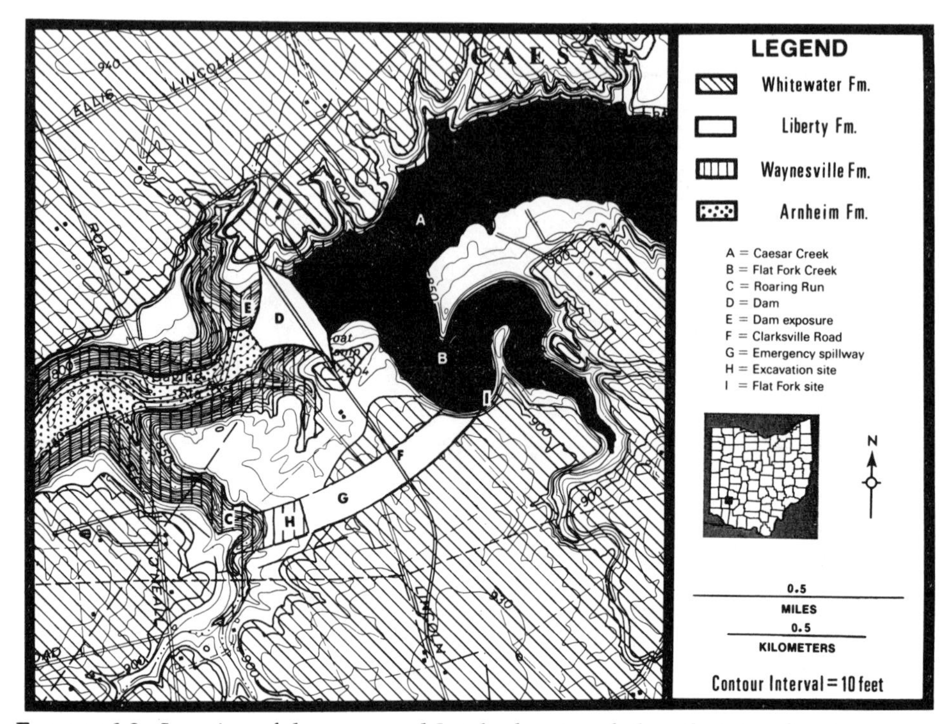

Figure 6.2. Location of the excavated *Isotelus*-bearing shale, other significant outcrops and landmarks, and a geologic map of the dam and spillway region of Caesar Creek State Park, Warren County, Ohio. Note: The dam for Caesar Creek Lake is constructed of bedrock excavated from the emergency spillway and does not represent the Liberty Formation as the legend shows.

to stimulate additional investigations of this *Isotelus*-bearing shale and other "trilobite shales" in the upper Waynesville Formation.

Materials and Methods

Well-preserved trilobites and other fossils were collected from the excavation of a 1-m^2 pit located in the middle of the Caesar Creek emergency spillway. The pit was located approximately 450 m west-southwest of the midpoint where Clarksville Road crosses the Caesar Creek emergency spillway (fig. 6.2; appendix 1). Table 6.1 and figures 6.3 and 6.4 summarize the field methods used in the excavation of the lower *Isotelus*-bearing shale and table 6.2 outlines the laboratory methods employed. Laboratory analyses are summarized in tables 6.3 and 6.4.

The data collected from the lower *Isotelus*-bearing shale were supplemented with observations from the Roxanna Road (fig. 6.1, site 2) and Flat Fork (fig. 6.2, site 1) localities. Because of time constraints, detailed excavations were not attempted at these localities.

Three continuous cores were measured and described in a manner consistent with the methods used for the excavated bed (table 6.1). Shale percentage logs were generated for each core using the methodology of Sweet et al. (1974).

Geologic Setting and Paleoenvironmental Setting

In the Late Ordovician the majority of eastern North America was inundated by a broad, shallow epeiric sea (Elias 1982; Weir et al. 1984; Frey 1987a). Southwestern Ohio was located between 15 and 20 degrees south latitude (Scotese and McKerrow 1990) within the tropical to subtropical climate zone (Spjeldnaes 1981). The interbedded limestone and shale characteristic of the stratigraphic section at Caesar Creek were deposited on a gently eastward-sloping, mixed siliciclastic-carbonate ramp that originated in east-central Indiana and extended into west-central Ohio (Gray 1972; Schumacher, in press). Estimated water depths along this ramp ranged from less than 1 m in supratidal to high intertidal environments to 20–50 m in subtidal offshore environments (Hatfield 1968; Weir et al. 1984; Frey 1987a, 1987b; Schumacher, in press). Hurricanes

TABLE 6.1 *Field Methods Employed*

Excavation of 1 m^2 of interbedded limestone and shale to depth of 49 cm*
Observations recorded for each limestone and shale bed
- Rock type according to Dunham (1962) for limestones and Potter et al. (1980) for shales
- Thickness
- Color according to Goddard et al. (1951)
- Bed induration
- Bedding style
- Nature of bed contacts
- Primary and secondary sedimentary structures
- Limestone joint orientations
- Shale fracture and parting characteristics according to Potter et al. (1980)

Scale mapping of body and trace fossils within the *Isotelus*-bearing shale using standard 1-m and 30-cm grids (fig 6.3). Examples of mapped horizons are shown in figure 6.4.
Fossil data collected
- Tentative identification
- Preferred azimuth orientation
- Orientation relative to bedding
- Dorsal/ventral orientation of trilobite carcasses, moult remains, and aggregates of sclerites†
- Enrolled/outstretched trilobites
- Type and length of trace fossils

Sampling for laboratory analysis
- Overlying limestone beds (fig. 6.5, units 2 and 4)
- *Isotelus*-bearing shale
  - Selected body fossils with encasing shale or supporting plaster mold
- Fourteen bulk samples of approximately 2 kg for X-ray diffraction, textural analysis, and faunal composition analysis (fig. 6.3)
- Thirteen vertically oriented samples representing the stratigraphic interval excavated for thin-section analysis (fig. 6.3)
- Grab samples at the 6–7-, 8–9-, and 11–12-cm stratigraphic intervals in order to study the clumps of *Zygospira modesta* (Say)

*At 49 cm the slow influx of ground water forced abandonment of the excavation.
†Dorsal/ventral data were not collected for the numerous individual trilobite sclerites excavated.

TABLE 6.2 *Laboratory Methods Employed*

Collected limestone beds slabbed and described in detail.
Observations recorded from each limestone bed
 Rock type(s) according to Dunham (1962)
 Thickness
 Color according to Goddard et al. (1951)
 Bed induration
 Primary and secondary sedimentary structures
 Fossil orientation, preservation, and disarticulation
One sample of 50 g ± 3 g was removed from each of the 14 shale samples.
 Percentage of sand-sized, silt-sized, and clay-sized sedimentary grains was determined for each 50 g ± 3 g sample using the textural analysis method of Folk (1974). Results are listed in table 6.3.
 Statistical analysis of textural analysis using T-test following methods outlined in Ausich and Kammer (1991).
 Each 1,000 g ± 3 g sample was disaggregated using kerosene and water, then wet-sieved through 350-μm, 250-μm, and 125-μm sieves. Residue of each size fraction was dried and weighed. Results are listed in table. 6.4.
Grab samples of the clumps of *Zygospira modesta* were disaggregated using kerosene and water, then wet-sieved through 350-μm and 250-μm sieves.
 The pedicle valve of each *Z. modesta* was measured to a maximum error of ±0.05 mm from the pedicle to the commissure along the midline of the specimen using a binocular microscope.
Thin sections of the 13 vertically oriented shale samples were produced following methods of Martin et al. (1979). These thin sections were not point counted.
X-ray diffraction analyses were conducted on 8 of the 14 bulk samples collected from the *Isotelus*-bearing shale.
The clay mineralogy of this bed was determined using standard techniques and identification procedures of Starkley et al. (1985).
Material collected and analyzed in this study is curated at the U.S. National Museum in Washington, D.C.

TABLE 6.3 *Textural Analysis Data and Percentages from Standard Samples of the Isotelus-Bearing Shale Bed (All Weight in Grams)*

Sample	Interval (cm)	Gross weight of sample*	Sand	% sand	Silt	% silt	Clay	% clay
1	0–1	48.51	1.37	2.8%	20.67	42.6%	26.47	54.6%
2	1–2	49.82	0.43	0.9%	17.86	35.8%	31.53	63.3%
3	2–3	49.98	0.24	0.5%	13.95	27.9%	35.79	71.6%
4	3–4	50.31	0.26	0.5%	16.19	32.2%	33.86	67.3%
5	4–5	50.63	0.21	0.4%	17.41	34.4%	33.01	65.2%
6	5–6	51.16	0.20	0.4%	18.61	36.4%	32.35	63.2%
7	6–7	50.79	0.08	0.2%	18.91	37.2%	31.80	62.6%
8	7–8	51.33	0.82	1.6%	23.02	44.8%	27.49	53.6%
9	8–9	50.83	0.43	0.8%	16.51	32.5%	33.89	66.7%
10	9–10	52.52	0.44	0.8%	20.26	38.6%	31.82	60.6%
11	10–11	51.32	0.28	0.5%	20.94	40.8%	30.10	58.7%
12	11–12	51.24	0.19	0.4%	16.02	31.3%	35.03	68.4%
13	12–13	49.69	0.31	0.6%	20.36	41.0%	29.02	58.4%
14	13–14	49.79	0.41	0.8%	18.82	37.8%	30.56	61.4%
15	14–15	50.83	0.21	0.4%	16.67	32.8%	33.95	66.8%

*Weight of gravel—sized particles has been subtracted.

TABLE 6.4 *Bulk Sample Weights and Amount of Fossil Residue from Standard Samples of the Isotelus-Bearing Shale Bed (All Weight In Grams)*

Sample number	Sample interval (cm)	Dry weight	Used weight	350*μ weight	250*μ weight	125*μ weight	Total residue
1	0–1	2,049.2	1,000.0	24.225	0.730	2.750	27.705
2	1–2	1,120.9	1,000.0	9.860	0.600	1.870	12.330
3	2–3	1,266.0	1,000.0	31.565	0.500	1.690	33.755
4	3–4	2,554.0	1,000.0	2.500	0.250	0.810	3.560
5	4–5	1,095.0	1,000.0	2.710	0.200	0.650	3.560
6	5–6	1,616.3	1,000.0	1.835	0.220	0.745	2.800
7	6–7	1,140.2	1,000.0	4.535	0.660	5.455	10.650
8	7–8	1,935.2	1,000.0	31.945	1.580	7.210	40.735
9	8–9	1,536.0	1,000.0	2.605	0.385	1.245	4.235
10	9–10	1,576.8	1,000.0	10.230	0.655	1.900	12.785
11	10–11	2,424.4	1,000.0	16.575	0.945	3.700	21.220
12	11–12	1,260.8	1,000.0	6.130	0.650	3.240	10.020
13	12–13	1,863.5	1,000.0	6.150	0.490	2.425	9.065
14	13–14	2,046.9	1,000.0	15.045	2.255	4.915	22.215

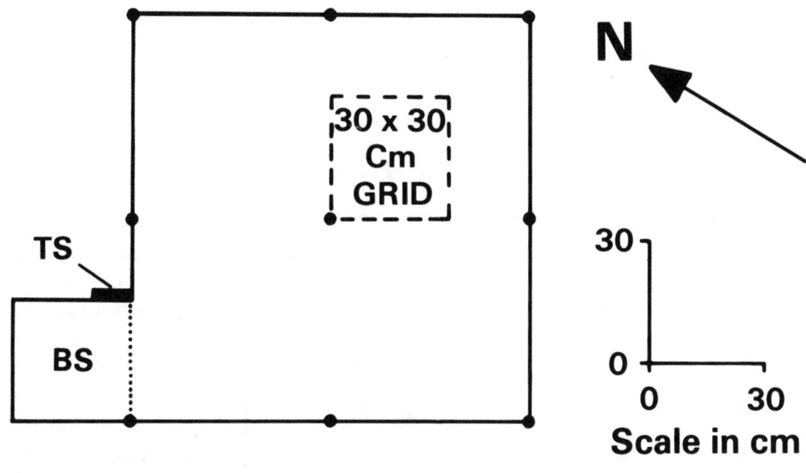

FIGURE 6.3. Schematic diagram of the excavation pit illustrating the methods used to record the position of fossils and locations of bulk sample area and area sampled for thin-section analysis. Fossils were mapped using 1- × 1-cm squares within the 30- × 30-cm grid. Fossils of more than 1 cm were plotted to scale (1 cm = 8 cm) and fossils of less than 1 cm were recorded with a dot.

periodically moved across the region, producing numerous tempestites (Schumacher and Ausich 1983; Shrake et al. 1988; Frey 1987a, 1987b; Schumacher and Shrake 1989).

Distinct lithofacies developed along this ramp as a function of varying water depth, storm sedimentation, regional and local tectonics, and biological produc-

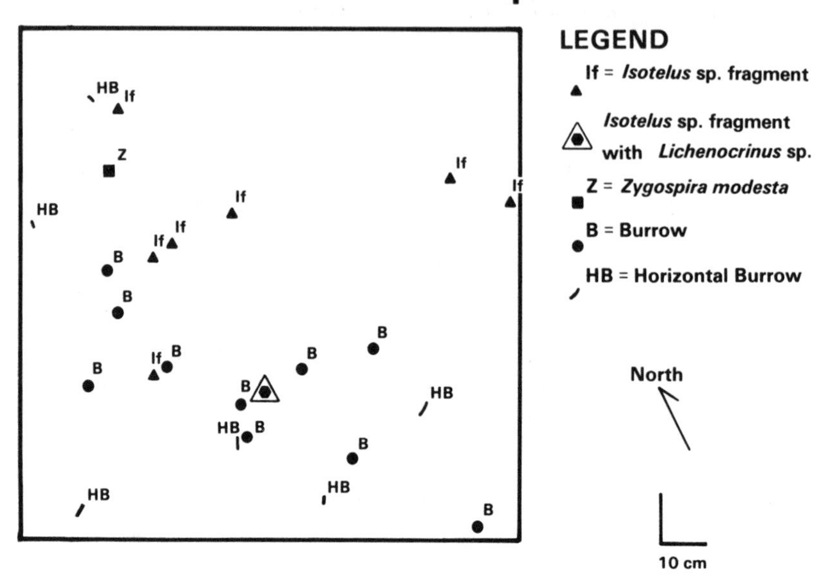

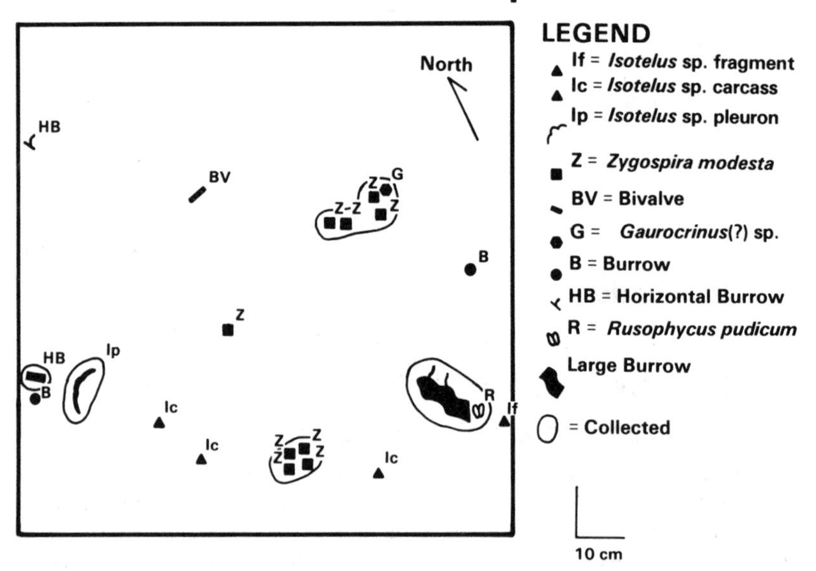

FIGURE 6.4. Mapped positions of the fossils encountered at the 2–3- and 6–7-cm stratigraphic intervals of the lower *Isotelus*-bearing shale. Grab samples and fossil specimens collected for laboratory analysis were taken within the circled areas of the 6–7-cm interval.

tivity (Anstey and Fowler 1969; Hay et al. 1981; Tobin 1982; Schumacher, in press; Holland 1993). The Caesar Creek emergency spillway section is characterized by a shoaling-upward succession of lithofacies ranging from the upper shoreface environment to the subtidal offshore environment (fig. 6.5). The upper shoreface environment is represented by the wavy- to nodular-bedded limestone-dominant lithofacies of the Whitewater Formation. The transitional

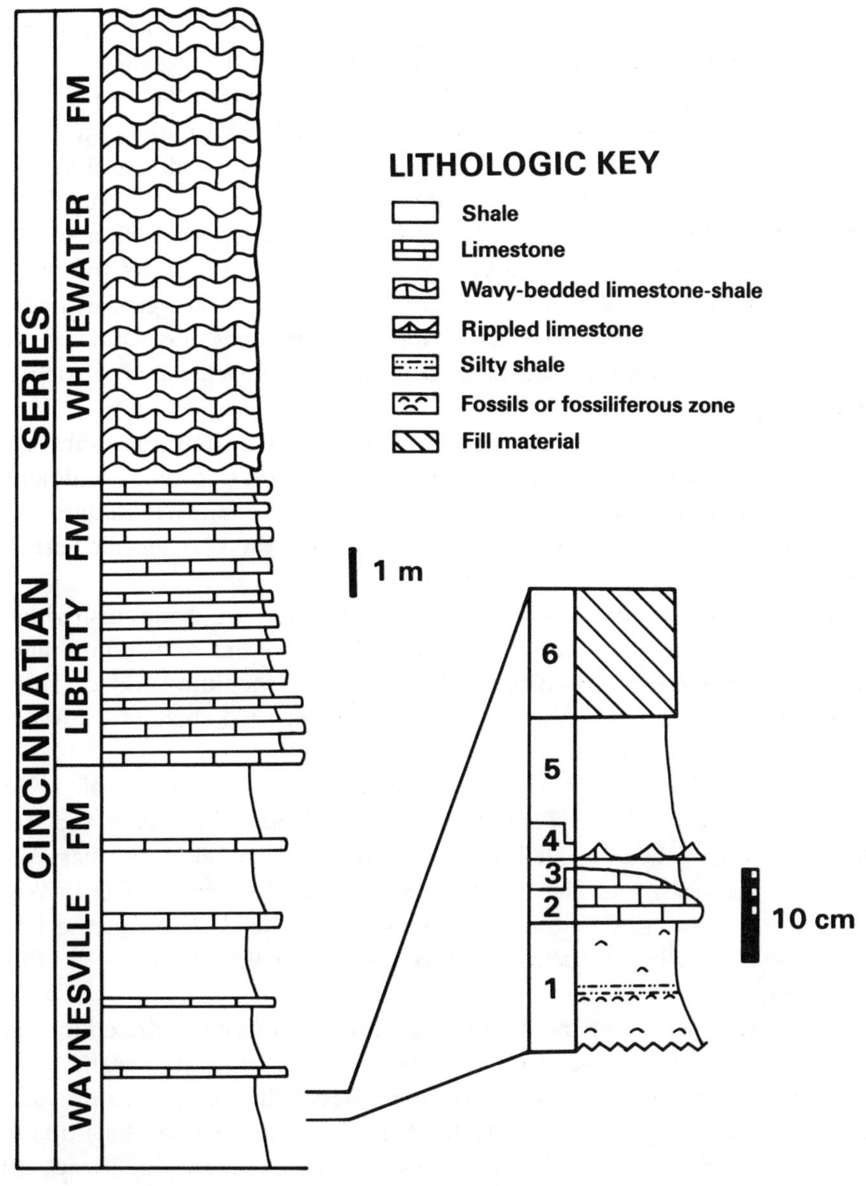

FIGURE 6.5. Stratigraphic section exposed at the emergency spillway of Caesar Creek State Park, Warren County, Ohio (modified from Schumacher and Ausich 1983). The stratigraphic section of the excavated interval is shown in the inset.

environment between the shoreface and offshore environment is characterized by the planar- to irregular-bedded mixed limestone and shale lithofacies of the Liberty Formation. The offshore environment contains the planar- to lenticular-bedded shale-dominant lithofacies of the Waynesville Formation.

Detailed Stratigraphy

The lower *Isotelus*-bearing shale was deposited in the offshore environment of the Waynesville Formation and occurs 7 m below the Waynesville/Liberty formational contact (fig. 6.5, unit 1). The upper 7 m of the Waynesville Formation averages 75% shale and 25% limestone. Shale beds are medium- to thick-bedded, sparsely fossiliferous, and dolomitic. Limestone beds are thin- to medium-bedded and abundantly fossiliferous (Schumacher and Ausich 1983; Tobin 1986). Many limestone beds were deposited by the action of storm-generated currents. Evidence supporting this conclusion includes the presence of small-scale hummocky cross-stratification, crude upward-graded bedding, sole and tool marks, gutter casts, shale intraclasts in limestone beds, imbricated brachiopod valves, and current-aligned fossils (Schumacher and Ausich 1983; Tobin 1986; appendix 2).

The Liberty Formation averages 57% shale and 43% limestone. Shale beds are thin- to medium-bedded, sparsely fossiliferous and calcareous to dolomitic. Limestone beds are thin- to medium-bedded, are abundantly fossiliferous, and display sedimentary structures similar to the beds of the Waynesville Formation (Schumacher and Ausich 1983; Tobin 1986).

The Waynesville Formation and Liberty Formation are distinguished by the average percentage of shale (75% versus 57%) and the reduction shale bed thickness from medium- to thick-bedded to thin- to medium-bedded. The contact is placed at the top of the uppermost, thick-bedded shale of the Waynesville Formation.

The second *Glyptorthis insculpta* brachiopod zone of Caster et al. (1955) is present in the basal Liberty Formation (Wolford 1930; Schumacher and Ausich 1983). The *Isotelus*-bearing shales occur 1 m and 7 m below the base of the second *G. insculpta* zone. This zone is a reliable stratigraphic marker in Warren and Greene counties, Ohio, and is useful in locating the base of the Liberty Formation in weathered sections where average shale percentage is difficult to determine.

Shale-percentage logs generated for three continuous cores drilled in Warren and Clinton counties, Ohio, show a distinct increase in shale percent 0–2 and 6–8 m below the top of the Waynesville Formation (fig. 6.6), thus providing a simple diagnostic means to correlate these intervals. These shale-dominant intervals have been observed at the Caesar Creek emergency spillway section and the Flat Fork and Roxanna Road localities by the senior author (fig. 6.7).

The *Isotelus*-bearing shale-dominant intervals present at the Caesar Creek

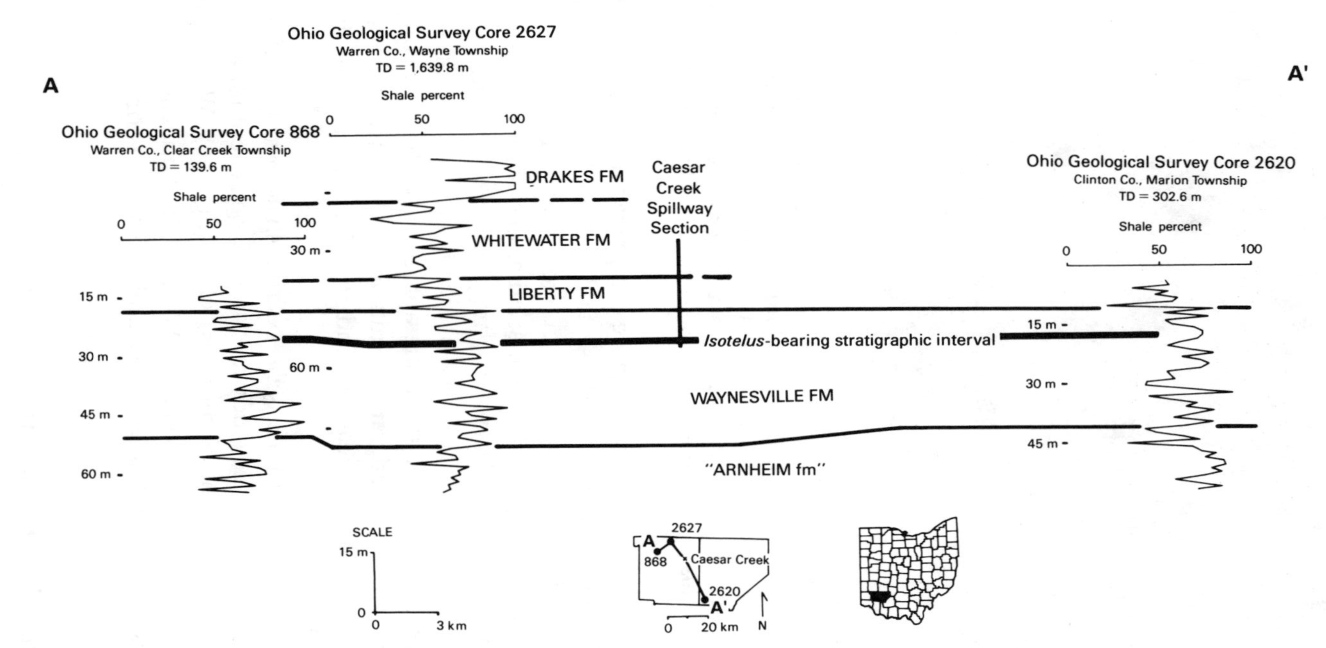

FIGURE 6.6. Lateral correlation of the lower *Isotelus*-bearing interval and the lithostratigraphy in Warren and Clinton counties, Ohio, using shale-percentage logs. Datum is the top of the Waynesville Formation.

FIGURE 6.7. Caesar Creek emergency spillway section illustrating the Waynesville-Liberty formational contact as exposed in 1980. The upper *Isotelus*-bearing, shale-dominant stratigraphic interval of Foerste (1919) and Wolford (1930) is well exposed in the upper 2 m of the Waynesville Formation (area A). Note the increase in limestone below the *Isotelus*-bearing interval (area B). The approximate stratigraphic position of the lower *Isotelus*-bearing shale of this study is the floor of the spillway visible in the foreground.

emergency spillway are correlative to those recognized in the cores: (1) *Isotelus*-bearing shale-dominant intervals at 0–2 and 6–8 m below the top of the Waynesville Formation occur in both outcrop and core; (2) thick-bedded claystone lithology with large *Isotelus* sp. sclerites and abundant *Chondrites* sp., which decrease in abundance downward, are present in both outcrop and core; (3) the second *G. insculpta* zone of the basal Liberty Formation is present in both outcrop and core; and (4) previous stratigraphic studies have documented the lateral persistence of thin stratigraphic units in the Waynesville Formation of southwestern Ohio and southeastern Indiana (Hay 1981; Frey 1987b; and Shrake et al. 1988).

Our evidence indicates that the *Isotelus*-bearing shales recognized by Foerste, Austin, Wolford, Frey, and this study are laterally persistent for at least 40 km throughout Warren and Clinton counties, Ohio (fig. 6.6). We suspect these intervals can be traced farther along the Waynesville Formation outcrop belt. Future field work supplemented with core drilling should verify our suspicions.

Autecology and Biostratinomy

A low-diversity fauna consisting of more than 15 genera of macrofauna, more than five genera of microfauna, and three varieties of trace fossils was identified

TABLE 6.5 *Macrofauna, Microfauna, and Ichnofossils Present in the* Isotelus*-Bearing Shale Bed (Listed in Order of Relative Abundance)*

Macrofauna	
Zygospira modesta (Say)	Abundant
Cornulites corrugatus (Nicholson)	Common
Isotelus sp. (Dekay)	Common
Flexicalymene meeki (Foerste)	Common
Caritodens demissa (Foerste)	Rare
Lichenocrinus sp. (Hall)	Rare
Gaurocrinus(?) sp. (Hall)	Rare
Cincinnaticrinus sp. (Warn and Strimple)	Rare
Xenocrinus(?) sp. (S. A. Miller)	Rare
Leptaena richmondensis (Foerste)	Rare
Orthodesma sp. (Hall and Whitfield)	Rare
Megalograptus welchi (Miller)	Rare
Encrusting bryozoans	Rare
Cephalopod cast	Rare
Microfauna	
Ostracodes	Common
Chitinozoans	Common
Ramose bryozoans	Common
Conodonts	Rare
Scolecodonts	Rare
Ichnofossils	
Chondrites sp. (Sternberg)	Abundant
Large horizontal feeding traces	Common
Rusophycus pudicum (Hall)	Rare

from the *Isotelus*-bearing shale 7 m below the top of the Waynesville Formation (table 6.5). The preservation of the fauna ranged from complete, articulated specimens to fragmented skeletal remains (table 6.6). Our discussion of the autecology and biostratinomy of the *Isotelus*-bearing shale fauna will be restricted to those taxa providing the maximum taphonomic information about the variability of sedimentation rates within this bed (for summary see table 6.9). Detailed notes describing the autecology and preservation of the fauna of this bed are available from the authors and have been filed with the fossil specimens curated at the U.S. National Museum.

Bivalvia

Caritodens demissa (Foerste) and *Orthodesma* sp. (Hall and Whitfield) are present in the *Isotelus*-bearing shale (table 6.5). The bivalve specimens range in size from 3 to 20 mm, representing juveniles to young adults according to the ontogenetic growth study of Frey (1983).

Bivalve preservation ranges from articulated specimens with calcareous shell material to external molds of single, unfractured valves. Most specimens are preserved as external molds. However, at the 9–10-cm level of the *Isotelus*-bearing shale, a single, well-preserved, articulated specimen of *C. demissa* was found in possible life position with the valves parallel to bedding. Also at this

TABLE 6.6 *Census and Mode of Fossil Preservation of the Isotelus-Bearing Shale Fauna of the 1-m^2 Excavation and the 900 cm^2 Bulk Samples*

		Stratigraphic interval (cm)													
		0–1	1–2	2–3	3–4	4–5	5–6	6–7	7–8	8–9	9–10	10–11	11–12	12–13	Totals
Articulated carcases	*Caritodens demissa*	—	—	—	—	—	—	—	—	—	1	—	—	—	1
	Orthodesma sp.	—	—	—	—	—	—	—	—	—	1	—	—	—	1
	Zygospira modesta	3	—	1	1	4	10	192	5	344	2	9	43	11	625
	Chitinozoans	C	C	C	A	A	A	C	R	A	C	R	C	C	0
	Ostracodes	A	R	A	A	A	C	C	A	A	C	C	C	C	0
	Flexicalymene meeki	1	—	—	—	1	4	—	—	1	2	2	1	3	15
	Isotelus sp.	—	1	—	—	1	—	3	1	1	—	3	—	1	11
	Cornulites corrugatus	—	—	—	—	—	—	—	—	44	—	—	70	—	114
Partially articulated skelatal remains	*Caritodens demissa*	—	—	—	—	—	—	—	—	3	1	1	1	—	6
	Leptaena richmondensis	—	—	—	—	—	—	—	—	1	—	—	—	—	1
	Encrusting bryozoans	—	—	—	—	1	—	—	—	2	—	—	1	—	4
	Chitinozoans	C	C	C	R	R	R	R	R	R	R	C	R	R	0
	Conodonts	R	R	R	R	R	R	R	R	R	R	R	R	R	0
	Crinoids	—	—	3	—	—	—	1	—	3	1	—	—	—	8

	Ostracodes	R	R	R	R	R	R	R	R	R	R	R	R	—	0
	Scolecodonts	N	R	R	R	R	R	N	R	R	N	R	N	N	0
	Flexicalymene meeki	—	—	—	—	—	2	—	3	—	—	—	—	—	5
	Isotelus sp.	—	—	—	1	4	3	1	2	2	—	1	2	4	20
Skeletal fragments	Bivalves	N	N	N	N	N	N	N	N	N	N	N	N	R	0
	Brachiopods	R	C	C	R	C	R	R	R	R	R	R	R	R	0
	Bryozoans	C	C	C	C	C	R	R	N	R	N	N	N	N	0
	Chitizozoans	C	R	R	R	R	R	R	R	R	R	R	R	R	0
	Conodonts	R	R	R	N	R	R	R	R	R	R	R	R	R	0
	Crinoids	R	R	N	N	N	R	N	N	N	N	R	N	R	0
	Eurypterids	N	N	N	N	N	N	N	N	N	N	N	N	R	0
	Ostracodes	R	R	R	R	R	R	R	R	R	R	R	R	R	0
	Trilobites	C	C	C	C	A	A	A	R	A	A	A	R	R	0
	Worm tubes	N	N	N	N	N	N	N	N	N	N	N	N	N	0
	Scolecodonts	N	N	R	N	N	N	N	N	R	N	R	N	N	0

Note: *A* = *Abundant* *C* = *Common* *R* = *Rare* *N* = *Not Present* level, a single external mold of an articulated specimen of *Orthodesma* sp. was found in probable life position oriented perpendicular to bedding (Frey 1983, 1987a).

The variability in bivalve preservation provides some interesting insights as to the relative sedimentation within the 8–12-cm interval of the *Isotelus*-bearing shale. Actualistic studies of the disarticulation rates of modern bivalves show that within a few days of death the valves of bivalves will gap open and then rapidly disarticulate (Schäfer 1972). Thus the presence of well-preserved, articulated bivalves at the 9–10-cm level suggests they were buried alive or died shortly before rapid burial.

BRACHIOPODA

Zygospira modesta (Say) and *Leptaena richmondensis* (Foerste) are the brachiopod taxa present within the *Isotelus*-bearing shale (tables 6.5 and 6.6). *Z. modesta* occurs as single articulated specimens or as clumps (*sensu* Kidwell et al. 1986) of a few articulated individuals to more than 300 articulated specimens. All specimens are well preserved, displaying no significant abrasion or fracturing, and ranged in size from 0.4 to 4.4 mm (fig. 6.8). A few specimens of *Z. modesta* examined from the 8–9-cm level were encrusted with small specimens of the worm tube *Cornulites corrugatus* (Nicholson).

The preservation of large numbers of articulated specimens of *Z. modesta* would suggest rapid burial of living or recently dead individuals. However, *Z. modesta* possesses a cyrtomatodont hinging mechanism, which consists of individual sockets that grow around the bulbous teeth, thus locking the valves together. This dentition causes a bias toward preservation of articulated *Z. modesta* specimens and potentially limits the usefulness of *Z. modesta* in determining the sedimentologic history of this bed.

However, the preservation of clumps of *Z. modesta* with many individuals oriented with the pedicle opening adjacent to trilobite sclerites, trace fossils, or other fossil fragments suggests preservation in life position (Richards 1972). Clumps of *Z. modesta* preserved in apparent life position occurred at the 6–7-cm, 8–9-cm, 10–11-cm, and 11–12-cm intervals. Preservation of brachiopods in life position suggests that these clumps were rapidly buried during a period of high sedimentation.

Size-frequency analysis of the two clumps of *Z. modesta* collected from the 6–7-cm and 8–9-cm intervals may suggest the presence of a single population of brachiopods or multiple populations of brachiopods that flourished for a considerable time (fig. 6.8). Regardless of which interpretation is correct, these clumps represent an indeterminate period of low sedimentation during which the clumps accumulated.

Zygospira modesta provides valuable information regarding the sedimentation rates within the *Isotelus*-bearing shale. The development of brachiopod clumps required periods of low sedimentation to allow the population(s) of brachiopods

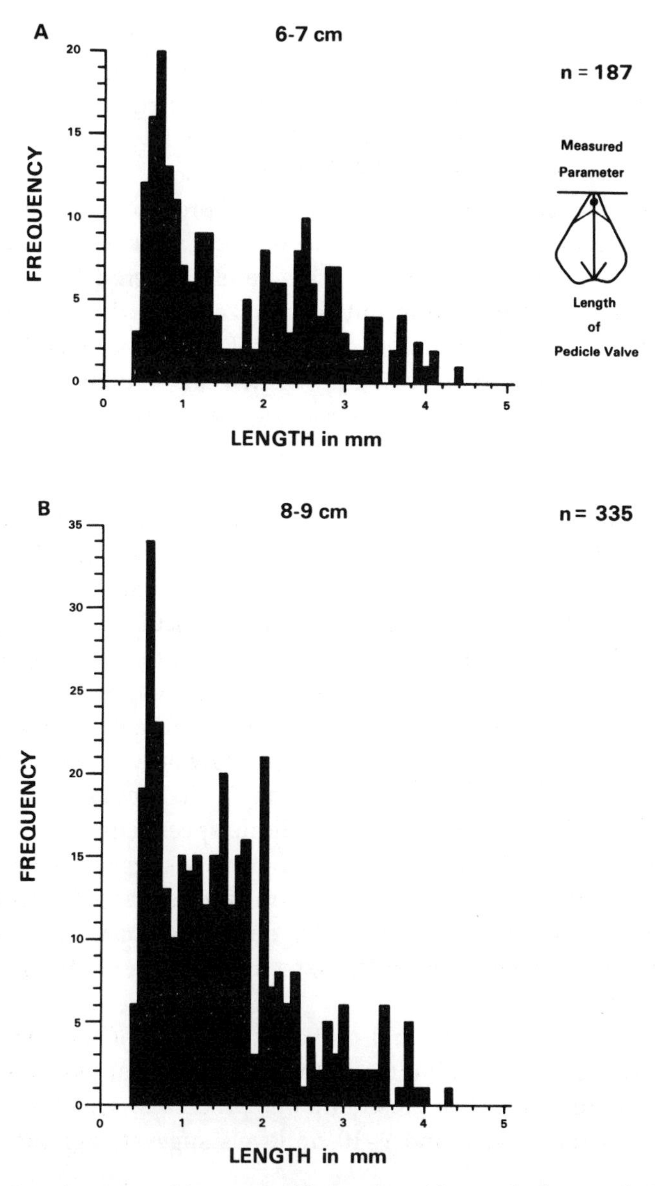

FIGURE 6.8. Size-frequency histograms illustrating either a single population or multiple populations of *Zygospira modesta* from grab samples collected from the 6–7- and 8–9-cm intervals of the lower *Isotelus*-bearing shale.

to develop, whereas the preservation of brachiopods in apparent life position indicates relatively rapid sedimentation and burial of the thriving brachiopods.

BRYOZOA

A number of the larger *Isotelus* sp. sclerites were encrusted by well-preserved bryozoans, generally located on only one side of the sclerite (table 6.6). The bryozoan colonies range in diameter from 1 mm to more than 1 cm. One anomalously large colony is present on one *Isotelus* sp. sclerite encountered at the 8–9-cm level. The sclerite is completely encrusted on the upper surface (relative to bedding) and approximately 75% of the lower surface. The area of encrustation is estimated to be 35 cm^2 for both surfaces.

Encrusting bryozoans on *Isotelus* sp. sclerites provide evidence of low sedimentation at specific intervals throughout the *Isotelus*-bearing shale (Alexander and Scharpf 1990). Bryozoan encrustation rarely covered the entire trilobite sclerite, indicating that time from initial colonization to destruction of an individual colony was short. Alexander and Scharpf (1990) speculated that this type of bryozoan preservation represents only a few years' growth of the colony.

CRINOIDEA

Crinoids identified from the *Isotelus*-bearing shale are *Cincinnaticrinus* sp. (Warn and Strimple), *Lichenocrinus* sp. (Hall), *Xenocrinus*(?) sp. (S. A. Miller), and juvenile specimens of *Gaurocrinus*(?) sp. (Hall) (table 6.5). Crinoid preservation ranged from partially articulated specimens to disarticulated remains (table 6.6). Partially articulated specimens were recovered from the 2–3-cm, 6–7-cm, 8–9-cm, and 9–10-cm levels and consist of five crinoid holdfasts, two nearly complete crowns of *Gaurocrinus*(?) sp., and a few centimeters of an articulated *Xenocrinus*(?) column. The *Gaurocrinus*(?) sp. crowns were probably complete prior to excavation; however, the excavation of the *Isotelus*-bearing shale destroyed most of the arms of these small, delicate specimens.

The presence of nearly complete crinoid crowns suggests a period of rapid sedimentation at the 6–7-cm and the 8–9-cm levels. Crinoids must be rapidly buried or the biostratinomic processes of decay and scavenging will quickly disarticulate their multielement skeleton, commonly in a matter of days (Meyer 1971; Liddell 1975).

Disarticulated crinoid debris and partial articulation of resistant portions of the crinoid skeleton (short column lengths and holdfasts) indicate longer periods of exposure to destructive biostratinomic processes. This type of preservation observed at the 2–3-cm and 9–10-cm levels suggests periods of low sedimentation.

OSTRACODA

Ostracodes were common fossils in the sieved residues from the *Isotelus*-bearing shale (table 6.6). The majority of specimens examined were articulated, uncrushed individuals; in some cases, detailed surface ornamentation was preserved.

The abundance of articulated individuals suggests high sedimentation rates throughout much of the shale's deposition. One possibility is that a period of rapid sedimentation buried the ostracodes before they could disarticulate. A second possibility is that most ostracodes collected were burrowing organisms that were overwhelmed by an episodic high sedimentation event. A third possibility is that upon death the ostracode specimens sank far enough into the relatively fluid muds of the Waynesville seafloor so that the encasing sediment preserved them intact. We are uncertain as to which of these possibilities applies to the articulated ostracode carcasses of this bed.

Trace Fossils

Chondrites sp. (Sternberg), unidentified feeding traces, and *Rusophycus pudicum* (Hall) are the trace fossils occurring in the *Isotelus*-bearing shale (table 6.5). The distribution of trace fossils is sparse (fig. 6.4), and the abundance decreases to zero below the 12-cm level of the bed. *Chondrites* sp. occurs as vertical to subparallel traces cross-cut the upper 12 cm of the bed. The large horizontal feeding traces and *R. pudicum* occur at the 3–4-cm, 5–6-cm, 6–7-cm, and 10–11-cm levels. *Rusophycus pudicum* has been interpreted as the resting trace of the trilobite *Flexicalymene* (Osgood 1970) and was produced at the sediment/water interface through shallow excavation of sediment by the trilobite's legs. Thus within the 6–7-cm level of the *Isotelus*-bearing shale a break in sedimentation occurred, allowing *R. pudicum* to be produced.

Large horizontal feeding traces are associated with *R. pudicum* at the 6–7-cm level of the *Isotelus*-bearing shale, suggesting that the large horizontal trace fossils were produced at or slightly below the sediment/water interface. These traces represent periods of low sedimentation for a horizon slightly above the position of these traces.

Trilobita

Isotelus sp. (DeKay) and *Flexicalymene meeki* (Foerste) represent the two genera of trilobites present in the *Isotelus*-bearing shale. Trilobite preservation ranged from well-preserved, articulated carcasses or moult remains to fragmented, encrusted sclerites (table 6.6). Tables 6.7 and 6.8 provide additional details concerning the preservation and orientation of articulated trilobite carcasses and partially articulated trilobite remains.

Trilobite sclerites were a common to abundant constituent in most intervals within the bed (table 6.6). Sclerite preservation ranged from individual cephalic, pygidial, and thoracic segments; genal spines; cranidia; and librigenae to fragments that could not be assigned to any particular portion of a trilobite specimen. Sclerites were generally not encrusted with epibionts, although some specimens did contain encrusting bryozoans and *Cornulites corrugatus*.

The preservation of articulated trilobite carcasses and moult remains and of partially disarticulated carcasses and moult remains indicates that much of the bed accumulated during periods of high sedimentation. Schäfer (1972), Plotnick (1986), and Allison (1986, 1988) have demonstrated, through actualistic studies,

TABLE 6.7 *Summary of Biostratinomic Data Collected from Articulated Trilobite Carcasses Excavated from the* Isotelus-*Bearing Shale Bed. (Sample Size = 26.)*

Stratigraphic interval (in cm)	Trilobite specimen	Azimuth* orientation	Outstretched vs. enrolled	Dorsal vs. ventral position related to bedding
0–1	*Flexicalymene meeki*	N 5 W	Outstretched	Dorsal up
1–2	*Isotelus* sp.	N 70 E	Outstretched	Dorsal up
4–5	*F. meeki*	N 45 W	Outstretched	Ventral up
4–5	*Isotelus* sp.	N 70 E	Enrolled	Dorsal up
5–6	*F. meeki*	N 10 W	Enrolled	Dorsal up
5–6	*F. meeki*	N 10 E	Enrolled	Dorsal up
5–6	*F. meeki*	N 10 E	Outstretched	Ventral up
5–6	*F. meeki*	N 10 E	Outstretched	Ventral up
6–7	*Isotelus* sp.	N 35 E	Outstretched	Ventral up
6–7	*Isotelus* sp.	N 10 E	Outstretched	Ventral up
6–7	*Isotelus* sp.	N 10 W	Outstretched	Dorsal up
7–8	*Isotelus* sp.	N 70 E	Enrolled	Dorsal up
8–9	*F. meeki*	ND†	Enrolled	Dorsal up
8–9	*Isotelus* sp.	N–S	Outstretched	Dorsal up
9–10	*F. meeki*	N 45 E	Outstretched	Dorsal up
9–10	*F. meeki*	N 80 W	Outstretched	Dorsal up
10–11	*Isotelus* sp.	N 30 W	Enrolled	Dorsal up
10–11	*Isotelus* sp.	N 55 W	Enrolled	Dorsal up
10–11	*Isotelus* sp.	S 35 E	Outstretched	Dorsal up
10–11	*F. meeki*	N–S	Outstretched	Dorsal up
10–11	*F. meeki*	N 80 E	Enrolled	Dorsal up
11–12	*F. meeki*	N 5 W	Enrolled	Dorsal up
12–13	*Isotelus* sp.	N 45 E	Outstretched	Dorsal up
12–13	*F. meeki*	N 12 E	Outstretched	Dorsal up
12–13	*F. meeki*	S 15 W	Enrolled	Dorsal up
12–13	*F. meeki*	N 45 W	Outstretched	Dorsal up

*Preferred azimuth orientations taken along the midline of a trilobite specimen from anterior to posterior.

†Not determined.

that modern arthropods disarticulate in a matter of a few days to a few months after death as the result of decay and scavenging. Thus the preservation of articulated arthropods in the fossil record is considered proof of rapid sediment influx and final burial in nearly all environments (Brett and Baird 1986).

In contrast, trilobite sclerites that are encrusted with bryozoans and *C. corrugatus* indicate long periods of seafloor exposure. The presence of both types of trilobite preservation in a given stratigraphic interval suggests the bed was deposited either in one episode of high sedimentation that incorporated both living and disarticulated remains of trilobites or in a number of high-sedimentation events punctuated by periods of little or no sedimentation.

WORM TUBES

Cornulites corrugatus, presumed annelid worm tube, were recovered from the 8–9-cm and 11–12-cm levels of the *Isotelus*-bearing shale (table 6.6). These encrusting organisms were preserved (1) in the plication furrows of *Zygospira modesta* with the anterior end of the tube oriented toward the brachiopod's

TABLE 6.8 *Summary of Biostratinomic Data Collected from Partially Articulated Trilobite Remains Excavated from the* Isotelus-*Bearing Shale Bed. (Sample Size = 25.)*

Stratigraphic interval (in cm)	Trilobite specimen	Skeletal remains	Azimuth* orientation	Dorsal vs. ventral orientation†	Outstretched vs. enrolled
3–4	*Isotelus* sp.	Pygidium			
4–5	*Isotelus* sp.	Thoracopygon	N 60 W	Ventral up	Outstretched
4–5	*Isotelus* sp.	Thoracic segments		Ventral up	Outstretched
4–5	*Isotelus* sp.	Pygidium	N 80 W		
4–5	*Isotelus* sp.	Thoracic segments			Outstretched
5–6	*Flexicalymene meeki*	Thoracopygon	N 15 E	Dorsal up	Outstretched
5–6	*F. meeki*	Cephalon			
5–6	*Isotelus* sp.	Pygidium	N 40 E		
5–6	*Isotelus* sp.	Pygidium			
5–6	*Isotelus* sp.	Cephalon		Dorsal up	
6–7	*Isotelus* sp.	Cephalothorax	N 10 W	Dorsal up	Outstretched
7–8	*F. meeki*	Thoracopygon	N 40 W	Dorsal up	Outstretched
7–8	*F. meeki*	Thoracopygon	N 85 W	Dorsal up	Outstretched
7–8	*F. meeki*	Cephalon	N 15 W		
7–8	*Isotelus* sp.	Pygidium	N 50 W		
7–8	*Isotelus* sp.	Cephalon, Pygidium	Random	Mixed	Outstretched
8–9	*Isotelus* sp.	Pygidium	N 35 W		
8–9	*Isotelus* sp.	Cephalon	N 40 E		
10–11	*Isotelus* sp.	Pygidium	N 40 E		
11–12	*Isotelus* sp.	Cephalon	N 60 W		
11–12	*Isotelus* sp.	Cephalon	S 73 E		
12–13	*Isotelus* sp.	Thoracopygon	N 80 E		Outstretched
12–13	*Isotelus* sp.	Pygidium	N 70 W		
12–13	*Isotelus* sp.	Pygidium	N 73 W		
12–13	*Isotelus* sp.	Pygidium			

*Azimuth orientation was not determined on all specimens.
†Dorsal/ventral data were not collected on isolated trilobite sclerites.

commissure; (2) as randomly oriented, isolated individuals on *Isotelus* sp. sclerites; or (3) as rosettes of individuals encrusting *Isotelus* sp. sclerites.

The colonization of *Z. modesta* and *Isotelus* sp. sclerites by *C. corrugatus* indicates that low sedimentation rates occurred at these levels. High rates of sedimentation would bury the necessary hard substrates and preclude these encrusting organisms.

Synecology

The Waynesville seafloor throughout the deposition of the *Isotelus*-bearing shale was a relatively barren, soft mud containing rare silt-rich animal traces and skeletal fragments. This substrate was inhabited by a low-diversity fauna of sedentary, semi-infaunal, endobyssate, suspension-feeding bivalves; epifaunal and infaunal ostracodes; vagrant, epifaunal, and infaunal deposit-feeding trilobites; and a host of burrowing infaunal, soft-bodied organisms. The skeletal fragments and silt-rich feeding traces were colonized by sessile, epifaunal, suspension-feeding brachiopods, bivalves, bryozoans, crinoids, and test-form-

Table 6.9 *Summary of Autecology and Biostratinomic Data Used to Determine Relative Rates of Sedimentation in the* Isotelus-*Bearing Shale*

Depth (in cm)	Low sedimentation rates	High sedimentation rates
0–1	Disarticulated skeletal elements (DSE)	Articulated ostracode and trilobite carcasses (AC)
1–2	DSE	AC
2–3	DSE, crinoid holdfasts on *Isotelus* sp. sclerites	Evidence not observed
3–4	DSE, horizontal feeding traces	Evidence not observed
4–5	DSE, encrusting bryozoans on *Isotelus* sp. sclerite	AC
5–6	DSE, horizontal feeding traces	AC
6–7	DSE, trilobite resting traces, *Zygospira modesta* clumps	AC, nearly complete crinoid crown, *Z. modesta* in life position
7–8	DSE	AC
8–9	DSE, encrusting bryozoans on *Isotelus* sp. sclerite, *Cornulites corrugatus* encrusting *Z. modesta*, *Z. modesta* clumps	AC, nearly complete crinoid crown, *Z. modesta* in life position
9–10	DSE	AC, bivalves in life position
10–11	DSE, horizontal feeding traces	AC, *Z. modesta* in life position
11–12	DSE, encrusting bryozoans and *C. corrugatus* on *Isotelus* sp. sclerites, *Z. modesta* clumps	AC, *Z. modesta* in life position
12–13	DSE	AC

ing worms. Eurypterids, cephalopods, conodont animals, and scolecodont-bearing worms were the preserved predators of this bed.

Sessile, epifaunal, suspension-feeding organisms were restricted to small colonies or clumps associated with the scarce, isolated areas of firm or hard substratum during times of low sedimentation (table 6.9). Monotypic and polytypic fossil concentrations developed at the 2–3-, 4–5-, 6–7-, 8–9-, 10–11-, and 11–12-cm levels of the *Isotelus*-bearing shale as the suspension-feeding organisms colonized the limited areas of suitable substrate. The fossil concentrations present at the 6–7-, 8–9-, 10–11-, and 11–12-cm levels contain fossils preserved in life position and represent autochthonous fossil assemblages (*sensu* Kidwell et al. 1986) (table 6.9).

Fossil Diagenesis

Skeletal dissolution and pyrite mineralization of fossils occurred following the deposition of the *Isotelus*-bearing shale. The single specimen of *Orthodesma* sp. was preserved as a composite mold as the result of the dissolution of the inferred aragonitic valves. Burial of shell material in organic-rich sediments typically accelerates dissolution because of the buildup of CO_2 and H_2SO_4 in the course of anaerobic decay (Hecht 1933; Brett and Baird 1986). This process reduces the pH of the sediment, resulting in dissolution of shell carbonates. The specimen of *Orthodesma* sp. was likely subjected to reducing conditions after burial, which resulted in dissolution of the shell.

Additional evidence supporting the development of reducing conditions is the presence of pyrite as thin patches or patinas on ostracode valves and trilobite sclerites, the occurrence of the trace fossil *Chondrites* sp., and the lack of feeding traces deeper than 12 cm. Rapid burial of organisms in organic-rich, sulfide-rich muds, under reducing conditions, will produce thin pyritic skins or patinas on skeletons (Fisher and Hudson 1985; Dick and Brett 1986). Bromley and Ekdale (1984) documented that as oxygen levels decline in marine sediments the diversity of trace fossils is reduced until only *Chondrites* is present in nearly anoxic sediments. *Chondrites* sp. traces cross-cut bedding and other traces, indicating that it was the last trace fossil produced in this bed. This relationship suggests that the oxygen content of *Isotelus*-bearing shale was low when *Chondrites* sp. traces were formed.

Discussion

The trilobite-bearing shales discussed in this study are not unique in the Waynesville Formation or the Late Ordovician. Frey (1987b) traced the *Treptoceras duseri* shale for approximately 40 km in southwestern Ohio and recognized a temporally equivalent shale unit that did not contain abundant cephalopods. This unit, termed the *trilobite shale,* was traced over 40 km in west-central Ohio and southeastern Indiana. Hence the "trilobite shale" and the *T. duseri* shale bed represent a trilobite- and cephalopod-rich bed traceable for nearly 100 km. Brandt Velbel (1985) documented the occurrence of five trilobite shales in the Late Ordovician rocks of southwestern Ohio and southeastern Indiana. These beds were not individually named, nor was the lateral distribution of these beds addressed. Baird and Brett (1983), Brett et al. (1986), and Babcock and Speyer (1987) recognized five laterally persistent trilobite beds from the Middle Devonian Hamilton Group of New York. These beds were traced for a minimum of 50 km to a maximum of 120 km along the outcrop belt of the Hamilton Group of western New York.

These Late Ordovician and the Middle Devonian "trilobite shales" were deposited in shallow, epeiric seas and represent distal, shelf-to-slope facies below the normal wave base but within the storm wave base (Brett et al. 1986; Frey 1987a, 1987b). Water depths of these "deeper-water" facies have been estimated to be between 20 and 100 m (Frey 1987a, 1987b; Vogel et al. 1987).

Frey (1987b) suggested that the *T. duseri* shale bed and trilobite shale bed represent the influx of fine-grained clastic sediments in a low-energy, quiet-water environment subjected to episodic disturbance by storms. Brandt Velbel (1985) indicated that ichnologic, taphonomic, and sedimentologic evidence observed in Late Ordovician trilobite shale beds pointed to rapid deposition of silt and clay by an undetermined sedimentary process. Brett et al. (1986) concluded that the repeated occurrence of widespread, persistent trilobite beds in the Middle Devonian suggested that single, powerful, large storms buried each of these beds. They envisioned storm-generated currents eroding and resuspending sediments from penecontemporaneous, "shallow-water," upslope facies. The resulting cloud of sediment would move offshore, possibly by a high-density, water-saturated sediment flow (Speyer and Brett 1991). The resulting storm deposit would grade laterally from coarse, fragmented skeletal debris through calcisiltites to mud layers. The trilobite beds of the Hamilton Group presumably resulted from rapid smothering of large areas of the seafloor by large storm events.

The model proposed by Brett et al. (1986) is applicable to the formation of the *Isotelus*-bearing shale excavated at Caesar Creek. The rapid influx of a thick blanket of mud smothered the prestorm benthic community. The "deeper-water" offshore depositional setting favored sedimentation over erosion (Speyer and Brett 1988), producing the stacking of multiple mud-dominated tempestites as observed in this study and by Frey (1987b).

The lower Caesar Creek *Isotelus*-bearing shale contains at least seven muddy tempestites based on our analysis of the lithologic, paleoecological, and taphonomic data (fig. 6.9). These muddy tempestites lacked (1) an erosional base with sole and tool marks, (2) gutter casts, (3) upward-graded bedding, (4) small-scale hummocky cross-stratification, (5) shale intraclasts in limestone beds, and (6) imbricated brachiopods and current-aligned fossils that are common in many tempestites present in the upper part of the Waynesville Formation. The muddy tempestite between 6 and 8 cm proved to be the exception because this interval was an upward-graded sequence. Graded bedding, although not diagnostic, is one of the characteristics of tempestite sedimentation (Kreisa 1981; Aigner 1985; Brett et al. 1986), and the preservation of articulated ostracode and trilobite carcasses provides convincing evidence of rapid sedimentation for this stratigraphic interval (Brett and Baird 1986).

Paleoecological and taphonomic analyses provided the evidence required to recognize the remaining six tempestites. The boundaries of each tempestite were determined using the stratigraphic position of autochthonous fossil assemblages or other evidence from fossils of periods of low sedimentation (table 6.9). The exact contact of these tempestites could not be determined because the contact between each storm-generated blanket of mud was indistinguishable by the methods employed in this study (see appendix 2). Articulated ostracode and trilobite carcasses, nearly complete crinoid crowns, and fossils preserved in life position provide the evidence to interpret the remaining inferred tempestites (table 6.9; fig. 6.9).

We were able to compare the taphonomy of the *Isotelus*-bearing shale to the less rigorously defined taphofacies models of Speyer and Brett (1991). This unit displays intervals with taphonomic and sedimentologic features that are

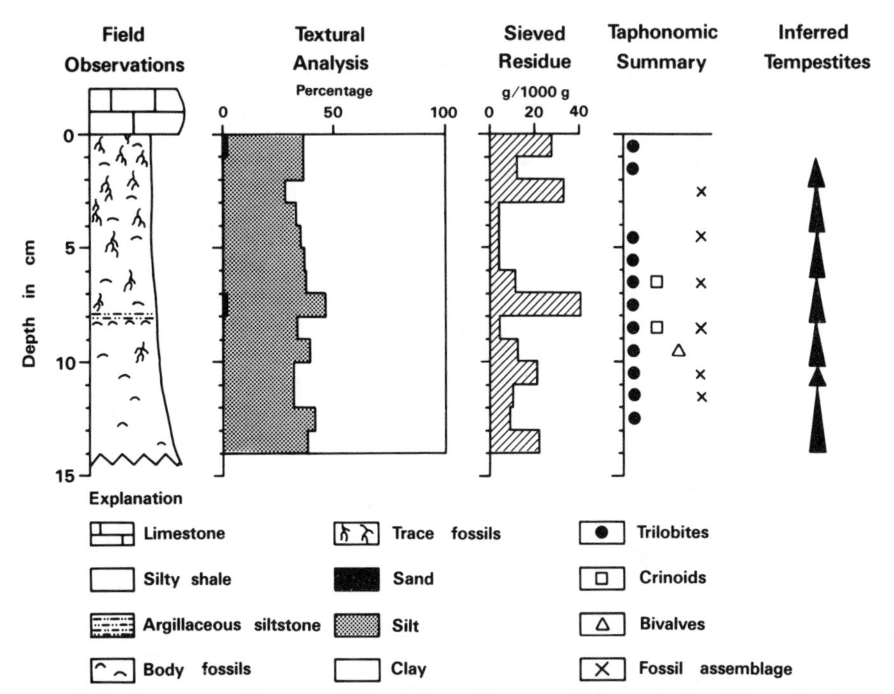

FIGURE 6.9. Summary diagram of the inferred tempestites recognized from the lower *Isotelus*-bearing shale on the basis of field observations, textural analysis, sieved residues, and comparative taphonomy. The stratigraphic position of articulated trilobites and bivalves, nearly complete crinoids, and autochthonous fossil assemblages shown in the taphonomic summary column.

characteristic of taphofacies 4 of Speyer and Brett (1988) or 4b of Speyer and Brett (1991). Moreover, this unit contains mudstones and claystones that have articulated fossils commonly in life position mixed with disarticulated fossil debris characteristic of taphofacies 5 of Speyer and Brett (1988, 1991).

The following conclusions can be drawn from the preceding:

1. Two thin shale-dominant, stratigraphic intervals containing large abundant specimens of the trilobite *Isotelus* occur in the upper 7 m of the Waynesville Formation of Warren and Clinton counties, Ohio. These intervals are laterally correlative for more than 40 km and contain similar lithologic and paleontological characteristics over this range in outcrop and core.
2. The *Isotelus*-bearing shales were deposited in the offshore environment below normal wave base but within storm wave base. These shales were formed by episodic deposition of muddy sediment resuspended and transported by storm-generated processes. Each mud blanket was separated by periods of low sedimentation.
3. Seven tempestites were recognized from the lower *Isotelus*-bearing shale excavated at Caesar Creek emergency spillway. The base of each tempestite contains an autochthonous fossil assemblage or taphonomic evidence indic-

ative of periods of low sedimentation. The fossil assemblages were generally overlain by sediments containing articulated, well-preserved ostracode and trilobite carcasses and bivalves preserved in life position indicative of rapid sedimentation.

4. The fauna present in the lower *Isotelus*-bearing shale is a diverse collection of vagrant deposit feeders; sessile, semi-infaunal suspension feeders; and predators. Vagrant deposit feeders and sessile, semi-infaunal suspension feeders inhabited the soft-bottom environments. Epifaunal suspension feeders were restricted to firm or hard substrates, and predators inhabited the water column and the soft- and hard-bottom environments.
5. The lower *Isotelus*-bearing shale displays lithologic and taphonomic features comparable to the trilobite shales studied by Brandt Velbel (1985) and Frey (1987b), and the taphofacies models defined by Speyer and Brett (1988, 1991).

Appendix 1

Locality information for *Isotelus*-bearing shale in Greene and Warren counties, Ohio. The Roxanna Road, Caesar Creek, and Flat Fork localities are indicated by sites 2, 3, and 4 on figure 6.2.

Site 2, Roxanna Road, is located approximately 1.3 km north of the Warren-Greene county border on U.S. Route 42, on the west side of the road. The approximate X-Y coordinates are: X = 1,568,250, Y = 580,750 (Ohio Coordinate System, south zone), located on the U.S. Geological Survey Waynesville, Ohio, 7.5- minute quadrangle map.

Site 3, Caesar Creek, is located in the northeastern corner of Warren County and adjacent portions of Clinton and Greene counties. The emergency spillway is located approximately 4.8 km southwest of Waynesville, Ohio, within Caesar Creek State Park. The approximate X-Y coordinates of the excavation are: X = 1,559,100, Y = 542,150 (Ohio Coordinate System, south zone), located on the U.S. Geological Survey Oregonia, Ohio, 7.5-minute quadrangle map.

Site 4, Flat Fork enters Caesar Creek Lake slightly northeast of the dam across Caesar Creek. The approximate X-Y coordinates of the Flat Fork locality are: X = 1,561,500, Y = 543,600 (Ohio Coordinate System, south zone), located on the U.S. Geological Survey Oregonia, Ohio, 7.5-minute quadrangle map.

Appendix 2

Description of the excavated stratigraphic section from the emergency spillway at Caesar Creek State Park (figure 6.5, units 1–6). Section is presented in descending order.

Unit 6 (0–14 cm) is 14 cm of randomly oriented, weathered, dark-yellowish-brown (5 YR 4/2) shale and limestone fragments held loosely together by modern plant roots. This unit is fill material deposited in the course of final grading of the floor of the Caesar Creek Reservoir emergency spillway. The contact with unit 5 is sharp.

Unit 5 (14–28 cm) is 14 cm of calcareous, silty, sparsely fossiliferous, medium-gray (N5) shale. The bed displayed platy to flaggy partings and conchoidal fracture. Small limestone nodules up to 5 cm in diameter were randomly distributed throughout this bed. Pyrite-filled burrows were encountered in the course of excavation. The contact with unit 4 is sharp.

Unit 4 (28–29 cm) is 0–2 cm of light-olive-gray (5 Y 6/1) to medium-dark-gray (N4) limestone. This bed is a starved ripple with an amplitude of 2 cm consisting of skeletal fragments, argillaceous material, spar cement, and micrite. Sedimentary structures exhibited by this bed are crude graded bedding, rare rounded shale clasts, and indistinct, subparallel cross-lamination. Fossil preservation ranged from complete to fragmented. Minor bioturbation was present on the top of the bed. Joints were not noted in this bed. The contact with unit 3 is sharp.

Unit 3 (29–30 cm) is 1 cm of calcareous, silty, sparsely fossiliferous medium-gray (N5) shale. The bed exhibited platy to flaggy partings and conchoidal fracture. The contact with unit 2 is gradational over a few millimeters.

Unit 2 (30–35 cm) is 3–6 cm of light-olive-gray (5 Y 6/1)to dark-gray (N3) limestone. This bed is a lenticular-bedded packstone consisting of skeletal fragments, argillaceous material, spar cement, rare pyrite, and micrite. Sedimentary structures include rare rounded shale clasts, crude grading of the upper few millimeters of the bed, and shale envelopes on fossil fragments. Geopetal fabric with vug-filling spar cement is present under large brachiopods. Load structures were present on the base of the bed. Preservation of fossils ranged from complete, articulated specimens to fragmental, disarticulated specimens. Elongate fossil fragments were oriented parallel to bedding. Trace fossils were present in the upper surface of the bed. Prominent joints occur in a bimodal pattern oriented north-south and N 77° E. The contact with the *Isotelus*-bearing shale (unit 1) is sharp.

Unit 1 (35–49 cm), the *Isotelus*-bearing shale, consists of more than 14 cm of medium-gray (N5) shale. Thin-section analyses indicate that the bed is composed of clay minerals, calcite cement, dolomite, pyrite, and fossil debris. The fining-upward sequences observed in the "trilobite shales" studied by Brandt Velbel (1985) were not observed in this bed. Illite, chlorite, kaolinite(?), and quartz are the dominant clay minerals composing the clay fraction of this bed. Statistical analysis indicates that no statistically significant variation in the texture occurs in this bed. Thus the recognition of the tempestites of this bed was not supported by statistical analysis. The bed displays platy to flaggy parting and conchoidal fractures. The lower contact of this bed was not excavated.

ACKNOWLEDGMENTS

The authors express their gratitude to the U.S. Army Corps of Engineers and Thomas T. Johnson for their support and cooperation in the course of this study. The authors appreciate the helpful suggestions from William I. Ausich, Michael C. Hansen, Thomas T. Johnson, and David L. Meyer concerning many of the methods used. William I. Ausich supplied the various grids used, helped with crinoid identification, and, through The Ohio State University computer center, conducted the statistical analysis of the textural data collected from the lower *Isotelus*-bearing shale. Robert C. Frey helped with the identification of *Orthodesma* sp. Richard W. Carlton identified the clay minerals of this bed. Kim E. Vorbau processed the samples and conducted the textural analysis. We thank Loren E. Babcock for his thoughtful discussions of trilobite taphonomy and ecdysis. The authors also thank Jean M. Lesher, Michael R. Lester, Edward V. Kuehnle, and Lisa Van Doren for their assistance in preparing the figures for this paper. The authors are also appreciative of the constructive criticism of this paper by Gordon C. Baird, Thomas M. Berg, Carlton E. Brett, Merrianne Hackathorn, Dennis N. Hull, David Lehmann, E. Mac Swinford, and Danita Brandt Velbel, which greatly improved earlier versions of this paper. Financial support for our field expenses was provided by the Ohio Department of Natural Resources, Division of Geological Survey.

REFERENCES

Aigner, T. 1985. *Storm Depositional Systems: Dynamic Stratigraphy in Modern and Ancient Shallow Marine Sequences. Lecture Notes in Earth Sciences,* vol. 3. Berlin: Springer-Verlag.

Alexander, R. R. and C. D. Scharpf. 1990. Epizoans on Late Ordovician brachiopods from southeastern Indiana. *Historical Biology* 4:179–202.

Allison, P. A. 1986. Soft-bodied animals in the fossil record: The role of decay in fragmentation during transport. *Geology* 14(12):979–81.

———. 1988. The role of anoxia in the decay and mineralization of proteinaceous macro-fossils. *Paleobiology* 14(2):139–54.

Anstey, R. L. and M. L. Fowler. 1969. Lithostratigraphy and depositional environment of the Eden Shale (Ordovician) in the tri-state area of Indiana, Kentucky, and Ohio. *Journal of Geology* 77(6):668–82.

Ausich, W. I. and T. W. Kammer 1991. Late Osagean and Meramecian *Actinocrinites* (Echinodermata: Crinoidea) from the Mississippian Stratotype region. *Journal of Paleontology* 65(3):485–99.

Austin, G. M. 1927. Richmond faunal zones in Warren and Clinton counties, Ohio. *Proceedings of the United States National Museum* 70(2671), Art. 22:1–18.

Babcock, L. E. and S. E. Speyer 1987. Enrolled trilobites from the Alden Pyrite Bed, Ledyard Shale (Middle Devonian) of western New York. *Journal of Paleontology* 61(3):539–48.

Baird, G. C. and C. E. Brett. 1983. Regional variation and paleontology of two coral beds in the Middle Devonian Hamilton Group of western New York. *Journal of Paleontology* 57(3):417–46.

Brandt Velbel, D. 1985. Ichnologic, taphonomic, and sedimentologic clues to the deposition of the Cincinnatian shales (Upper Ordovician), Ohio, U.S.A. In H. A. Curran, ed., *Biogenic Structures: Their Use in Interpreting Depositional Environments,* pp. 299–307. SEPM Special Publication 35.

Brett, C. E. and G. C. Baird. 1986. Comparative taphonomy: A key to paleoenvironmental interpretation based on fossil preservation. *Palaios* 1(3):207–27.

Brett, C. E., G. C. Baird, and S. E. Speyer. 1986. Storm-generated sedimentary units:

Tempestite proximity and event stratification in the Middle Devonian Hamilton Group of New York. In C. E. Brett, ed., *Dynamic Stratigraphy and Depositional Environments of the Hamilton Group (Middle Devonian) in New York State, Part 1, New York State Museum Bulletin* 457:129–56.

Bromley, R. G. and A. A. Ekdale. 1984. *Chondrites:* A trace fossil indicator of anoxia in sediments. *Science* 224:872–74.

Caster, K. E., E. A. Dalvé, and J. K. Pope. 1955. *Elementary Guide to the Fossils and Strata of the Ordovician in the Vicinity of Cincinnati, Ohio,* Cincinnati Museum of Natural History.

Dick, V. B. and C. E. Brett. 1986. Petrology, taphonomy, and sedimentary environments of pyritic fossil beds from the Hamilton Group (Middle Devonian) of western New York. In C. E. Brett, ed., *Dynamic Stratigraphy and Depositional Environments of the Hamilton Group (Middle Devonian) in New York State, Part 1. New York State Museum Bulletin* 457:102–28.

Dunham, R. J. 1962. Classification of carbonate rocks according to depositional texture. In W. E. Ham, ed., *Classification of Carbonate Rocks, American Association of Petroleum Geologists Memoir* No. 1, pp. 108–21.

Elias, R. J. 1982. Latest Ordovician solitary rugose corals of eastern North America. *Bulletins of American Paleontology* 81(314):1–116.

Fisher, I. St. J. and J. D. Hudson. 1985. Pyrite geochemistry and fossil preservation in shales. In H. B. Whittington and S. C. Morris, eds., *Extraordinary Fossil Biotas: Their Ecological and Evolutionary Significance, Philosophical Transactions of the Royal Society of London B* 311:167–69.

Foerste, A. F. 1919. Notes on *Isotelus, Acrolichas, Calymene,* and *Encrinus, Scientific Laboratories of Denison University Bulletin* 19:65–82.

Folk, R. L. 1974. *Petrology of Sedimentary Rocks.* Austin, Texas: Hemphill.

Frey, R. C. 1983. The paleontology and paleoecology of the *Treptoceras duseri* shale (Late Ordovician, Richmondian) of southwestern Ohio, Ph.D. diss., Miami University.

———. 1987a. The occurrence of pelecypods in Early Paleozoic epeiric-sea environments, Late Ordovician of the Cincinnati, Ohio, area. *Palaios* 2(1):3–23.

———. 1987b. The paleoecology of a Late Ordovician shale unit from southwestern Ohio and southeastern Indiana. *Journal of Paleontology* 61(2):242–67.

Goddard, E. N., P. D. Trask, R. K. DeFord, O. N. Rowe, J. T. Singewald, and R. M. Overbeck. 1951. *Rock-color chart,* Geological Society of America, Boulder, Colorado.

Gray, H. H. 1972. *Lithostratigraphy of the Maquoketa Group (Ordovician) in Indiana, Indiana Geological Survey Special Report* No. 7.

Hatfield, C. B. 1968. *Stratigraphy and Paleoecology of the Saluda Formation (Cincinnatian) in Indiana, Ohio, and Kentucky. Geological Society of America Special Paper* No. 95.

Hay, H. B. 1981. Lithofacies and formations of the Cincinnatian Series (Upper Ordovician), southeastern Indiana and southwestern Ohio, Ph.D. diss., Miami University.

Hay, H. B., J. K. Pope, and R. C. Frey. 1981. Lithostratigraphy, cyclic sedimentation, and paleoecology of the Cincinnatian Series in southwestern Ohio and southeastern Indiana. In T. G. Roberts, ed., *Geological Society of America, 1981 Field Trip Guidebooks, 1: Stratigraphy, Sedimentology,* pp. 73–86.

Hecht, F. 1933. Der Verbleib der organischer substanz der Tiere bei meerischer Einbettung: *Senckenbergiana* 15:165–249.

Holland, S. M. 1993. Sequence stratigraphy of a carbonate-clastic ramp: The Cincinnatian Series (Upper Ordovician) in its type area. *Geological Society of America Bulletin* 105(3):306–22.

Kidwell, S. M., F. T. Fürsich, and T. Aigner. 1986. Conceptual framework for the analysis and classification of fossil concentrations. *Palaios* 1(3):228–38.

Kreisa, R. D. 1981. Storm-generated sedimentary structures in subtidal marine facies with examples from the Middle and Upper Ordovician of southeastern Virginia. *Journal of Sedimentary Petrology* 51(3):823–48.

Liddell, W. D. 1975. Recent crinoid biostratinomy. *Geological Society of America Abstracts with Programs* 7(7):1169.

Martin, R., P. E. Litz, and W. D. Huff. 1979. A new technique for making thin sections of clayey sediments. *Journal of Sedimentary Petrology* 49(2):641–43.

Meyer, D. L. 1971. Post-mortem disarticulation of crinoids and ophiuroids under natural conditions. *Geological Society of America Abstracts with Programs* 3(7):645.

Osgood, R. G., Jr. 1970. Trace fossils of the Cincinnati area. *Palaeontographica Americana* 6(41):281–444.

Plotnick, R. E. 1986. Taphonomy of a modern shrimp: Implications for the arthropod fossil record. *Palaios* 1(3):286–93.

Potter, P. E., J. B. Maynard, and W. A. Pryor. 1980. *Sedimentology of Shale: Study Guide and Reference Source.* New York: Springer-Verlag.

Richards, R. P. 1972. Autecology of Richmondian brachiopods (Late Ordovician of Indiana and Ohio). *Journal of Paleontology* 46(3):385–405.

Schäfer, W. 1972. *Ecology and Paleoecology of Marine Environments.* Chicago: University of Chicago Press.

Schumacher, G. A. In press. A new look at the Cincinnatian Series from a mapping perspective. In R. A. Davis and R. J. Cuffey, eds., *Sampling the Layer Cake that Isn't: The Stratigraphy and Paleontology of the "Type Cincinnatian," Ohio Division of Geological Survey Guidebook* 13, Ohio Division of Geological Survey, Columbus, Ohio.

Schumacher, G. A. and W. I. Ausich. 1983. New Upper Ordovician echinoderm site: Bull Fork Formation, Caesar Creek Reservoir (Warren County, Ohio). *Ohio Journal of Science* 83(1):60–64.

Schumacher, G. A. and D. L. Shrake. 1989. The *Isotelus gigas* (trilobite) bed: revisited. *Geological Society of America Abstracts with Programs* 21(2):65.

Scotese, C. R. and W. S. McKerrow. 1990. Revised world maps and introduction. In W. S. McKerrow and C. R. Scotese, eds., *Paleozoic Paleogeography and Biogeography,* pp. 1–21. *The Geological Society of London Memoir* No. 12.

Shrake, D. L., G. A. Schumacher, and E. M. Swinford. 1988. The stratigraphy, sedimentology, and paleontology of the Upper Ordovician Cincinnati Group of southwestern Ohio, *SEPM Field Trip Guidebook, 5th Midyear Meeting,* Columbus, Ohio.

Speyer, S. E. and C. E. Brett. 1988. Taphofacies models for epeiric sea environments: Middle Paleozoic examples. In A. K. Behrensmeyer and S. M. Kidwell, eds., *Ecological and Evolutionary Implications of Taphonomic Processes. Palaeogeography, Palaeoclimatology, Palaeoecology,* Special Issue 63(1–3):225–62.

Speyer, S. E. and C. E. Brett. 1991. Taphofacies controls: Background and episodic processes in fossil assemblage preservation. In P. A. Allison and D. E. G. Briggs, eds., *Taphonomy: Releasing the Data Locked in the Fossil Record.* London: Plenum Press.

Spjeldnaes, N. 1981. Lower Paleozoic paleoclimatology. In C. H. Holland, ed., *Lower Paleozoic of the Middle East, Eastern and Southern Africa, and Antarctica.* London: John Wiley .

Starkley, H. C., P. D. Blackman, and P. L. Hauff. 1985. The routine mineralogical analysis of clay-bearing samples. *U.S. Geological Survey Bulletin* No. 1563.

Sweet, W. C., H. Harper, Jr., and D. Zlatkin. 1974. The American Upper Ordovician Standard XIX: A Middle and Upper Ordovician reference standard for the eastern Cincinnati region. *Ohio Journal of Science* 74(1):47–54.

Tobin, R. C. 1982. A model for cyclic deposition in the Cincinnatian Series of southwestern Ohio, northern Kentucky, and southeastern Indiana, Ph.D. diss. (unpub.), University of Cincinnati.

———. 1986. An assessment of the lithostratigraphic and interpretive value of the traditional "biostratigraphy" of the type Upper Ordovician of North America. *American Journal of Science* 286:673–701.

Vogel, K. K., S. Golubic, and C. E. Brett. 1987. Endolithic associations and their relation to facies distributions in the Middle Devonian of New York State. *Lethaia* 20:263–90.

Weir, G. W., W. L. Peterson, and W C Swadley. 1984. Lithostratigraphy of Upper Ordovician strata exposed in Kentucky, U.S. Geological Survey Professional Paper 1151–E.

Wolford, J. J. 1930. The stratigraphy of the Oregonia-Fort Ancient region, southwestern Ohio. *Ohio Journal of Science* 30(5):301–308.

7

Silurian Event Horizons Related to the Evolution and Ecology of Pentamerid Brachiopods

Markes E. Johnson

ABSTRACT

The big-shell communities dominated by pentamerid brachiopods formed some of the most simple but spatially prolific Silurian associations. They attained a broad intercontinental distribution and thrived equally well in carbonate, mixed carbonate–clastic, and siliciclastic settings. Species diversity is variable but often verges on monotypic. The Llandovery or Wenlock species *Borealis borealis, Virgiana decussata, Pentamerus oblongus, P. dorsoplanus,* and *Pentameroides subrectus* frequently occur in large numbers preserved in growth position. Late Wenlock to Ludlow forms such as *Kirkidium* sp. contributed to the fabric of bioherms but are also known from biostromes in growth position. What constitutes a fossil community is difficult for paleontologists to agree on, but genuine pentamerid populations are well documented. Pentamerid ecology is summarized on the basis of North American, northern European, Siberian, and Chinese material, with emphasis on a variety of large to small-scale event horizons related to evolutionary novelty, biogeography, fluctuations in sea level, local seafloor topography, and generation of tempestites.

The *Pentamerus* community and closely related *Pentameroides* community originally were defined by Ziegler (1965) as the middle of five Silurian marine communities in Wales and the Welsh Borderland controlled by factors related to water depth and distance from shore. Subsequently, Ziegler et al. (1968, Pl. 121) illustrated a layer from the Lower Silurian Red Mountain Formation of Alabama composed exclusively of *Pentamerus oblongus* preserved in growth position. Later, the *Pentamerus* community from a clastic setting in the Welsh

Borderland was defined as consisting of 18 species, with 68% of the membership dominated by *P. oblongus* (Ziegler et al. 1968:13). These statistics were used to construct a popular block diagram of the community (Ziegler et al. 1968:15). Similar statistics were upheld by Cocks and McKerrow (1984) in their reevaluation of the Welsh data. In view of the Alabama material so exquisitely preserved in growth position, the composition and diversity of the *Pentamerus* community derived from the Welsh material was considered inflated by Boucot (1975 p. 208). He suggested that the slope of the Welsh seafloor promoted biological contamination by slumping across narrow community zones. Johnson (1980, table 4) collected diversity and evenness data for a variety of pentamerid communities, including *Pentamerus oblongus, Harpidium maquoketa,* and *Pentameroides subrectus* from platform carbonates in the Lower Silurian Hopkinton and Scotch Grove formations of eastern Iowa. The Iowan *Pentamerus* community, for example, included a maximum of seven species dominated up to 94% by *P. oblongus.* These figures are much closer to the monotypic nature of the in situ Alabama material (Ziegler et al. 1966). On the other hand, calculations show that the various Iowan pentamerid communities often approached or matched the diversity and evenness of the type Welsh community (compare tables 4 and 7 in Johnson 1980).

The precise composition of a pentamerid community is open to dispute. Paleoecologists are unable to agree whether fossil communities are discernible in the stratigraphic record, but there is no question that large, genuine populations of pentamerid siblings were frequently preserved intact. The purpose of this contribution is to summarize what is known about the ecology of pentamerid populations, especially in relationship to the formation of event horizons. A wide range of stratigraphically distinctive events associated with these prolific brachiopods may be related to the effects of storms, bathymetric relief, sea-level fluctuations, paleogeography, and the introduction of evolutionary novelty.

Evolutionary Events

Typically, robust pentamerid brachiopods were first described and illustrated by Sowerby (1813, plates 28 and 29) from the Silurian of England. The Pentamerida is a distinctive order differentiated on the basis of a fivefold division in the shell: The larger pedicle valve is partitioned into two sections by a median septum, and the smaller brachial valve is divided into three sections by a pair of outer plates. These internal skeletal features are easily observed in lateral cross sections through the shell or by way of internal molds (fig. 7.1).

At least three genera are closely linked through direct evolutionary descent. Working in the Oslo region of Norway, Kiaer (1908) was the first to draw attention to the *Borealis/Pentamerus/Pentameroides* lineage. He referred to this lineage as a "morphological series" extending from "*Pentamerus borealis* to *P. oblongus* to *P. gothlandicus.*" Subsequent taxonomic revisions have seen the

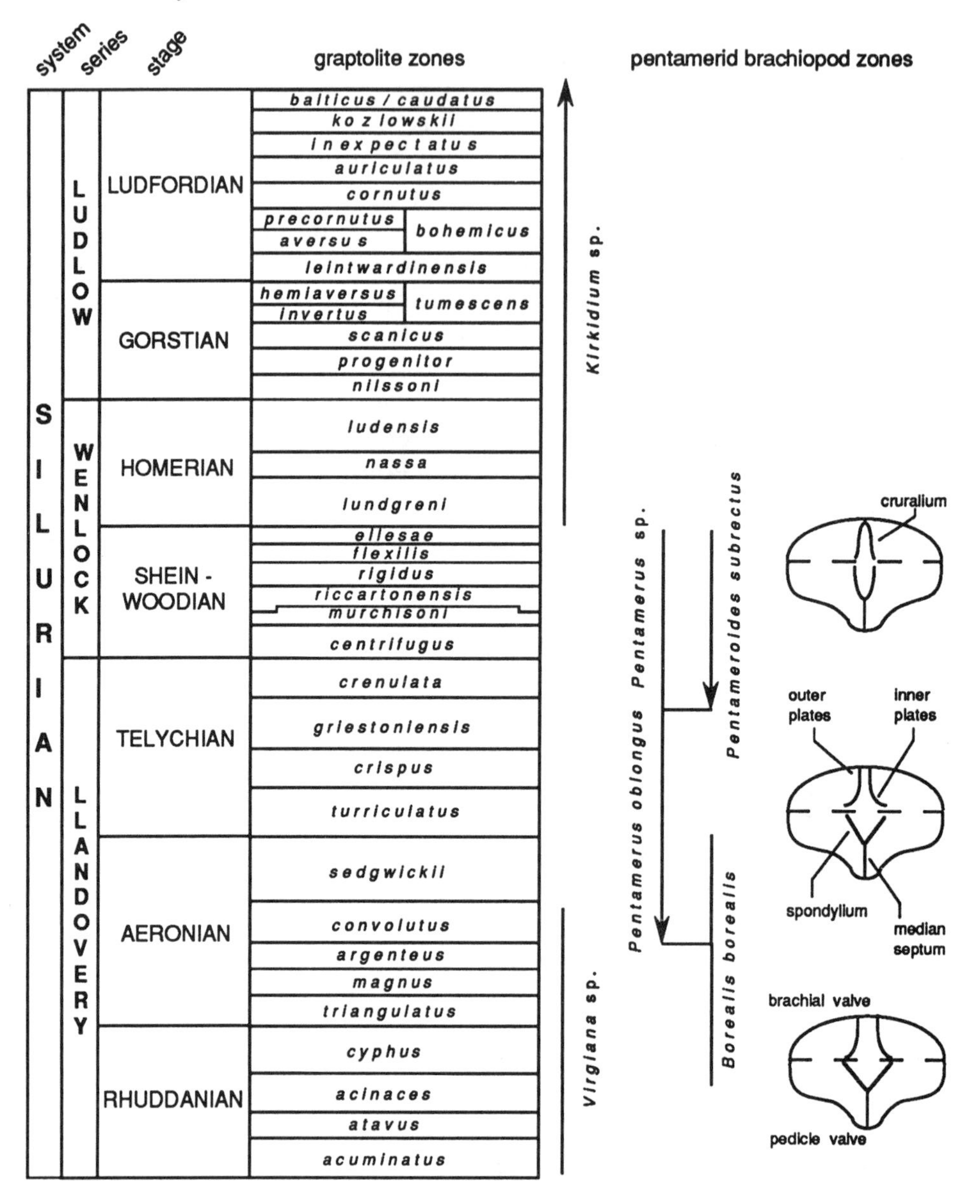

FIGURE 7.1. Stratigraphic range and morphology of some selected Silurian pentamerid bachiopods (the last, or Pridoli Series, is not shown).

erection of the genus *Borealis* for the ancestral form and a recognition that the descendant form called *Pentamerus gothlandicus* in the Oslo region actually belongs to the genus *Pentameroides* (see Baarli and Johnson 1988). *Borealis* gives rise to *Pentamerus* in the Lower Aeronian Stage, and *Pentamerus* gives rise to *Pentameroides* in the middle Telychian Stage (fig. 7.1). In any case, Kiaer (1908:499–500) correctly noted the gradual lengthening of the median septum from one lineage member to the next.

Also working in the Oslo region, St. Joseph (1938:286) first drew attention to the gradual closure of the outer plates in *Pentamerus* until they merge into an elevated cruralium diagnostic of *Pentameroides* (fig. 7.1). Careful measurements on the changing width of the outer plates in stratigraphically successive populations of *Pentamerus oblongus* confirm that this gradualistic trend occurred widely in North America (Johnson 1979; Johnson and Colville 1982). Only when the two outer plates actually reached a point of merger was the comparatively rapid evolution of a cruralium physically possible. Such a pattern fits the concept of a functional threshold punctuating gradual morphological change (Jaanusson 1981). Although some *Pentamerus* persisted for a time, the coeval appearance of *Pentameroides* during mid-Telychian time in Baltica and Laurentia (North America) represents a well-defined morphological event horizon. Individuals from the *Pentamerus* population illustrated by Ziegler et al. (1966) actually represent a transitional stage between *Pentamerus* and *Pentameroides* signified by the presence of a primitive cruralium at the anterior end of vestigial outer plates (see Johnson and Colville 1982, fig. 7).

Fusion of the outer plates in the brachial valve of *Pentamerus* may have imitated a similar process leading to the much earlier development in ancestral stock of the pedicle valve's median septum and spondylium. In this regard, the sudden appearance of such Late Ordovician pentamerids as *Holorhynchus, Eoconchidium, Proconchidium,* and *Tscherskidium* represents equivalent evolutionary event horizons (Rong et al. 1989).

In his review of the *Borealis/Pentamerus* relationship in Norway, Mørk (1981) found evidence for several gradualistic trends, but one punctuated change also was recognized at the transition between the two. The lamellar shell layer associated with the inner and outer plates in the cardinalia of *Borealis borealis* is abruptly absent in *Pentamerus oblongus*. This diagnostic change in shell structure marks another Silurian event horizon of an evolutionary nature.

Influence of Paleogeography

The timing of the *Borealis/Pentamerus/Pentameroides* lineage shown in fig. 7.1 is accurate for the paleocontinent of Baltica (including present-day Norway, Estonia, and the island of Novaya Zemlya) and partly for the separate paleocontinent of Siberia. Occupation of the pentamerid niche apparently followed a somewhat different history with delayed timing on the large paleocontinent of Laurentia (North America). After the Late Ordovician mass extinctions, the depleted endemic faunas of Laurentia were largely replaced by waves of brachiopod immigration (Sheehan 1975).

Borealis is found in the Lower Silurian of Baltica and Siberia (Boucot et al. 1969), but it is absent altogether from Laurentia (fig. 7.2). Instead the pentamerid niche was adopted there by an early immigrant of meristellid stock belonging to *Cryptothyrella*. At home in Baltica during the Late Ordovician, this brachiopod was a reef dweller (Sheehan 1975), but during Early Silurian (Rhuddanian) time in Laurentia it successfully diversified to an open-shelf biostromal habitat. During Late Rhuddanian to Early Aeronian time, *Cryptothyrella* was ecologically replaced in Laurentia by a true pentamerid immigrant, *Virgiana*. *Virgiana* and *Borealis* coexisted in Baltica and Siberia (Boucot et al. 1969), but *Virgiana* alone dominated the pentamerid niche in Laurentia (fig. 7.2). During Middle Aeronian time, *Virgiana* continued to dominate in Laurentia, while *Borealis* gave way to its descendant *Pentamerus* in Baltica and Siberia. Only in Late Aeronian time did *Pentamerus* immigrate to Laurentia and ecologically replace *Virgiana*. Finally in middle Telychian time, *Pentamerus* gave way to its descendant *Pentameroides* uniformly on at least two paleocontinents, but it is unknown in Siberia (fig. 7.2).

The evolutionary event horizon represented by the transition from *Pentamerus* to *Pentameroides* was intercontinental in nature because the ancestor was already intercontinental in its range. That is, *Pentameroides* developed in place (except in Siberia) and was not an invader. Other major event horizons shaped by postimmigration domination, such as *Pentamerus* over *Virgiana*, were more limited but still continental in extent.

Pentamerid Brachiopod Ecology

Boucot (1975) devised a bathymetric scheme whereby fossil communities or associations could be assigned to one of five contiguous "benthic-assemblage zones." Pentamerid communities or populations are assigned to BA3, bracketed by shallower shelly faunas in BA1–2 (such as the *Lingula* and *Eocoelia* communities) and deeper faunas in BA4–5 (such as the *Stricklandia* and *Clorinda* communities). In their treatment of the family Pentameridae, Boucot and Johnson (1979) briefly discussed the ecology of Silurian forms. They attributed pentamerids to a range of conditions from rough-water to calm-water, level-bottom environments restricted to BA3.

This review of additional information on pentamerid paleoecology focuses on the nonreefal species that commonly developed immense populations during Llandovery to Wenlock times. Although not widely recognized, some of the Late Ordovician pentamerids (Rong et al. 1989, table 1) and Late Silurian pentamerids (Boucot and Johnson 1979, fig. 1) formed comparable biostromal deposits in growth position.

Population Growth

From the early descriptions of in situ pentamerids by Ziegler et al. (1966), crowding was found to have been intense enough to produce growth deformities and the immurement of small individuals within cavities below the umbonal regions of larger individuals. Small and large individuals alike were often

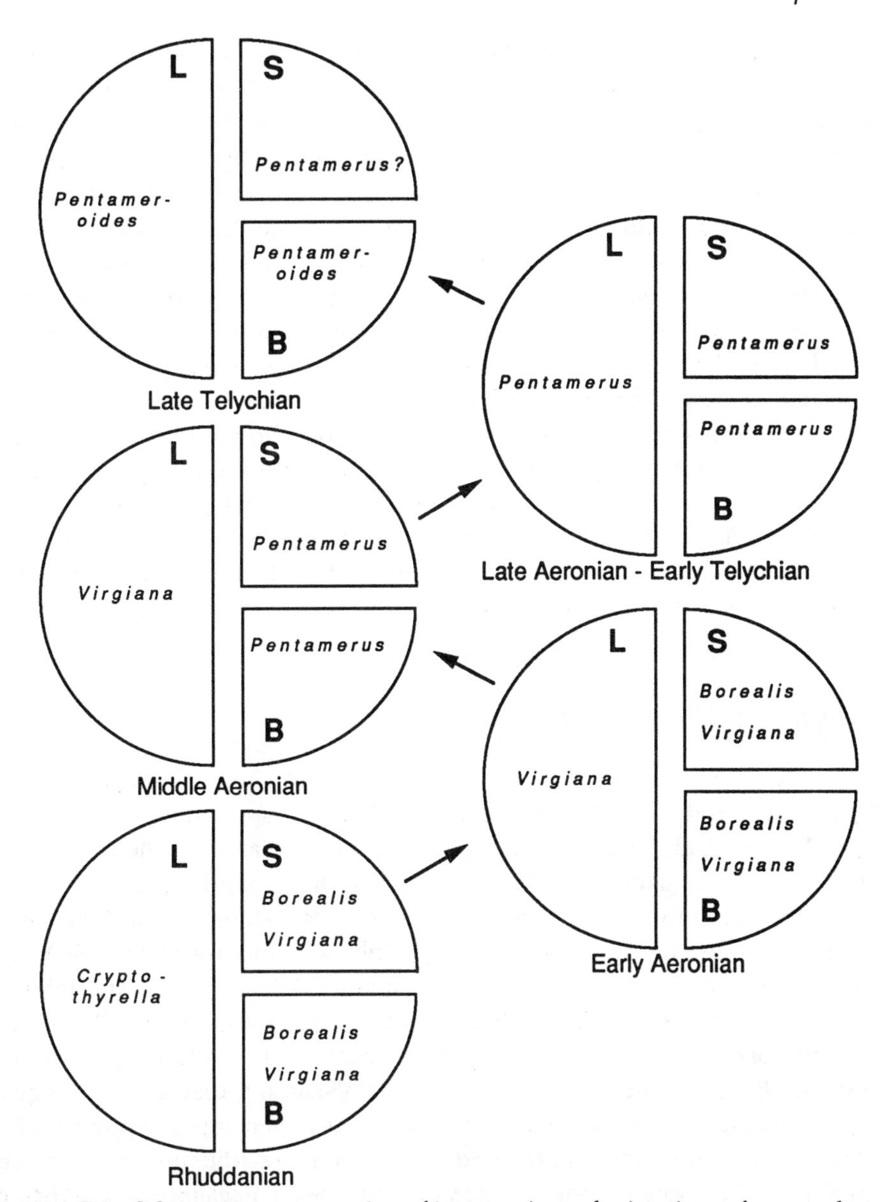

FIGURE 7.2. Schematic representation of intercontinental migrations of some selected pentamerid brachiopods: L = Laurentia; B = Baltica; S = Siberia.

siblings from the same spat fall. The main basis for selection was probably a slight variability in growth rates. Small individuals had a functional pedicle that served to hold them in place on the seafloor. Based on the closure of the delthyrial opening in large individuals, however, this feature was lost with maturity. Crowding was the primary factor that served to hold large populations together.

High mortality with the successful maturation of only 20% of a population is well illustrated by a remarkable block of *Pentamerus oblongus* from eastern Iowa (Johnson 1977:87–88). Articulated shells ranging in cross-sectional diameter from 1 to 5 cm are preserved in a biologically graded bed having a thickness of only 10 cm. Approximately 50 individuals per 100 cm^2 form the bottom of this bed; only eight or nine individuals fill the same space at the top of the bed. Traced laterally on both sides of a highway outcrop, the bed covers a minimum area of 1,125 m^2. This means that the traceable population included over 5.5 million members but ultimately declined to just over 1 million members.

Although the full spatial extent of this particular example is unknown, pentamerid populations were patchy even within a benign environment of vast dimensions. At any given time, neighboring populations were at various stages of ontological development between colonization (on hardgrounds or shell pavements) and the demise of adult clusters due to local conditions. Ziegler et al. (1968:25) claimed that the Early Silurian brachiopod-dominated communities depended more on hydrographic factors than on the sedimentological nature of the substrate. This seems especially true for pentamerid populations, in the sense that muddy substrates (whether carbonate or clastic) were transformed into viable living spaces through the development of shell pavements that often included disarticulated pentamerid shells (Johnson 1989, fig. 3.1).

Extent and Nature of Habitat

After many years of field experience in Michigan's Upper Peninsula, Ehlers (1973:132) reported, "The lower and upper *Pentamerus* dolomite and the thin, even-bedded gray dolomites between them are remarkable for their continuity from St. Martin's Island, Big Bay de Noc, to Drummond Island." In terms of present stratigraphic nomenclature, these beds belong to the Schoolcraft Formation within the Manistique Group (Johnson and Campbell 1980); the linear distance traced by the outcrop belt is about 200 km. Colville and Johnson (1982:969) regarded this pair of *Pentamerus* biostromes as time-parallel marker beds and correlated them from northern Michigan to Ontario's Manitoulin Island, the Bruce Peninsula, and the Lake Timiskaming district. A triangular region enclosing these outcrop areas is calculated to encompass approximately 85,000 km^2 of level-bottom pentamerid habitat. Making adjustments for patchy distribution, the monotonous proliferation of coeval *Pentamerus* populations must still be considered "astronomical" in terms of occupied space and the enormous numbers of individuals involved. In addition to the *Pentamerus* beds recognized by Ehlers (1973), the older *Virgiana* and younger *Pentameroides* are equally well represented by beds signifying event horizons in the Michigan Basin (fig. 7.3). Research by Beadle and Johnson (1986) on ecotypic size variation in the green alga *Cyclocrinites* confirms that *Pentamerus* communities in North America, the British Isles, and Norway thrived in shallower, more brightly lit waters than the stricklandiid communities. The maximum thallus size of green algae from a pentamerid community is on average a third as large as those less abundant individuals from a stricklandiid community. Carbonate sequences typically may feature pentamerid layers bracketed between stricklan-

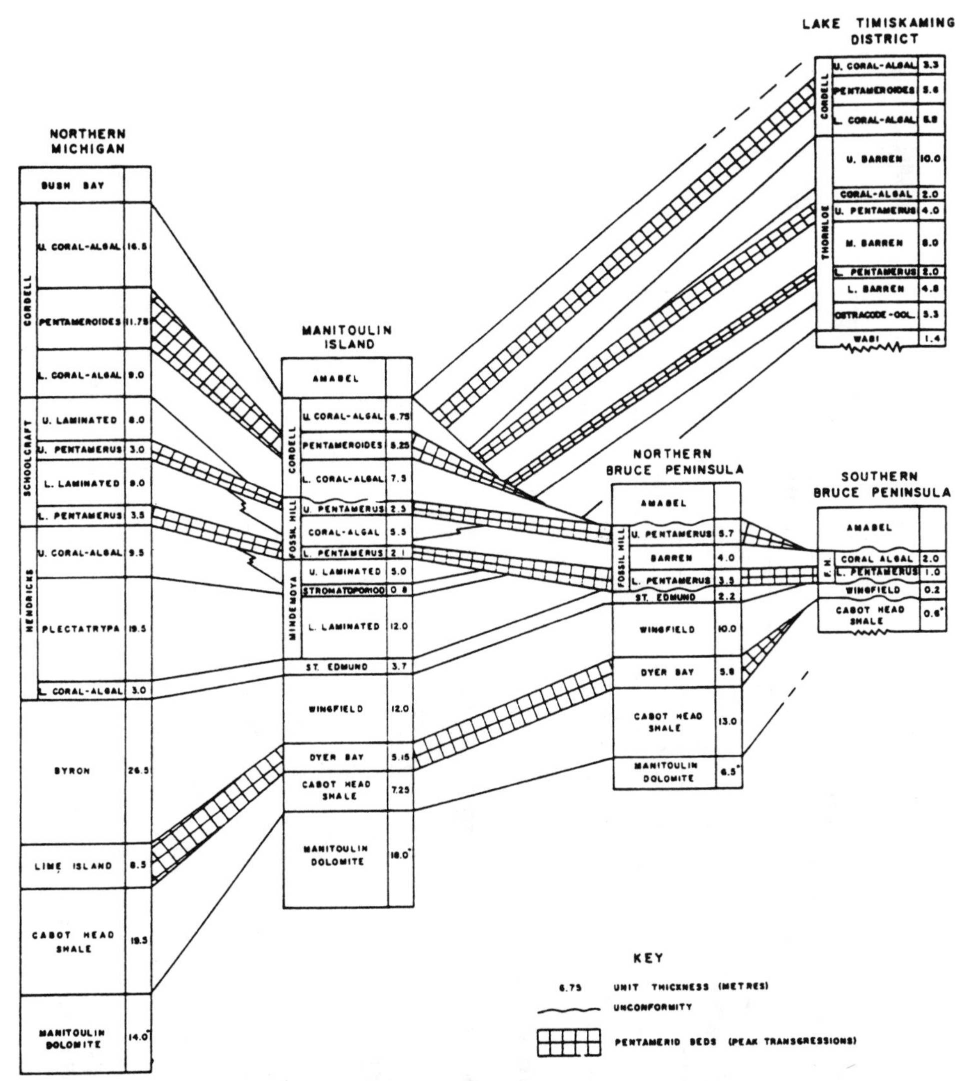

FIGURE 7.3. Fence diagram showing the correlation of Lower Silurian strata throughout the northern area of the Michigan Basin; the checked pattern involves regional event horizons marked by major pentamerid beds following peak highstands in sea level (after Colville and Johnson 1982).

diid-rich layers of deeper-water origin (BA4) and wave-worked, coral-rich layers of shallower-water origin (BA2).

Bathymetry and Sea-Level Changes

In contrast to the American platform seas, with their low bathymetric relief, the narrow community belts mapped by Ziegler (1965) in Wales reflect a pronounced paleoslope. Conspicuously parallel to the paleoshore, the Welsh pen-

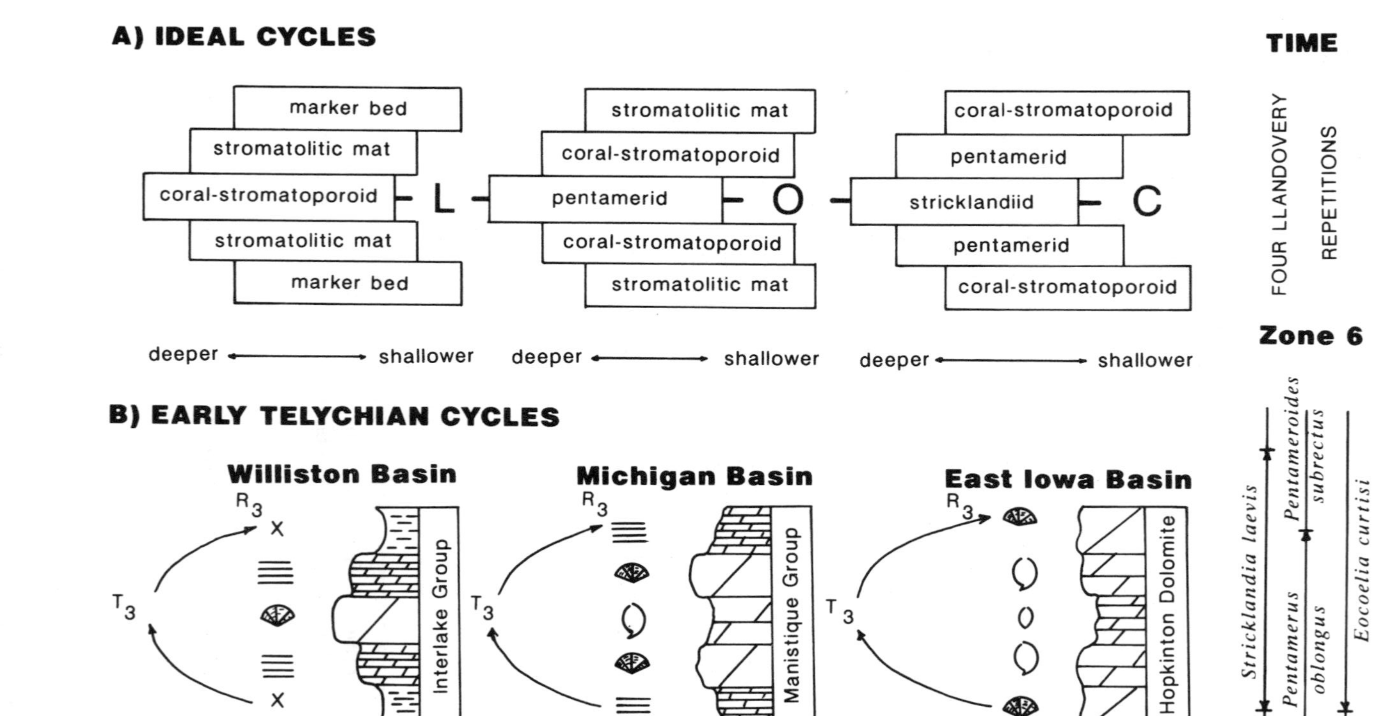

FIGURE 7.4. Correlation of typical sea-level cycles from the Lower Silurian of the Williston, Michigan, and East Iowa basins: (A) Ideal cycles (L-O-C denotes line of correlation linking maximum transgressive peaks in each region), (B) Example of the third of four transgressive-regressive cycles of Early Telychican age. Note that highstands in sea level are recorded by pentamerid communities only in the Michigan Basin area; the Williston Basin area was consistently shallower in water depth, and the East Iowa Basin area was deeper.

tamerid belt was no more than 15–30 km in width. A similar pentamerid belt developed on the flanks of a troughlike basin in the Oslo region (Johnson 1989). Pentamerid distribution basically was regulated by multiple physical factors related to water depth and seafloor topography. Local water depth sometimes changed because of local conditions, but a series of at least four cycles in sea-level fluctuation during the Llandovery Epoch was global in nature (Johnson et al. 1991). When the sea level rose or fell in the region of the Welsh Basin, pentamerid communities shifted only a short distance up or down slope tracking their preferred habitat.

Much larger areas in Laurentia were affected by these eustatic fluctuations. During each Llandovery highstand in sea level, pentamerid communities dominated the region of the Michigan Basin (BA3), while deeper-water stricklandiid communities usually were favored in the area of the East Iowa Basin (BA4) and shallower-water coral-stromatoporoid communities usually thrived in the area

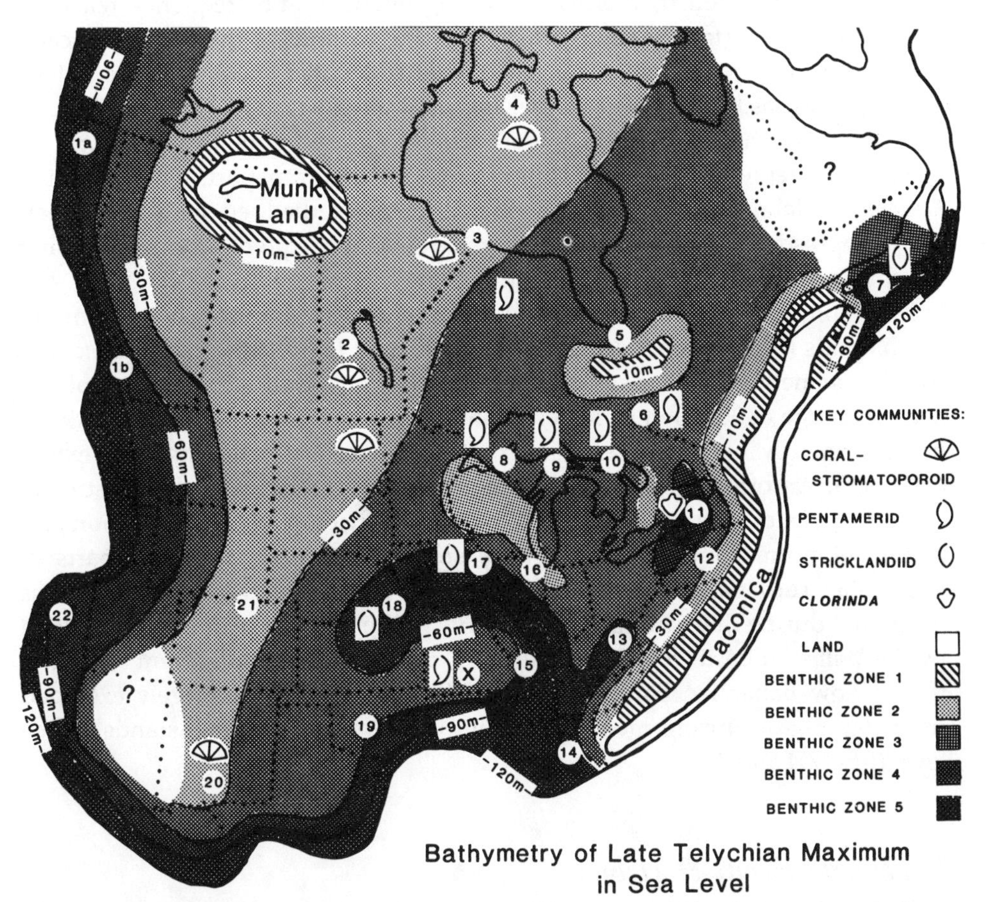

FIGURE 7.5. Bathymetric map reconstruction of the ancestral North American platform (Laurentia) during the Early Telychian highstand in sea level. The O and W refer to the Ozark and Wisconsin domes, respectively. Prime pentamerid habitat is denoted by tghe shading for benthic zone 3 (after Johnson 1987).

of the Williston Basin (BA2). The ideal biostratigraphic cycles characteristic of these basins are correlated with one another in fig. 7.4A; a specific Early Telychian cycle is correlated in fig. 7.4B. During low stands in sea level, pentamerid communities were forced to abandon the region of the Michigan Basin and migrate to the East Iowa Basin and other down-gradient areas (Johnson 1987). A biogeographic map of Laurentia for Early Telychian time gives some perspective on the bathymetry of the paleocontinent and the way it influenced community distribution (fig. 7.5).

Generation of Tempestites

In a study on the derivation of Silurian tempestites in Norway (Johnson 1989), it was possible to show that *Pentamerus* populations lived across a range of BA3 water depths affected by storms of varying intensity and frequency reaching below the normal (fair-weather) wave base. Populations at the bathymetrically lower fringes of habitation were sometimes suffocated by fine clastic fallout stirred by storms but otherwise never mechanically disturbed (fig. 7.6). Populations living in shallower settings were liable to more frequent disruptions involving progressively more mechanical breakage and transport. Finely crushed *Pentamerus* debris (from immature populations) interbedded with thin green shales represent local event horizons generated by frequent storms sweeping the shallow upper end of the habitat (fig. 7.6).

Environmental constraints indicate that the preferred habitat of pentamerid brachiopods (BA3) extended from at or just below fair-weather wave base to a position somewhat below storm wave base but still well within the euphotic zone. A range of 30–60 m is one estimate for this benthic zone (Johnson 1987), based primarily on the observation that shallower-water depths over extant carbonate platforms such as the Bahamian Banks do not engender biostromal developments of lateral uniformity (Beach and Ginsburg 1980). Circulation and mixing of normal marine waters were sufficient enough over vast parts of Laurentia's carbonate platform to promote the prolific pentamerid populations preserved during Silurian time. Other parts of the Silurian platform, as typified by the Williston Basin region (Roehl 1967), are considered an ancient analog to the shallow Bahamian Banks. Pentamerid populations were largely excluded from this region during most of the Silurian, even during highstands in sea level (figs. 7.4 and 7.5).

Ephemeral Event Horizons

Pentamerid brachiopods were prone to take advantage of calm waters, which occurred under some circumstances in protected settings shallower than BA3. Fürsich and Hurst (1980) suggested that monotypic assemblages of *Virgiana*

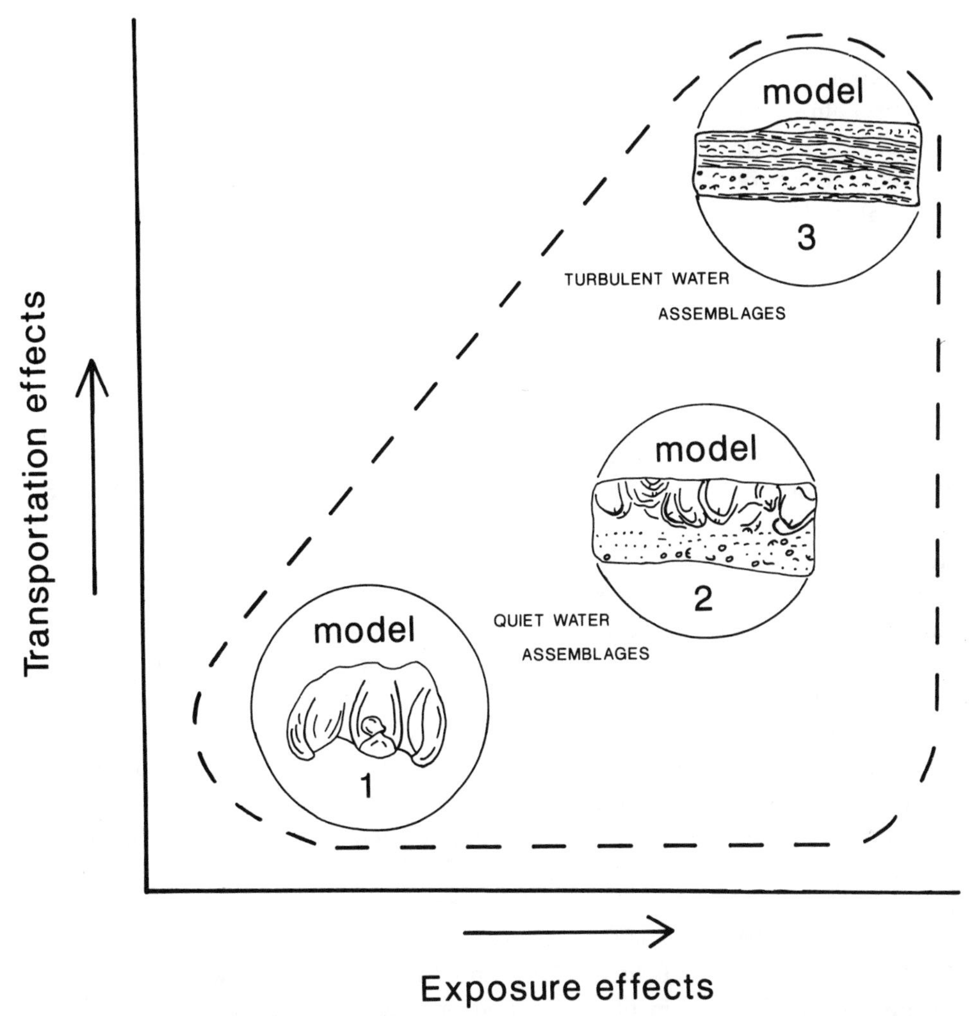

FIGURE 7.6. Models for the development of tempestites generated from pentamerid shells, based on extremes in postmortem exposure and transport (after Johnson 1989).

from the Lower Silurian of Greenland represent marginal marine or slightly hypersaline conditions. Such conditions undoubtedly prevailed during the deposition of bituminous, algal-laminated dolomites (see fig. 2 of Fürsich and Hurst 1980:305), but pentamerid shell beds stratigraphically interspersed among these laminated dolomites (see fig. 1 of Fürsich and Hurst 1980:304) do not prove that the brachiopods lived under the same conditions. Dominance and lack of size reduction are further cited by Fürsich and Hurst (1980:306) as evidence that *Virgiana* was "well adapted for living under raised salinity conditions."

Johnson and Campbell (1980:1049) recovered a population of *Pentamerus oblongus* preserved in growth position within one of the "even-bedded gray dolomites" with carbonaceous partings and occasional salt-crystal casts previously described by Ehlers (1973) in Michigan. In this case, the pentamerids are very small (approximately 50 per 100 cm^2). Pentamerid brachiopods probably were not euryhaline to the extent imagined by Fürsich and Hurst (1980), but they were able to establish ephemeral populations in protected, shallow-water environments with modestly raised salinities. Size reduction of populations in situ within algal dolomites suggests that those capabilities were not normal.

Listing of Cited Populations

A diverse body of literature contains references to pentamerid populations preserved in growth position. The first reports of *Pentamerus* were treated as something of a novelty, but this characteristic was soon widely recognized. In situ populations are important because they allow paleoecologists to attribute bathymetric significance to stratigraphic replacement patterns in fossil communities or associations. It is also important to trace this characteristic to allied species or genera. In addition to *Pentamerus oblongus*, clusters of *P. dorsoplanus*, the older *Borealis borealis*, and *Virgiana decussata* as well as the younger *Pentameroides subrectus* have been observed in growth position. *Kirkidium* sp. is found in biohermal settings but did form comparable biostromal beds as well.

Table 7.1 lists information on collected blocks of fossil material illustrated or described in the literature. These materials are curated at the following institutions: Field Museum of Natural History in Chicago (FMNH; UC denotes collections linked to the University of Chicago), United States National Museum, Smithsonian Institution in Washington, D.C. (USNM), the type fossil collection at the Geological Survey of Canada in Ottawa (GSC), the Paleontologisk Museum in Oslo (PMO), and the Geology Department of Williams College (WCGD). Table 7.2 lists field localities known to have pentamerids in growth position, which have been photographed or otherwise referred to in the literature.

Table 7.1 includes two references to *Pentameroides subrectus*, which are treated herein. The first photograph recording a block of *Pentameroides* preserved in growth position is shown in fig. 7.7. The material was collected from the interreef facies of the Johns Creek Quarry Member belonging to the basal part of the Scotch Grove Formation in eastern Iowa. This is the earliest stratigraphic occurrence of *Pentameroides* in Iowa and is considered to be Late Llandovery in age. The collecting site is the Andrew Quarry, located 2.5 km west of the village of Andrew (Cen. NW 1/4, Sec. 24, T85N, R3E) in Jackson County, Iowa. The cluster shows some disruption by bottom currents, but all the individuals are articulated and still in growth position. This block is part of the Johnson collection reposited with the Field Museum of Natural History.

Another block with *Pentameroides* in growth position from this collection (table 7.1) comes from a stratigraphically higher level in the same formation (Buck Creek Quarry Member) and is Early Wenlock in age. The material was

TABLE 7.1 *Collected Blocks with Material in Growth Position*

Species	Place	Curation	Reference
Pentamerus oblongus	Michigan	unknown	Ehlers (1973, plate 12, fig. 10)
	Michigan	WCGD	Johnson & Campbell (1980: 1049)
	Alabama	USNM 140428	Ziegler er al. (1966, plate 121)
	Alabama	FMNH PE 53958	Baarli et al. (1992, Plate 2)
	Iowa	WCGD	Johnson (1975, plate 1, fig. 1)
	Iowa	FMNH UC 58837	Johnson (1977:87, fig. 4)
	Iowa	FMNH UC 58840	Johnson (1977:86, fig. 3)
	Norway	PMO 117,066	Johnson (1989, figs. 3,2)
Virgiana decussata	Michigan	FMNH PE 25915 & 25916	Johnson & Campbell (1980, plate 3)
	Manitoba	GSC 80481	Johnson & Lescinsky (1986, fig. 4)
Pentameroides subrectus	Iowa	FMNH UC 63682	herein
	Iowa	FMNH UC 65919	herein

TABLE 7.2 *Field Notes on Material in Growth Position*

Species	Place	Data	Reference
Borealis borealis	Estonia	text note	Herein
Pentamerus oblongus	Ontario	photo	Johnson & Colville (1982, Fig. 2)
	Ontario	photo	Johnson, Rong, & Yang (1985, Fig. 4C)
	Quebec	text figure	Johnson et al. (1981, Fig. 3)
	Wales	text note	Ziegler et al. (1966, p. 1033)
	Estonia	text note	herein
Pentamerus dorsoplanus	South China	photo	Johnson, Rong, & Yang (1985, Fig. 4D)
		photo	Rong, Johnson, & Yang (1984, Plate 3, Fig. 5)
Kirkidium sp.	Indiana	text note	herein

collected in the Hanken Quarry (abandoned), located 5 km southeast of Monticello on Iowa Highway 38 and 2 km east on a cross road (NW 1/4, NE 1/4, Sec. 6, T85N, R2W) in Jones County, Iowa. This site has been designated the type locality for the *Pentameroides* community. Table 7.2 includes three references to field observations on pentamerids, also discussed herein. Two of these concern the Lower Silurian of Estonia. *Borealis* coquinas of disarticulated shells (mostly pedicle valves) are very common in northeast Estonia, as typified by the 6.0-m massive coquina in Karinu Quarry (Kaljo and Nestor 1990:155–56). Other localities in western Estonia yield horizons with articulated individuals in growth position. A type locality for the *Borealis borealis* community (M. Rubel, personal communication, 1990) with in situ populations of these brachiopods is the Kirimaì Quarry (near Kiltsi) in the Tammiku Member of the Tamsalu Formation.

FIGURE 7.7. A cluster of *Pentameroides subrectus* preserved in growth position from the Upper Llandovery Johns Creek Quarry Member in the basal Scotch Grove Formation of eastern Iowa (FMNH UC 63682). Arrows point to individuals where the distinctive cruralium of the brachial valve is exposed. The dimensions of this block are 44 by 30 cm.

In the Rumba Formation, the oval outlines of truncated *Pentamerus* preserved in growth position may be observed directly on bedding planes on the floor of the Pari Quarry (located 5 km southwest of the village of Kullamaa and 1 km northwest of the Tallinn/Virtsu highway). Submarine abrasion scoured the anterior portions of most of the shells in their upright position, but some complete *Pentamerus* shells that fell into a horizontal position may be excavated from below the plane of abrasion. The locality also is very rich in tabulate

FIGURE 7.8. Bedding surface from the Ludlow Wabash Formation in Fort Wayne, Indiana, showing several *Kirkidium* sp. in growth position.

corals, indicating that strata were deposited under conditions locally oscillating between BA2 and BA3.

Figure 7.8 illustrates a bedding surface from the upper part of the Wabash Formation (Ludlow in age) with several articulated individuals of *Kirkidium* sp. in growth position. This field photo was taken in the Ardmore Avenue Quarry of the May Stone & Sand Company, in Fort Wayne, Indiana (NE 1/4, Sec. 29, T30N, R12E, Allen County). While preservation is imperfect, it is clear that some of the *Kirkidium* deposits in the so-called Fort Wayne Bank are biostromal in nature and include material in growth position. Similar relationships should be expected in some of the many other Ludlow/Pridoli pentamerids and in the much older Ashgill (Late Ordovician) pentamerids.

This paper provides a summary of all published material with references to Silurian pentamerid brachiopods preserved in growth position. The primary living conditions of *Virgiana decussata, Borealis borealis, Pentamerus oblongus, P. dorsoplanus,* and *Pentameroides subrectus* were in environments below fair-weather wave base (BA3), where large populations were typically so crowded together that individual shells matured while "standing" on the tip of the pedicle valve. Stratigraphically younger pentamerids, such as *Kirkidium* sp., sometimes acquired the same habit but also lived in protected biohermal settings. Although the preservation of natural populations is emphasized in this survey, pentamerid brachiopods also formed major bank deposits of the sort called "parautochthonous and allochthonous" fossil concentrations by Kidwell et al. (1986).

The information reviewed here is based on data from North America, northern Europe (British Isles, Norway, and Estonia), Siberia, and South China. During Silurian time, these regions formed part of what Boucot (1975) called the North Silurian Realm, as distinguished from the colder Malvinokaffric Realm, which is mainly composed of the Gondwanan continents. Pentamerid brachiopods are reported from New South Wales in Australia (Jenkins 1978) and from northern India (Goel et al. 1987) on the fringes of Gondwana. Presumably, these outlying areas were bathed by milder waters more associated with the North Silurian Realm. There is no record of Australian or Indian material preserved in growth position, but it is probable that such material will be recognized through future work.

ACKNOWLEDGMENTS

It is commonly said that Americans relish big, flashy things of all kinds. I cannot deny my attraction to the big-shelled pentamerid brachiopods. Most of my research proposals to study in various places where the pentamerids are found and most of the resulting articles I have authored on this subject were reviewed in one form or another by Arthur J. Boucot (University of Oregon). I am most grateful for his long-standing help and interest in these affairs and for his encouragement to complete this summary.

REFERENCES

Baarli, B. G. and M. E Johnson. 1988. Biostratigraphy of key brachiopod lineages from the Llandovery Series (Lower Silurian) of the Oslo Region. *Norsk Geologisk Tidsskrift* 68:259–74.

Beach, D. K. and R. N. Ginsburg. 1980. Facies succession of Pliocene-Pleistocene carbonates, Northwestern Great Bahama Bank. *American Association Petroleum Geologists Bulletin* 64:1634–42.

Beadle, S. C. and M. E. Johnson. 1986. Palaeoecology of Silurian cyclocrinitid algae. *Palaentology* 29:585–601.

Boucot, A. J. 1969. Stratigraphic range of the Early Silurian Virgianiinae (Brachiopoda). *Eesti NSV Teaduste Akadeemia Toimetised, Keemia Geoloogia* 1:76–79.

———. 1975. *Evolution and Extinction Rate Controls.* New York: Elsevier.

Boucot, A. J. and J. G. Johnson. 1979. Pentamerinae (Silurian Brachiopoda). *Palaeontographica A* 163:87–129.

Cocks, L. R. M. and W. S. McKerrow. 1984. Review of the distribution of the commoner animals in Lower Silurian marine benthic communities. *Palaeontology* 27:663–70.

Colville, V. R. and M. E. Johnson. 1982. Correlation of sea-level curves for the Lower Silurian of the Bruce Peninsula and Lake Timiskaming District (Ontario). *Canadian Journal Earth Sciences* 19:962–74.

Ehlers, G. M. 1973. Stratigraphy of the Niagaran Series of the Northern Peninsula of Michigan. *Museum of Paleontology (University of Michigan) Papers on Paleontology* No. 3.

Fürsich, F. T. and J. M. Hurst. 1980. Euryhalinity of Palaeozoic articulate brachiopods. *Lethaia* 13:303–12.

Goel, R. K., M. Kato, A. K. Jain, and S. S. Srivastavea. 1987. Fauna from the "Muth Quartzite," Garhwal Himalaya, India. *Journal Faculty of Science, Hokkaido University* 22:247–57.

Jaanusson, V. 1981. Functional thresholds in evolutionary progress. *Lethaia* 14:251–60.

Jenkins, C. J. 1978. Llandovery and Wenlock stratigraphy of the Panuara Area, central New South Wales. *Proceedings Linnean Society New South Wales* 102:109–30.

Johnson, M. E. 1975. Recurrent community patterns in epeiric seas: The Lower Silurian of eastern Iowa. *Proceedings Iowa Academy of Science* 82:130–39.

———. 1977. Succession and replacement in the development of Silurian brachiopod populations. *Lethaia* 10:83–93.

———. 1979. Evolutionary brachiopod lineages from the Llandovery Series of eastern Iowa. *Palaeontology* 22:549–67.

———. 1980. Paleoecological structure in early Silurian platform seas of the North American midcontinent. *Palaeogeography, Palaeoclimatology, Palaeoecology* 30:191–216.

———. 1989. Tempestites recorded as variable *Pentamerus* layers in the Lower Silurian of southern Norway. *Journal of Paleontology* 63:195–205.

Johnson, M. E. and G. T. Campbell. 1980. Recurrent carbonate environments in the Lower Silurian of northern Michigan and their inter-regional correlation. *Journal of Paleontology* 54:1041–57.

Johnson, M. E., L. R. M. Cocks, and P. Copper. 1981. Late Ordovician-Early Silurian fluctuations in sea level from eastern Anticosti Island, Quebec. *Lethaia* 14:73–82.

Johnson, M. E. and V. R. Colville. 1982. Regional integration of evidence for evolution in the Silurian *Pentamerus-Pentameroides* lineage. *Lethaia* 15:41–54.

Johnson, M. E., D. Kaljo, and J.-Y. Rong. 1991. Silurian eustasy. In M. G. Bassett, P. D. Lane, and D. Edwards, eds., *The Murchison Symposium: Proceedings of an International Conference on the Silurian System. Special Papers in Palaeontology* No. 44, pp. 145–63.

Johnson, M. E. and H. L. Lescinsky. 1986. Depositional dynamics of cyclic carbonates from the Interlake Group (Lower Silurian) of the Williston Basin. *Palaios* 1:111–21.

Johnson, M. E., J.-Y. Rong, and X.-C. Yang. 1985. Inter-continental correlation by sea-level events in the Early Silurian of North America and China (Yangtze Platform). *Geological Society America Bulletin* 96:1384–97.

Kaljo, D. and H. Nestor, eds. 1990. *Field Meeting Estonia 1990, An Excursion Guidebook.* Tallinn: Institute of Geology, Estonian Academy of Sciences.

Kiaer, J. 1908. Das Obersilur im Kristianiagbiete. *Skrifter Det Norske Videnskaps-akademi i Kristiania, Matematisk-Naturvidenskapelig Klasse 1906* (2) *Kristiania,* 595p.

Kidwell, S. M., F. T. Fürsich, and T. Aigner. 1986. Conceptual framework for the analysis and classification of fossil concentrations. *Palaios* 1:228–38.

Mørk, A. 1981. A reappraisal of the lower Silurian brachiopods *Borealis* and *Pentamerus. Palaeontology* 24:537–53.

Roehl, P. O. 1967. Stony Mountain (Ordovician) and Interlake (Silurian) facies analogs of recent low-energy marine and subaerial carbonates, Bahamas. *American Association Petroleum Geologists Bulletin* 51:1979–2032.

Rong, J.-Y., M. E. Johnson, and X.-C. Yang. 1984. Early Silurian (Llandovery) sea-level changes in the Upper Yangtze Region of central and southwestern China. *Acta Palaeontologica Sinica* 23:672–93.

Rong, J.-Y., B. Jones, and F. W. Nentwich. 1989. *Proconchidium* from late Ordovician strata of Brodeur Peninsula, Baffin Island, Arctic Canada. *Journal of Paleontology* 63:25–33.

Sheehan, P. M. 1975. Brachiopod synecology in a time of crisis (Late Ordovician–Early Silurian). *Paleobiology* 1:205–12.

Sowerby, J. 1813. *The Mineral Conchology of Great Britain* 1:1–234 .

St. Joseph, J. K. S. 1938. The Pentameracea of the Oslo Region. *Norsk Geologisk Tidsskrift* 17:225–336.

Ziegler, A. M. 1965. Silurian marine communities and their environmental significance. *Nature* 207:270–72.

Ziegler, A. M., A. J. Boucot, and R. P. Sheldon. 1966. Silurian pentameroid brachiopods preserved in position of growth. *Journal of Paleontology* 40:1032–36.

Ziegler, A. M., L. R. M. Cocks, and R. K. Bambach, 1968. The composition and structure of Lower Silurian marine communities. *Lethaia* 1:1–27.

8

The Homocrinus *Beds: Silurian Crinoid Lagerstätten of Western New York and Southern Ontario*

Carlton E. Brett and Wendy L. Taylor

ABSTRACT

Obrution deposits, reflecting nearly instantaneous burial of communities, provide brief "snapshots" of ancient seafloor conditions. Such obrution horizons may occur nonrandomly in restricted transitional portions of transgressive/regressive depositional cycles.

The middle Lewiston Member of the Rochester Shale (Silurian, Wenlockian) in western New York and southern Ontario contains several well-developed obrution-bearing intervals of echinoderm *Lagerstätten* (*Homocrinus* beds) that are traceable along depositional strike for more than 80 km. These intervals display an abundance of completely articulated fossil crinoids, cystoids, edrioasteroids, starfish, trilobites, and other delicate fossils. The lower *Homocrinus* interval comprises a 0.5-m transitional zone between bryozoan-rich fossiliferous mudstones and packstones and the sparsely fossiliferous middle Lewiston Member shales. Upper, 1–2-m-thick obrution-bearing intervals, with similar faunas and nearly identical taphofacies, occur at the opposite transition from barren shales to richly fossiliferous bryozoan-rich mudstones. Fossils are preserved at the bases of thin (up to 10 cm) siliciclastic mudstone and carbonate silt beds representing distal tempestites. The mudstones, which comprise the bulk of Rochester Shale thickness, accumulated in brief pulses spanning hours to days. Muds were deposited rapidly as flocculated aggregate grains. Disturbance was minimal but included toppling and inversion of shells and minor disarticulation prior to burial. Intervening thin shell pavements record "background" periods of time-averaged skeletal accumulation and taphonomic feedback.

The constrained position of the obrution layers within the lithofacies and biofacies spectrum (*Striispirifer* brachiopod community) suggests coincidence of ecological and taphonomic effects. The fauna represents a distinctive ecotonal community occurring between optimal shallow water conditions and deeper, more dysaerobic, turbid water settings where few organisms lived. The most favorable conditions for burial of these communities existed near the lower end of storm wave base, where relatively thick distal mud flows were frequently generated but where energy conditions precluded later reworking. The obrution horizons are particularly well developed in the lower transgressive interval, consisting of a condensed section recorded by only 5–7 stacked burial layers. The upper *Homocrinus* interval is more diffuse, indicative of more persistent background sedimentation.

Fossil *Lagerstätten* have received considerable attention in recent literature (see, for example, papers in Whittington and Conway Morris 1985; Allison 1986, 1988a, 1988b, 1990; Gould 1989; Butterfield 1990; Allison and Briggs 1991). Obrution deposits, reflecting nearly instantaneous burial of organism communities, are among the most abundant of fossil *Lagerstätten,* but until recently these deposits have been less well documented than the famous soft-bodied occurrences (see Brett and Seilacher [1991] for a summary). Although most obrution layers do not preserve soft parts of organisms, they commonly reveal exquisite details of multielement skeletons (e.g., echinoderms, arthropods, or vertebrates) that are rarely seen fully articulated. Beds of intact arthropods and echinoderms are among the classic obrution deposits (see Seilacher et al. 1985; Speyer and Brett 1985; and Speyer 1988). For example, the famous Mississippian Crawfordsville crinoid beds of the North American mid-continent have been studied extensively and have revealed important aspects of community paleoecology and paleoautecology of crinoid-dominated communities (Lane 1963, 1969, 1973; Ausich et al. 1979; Ausich 1980). Such horizons provide exceptionally well-resolved "snapshots" of skeletonized components of seafloor communities and thus are of considerable paleobiological importance, although in most cases their full potential has not been realized. Smothered bottom assemblages represent census samples of particular facies whose type and mode of formation can be predictably related to sedimentary environments. Furthermore, obrution deposits are instantaneous event horizons that have potential utility as refined markers for high-resolution event stratigraphy (see Kauffman 1986, 1988). In this regard, obrution *Lagerstätten* have received very little attention. Our inherent bias would lead us to suspect that extraordinary burial circumstances might be spatially restricted, limited to small patches of seafloor that happen to be rapidly covered by sediment, and this is partially correct. Most obrution deposits are patchy on a scale of meters within outcrops. However, if a large enough outcrop is available, it is typically observed that the patches are parts of a larger, nearly tabular unit of overall well-preserved fossil remains. If obrution deposits are the result of particularly "high-intensity" rare events, we might expect that they would have widespread signatures. In a few cases this has been documented. Parsons et al. (1988), for example, report Devonian shell and crinoid-rich layers that appear to be widespread on a scale of several tens of kilometers. Brett et al. (1986) reported a trilobite-rich layer, in which all trilobite individuals showed contorted orientation, that was traceable for about 100 km in the Devonian of New York State. This persistence of horizons, of course, leads to questions of sedimentary and biological processes that would be responsible for the formation of a rather uniform layer over a broad region. How much spatial and temporal resolution may be preserved in individual obrution layers remains an open question.

A last point that needs further study is the relationship of obrution deposits to epiboles or widespread thin horizons characterized by an anomalous abundance of otherwise rare or absent fossil species. Clearly, in some instances the two may coincide almost perfectly. Fragile skeletons may be present, at least as identifiable remains, only under extraordinary burial conditions of the sort that lead to obrution *Lagerstätten.* Brett, Miller, and Baird (1990) coined the term *taphonomic* epiboles to describe this sort of occurrence. In other cases, however, the general environments that favor the development and/or preservation of obrution layers may also possess distinctive combinations of physical and chemical factors that facilitate colonization by normally rare organisms. These would be true ecological epiboles (see Brett, Miller, and Baird 1990).

Historical Overview

Herein we describe extraordinary fossil beds from the Silurian Rochester Shale of New York and Ontario that appear to reflect a limited number of widespread burial events. Although he did not state precisely the horizon of his fossils, it is evident that James Hall (1852, 1860) was among the first to collect from layers about 5 m above the base of the Rochester Shale in the Erie Canal locks at Lockport, N.Y., which were later designated the *Homocrinus* band (Ringueberg 1888). Hall described forms such as *Homocrinus parvus* and *Hemicystites parasiticus,* which are virtually restricted to this level. Dr. E. N. S. Ringueberg, a contemporary of James Hall and a Lockport physician, was the first person to attempt faunal subdivision of the Rochester Shale. At Lockport Gulf (the gully of 18-Mile Creek) he recognized three unequal divisions, which he later termed the *lower, middle,* and *upper thirds,* each characterized by a distinctive suite of fossils. Of these the lower two correspond roughly to the Lewiston Member of the Rochester, while the upper third represents the Burleigh Hill Member (Brett 1983a). Ringueberg (1888) placed the break between the lower and middle third at the top of a fossiliferous horizon, which he termed the *Homocrinus* band because of the abundance of the inadunate crinoid *Homocrinus parvus* Hall at this level. It is not entirely clear whether Ringueberg meant to imply a single bedding plane or several layers by the term *band*. But he does mention that well-preserved fossils occurred "within a few feet in either direction from the *Homocrinus* band." Ringueberg (1888, pp. 268–70) described the *Homocrinus* band fauna in considerable detail and listed 12 species of echinoderms that characterize this horizon: *Homocrinus parvus*, four species of "*Lecanocrinus*" (includes *Asaphocrinus*, two of the species have subsequently been synonymized with *A. ornatus* (Hall)), *Caryocrinites ornatus* Say, three stelleroids and the edrioasteroid *Hemicystites parasiticus* Hall (fig. 1). He also noted the abundance of the brachiopod *Striispirifer niagarensis* (Conrad), which he found to be "gregarious, living in small-sized colonies so that specimens at times completely cover slabs of shale, with comparatively few individuals in the intervening spaces."

Further extensive collecting from the Lockport Gulf crinoid beds was undertaken by Frederick Braun, who was employed by the great crinoid taxonomist Frank Springer in the summers of 1911 and 1914. To obtain additional specimens of crinoids and cystoids, Braun used dynamite to excavate quarried openings at the site of Ringueberg's earlier discoveries and at two other locations farther downstream. Braun's operations resulted in the discovery of numerous well-preserved echinoderm specimens that are now deposited in the Springer Collection of the U.S. National Museum. A greater number of specimens was obtained by Braun than had been amassed by the combined efforts of Hall, Ringueberg, and other Lockport collectors. However, aside from detailed work on the myelodactylids and *Asaphocrinus* specimens (Springer 1920, 1926), relatively little has been published on this material.

The *Homocrinus*-bearing horizons contain exquisitely preserved cystoids, edrioasteroids, starfish, trilobites, graptolites, and other fragile fossils. The general state of preservation for these fossils is remarkable, given that the same species from other horizons rarely, if ever, show completely articulated conditions. For example, in more than two decades of study of the other beds of the Rochester Shale, the senior author has observed only some 20 specimens out of more than 1,000 of the cystoid *Caryocrinites* with intact columns and still fewer with articulated arms (see fig. 8.1). Yet absolutely complete specimens are abundant and even the rule in some layers of the *Homocrinus* intervals. In the sense of comparative taphonomy, the *Homocrinus* intervals represent a distinctive taphofacies of the Rochester Shale characterized by a very high proportion of articulated and in situ fossils with little breakage and minor evidence for reworking. Other fossil beds in the Rochester Shale appear to represent more standard "disturbed neighborhood assemblages" that have relatively high proportions of disarticulated shells and multielement skeletons. These beds display evidence of more breakage, abrasion, and encrustation, and they possess biased ratios of brachiopod valves and trilobite tagmata. This evidence suggests recurring conditions, briefly developed during the deposition of the *Homocrinus* intervals, that were more favorable to fossil preservation than during the accumulation of other parts of the Rochester Formation.

In 1978 the senior author rediscovered the site of the original Ringueberg

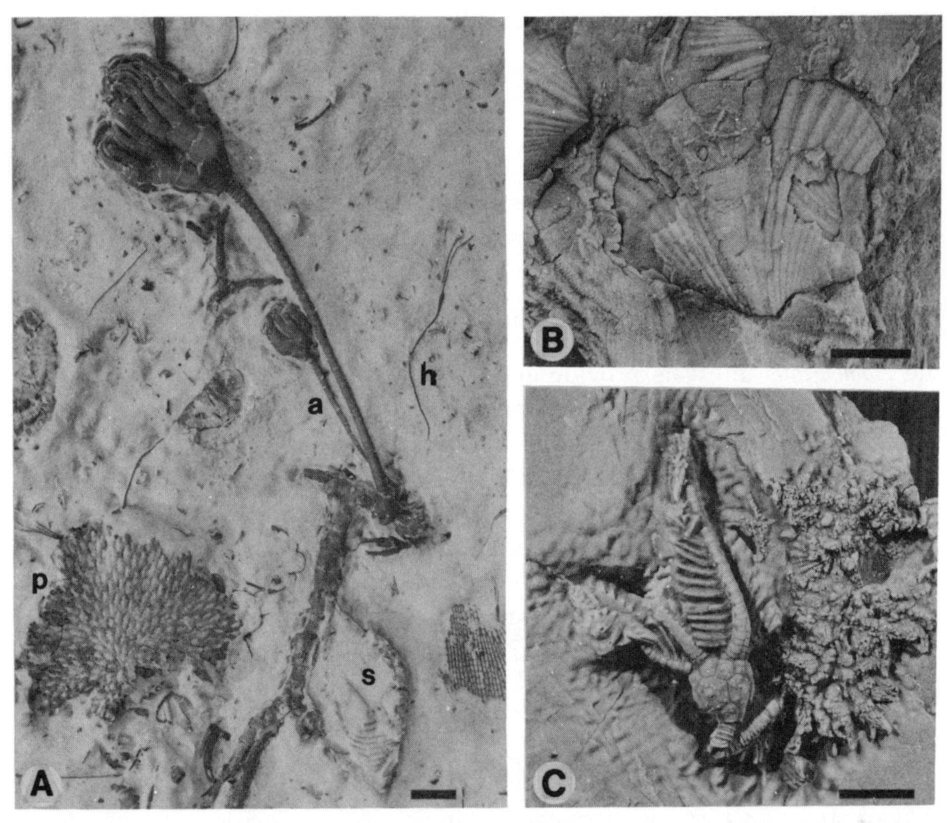

FIGURE 8.1. Representative fossils from the Rochester Shale lower *Homocrinus*-bearing obrution interval; (A) upper surface of a slab displaying intact paleocommunity with two complete specimens of *Asaphocrinus ornatus* (Hall) (a) of differing size attached by discoidal holdfasts to a ramose bryozoan, (p) an intact colony of the bryozoan *Phylloporina asperato-striata* (Hall), (s) truncated but in situ, articulated *Striispirifer niagarensis* (Conrad), note intact spiralia, (h) crown and partial stem of *Homocrinus parvus* Hall, ROM 48069; (B) group of *Striispirifer niagarensis* valves, note specimen of the edrioasteroid *Hemicystites parasiticus* Hall attached to the valve in center; (C) small individual of the rhombiferan *Caryocrinites ornatus* Say, note the "arms" of differing length with intact brachioles, USNM 449447. Scale bars are 1 cm.

and Braun excavation at Lockport Gulf and, with James Eckert, collected a considerable amount of new material, now mainly housed in the Royal Ontario Museum. Subsequently, several additional outcrops of the lower *Homocrinus*-bearing beds were found at the Lewiston B/C submember boundary. This extended the range beyond the original site by more than 50 km to the west and 80 km to the east. The most notable eastern outcrop at Middleport was collected by James Eckert, Denis Tetreault, James Nardi, and others. In the summer of 1990 it was systematically excavated in a joint study involving the University of Rochester and the Smithsonian Institution. Furthermore, in 1989 two additional higher *Homocrinus*-bearing intervals were discovered in the Rochester Shale at three newly excavated shale pits in Orleans and Monroe counties. These upper *Homocrinus* intervals bear a similar fauna that occurs at the opposite transition from barren shales of the C submember to richly fossiliferous bryozoan-rich mudstones (D submember). The recurrent *Homocrinus* intervals provide an opportunity to assess what is general and what is truly unique to these tapho- and biofacies.

Detailed microstratigraphy of the *Homocrinus* intervals at several localities along the generally east-to-west-trending Silurian outcrop belt in New York and Ontario reveals a very similar succession of beds. While there are local differences in the appearance of the beds, their general pattern is so similar that we can conclude that the same events may be represented at many localities. Thus the *Homocrinus* intervals provide an exceptionally well-resolved time line for local stratigraphy. This leads us also to speculate on the possible causes for widespread occurrence of obrutionary layers.

In the following sections we describe the regional stratigraphy, fauna, taphonomy, and paleoecology of the original and recurrent *Homocrinus* beds. We then present a detailed interpretation of seafloor conditions at the time of deposition and a qualitative model for the development of widespread obrution-rich taphofacies. This study serves to underscore the paleobiological, taphonomic, and time stratigraphic significance of obrution layers and epiboles. Similar approaches should yield valuable new data when applied to other classic *Lagerstätten*.

Study Area and General Stratigraphy

The obrution deposits described herein were studied at a total of 10 localities along the east-to-west-trending Silurian outcrop belt (Niagara escarpment). Sections were located in Monroe, Orleans, and Niagara counties of western New York State and in Niagara and Wellington counties in adjacent Ontario, Canada (fig. 8.2). The lower horizon has been identified and studied primarily in the western localities (locations 4–9). Unfortunately, the interval containing this band is very poorly exposed or inaccessible east of Niagara County. The opposite is true for the upper horizons, which are well represented east of Niagara County, but generally occurs in cliff exposures to the west and so is poorly known there. Consequently, three levels can be studied in detail together only in the Middleport area. However, detailed physical stratigraphic correlation of lower and upper intervals is possible at least from Monroe County, New York, westward to near St. Catharines, Ontario.

In the study area the Rochester Shale consists of 15–25 m of medium gray, calcareous to dolomitic mudstone with numerous thin limestone interbeds. It is the upper portion of the Upper Clinton Group and has recently been interpreted as a third-order depositional sequence (Sequence V) by Brett, Goodman, and LoDuca (1990). The Rochester Shale in western New York is subdivisible into two members, a lower or Lewiston Member and an upper Burleigh Hill

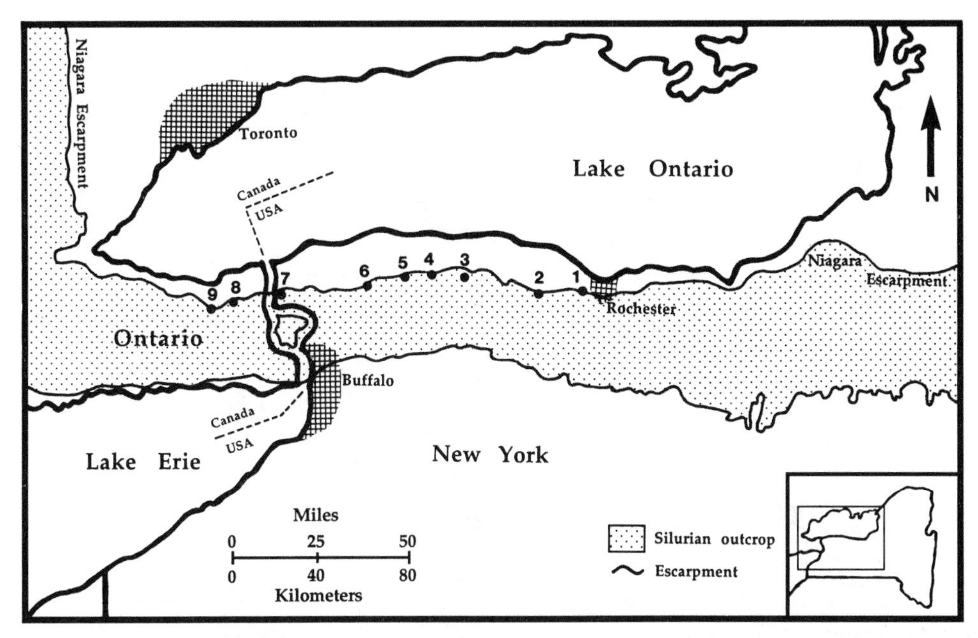

FIGURE 8.2. Locality map of southern Ontario and western New York showing major study locations (numbered) along the Silurian outcrop belt (stippled). See appendix for locality descriptions.

Member. All *Homocrinus* horizons discussed herein occur within the Lewiston Member of the Rochester Shale defined by Brett (1983a). This member consists of medium gray, calcareous mudstones with intervals of closely interbedded bryozoan- and brachiopod-rich mudstone or limestone; beds range from less than 1 to 10 cm thick. Other intervals contain barren shales with laminated calcisiltite up to 10 cm thick. In some intervals, particularly near the base and at the top of the Lewiston Member, bryozoan-rich coquinites are abundant. Brett (1983a) subdivided the Lewiston Member into five informal submembers designated A–E (fig. 8.3). A thin (0.5–1.0-m), basal brachiopod-rich hash, transitional with the underlying Irondequoit Limestone, comprises the A submember. Unit B is a thicker interval, up to 6 m thick, containing abundant lenses of ramose bryozoan, brachiopod, and pelmatozoan debris interbedded with fossiliferous mudstone. The middle portion of the Lewiston Member comprises an interval of nearly barren shale with a few thin calcisiltite bands, designated the C submember; this interval ranges from 2 to 6 m in thickness. A variable upper interval 3–5 m thick, somewhat resembling the B submember, comprises the D interval and is characterized by bryozoan-rich mudstone beds. Finally, an interval of bryozoan- and crinoid-rich packstone ("bryozoa beds" of Grabau 1901) comprises the Lewiston E, which caps the Lewiston Member and typically displays a sharp upper contact with the overlying Burleigh Hill barren shales.

The *Homocrinus*-bearing beds described in this paper occur, respectively, at the base and near the top of the sparsely fossiliferous middle C submember of

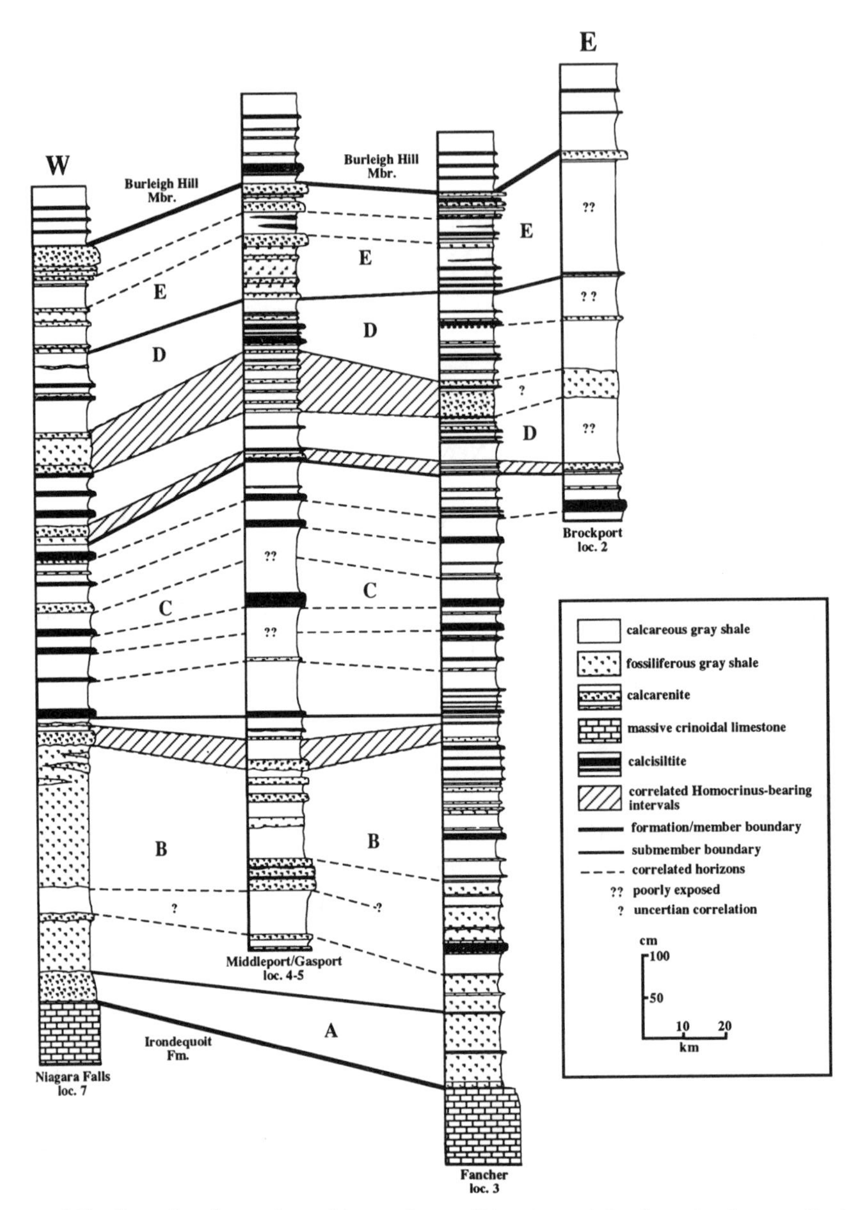

FIGURE 8.3. Correlated stratigraphic sections of Lewiston Member, Rochester Shale for selected localities in western New York. Location of sections shown in figure 8.1.

the Lewiston Member. The lower and better-known horizon, described originally by Ringueberg (1888), occupies 15–30 cm of transitional strata overlying the highest bryozoan-rich biostromal layer of the B submember. The newly discovered upper *Homocrinus* intervals are less well defined but appear to comprise about 1–2.5 m of transitional shale in the upper portion of the C submember, extending up to the base of *Striispirifer* shell beds of the D submember.

Geologic and Paleoenvironmental Setting

The paleoenvironmental setting of the Rochester Shale was discussed in detail by Brett (1983b). The siliciclastic muds derived from the erosion of tectonic highlands to the southeast in New York State were deposited in the northern, apparently blind, end of the Appalachian Foreland Basin whose axis trended approximately from northeast to southwest. These sediments accumulated in a shallow subtropical muddy sea approximately 15–20 degrees south of the paleoequator (Ziegler et al. 1977; Witzke 1990). Brett concluded that the local facies strike in western New York and Ontario was approximately east to west, paralleling the modern Niagara escarpment outcrop belt. This gives the appearance of "layer cake" stratigraphy within the escarpment outcrop belt. However, fossiliferous facies of the Lewiston B and D intervals generally pass southward along the Niagara Gorge into barren shales with only a few minor fossiliferous beds. This facies transition, which occurs within about 8 km, suggests the presence of a southward-dipping ramp during deposition of the Lewiston Member in this area. The shallower portions of the Rochester Shale were deposited in a muddy outer shelf, in perhaps 20–50 m of water, below normal wave base but strongly influenced by episodic storms. The interbedded limestones within the Rochester Shale appear to be tempestites or storm deposits (Cant 1980; Aigner 1982, 1985; Wu 1982; DeCelles 1987; Baarli 1988; and Duke 1990; Jennette and Pryor 1993). Winnowed bryozoan-rich coquinites were generated as proximal deposits by the aggradation and winnowing of skeletal material in areas affected directly by storm waves. Calcisiltites and muddy siltstone beds appear to be distal silt layers that were deposited by gradient or density currents slightly below storm wave base. The occurrence of small-scale hummocky cross-stratification on the tops of some of the thicker calcisiltites suggests that these areas were also influenced by the deepest parts of storm waves. Obrution layers described herein are interpreted generally as the result of distal mud tempestites similar to those described in the deeper shelf environment (greater than 30 m) of the modern Helgoland Bight by Aigner and Reineck (1982).

The Rochester Shale was deposited during a relatively major transgressive/regressive cycle that spans from the Irondequoit Limestone through the Rochester mudstones and back again to high-energy grainstones of the lower Lockport Group. This is interpreted as a third-order sequence, with the Irondequoit representing transgressive systems tract deposits and the Rochester Shale representing highstand deposits (Brett, Goodman, and LoDuca 1990). The Irondequoit through Lewiston Member represents a smaller-scale subsequence. In turn, the submember divisions of the Lewiston (A–E) are interpreted by Brett, Goodman, and LoDuca (1990) as the product of still smaller-scale, perhaps fifth-order, cycles of relative sea-level change. The middle Lewiston barren shales presumably reflect maximum water depths in a lower aerobic to dysaerobic muddy basinal setting below storm wave base (Brett 1983; Tetreault 1994).

Fossiliferous, bryozoan-rich mudstones and packstones that border this central barren zone reflect shallower water deposits formed during early and later parts of the transgressive/regressive cycle (Tetreault 1994). The lower *Homocrinus* beds are interpreted as the result of episodic deposition during a phase of relatively rapid deepening into the C submember. There shales overlie a slightly condensed interval represented by the densely packed bryozoan and cystoid layers that cap the B submember.

The upper *Homocrinus* beds occupy an approximately mirror-image position reflecting gradual shallowing into the D submember. This is associated with a later relative highstand interval of Rochester deposition and may reflect a combination of real shallowing and/or progradation of muds. This interval is similar to but somewhat less condensed than the lower one. Both are characterized by an abundance of the brachiopod *Striispirifer niagarensis* and by the occurrence of obrution layers that provide exceptional preservation.

Detailed examination of lower, middle, and upper *Homocrinus*-bearing intervals reveals a remarkable similarity between outcrops separated by tens of kilometers. The nearly identical stacking order of three types of beds (shell debris beds, calcisiltites, and barren mudstones), similarities in relative thickness and spacing of intervals, consistency of appearance of obrution layers with unique taphonomic features, and position in larger-scale cyclic successions all support a correlation of individual shelly layers (at least the thickest ones) and intervening event beds (mudstones and calcisiltites) over distances of up to 80 km. This finding is consistent with observations of Brett et al. (1986) and Parsons et al. (1988) of wide lateral extent of shell beds and mudstones in the Devonian Hamilton Group. However, it should also be noted that most of these layers are only known from the main east–west-trending Niagara Escarpment, which is apparently oriented nearly parallel to depositional strike for the Rochester Shale (i.e., perpendicular to southward-sloping submarine ramp). Attempts to trace the beds across depositional strike have not been as successful. For example, presumably deeper-water facies in the southern end of Niagara Gorge (about 8–10 km south of the escarpment) or in the thickened sections associated with the Rochester Shale depocenter in Monroe County have not yielded these layers to date.

Lower *Homocrinus* Interval

Figure 8.4 illustrates the microstratigraphy of the lower *Homocrinus* interval at four sections in Ontario and western New York. Subsequent to their rediscovery at Lockport, the lower *Homocrinus* beds have been identified at five other localities both to the east and west of this type area in Ontario and western New York.

The westernmost probable location of the lower *Homocrinus* interval is at Jordan, Ontario (16 Mile Creek). Brett and Eckert (1982) described a horizon 1.4 m above the base of the Rochester Shale at this outcrop. A bed approximately

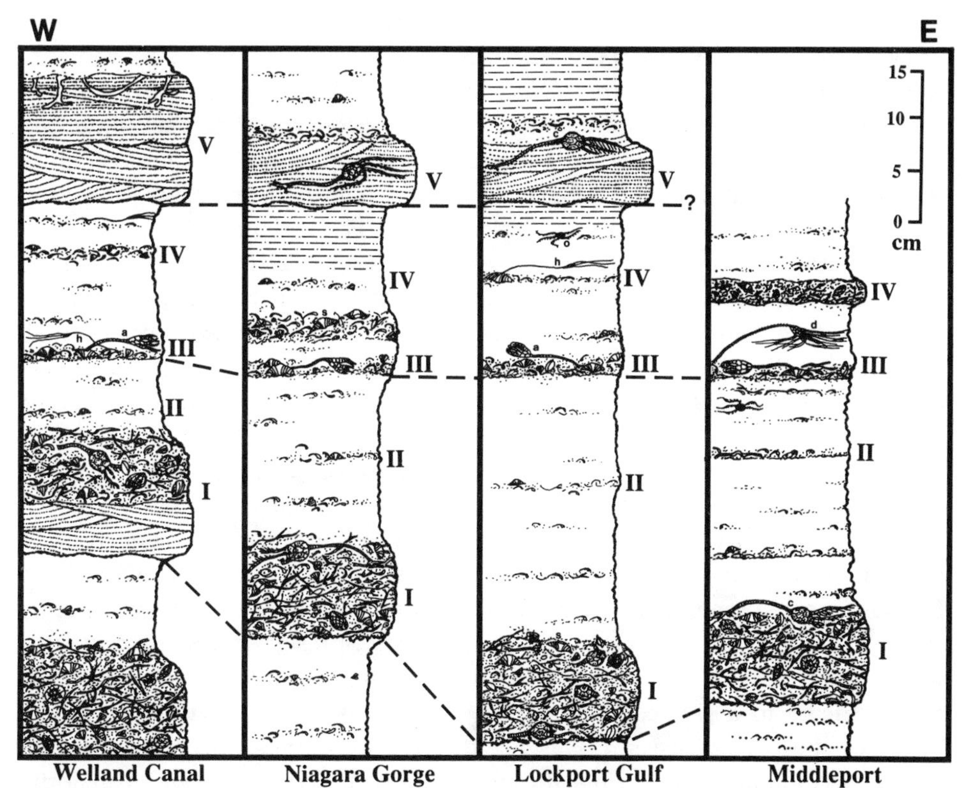

FIGURE 8.4. Microstratigraphy of the lower *Homocrinus* interval at four selected locations in western New York; the following characteristic fossils are denoted as *Asaphocrinus* (a), *Caryocrinites* (c), *Dendrocrinus* (d), *Homocrinus* (h), ophiuroid (o) and *Striispirifer* (s). Note the nearly identical succession of beds.

1–2 cm thick contains a layer with numerous complete *Eucalyptocrinites, Dimerocrinites,* and *Homocrinus* individuals. A single complete specimen of *Asaphocrinus* and one of *Dendrocrinus longidactylus* were obtained 2.16 m and 1.68 cm, respectively, above the *Eucalyptocrinites*-bearing horizon.

The lower *Homocrinus* interval is well exposed in bedding planes along the floor of the Welland Canal and at DeCew Falls Hydroelectric plant at St. Catharines, Ontario (loc. 9), where *Homocrinus parvus* was obtained from shale about 3.4 m above the base of the Lewiston Member at the boundary between bryozoan-rich biostromes (submember B) and sparsely fossiliferous mudstones with thin calcisiltites (submember C). *Striispirifer* shell pavements with numerous complete *Asaphocrinus, Macrostylocrinus, Caryocrinites,* and other pelmatozoans occur on two bedding planes a few centimeters above the highest relatively thick bryozoan biostrome. A 10-cm-thick bed composed of two amalgamated calcisiltites occurs about 30 cm above the bryozoan-rich layer. A few specimens of *Homocrinus* occurred on the base of and just above the calcisiltite.

Strangely, the *Homocrinus* fauna is poorly represented in the extensive exposures of Rochester Shale at the north end of Niagara Gorge. Hints of the fauna are known. Rare specimens of the crinoid *Homocrinus* and the extraordinarily well-preserved *Icthyocrinus* and *Caryocrinites* have been collected from shales near the Lewiston B/C boundary at this section. A compact packstone bed comprised mainly of *Striispirifer* valves occurs at approximately the position of the Welland Canal *Striispirifer* beds. Again, this bed is overlain by shale with very thin fossil hash layers, and this in turn is overlain by a 10-cm-thick amalgamated bed that clearly resembles the marker calcisiltite of the Welland Canal. No extensive beds of *Homocrinus* and *Asaphocrinus* have been located at Niagara Gorge. However, the occurrence of nearly identical successions of fossil beds with the same rare fossils both west (St. Catharines, Ontario) and east (Lockport) of the Niagara Gorge suggests that the interval was once continuous through the Niagara region.

The lower *Homocrinus* interval is classically exposed at Lockport, along 18 Mile Creek and in the Erie Canal Locks cut (Ringueberg, 1888). As at St. Catharines, the lower *Homocrinus* fauna occurs at the transition between bryozoan-rich fossiliferous lower mudstones of the B submember and the sparsely fossiliferous C submember of the Lewiston Member. Detailed microstratigraphy of the Lockport Gulf reveals assemblages of well-preserved fossils in at least four discrete levels within the 30–40 cm B/C transition interval; *Homocrinus* is not restricted to a single bedding plane but occurs throughout this interval. Three of these fossil assemblages are directly associated with horizons containing accumulations of *Striispirifer* shells and scattered bryozoans. A majority of the characteristic *Homocrinus* fauna fossils, including *Homocrinus parvus, Asaphocrinus, Hemicystites* and stelleroids, occur on two bedding planes in a 25–30-cm-thick interval that underlies a 6-cm marker calcisiltite resembling that seen at Niagara (unit V of fig. 8.4) and St. Catharines. Fossils occur in a tough medium-gray, silty mudstone. The base of this lower *Homocrinus*-bearing interval is marked by a very fossiliferous bedding plane (top of B submember) containing densely packed bryozoans and spiriferid and atrypid brachiopods.

The upper surface of the calcisiltite appears slightly gradational through about 0.5 cm into the overlying shale. Few fossils occur within the calcisiltite, and none were observed on the lower surface of this bed, but three well-preserved *Caryocrinites* were collected from the upper surface of the layer. Single specimens of *Caryocrinites, Homocrinus,* and *Asaphocrinus* were obtained from this layer; otherwise the fauna consists of brachiopods. Higher beds are largely covered by slumped debris but appear to be barren mudstone.

East of Lockport, poor exposures at Cottage Road Creek near Gasport, N.Y. (loc. 5), display a microstratigraphy similar to that seen at Lockport and elsewhere. Again, the uppermost bryozoan biostrome of the Lewiston B submember is overlain by about a 30-cm succession of more sparsely fossiliferous beds containing an abundance of the brachiopod *Striispirifer.* Several complete specimens of the trilobite *Calymene* and abundant fragments of *Arctinurus* have been found.

Finally, a similar succession of beds is observed at Middleport, N.Y. (loc. 4), where the *Homocrinus*-bearing strata occur about 6 m above the base of the Rochester Shale. Once again, a basal unit consists of a 10–12-cm-thick layer packed with ramose bryozoans, atrypids, and spiriferids. Its upper surface is overlain by 5–30 cm of sparsely fossiliferous mudstone that contains, near the base, exceptionally well-preserved specimens of *Caryocrinites,* many of which are attached to bryozoans in the underlying biostrome. This *Caryocrinites* bed is overlain by an interval of 25–30 cm of sparsely fossiliferous mudstone. This interval apparently constitutes two or three thick mud tempestites. Thin (1–2 mm) sole-marked calcisiltite laminae, overlain by skeletal debris, are noted 5 cm and 15 cm above the top. These levels have yielded nearly complete *Arctinurus* trilobites. The sparsely fossiliferous interval is capped by a complex shell bed displaying slightly undulatory load-casted base. The upper surface had yielded well-preserved crinoids, including *Asaphocrinus, Macrostylocrinus,* and *Homocrinus;* well-preserved *Dendrocrinus* occur in the overlying mudstone (see fig. 8.4). This level may correspond to the main *Striispirifer-* and *Asaphocrinus*-bearing bed at Lockport, although it does not display the high-density patches of *Striispirifer.* Higher beds are not exposed at Middleport, and the calcisiltite has not been identified.

Although the lower *Homocrinus* interval is poorly exposed east of Middleport, a drill core recently obtained from the region of Fancher, Orleans County, N.Y. (Howard Farms Inc. pit), displays a very similar succession of beds at the top of the B submember of the Lewiston Member. These beds occur at 5.6 m above the Irondequoit Limestone, a position very similar to that measured at Lockport (5.15 m) and Middleport (?5.15–5.45 m) to the west.

Thus, in summary, a consistent microstratigraphy typifies the lower *Homocrinus* beds from at least St. Catharines, Ontario, to Middleport, N.Y., and perhaps to Fancher, N.Y., a lateral distance of 80–120 km along the Niagara Escarpment. In all these areas the succession includes the following:

1. A relatively thick bed of brachiopod, bryozoan, and pelmatozoan debris, locally developed into biostromes of ramose and fenestrate bryozoans and everywhere having an abundance of the rhombiferan *Caryocrinites.*
2. A silty mudstone up to 10 cm thick containing at its base extraordinarily well-preserved *Caryocrinites* and other pelmatozoans, and at its top a very thin skeletal debris layer that has yielded exceptionally well-preserved *Arctinurus* trilobites at Middleport and Lockport.
3. A silty mudstone up to 15 cm thick split by another thin shelly debris layer at Middleport.
4. A thicker (up to 5 cm thick) layer typically with a sharp erosional to burrowed base and carrying a great abundance of *Striispirifer* together with localized bryozoan thickets and diverse fossils. The buried top of this bed has yielded some of the best crinoids at Middleport and St. Catharines. In particular, over this entire distance this bed yields completely articulated *Asaphocrinus, Macrostylocrinus,* and others. Complete *Dendrocrinus* and ophiuroids occur in the mudstone (up to 6 cm thick) that blankets this bed.

5. An upper *Striispirifer*-bearing hash bed that at Lockport has yielded some of the best *Asaphocrinus* at its top. It is a hash layer (or thin calcarenite) at Middleport but a thin *Striispirifer* bed at Niagara and St. Catharines.
6. An upper barren silty mudstone that has yielded scattered specimens of ophiuroids and *Homocrinus* at all localities.
7. Double calcisiltite beds that yield rare cystoids.
8. One or two thin hash beds with *Homocrinus* and other fossil debris in the 10 cm overlying the calcisiltite.

MIDDLE *HOMOCRINUS* INTERVAL

Until recently, the lower *Homocrinus*-bearing interval described in the preceding section was thought to be the only occurrence of this minute crinoid in the New York area. However, late in 1989, excavations for a supermarket (Wegman's Plaza) at Brockport, N.Y. (loc. 2), revealed two new and apparently higher occurrences of *Homocrinus parvus* and a large number of other fossil taxa shared with the lower *Homocrinus* beds (table 8.1). These fossils, particularly the echinoderms and trilobites, were well preserved in an exquisitely articulated condition. Indeed, the taphonomic conditions as well as the paleoecology of the upper horizons appear to have been extremely similar to the lower, although there are significant differences.

The interval occurs within the upper half of the Lewiston C submember, which appears to be somewhat gradational upward into the overlying, more richly fossiliferous D interval. In general, the beds are sparsely fossiliferous. However, certain bedding planes throughout the interval contain pavements of the brachiopod *Striispirifer niagarensis*. Subsequent field study revealed a nearly identical section with *Homocrinus* beds at two other locations to the west, near

TABLE 8.1 *Species composition of the lower* Homocrinus *interval sampled at Lockport, N.Y. (loc. 6) and the upper* Homocrinus *interval sampled at Brockport, N.Y. (loc. 2).*

Taxa	Lower homocrinus interval Lockport, New York	Upper homocrinus interval, Brockport, NY
Arthropoda		
Arctinurus boltoni	2	28
Bumastus ioxus	1	13
Calymene niagarensis	15	26
Dalmanites limulurus	13	118
Proetus? sp.	1	1
Trimerus delphinocephalus	2	5
Brachiopoda		
Amphistrophia sp.	15	16
Atrypa reticularis	49	7
Atrypina rugosa	10	1
"Camerotoechia" sp.	0	10
Coolinia subplana	9	17
Dalejina sp.	1	1
Dicoelosia biloba	19	0

Taxa	Lower homocrinus interval Lockport, New York	Upper homocrinus interval, Brockport, NY
Dictyonella corallifera	15	0
Eoplectodonta transversalis	0	1
Eospirifer radiatus	2	5
Leptaena rhomboidalis	73	56
Parmorthis elegantula	1	5
Pholidops squammiformis	15	22
Plectatrypa nodostriata	1	31
Resserella sp.	2	23
Rhipidomella hybrida	15	0
Rhynchotreta americana	0	10
Stegerhyncus sp.	9	59
Strüspirifer niagarensis	214	370
Trematospira camura	10	2
Whitfieldella nitida	5	14
W. oblata	0	8
Bryozoa		
ramose bryozoans	42+	275+
Fenestella sp.	26	56
Lichenelia concentrica	15	41
Phylloporina asperato-striata	1	0
Cnidaria		
Enterolasma calculum	0	26
Favosites parasiticus	11	24
Conulariida		
Conularia sp.	0	19
Echinodermata		
Asaphocrinus ornatus	102	0
Calceocrinus chrysalis	2	0
Callocystites jewetti	0	1
Caryocrinites ornatus	10	1
Crinobrachiatus brachiatus	X	0
Dendrocrinus longidactylus	0	3
Dimerocrinites sp.	X	0
Eucalyptocrinites caelatus	0	10
Homocrinus parvus	155	46
Icthyocrinus laevis	1	0
Macrostylocrinus ornatus	4	0
Paleaster niagarensis	2	0
Protaster stellifer	23	6
Saccocrinus speciosus	0	5
Hemichordata		
Dictyonema retiforme	4	29
Inocaulus sp.	0	3
Mollusca		
Cornellites emaceratus	4	19
nautiloid cephalopods	1	11
platycerid gastropods	13	4

Numbers of individuals/colonies for each species were determined as follows: Bryozoans and graptolites were recorded as number of colonies and/or colony fragments; bivalves were recorded as total number of valves; brachiopods were recorded as number of articulated valves plus maximum number of either pedical or brachial valves; trilobites were recorded as the number of articulated individuals plus maximum number of pygidia and/or cephala; pelmatozoans were recorded as total number of thecae or calyces; annelids, conulariids, gastropods, rugosans and stelleroids were recorded as total number of individuals; X indicates species was observed.

Fancher, Orleans County, and Shelby/Middleport, Niagara County. Similar stratigraphic successions have been measured westward at Niagara Gorge and St. Catharines, but the nearly vertical cliff exposures at these locations preclude detailed paleontological study.

Microstratigraphically, the middle interval can be subdivided readily into a series of thin skeletal-rich zones dominated by *Striispirifer* and relatively thick calcisiltites separated by more sparsely fossiliferous mudstones. Again the main *Homocrinus*-bearing horizon occurs at a transition between the fossiliferous *Striispirifer*- and/or bryozoan-rich beds and the barren intervals. The upper transition of the Lewiston submember C and submember D ranges from about 1.5 to 2.0 m and bears a series of four or five thicker (5–15 cm) calcisiltite marker beds, several of which appear to be amalgams of two separate depositional events. Again, the shell and calcisiltite beds demarcate apparently correlative intervals that can be described as follows:

1. A relatively thick (5–10 cm) calcisiltite has been used to mark the base of the middle *Homocrinus*-bearing interval. The bed is vaguely graded with a thin (1–5 mm) skeletal hash bed at its base. At an outcrop near Middleport (loc. 4), this basal hash contains abundant fragments of *Calymene* trilobites and abundant complete crowns and columns of *Homocrinus*. The apparently correlative bed at Fancher (loc. 3) has yielded *Calymene* fragments and probable columns of *Homocrinus*. A similar hash, probably equivalent, occurs at Brockport; it was not found to contain *Homocrinus*, but it underlies an interval that does yield these crinoids (see later). A possibly correlative bed has been found westward at least to Niagara Gorge.
2. A sparsely fossiliferous shale at Brockport overlying the calcisiltite bed has yielded abundant current-aligned specimens of *Homocrinus* and other rare crinoids, including complete individuals of *Saccocrinus* (fig. 8.5) and *Dimerocrinites*, nearly all of which had attached *Naticonema* gastropods. Thus far this crinoid horizon has not been traced to the west, although the "barren" shale zone is traceable to Middleport.
3. One of the best markers, a very distinctive 0.5-m interval, displays two or three fossil-rich beds composed of thin calcisiltites (locally pinching out), overlain by a skeletal hash dominated by fragmentary valves as well as articulated *in situ* specimens of *Striispirifer*. This horizon is readily traceable from Brockport westward to Fancher (loc. 3) and Middleport (loc. 4), and an apparently correlative shell-rich bed occurs at Niagara Gorge (loc. 7) and St. Catharines (loc. 8). At Brockport the sparsely fossiliferous shale overlying this bed was found to yield an obrution *Lagerstätten* with intact *Dalmanites*; the crinoids *Homocrinus, Macrostylocrinus, Dimerocrinites*; rare *Eucalyptocrinites*; and the ophiuroid *Protaster*. Although this horizon has not been extensively investigated at Middleport or Fancher, a single specimen of *Striispirifer*, apparently from this level at the latter locality, had the edrioasteroid *Hemicystites* attached. Loose rubble at Niagara Gorge derived from approximately this level also produced a *Hemicystites*. Thus far these are the only occurrences of this edrioasteroid above the *Striispirifer-Asaphocrinus* bed(s) of the lower *Homocrinus* interval. This points up the extraordinary

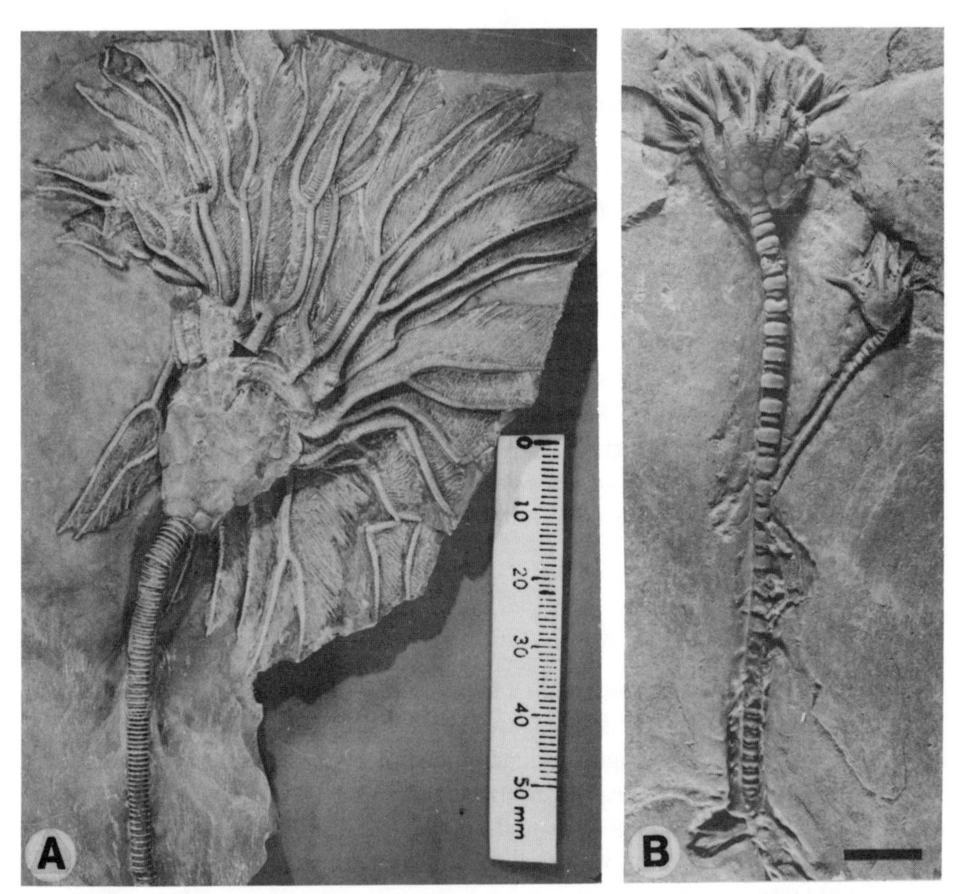

FIGURE 8.5. Crinoids from the upper *Homocrinus* interval; (A) *Saccocrinus speciosus* Hall, complete crown, USNM 454591; (B) two small individuals of the camerate *Eucalyptocrinites caelatus* (Hall), the larger displays small radicular holdfast, the smaller juvenile is attached to the column of the former, photo courtesy of S. T. LoDuca. Scale bars 1 cm or as shown.

degree of resemblance of the two horizons. However, *Asaphocrinus*, which is the most abundant crinoid in the lower beds, is absent in the middle, whereas *Saccocrinus* and *Eucalyptocrinites*, which are rare or absent in the lower interval, are more common in the middle one.

This shell-rich *Striispirifer* band is overlain by sparsely fossiliferous shales and calcisiltites that have yielded more *Dalmanites* and *Homocrinus* at Brockport. That shale passes upward again through *Striispirifer*-rich shell beds into an interval of 0.5–0.75 m thick with abundant fossil hash beds (D submember). As far as is known no major obrution layers are associated with these transitions. However, rare specimens of complete *Eucalyptocrinites* are found in the sparsely fossiliferous mudstone overlying this bed at Fancher. This may be a zone with *Lagerstätten*, but as yet it has not been fully investigated.

Upper *Eucalyptocrinites-Homocrinus* Interval

The upper *Homocrinus*-bearing interval is poorly known as yet; it has been identified tentatively at Middleport, Fancher, and Brockport. At all these localities it is characterized by crowns of the camerate *Eucalyptocrinites* and abundant specimens of the trilobite *Calymene*. It occurs about 2.5–3.0 m above the middle *Striispirifer—Homocrinus* beds. This horizon consists again of a bed rich in *Striispirifer* but also with scattered ramose and fenestrate bryozoans and diverse brachiopods. The most notable feature of this horizon is the presence of complete small individuals of *Saccocrinus* and *Eucalyptocrinites* (fig. 8.5). It is overlain by 0.75–1.0 m of sparsely fossiliferous shale or undulating very thin bryozoan-rich calcisiltites (at Fancher). Rare specimens of *Homocrinus* have been observed from this shale at Brockport only. Again, it is notable that this *Lagerstätten* horizon occurs at the interface between a moderately shell-rich interval and barren mudstone and is associated with a *Striispirifer*-dominated bed.

Sedimentology

Homocrinus intervals are composed of three distinct categories of beds: skeletal debris layers, laminated calcisiltites, and sparsely fossiliferous mudstones. Biostratinomic and sedimentologic evidence indicates that the skeletal debris layers represent long-term "background accumulations," in some cases modified by winnowing events, whereas the mudstone and calcisiltite beds reflect rapid depositional events (cf. Parsons et al. 1988). Each of these beds is discussed in more detail later.

Skeletal Debris Layers

Skeletal debris layers comprise intervals a few millimeters to more than 10 cm thick, characterized by an abundance of fossil material. At the simplest these are thin stringers of disarticulated pelmatozoan and trilobite debris, whole and fragmented valves of brachiopods, and rare bryozoan fragments. Such layers may appear relatively continuous for several meters along outcrops, or they may be somewhat disrupted and "blended" downward into underlying mudstones by bioturbation. In some cases skeletal debris may be scattered through a few millimeters of mudstone with indistinct lower boundaries. These (1–5 mm) nearly flat pavements of skeletal hash may be sharply defined and appear to represent shelly debris concentrated by minor winnowing events. Completely articulated skeletons may occur at the tops of these beds.

Thicker skeletal layers, up to several centimeters thick, are more complex and may display local discontinuous lenses of sparsely fossiliferous mudstone. Skeletons range from disarticulated, comminuted fragments of shell and crinoid ossicles to relatively complete, branching colonies of ramose and fenestrate bryozoans. Articulated brachiopods and portions of columns and thecae of pelmatozoans such as *Stephanocrinus* and *Caryocrinites* may occur within these layers, particularly, but not exclusively, at their upper surfaces. Fossils are packed in a silty, calcareous mudstone matrix, generally in random orientations. Fabrics range from wackestone to packstone. Nearly in situ thickets of bryozoans, bordering on biostromes, occur in local patches.

Examination of the silty mudstone matrix of skeletal beds by scanning electron microscopy (SEM) reveals the presence of compressed clay floccule fabrics and quartz silt grains that resemble those of intervening sparsely fossiliferous mudstone (O'Brien, Brett, and Taylor 1992, 1994). Skeletal debris beds may have sharp contacts on lower surfaces, upper contacts, or both. Sharp bases characterize beds that also display the greatest proportion of fragmented comminuted skeletal debris and tightly packed partially winnowed fabrics. Sharp bases are typically slightly irregular and may even be erosional with gutterlike scours. In other cases vague molds of burrows are present on bed soles; skeletal debris may fill cone-shaped depressions, probably attributable to the mining(?) trace *Conostichus*.

Where beds display sharp planar tops, the upper few millimeters typically show evidence of winnowing, including pack- or thin grainstone fabrics and preferential convex-upward orientation of brachiopod valves. These layers may be sharply overlain by barren mudstone or calcisiltite beds. Beds with sharp tops and, in some instances, sharp bases commonly occur at important boundaries between distinctive lithologic packages. For example, these commonly mark the tops of intervals of abundant skeletal debris layers and are overlain by intervals of more sparsely fossiliferous mudstones and/or calcisiltites. This suggests that such beds may represent slightly condensed intervals associated with sediment starvation at the tops of small-scale upward shallowing cycles or parasequences. Sharpness of the tops may again reflect concentration of skeletal hash by winnowing events and/or nondeposition of muds. Sharp bases, on the other hand, suggest minor episodes of erosion, probably associated with particularly strong storms that reworked the entire blanket of skeletal debris rather than just the tops of the accumulation. The sharply delineated burrows and scour features in these bed bases suggest that surficial muds had been removed down to a level at which more compact sediment was encountered. In such beds the skeletal debris may have accumulated over a period of hundreds or even a few thousand years, but the final deposition of this debris may record a single event. In some cases, however, this was clearly not the final event to affect the beds, as the tops of sharp-based beds may display perfectly preserved in situ fossils that represent smothered bottom assemblages not reworked by waves or currents.

Calcisiltites

A common type of interbedded lithology in the *Homocrinus*-bearing intervals consists of tabular to undulatory, sheetlike beds of laminated carbonate silt or fine sand. Such beds range from wispy stringers a few millimeters thick to layers exceeding 15 cm. In outcrop these beds are resistant and weather as ledges. They are typified by alternations of millimeter-scale dark and light laminae representing more or less clay-rich layers. Millimeter-scale pyrite blebs are common in some calcisiltites. They display planar to low-angle irregular cross-laminations, some of which display complex cross-cutting relationships and slightly convex-upward surfaces typical of small-scale hummocky cross-stratification (ssHCS, Jennette and Pryor 1993).

Bases of the calcisiltite beds are invariably sharp, but tops may be more difficult to define, particularly in drill cores, where the beds appear to grade upward into mudstones. In a few instances bases of calcisiltites display a concentration of fine skeletal debris, especially crinoid ossicles, trilobite segments, and small, thin brachiopod shells. This debris is probably locally derived and may be obviously scoured from underlying skeletal-rich layers; however, it is more obviously sorted and may be current aligned. Complete small crinoids occur at the bases of a few calcisiltites (see later). Sharply defined scour and groove casts indicate minor erosion of underlying mudstone and dragging of tools, probably skeletal grains along the base of the flow. In some instances load casts also indicate deposition on fluid muds.

Most calcisiltites are heavily burrowed, particularly in their upper portions, and some display an irregularly burrowed top. Most typical are *Chondrites* and oblique (30-degree angle) burrows that appear to be associated with *Teichichnus* spreiten. In many instances burrows appear to emanate from overlying mudstones rather than from the highest silt-sized sediment. Burrow fills invariably are lighter in color, often displaying distinct pelletal fabrics, and may be incomplete with geopetal upper parts commonly infilled with secondary anhydrite or gypsum. Hence the calcisiltites display a classic "lam-scram" fabric (Ekdale et al. 1984) typical of storm silt layers. The tops of some calcisiltites are marked by minor skeletal hash layers, typically in burrowed mudstone matrix. This indicates a return to background conditions and recolonization of the muddy seafloor, but only after much of the burrowing of the carbonate silts had taken place. Thus the producers of *Teichichnus* and *Chondrites* appear to have behaved as opportunistic pioneers following silt deposition (see Vossler and Pemberton 1988).

Sparsely Fossiliferous Mudstone Layers

Brett (1983b) noted a third type of event bed in the Rochester Shale, represented in drill core or outcrop as sparsely fossiliferous to barren layers of silty mudstone. In rare instances, these beds may display fine laminae of carbonate silt near their bases, but they are typically structureless, and SEM confirms a homogeneous vertical texture to these beds. Nonetheless they are typically not completely bioturbated. The thickness of the layers ranges from 0.5 to more than 10 cm.

Scanning electron microscopy reveals that these mudstone layers, at least within the lower *Homocrinus* interval, are composed of a mixture of clay floccules with both face-to-face and edge-to-face contacts of illite crystals and quartz and carbonate silt (fig. 8.6). In some cases, quartz and silt grains are enveloped by clay floccules up to 30–40 μm in diameter. Evidently, these are aggregate grains that formed by electrostatic attraction of clays and silt; floccules are typically, but not invariably, flattened by compaction (O'Brien, Brett, and Taylor 1992, 1994). There is no evidence of grading in most mudstone layers, although they appear to be similar to the gradational muddy fabric of calcisiltites. The lack of grading or lamination is probably attributable to the absence of large particle size. Rather, the beds appear to have been deposited rapidly as a mix of medium to coarse silt and clay floccules. The observation of floccules provides an explanation for rapid deposition. In fact, all the sediment behaved as silt rather than as clay, which would settle much more gradually from suspension.

As with calcisiltite layers, bases of mud layers are usually sharply demarcated, particularly where they overlie skeletal debris beds. These junctions yield the bulk of the extraordinarily well-preserved fossil assemblages in the *Homocrinus* intervals, although the processes are clearly analogous to those described earlier for the somewhat coarser calcisiltite obrution layers. Details of taphonomy strongly support a nearly instantaneous deposition of many mudstone beds. This evidence will be considered next.

Biostratinomy

Analysis of fossil preservation provides insight into the history of these mud-bottom communities and paves the way for more detailed paleoecological studies. The relationships between the shell-rich horizons and the under- and overlying units are of key importance in the interpretation of the taphonomy and paleoecology of the *Homocrinus* intervals.

Background Shell Beds

Several lines of evidence suggest that shell-rich surfaces (e.g., those designated as horizons I–V) are bedding planes representing minor discontinuity surfaces (see fig. 8.4). First, the shell layers are fairly continuous within the outcrop and are more or less planar and parallel. Second, some beds of concentrated *Striispirifer* shells sharply overlie unfossiliferous mudstone. Where the shells are absent the surface may be difficult to discern because of the similarity of the mudstone on either side, as is typical of diastems occurring within mudstones and shales. Third, the horizons contain abundant well-preserved fossils, some of which appear to be preserved in life position. Finally, there is evidence that the fossil layers contain remains of several distinct generations of organisms that accumulated over a span of years.

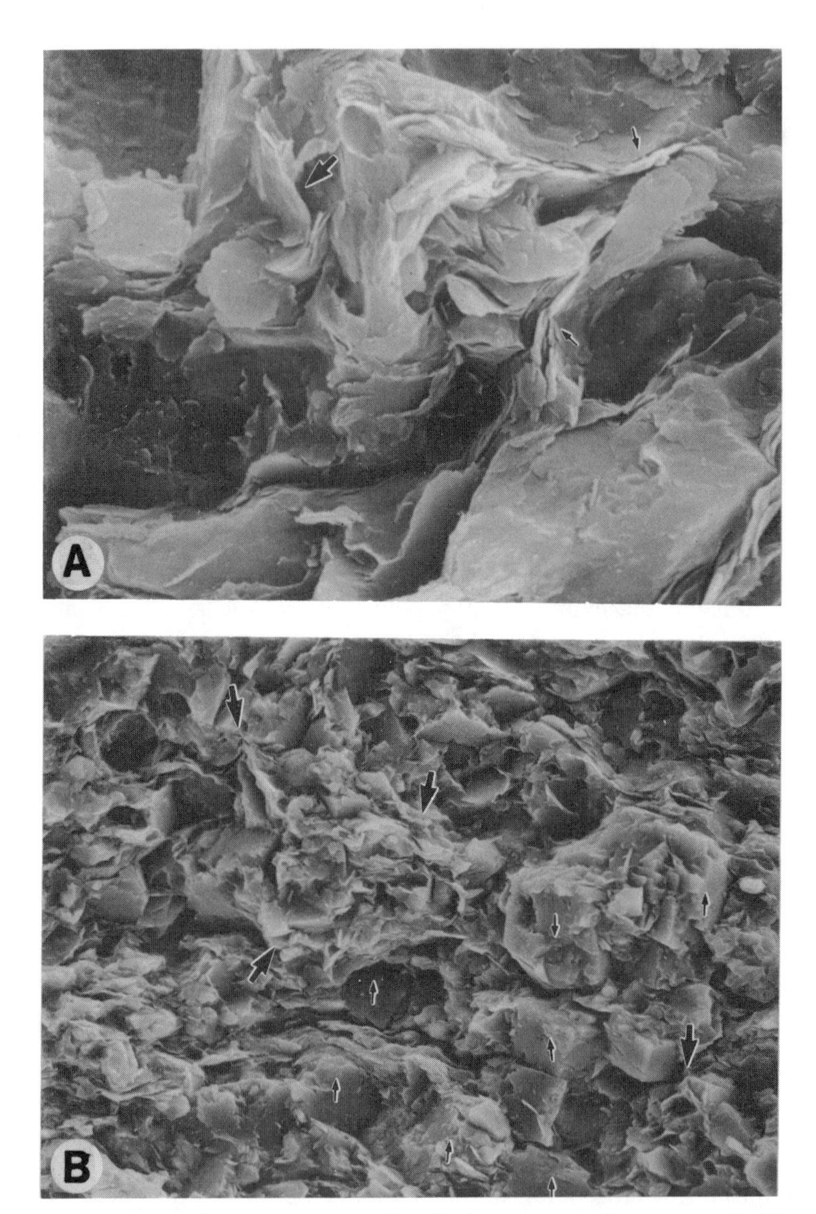

FIGURE 8.6. Scanning electron micrographs of calcareous mudstone sampled from broken surfaces perpendicular to bedding, lower *Homocrinus* interval at Middleport, N.Y. (loc.4); (A) mudstones typically have a flocculated fabric composed of a mixture of clay floccules with both face-to-face (small arrows) and edge-to-face contacts (large arrows) and quartz and carbonate silt, ×4500; (B) often quartz and silt grains are enveloped by clay floccules up to 30–40 microns in diameter, small arrows point to silt/clay size quartz grains and large arrows point to clay aggregates, ×700, photos courtesy of Neal O'Brien.

An intimate admixture of well-preserved and fragmentary fossils is characteristic of the shell-rich assemblages. Valves of *Striispirifer,* bryozoan fragments, and trilobite segments clearly represent remains of earlier generations of organisms that had disintegrated on the seabottom prior to final burial. Furthermore, the broken condition of many of these fossils indicates earlier periods of reworking. The mechanism of breakage is somewhat problematic. Similarities between the broken and the well-preserved fossil assemblages mitigate against transport of the debris form distant shoal environments as opposed to *in situ* accumulation. Aside from nautiloid cephalopods, which are very rare in these horizons, no known predaceous or scavenging organisms had jaws capable of breaking fresh skeletal hard parts. However, disintegration of the skeletal material may have been facilitated by the action of microborers, such as endolithic algae, that are known to be important agents in the destruction of Recent skeletal materials (Warme 1975). Once shells were weakened by such microborers, the activities of scavengers or of weak bottom currents may have been sufficient to cause breakage. The occurrence of well-preserved fossils (e.g., edrioasteroids, bryozoans, and crinoid holdfasts encrusting the fragmentary fossils) is evidence that at least some of the breakage occurred prior to, rather than coincident with, the final burial event. Some of the material may, in fact, represent a storm lag of material reworked from a previous burial cycle.

Biostratinomic evidence indicates that the seafloor was relatively undisturbed for a considerable timespan preceding the disturbances. The attachment of sessile epizoans, such as bryozoans and mature edrioasteroids onto the interiors of *Striispirifer* valves, proves that these shells were exposed in unstable concave-up positions on the seabottom for spans of months to years, up to the time of burial. Certain crinoids, edrioasteroids, and numerous bryozoans were also cemented to extremely small and lightweight skeletal fragments. For example, one cephalon of the trilobite *Trimerus* possesses at least nine minute holdfasts of *Homocrinus* on its upper surface. It is improbable that crinoid larvae settled on a living trilobite; instead this probably represents a molted cephalic shield. Judging from modern arthropod exoskeletons, this would have been very lightweight and readily moved by a current (Kidwell and Bosence 1991). Successful colonization of such an unstable substrate would therefore require undisturbed bottom conditions. Certain crinoids that attached to small bits of shell or bryozoan detritus were obviously uprooted and moved at the time of burial; these examples, again, argue for a normally quiet water environment (see fig. 8.7).

Obrutionary Calcisiltites

Carbonate silt beds are typically unfossiliferous, but in some instances well-preserved skeletons are found within them. One bed from Shelby displays abundant articulated *Homocrinus* on its base. At Fancher, Orleans County (Howard Farms shale pit) a layer of nearly complete *Eucalyptocrinites* crowns with articulated columns was buried by a thin lenticular calcisiltite. Significantly,

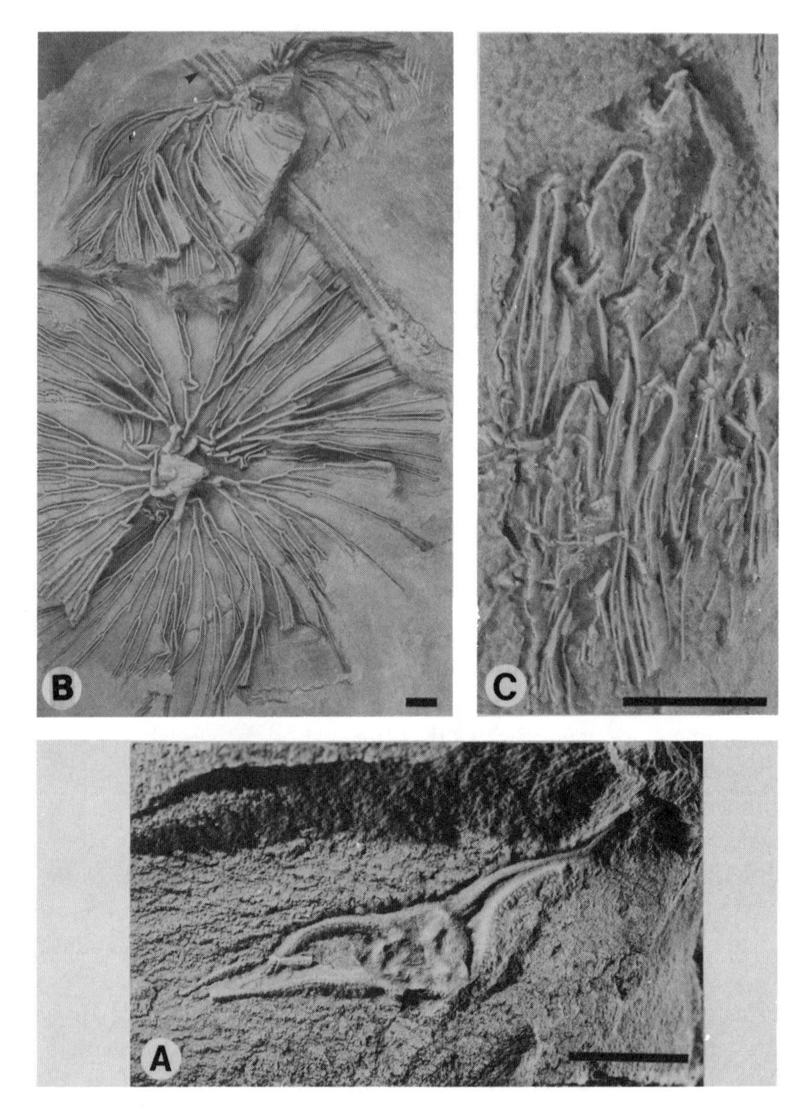

Figure 8.7. Fossils indicative of extremely rapid burial or obrution, collected from sparsely fossiliferous mudstones of the lower *Homocrinus* interval at Middleport, N.Y.; (A) obliquely embedded ophiuroid *Protaster stellifer* Ringueberg found in otherwise barren mudstone, note the contorted rays suggestive of escape orientation, USNM 459650; (B) complete crown and nearly complete individual of the inadunate *Dendrocrinus longidactylus* Hall, note the radially splayed arms and partially exposed anal sac (arrow), USNM 459643; (C) cluster of twenty individuals of the tiny crinoid *Homocrinus parvus*, note the alignment of crowns. Scale bars are 1 cm.

these crinoids display a partial disarticulation of the upper parts of the crowns (although lower surfaces are largely intact). This indicates a brief period of decay (hours?) of the dead crinoids prior to calcisiltite deposition. As in many cases, the death of these organisms preceded final burial, but only by a short interval, which indicates that mortality and burial were related to (different phases of) a single major disturbance. Body fossils are uncommon within the calcisiltite beds except at their bases. However, rarely, articulated specimens of trilobites or pelmatozoans may occur along laminae within these beds. A distinctive 6–15-cm-thick amalgamated marker calcisiltite that occurs near the top of the lower *Homocrinus* interval has yielded articulated *Caryocrinites* at two localities (Niagara Gorge and Lockport). Specimens embedded obliquely through much of the thickness of the upper half of this bed demonstrate that the calcisiltite sediment accumulated as a single event within a few hours. Such fossils may represent carcasses that were transported at least short distances in the turbid flow.

Obrutionary Mudstones

The bases of mudstones immediately above shelly beds display exquisite preservation of intact crinoids, cystoids, and trilobites. Furthermore, these articulated fossils may extend upward for up to several centimeters into the overlying "barren mudstone." Mudstone from 1 to 2 cm above the main layers of shell concentration contains occasional large, articulated specimens of *Striispirifer*. Most of the pelmatozoan crowns have been found in these zones. Furthermore, certain fossils, particularly planar shells of *Amphistrophia* and *Striispirifer*, are embedded edgewise. Some crinoid and cystoid crowns occur well above the shell beds, but the columns run obliquely or in zigzag fashion downward through the mudstone to holdfasts that are directly cemented to objects in the main layer; similarly, some ramose bryozoans run through several centimeters of the mudstone. At Middleport a 6-cm mudstone overlying horizon III displays abundant articulated specimens of *Dendrocrinus longidactylus*, many of which have radially splayed arms (fig. 8.7). In some instances these crinoids are obliquely embedded in the otherwise barren mudstone and may have columns attached to the underlying shell bed. We infer that these long-stemmed (up to 50 cm) crinoids were a part of the community preserved at the base of the mudstone. However, because of their elevated position they were killed and/or buried later than lower-tier organisms and became entombed in upper portions of the mud layers.

Evidence for rapid burial is present in all mudstone horizons of the *Homocrinus* intervals. The most obvious feature of the *Homocrinus* beds is the exceptional preservation of multielement fossils. Ringueberg (1888: 269) aptly pointed out that the "fine state of preservation of many fossils occurring in this band and the layers immediately above and below it is noteworthy. This is especially true of species as a rule found only in a disconnected condition." For example, nearly all specimens of *Caryocrinites ornatus* from the *Homocrinus* layers (including many specimens in the Springer Collection) preserve both columns and complete arms. In contrast, specimens of this cystoid from other horizons in the Rochester Shale consist of thecae with, at best, short sections of stem and no arms preserved. Even in beds where *Caryocrinites* thecae are common,

Asaphocrinus is usually represented only as disarticulated plates; evidently, this flexible crinoid was more rapidly disarticulated than the cystoids. Yet in the *Homocrinus* intervals both forms are equally well preserved.

In addition, the *Homocrinus* beds preserve several forms that are extremely rare or completely unknown in the rest of the Rochester Shale, including *Hemicystites* and various stelleroids. Occasional fragments of these fossils found in washings from other horizons prove that they were present in many areas but were too fragile to be commonly preserved as intact fossils.

The preservation of intact vagrant benthic organisms, such as stelleroids and trilobites, provides further evidence for rapid and deep burial of the bottom fauna. These include both stelleroids and trilobites. Modern ophiuroids are relatively active organisms capable of shallow burrowing by sinuous motions of the slender rays. Brittle stars covered by less than 6 cm of sand are able to dig themselves out (Schäfer 1972), although considerably thinner layers of mud may be lethal because the fine-grained sediment interferes with the water vascular systems of ophiuroids (Rosenkranz 1971). Schafer noted that assemblages of brittle stars are occasionally buried alive in the North Sea (1972, pp. 98–99): "The time when the water calms down after a gale is particularly dangerous, because then all the suspended sediment settles back to the seafloor."

Many ophiuroids from the lower *Homocrinus* beds at Lockport and Middleport were obtained from horizon III in mudstone 0.5–2.0 cm above the main shell beds. Specimens occurred in a variety of orientations both upright and inverted, and with rays variously curved as during crawling. Rarely, specimens occurred on fracture surfaces split across the presumed bedding direction, indicating oblique embedment, as might be expected if the organisms were moving upward through the sediment (fig. 8.7). These observations suggest that the ophiuroids attempted to escape burial but that the sedimentation rate was too rapid to permit this. No ophiuroids were found above most fossiliferous horizons that were covered by thinner layers of mudstone, suggesting shallower burial. On the other hand, some ophiuroid specimens show evidence for incipient decay, such as smearing of the disk, and passive orientation of rays by currents. These individuals evidently died prior to final burial.

Well-preserved trilobites are also found in mudstone overlying shell layers in both levels. The small size of most of these specimens may be significant. Nonetheless, abundant disarticulated remains (exuviae) of larger trilobites (*Arctinurus, Trimerus*) suggest that they were originally common on the seabottom. The larger trilobites may have been able to escape most burial events. This seems particularly likely for forms that are interpreted on morphological grounds to have been burrowing scavengers (e.g., *Trimerus*). Nonetheless, at least two horizons in the lower *Homocrinus* interval and one or more in the upper yielded relatively abundant and perfectly articulated specimens of the large lichid trilobite *Arctinurus boltoni* (fig. 8.8). Evidently in these cases the organisms were killed shortly prior to or during burial and so were preserved intact (see Tetrault 1992).

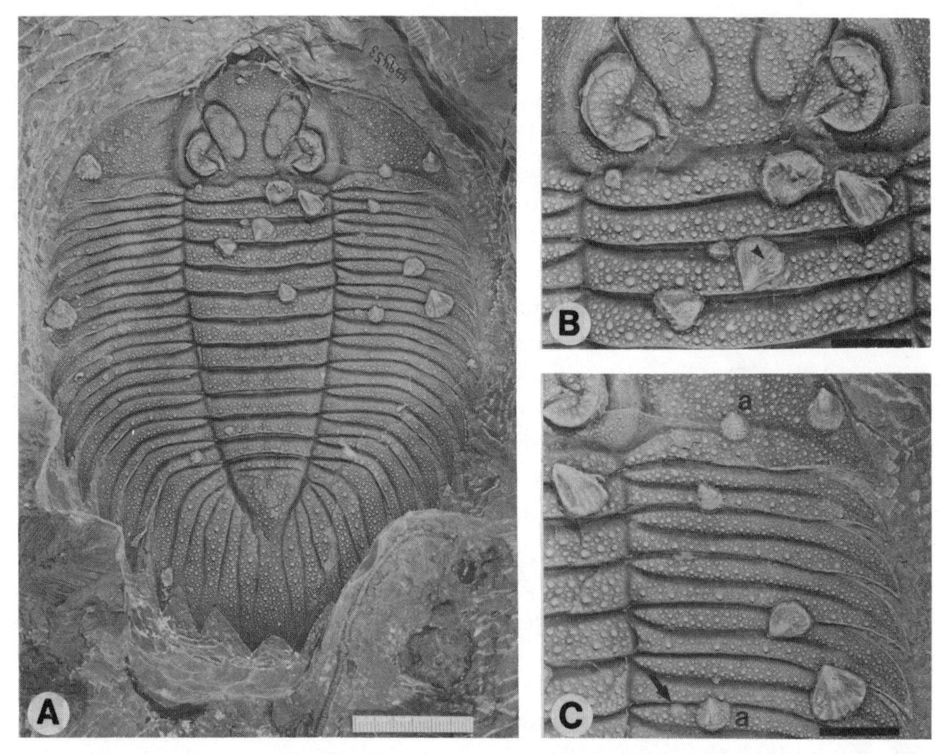

FIGURE 8.8. Large lichid trilobite *Arctinurus boltoni* from Middleport, N.Y., USNM 449453; (A) *Arctinurus* with a microcommunity of encrusters including, *Cornulites*, bryozoans and aligned rhynchonellid and atrypid brachiopods; (B) cluster of *Stegerhynchus* sp. showing several growth stages; many of the larger individuals were themselves encrusted by a minute unidentified bryozoan? (arrow); (C) closeup showing two individuals of *Atrypa* sp. (a), a small bryozoan colony (arrow) and five individuals of *Stegerhynchus* sp. Scale bars are 1 cm or as shown.

Evidence for disturbance of the seafloor immediately prior to the final burial events was found in the Lockport section. Brachiopod valves were evidently overturned at this time, as inverted (concave-up) specimens with attached edrioasteroids were found. Obviously, the edrioasteroids could not have lived in this position, and if overturning of the shell had not immediately preceded burial, it is unlikely that the fragile echinoderms would have been well preserved. Crinoid holdfasts were often preserved on the undersides of inverted brachiopods; several also have the stem and complete crown attached, again proving disturbance just prior to burial.

The fact that some brachiopod valves were turned over from hydrodynamically stable (concave-down) to unstable (concave-up) orientations suggests that the bottom was not swept by a persistent current, as even slight (horizontal) current action will tend to orient all valves in the concave-down position (Brenchly and Newall 1967; Kelling and Williams 1967; Nagle 1967). Instead the shells were probably lifted from the seafloor slightly and then resettled. Such reorientation may have taken place during brief scouring of the seafloor by storm waves.

A few echinoderms show signs of very slight decay prior to burial. For instance, certain *Asaphocrinus* specimens have collapsed crowns and stems or are missing terminal portions of the arms. In several instances the column was preserved as other associated but disarticulated segments that have been shifted slightly in relation to each other. This is best explained as due to slight shifting of a partly decayed specimen along the seafloor. Such crinoids must have died and started to decay before the final burial. The very slight amount of decay may have taken place within hours after the death of these crinoids (Meyer 1971; Liddell 1975). Changes in salinity or other perturbations of the environment may have occurred immediately prior to the rapid deposition of sediments; if so, these changes were probably related to the larger environmental disturbances (i.e., storm) that, in turn, induced rapid burial. Many of the severely collapsed *Asaphocrinus* in horizon III at Lockport are relatively small specimens that might have been especially sensitive to disturbances.

Fossils in horizon II at Lockport provide evidence for a slightly longer period of disturbance prior to burial. Here, several *Asaphocrinus* specimens occurred as isolated calyces or even partial calyces that had apparently decayed for a longer time before being covered.

In each of the fossil horizons, fragmentary fossils become increasingly abundant upward from the main shell layer. The complete/fragmented valve ratio for the shell layer of horizon III is approximately 2.3, while the ratio for mudstone overlying this layer is 0.56. Trilobite thoracic segments and chips of *Striispirifer* valves are the most common fossils in the upper unit. Presumably, the lighter fragments were picked up from the seafloor during scouring and then later resedimented along with the flocculated mud. Fragments are closely associated with well-preserved fossils of small size and thus light weight, such as *Homocrinus* and ostracodes. Consequently, assemblages of small-sized fossils are thought to represent a crudely size-sorted fraction of the original biocoenoses.

Fossil Diagenesis

Following the deposition of mud layers, most of the *Homocrinus* fossil assemblages suffered little further disturbance. As noted previously, burrows are rather infrequent but do penetrate at least into the upper layers of mudstones. In rare instances burrows slightly disrupted the articulation of crinoid arms, but in most cases the carcasses suffered no further disturbance. In part this is due to the thickness of the sediment layer overlying the majority of the fossils. Even discounting the effects of compaction (which almost certainly took place) most fossils are covered by 2–3 cm of sediment. Only a few remains occurred in the upper portions of the mudstone. Apparently, feeding burrows generally did not penetrate deeply enough to affect the skeletal remains. Intense bioturbation of the sediment apparently did not take place.

Compaction of the sediment and entombed fossils was fairly intense, as indicated by the strongly compressed nature of most of the brachiopods. In part, the flattening of articulated specimens may be due to the fact that they were relatively thin-shelled and were not filled with sediment but were essentially empty after decay of soft tissue. Crinoids have suffered less compaction, possibly because of their rigid construction, but some of the calyces are broken open and collapsed perpendicular to the presumed vertical stress. A few specimens that were buried with the crown oriented perpendicular to bedding are uncrushed and spar-filled. Shells of cephalopod and other mollusks are poorly preserved because of dissolution of original aragonite; they have been severely compressed and some are scarcely recognizable.

Pyrite is not common and no pyritic steinkerns are known. However, a few of the complete *Asaphocrinus* specimens from horizon III at Lockport (including one with full stem and holdfast attached to a brachiopod) show crusts of finely crystalline pyrite adhering to the plate surfaces. Numerous brachiopods in the shell layer also have rusty limonitic coatings. This suggests that local reducing environments existed around certain fossils as a result of anaerobic decay of the organic matter after burial.

Homocrinus *Intervals as Epiboles*

In a sense, the *Homocrinus* beds also represent epiboles (see Brett, Speyer, and Baird 1986; Brett, Miller, and Baird 1990). These thin intervals are typified by an abundance of several species that are otherwise very rare or absent in the Rochester Shale or other Silurian formations. These consist primarily of echinoderms but also include trilobites, graptolites, and brachiopods. The distinctive elements of the *Homocrinus* fauna include *Homocrinus parvus* itself, the flexible crinoid *Asaphocrinus ornatus* (lower beds only), the edrioasteroid *Hemicystites parasiticus,* the ophiuroid *Protaster stellifer,* trilobites *Radnoria stokesi* and *Arctinurus boltoni,* and the brachiopod *Striispirifer niagarensis.* In part, this occurrence surely reflects a taphonomic effect. All these fossils, except for the brachiopods, are fragile and easily disarticulated. This is particularly true in the case of the minute *Homocrinus,* starfish, and edrioasteroids. Furthermore, the occurrence of abundant and previously unrecognized ossicles of these echinoderms in washings from more typically preserved Rochester Shale assemblages (Derstler, personal communication, 1980) indicates that these organisms may indeed have been present or abundant in other environmental settings. They are simply not recognized because of their complete disarticulation. In this sense the beds are partly artifacts of appropriate biostratinomic conditions ("*taphonomic epiboles*").

On the other hand, certain elements of this fauna, such as the crinoid

Asaphocrinus and the trilobite *Arctinurus*, are preservable as partial remains that are quite identifiable. Indeed, these are found in other parts of the Rochester Shale, but in much lower abundance than in the case of the *Homocrinus* beds. As noted earlier, the lower *Homocrinus* interval represents an epibole for the crinoid *Asaphocrinus ornatus*. In no other layer of the Rochester Shale is this crinoid at all common. Yet between St. Catharines, Ontario, and Lockport, NY, this species is the first or second most common echinoderm in bed III. Conversely, the upper *Homocrinus* intervals constitute epiboles for the camerate crinoids *Eucalyptocrinites caelatus* and *Saccocrinus speciosus*, which are rare in most of the Rochester Shale. The precise causes of these differences, given the overall resemblance of the three intervals, is difficult to account for. Presumably, slight environmental differences existed that are not otherwise manifest in the sedimentary record. These would represent *ecological epiboles* in the sense of Brett, Miller, and Baird (1990).

Sedimentologic and Paleontological Model for Obrution Intervals

GENESIS OF OBRUTION DEPOSITS

Sedimentologic and taphonomic evidence indicates that the Rochester Shale echinoderm *Lagerstätten* represent episodes of widespread mass mortality of benthic communities coincident with or immediately followed by burial. The lateral persistence of these layers in outcrop and apparently along depositional strike for distances of tens of kilometers reveals that blanketing of tracts of the seabottom by fine-grained sediment was relatively widespread and uniform. Most of the obrution layers overlie thin skeletal debris layers that are interpreted as background accumulations; in many instances mixtures of whole and fragmentary shell material support the notion of multiple generations of debris in the layers (ranging from "old," corroded to freshly buried shell). Mixing within these layers was undoubtedly mediated to some extent by bioturbation. Sharp basal concentrations indicate brief intervals of scouring and winnowing, probably by deep storm waves and/or gradient currents. Sharp-topped shell beds indicate abrupt smothering by sediment and termination of debris accumulation. In some instances this followed a minor winnowing event.

Interbedded calcisiltite beds bear strong resemblance to "graded rhythmites" described by Reineck and Singh (1972) from the modern North Sea. These are interpreted as layers of pelletal carbonate and siliciclastic fine sand and silt deposited by gradient currents. Sharply scoured bases with tool marks, and planar to small-scale hummocky cross-lamination support these interpretations. Mudstone caps of calcisiltites appear to represent final depositional phases when somewhat finer muds settled rapidly from suspension. The well-preserved fossils, such as graptolites and rarely cystoids, that occur infrequently in these upper portions may represent organisms that were caught up in the gradient currents that transported sediments to their depositional site. The source area of the carbonate silts is poorly understood; carbonate shoals lying to the north of the present outcrop belt represent a possible explanation. Such carbonate shoals (Amabel Formation) are known to have existed to the north of the Rochester Shale basin in Ontario, but direct evidence of shoals north of New York State, if such existed, has been removed by erosion. The apparent lateral persistence of at least some thicker beds suggests that these layers were produced by infrequent, exceptionally large storm events. The occurrence of such beds overlying some diastems (flooding surfaces) suggests that shallow-water sediment reservoirs built up over periods of time may have been intermittently emptied and the sediments redistributed basinally as thin widespread sheets.

Obrution mudstone layers may represent similar processes of rapid flushing and resedimentation of muds. Their intimate association with and rare gradation into calcisiltite beds suggest that the two types of deposits are genetically related. The absence of calcisiltite beds may reflect a more distal position for a given event (i.e., deeper than the area in which coarser carbonate silt dropped out of suspension). Alternatively, during some times the source of coarser carbonate silt may have been unavailable. In any case, it is obvious from SEM analysis that the mudstone layers were deposited as fine to medium silt-sized grains composed of quartz silt and clay floccules (see fig. 8.6), hence permitting the very rapid sedimentation that was required for intact preservation of echinoderms and other multielement skeletons.

Whether the clay floccules were derived directly from terrigenous sources, as during storm flood runoff, or resedimented from previously deposited marine muds remains uncertain. However, the association with obviously marine-derived carbonate silts and pellets suggests the latter possibility.

Causes of mortality of benthic communities are obscure, but in some instances it is clear that mortality preceded burial itself and so was probably related to other factors, such as physical disturbance, turbidity, and chemical changes of the bottom water. There is also some evidence for uprooting of crinoids and reorientation of shells, probably by waves, prior to burial, and this type of disturbance may have been lethal to some organisms.

Thus the obrution layers are interpreted to represent burial events triggered by major storms that caused widespread mortality because of disturbances of benthic communities, produced gradient currents that winnowed more proximal areas of seafloor, and imported large quantities of probably resuspended sediment that blanketed extensive tracts of seabottom together with living, dead, and/or moribund organisms.

Brett and Seilacher (1991) recognized several distinctive types of obrutionary deposits. The *Homocrinus* beds described herein belong to the "Hamilton-type" obrution bed category with thin shell beds blanketed by mudstone or siltstone layers. The position for all these accumulations is interpreted as moderate depth (perhaps more than 50 m) well below normal wave base and close to the lower limit of the most severe storm waves. They were areas in which distal gradient currents flowing basinward slowed to allow abrupt deposition of fine to coarse silt-sized sediment.

Nonrandom Distribution of Obrution Deposit Intervals in Facies: Environmental Gradients

It is apparent that echinoderm *Lagerstätten* in the Silurian Rochester Shale occur nonrandomly. Brett (in press) has recognized about 10 distinctive fossil associations in the Rochester Shale, ranging from high-diversity (60–100 species) bryozoan-rich assemblages of the *Atrypa*-bryozoan community to sparse, low-diversity assemblages (10–15 species) of the *Amphistrophia-Dalmanites* community (see Tetreault 1994). Relatively well-preserved calyces, or very rarely, crowns of pelmatozoans are known from several of these biofacies, indicating that moderately rapid burial of echinoderm remains occurred infrequently in most Rochester Shale environments. However, preservation of complete echinoderms is unknown over most of the biofacies spectrum.

All beds displaying completely articulated pelmatozoans and ophiuroids occur in transitional facies characterized by moderate diversity (30–40 species), *Striispirifer* brachiopod-dominated associations. The lithofacies are typified by a bioturbated mixture of thin shell-bryozoan hash beds, calcisiltites, and sparsely fossiliferous mudstones. The majority of sparsely fossiliferous mudstones and of shelly or bryozoan-rich intervals do not yield recognizable obrution beds.

Several successive obrutionary layers may occur within relatively narrow intervals (0.3–1.5 m), indicating that, at times, conditions responsible for obrution deposition occurred repetitively. Yet these "bundles" of obrution beds are separated from one another by thicker intervals of up to several meters in which no obrution layers have been recognized. Obrution beds occur nearly symmetrically arranged in analogous transitional facies on both deepening and regressive phases of a large-scale (approximately 0.5–1.0 m.y. duration) cycle in the lower half of the Rochester Shale. However, examined at a finer scale, both the lower (transgressive) and upper (regressive) *Homocrinus* intervals appear to occur at or near the caps of meter scale shallowing-upward cycles.

Finally, and perhaps most surprisingly, detailed microstratigraphy indicates that individual obrution layers and groups thereof are traceable as much as 80 km along depositional strike but only a few kilometers (perhaps as little as 2–5 km) across strike.

The nonrandom distribution of *Lagerstätten* in these transitional facies probably reflects both ecological and taphonomic factors. Obviously, these obrution layers will only be recognizable where at least moderate abundance of echinoderms, trilobites, or other fragile organisms were subject to periodic mass mortalities. Deeper-water facies of the Rochester Shale are represented by very sparsely fossiliferous, dark gray mudstone. The rarity of pelmatozoans or other organisms with multielement skeletons certainly limits our ability to recognize obrution deposits. Scattered complete specimens of the trilobite *Dalmanites* prove that rapid burial did, in fact, occur in these settings, but most bedding planes display comminuted trilobite and brachiopod debris. This taphonomic evidence suggests that rapid blanketing of seafloor environments was less common in these settings, perhaps because they were most distal to areas of sediment resuspension.

The absence of traceable obrution layers (as opposed to scattered completely articulated fossils) in the apparently shallower-water shelly bryozoan-rich facies surely does not reflect the absence of indicator organisms. Pelmatozoans and trilobites of several species occur in these facies. Indeed, they are typically more diverse and abundant than in the *Homocrinus* biofacies. The fact that certain taxa that are restricted to these more shelly facies (i.e., the coronoid *Stephanocrinus* and the flexible crinoid *Lecanocrinus*) are completely unknown from articulated intact skeletons further attests to the rarity or absence of conditions conducive to obrution in these environments. Obviously, taphonomic, not ecological conditions are responsible for the absence of obrution *Lagerstätten* in these settings. The very conditions that were favorable to colonization of the seafloor by diverse suspension feeding benthos may have limited preservation potential (i.e., well-oxygenated bottom waters would favor rapid bacterial decay). Turbulence aided in disarticulation and disinterment of once-buried remains, and winnowing and bypass of suspended fine-grained sediment would prevent rapid or permanent burial of remains. Moreover, organisms in this setting may not have been as subject to rapid mass mortality. Upslope communities were certainly more subject to direct physical stresses (e.g., uprooting and breakage during storms) but may not have been as susceptible to the effects of increased turbidity or sediment blanketing. Furthermore, shallower-water organisms may have been adapted to withstand episodic stresses.

Incipient decay in some of the carcasses in obrution beds indicates preburial mortality. The conditions that led to this situation are of course difficult to ascertain but may well have included physical disturbance, increased turbidity, and/or possible toxicity of water associated with input of sediment-charged gradient currents. Furthermore, these communities evidently lived under normally stable low-energy conditions and so were more susceptible to disruption by minor disturbances.

Position of Obrutionary Intervals in Sedimentary Cycles

More enigmatic is the fact that the obrution intervals occur selectively in what appear to be relatively condensed deposits. Thick successions of shale and siltstones in the Rochester Shale depocenter (Monroe and Wayne counties) have not yielded abundant or distinctive obrution beds. Nor do thicker regressive successions of Rochester strata in western sections yield these deposits. Rather, the layers are "bundled" above the caps of minor shallowing cycles associated with marine flooding events.

Again, at least two factors may explain this paradox. First, thicker successions, reflecting higher net rates of sedimentation, are not characterized by an abundance of suspension-feeding organisms, such as pelmatozoans, which comprise the most recognizable parts of obrution deposits. Second, although horizons of well-preserved fossils are known to occur in the thicker mudstone successions, these are much more widely spaced and are separated by a far greater number of layers that display poorer preservation associated with background conditions of more gradual or thinner burial layers. Thus the relative rarity of obrution beds in thicker, less condensed sections may, in fact, represent a "dilution" factor. The higher rate of background sedimentation means that episodic major event beds are more widely separated. This, together with the sparser fauna, leads to the impression of only "scattered complete" fossils rather than *Lagerstätten*. Conversely, in areas where sediment starvation prevails, background conditions may be condensed into thin reworked fossil hash beds, such as those that underlie the obrution beds described herein. Only the most major storm events were sufficient to bring in large quantities of sediment.

This latter deduction leads us to an explanation of one of the remaining observations, the lateral persistence of obrution beds with the condensed *Homocrinus* intervals. During exceptionally major storms or flooding events large amounts of sediment, which may have accumulated over large upslope areas during long periods of time, were abruptly resuspended and transported basinward by gradient currents. These suspended muds uniformly blanketed areas below wave base, leading to deposition of sheetlike layers over thin condensed hash beds that represent vastly longer periods of minimal accumulation. It is certainly possible that layers of this regional extent also occur in thick, noncondensed intervals, but they are difficult or impossible to discriminate in the absence of concentrated obrution deposits. In contrast, during generally sediment-starved time intervals, as would be associated with rapid minor transgressive phases (marine flooding surfaces), practically the only sediments that were deposited were those that accumulated during extremely major large storms.

Thus, for example, the major transition from bryozoan biostromes at the top of the lower Lewiston (B submember) to the sparsely fossiliferous deeper-water facies of the C submember is represented in the 30–50-cm lower *Homocrinus* interval. Here it is only recorded at all because of a series of perhaps five to seven major burial events. In the absence of these events, the transition would be unresolved as a thin skeletal hash bed. Each successive bed appears to provide a "snapshot" of a somewhat lower-diversity, higher-dominance biota along a transition from the B to C submember facies.

It is certainly noteworthy and counterintuitive that event beds during a generally sediment-starved interval should be as thick or thicker than those during times of greater net sedimentation, but this seems to be the case. Taphonomic evidence proves that layers of up to 10 cm of carbonate silt or mudstone accumulated as single events. As these layers are continuous along outcrop for up to 100 km and up to at least 1–2 km (and extend possibly much further) perpendicular to strike, the volumes of sediment transported and deposited during such events are very large (up to 10^{14} cm^3). Nonetheless, similarly widespread layers are known to accumulate during modern hurricanes (Hayes 1967; Davis et al. 1989). The general sediment starvation conditions that prevailed because of entrapment of sediment in near-shore areas were briefly interrupted by severe storm events that transported large quantities of temporarily stored sediment into offshore areas. Isopach maps of the Rochester Shale (Goodman and Brett, in prep.) indicate a local input source direction of terrigenous sediment to the northeast of central New York State. This area could have provided intermittent pulses of siliciclastic mud and silt.

Discussion

The generalizations regarding distribution patterns of obrution deposits in the Rochester Shale appear to have broader application. Many of the best-known middle Paleozoic (Ordovician to Devonian, table 8.2) echinoderm *Lagerstatten* bear close facies resemblance to those of the *Homocrinus* beds. Complete crinoids most commonly occur in fine siliciclastic muds overlying thin shell beds rich in ramose and fenestrate bryozoans and spiriferid, strophomenid, and/or atrypid brachiopods. Underlying shell beds commonly display features suggestive of storm reworking and amalgamation and may show indications of sedimentary omission, such as firmgrounds.

The typical lithology is thin-bedded shell-rich carbonates with interbedded shales. Furthermore, the stratigraphic context in most cases is in transgressive intervals overlying pelmatozoan shoal packstones and grainstones. The depositional environments in these cases can be characterized as normally low energy, below wave base, and near the lower end of storm wave base. Burial sediment is invariably resuspended siliciclastic mud; it is not yet known whether floccules are present in all such cases.

It is evident in several cases that obrution deposits occur in clusters within relatively shaley intervals. It is less easy to ascertain from published accounts whether the obrution layers are associated with condensed transgressive parts of small-scale cycles. However, personal observations of several occurrences suggest that this may be the case. For example, in the middle Ordovician Trenton Group of New York and Ontario, the best obrution deposits appear in beds immediately overlying hardgrounds or firmgrounds that cap small-scale cycles.

One implication of these generalizations is that most pelmatozoan *Lagerstätten* form in a relatively narrow portion of the facies spectrum. This aids in prospecting for new occurrences and has been used successfully to locate new echinoderm *Lagerstätten* in the Paleozoic of New York. On the other hand, it also implies that the best-known echinoderm occurrences reflect an inherently

Table 8.2 *Key middle Paleozoic formations that contain echinoderm* Lagerstätten *deposits.*

Age/formation	Location	References	Lithology/BA
Ordovician			
Benbolt limestone	Tennessee	Brower and Veinus 1974	1 BA-3/4
Bromide Formation	Oklahoma	Sprinkle and Longmann 1977; Sprinkle 1982	1 BA-3/4
Corryville Formation	Kentucky, Ohio	Meyer, 1990; Meyer et al. 1981	1,2 BA-4
Denley Formation	New York	Beecher 1894; Walcott 1875, 1881	1 BA-4
Girardeau Formation	Missouri	Brower 1973	1 BA-3
Lebanon limestone	Tennessee	Guensberg 1992	1 BA-3/4
Martinsburg Formation	Pennsylvania	Cramer 1957	3 BA-4/5
Verulam Formation	Ontario	Springer 1911	1 BA-2?, 3/4
Silurian			
Eke beds	Gotland	Franzen 1982	1 BA-3/4
Keyser Formation	West Virginia	Schuchert 1904, 1913	1,2 BA-3
Medina group	N.Y., Ontario	Eckert 1984	2,3 BA-2/3
Much Wenlock Fm.	England	Calef and Hancock 1974; Hurst 1975; Watkins and Hurst 1977	1 BA-3/4
Rochester Formation	N.Y., Ontario	Brett and Eckert 1982	2,3 BA-4
Devonian			
Arkona Formation	Ontario	Kesling 1969, 1970, 1971	2,3 BA-4
Hamilton group Moscow Formation	N.Y.	Parsons et al. 1988; Brett and Seilacher 1991	2,3 BA-3/4
Hunsruck Slate	Germany	Schmidt 1934, 1941; Sturmer 1970	3 BA-4/5
Silica Formation	Ohio	Kier 1952; Kesling and Chilman 1975	2,3 BA-4
Carboniferous			
Banff Formation	Alberta	Laudon et al. 1952	1 BA-3/4
Borden group	Indiana	Lane 1973	3 BA-3/4
Fort Payne Formation	Kentucky, Tenn.	Meyer et al. 1989	1 BA-3/4
Hampton Formation	Iowa	Laudon and Beane 1937	1 BA-3/4
LaSalle limestone	Illinois	Strimple and Moore 1971	1 BA-3/4

Lithology represented by the following code: 1 = carbonates; 2 = interbedded carbonates and shales; 3 = shales; and BA = probable Benthic Assemblage (or range) for the deposit.

biased sample of a narrow belt of facies. Unfortunately, these facies probably represent suboptimal environments for crinoids, and so the well-preserved faunas usually feature relatively eurytopic, generalized crinoids (e.g., *Dimerocrinites* and *Dendrocrinus*) while other taxa, represented by abundant disarticulated material in other taphofacies, is rarely seen in obrution beds.

In terms of event stratigraphy, obrution *Lagerstätten*, such as the *Homocrinus* intervals, may be very useful as precise local time lines. Because of unique taphonomic and taxonomic conditions certain beds may be relatively easy to identify and correlate precisely. In this sense, obrution beds may be even more valuable to local stratigraphy than ash layers.

Finally, obrution beds provide unique glimpses into ancient benthic communities and sedimentary dynamics of otherwise monotonous-appearing mudstones. In this sense they provide taphonomic windows for paleoenvironmental study.

Conclusions

1. *Homocrinus*-bearing intervals of the medial Silurian (Wenlockian) Rochester Shale in Ontario and western New York State represent bundles of obrutionary deposits in which fragile echinoderms, trilobites, and other fossils are exceptionally well articulated.
2. Groups of obrution beds and individual layers occur consistently at three distinct levels within the lower Rochester Shale, Lewiston Member at the transitions from lower bryozoan-rich B submember to the sparsely fossiliferous C submember, and again at the transition from C to the overlying fossiliferous D submember.
3. Groups of obrution layers and apparently distinctive individual beds are traceable for 80 to perhaps more than 120 km along depositional strike but for only a few kilometers across strike (e.g., south in Niagara Gorge).
4. Within the intervals three distinctive lithologies are interbedded: (a) shell- and bryozoan-rich debris layers (0.2–10 cm), (b) laminated calcisiltites, and (c) sparsely fossiliferous mudstone layers. The shell hash beds and bryozoan thickets represent amalgamated debris accumulated during background conditions; calcisiltite and mudstone layers represent event deposits.
5. *Homocrinus* obrution deposits reflect very rapid deposition of laterally extensive (tens of kilometers) sheets of carbonate silt and flocculated clays (calcisiltites) that were suspended and transported basinward by gradient currents during storms. Mass mortality of benthic organisms briefly preceded or coincided with mud and silt deposition. Communities were buried largely in situ, although local transport of carcasses may have occurred in carbonate silt flows.
6. The *Homocrinus* obrution beds occur nonrandomly in a biofacies spectrum. They are associated with a distinct ecotonal community type (*Striispirifer* community) that is gradational between highly diverse brachiopod and bryozoan assemblages in shallow-water facies and low-diversity, deeper, mud-bottom *Amphistrophia-Dalmanites* associations.
7. *Homocrinus* beds represent both taphonomic and ecological epiboles (i.e., anomalous abundances of certain species, because of taphonomic and ecological factors).
8. Obrution deposits typify a particular suite of sedimentary facies. They are preserved most frequently in facies that were deposited near the lower limit of storm wave base. Here there were abundant benthic organisms that lived under lower-energy, rarely disturbed bottom conditions. These organisms were vulnerable to episodic disturbances. Mud tempestites were deposited preferentially in these settings and were not reworked.
9. Rochester Shale obrution deposits appear to be *concentrated* in bundles overlying marine flooding surfaces in small-scale cycles. During these inter-

vals there was relatively low *net* sediment accumulation, perhaps as a result of rapid relative sea-level rise. However, severe storm events still redistributed large quantities of sediment into basinal areas. Thus thin transgressive intervals are relatively condensed but may be recorded as a series of discrete widespread depositional episodes.

Appendix: Register of Localities

1. Exposures of upper Lewiston and Gates members of Rochester Shale on NE and SW sides of New York State Barge Canal beginning about 0.5 km SE on Long Pond Road and extending to the overpass of Rt. 390 over canal, Towns of Gates (southwest side) and South Greece (on east side), Monroe County, N.Y. (Rochester West 7.5 Quadrangle).
2. Temporary exposures of upper Lewiston Member, excavated in fall of, 1989, during construction of Wegman's Plaza immediately north of N.Y. Rt. 31A, 0.5 km west of junction of N.Y. 19/31 and just east of Hartshorne Place, Brockport, Monroe County, N.Y. (Brockport 7.5 Quadrangle).
3. Exposures of the middle Lewiston Member in shale pit (property of Howard Farms, Inc.), 1.8 km north of N.Y. Rt. 31, just east of Hindsburg Road, and 1. 0 km west of Fancher, Orleans County, N.Y. (Holley 7.5 Quadrangle).
4. Exposures of the Lewiston Member (about 0.4 m above the base of the formation) along a north-flowing tributary of Jeddo Creek, about 0.5 km south of N.Y. Rt. 31, and 1.4 km east of Freeman Road, Middleport, Niagara County, N.Y. (Medina 7.5 Quadrangle).
5. Discontinuous outcrops of middle and upper Rochester Shale along an unnamed, north-flowing creek 0.1 km west of Cottage Road north to N.Y., Rt. 31, 3. 2 km northwest of Gasport, Niagara County N.Y. (Gasport 7.5 Quadrangle).
6. Exposures of Rochester Shale along the west branch of 18 Mile Creek at Lockport Gulf, 0.8 km north of N.Y. Rt. 31 and 0. 3 km west of N.Y. Central railroad tracks, Lockport, Niagara County, N.Y. (Lockport 7.5 Quadrangle).

7a. Exposure of Rochester Shale (lowest 0.5 m), in the Fish Creek Gully, just below (west of) Lewiston Heights and 0.8 km south of Artpark, Lewiston, Niagara County, N.Y. (Lewiston 7.5 Quadrangle).

7b. Exposures and talus from the Rochester Shale, east wall Niagara Gorge directly below to 0.4 km south of power lines crossing gorge from U.S. to Canadian side, 0.2 km north of Robert Moses Hydroelectric Generating Plant and 2.5 km south of Lewiston, Niagara County, N.Y. (Lewiston 7.5 Quadrangle).

8. Exposures of middle Lewiston Member in the bottom of lock no. 3, Welland Canal, Thorold, Ontario (NTS 1:25,000 Series, 30M/3g; St. Catharines Sheet).
9. Exposure of Rochester Shale in cliff along north face of Niagara Escarpment at DeCew Falls Hydroelectric Generating Plant, St. Catharines, Ontario (NTS 1:25,000 Series, 30M/3c; Fonthill Sheet).

ACKNOWLEDGMENTS

This paper was critically reviewed by David L. Meyer, Gordon C. Baird, and George C. McIntosh, who provided numerous helpful comments and suggestions. Neal O'Brien (SUNY College at Potsdam) studied fabrics of Rochester obrutionary mudstones and kindly provided SEM photomicrographs for fig. 8.6; a more thorough study of Silurian mudrock fabrics is in preparation. Talia Sher aided with figure and text preparation, and James Mott provided special assistance with measurements in the field. Excavation of the Middleport, N.Y., site was greatly assisted by Mr. James Nardi and a party from the U.S. National Museum coordinated by Frederick J. Collier. Mr. Nardi also generously donated numerous specimens for study. James Eckert and Denis Tetreault collected some of the material from Lockport used in this study. Research for this study was funded by grants from the Geological Society of America, Sigma Xi, New York State Honorarium, and grant 21987-AC8 from the donors to the Petroleum Research Fund of the American Chemical Society. Specimens illustrated in this paper are reposited in collections of the Royal Ontario Museum (ROM) and U.S. National Museum (USNM).

REFERENCES

Aigner, T. 1982. Calcareous tempestites: Storm-dominated stratification in Upper Muschelkalk limestones (Middle Triassic, southwestern Germany). In G. Einsele and A. Seilacher, eds., *Cyclic and Event Stratification*, pp. 180–98. New York: Springer-Verlag.

———. 1985. *Storm Depositional Systems: Dynamic Stratigraphy in Modern and Ancient Shallow-Marine Sequences*. Lecture Notes in Earth Sciences, vol. 3. New York: Springer-Verlag.

Aigner, T. and H. E. Reineck 1982. Proximality trends in modern storm sands from the Helgoland Bight (North Sea) and their implications for basin analysis. *Senckenbergiana Maritimia* 14:183–215.

Allison, P. A. 1986. Soft-bodied animals in the fossil record: The role of decay in fragmentation during transport. *Geology* 14:979–81.

———. 1988a. *Konservat–Lagerstätten:* Cause and classification. *Paleobiology* 14:331–44.

———. 1988b. The role of anoxia in decay and mineralization of proteinaceous macrofossils. *Paleobiology* 14:139–54.

———. 1990. Variation in rates of decay and disarticulation of Echinodermata: Implications for the application of actualistic data. *Palaios* 5:432–40.

Allison, P. A. and D. E. G. Briggs, eds. 1991. *Taphonomy: Releasing the Data Locked in the Fossil Record*. New York: Plenum Press.

Ausich, W. I. 1980. A model for niche differentiation in Lower Mississippian crinoid communities. *Journal of Paleontology* 54:273–88.

Ausich, W. I., T. W. Kammer, and N. G. Lane. 1979. Fossil communities of the Borden (Mississippian) Delta in Indiana and northern Kentucky. *Journal of Paleontology* 53:1182–96.

Baarli, B. G. 1988. Bathymetric co-ordination of proximality trends and level-bottom communities: A case study from the lower Silurian of Norway. *Palaios* 3:577–87.

Beecher, C. E. 1894. On the mode of occurrence and the structure and development of *Triarthrus becki*. *American Geologist* 13:38–43.

Brenchly, P. J. and G. Newall 1967. Flume experiments on the orientation and transport of models and shell valves. *Palaeogeography, Palaeoclimatology, and Palaeoecology* 7:185–220.

Brett, C. E. 1983a. Stratigraphy and facies relationships of the Silurian Rochester Shale (Wenlockian; Clinton Group) in New York State and Ontario. *Proceedings of the Rochester Academy of Science Inc.*, Centennial Colloquium Issue 15:118–40.

———. 1983b. Sedimentology, facies and depositional environments of the Rochester Shale (Silurian; Wenlockian) in western New York and Ontario. *Journal of Sedimentary Petrology* 53:947–71.

———. (in press). Wenlockian fossil communities in New York State and adjacent areas: Paleontology and paleoecology. In A. J. Boucot and J. Lawson, eds., *Paleoecology of Silurian to Early Devonian Communities,* Final Report of Project Ecostratigraphy, Cambridge University Press.

Brett, C. E. and J. D. Eckert 1982. Paleoecology of a well-preserved crinoid colony from the Silurian Rochester Shale in Ontario. *Royal Ontario Museum Life Sciences Contributions* 131.

Brett, C. E., S. E. Speyer, and G. C. Baird. 1986. Storm-generated sedimentary units: Tempestite proximality and event stratigraphy in the Middle Devonian Hamilton Group of New York. In C. E. Brett, ed., *Dynamic Stratigraphy and Depositional Environments of the Hamilton Group (Middle Devonian) in New York State,* part 1, pp. 129–54, *New York State Museum Bulletin* 457.

Brett, C. E., W. M. Goodman, and S. T. LoDuca. 1990. Sequences, cycles, and basin dynamics in the Silurian of the Appalachian Foreland Basin. *Sedimentary Geology* 69:191–244.

Brett, C. E., K. B. Miller, and G. C. Baird. 1990. A temporal hierarchy of paleoecological processes in a Middle Devonian epieric sea. In W. Miller III, ed., *Paleocommunity Temporal Dynamics: The Long-Term Development of Multispecies Assemblages.* Paleontological Society Special Paper No. 5, pp. 178–209.

Brett, C. E. and A. Seilacher. 1991. Fossil Lagerstätten: A taphonomic consequence of event sedimentation. In G. Einsele, W. Ricken, and A. Seilacher, eds., *Cycles and Events in Stratigraphy,* pp. 283–97. New York: Springer-Verlag.

Brower, J. C. 1973. Crinoids from the Girardeau Limestone (Ordovician). *Palaeontographica Americana* 7(46):259–499.

Brower, J. C. and J. Veinus. 1974. Middle Ordovician crinoids from southwestern Virginia and eastern Tennessee. *Bulletins of American Paleontology* 66:283–300.

Butterfield, N. J. 1990. Organic preservation of non-mineralizing organisms and the taphonomy of the Burgess Shale. *Paleobiology* 16:272–86.

Calef, C. E. and N. J. Hancock. 1974. Wenlock and Ludlow marine communities of Wales and the Welsh Borderland. *Palaeontology* 17:779–810.

Cant, D. J. 1980. Storm dominated shallow marine sediments of the Arisaig Group (Silurian-Devonian) of Nova Scotia. *Canadian Journal of the Earth Sciences* 17:120–31.

Cramer, H. R. 1957. Ordovician starfish from the Martinsburg Shale, Swatara Gap, Pennsylvania. *Journal of Paleontology* 31:903–907.

Davis, R. A., Jr., S. C. Knowles, and M. J. Bland. 1989. Role of hurricanes in the Holocene stratigraphy of estuaries: Examples from the Gulf Coast of Florida. *Journal of Sedimentary Petrology* 59:1052–61.

DeCelles, P. G. 1987. Variable preservation of middle Tertiary, coarse-grained nearshore to outer-shelf storm deposits in southern California. *Journal of Sedimentary Petrology* 57:250–64.

Duke, W. L. 1990. Geostrophic circulation or shallow marine turbidity currents? The dilemma of paleoflow patterns in storm-influenced prograding shorelines. *Journal of Sedimentary Petrology* 60:870–83.

Eckert, J. D. 1984. Early Llandovery crinoids and stelleroids from the Cataract Group (lower Silurian) in southern Ontario, Canada. *Royal Ontario Museum Life Sciences Contributions* 137.

Ekdale, A. A., R. G. Bromley, and S. G. Pemberton. 1984. Ichnology: Trace fossils in sedimentology and stratigraphy. Society of Economic Paleontologists and Mineralogists, Short Course, Number 15.

Franzen, C. 1982. A Silurian crinoid thanatotope from Gotland. *Geologiska Foreningens i Stockholm Forhandlingar,* 103(4):469–90.
Goodman, W. M. and C. E. Brett. 1994. Tectonics vs. eustatic controls on the stratigraphic architecture of the Silurian northern Appalachian basin. *Society for Sedimentary Geology (SEPM), Concepts in Sedimentology and Paleontology,* vol. 4, pp. 147–69.
Gould, S. J. 1989. *Wonderful Life: The Burgess Shale and the Nature of History.* New York: Norton.
Grabau, A. W. 1901. Guide to the geology and paleontology of Niagara Falls and vicinity. *Buffalo Society of Natural Sciences Bulletin,* vol. 1.
Guensburg, T. E. 1992. Paleoecology of hardground encrusting and commensal crinoids, Middle Ordovician, Tennessee. *Journal of Paleontology* 66(1):129–47.
Hall, J. 1852. Containing descriptions of the organic remains of the lower middle division of the New York System. *Paleontology of New York* 2:i-vii.
———. 1860. Observations upon a new genus of crinoidea: *Cheirocrinus. Contributions to Paleontology 1858 and 1859, New York State Cabinet Natural History Annual Report,* vol. 13.
Hayes, M. O. 1967. Hurricanes as geological agents: Case studies of Hurricane Carla, 1961, and Cindy, 1963. University of Texas, Bureau of Economic Geology, Report of Investigation, Number 61.
Hurst, J. M. 1975. Wenlock carbonate level-bottom brachiopod dominated communities from Wales and the Welsh Borderland. *Palaeogeography, Palaeoclimatology, Palaeoecology* 17:227–55.
Jennette, D. C. and W. A. Pryor, 1993. Cyclic alternations of proximal and distal storm facies on a prograding ramp: Examples from the Kope and Fairview Formations (Upper Ordovician), Ohio and Kentucky. *Journal of Sedimentary Petrology* 63:183–203.
Kauffman, E. G. 1986. High-resolution event stratigraphy: Regional and global Cretaceous bio-events. In O. Walliser, ed., *Global bio-events,* pp. 279–337. New York: Springer-Verlag.
———. 1988. Concepts and methods of high-resolution event stratigraphy. *Annual Review of the Earth and Planetary Sciences* 16:605–54.
Kelling, N. and N. Williams. 1967. Flume studies of orientation of pebbles and shells on the continental shelf. *Journal of Sedimentary Petrology* 38:1264–69.
Kesling, R. V. 1969. A new brittle-star from the Middle Devonian Arkona Shale of Ontario. *Contributions from the Museum of Paleontology, University of Michigan,* 23(2):37–51.
Kesling, R. V. 1970. *Drepanaster Wrighti,* a new species of brittle-star from the Middle Devonian Arkona Shale of Ontario. *Contributions from the Museum of Paleontology, University of Michigan* 23(4):73–79.
———. 1971. *Proctothylacocrinus berryorum,* a new crinoid from the Middle Devonian Arkona Shale of Ontario. *Contributions from the Museum of Paleontology,* University of Michigan 23(21):343–47.
Kesling, R. V. and R. B. Chilman. 1975. Strata and megafossils of the Middle Devonian Silica Formation. *University of Michigan, Museum of Paleontology Papers on Paleontology,* Number 8.
Kidwell, S. and D. W. J. Bosence. 1991. Taphonomy and time-averaging of marine shelly faunas. In P. A. Allison and D. E. G. Briggs, eds., *Taphonomy: Releasing the Data Locked in the Fossil Record,* pp. 115–209. New York: Plenum Press.
Kier, P. M. 1952. Echinoderms of the Middle Devonian Silica Formation of Ohio. *Contributions from the Museum of Paleontology, University of Michigan* 10(4):59–81.
Lane, N. G. 1963. The Berkeley crinoid collection from Crawfordsville, Indiana. *Journal of Paleontology* 37:1001–1008.
———. 1969. Crinoids and reefs. *Proceedings of the North American Paleontological Convention* 1:1430–43.

———. 1973. Paleontology and paleoecology of the Crawfordsville fossil site (upper Osagian, Indiana). *University of California Publications in Geological Sciences* 99:1–141.

Laudon, L. R. and B. H. Beane. 1937. The crinoid fauna of the Hampton Formation at LeGrand Iowa. *University of Iowa Studies in Natural History* 17:227–72.

Laudon, L. R., J. M. Parks, and A. C. Spreng. 1952. Mississippian crinoid fauna from the Banff Formation, Sunwapta Pass, Alberta. *Journal of Paleontology* 26(4):544–75.

Liddell, W. D. 1975. Recent crinoid biostratinomy. *Geological Society of America Abstracts with Programs* 7:1169.

Meyer, D. L. 1971. Post-mortem disarticulation of Recent crinoids and ophiuroids under natural conditions. *Geological Society of America Abstracts with Programs* 3:645.

———. 1990. Population paleoecology and comparative taphonomy of two edrioasteroid (Echinodermata) pavements: Upper Ordovician of Kentucky and Ohio. *Historical Biology* 4:155–78.

Meyer, D. L., R. C. Tobin, W. A. Pryor, W. B. Harrison, and R. G. Osgood. 1981. Stratigraphy, sedimentology, and paleoecology of the Cincinnatian Series (Upper Ordovician) in the vicinity of Cincinnati, Ohio. In T. G. Roberts, ed., *Geological Society of America, Cincinnati, 1981 Field Trip Guidebooks,* vol. 1, pp. 31–71, American Geological Institute, Alexandria, Virginia.

Meyer, D. L., W. I. Ausich, and R. E. Terry. 1989. Comparative taphonomy of echinoderms in carbonate facies: Fort Payne Formation (Lower Mississippian) of Kentucky and Tennessee. *Palaios* 4:533–52.

Nagle, N. 1967. Wave and current orientation of shells. *Journal Sedimentary Petrology* 37:1124–38.

O'Brien, N. R., C. E. Brett, and W. L. Taylor. 1992. Taphonomy and microfabric analysis: Keys to the recognition of sedimentary processes in Silurian shales. *Abstracts with Programs, Geological Society of America Annual Meeting,* Cincinnati, Ohio, A345.

———. 1944. The significance of microfabric and taphonomic analysis in determining sedimentary processes in marine mudrocks: Examples from the Silurian of New York, *Journal of Sedimentary Research, Part A: Sedimentary Petrology and Processes,* vol. A64, pp. 847–52.

Parsons, K. M., C. E. Brett, and K. B. Miller. 1988. Taphonomy and depositional dynamics of Devonian shell-rich mudstones. *Palaeogeography, Palaeoclimatology, and Palaeoecology* 63:109–34.

Reineck, H. E. and I. B. Singh. 1972. Genesis of laminated sand and graded rhythmites in storm sand layers of shelf mud. *Sedimentology* 18:123–28.

Ringueberg, E. N. S. 1888. Niagara shales of western New York: A study of their origin and of their subdivisions and faunas. *American Geologist* 1:264–72.

Rosenkranz, D. 1971. Zue sedimentologie und ökologie von echinodermen-Lagerstätten. *Neues Jahrbuch für Geologie und Paläontologie Abhandlungen* 138:221–58.

Schäfer, W. 1972. *Ecology and paleoecology of marine environments.* Chicago: University of Chicago Press.

Schmidt, W. E. 1934. Die Crinoideen des rheinischen Devons, Teil I, Die Crinoideen des Hunsrückschiefers. *Abhandlungen der Preussischen Geologischen Landesanstalt (Neue folge)* 163:1–149.

Schmidt, W. E. 1941. Die Crinoideen des rheinischen Devons, Teil II, A. Nachtrag zu: Die Crinoideen des Hunsrückschiefers. B. Die Crinoideen des Unterdevons bis zur Cultrijugatus-Zone (mit Ausschluss des Hunsrückschiefers). *Abhandlungen der Reichsstelle fur Bodenforschung (Neue folge)* 182:113–230.

Schuchert, C. 1904. On Siluric and Devonic cystidea and *Camarocrinus. Smithsonian Miscellaneous Collections,* 47(2):201–72.

———. 1913. Systematic paleontology of the Lower Devonian deposits of Maryland; Cystoidea. *Maryland Geological Survey, Lower Devonian,* pp. 227–48.

Seilacher, A., W. E. Reif, and F. Westphal. 1985. Sedimentological, ecological and temporal patterns of fossil Lagerstätten. *Philosophical Transactions of the Royal Society of London* B311:5–23.

Speyer, S. E. 1988. Comparative taphonomy and paleoecology of trilobite Lagerstätten. *Alcheringa* 11:205–32.

Speyer, S. E. and C. E. Brett. 1985. Clustered trilobite assemblages in the Middle Devonian Hamilton Group. Lethaia 18:85–103.

Springer, F. 1911. On a Trenton echinoderm fauna at Kirkfield, Ontario. Canada, Department of Mines and Resources, Geological Survey, Memoir 15–P.

———. 1920. The crinoidea Flexibilia. *Smithsonian Institution Publication* 2501.

———. 1926. Unusual forms of fossil crinoids. *United States National Museum, Proceedings*, vol. 67, article 9.

Sprinkle, J. 1982. Echinoderm faunas from the Bromide Formation (Middle Ordovician) of Oklahoma. *University of Kansas Paleontology Contributions*, Monograph 1.

Sprinkle, J. and M. W. Longmann. 1977. Echinoderm faunas and paleoecology of the Bromide Formation (Middle Ordovician) of Oklahoma. *Journal of Paleontology* 51:26.

Strimple, H. L. and R. C. Moore. 1971. Crinoids of the LaSalle Limestone (Pennsylvanian) of Illinois. *University of Kansas, Paleontological Contributions*, article 55 (Echinoderms 11).

Sturmer, W. F. 1970. Soft parts of cephalopods and trilobites: Some surprising results of X-ray examination of slates. *Science* 1170:1300–1302.

Tetreault, D. 1992. Paleoecologic implications of epibionts on the Silurian Liichid Trilobite, *Arctinurus:* Fifth North American Paleontologica Convention, Abstracts and Programs, p. 289.

———. 1994. Brachiopod and trilobite biofacies of the Rochester Shale (Siluricum, Wenlockian Series) in western New York. In E. Landing, ed., *Studies in Stratigraphy and Paleontology in Honor of Donald W. Fisher*. New York: New York State Museum, Bulletin No. 481, pp. 347–61.

Vossler, S. M. and S. G. Pemberton. 1988. Superabundant *Chondrites:* A response to storm-buried organic material? *Lethaia* 21:94.

Walcott, C. D. 1875. Notes on *Ceraurus pleurexanthemus* Green. *Annals of the Lyceum of Natural History* 11(7–8):18–20.

———. 1881. The Trilobite: New and old evidence relating to its organization. *Bulletin of the Museum of Comparative Zoology* 10:191–224.

Warme, J. W. 1975. Borings as trace fossils, and the processes of marine bioerosion. In R. W. Frey, ed., *The Study of Trace Fossils*, pp. 181–227. New York: Springer-Verlag.

Watkins, R. and J. M. Hurst. 1977. Community relations of Silurian crinoids at Dudley, England. *Paleobiology* 3:207–17.

Whittington, H. B. and S. Conway Morris, eds., 1985. Extraordinary fossil biotas: Their ecological and evolutionary significance. *Philosophical Transactions of the Royal Society of London, Series B: Biological Sciences* 311:1–192.

Witzke, B. 1990. Palaeoclimatic constraints for Palaeozoic palaeolatitudes of Laurentia and Euramerica. In W. S. McKerrow and C. R. Scotese, eds., *Palaeozoic Palaeogeography and Biogeography*, pp. 57–73, Geological Society Memoir No. 12.

Wu, X. 1982. Storm-generated depositional types and associated trace fossils in lower Carboniferous shallow-marine carbonates of Three Cliffs Bay and Ogmore-By-Sea, South Wales. *Palaeogeography, Palaeoclimatology, Palaeoecology* 39:187–202.

Ziegler, D. H., K. S. Hanson, M. E. Johnson, M. A. Kelly, C. R. Scotese, and R. van der Voo. 1977. Silurian continental distributions, paleogeography, climatology and biogeography. *Tectonophysics* 40:13–51.

9

Middle Devonian (Givetian) Trilobite Clusters from Eastern Iowa and Northwestern Illinois

William J. Hickerson

ABSTRACT

Middle Devonian (Late Givetian) trilobite cluster horizons are known from six discrete beds within the basal 4.5 m of the Little Cedar Formation, Cedar Valley Group, in Rock Island County, Illinois, and adjacent eastern Iowa. Three geographically restricted cluster horizons (S1, S2, and S3) are present in the Solon Member (coralline calcarenites), two within the basal 50 cm in Rock Island County; a third, poorly studied horizon is known from the upper 25 cm of the Solon Member in Johnson County, Iowa. Another three laterally continuous (120–160 km) cluster beds (R1, R2, and R3) are present in the basal 1 m of the more argillaceous Rapid Member (brachiopod/bryozoan calcilutites). Two biostratigraphically exclusive trilobite assemblages occur, a highly diverse (17 species) cosmopolitan Solon fauna and a less diverse (four species) basal Rapid association. The cluster horizons are analyzed in terms of (1) trilobite species present, (2) lithology and environment of deposition, (3) associated macrofauna, and (4) trilobite taphonomy. Data are compared to trilobite cluster beds from the Hamilton Group (Givetian) of New York, which has a similar trilobite fauna. Trilobites from the lower five horizons were buried in sediments deposited during a bottom-smothering event. Trilobites from cluster horizon R3 were hydrodynamically concentrated as a lag at the terminations of large infill burrows.

Clustered Middle Devonian (Givetian) trilobite assemblages have been reported by Speyer and Brett (1985) and Speyer (1987) from the Hamilton Group of New York. Similar assemblages are known from the Little Cedar Formation (Late Givetian) of the Cedar Valley Group (L. Givetian/E. Frasnian), in Rock Island

County, Illinois (fig. 9.1) (Hickerson et al. 1990; Hickerson 1992b). Speyer and Brett (1985) define a cluster as "a group of three or more trilobites along a single bedding plane in which adjacent individuals are no more than two centimeters from one another." Groups of complete individuals are referred to as body clusters, whereas groups of exuviae are termed *moult clusters.* Body clusters occur within six distinct beds at the base of the Little Cedar Formation.

Little Cedar trilobite clusters contain complete specimens (enrolled, reflexed, and outstretched), disarticulated individuals, as well as isolated sclerites that may represent reworked exuviae. In Rock Island County two body cluster horizons occur directly above the base of the Solon Member and three trilobite cluster horizons are known from the basal 2 m of the Rapid Member (fig. 9.2). Moult clusters are difficult to recognize; most Little Cedar body clusters contain isolated sclerites and slightly disarticulated exoskeletons. Two interpretations are possible: These are (1) moult assemblages in various stages of disarticulation or (2) body clusters that have been partially disarticulated by scavengers. The problem of differentiating moulted exoskeletons from trilobite carcasses is apparent in the trilobite cluster illustrated in plate 9.1, number 7. Three complete specimens of *Dechenella? elevata* are apparently preserved here. However, detached hypostomes lie near the cephalon of each outstretched specimen, which is strong evidence that these are actually moulted exoskeletons. Without the fortuitous close association of detached hypostomes this assemblage would be mistaken for a body cluster. Rare monospecific moult clusters are present in the

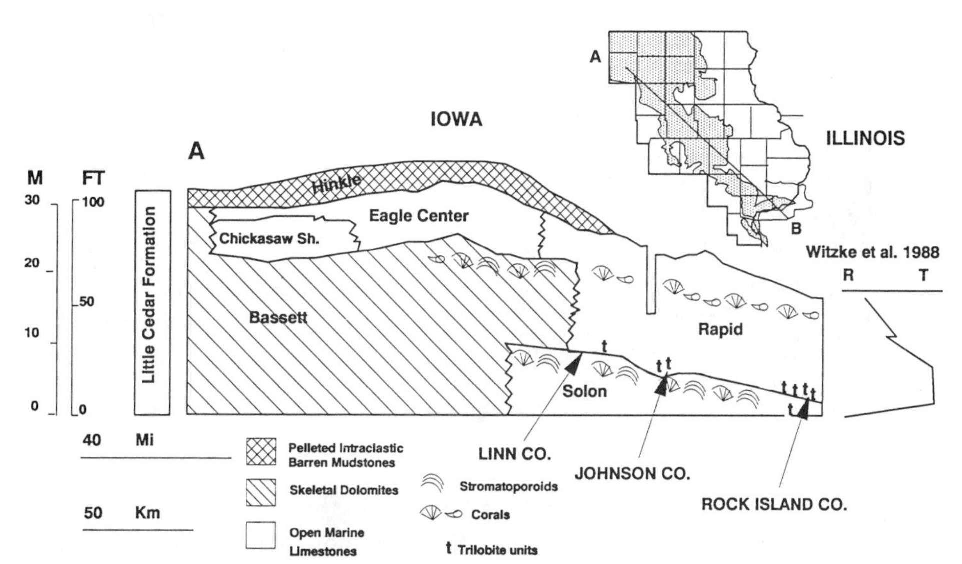

FIGURE 9.1. Generalized regional lithostratigraphic cross-section from north central Iowa to northwestern Illinois. Stippled area on map shows Devonian outcrop belt. A sea-level curve for Little Cedar Formation is at the far right. Dominant lithologies, trilobite cluster horizons, and coral/stromatoporoid biostromes are represented. Modified from Witzke et al. (1988) (Hickerson et al. 1990).

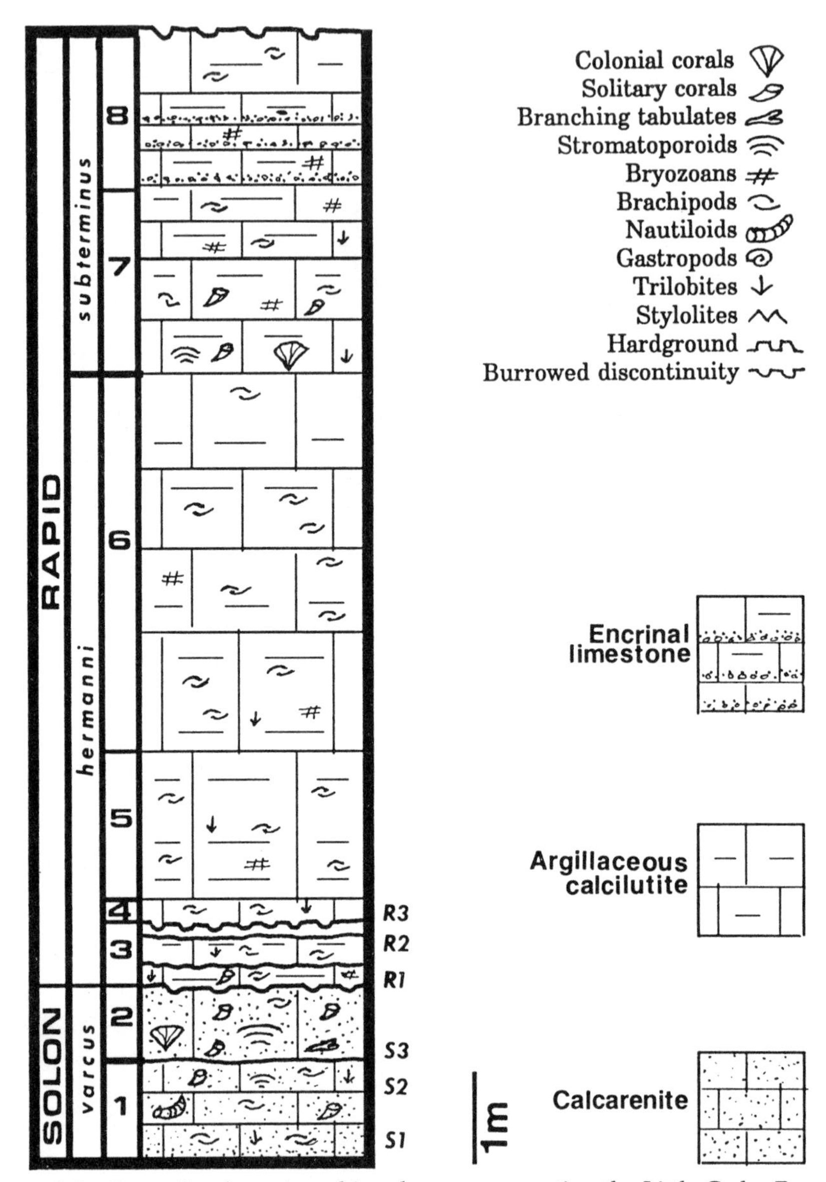

FIGURE 9.2. Generalized stratigraphic column representing the Little Cedar Formation of Rock Island County, Illinois, showing the position of trilobite cluster beds. Conodont zonation and stratigraphic unit numbers are shown.

overlying Middle Rapid Member (Units 5 and 6); in this more argillaceous strata individual moults (plates 9.2, 9.3) are fairly common, while moult clusters of *Phacops norwoodensis* and *Neometacanthus barrisi* (plate 9.3, number 9) are rare. Trilobite remains from the higher-energy, crinoid-rich Upper Rapid (Units 7 and 8) are almost exclusively isolated sclerites or moult ensembles (plate 9.3, number 4). Complete exoskeletons collected from Units 7 and 8 (plate 9.3,

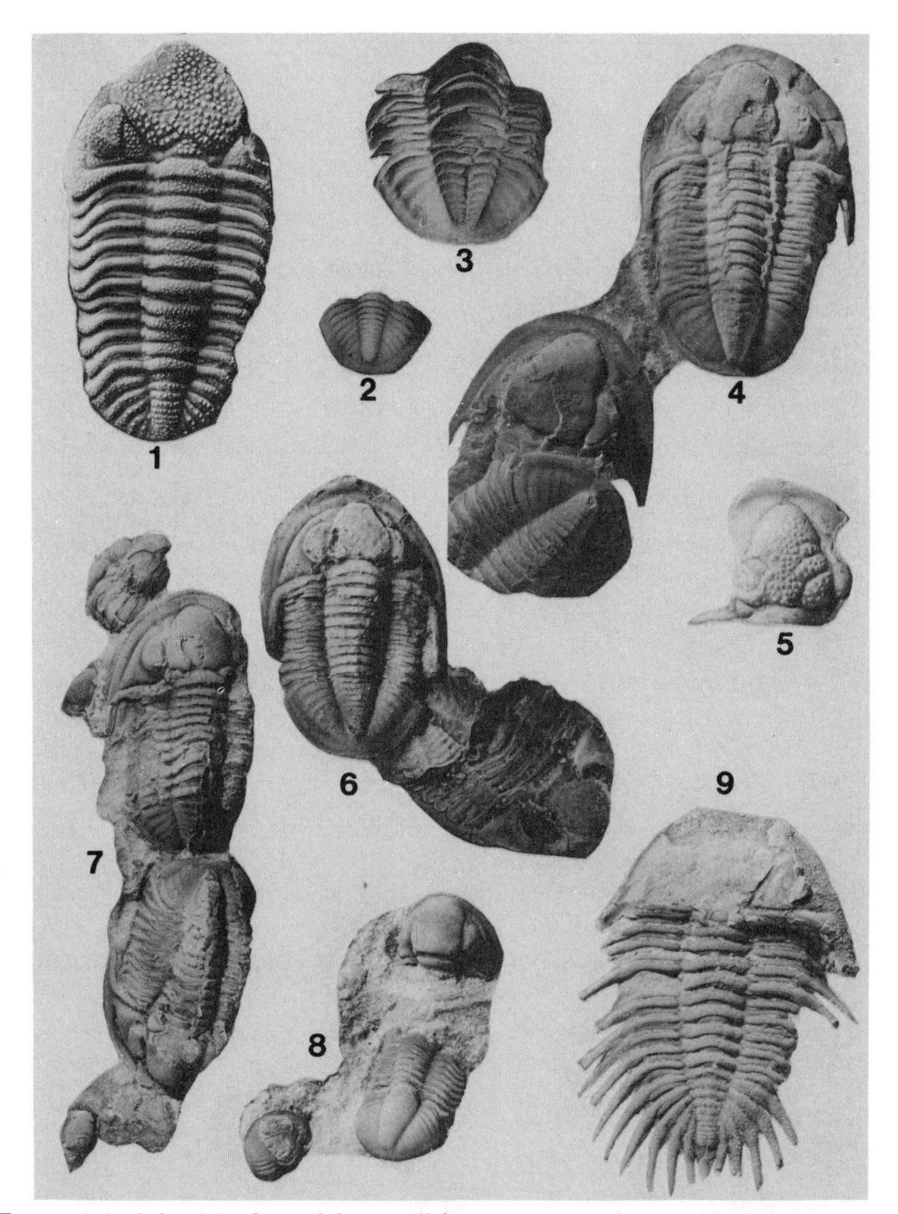

PLATE 9.1. Solon Member trilobites. All figures $\times 2$. (1) *Phacops iowensis iowensis*, SUI 9266, latex cast of holotype, Linn Junction, Linn Co., Iowa; (2) *Eudechenella haldemani*, SUI 9096, Linn Junction, Linn Co., Iowa; (3) *Dechenella? elevata*, locality MRL, Rock Island Co., Illinois; (4) *D.? elevata*, SUI 80690, locality MRL, Rock Island Co., Illinois; (5) *Dechenella nortoni*, SUI 9098, Linn Junction, Linn Co., Iowa; (6) *Dechenella? prouti*, SUI 80687, locality MRL, Rock Island Co., Illinois; (7) *D.? elevata*, locality MRL, Rock Island Co., Illinois; (8) *Crassiproetus arietinus*, SUI 80676, confluence of Rock and Mississippi Rivers, Rock Island Co., Illinois; (9) *Neometacanthus* sp., SUI 80665, locality MRL, Rock Island Co., Illinois.

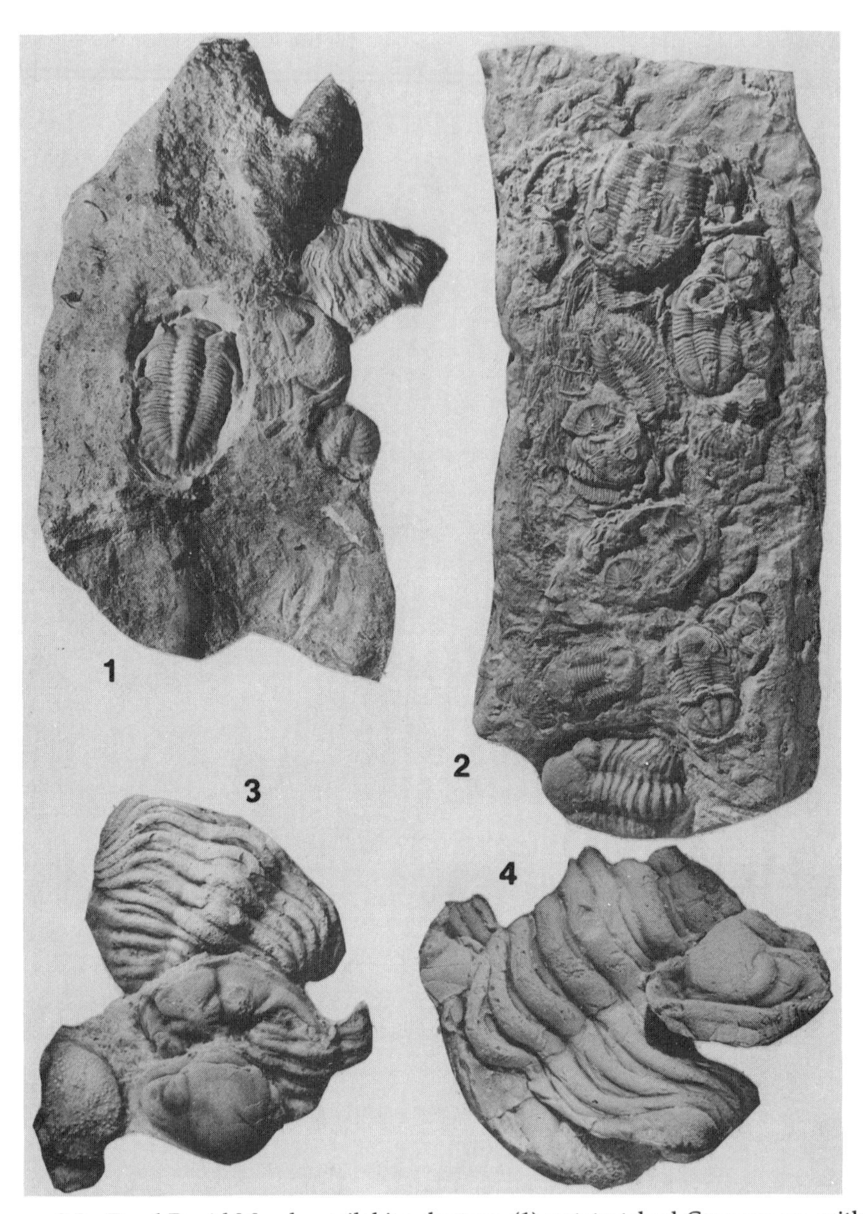

PLATE 9.2. Basal Rapid Member trilobite clusters. (1) outstretched *Greenops* sp. with three enrolled/reflexed *Phacops norwoodensis*, *Cruziana* bed, Unit 4, locality MRL, Rock Island Co., Illinois, ×1; (2) *Rusophycus* termination of a *Cruziana* trace containing sclerites, enrolled, and outstretched *Greenops* sp. as well as three outstretched *Crassiproetus* sp. B, a reflexed *P. norwoodensis* and partially articulated blastoid *(Nucleocrinus melonformis?)* with brachioles preserved. SUI 80682, locality MRL, *Cruziana* bed, Unit 4, Rock Island Co., Illinois, ×1; (3) *P. norwoodensis* (enrolled and cephalon), *Greenops* sp. (cephalon), and *Crassiproetus* sp. B (cephalon), SUI 80689, *Cruziana* bed, Unit 4, locality MRL, Rock Island Co., Illinois, ×2; (4) enrolled *Crassiproetus* sp. B on enrolled *P. norwoodensis*, Unit 3, locality MSQ, Rock Island Co., Illinois, ×2.

PLATE 9.3. Rapid Member trilobites. All specimens ×2. (1) *Crassiproetus occidens*, SUI 80621, *Cruziana* bed, Unit 4, locality MC, Rock Island Co., Illinois; (2) *Neometacanthus barrisi*, Unit 5, locality MRL, Rock Island Co., Illinois; (3) *N. barrisi*, SUI 80662, Unit 6, locality ASQ, Rock Island Co., Illinois; (4) *Crassiproetus bumastoides*, SUI 80687, Unit 8, Buffalo, Scott Co., Iowa; (5) *C. bumastoides*, SUI 80688, Unit 8, Milan, Rock Island Co., Illinois; (6) *Greenops* sp., SUI 80661, Unit 3, locality MSQ, Rock Island Co., Illinois; (7) *Dechenella?* s. B, SUI 80668, Unit 7, locality DCQ, Scott Co., Iowa; (8) *Crassiproetus* sp. B, *Cruziana* bed, Unit 4, locality MRL, Rock Island Co., Illinois; (9) Moult assemblages of *Neometacanthus barrisi*, SUI 80660, Unit 5?, Muscatine Co., Iowa.

number 7) number less than 10, whereas more than 250 complete specimens were recovered from the lower Solon and Rapid members.

The Hamilton Group is slightly older than the Little Cedar Formation (fig. 9.3); however, most of the same genera occur in both (*Phacops, Greenops, Neometacanthus, Dechenella?, Eudechenella, Cyphaspis?, Mystrocephala,* and *Scutellum*). Trilobites species found in the Solon Member are similar or in some cases conspecific with Hamilton species. Possible Hamilton Group trilobite species present in the Solon Member are *Phacops rana, Eudechenella haldemani,* and *Dechenella? rowi.* In addition, *Scutellum depressum* is morphologically similar to *S. tullium* from the Tully Limestone, and *Cyphaspis?* sp. may prove to be conspecific with *C.? craspedota.* This trilobite faunal overlap allows close comparison with taphonomic and cluster data published by Speyer and Brett (1985, 1986) and Speyer (1987).

Material and Localities

More than 800 trilobite specimens, ranging from isolated sclerites to complete individuals, were collected from localities in Rock Island County, Illinois. Additional specimens come from Johnson (75 specimens) and Scott (120 specimens)

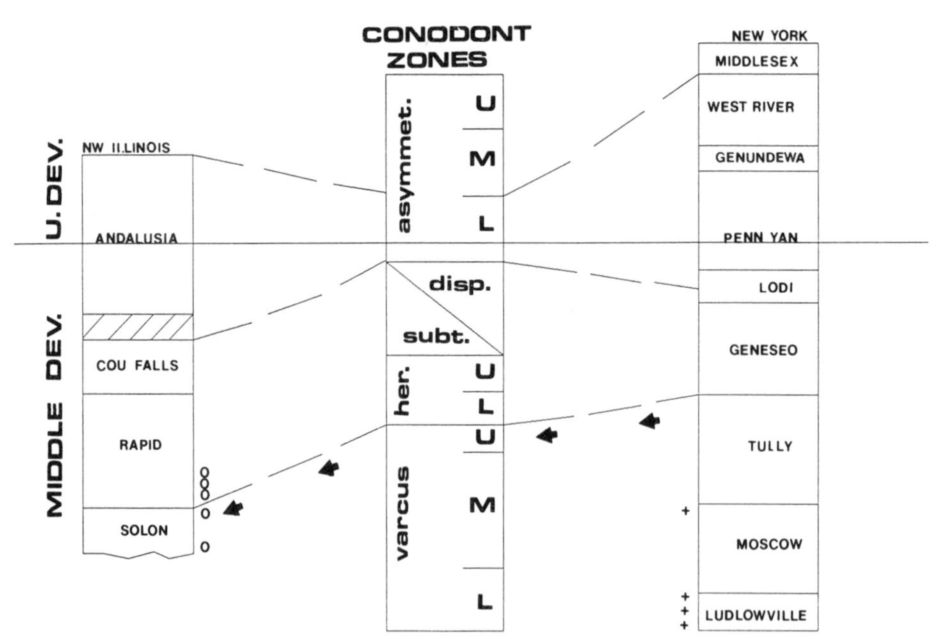

FIGURE 9.3. Diagram showing the relationship between the studied interval in northwestern Illinois and equivalent strata in New York, based on conodont biostratigraphy. Large black arrows indicate approximate position of Taghanic Onlap (cycle IIa-2, Johnson et al. 1985). O = Little Cedar trilobite cluster horizons and + = Hamilton group trilobite cluster beds (Kralick in Hickerson et al. 1990).

counties, Iowa (fig. 9.1). Rock Island County collecting localities include Milan Stone Quarry (MSQ), Milan; Allied Stone Quarry (ASQ), Rock Island; natural exposures and a small abandoned quarry along Mill Creek (MC); and transported slabs from the levee along the Mississippi River (MRL) between the cities of Rock Island and Moline. Several localities west and northwest of Rock Island, in eastern Iowa, were visited to determine the geographic distribution of the trilobite cluster beds. Eastern Iowa localities include Davenport Cement Company Quarry (DCCQ), Buffalo, Scott County; River Products Conklin Quarry (RPCQ), Coralville, Johnson County; Lake Macbride Spillway (LMS), Johnson County; and Four Counties Quarry (FCQ), Johnson County. Supplemental specimens from the Fryxell Geology Museum, Augustana College, Rock Island; Putnam Museum, Davenport, Iowa; and the fossil repository at the University of Iowa (SUI), Iowa City, were examined.

Trilobites collected from the cluster horizons were prepared using an air abrasive unit, identified, and then analyzed for taphonomic and biostratigraphic implications. Most trilobite specimens from the Solon Member were found by spitting large slabs and then carefully examining each broken piece for the smallest sclerites. Because the slabs are more likely to split around larger fossils, sclerites are probably more abundant than indicated by the sample counts from this member. Trilobite specimens from the basal Rapid Member cluster horizons (R1 and R2) were collected from the surface of weathered bedding planes, so the data may be somewhat biased because of preferential collection of complete specimens. Sclerites are certainly more abundant than indicated. Alternatively, counts from the *Cruziana* bed (base of unit 4) are the most accurate in this horizon because of the known trace fossil disposition relative to the overlying interbed; trilobite exoskeleton orientation can be recorded from even loose blocks. This highly burrowed surface is quite distinctive and is stratigraphically unique, allowing transported blocks to be used in this study.

Little Cedar Formation Stratigraphy and Environments of Deposition

The Devonian outcrop belt trends northwest to southeast across eastern Iowa into northwestern Illinois (fig. 9.1). Little Cedar Formation lithologies in Rock Island County are represented by skeletal-rich calcilutites (mudstone, wackestone) and calcarenites (packstone, rare grainstone). Two members are present, the coral/stromatoporoid-dominated, calcarenitic Solon Member ranges from 1.5 to 3 m in thickness, and the overlying argillaceous Rapid Member calcilutites are approximately 17 m thick. The Little Cedar Formation is underlain by limestone breccia and sublithographic calcilutites of the Davenport Member of the Pinicon Ridge Formation and overlain by coralline calcarenites of the Coralville Formation.

Traditionally, the Solon Member has been divided into two faunal zones (Stainbrook 1935, 1941), with the lower half dominated by atrypid brachiopods (*Independatrypa independensis* Zone) and an upper colonial coral/stromatoporoid-dominated interval (*Hexagonaria profunda* Zone). Solon Member lithologies consist of fine- to medium-grained packstones with minor shaley partings. Bedding is massive, irregular, or blocky and is brecciated in places. At most localities a hardground surface or burrowed discontinuity separates the Solon from the overlying Rapid Member. Conodont elements recovered from the Solon Member are diagnostic of the upper middle varcus Subzone (Witzke et al. 1985; fig. 9.3). Solon Member deposition occurred during the initial phases of a major marine transgression (Taghanic Onlap) and is characterized by normal marine salinities in a subtidal setting (Witzke et al. 1985).

Lower Rapid Member lithologies are dominated by brachiopod/bryozoan-rich argillaceous calcilutites (wackestone) separated by sparsely fossiliferous to barren burrowed calcilutite (mudstone) interbeds (fig. 9.2). The basal 1.5 m (units 3 and 4) differ faunally and lithologically from the rest of the Lower Rapid in being less argillaceous and containing common solitary rugose and tabulate corals, which are rare or absent in the overlying 10 m (Udden 1897; Savage and Udden 1921). Upper Rapid Member lithologies consist of crinoidal calcarenites (packstone) with argillaceous calcilutite (wackestone) interbeds. A coral biostrome is present locally, approximately 4 m below the burrowed discontinuity at the base of the overlying Coralville Formation. Rapid Member strata below the coral biostrome are assigned to the hermanni conodont subzone, while conodonts recovered from the biostromal and upper crinoid-dominated beds indicate the Lower Subterminus Subzone (Witzke et al. 1985). The Lower Rapid Member was deposited as the seaway deepened following Solon Member deposition, thereafter shallowing upward into higher-energy crinoid and coral-dominated facies; a minor deepening event is recorded near the top of the member (Witzke et al. 1988).

Little Cedar Formation Trilobite Cluster Horizons

At least six trilobite cluster horizons occur in the basal 5 m of the Little Cedar Formation in Rock Island County (fig. 9.4). Three are present at the base of the Solon Member (S1, S2, and S3), and three (R1, R2, and R3) occur at the base of the Rapid Member. Another poorly sampled trilobite cluster horizon is present at the top of the Solon Member in Rock Island County and at locality LMS in Johnson County, Iowa. Trilobite clusters from two distinctly different trilobite faunas are well represented; there is no overlap of trilobite species between the Solon and Rapid members (fig. 9.5). The reason for the segregation of the two trilobite faunas may be related to a missing time interval represented by the hardground/burrowed discontinuity that divides the two members, or it may be related to faunal changes along an environmental gradient. Faunally, the uppermost Solon Member is similar to the lowest Rapid Member. Both units contain common spiriferid brachiopods (*Orthospirifer iowensis*), solitary rugose/tabulate corals, and the trilobite genera *Phacops* and *Crassiproetus*. Lithologically, the upper Solon Member is a fine-grained calcarenite, while the basal Rapid Member is argillaceous calcilutite (wackestone). In addition, the trilobites differ at the species level, *Phacops rana?* is found in the Upper part of the Solon, while *P. norwoodensis* is present in the lowest Rapid bed. The upper Solon Member also contains common stromatoporoid grains in petrographic thin section, while stromatoporoids are rare in the Lower Rapid Member. Thus the environments of deposition for the upper Solon and Lower Rapid are distinct.

Givetian-Frasnian Trilobites from Eastern Iowa and Northwestern Illinois

FRASNIAN
- LIME CREEK FORMATION
  - Owen Member
    - *Scutellum thomasi* (Walter 1923)
  - Independence Shale
    - *Harpidella? brandonensis* (Walter 1923)

GIVETIAN
- CORALVILLE FORMATION
  - COU Falls Member
    - *Crassiproetus searighti* (Walter 1923)
    - *Mystrocephala raripustulosa* (Walter 1923)
    - *Dechenella?* sp. C
- LITTLE CEDAR FORMATION
  - Rapid Member
    - *Phacops norwoodensis* (Stumm 1953)
    - *Neometacanthus barrisi* (Hall 1888)
    - *Greenops* sp.
    - *Neometacanthus? fitzpatricki* (Walter 1923)
    - *Crassiproetus occidens* (Hall 1861)
    - *Crassiproetus bumastoides* (Walter 1923)
    - *Crassiproetus* sp. B
    - *Dechenella?* sp. B
  - Solon Member
    - *Phacops iowensis iowensis* Delo 1935
    - *Phacops rana?* (Green 1838)
    - *Neometacanthus* sp.
    - *Crassiproetus arietinus* (Walter 1923)
    - *Crassiproetus* sp. A
    - *Dechenella? prouti* (Shumard 1863)
    - *Dechenella? elevata* Cooper and Cloud 1938
    - *Dechenella? rowi?* (Green 1838)
    - *Dechenella?* sp. A
    - *Dechenella nortoni* (Walter 1923)
    - *Eudechenella haldemani* (Hall 1861)
    - *Eudechenella* sp. A
    - *Eudechenella* sp. B
    - *Mystrocephala pulchra* Cooper and Cloud 1938
    - *Cyphaspis* sp.
    - *Proetus?* sp.
    - *Scutellum depressum* Cooper and Cloud 1938

FIGURE 9.4. Trilobites occurring in Late Givetian/Late Frasnian strata of eastern Iowa and northwestern Illinois.

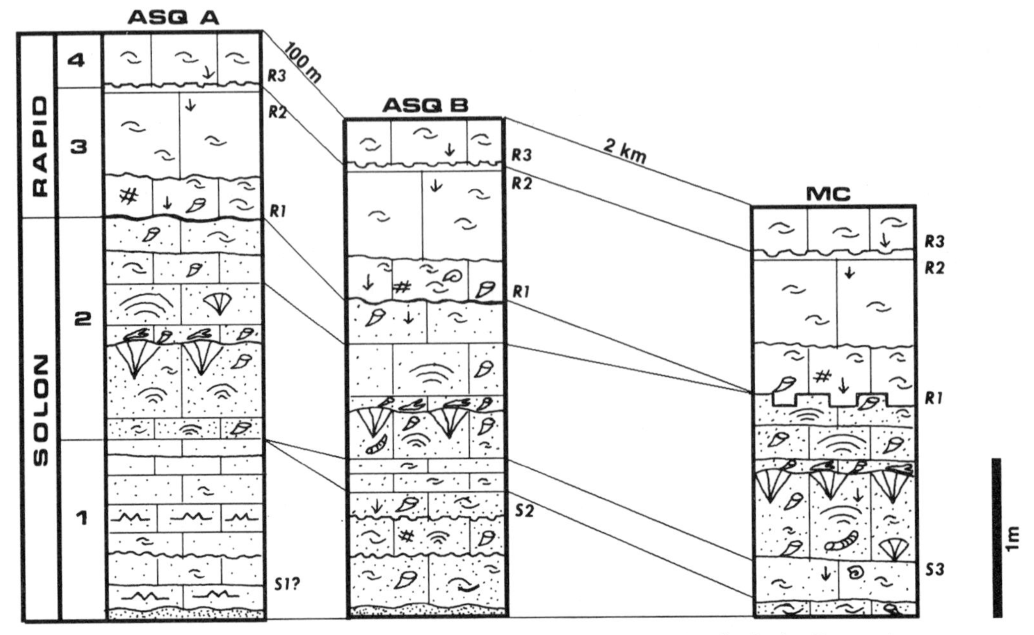

FIGURE 9.5. Measured stratigraphic sections of the Lower Little Cedar Formation correlated from Allied Stone Quarry (ASQ A and B) south (2 km) to Mill Creek in Rock Island County, Illinois. Solon Member (S1 and S2) and Rapid Member (R1, R2, and R3) trilobite cluster beds. Stratigraphic unit numbers are indicated.

In the following section each trilobite cluster bed is discussed with respect to (1) trilobite species present, (2) lithology of associated sediment, (3) associated shelly macrofauna, and (4) trilobite taphonomy.

BASAL SOLON MEMBER TRILOBITE CLUSTER HORIZONS

Trilobite Association *Dechenella? elevata* (plate 9.1, numbers 3, 4, and 7) and *Phacops iowensis iowensis* (plate 9.1, number 1) dominate this diverse trilobite fauna (Hickerson 1992). *Crassiproetus arietinus* (plate 9.1, number 8), *Dechenella? prouti* (plate 9.1, number 6), and *Neometacanthus* sp. (plate 9.1, number 9) are common locally. *Eudechenella* sp. B occurs rarely at several localities in which trilobite body clusters are present, while pygidia of *E. haldemani* (plate 9.1, number 2 and *E.* sp. A are restricted to an argillaceous calcarenite developed at the base of the Solon Member in Scott County, Iowa (DCCQ). *Mystrocephala pulchra* does not occur in the cluster beds. This trilobite is restricted to a 12-cm-thick bed within the lower part of the profunda Zone, associated with terebratulid brachiopods (*Cranaena*), rostroconchs, laminar stromatoporoids, and solitary rugose corals. *Phacops rana? Cyphaspis?* sp., *Proetus?* sp., *Crassiproetus* sp. A, *Eudechenella* sp. B, and *Scutellum depressum* occur rarely in the cluster bed. In addition, *Dechenella nortoni* (plate 9.1, number 5) and *Dechenella?* sp. B are known from the Solon Member in

Johnson County but have not been noted in Rock Island County. The distribution and relative abundance of preserved Solon trilobites is dependent on the lithofacies present at a particular locality. Species of *Dechenella?*, *Crassiproetus*, *Proetus?*, and *Mystrocephala* are associated with strata that contain either diverse mixed or coralline faunas, whereas *Eudechenella*, *Dechenella*, *Neometacanthus*, *Scutellum*, and *Cyphaspis?* prefer brachiopod-dominated argillaceous facies. Species of *Phacops* are found in all facies, with *P. iowensis iowensis* restricted to the brachiopod-dominated independensis Zone and *P. rana?* associated with the coralline fauna of the upper Solon Member.

To avoid confusion, a brief discussion concerning the taxonomic position of the genus *Dechenella* is in order. In past papers by various authors (e.g., Cooper and Cloud 1938; Stumm 1953) three distinctly different trilobite genera have been referred to the genus *Dechenella*. In this paper two "true" dechenellid genera are recognized. *Eudechenella* is characterized by species that are dorsally/ventrally flattened and possess a short, pear-shaped glabella with incised glabellar furrows and an elongated pygidium with 14–16 axial rings (plate 9.1, number 2). *Dechenella nortoni* has an exoskeleton that is more convex than *Eudechenella* and a highly tuberculate cephalon with deeply incised glabellar furrows (plate 9.1, number 5). *Dechenella?* includes forms that were mistakenly assigned to *Dechenella* (*Basidechenella*) by Stumm 1953. Species of this genus have a subconical-shaped glabella and lack deeply incised glabellar furrows (plate 9.1, figs. 9.4, 9.6, and 9.7), although on well-preserved specimens the position of the furrows is indicated by superficial color markings. The proportionally shorter, wider pygidium (plate 9.1, number 3) differs from *Dechenella* and *Eudechenella*. *Dechenella?* has fewer axial rings (10–13), and axial nodes are present on most species. It is probable these forms, exemplified by *D.? rowi* and *D.? elevata* are not dechenellids but represent a new subfamily within the Proetidae. Ormiston (1967) comments on the similarities and relationships of *Dechenella?* and the Proetidae.

Lithology A wide variety of lithologies is present in the Lower Solon Member, including calcareous sandstone, sandy calcarenite (grainstone), argillaceous fine-grained brachiopod calcarenite, fine- to medium-grained diverse fauna calcarenite, unfossiliferous fine-grained calcarenite, and minor shaley partings. Trilobite body clusters are restricted to the finer-grained/argillaceous brachiopod/diverse mixed fauna lithofacies that occurs in the lower part of the member (fig. 9.5). Numerous laterally discontinuous hardground surfaces and burrowed discontinuities add to the complexity and make correlation of trilobite-bearing bedding planes over distances of 2 km difficult. Two distinct trilobite cluster horizons can be recognized (S1 and S2). More than 90% of the trilobite clusters from the Lower Solon Member (S1) were collected from loose blocks dumped along the Mississippi River Levee. The massive nature of Solon lithologies has impeded in situ faunal studies. Blocks can only be split only after intense weathering. These blocks were dredged out of the Mississippi River bed just north of Buffalo, Scott County, Iowa, during channel-deepening operations. Lithologically and faunally, these slabs display the following characteristics: (1) fine-grained calcarenite, faintly laminated in places; (2) common dissolution surfaces and stylolites; (3) a prominent burrowed discontinuity (*Paleophycus*); (4) bivalves preserved in life position and large euomphalid gastropods; (5) sparsely fossiliferous slabs containing only rare trilobites and brachiopods, or highly fossiliferous slabs with groups of trilobites and large brachiopod clusters (25–100 specimens) dominated by *Independatrypa*.

Solon Member exposures at localities ASQ and MC will serve as reference sections. At locality ASQ, stratigraphic Section A (fig. 9.5), located in the northwest corner of the quarry, exposes approximately 3 m of Solon. Here the Lower Solon consists of sparsely fossiliferous, thinly bedded, fine-grained calcarenite separated by dissolution surfaces and burrowed discontinuities. The only fossils noted were a few small atrypid brachiopods. Section B approximately 100 m to the south, in the southwest corner of the quarry, exposes 68 cm of Lower Solon. Here the Basal Solon is coarser grained, with thicker beds separated by burrowed discontinuities, and is abundantly fossiliferous with large atrypids (*Independatrypa*), nautiloids, bryozoans, solitary rugosans, tabulate corals, stromatoporoids, placoderms, crinoidal debris, and trilobites. Section MC, 2 km south of ASQ, exposes a reduced Lower Solon interval (10 cm), with a 4-cm-thick bed containing large brachiopods (*Independatrypa, Orthospirifer)* at the base. Common stylolites and dissolution surfaces are present in the basal 15 cm. Stratigraphic placement of the trilobite clusters (S1) from transported slabs dumped along the Mississippi River Levee is attempted using lithologic and biostratigraphic evidence. *Independatrypa* is invariably associated with S1 clusters and is restricted to the Lower Solon. The highest in situ occurrence noted is 60 cm up from the base of the Solon at Section ASQ B. Clusters of small *Independatrypa* identical to those occurring with trilobite clusters from locality MRL are found only in the lowest bed at Section ASQ B. The top of this bed is covered with *Paleophycus* traces, at locality MRL loose slabs displaying *Paleophycus* traces uncommonly yield trilobite clusters when split. At Section ASQ A, this interval is sparsely fossiliferous with dissolution surfaces, again characteristic of slabs from MRL. At locality MC a trilobite body cluster was collected in situ 25 cm above the base of the Solon, just below a discontinuity surface; this horizon also displays bivalves in life position and contains euomphalid gastropods. The body cluster included *Phacops iowensis iowensis* and *Dechenella? elevata,* two species that dominate trilobite clusters from MRL. Based on this evidence, Solon trilobite cluster horizon S1 is placed with the basal 25 cm at localities ASQ and MC.

A second Lower Solon trilobite body cluster bed (S2) is present at locality ASQ in stratigraphic Section B. Articulated trilobites, inadunate crinoids (crowns), and a blastoid (with brachioles) were collected from the burrowed discontinuity surface at the base of Bed 3. *Phacops iowensis iowensis* dominates the trilobite fauna (more than 80% of trilobites) in this bed. Lithologically, this bed is coarser-grained than horizon S1 and differs faunally as well. Associated with the trilobites and echinoderms in S2 are large brachiopods (*Independatrypa*), tabulate and solitary rugose corals, and bulbous stromatoporoids. These beds represent a biostromal buildup within the independensis Zone.

The base of the Solon is sandy and varies locally from isolated sand grains to 3–6-cm-thick sandy calcarenite to 15–25-cm calcareous sandstone. Strata directly above this sandy interval are generally fine grained (*Independatrypa* beds) and coarsen upward into prominent coral/stromatoporoid biostromes in the upper part of the member. Above the biostrome the grain size is generally finer. Periodically, the Solon seafloor was swept by strong currents, as evidenced by the lack of carbonate mud and overturning of large colonial coral and stromatoporoid heads with the biostrome.

Associated Fauna

Atrypid (*Independatrypa* and *Spinatrypa*), spiriferid (Orthospirifer), and pentamerid (*Gypidula*) brachiopods are abundant in the lower part of the Solon Member at most localities. Additional common fossils associated with trilobite clusters (S1 and S2) include bivalves, euomphalid gastropods, nautiloids, and branching auloporids?. Solitary rugose corals, colonial corals, bulbous stromatoporoids, and crinoid debris are common locally.

This brachiopod association (Spiriferid/Atrypid community, Koch 1978) implies that the Lower Solon (independensis Zone) was deposited in a more offshore setting than overlying coral/stromatoporoid-dominated facies (profunda Zone). Lower profunda Zone brachiopod faunas (S3), dominated by several species of the terebratulid *Cranaena* with pentamerids, rhynchonellids, and strophomenids, are associated with shallower, higher-energy coralline facies (*Pentamerella/Orthospirifer/Cranaena* community, Koch 1978).

Trilobite Taphonomy

Trilobites occur as isolated sclerites, as moult ensembles, and as complete enrolled, reflexed, and outstretched exoskeletons. Body clusters from this horizon are typically monospecific (57%, table 9.1). Monospecific body clusters continuing *Dechenella? elevata* (plate 9.1, number 4) and *Phacops iowensis iowensis* are most common (five clusters of each), followed by *Neometacanthus* sp. (three clusters) and *Dechenella? prouti* (one cluster, plate 9.1, number 6). Mixed-species body clusters are generally dominated by one species, *P. iowensis iowensis* (two mixed clusters), *Neometacanthus* sp., *D.? elevata*, and *Crassiproetus arietinus* (one mixed cluster each). *Neometacanthus* sp. rarely occurs as individual specimens. More than 80% of known specimens are from clusters (S1), illustrating the gregarious nature of this trilobite.

Body clusters generally have isolated sclerites associated and include com-

TABLE 9.1 *Trilobite Cluster Data*

Cluster bed	T.C.	Bo.	Mou.	Pct mon.	Pct mix.	# spec.	Avg.
S1	14	10	4	57.1	42.9	84	6.0
S2	2	2	—	100.0	00.0	10	5.0
S3	1	1	—	00.0	100.0	3	3.0
S4	2	1	1	100.0	00.0	6	3.0
Solon total	19	14	5	63.2	36.8	103	5.4
R1	9	7	2	55.0	45.0	47	5.2
R2	5	4	1	60.0	40.0	20	4.0
Unit 3 total	14	11	3	57.1	42.9	67	4.8
TOTAL	33	25	8	60.6	39.4	170	5.2
R3	34	—	—	47.7	52.9	180	5.3

plete individuals as well as specimens in various stages of disarticulation. The number of specimens in a cluster ranges from three to 22, with an average of 5.7 (table 9.1). One large mixed cluster includes 15 *Dechenella? elevata* (two complete outstretched, four enrolled, two incomplete, five pygidia, and two librigenae), four *Phacops iowensis iowensis* (one enrolled, two cephala, one pygidium), and a *Scutellum depressum* pygidium. The ratio of inverted to convex up exoskeletons appears to be nearly equal for specimens from this horizon.

Biostratinomic indicators within the basal Solon cluster bed (S1) include articulated trilobites, trilobite moult ensembles, bivalves preserved in life position, burrowed surfaces, and brachiopod/trilobite clusters. Rapid burial of fauna with little or no postmortem displacement during a bottom-smothering event is evident. Multielement skeletons of arthropods can only be preserved intact if individuals are buried very rapidly, usually within hours (Speyer 1987). Arthropods can escape rapid burial, suggesting that the trilobites were killed or immobilized prior to burial (Speyer and Brett 1985; Speyer and Brett 1986).

BASAL RAPID MEMBER CLUSTER HORIZONS

Trilobite Association Two distinct trilobite cluster horizons (R1 and R2) are discussed together because of their close stratigraphic association, similar brachiopod fauna, and identical trilobite assemblage. *Phacops norwoodensis* (plate 9.2, figs. 9.1–9.4) dominates the trilobite fauna (less than 80%). *Crassiproetus* sp. B (plate 9.2, figs. 9.2–9.4; plate 9.3, number 8) and Greenops sp. (plate 9.2, figs. 9.1–9.3; plate 9.3, number 6) are also present.

LITHOLOGIES

Cluster horizon R1 is characterized by argillaceous mixed skeletal wackestone overlying the discontinuity surface developed at the top of the Solon Member. Reworked corals, phosphatic clasts, lithoclasts, and ptyctodont dental plates are present as a lag within the basal 6 cm. The basal 25 cm of Unit 3 is more argillaceous and weathers to a rusty brown color. The base of this unit represents a major transgressive event, as evidenced by the change from higher-energy, calcarenitic coral/stromatoporoid depositional environments of the upper Solon to a more offshore, lower-energy brachiopod-rich (spiriferids and atrypids) argillaceous calcilutite.

Cluster horizon R2 in a mixed-fauna wackestone interbed, separated from cluster bed R1 by approximately 15 cm of brachiopod-rich (atrypids) argillaceous wackestone. The bottom of this bed is burrowed, with trilobites occurring between the traces on the bedding plane and within the bed.

Associated Fauna

Solitary rugose corals and brachiopods comprise more than 80% of the macrofauna associated with cluster horizon R1. Other fauna includes common tabulate corals (*Favosites*), bryozoans, and platycerid gastropods; uncommon sponges (*Astreospongia*); nautiloids; placoderm dermal plates; and rare crinoid calyces (*Megistocrinus, Melocrinites?*). The diverse brachiopod fauna is dominated by spiriferids (*Orthospirifer, Tylothyris*), atrypids (*Neatrypa, Spinatrypa*), orthids (*Schizophoria*), and strophodontids (*Strophodonta, Protoleptostrophia*).

Trilobite bed R2 has a less diverse fauna, dominated by atrypid brachiopods (*Spinatrypa* and *Pseudoatrypa*) and lacks the coralline component present in the previous bed. The burrowed discontinuity present at the bottom of this bed and the less diverse associated fauna, lacking corals, indicates further deepening of the Little Cedar seaway.

Trilobite Taphonomy

Trilobite body clusters are less common in Unit 3 than in the basal Solon Member (S1). Trilobites occur more typically as isolated sclerites or enrolled/reflexed individuals. Half of the body clusters collected are monospecific (table 9.1). Three of the six mixed-species body clusters were dominated by *Phacops norwoodensis*. *Crassiproetus* sp. B and *Greenops* sp. dominated one each, and the other contained equal numbers of *P. norwoodensis* and *Crassiproetus* sp. B. The number of trilobites within clusters ranges from three to 14, with an average of 5.2. As with the Solon cluster bed S1 the ratio of inverted to convex-up specimens is nearly equal. More than 80% of the complete specimens are enrolled or reflexed. All enrolled specimens are sediment filled and crushed, which suggests that sediments were compacted prior to pore space preservation (Speyer and Brett 1985).

The large number of enrolled/reflexed trilobites may reflect a response by the trilobites to adverse ecological conditions. Speyer (1987) discusses enrolled trilobite beds from the Hamilton Group that are interpreted to have formed under storm event smothered-bottom conditions. The majority of enrolled P. *norwoodensis* exhibit imploded pygidia. It has been suggested that the implosion phenomenon is caused by gas buildup caused by decay within the enrolled trilobite combined with pressure resulting from compaction, forcing a disjunction at the thoraco-pygidial suture (Speyer 1987).

CRUZIANA BED CLUSTER HORIZON

Trilobite Assemblage *Phacops norwoodensis* is the most common trilobite (more than 45%), but *Greenops* sp. (29%) and *Crassiproetus* sp. B (23%) occur more commonly here than in Unit 3. *Crassiproetus occidens* (plate 9.3, number 1), the dominant proetid in overlying beds (Units 5–7), is first noted in this bed. *Greenops* sp. and *Crassiproetus* sp. B are not known to occur in strata above Unit 4 in Rock Island County, but do occur stratigraphically higher in Johnson County, Iowa.

LITHOLOGY

Hypichnial traces extend down into underlying argillaceous bed; mixed skeletal wackestone identical to sediments overlying the traces infill the burrows and trails. Phosphatic clasts occur as a lag in the bottom of the traces. Clasts are elongate or subspherical and structureless, perhaps originally fecal pellets. Commonly, the terminations of large burrows are packed with brachiopods (detached valves or complete specimens) and trilobites (commonly articulated). Partially articulated echinoderms (crinoids and blastoids) rarely occur in burrow terminations (plate 9.2, number 2). The presence of articulated trilobites and echinoderms implies the fauna was not hydrodynamically transported long distances.

ASSOCIATED FAUNA

From the standpoint of fauna this horizon is less diverse than the basal Solon and Rapid trilobite cluster beds. The orthid brachiopod *Schizophoria iowensis* dominates the macrofauna (more than 75%). *Spinatrypa bellula* and *Orthospirifer iowensis* are the only other brachiopods commonly noted. Other uncommon faunal elements include small stereolasmid corals, fenestellid bryozoans, tentaculitids, large ostracods, crinoids (*Megistocrinus*), and blastoids (plate 9.2, number 2). The orthid/spiriferid/atrypid brachiopod fauna in association with common phacopid trilobites, and rare solitary rugose corals, indicates a low-energy offshore environment. The burrowed discontinuity may represent maximum transgression of the Little Cedar Seaway.

Ichnogenera abundantly preserved on the discontinuity surface include *Cruziana, Rusophycus, Rhizocorallium, Paleophycus, and Planolites.* Similar burrowed discontinuity surfaces are present in Givetian strata of the Appalachian Basin (e.g., the Silica Formation, Northwestern Ohio [Kesling and Chilman 1975:47] and Hungry Hollow Formation, Southwestern Ontario [Landing and Brett 1987]).

TRILOBITE TAPHONOMY

This horizon is unique in that complete outstretched and enrolled trilobite exoskeletons, disarticulated exoskeletons, and isolated sclerites are associated with, and commonly occur within, traces that have been interpreted as being produced by trilobites or other arthropods (plate 9.2, figs. 9.1–9.3; plate 9.3, number 8). Trilobite body clusters are commonly preserved in the *Rusophycus*-like terminations of large, deep, arthropod excavations similar to *Cruziana transversa* (Landing and Brett 1987). Other arthropod traces exhibiting scratch marks resemble *Cruziana reticulata,* another arthropod trace from the Hungry Hollow (Landing and Brett 1987). In general, the trilobites are much smaller than the traces in which they occur, but rare, solitary enrolled specimens of *Phacops norwoodensis* are present in *Rusophycus* terminations of *Cruziana* traces of similar widths to the associated trilobites.

Trilobite body clusters preserved within traces tend to be mixed (more than 50%) and commonly contain all three genera (plate 9.2, figs. 9.1–9.3). Monospecific body clusters, individual complete trilobites, and sclerites are rarely found between traces. More than 90% of trilobite exoskeletons tabulated from this horizon are inverted. The fact that the trilobites are usually found jumbled together at the terminations of long burrows, combined with their inverted orientation, suggests that they were swept into open burrows on the seafloor by currents, perhaps related to a storm event.

Complete cephala of *Crassiproetus* are common, but isolated cranidia and librigenae are rare (plate 9.2, number 3; plate 9.3, figs. 9.5 and 9.8). This is peculiar, as proetids generally shed their librigenae during exuviation (plate 9.3, number 4). Thus either these specimens represent complete exoskeletons that were "decapitated" by currents or scavengers (plate 9.1, number 8) prior to burial or these particular proetids had fused facial sutures and did not shed librigenae during the moulting process. *Crassiproetus* and *Greenops* often occur as complete outstretched or slightly reflexed specimens, whereas *Phacops norwoodensis* exoskeletons are invariably highly reflexed or enrolled.

Discussion

Speyer and Brett (1987) noted that trilobite body clusters from the Hamilton Group are characterized by (1) segregation of species, (2) size segregation, (3) predominantly inverted specimens, and (4) restriction to narrow beds with wide geographic disposition. The most striking difference between Hamilton and Little Cedar body clusters relates to species segregation. Approximately 44% of Little Cedar clusters are mixed-species assemblages (table 9.1). This may indicate mixing of discrete Little Cedar monospecific clusters prior to burial, particularly in the *Cruziana* bed. All other Little Cedar cluster horizons are within sediments that were deposited during bottom-smothering events and exhibit no hydrodynamic shorting before burial. The presence of isolated sclerites and disarticulated exoskeletons within Little Cedar body clusters is attributed to scavenging of trilobite carcasses or discarded exuviae. It has also been suggested (Speyer and Brett 1987) that Hamilton trilobites congregated into monospecific groups to moult. Large monospecific moult clusters are present on the same bedding planes as monospecific body clusters within Hamilton cluster beds. This hypothesis is not supported by Little Cedar cluster data. It is doubtful that trilobites differing at the generic level would congregate in order to moult. It is conceivable that Little Cedar mixed-species clusters were originally closely spaced monospecific clusters that were mixed by scavenging prior to burial, but this still implies that different trilobite genera congregated in a relatively small area of the seafloor and moulted in adjacent groups. Another possibility explaining the mixing of genera is a congregating response by the trilobites to the unstable ecological conditions (increased turbidity, clouds of suspended sediment) preceding a bottom-smothering event.

Little Cedar clusters tend to be size-sorted, but some size-sorted trilobite clusters include a specimen that is significantly smaller than the bulk of exoskeletons in that group (plate 9.2, number 4). In a few clusters a range in size of the same species is evident. Many of the mixed-species clusters include similarly sized specimens of one species associated with a very small specimen of another genus. Plate 9.2, number 4 illustrates this peculiar association. This example is from cluster bed R1 and shows a very small enrolled *Crassiproetus* sp. B compressed on top of an average-sized enrolled *Phacops norwoodensis*. Another example from the basal Solon cluster bed (S1) includes three enrolled *Phacops iowensis iowensis* of average size (35 mm) and an outstretched *Dechenella? prouti*, the smallest complete specimen (17.5 mm) of this species known. A similar association was documented in Hamilton Group moult clusters (Speyer and Brett 1987). Lone complete specimens of *Greenops boothi* rarely occur within *Phacops rana* moult assemblages. This may reflect scavenging activity by *Greenops* that, along with the cluster being scavenged, was buried by rapid mud accumulation (Speyer and Brett 1987).

Inverted exoskeletons dominate (more than 90%) the *Cruziana* cluster bed (R3), while the inverted/convex-up ratios for the basal Rapid (R1, R2) and Lower Solon (S1) are approximately equal. Trilobite inversion has been interpreted as (1) a reflection of trilobite swimming orientation (upside down), (2) a response to postmortem gas buildup and subsequent flipping by gentle currents or scavengers, and (3) a function of moulting and apolysis (Speyer 1987). It is clear that *Cruziana* bed clusters were hydrodynamically concentrated within large infill burrows. This accounts for the large percentage of inverted specimens. The other basal Rapid and Solon trilobite clusters were buried with little or no postmortem movement, suggesting that occurrence of inverted exoskeletons in these clusters is related to scavenging, gas buildup, or moulting processes. Evidence of scavenging is apparent in most body clusters, indicated by the large number of disarticulated exoskeletons and isolated sclerites occurring alongside complete specimens.

The preservation of multiple trilobite cluster beds at the base of the Little Cedar Formation is attributed to rapid sedimentation rates resulting from episodic storm events. Evidence includes (1) preservation of bivalves and solitary rugose corals oriented in life position; (2) preservation of multielement skeletons (trilobites, echinoderms, and placoderms); and (3) preservation of trilobite moult ensembles. The mechanics and paleoecological significance of storm-generated deposits (distal tempestites) is discussed in detail by Miller et al. (1988).

Geographic Extent of Little Trilobite Cedar Cluster Beds

Exposures of the lower part of the Little Cedar Formation in Johnson County (RPCQ, LMS, and FCQ), Iowa, were examined to determine the geographic continuity of trilobite beds.

At LMS a body cluster containing three outstretched *Dechenella? prouti* and a *Phacops rana?* moult assemblage were collected approximately 25 cm below the top of the Solon Member. This indicates that a poorly sampled trilobite cluster bed is present in the upper Solon. No clusters or complete trilobites were found in the Lower Solon (independensis Zone). This is not surprising, given the rapid facies changes that occur over a small area in Rock Island County.

Two trilobite cluster horizons are present at all three Johnson County localities within the basal 2 m of the Rapid Member. The lower horizon (basal 1 m) carries the same trilobite and associated fauna as cluster bed R1 in Rock Island County. This bed can be recognized at all three localities by its stratigraphic position (lowest Rapid bed) and common occurrence of *Orthospirifer* and *Favosites* with enrolled *Phacops norwoodensis*. A monospecific cluster of *P. norwoodensis* was collected at RPCQ, and a mixed-species cluster with *P. norwoodensis* and *Crassiproetus* sp. B was found at FCQ in this bed. This evidence suggests that cluster bed R1 is continuous from Rock Island County to the northwest corner of Johnson County, a distance of approximately 160 km.

Correlation of the upper trilobite bed present in Johnson County with cluster beds R2 and R3 in Rock Island County is problematic. The exact stratigraphic position of this trilobite bed is unknown, though it is confined to the basal 2 m of the Rapid Member. Further field study is needed to locate this horizon in situ.

The *Cruziana* discontinuity surface (R3) found at all localities in Rock Island and Scott counties may be present at RPCQ in Johnson County. A burrowed discontinuity that includes abundant *Cruziana* traces and its distinctive associated ichnofauna is present at approximately the same stratigraphic position as R3. However, it lacks the abundant *Schizophoria* and trilobite fauna that characterizes this bed in Rock Island County. In addition, the burrows display less relief and are not infilled with macrofossils. These faunal and depositional differences probably reflect lateral changes in biofacies and taphonomic processes.

Trilobite clusters occur within at least five separate beds in the lower part of the Little Cedar Formation in Rock Island County, Illinois, and adjacent Scott County, Iowa. In addition, another poorly sampled bed occurs in the upper Solon, Johnson County, Iowa. Solon Member cluster beds pinch out within 1–2 km or are developed locally. Basal Rapid cluster horizons and lithologies are regionally continuous and in some instances can be traced 160 km to the northwest.

Little Cedar trilobite body clusters are characterized by both monospecific (56%) and mixed-species (44%) accumulations. The percentage of monospecific trilobite clusters is higher in Lower Solon (S1 and S2) and basal Rapid (R1 and R2) beds, which were deposited during smothered-bottom events. The predominantly mixed-species trilobite clusters, characterized by inverted trilobite exoskeletons collected from the *Cruziana* bed (R3) were transported hydrodynamically or suspended within mud and deposited in the terminations of large infill burrows.

Uncommon trilobite moult assemblages are present within all cluster beds. Moult clusters may be more common than noted because of problems in differentiating complete moulted exoskeletons from complete trilobite carcasses and problems in determining whether partially articulated trilobite exoskeletons are moults or carcasses that have been scavenged. Within the cluster beds complete cephala of proetids far outnumber isolated cranidia. Two explanations for this occurrence are possible: (1) Complete cephala represent carcasses that were disarticulated by scavengers or (2) not all proetids shed their librigenae during exuviation.

ACKNOWLEDGMENTS

Most of the information presented in this paper is from my M.S. thesis completed at the University of Iowa. I thank my thesis adviser, Brian Glenister, and committee members Brian Witzke and Gilbert Klapper for reviewing that work. Carlton Brett (University of Rochester), Steven Speyer (Arizona State University), and Richard Anderson (Augustana College) reviewed the manuscript and offered many helpful suggestions. Julia Golden, curator of the fossil repository at the University of Iowa, helped me track down many specimens. Greg Castrey, geologist for Moline Consumers Co., and Fred Bagnell, geologist for Martin Marietta Co., allowed entry and accompanied me in the field in quarries owned by their respective companies. Key specimens were collected by University of Iowa graduate students Orrin Plocher (plate 9.3, number 9) and Jay Woodson (plate 9.3, number 7). Other important specimens were collected and donated by local fossil collectors David Sivill (plate 9.1, number 9) and Randy Meyers (plate 9.1, number 3). Finally, thanks to my wife, Elizabeth, for taking time to type the manuscript.

REFERENCES

Brett, C. E. and G. Baird. 1986. Comparative taphonomy: A key to interpretation of sedimentary sequences. *Palaios* 1:207–27.

Cooper, G. A. and P. E. Cloud. 1938. New Devonian fossils from Calhoun County, Illinois. *Journal of Paleontology* 12:444–60.

Hickerson, W. J. 1992a. Trilobites from the Late Givetian Solon Member, Little Cedar Formation of eastern Iowa and northwestern Illinois. In *The Stratigraphy, Paleontology, Depositional and Diagenetic History of the Middle-Upper Devonian Cedar Valley Group of Central and Eastern Iowa. Iowa Department of Natural Resources Guidebook* 16:123–41.

———. 1992b. "Trilobites from the Little Cedar Formation of Eastern Iowa and Northwestern Illinois." M.S. thesis, University of Iowa.

Hickerson, W. J., O. W. Plocher, and J. Kralick. 1990. Middle Devonian trilobite clusters from the Rapid Member of the Little Cedar Formation, Rock Island County, Illinois. *Geological Society of America Abstracts with Programs,* 24th annual meeting north-central section, p. 13.

Johnson, J. G., G. Klapper, and C. A. Sandberg. 1985. Devonian eustatic fluctuations in Euramerica. *Geological Society of America Bulletin* 96:567–87.

Kesling, R. V. and R. B. Chilman. 1975. Strata and megafossils of the Middle Devonian Silica Formation. *University of Michigan Museum of Paleontology, Papers on Paleontology,* vol. 8.

Koch, W. F. 1978. "Brachiopod paleoecology, paleobiogeography, and biostratigraphy in the Middle Devonian of eastern North America—an ecofacies model for the Appalachian, Michigan, and Illinois basins." Ph.D. dissertation, Oregon State University.

Landing, E. and C. E. Brett. 1987. Trace fossils and regional significance of a Middle Devonian (Givetian) disconformity in southwestern Ontario. *Journal of Paleontology* 61:205–30.

Miller, K. B., C. E. Brett, and K. M. Parsons. 1988. The Paleoecologic significance of storm-generated disturbance within a Middle Devonian Epeiric Sea. *Palaios* 3:35–52.

Ormiston, A. R. 1967. Lower and Middle Devonian Trilobites of the Canadian Arctic Islands. *Geological Survey of Canada Bulletin* 153:1–148.

Savage, T. E. and J. A. Udden. 1921. The geology and mineral resources of the Edgington and Milan Quadrangles. *Illinois Geological Survey Bulletin* 38:96.

Speyer, S. E. 1987. Comparative taphonomy and paleoecology of trilobite lagerstatten. *Alcheringa* 11:205–32.

Speyer, S. E. and C. E. Brett. 1985. Clustered trilobite assemblages in the Middle Devonian Hamilton Group. *Lethaia* 18:85–103.

———. 1986. Trilobite taphonomy and Middle Devonian taphofacies. *Palaios* 1:312–27.

Stainbrook, M. A. 1935. Stratigraphy of the Devonian of the upper Mississippi Valley. *Kansas Geological Society Guidebook, 9th Annual Field Conference,* pp. 248–60.

———. 1941. Biotic analysis of Owen's Cedar Valley Limestones. *Pan-American Geology* 75:321–27.

Stumm, E. C. 1953. Trilobites of the Devonian Traverse Group of Michigan. University of Michigan, *Museum of Paleontology Contributions* 10:101–36.

Udden, J. A. 1897. A brief description of the section of Devonian rocks exposed in the vicinity of Rock Island, Illinois, with a statement of the nature of its fish remains. *Journal of the Cincinnati Society of Natural History* 19:93–98.

Walter, O. T. 1923. Trilobites of Iowa and some related Paleozoic forms. *Iowa Geological Survey Annual Report* 31:167–390.

Witzke, B. J., B. J. Bunker, and G. Klapper. 1985. Devonian Stratigraphy in the Quad Cities Area, eastern Iowa–northwestern Illinois. In Devonian and Pennsylvanian stratigraphy of the Quad-Cities Region, Illinois–Iowa. In W. R. Hammer, R. C. Anderson, and D. A. Schroeder, eds., *Great Lakes Section, Society of Economic Paleontologists and Mineralogists 15th Annual Field Conference Guidebook*, pp. 19–64.

Witzke, B. J., B. J. Bunker, and F. S. Rogers. 1988. Eifelian through Frasnian stratigraphy and deposition in the Iowa area, central midcontinent, U.S.A. In Devonian of the World. *Canadian Society of Petroleum Geologists Memoir*, 14:221–50.

PART TWO

Epiboles and Longer-Term Bioevents

10

Epiboles, Outages, and Ecological Evolutionary Bioevents: Taphonomic, Ecological, and Biogeographic Factors

Carlton E. Brett and Gordon C. Baird

ABSTRACT

The Phanerozoic marine sediment record is characterized by a hierarchy of paleontological events of varying temporal and spatial magnitudes. These are generally isochronous or nearly so, with some events being potentially extrabasinal or global in extent. Study of these events in the context of paleogeography and both taxon- and community-scale evolution offers the potential for the establishment of a successional history of biotic events that is the paleontological counterpart (and possibly the paleontological expression) of patterns observed in sequence stratigraphy. Herein we advance a redefinition of an old term, *epibole,* as a concept to denote an unusual abundance of a taxon that is normally rare to absent. Epiboles are classified as taphonomic, ecological, and incursion related. Taphonomic epiboles are local to regional events leading to selective preservation of taxa usually unrecorded. Ecological epiboles include environment-controlled population bursts (proliferations) and colonization events; such events may be induced by substrate changes and include redox-related and diastem (hardground) colonization events as well as persistent biostrome or bioherm-bearing intervals. Incursion epiboles represent brief invasions of a species or group of species across basin or realm boundaries followed by a rapid die-off or emigration of these organisms. Because the causes of most epiboles remain problematic, the term *epibole* should be used in a flexible, nongenetic manner beyond the preceding category designations. Herein we advance the term *outage* (" antiepibole") for the absence of a taxon normally present in tracking communities. Ecological and incursion epiboles may be reciprocally linked to outages, thus providing critical insight to the ecological structure of paleocommunities.

Longer-term paleontological events include mass extinctions, evolutionary radiations, emigrations/immigrations, and episodes of community reorganization. Alternating episodes of community reorganization events and longer intervals of evolutionary stasis with community stability between these events may ultimately be linked to the hierarchy of climate-forcing and eustatic events that account for patterns in sequence stratigraphy.

Phanerozoic sedimentary deposits, particularly marine facies, record a rich spectrum of paleontological events ranging from nearly instantaneous events up to the scale of infrequent global community-realm-level reorganizations, extinctions, and originations within the biosphere. These events offer the potential for extremely precise stratal correlations both within basins and between them. Understanding the regional/temporal scale of paleontological events, their cyclic recurrence where developed, and the causal mechanisms behind them provides insight into the connection between fossil occurrences and global changes in sea level, climate, and other factors.

In the present paper we review differing types of paleontological events of local-to-regional scale, herein designated *epiboles,* which are expressed as beds yielding an unusual abundance of certain taxa or groups of taxa. Moreover, we also review events wherein taxa, normally present or abundant, are anomalously absent from communities. We believe that these paleontological absenteeisms, which we term *outages,* are no less important than epiboles in the study of long-ranging fossil communities. We also examine longer-term paleontological event periods, herein referred to as *"ecological-evolutionary subunits"* characterized by stable community structure. These appear to alternate with episodes of community reorganization that involve extinction, minor variations, and the shift of the ecological roles of some organisms. Ecological evolutionary subunits in the paleontological record vary in temporal length and discreteness, but some appear to correspond to stages in traditional biostratigraphic parlance.

The term *epibole* was coined by Trueman (1923) to imply an "acme" or maximum abundance of a particular species in strata. Epiboles were initially considered to have time-stratigraphic significance as formal biostratigraphic zones. Initial use of *epibole* in a time-stratigraphic sense led to miscorrelations until it was realized that ecological factors played an important role in the spatial distribution of organisms (see later). Early enthusiasm in using *epiboles,* or hemeras, for correlation waned and from the 1920s onward, these terms fell into disuse. Part of this decline also reflected a greater interest in application of the facies concept and waltherian principles and a movement away from "layer cake" biostratigraphy.

In the last 15 years there has been an important revival of stratigraphy as a scientific enterprise, particularly global-scale chronostratigraphy. Recognition of several scales of sedimentary cyclicity within basins, plus the development of the sequence stratigraphic paradigm, shows that high "lithochronostratigraphic" resolution within and between basins is achievable. Moreover, stratigraphy-oriented paleontologists (Kauffman 1986, 1988; Sageman and Johnson 1985; Sageman et al. in this volume) have long recognized the correlational significance of paleontological units and events below the scale of zone. The present authors have documented numerous paleontological event beds that appear to be widespread and isochronous in the middle Paleozoic of the northern Appalachian Basin (see figs. 10.1–10.3). These events, which are of varying temporal magnitude and cause (see below), convince us that facies patterns are commonly an overprint on the paleontological "layer cake." They also indicate that a paleontological event-bed chronostratigraphy with resolution far greater than that resolvable through biostratigraphy is not only achievable but demonstrable. It is with this perception that we advance the terms *epibole* and *outage* as stratigraphic concepts and correlation tools.

Epiboles as Paleontological Events

History of the Epibole Concept

In a series of papers Buckman (1893, 1898, 1902–1903) defined *hemera* as a time unit corresponding to the acme of a given zonal species. In essence, it was to be the smallest unit of consecutive time units that one could discern in a hypothetically complete stratigraphic section. Following from this, Trueman (1923) defined the *epibole* as a time-stratigraphic term to encompass the expression of the hemera in a section. Subsequent to this, British workers (Lang 1924; Tutcher and Trueman 1925) endeavored to compile hemeral charts for the Jurassic and listed the sequences of corresponding epiboles for sections in various regions. Buckman (1893, 1898) idealized the hemera to include a period of upward-increasing abundance of the given species followed by peak abundance (acme) and decline. Following from this, Trueman (1923) envisioned the epibole as being the time-rock expression of the Remera. Natural patterns are rarely so simple. Not only is the hemera implicitly diachronous, owing to paleoecology and migration timing, but its epibole (in the sense of Trueman 1923 and Arkell 1933) might not match with a single species "acme" in sections. Many taxa do not show a simple symmetrical buildup to an acme of abundance followed by gradual decline; a fossil species actually may be abundant near the base or top of its stratigraphic range or common on only a few bedding planes. We have observed many mid-Paleozoic taxa to be very rare or absent from almost all bedding planes within their long stratigraphic ranges with the exception of a handful of thin but very widespread beds in which the fossil is densely concentrated. These mappable horizons of anomalous abundance correspond closely to units that have been termed *epiboles*.

In a trenchant critique of the hemera concept, Arkell (1933) showed that hemeral successions varied for different basins with notable reversals or alternations of epiboles that were supposed to be strictly consecutive according to biological succession. Arkell (1933) deduced that the acme of a zonal species is controlled by such dynamic variables as migration rates, paleoecological factors, and structural barriers such that it could never be consistently used in a time sense. Vagaries of migration would allow the observed epibole of a given species to occur stratigraphically below another in one region but above it elsewhere (Arkell 1933:33, fig. 1).

One source of such discrepancies is that the inferred acme range of a species could be only a small fraction of the hypothetical total time of its evolutionary existence (the species *biochron* of H. S. Williams [1901]). The smaller the proportion of hemeral time to species biochron time, the greater the potential for reversals of epiboles. This implied diachroneity of the hemera at the global scale led Arkell (1933) to downgrade it as of little value to chronostratigraphy; although Arkell retained the word *epibole* as a term to denote the acme interval of a species in local sections. Thus redefined, the *epibole* was considered a reflection of local conditions that allowed that organism to flourish during some subset of time of its evolutionary existence.

Because the concept of the acme is mainly useful at the local and, perhaps, regional level only, its application to chronostratigraphy waned in importance. Both Dunbar and Rogers (1957) and Shaw (1964) viewed it primarily as an ecological effect that was local with little potential use in correlation. Although the term *epibole* dropped from common usage, other time-rock terms also denoting exceptional developments of taxa in sections were later advanced; Hedberg (1972) proposed *acme zone* for a body of strata representing the acme or maximum development of a particular taxon. Likewise, Ernst and Seibertz (1977) advanced the term *peak zones* for periods of maximum abundance of guide fossils. Both *acme zone* and *peak zone* are understood to be of use in local correlation but not in global chronostratigraphy because of the constraints cited by Arkell (1933); however, we herein find *epiboles* to be highly useful in intrabasinal correlations. Therefore we herein redefine and classify epiboles.

Redefinition of Epiboles

Epiboles, as defined herein, represent thin intervals (rarely single bedding planes) ranging in thickness from a few centimeters to a few meters, characterized by an unusual abundance of a fossil taxon or taxa that are normally rare or absent in a particular facies or set of interfingering facies. As such, we hope to establish the term, in its broadest sense, for objectively recognizable fossil concentrations *without* regard to their mode of origin. These anomalous abundances stand out against the background of persistent "tracking" biofacies. More than one epibole may be observed for a particular species that is rare within most of its range. Epiboles apparently represent several distinct phenomena. Consequently, we further classify them as *taphonomic, ecological,* and *incursion epiboles* (see table 10.1). It should be noted that these epiboles are the

TABLE 10.1 *Terminology, Temporal Scales and Inferred Processes Involved in Formation of Taphonomic, Ecological, Biogeographic, and Evolutionary Event Strata Discussed in This Essay*

Stratal term (this paper)	Bio-event term (Kaufmann 1986)	Temporal scale	Spatial scale/processes
I. *Taphonomic epiboles*	—	(10^{-1}–10^5 yrs)	Unusual taphonomic conditions lead to preservation of normally rare taxa
A. *Obrutionary epiboles*	Mass mortality	(minutes to days)	Local to regional mortality and burial preserves delicate multielement forms
B. *Diagenetic epiboles*	—	10^2–10^4 yrs.	Unusual early diagenetic mineralization commonly combined with obrution preserves normally rare species
II. *Ecological epiboles*	Positive bioevents	10^1–10^5 yrs.	Unusual environmental conditions lead to relatively brief colonization and/or regional scale proliferation of normally present, but rare species
A. *Colonization epiboles*	Colonization bioevents	10^0–10^5 yrs.	Unusual substrate conditions lead to short-term regional colonization of normally rare (and isolated) species
1. Redox related	—	1–10^2 yrs.	Colonization due to brief oxygenation events
2. Event deposit	—	1–10^2 yrs.	Colonization due to event deposits
3. Diastemic	—	10–10^3 yrs.	Colonization of firm- to hardground diastem surface
4. Biostrome	—	10^2–10^5 yrs.	Colonization of biostrome or bioherm-forming taxa
B. *Proliferation epiboles*	Population bursts	10^2–10^5 yrs.	Intermediate scale (thousands of years) fluctuations in environment permit relatively brief proliferation of normally rare species over broad geographic area
III. *Outages*	Emigration events (in part)	10^3–10^5 yrs.	Unfavorable environments or selective negative factors inhibit a normally common taxon in a given region
IV. *Incursion epiboles*	Immigration events (in part)	10^3–10^5 yrs.	Connection of normally isolated biogeographic provinces and/or unusual environmental factors (e.g., influx of atypical water masses) permits influx of taxon or fauna) from outside of basin
V. *Ecological evolutionary unit/subunit boundaries*	Punctuational evolutionary bioevents (mass extinctions)	10^5–10^6 yrs.	Regional to global scale environmental changes lead to mass extinctions, collapse and restructuring of ecosystems; these major bioevents punctuate much longer intervals (up to 10^7 yrs.) of relative stability (coordinated stasis)

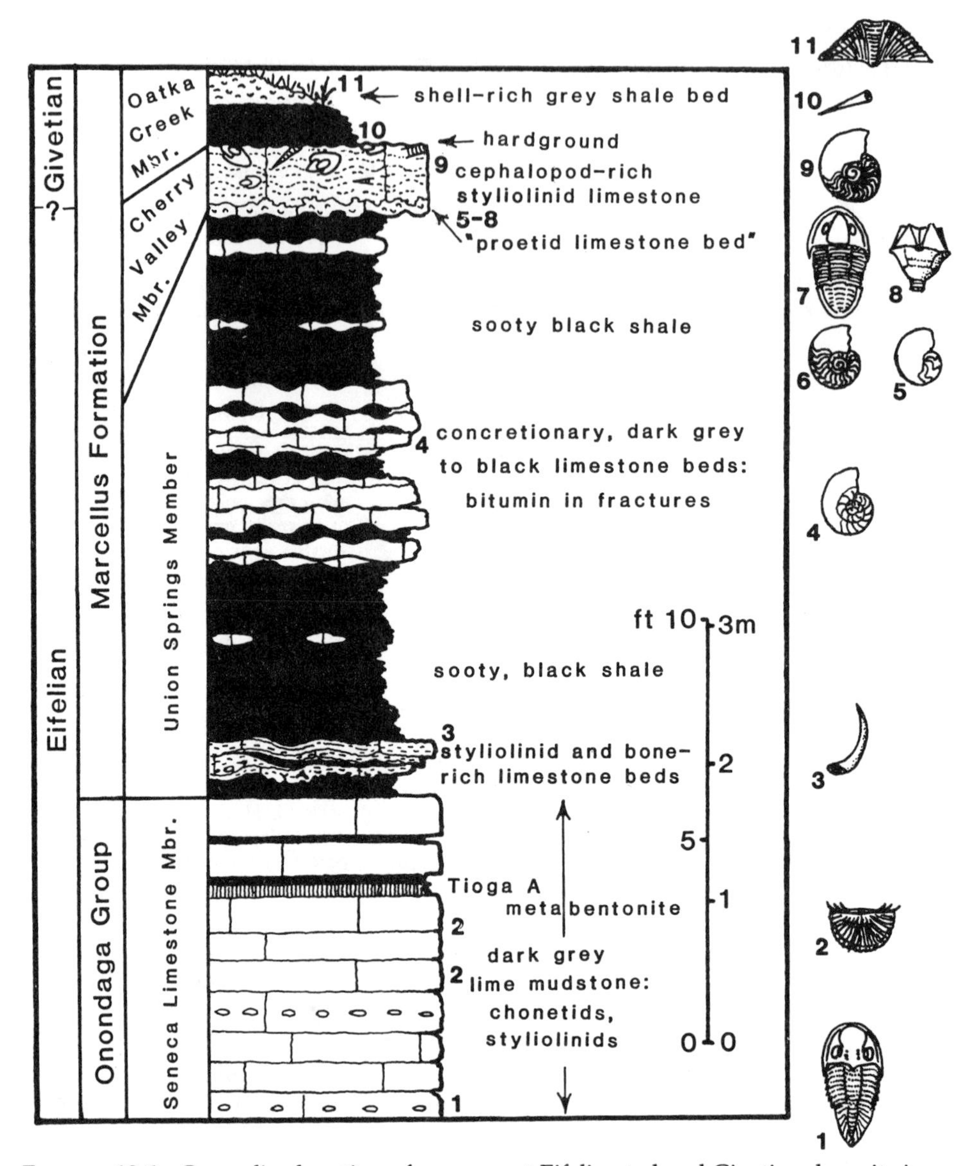

FIGURE 10.1. Generalized section of uppermost Eifelian to basal Givetian deposits in western New York showing major faunal turnover that corresponds to Kacak-*otomari* global bio-event of European workers (House 1985; Tryols-Massoni et al. 1990; Walliser 1992a). This quarry section in west-central New York State records the following widespread paleontological event horizons within condensed transgressive-highstand facies: (1) beds rich in the trilobite *Odontocephalus selenurus;* (2) beds of pink-colored *Hallinetes lineatus* chonetid brachiopods; (3) bone bed, rich in teeth of the crossopterygian fish *Onychodus;* (4) concretionary beds with *Cabrieroceras plebeiforme* (taphonomic and incursion epibole); (5–8) Chestnut Street bed (see Griffing and Ver Straeten 1991) with short-lived incursion fauna, including (5) goniatite *Subanarcestes* cf. *S. macromphalus,* (6) goniatite *Agoniatites nodiferus,* (7) trilobite *Dechenella haldemani,* and (8) crinoid *Haplocrinites clio;* this bed corresponds approximately with the Kacak-*otomari* event as defined by Tryols-

stratigraphic signature of varied local-to-regional biological events (or bioevents, sensu Walliser 1986b). Epiboles usually are not global events—at least, not demonstrably global—but regional, basin-scale to extrabasinal events that are defined through stratigraphic mapping to be pervasive over hundreds to thousands of square kilometers (e.g., West et al. in this volume). However, some epiboles (e.g., *Palaeocyclus* coral epibole, *Foerstia* epibole, see later) do appear to reflect global bioevents. They are generally isochronous, at least within the resolution of biostratigraphy.

We favor the retention of the term *epibole* as a general term for a class of paleontological stratal phenomena that have long been recognized by many stratigraphers. Our main reasons for retaining the term rather than proposing a new one are that it is familiar to many biostratigraphers and that the class of biostratigraphic horizons we denote with this term includes many horizons that have been traditionally called "epiboles" (e.g., the *Dicoelosia* epibole in the Lower Devonian Helderberg Group [Goodwin and Anderson 1986]). Thus we hope to emphasize that the intervals recognized by earlier stratigraphers are real, objective units and that they commonly do have a time-stratigraphic significance.

Kauffman (1986, 1988), in developing a high-resolution event stratigraphy for the midcontinent interior Cretaceous seaway, recognizes a variety of biological "event-markers" in his stratigraphic sections (table 10.1). In particular, he refers to "population bursts," which denote major population blooms of particular taxa, especially planktonic and nektonic forms, resulting from favorable environmental conditions. Although he also links epiboles to the acme zone concept of earlier stratigraphers (see Kauffman 1988:632), there is some ambiguity as to whether he views the epibole acme as a single or multiple abundance peak(s) within the range of the taxon. The wording suggests that the acme is controlled not only by evolution, but also by temporal vagaries of ecology and other factors. Using the expanded definition of *epibole* favored herein, several of Kauffman's (1986) regional bioevents—including population bursts, productivity, immigration/emigration, regional colonization, and mass mortality events—may be manifested in the stratigraphic record as *epiboles* (table 10.1; column b).

In defining epibole units it must be noted that many, if not most, biological event horizons of epibole scale are of multiple, not single, causes. The *Agoniatites vanuxemi* epibole (fig. 10.1) in a thin, widespread limestone (latest Eifelian Cherry Valley Limestone) of the northern Appalachian basin is nearly coinci-

Massoni et al. (1990) or *otomari* "a" horizon of Walliser (1990a); 9–10 Cherry Valley ("*Agoniatites*") Limestone packed with (9) *Agoniatites vanuxemi*, and (10) nautiloid *Striacoceras typum*; this level corresponds to the lower Kacak horizon of House (1985) or *otomari* "b" horizon of Walliser (1990a); (11) brachiopod *Mediospirifer* cf. *M. audaculus* epitomizes the incursion of a typical Hamilton Group (Givetian) brachiopod fauna in a thin gray shale (Halihan Hill-LeRoy bed) in lowermost Oatka Creek Shale; this level corresponds to *otomari* horizon "c" of Walliser (1990a).

dent with the Kacak global bioevent of European workers (House 1985; Chlupac and Kukal 1986; Tryols-Massoni et al. 1990) and is probably a result of biogeographic incursion of "Old World" ammonoids resulting from transgression. However, it is also an ecological event that probably reflects distinctive water mass and nutrient conditions owing to the same transgressive event. Eventual return of "normal" water conditions would also explain the abrupt disappearance of the epibole taxa. Finally, this goniatite bed reflects unique taphonomic conditions of episodic burial within a setting of general sediment starvation and near-surface sediment cementation (Griffing and Ver Straeten 1991). We agree with Walliser (1986b, 1990) that our event terminology should be sufficiently flexible to mirror the real dynamics of bioevents; epibole categories thus should never be used as exclusive explanatory labels for events whose causes are poorly known. Nonetheless we propose a preliminary subdivision of epiboles based on their *dominant* mode of origin. Taphonomic, ecological, and incursion epiboles are defined and discussed in the following sections.

Taphonomic Epiboles

Taphonomic epiboles are partly an artifact of unusual burial and/or early diagenetic conditions that favor preservation of fossil species that are not normally preserved, at least not as specifically identifiable fossils. They overlap with Kauffman's (1986, 1988) mass mortality bioevents and obrution deposits (rapidly buried assemblages) discussed by Brett et al. (essay 1 in this volume; see table 10.1). Typically, taphonomic epiboles involve organisms with easily disarticulated skeletons, such as those of trilobites and echinoderms or those of chitinous or aragonitic skeletons that do not normally preserve well. Outstanding examples of this type of *obrutionary epibole* include regionally persistent algal-, starfish-, and crinoid beds, such as those reported from the Ordovician, Silurian, and Devonian of the Appalachian basin (see LoDuca and Brett in this volume; Brett and Taylor in this volume). Such examples are a special type of obrution deposits, but they may also be termed epiboles in that they contain faunal elements that normally are unrecognized. Nonetheless, taphonomic epiboles generally are not confined to single obrutionary layers; instead they represent intervals in which unusual burial conditions recurred over an interval of time. With this definition, intervals of a few meters thickness yielding extraordinary biotas in a series of obrution horizons could be termed *taphonomic epiboles*. An example would be the Burgess Shale *Marella* beds of the Walcott Quarry (see Conway-Morris 1986).

A second type of taphonomic epibole reflects unusual early *diagenetic effects*, such as beds of concretionary or pyritized fossils that selectively preserve aragonitic shells (table 10.1). For example, ammonoid cephalopods are abundant in many horizons in Middle and Upper Devonian siliciclastic deposits in New York State (Kirchgasser and House 1981) but are generally so poorly preserved as composite internal/external molds that they are not recognizable to species or generic level. However, these same fossils are usually recognizable to species level when pyritized or when encapsulated within carbonate concretions that have formed prior to sediment compaction. These form widely recognized horizons that have proven extremely useful in stratigraphic correlation (see Arkell 1933; House 1962; Kirchgasser and House 1981; fig. 10.2).

It could be argued that taphonomic events are merely the effects of anomalous mechanical and chemical processes. However, as defined herein, *epibole* only implies a stratal pattern of anomalous abundance. Furthermore, major, taphonomic events (as opposed to most short-term processes) *are* usually associated with some biological changes, most notably mass mortality. Many taphonomic features are facies-specific and hence are environmentally controlled. Major obrution-related *Konservat Lagerstätten* are associated only with certain biofacies (see Brett and Taylor in this volume). In addition, numerous taphonomic events record widespread, instantaneous ecologic disturbances that are circumbasinal and even extrabasinal in extent. Major volcanic ash falls, colossal tsunamis, and even outsized storms may produce epiboles if they decimate the benthos of an entire basin or set the stage for ecological overturn in addition to preserving organisms through rapid burial (see Kauffman 1986, and the discussion under "Outages").

Furthermore, in the context of sequence stratigraphy, many taphonomic epiboles can be linked to distinctive parts of sedimentary cycles, which are in turn linked to global climate-forcing and associated eustasy (Vail et al. 1991; Wilgus et al. 1988; Kidwell 1991a; Banerjee and Kidwell 1991). Thus within the range of a particular long-ranging goniatite, virtually all strata may be barren of the fossil except for a handful of widespread concretion layers strictly associated with the transgressive systems tracts of eustatic cycles (figs. 10.1 and 10.2). At these few levels in a sediment-starved, quiet-water basinal, obrution would have been followed by bacteria-mediated mineralization and winnowing to produce widespread beds packed with well-preserved cephalopods. Condensed cephalopod-rich limestone deposits (*Cephalopodenkalk*) in transgressive intervals provide excellent examples of this type of epibole (Wendt and Aigner 1985). Hence such epiboles are closely linked to secular paleoenvironmental changes that are extrabasinal and even global in extent.

Ecological Epiboles

Ecological epiboles reflect geologically brief pulses of expansion in particular species of probable environmental cause (table 10.1). They are identifiable as thin intervals, generally only a few millimeters to a few meters thick, in which a species that is normally well preserved but present in very low numbers is extraordinarily abundant. As was long recognized by stratigraphers, these "acme zones" may be quite widespread and may even crosscut facies. For

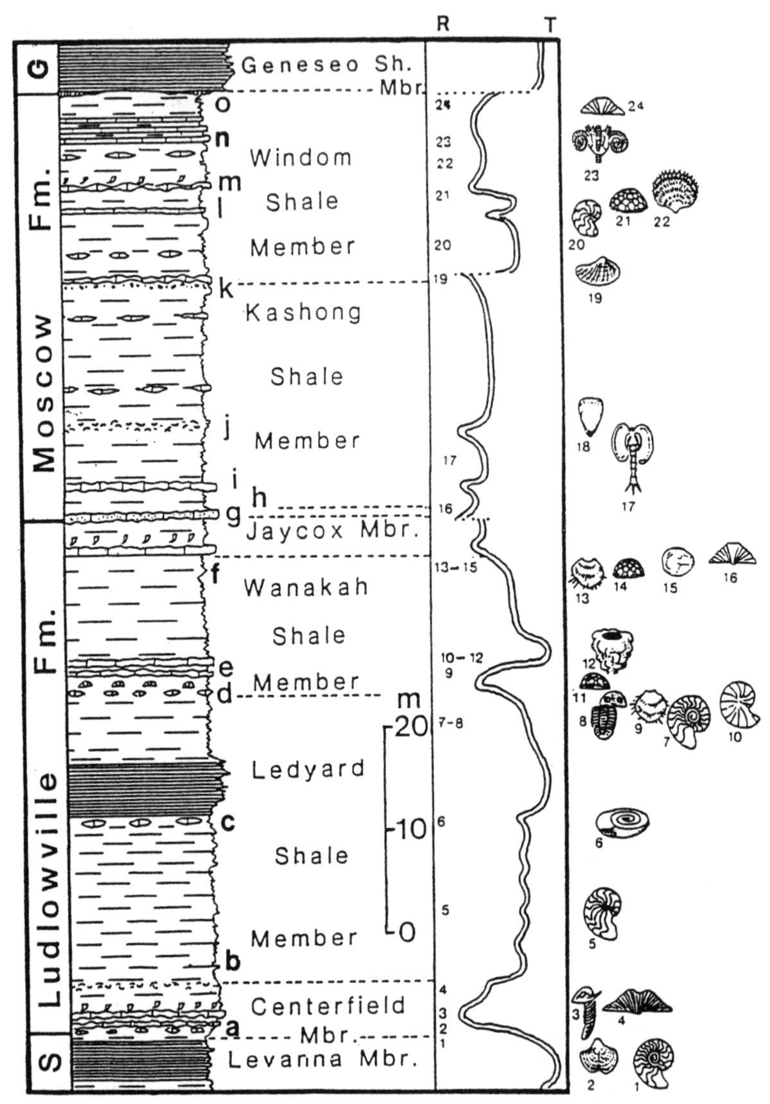

FIGURE 10.2. Generalized stratigraphic column for middle-upper Givetian deposits in western New York showing relationship of units to transgressive and regressive events, as well as succession of taphonomic (t), ecological (e), and incursion (i) epiboles recognized to date. For discussion of lettered stratigraphic units see Brett and Baird (1986); Brett et al. (1990). Horizons include (a) Peppermill Gulf bed; (b) Alden pyrite—Staghorn coral bed; (c) mid-Ledyard Shale-Joshua bed; (d) *"Truncalosia"* (Mt. Vernon) bed; (e) *"Pleurodictyum"* (Darien Center) bed; (f) *Strophodonta demissa* (Blasdell) bed; (g) Tichenor Limestone; (h) Deep Run Shale; (i) Menteth Limestone; (j) *Rhipidomella-Centronella* bed; (k) Barnes Gully phosphatic bed; (l) Bay View coral bed ; (m) Fall Brook Coral bed; (n) South Lansing coral bed; (o) *Allanella tullia* (Gage Gully) bed. Epibole species include the following: (1) *Agoniatites* cf. *A. unilobatus* (i); (2) brachiopod *Mucroclipeus eliei;* (3) proetid trilobite *Basidechenella rowi* obrution horizon (e,t); (4) brachiopod *Fimbrispirifer divaricatus* (e); (5) goniatite *Tornoceras uniangulare* (i, t) ; (6) gastropod *Straparollus rudis* (e), (7) goniatite *Agoniatites unilobatus* (i,t); (8) trilobite *Phacops rana* obrution horizon (t); (9) brachio-

example, the epibole species may occur in a range of lithologies, including those in which the taxon is normally absent (see lateral persistence of epiboles in fig. 10.3). These levels are isochronous, at least within the resolution of the most precise biostratigraphic zonations, and they parallel rather than crosscut other types of event marker beds such as ash layers (Kauffman 1986). Hence they appear to have value in local time-stratigraphic correlation. In contrast to the old view that such epiboles represent the unique "acme" in a "normal" progression of a species rise and fall, we find that certain species display multiple epiboles. This demonstrates that conditions that favored the species recurred, although infrequently. The stratigraphic boundaries of ecological epiboles may be sharply defined, especially in the case of the colonization events (see later). In such cases the lower surface may represent a discontinuity or sedimentary event bed. In other cases, the lower and upper boundaries are slightly more diffuse. The epibole taxa gradually increase or decrease in abundance, but generally only through intervals of a few centimeters.

Ecological epiboles obviously do not represent influxes of exotic faunal elements. Rather, they appear to reflect peculiar environmental conditions that allowed normally rare species or peculiar morphotypes to colonize and proliferate briefly. In some instances, the epibole interval may also display unusually low numbers or complete absence of a normally common species (see below under "Outages"). This suggests that biotic interactions may be related to the anomalous rises in abundance of certain species, although cause-and-effect relationships are difficult to judge. In other epiboles, colonization is clearly related to a particular type of substrate.

Such phenomena are difficult to explain but warrant substantially more investigation than they have received. Unlike many biofacies shifts, they appear to be noncyclic and to reflect unique combinations of ecological factors that permit success of a particular species for brief intervals of time in environments from which they normally are excluded or are relatively uncommon. Controlling variables for these events may never be understood. Nonetheless, these epiboles represent an important ecological phenomenon and provide unique time lines for basin analysis.

Kauffman (1986) identifies two types of biological events that we relate to ecological epiboles: *population bursts* and *regional colonization events* (Kauffman 1988:633). The former denote improved conditions causing biotic proliferation

pod *Truncalosia truncata* (i); (10) nautiloid *Nephriticeras magister* (e, t); (11) tabulate *Pleurodictyum americanum* (e); (12) pyritized hexactinellid sponge *Pseudohydnoceras* (t); (13) brachiopod *Truncalosia truncata* (i); (14) tabulate *Pleurodictyum americanum*; (15) problematic spherical organism (t); (16) brachiopod *Allanella tullia* (e); (17) obrution horizons yielding phyllocarids *Echinocaris* and *Rhinocaris* (t); (18) brachiopod *Centronella impressa* (i,e); (19) bivalve *Pholadella radiata(e)*; (20) goniatite *Tornoceras uniangulare* (e,t); (21) tabulate *Pleurodictyum americanum* (e); (22) brachiopod *Spinatrypa spinosa* (i); (23) obrution layer yielding the crinoid *Clarkeocrinus* (t); (24) brachiopod *Allanella tullia* (diminutive) (e).

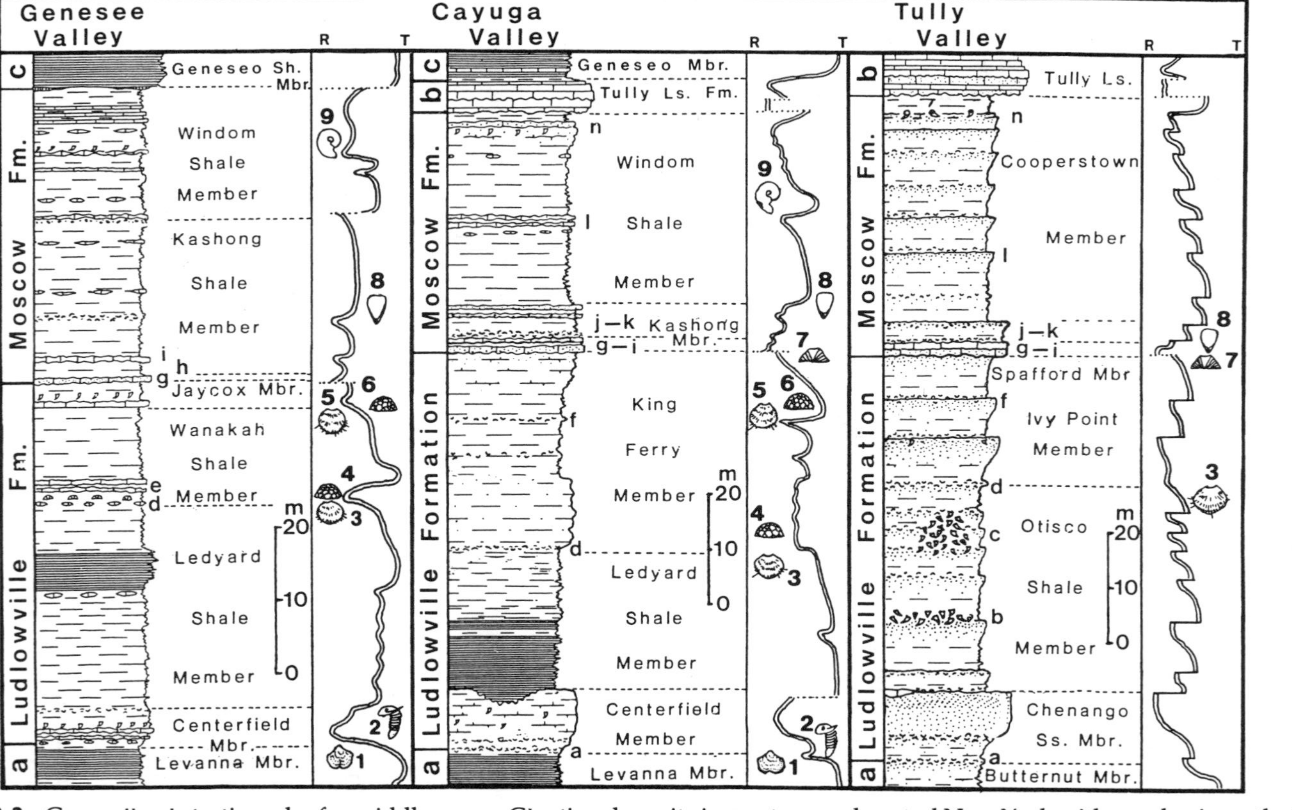

FIGURE 10.3. Generalized stratigraphy for middle-upper Givetian deposits in western and central New York with emphasis on the lateral persistence of the key epiboles from carbonate-rich mudstone facies in western New York into siltstone and even sandstone-rich proximal marine facies in central New York. Note also relationship of stratigraphic units and incursions to major transgression and regression events indicated by the relative sea-level curves (R = regression; T = transgression). Correlations extend over a total lateral distance of 150 km. from west (Genesee Valley) to east (Tully Valley). For lettered bed designations, see fig. 10.2; for epiboles, (e) = ecological; (i) = incursion; (t) = taphonomic. Numbered epiboles include (1) brachiopod *Mucroclipeus eliei*(i); (2) proetid trilobite *Basidechenella rowi* (t); (3,5) productid brachiopod *Truncalosia truncata* (i); (4,6) tabulate coral *Pleurodictyum americanum* (e); (7) spiriferid brachiopod *Allanella tullia* (e); (8) terebratulid brachiopod *Centronella impressa*(i); (9) goniatite *Maenioceras* sp. (i).

of groups of associated organisms; Kauffman's "productivity events" are related but normally are recorded as beds of biogenic sediment, such as coccolith oozes (chalks), rather than as epiboles.

Regional colonization events are unique associations of normally uncommon taxa that simultaneously colonize benthic environments due to abrupt changes in physicochemical parameters. Such colonization events—triggered by such factors as rapid changes in oxygenation or alteration of substrate due to widespread storm layers, turbidites, or ashfalls—may lead to epiboles. Although these various types of biotic events have differing underlying causes, the cause of population increases over broad areas in each case is fundamentally ecological. If short-lived (as is implied by Kauffman's term *events*), each of these phenomena could produce a thin interval displaying anomalous abundance of taxa, or *epibole* in the sense advocated herein. For this reason we favor the term *ecological epibole* as an umbrella for the stratigraphic expression of varying ecologically produced events. When possible, ecological epiboles should be further specified as being results of population bursts or colonization events (see tables 10.1a, 10.1b). These phenomena are discussed in the next sections.

Ecostratigraphic events (Kauffman 1986:633–634) mark widespread changes in community structure caused by cyclic changes in the paleoenvironment. Such broad-scale changes generally do not lead to thin stratigraphic intervals that would be recognized as epiboles. Rather, the wholesale change in community structure, in response to environmental shifts, is an example of abrupt community replacement or faunal tracking (Brett et al. 1990).

Colonization Epiboles

Certain widely distributed beds or intervals display in-place accumulations of skeletal remains. These appear to record widespread and nearly synchronous events of substrate colonization (Kauffman 1986). Simplest cases are in situ assemblages restricted to a single bedding plane that record short-lived colonization events. Thicker intervals of up to several meters (in bioherms) may be bounded at their bases by a distinct bedding plane that forms a colonization surface. Epifaunal colonizers occur above or directly attached to the colonization surface while infauna or their traces extend downward from it. The upper boundary of the colonization bed may be abrupt or gradational, depending upon conditions that terminated inhabitation by organisms. Colonization epibole strata may be laterally persistent for hundreds or thousands of square kilometers (see essay 16 in this volume).

We recognize four types of benthic colonization event beds, as discussed later: (1) redox, (2) event deposit, (3) diastem, and (4) biostrome/biohermal colonization events.

Redox Event Colonization Beds As noted in essay 1 of this volume (Brett et al.), colonization surfaces may occur as single bedding planes in black, laminated shales. These beds represent geologically brief colonization of an otherwise inhospitable bottom due to minor oxygenation events (see Wignall and Hallam 1991). The benthic organisms involved normally represent tolerant epifauna adapted to low oxygen conditions. These species may be specialized for life in dysaerobic or exaerobic environments, or they may be more generalized, opportunistic colonists (see Sageman et al. 1991 for further discussion of adaptive strategies). As such these species—including leiorhynchid and chonetid brachiopods in the Paleozoic and paper pectens, such as *Bositra*, in the Mesozoic—constitute a normal, recurring biofacies (Sageman et al. 1991). Thus, although colonization horizons contrast with a background of barren shale, they do not necessarily constitute examples of ecological epiboles. However, in certain cases an unusual faunal element may appear on colonization surfaces within a thin stratigraphic interval. These are a type of redox-related ecological epibole. An example would be thin dark shale intervals in the Middle Devonian Hamilton Group in which diminutive *Tropidoleptus* brachiopods occur in great abundance in bedding plane assemblages. This is an unusual facies association for *Tropidoleptus* that normally occurs in shallow-shelf, aerobic mudstones or siltstones. The brachiopod is not otherwise seen in dysoxic facies. Although we do not understand the parameters involved in this anomalous and probably abortive colonization, the small *Tropidoleptus* beds are persistent stratigraphic markers. This is similarly true for other small brachiopods in the Hamilton Group, such as *Truncalosia truncata* (fig. 10.2: epiboles 9 and 13) and a diminutive spiriferid *"Allanella" tullia* (fig. 10.2: epibole 24)

Event-Deposit Colonization Sedimentation event deposits, such as tempestites, turbidites, and ash falls may abruptly alter seafloor conditions, allowing for temporary colonization. For example, Vossler and Pemberton (1988) discuss distinctive suites of trace fossils that characterize storm layers. They attribute these traces to a community of opportunistic infaunal organisms that rapidly colonized new substrates to take advantage of buried organic matter. Follmi and Grimm (1990) discuss burrow galleries in turbidites from the otherwise nonburrowed (low-oxygen) Miocene Monterey Formation in California. The authors ascribe the burrows to temporary colonization events by "doomed pioneer" organisms imported by the turbidity current. Similarly, Kauffman (1986) discusses the short-term colonization, following storm sedimentation and ashfall events, of normally uninhabitable seafloors. He infers that the sediment or ash layers not only provided a new type of substrate but also blanketed an otherwise uninhabitable seafloor and sealed off the seepage of H_2S or other toxins.

Diastem-(Unconformity-) Related Colonization Epiboles Many colonization event beds display a *colonization surface* that coincides with a diastem or unconformity, such as a marine flooding surface at the top of a parasequence (see Kidwell 1991b), in which colonization may be favored by low rates of sedimentation. Commonly, such surfaces display evidence of altered substrate conditions. This is especially pronounced in the case of rockgrounds (see Johnson 1988 and references therein) and synsedimentary hardgrounds (see Brett 1988; Wilson and Palmer 1992, for reviews). Here colonization by epifaunal organisms is favored by the existence of a hard substrate upon which larvae could settle. Lag gravels, reworked concretions, and shell pavements may similarly induce colonization events, provided that they are exposed for a sufficient period of time in marine environments (Kidwell and Jablonski 1983; Kidwell 1991b). The upper contact of these colonization beds may be abrupt or gradational. A gradational upper contact may reflect a gradual change in conditions, such as resumption of sedimentation, which altered the substrate conditions. Taphonomic evidence commonly points to an abrupt termination of the colonization episodes. For example, many hardground assemblages display articulated fragile fossils, such as pelmatozoan echinoderms that were smothered by a pulse of sediment (Rosenkranz 1971; Brett and Liddell 1978). These are composite colonization events and obrution deposits (see Brett and Seilacher 1991 for further discussion).

In other cases, development of an unconsolidated, but stabilized, surface may permit colonization. Kauffman (1986) cites examples of colonization events of inoceramids and oysters on winnowed pelagic carbonates that form tops of cycles. Simple changes in the input of fine-grained siliciclastic sediments also alter the stability of the substrate. Particularly good examples of colonization events associated with condensed intervals are provided by thin but widespread layers of one or a few species of brachiopods or bivalves that occur at the tops of cephalopod limestones. For example, Havlicek and Storch (1990) and Kríẑ (1992) describe what could be called biostromes of densely packed articulated specimens of the brachiopod *Atrypoidea* at the top of a cephalopod-rich limestone from the Ludlovian Kopanina Formation in the Prague Basin. These colonization events, or epiboles, of *Atrypoidea* appear to be coincident with a flooding surface following maximum middle Ludlow regression. Low sedimentation rates combined with the skeletal substrate of the cephalopod aggregations may have stimulated colonization. These horizons appear to be traceable for several kilometers.

A similar case of a colonization event related to a condensed horizon is recorded in the remarkable Middle Devonian *pumilio* events of Europe and North Africa (Lottmann 1990). These are two thin (10–15-cm) intervals of brachiopod-rich (*"Terebratula" pumilio*) packstones, interbedded with cephalopod limestones, of Givetian age in Germany, southern France, and North Africa. They are exceptionally good marker horizons for detailed correlation across a distance of more than 3,000 km. The *pumilio* horizons are interpreted by Lottmann (1990) as single-event deposits resulting from mass transport of vast numbers of *"Terebratula" pumilio* by tsunamis. However, this interpretation raises many questions. For example, what could be the source and transport agent for such immense sheetlike accumulations? The upper "horizon" actually contains up to five graded units that have slight faunal differences (e.g., first and second *pumilio* layers, plus an *Ambocoelia*-rich layer). These layers contain both juveniles and rare adult specimens and thus are not sorted by size. Moreover, these beds carry a mixture of articulated and fragmentary valves. Such evidence is incompatible with the interpretation of the instantaneous transport of shells by geologic events. Rather, it is likely that these represent colonization bioevents, comparable to the *Atrypoidea* beds of the Prague Basin, or Kauffman's (1986) regional bivalve colonization events on winnowed skeletal layers. Although such shell beds do not represent absolutely isochronous markers, as suggested by Lottmann (1990), they may still reflect nearly simultaneous colonization under very temporary conditions and have great utility in event stratigraphy.

Colonization of diastemic surfaces by infaunal soft-bodied organisms may lead to formation of widespread burrowed horizons that commonly form important markers in otherwise unfossiliferous rocks (see Pemberton in this volume). Mortimore and Pomerol (1991) describe laterally persistent horizons typified by *Zoophycos, Thallasinoides,* or in some cases, extremely long vertical burrows of *Bathichnus paramoudrae* within the Upper Cretaceous Chalk of northern Europe. Certain horizons are accentuated by development of incipient hardgrounds or preferential early diagenesis of chert in and around burrow fillings. The authors attribute these persistent trace fossil horizons to minor fluctuations in sea level and attendant reduction in sedimentation rates. The upper terminations of these layers may reflect simply the resumption of sedimentation.

Biostromes and Bioherms Buildups of skeletal remains, such as biostromes, may record times of exceptional production of skeletons and/or entrapment of sediment by organisms. As an example, zones of distinctively shaped stromatolites are commonly traceable laterally for hundreds of square kilometers parallel to depositional strike. Grotzinger (1986) was able to trace individual stromatolite beds for more than 100 km in the well-exposed lower Proterozoic Rocknest carbonates of the Northwest Territories. Myro and Landing (1992) describe persistent algal horizons in the Lower Cambrian of Newfoundland, and Eagan and Liddell (essay 11 of this volume) discuss a similar case involving columnar stromatolites from the Middle Cambrian of Utah. In the Upper Silurian of the northern Appalachian basin region, thin (0.5–2.0-m) beds of thrombolitic algal mounds can be traced from Maryland to New York State along depositional strike. Perpendicular to depositional strike, these beds grade within tens of kilometers into somewhat larger thrombolites (1–2 m high), and these in turn grade into tabulate and stromatoporoid-bearing bioherms (Brett et al. 1990). Stromatolites and thrombolites thus are significant regional stratigraphic markers.

Bioherms of corals, stromatoporoids, and other heavily skeletonized organisms also may be persistent regionally. On the scale of outcrops, bioherms commonly appear to originate from a single bedding plane that obviously served as a pavement or substrate for the reef-building organisms (Walker and Alberstadt 1975). These levels reflect geologically brief intervals when conditions were particularly favorable for the buildup and preservation of algae and colonial skeletonized organisms. However, to date we are aware of only some general factors that might be involved in their formation. Abrupt climatic shifts, associated with reduced sedimentation and/or minor sea-level fluctuations, are potential explanations. It also should be noted that horizons of biohermal buildup commonly seem to be associated with marine flooding surfaces and thin intervals of relatively rapid deepening. During such times the seawater typically would be relatively free of suspended siliciclastic sediment, favoring reefal organisms. There also would be a greater chance of shell pavement development, which would favor reefal pioneer communities. Furthermore, a tendency toward vertical elevation of organism mounds might be driven by rising sea level (see Sarg 1988).

Thickets or biostromes are often associated with laterally persistent fossil-rich beds. Wolosz (essay 15 in this volume) discusses thicketing events in Middle Devonian reefs of New York, and West et al. (essay 16 in this volume) describes syringoporid epiboles or thickets in the Upper Carboniferous of Kansas that occur at consistent horizons for tens of kilometers in outcrop. At least in the second case, a temporary lowering of relative sea level may have induced the lateral spreading of opportunistic ("pioneer") coral thickets.

Altered substrate conditions may account for persistent small biostromes, such as the *Placunopsis* oyster banks described by Hagdorn (1982). The oyster buildups developed on hardgrounds in the Triassic Lower Muschelkalk of Germany. Similarly, Sageman and Johnson (1985) describe extensive lenticular biostromes of *Ostrea beloiti* immediately overlying the widespread x-bentonite in the Middle Cenomanian Greenhorn Formation. The biostromes extend from Wyoming and South Dakota to near the Texas/Mexican border and are evidently nearly coeval. In this case the development of the biostrome is related to the ashfall event, which may have sealed off seepage of toxic pore waters from lower anoxic sediments (Kauffman 1986).

The termination of biostromes or small bioherms may be related to gradual or abrupt pulses of increased sedimentation. However, larger bioherms and reefs are typically killed by one of two processes related to relatively rapid sea-level fluctuation. If deepening occurs rapidly, the reef may be unable to keep up in vertical growth and will drown. Conversely, major sea-level drop may expose portions of the reef, leading to production of an eroded or karstified upper surface that may be blanketed with a layer of reworked rubble. The latter also represents a type of taphonomic event bed.

Proliferation Epiboles

Kauffman's (1986) "population burst bioevents" correspond most closely to the traditional concept of epiboles (see table 10.1). Clearly, they represent an ecological phenomenon, but the precise causal agents are typically obscure. The term *population burst* is descriptive, but it should be noted that the phenomenon, designated herein as a type of ecological epibole; it records a geographically widespread event (hundreds to thousands of square kilometers) that probably occurred over periods of tens to perhaps tens of thousands of years. Hence these "bioevents" are a distinct phenomenon of greater temporal and spatial magnitude than a burst in abundance of a local population observed by a modern ecologist. These small-scale boom-and-bust cycles occur on a time scale of a few years—too short even to be recorded in most strata. Hence we substitute the more general term *proliferation* to replace *population burst.* Proliferation epiboles do show some features analogous to short-term population bursts. For example, many of the species involved are somewhat generalized and opportunistic in their distribution. The numbers of individuals may rise abruptly to exceptional levels in thin stratigraphic intervals and then abruptly drop in abundance ("crash") before disappearing in overlying strata.

Clearly, though, the environmental fluctuations responsible for the rise and collapse of these species actually involve many (up to thousands of) successive populations over extended periods. Thus, rather than reflecting the very short-term disturbances normally perceived by ecologists as causing population bursts, these proliferation epiboles reflect environmental perturbations of a much longer, although still geologically brief, time frame. It should be underscored that these fluctuations do not represent the same order of phenomena as ecological tracking (see Brett et al. 1990) that is recorded as nearly complete recurrence of normal fossil associations through sedimentary cycles. Rather, these ecological epiboles are superimposed upon the normal pattern of faunal assemblage cycles and stand out against a "background" of relatively stable tracking species.

One of the enigmatic features of these long-term proliferation epiboles is that commonly they are not associated with other obvious lithofacies or biofacies changes that would indicate unusual environments. Indeed, as noted, the epibole species may actually occur abundantly, more or less simultaneously in several lithofacies (fig. 10.3). This gives proliferation epiboles stratigraphic utility but makes ecological explanation more difficult. Apparently, environmental changes that favor the species do not cause obvious changes in sedimentary conditions. Plausible factors might include slight fluctuations of water temperature or salinity on a regional scale.

Whatever the factors, they seem to have had an effect on only one or a very few species and not on the majority. Such species specificity could be explained as the result of very sensitive physiological responses. Biotic processes, such as the outage of another species, might also be involved (see later).

In a previous paper (Brett et al. 1990) we provide examples of phenomena that might be included as proliferation epiboles. The following examples serve to illustrate the general concepts. The tabulate coral *Pleurodictyum* is a rare fossil in many assemblages in the Middle Devonian of the Appalachian Basin.

However, in certain thin calcareous layers, appropriately called "*Pleurodictyum* beds" (figs. 10.2 and 10.33), this coral is prolific (Grabau 1898; see also Brett and Cottrell 1982). Holland (this volume) describes similar anomalous occurrences of certain brachiopods, including *Thaerodonta,* within the Upper Ordovician Cincinnatian series of the mid-continent and Appalachian region. Likewise, Frey (this volume) describes beds composed almost exclusively of the brachiopod *Onniella* from the Waynesville Formation in Ohio, and Johnson (this volume) illustrates proliferation epiboles in *Pentamerus.* In these cases taphonomic factors cannot explain the anomalous occurrence, considering that these calcitic shells and corals have a good preservation potential and yet are not particularly common in most adjacent strata.

Another intriguing example involves the bryozoan *Reptaria,* normally a relatively rare encruster of Devonian nautiloids (see Baird et al. 1989). In certain horizons it occurs on a very high proportion of nautiloid specimens. Other layers with an equal abundance of nautiloids in similar facies display few or no encrustations. Some otherwise unrecorded factors enabled these organisms to proliferate at certain times.

A final peculiar example of a proliferation epibole is the occurrence of a fossil taxon in atypical facies. For example, the Devonian trilobite *Dipleura* is normally found in shallow marine siltstone facies. Yet, in one horizon, near the first appearance of this trilobite in the Hamilton Group, the species occurs in extreme abundance in dark gray offshore shales reflecting minimally oxic conditions. This unusual facies association is recorded from a thin bed for more than 200 km along the outcrop belt in New York State (see fig. 10.1: unit 11).

Outages ("Antiepiboles")

In some respects, the phenomena identified herein as "outages" represent the antithesis of ecological epiboles. Outages are thin stratigraphic intervals in which a species, normally abundant in a particular facies, is extremely rare or absent (table 10.1). The phenomenon was recognized by Brett et al. (1990), who pointed out examples in the Middle Devonian Hamilton Group in which common fossils are missing or are very rare in appropriate facies of particular cycles. Thus, for example, the common brachiopod *Pseudoatrypa* cf. *P. devoniana,* which is typically among the dominant fossils in diverse brachiopod/coral assemblages of calcareous mudstones, is virtually absent from these facies in two cycles in the middle part of the Hamilton Group (Ludlowville Formation). This "anomalous absenteeism" has been noted at all outcrops of these cycles across New York State and in their lateral equivalents in Pennsylvania. In contrast, most of the other diverse faunal elements of the coral bed biofacies do recur in these cycles, and the lithofacies are apparently identical to counterparts in cycles both above and below these outage levels in which *Pseudoatrypa* is absent (Brett et al. 1990).

Outage intervals, based as they are on negative evidence, have probably been dismissed by many previous researchers. However, they clearly record a phenomenon that demands further study. Outages demonstrate that faunal tracking as discussed by Brett et al. (1990) is not a completely adequate description of cyclic faunal change. Obviously, certain numerically important members of biofacies or communities fail to track environments at times. This is particularly anomalous in view of the fact that at least 90% of normally associated species do recur in the appropriate Hamilton facies. The fact that the species in question reappear abundantly in higher cycles indicates that they were "restocked" in particular communities following virtual local extinction. This observation further implies that reservoir populations of the species must continue to exist outside the region of local extermination. Outages persist for ecologically long spans of time, probably tens to hundreds of thousands of years. As is the case with ecological epiboles, the outage intervals are highly selective, seemingly involving only a single species. If ecological epiboles reflect a minor environmental shift that favors a normally rare taxon, then outages represent the opposite effect—an environmental parameter that decimates populations of a given species over a broad region. Again, as with epiboles, the factors responsible for the anomalous shift in abundance typically are not obvious from the enclosing strata or associated faunas.

In some instances outages of one species may be associated with epiboles of another taxon. For example, in the Devonian Hamilton Group certain thin (on the order of a decimeter to a few meters) intervals of dark gray shale display an anomalous abundance of diminutive (i.e., juvenile or paedomorphic) individuals of the orthid brachiopod *Tropidoleptus*. In these horizons morphologically similar small chonetid brachiopods that normally dominate these dysoxic facies are rare or absent. Cause-and-effect relationships are unclear: The epiboles of small *Tropidoleptus* may reflect ecological displacement or replacement of a taxon, such as chonetids, that has been decimated by external perturbations, or neither.

On the other hand, some outages, such as the *Pseudoatrypa* example mentioned earlier, are not linked to the differential success of any other species. Here, more clearly, other factors are responsible. At present we do not have any clear explanation for such selective outages. Specific bacterial or viral diseases could be suggested as a possible (albeit untestable) scenario. However, the relatively prolonged nature of outages (thousands of years, as compared to decades at most for known disease epidemics) raises doubts about this explanation.

In any case, outages represent an important and probably greatly underestimated type of ecological event related to epiboles. Outage intervals may have as much local stratigraphic utility as do epiboles. However, they must first be recognized as a real phenomenon worthy of study by stratigraphers and paleontologists.

Incursion Epiboles

Incursion epiboles are beds or thin (meters) intervals showing abrupt appearances of normally absent taxa in great numbers followed by equally abrupt disappearance of the taxa (table 10.1). In some cases the species may recur in a similar thin bed in another portion of the stratigraphic section. Incursion epiboles correspond closely to the "immigration-emigration bioevents" of Kauffman (1988, 634–638; see table 10.1).

One of the best illustrations of incursion epiboles is a series of thin horizons of goniatites and other fossils in the lower part of the Marcellus Shale and Cherry Valley Limestone in New York and Pennsylvania (fig. 10.1); in this stratigraphic interval each thin concretionary limestone bed contains a different goniatite species or series of species (Brett and Kloc 1988; Griffing and Ver Straeten 1991). The preservation of these goniatites is partially a taphonomic epibole phenomenon; they occur primarily in levels where concretionary carbonates or limestone beds prevented shell dissolution and compaction. However, this clearly is not the entire answer, as each successive bed contains a different ammonoid species. Furthermore, this succession of beds is traceable for hundreds of kilometers in the northern and central Appalachian Basin. From where did these taxa invade? Why were they viable for only brief intervals of time? Related species occur in the Old World Rhenish-Bohemian Realm of the Middle Devonian (Boucot 1990b). We conclude that these ammonoids briefly invaded into the Appalachian Basin from an outside source within the Old World Realm but were not viable for extended periods of time within it.

In the Devonian of the Appalachian region and elsewhere, the alga(?) *Foerstia* appears in profusion in a single, relatively narrow interval of the Upper Devonian Late Famennian (Murphy 1973; Schopf and Schwietering 1976). This abrupt, possibly global, appearance and disappearance event represents a very broad-scale epibole. Such large-scale events overlap substantially with traditional biostratigraphic units, but their mode of genesis is far from clear.

Another excellent example of an incursion epibole is the abrupt appearance of the small discoidal rugose coral *Palaeocyclus porpita* on both the Laurentian and Avalonian sides of the Early Silurian Iapetus Ocean (Scrutton and Deng 1994, and references therein; C. T. Scrutton, pers. comm., 1994). This coral occurs in a very narrow (typically 1 m to a few tens of meters) interval of Late Early Silurian (Llandovery C_6) of the northern Appalachians, Ontario, New York, Pennsylvania, and elsewhere. Similarly, the related genus, *Microcyclus,* appears only in a thin zone (less than 1 m) in the lower part of Middle Devonian Hamilton Group (Lower Givetian). Where did this lineage go during the intervening interval? How was this coral able to expand very rapidly with no record of an ancestral stock? It would appear in this case that the coral lineage descended from a stock that existed in otherwise unrecorded environments, such as the open oceanic realm. Periodically, corals of this lineage were able to advance rapidly into cratonic seas, perhaps because of favorable current patterns or breakdown of ecological or geographic barriers. In both cases, the coral incursions occur in offshore facies associated with a relatively major transgressive episode.

What is particularly enigmatic about these ephemeral faunas is that, within an epibole, the organisms may be extremely abundant and widespread, even cutting across facies (see fig. 10.3). Hence the organisms that characterize the incursion epibole appear not to have responded to typical facies-related environmental properties. Indeed, these occurrences appear to be an analog, on a large scale, of opportunistic behavior on the part of invading species. This may result from the fact that the epibole species were not integrated parts of normal tracking communities. They had a widespread larval dispersal and therefore could take over in many local communities, but only briefly. Given the widespread nature of these incursion taxa, their sudden disappearance is enigmatic. Perhaps local diseases or minor climatic shifts made the environments unfavorable to the invaders and caused local extinction. Certain epibole taxa recur at widely separated horizons. These cases seem to reflect multiple invasions of larvae into the basin at widely separated intervals and imply the existence of the "Lazarus" taxon in some other locality throughout the time interval.

In the case of incursion epiboles, repeated invasions of a taxon into a basin may reflect temporary opening of normally closed dispersal corridors. In particular, major highstands of sea level may permit influx of larvae into a local basin. Also, high sea level may allow temporary influx of unusual water masses and associated organisms into such basins. Along these lines of reasoning, Kloc (1986) pointed out that certain short-lived goniatite species appear in the Devonian of the Appalachian Basin in or above black shale tongues that represent major highstands (see fig. 10.1). Major transgressions may have linked normally separated faunal provinces and allowed the influx of transient taxa that were ultimately excluded from the Appalachian Basin. The invaders appear to have proliferated briefly over a broad range of environments, but ultimately they were excluded from the basin and did not become a permanent part of tracking communities.

Kauffman (1986) discussed "immigration" events from the Cretaceous of the Western Interior Seaway. Incursions here spanned a few thousand to about 100,000 years and the organisms dispersed over distances of 1,000–2,000 km. These events were apparently limited to organisms with mobile adults or planktotrophic larvae (Eicher and Diner 1985). Kauffman invokes breaching of geographic barriers (e.g., Transcontinental arch) to explain the abrupt immigration of subtropical biotas from the Gulf Coast in Colorado and of boreal bivalves into New Mexico. The causal agent for these immigrations was a major transgression.

Another instance, termed the *Thatcher bioevent,* marks the abrupt appearance of a diverse tropical fauna in a thin (0.25–1 m) concretionary bed within the Early Late Cretaceous Graneros Shale (Kauffman 1986). Surrounding Graneros shale carries only a low-diversity temperate fauna. Here Kauffman (1986) argues for a brief incursion of a subtropical water mass into the Western Interior Sea. This and other influxes of tropical species were associated with peak eustatic highstands (Eicher and Diner 1985). In relation to an event stratigraphic framework of bentonite beds, Kauffman (1986) demonstrates that these events are approximately simultaneous over regions of hundreds of square kilometers.

Thus incursion epiboles appear to be most effectively explained as a combination of breaching of biogeographic barriers and influx of water masses that have distinct properties. At least in these cases there is some type of linkage between local paleontological events and geologic processes. While this does not elucidate the causal mechanisms, it at least suggests that they relate to specific physiological tolerances. Possibilities might include slight differences in water temperature, salinity, or other chemical factors.

An intriguing example from our work is that of the small productid brachiopod *Truncalosia truncata*. This species is a common component of basinal dysoxic facies in the Eifelian Onondaga limestone and lower Hamilton Group of the Appalachian Basin (Koch 1981). It is extremely rare or absent from these facies in the overlying upper Hamilton Group but occurs as epiboles at two thin intervals (fig. 10.2: epiboles 9 and 13; fig. 10.3). An intriguing possibility presents itself in this case. Boucot (1990b) suggests that a major faunal changeover at the Eifelian-Givetian boundary Kacak-*Otomari* global bioevent of Walliser (1986a; Tryols-Massoni et al. 1990) may record global warming. The cool-water Malvinokaffric province faunas largely became extinct at about this time (Boucot 1988, 1990a, 1990b). Apparently, warm-water Rhenish-Bohemian faunas immigrated into the Appalachian basin, while recent discoveries suggest that some of the cooler-adapted Onondaga species occur mixed with the Malvinokaffric species (Boucot, pers. comm., 1992). Given this background, the *Truncalosia* epiboles could be explained as the result of incursion of slightly cooler bottom waters into the Appalachian Basin associated with transgressions.

Longer-Term Bioevents and Their Relation to Epiboles and Outages

To this point we have discussed bioevents of regional-to-global extent that operated over the range of hundreds of years to hundreds of thousands of years time. Such events produce discrete beds in sections yielding distinctive taxa over intervals of millimeters to several meters thickness. In contrast to ecological, migrational, and preservational events operating in this time frame are evolutionary punctuational events that recur at intervals of several million years or longer. These events, which are presumably global in magnitude, are expressed as mass extinctions, evolutionary radiations, emigrations/immigrations, and episodes of community reorganization. Such events include the major mass extinction and radiation events (see, for example, review volumes edited by Nitecki 1984; Elliot 1986; Walliser 1986a; Donovon 1989; and others), as well as minor mass extinctions and radiations (Boucot 1990a; Brett and Baird 1992). Following the cautionary notes of Walliser (1986b, 1990) to keep terminology regarding global bioevents relatively unconstrained and flexible, we strive herein to show that there may be a mesoscale evolutionary event or cycle that serves as the boundary for successive units of evolutionary stasis in the geologic record.

Early in the nineteenth century, Cuvier (1825) recognized a succession of "extinction-bounded" stratigraphic intervals in the Paris Basin, which were characterized by a predictable suite of organisms that had overlapping ranges within the given interval. Although many of these divisions were later found to be unconformity-bounded rather than extinction-bounded, later refinements by d'Orbigny (1849–1852, 1850–1852) showed the existence of chronostratigraphic units (stages) that were bounded by much narrower time intervals of extinction and origination. Although the catastrophist views of Cuvier and d'Orbigny are no longer held, the persistence of the stage as a dependable and important working unit in chronostratigraphy offers the possibility, if not the probability, that many stages are the record of an evolutionary rhythm affecting global ecosystems.

Eldredge and Gould (1972) advanced the view that the tempo of evolution for lower taxa is characterized by rapid transmutation events (punctuational speciation) followed by long episodes of little or no net evolutionary change through the range of the taxon (stasis). Since organisms in a community are often dependent upon one another, it stands to reason that punctuational events may occur collectively within communities as they are decimated and reorganized at times of minor mass extinction (Boucot 1990b). Thus if a hierarchy of cyclic or noncyclic global extinction events does exist for Earth, it should define a succession of community stasis units approximating the scale of stage and system. Boucot (1986, 1990b) has discussed the possibility of such a hierarchy of large- and small-scale intervals of community stability bounded by extinction events. Research to test this possibility is ongoing; we are currently endeavoring to document the existence of community-scale stasis or coordinated stasis, and second, to show that evolutionary stasis unit boundaries can be reorganized as mesoscale evolutionary bioevents (Brett and Baird 1995).

Taking the example with which we are most familiar, the Middle Paleozoic of the Appalachian basin, we can recognize a hierarchy of major and minor community reorganization events. This succession contains two of the five great mass extinction events: the Late Ordovician Ashgill crisis and the Late Devonian Frasnian and Famennian events (see Schindler 1990; McGhee 1990 and essay 18 in this volume). Unlike previously described boundaries, it is not always clear whether such extinctions are single sharp events or diffuse intervals including multiple step-wise extinctions. No iridium spikes are known from the Appalachian Basin, yet certain extinction signatures appear to be quite abrupt. For example, the disappearance of certain brachiopods occurs very abruptly at the Frasnian-Famennian boundary in New York State. However the late Frasmian extinction can be subdivided into a suite of lesser extinction events, as shown in Copper's (1986) study of atrypid brachiopods and ammonoids that show a series of extinction steps. Nonetheless, a potentially distinctive horizon marks the termination of each of these steps.

The interval between these extinction events is characterized by persistence of general community groups and thus comprises a single (or possibly two) ecological-evolutionary subunit (E-E unit V of Sheehan 1985, 1992; VI and V of Boucot 1983, 1990b).

We identify at least thirteen mesoscale community-level stasis units, *ecological-evolutionary* subunits (Boucot 1990b; Brett and Baird in press; fig. 10.4), which range in duration from 5 to 7 million years between the Middle Silurian (Llandoverian) and the Late Devonian (Frasnian-Famennian boundary); within these units, we find fossil assemblages to be nearly static for timespans of 5–7 million years.

For instance, Brett et al. (1990) discussed the oldest and youngest occurrence of coral beds within the Middle Devonian Hamilton Group and concluded that they are very similar on a species-by-species basis, with even similar rank abundances of various taxa over a period of about 6 million years. Against this backdrop of nearly perfect stasis in tracking communities (fig. 10.5), even minor stage-level extinction events stand out as times of major environmental punctuation. These represent perturbations severe enough to break down the equilibrium structure of stable communities (Brett and Baird 1995). What then happens is not entirely clear. Certain fossil taxa disappear abruptly within a single formation after persisting virtually unchanged for several millions of years. During the immediately ensuing time interval a number of in situ evolutionary changes take place, but these seem to be relatively minor species- or generic-level changes. Most community reorganization events also involve a rapid influx of new taxa from extrabasinal sources (fig. 10.5). In other cases, formerly rare taxa abruptly become common and certain ecological niches appear to remain unfilled. This type of change occurs typically within the first few hundred thousand years of an interval of stasis and then community structure appears to become locked at a nearly constant composition and diversity of species for any particular environment. Stratigraphic intervals recording community reorganization events are evidently of critical importance in the progress of life history, and it is at these particular punctuations that major changes in community structure occur due to accelerated evolution, immigration, and extinction.

What is the relationship of these large-scale evolutionary pulses in community structure to epibole and outage phenomena? Although this subject remains very poorly understood, a number of possible connections may exist between the two hierarchical levels of bioevents. First, in some senses incursion epiboles and outages are small-scale analogs of two key processes that are involved in the restructuring of many ecological/evolutionary faunas, namely immigration

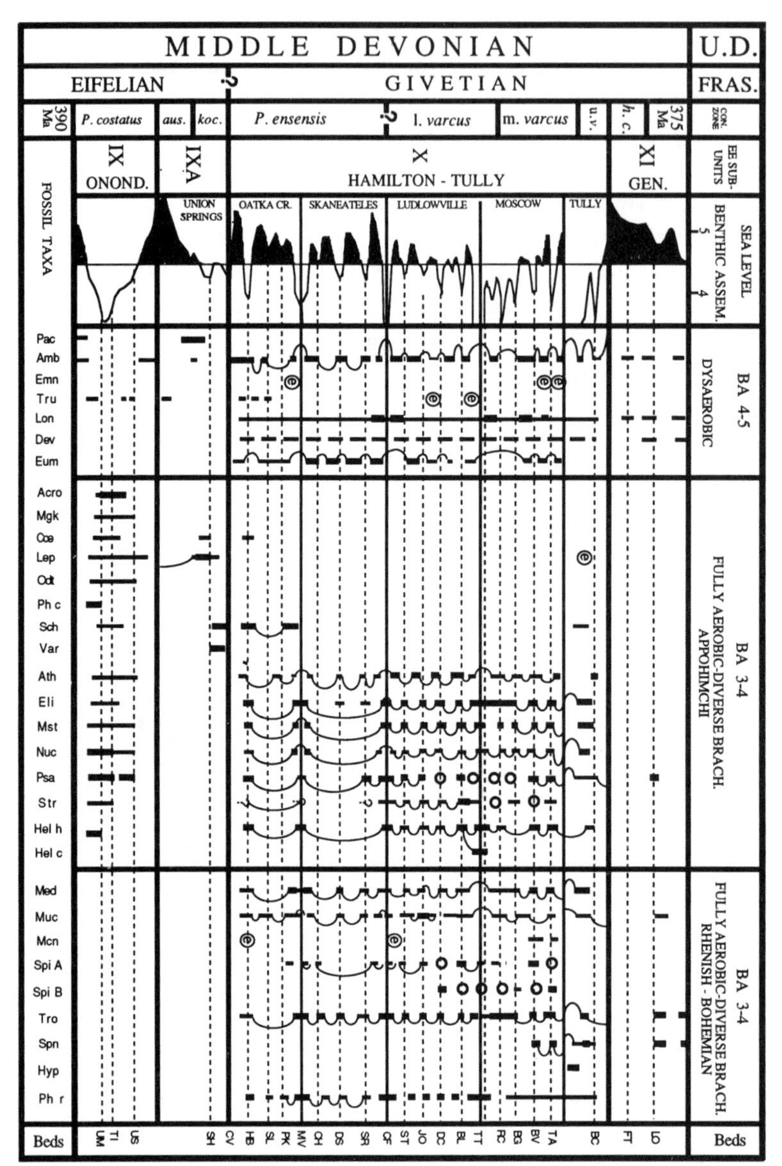

FIGURE 10.4. Stratigraphic ranges of Middle to earliest Late Devonian fossil species of two broadly defined biofacies in west-central New York State (Cayuga to Owasco Lake outcrops). Time-rock interval encompassed spans middle-upper Eifelian Stage (upper part of Onondaga Limestone and Union Springs Black Shale), Givetian Stage (remainder of Hamilton Group, Tully Limestone, and lower part of Genesee Group) and lowest Frasnian (remainder of Genesee Group). Abbreviations for biostratigraphic zones include *P. costatus: Polygnathus costatus costatus* zone; *aus: Polygnathus australis* zone; koc: *Tortodus kockelianus kockelianus* zone., *P. ensensus: Polygnathus xyleus ensensus* zone; 1., m., and u. *varcus:* lower, middle, and upper *Polygnathus varcus* subzones; h. c.: *Polygna-*

thus hermanni-Polygnathus cristatus zone. Absolute ages listed are in millions of years B.P. EE subunits are ecological-evolutionary "faunas" defined by Brett and Baird (1995); IX is the Onondaga fauna; IXA consists of a modified Onondaga assemblage plus unique elements of the Stony Hollow fauna; X is the Hamilton-Tully fauna; XI is the Genesee fauna. Relative sea-level curve is calibrated by fossil benthic assemblages as defined by Boucot (1975); note that only the deeper end of this scale is shown. BA 3.5 = coral-rich, diverse offshore communities; BA 4 = diverse brachiopod communities; BA 4.5 = ambocoeliid communities; and BA 5 = high-dominance lieorhynchid communities typical of dark gray to black shale. Center line subdivides dysoxic from fully oxic facies. Abbreviations for major marker beds include (for Onondaga) UM: uppermost Moorehouse Member grainstones; TI "Tioga-B" bentonite; US: upper Seneca Member; (for Hamilton-Tully) SH: Stony Hollow (thin limestone bed in western N.Y.); CV: Cherry Valley Limestone; HH: Halihan Hill fossil bed; SL: Solsville horizon; PK: Pecksport horizon; MV: Mottville (Stafford) Member; CH: Cole Hill fossil bed; DS: Delphi Station shell bed; SR: Slate Rock bed (cap of Pompey Member); CF: Centerfield (Chenango) Member; ST: Staghorn coral bed; JO: Joshua coral bed; DC: Darien Center coral bed; BL: Bloomer Creek fossil bed; TT: Tichenor Limestone; MT: Menteth Limestone; RC: *Rhipidomella-Centronella* fossil bed; BG: Barnes Gully phosphatic bed; BV: Bay View coral bed; TA: Taunton fossil beds; BC: Bellona coral bed; FT: Fir Tree fossil bed; LO: Lodi fossil bed. Fossil species are subdivided roughly by biofacies into two groups: (1) those that characterize deeper-water, dysoxic areas typified by small ambocoeliid, chonetid, and lieorhynchid brachiopods (BA 4.5–5.5); (2) those that typify diverse brachiopod biofacies of oxic shallower, calcareous gray mudstones or argillaceous limestones (BA 3.5–4). The latter are further subdivided into two groups: (a) those derived from the earlier Devonian Appohimchi Province (Appalachian Basin natives), (b) those that immigrated into the Appalachian Basin from the Rhenish-Bohemian area during late Eifelian to early Givetian time. Symbols for range lines are as follows: thick line: species common to dominant; thin lines: species present; arcuate curves to left imply species migrated temporarily from study area downslope (south); arcs to right indicate species migrated temporarily upslope (north) tracking favored environment; e = incursion epibole, brief incursion of species not normally present; o = outage (temporary absence of normally common species from appropriate biofacies.) Abbreviations for species include brachiopods (dysaerobic): Pac = *Pacificocoelia acutiplicata;* Ambo = *Ambocoelia* cf. *A. umbonata;* Emn = *Emanuella* cf. *E. praeumbona;* Tru = *Truncalosia truncata;* Lon = *Longispina mucronata* and/or *L. deflecta;* Dev = *"Devonochonetes" scitulus;* Eum = *Eumetabolotoechia* cf. *E. multicosta;* brachiopods (BA 3–4, Appohimchi): Acro = *Acrospirifer duodenaria;* Mgk = *Megakozlowskiella raricosta;* Coe = *Coelospira* cf. *C. camilla;* Lep = *Leptaena "rhomboidalis"*; trilobites: Odt = *Odontocephalus* sp.; Ph c = *Phacops cristata;* brachiopods: Sch = *Schizophoria* sp.; Var = *Variatrypa arctica* (Canadian Arctic affinities); Ath = *Athyris* cf. *A. cora* and *A. spiriferoides;* Eli = *Elita fimbriata;* Mst = *Megastrophia* cf. *M. concava;* Nuc = *Nucleospira* aff. *N. ventricosa* (Onondaga) and *N. concinna* (Hamilton); Psa = *"Pseudoatrypa"* sp. (Onondaga) and cf. *P. devoniana* (Hamilton-Tully); Str = *Strophodonta* cf. *S. demissa* (Onondaga); *S. demissa* (Hamilton); corals: Hel h = *Heliophyllum halli;* Hel c = *Heliophyllum confluens;* brachiopods (BA 3–4, Rhenish-Bohemian affinities): Med = *Mediospirifer* cf. *M. audaculus;* Muc = *Mucrospirifer mucronatus;* Mcn = *"Mucrospirifer" consobrinus;* Spi A = *Spinocyrtia* sp. A (no medial notch on fold); Spi B = *Spinocyrtia* sp. B (deep notch on fold)-note alternations of morphotypes; Tro = *Tropidoleptus carinatus;* Spn = *Spinatrypa spinosa;* Hyp = *Hypothyridina* cf. *H. cuboides;* trilobite Ph r = *Phacops rana.*

of "exotic" taxa and emigration/extinction of previously stable taxa (fig. 10.5). Both incursions and outages involve just a few species, or only one at a time, and both are in a sense abortive bioevents. This means that they do not lead to permanent changes in stable, tracking biotas that characterize a given E-E subunit (fig. 10.5). Yet in attempting to elucidate the processes that lead to temporary influx of an "exotic" species or absenteeism of a normally common form, we may arrive at a better understanding of processes that govern the evolutionary breakdown and restructuring of entire communities. Both of these small and large-scale bioevents point to the importance of biogeographic effects. In both cases the existence of external "reservoirs" of species is implied. The repeated incursion of epibole taxa, or for that matter, the reappearance of a normal indigenous form or related species following an outage interval, resembles the "Lazarus" effects observed at evolutionary time scales (Jablonski 1980, 1986). In both cases the reservoir area for the returning taxa is difficult to locate, particularly in Paleozoic examples. Oceanic island areas are a possible provenance of the holdover species (see also Jablonski 1985).

Second, there is some very preliminary evidence that similar widespread environmental shifts govern incursion epiboles as in wholesale overturn of faunas at E-E unit or subunit boundaries. For example, we have noted that a number of incursion events both in the Paleozoic of the Appalachian Basin and in the Cretaceous of the western interior are associated with major transgressive and anoxic events. This association also appears to be a common feature of many E-E subunit turnovers and larger extinction events (see papers in Walliser 1986a; Walliser 1990; Donovan 1989). In reviewing the E-E subunits of the Silurian to Middle Devonian of the Appalachian Basin (Brett and Baird 1992, 1995) we have noted that over half of the turnover events are associated temporarily with development of widespread black shale intervals. The Silurian Williamson Shale, and Devonian Esopus, Marcellus, and Genesee black shales are timed with major faunal turnovers. Both the Givetian-Frasnian and the Frasnian-Famennian mass extinctions occur during intervals of sea-level highstand and widespread anoxia. In some cases there is geochemical evidence for incursion of atypical water masses (e.g., seawater with warmer or cooler than normal temperatures, slightly elevated or reduced salinity) associated with intervals of incursion epiboles and larger-scale turnovers (see Kauffman 1986, for example). Such anomalies together with widespread anoxia could account for extermination/emigration of local faunas and introduction of exotics at least in areas above the oxycline (see the discussion of Ireviken event by Jeppsson [1987] in this volume). Also, highstands probably result in altered oceanic circulation patterns and breaking of biogeographic barriers to permit influx of planktotrophic larvae into a local basin. Careful attention to the details of larval paleoecology (e.g., Hansen 1980; Jablonski 1985, 1986) might permit a test of this postulate, at least for Mesozoic-Cenozoic events.

Third, both ecological epiboles and outages demonstrate that certain elements of seemingly stable biotas may be very sensitive to subtle environmental factors, including possible biotic effects such as pathogens and minor physicochemical perturbations. In both epiboles and outages these effects led only to a temporary change in a portion of the biota. However, it is possible that if such effects eliminated or promoted particular "keystone species" (Morris et al. 1992), they could trigger major collapse and reorganization of a community or ecosystem. On the other hand, most communities seem quite resilient to such turnover.

Fourth, perhaps the most important insight gained from study of outages and epiboles is that ecosystems are remarkably stable and capable of withstanding fairly major perturbations with little permanent effect. The very definitions of epiboles and outages reflect this: These are temporary, geologically short-lived bioevents. More than 90% of normally coexisting species reappear in the appropriate biofacies even during times of anomalous proliferation of a usually rare or absent taxon, or the complete absence of a normally dominant form. Thus epibole and outage events are not simply a mirror of major bioevents. Because they are abortive E-E events, epiboles and outages provide a natural test of the internal "homeostasis" of ecosystems; and they demonstrate a high degree of "buffering" or resilience against breakdown. A single new species introduction, even where that species proliferates to enormous and widespread population sizes, rarely leaves a lasting effect. In our studies of Appalachian basin faunas, at least, there are very few instances in which a single new taxon becomes established in a given cycle within the duration of a given E-E subunit and then becomes part of the tracking biota, reappearing in all subsequent cycles. Rather, almost all the restructuring and introduction of tracking forms occurs very near the beginning of E-E subunits. The fact that most subsequent incursions lead only to epiboles indicates that once community structure is reestablished it is refractory to the permanent introduction of even a few new species, for millions of years. The causes of this remarkable pattern of biofacies stability or "coordinated stasis" (Brett and Baird 1992) are beyond the scope of this paper and will form the subject of future reports (Morris et al. 1992; in press).

The observation that epiboles in some species coincide with rarity or outages of others indicates that some degree of shuffling of dominance positions of indigenous species may occur occasionally and temporarily; an exotic form may be accommodated in benthic ecosystems without major alteration of community structure. These "musical chair" patterns are a form of ecological noise, although they may point to a possible source of stability in diverse communities (Morris et al. 1992).

All this stresses the critical importance of large-scale ecological crises in producing the major bioevents that punctuate life history at least in benthic marine ecosystems (Boucot 1990a). Seemingly only very large-scale or excessively rapid environmental changes can knock ecosystems out of equilibrium sufficiently to produce sweeping changes in community organization. These infrequent events are accompanied by widespread, at least local, termination of long-ranging lineages, rapid evolution and the permanent influx of new taxa. Some epiboles may coincide with the large-scale turnover events (e.g., *Palaeocyclus* epibole is found near the base of the Middle Silurian, upper Clinton ecological-evolutionary subunit) but most of the newly appearing taxa become permanent components of tracking biofacies and persist to the end of the E-E subunits (Brett and Baird, in press).

Summary

Paleontological bioevents are herein classified either as short-term events in the form of epiboles and outages or as longer-term units in the form of community reorganization events separated by times of evolutionary community stasis and stability. Although paleontological events have long been studied on a case-study basis, there have been few attempts to determine how these events relate to one another in hierarchical magnitude or how they connect to the rapidly growing body of knowledge stemming from the study of sedimentary cyclicity within geological systems.

Of the paleontological event beds discussed herein, only some of the taphonomic epiboles are of strictly local spatial distribution within basins. All the other categories have the potential of being linked to extrabasinal processes such as eustasy and climate-forcing even though correlation is commonly less than global. More significantly, the existence of numerous large-scale episodes of accelerated evolution and community reorganization offers the possibility of a connection between these events and the traditional scheme of assemblage zones and stages now recognized. If this connection can be shown to be firm, this would suggest a further link to the hierarchy of processes that account for large-scale (third- to fifth-order) cyclical events now recognized as integral to the model of sequence stratigraphy.

The putative connections listed earlier argue that evolution of marine invertebrates and marine communities is pulsational and cyclical. Some epibole- and outage-scale events are controlled directly and indirectly by climatic cycles operating on scales corresponding to sixth-order cycles or less. By contrast, the pattern of alternating stasis periods and punctuation events may be ultimately linked to many larger-scale eustatic/climatic cycles. Hence paleontological event stratigraphy may be a critical link to chronostratigraphy.

ACKNOWLEDGMENTS

This manuscript was critically reviewed by Art Boucot, David Linsley, and Michael Savarese; Mary Nardi provided technical editing. The manuscript was processed by Fred Teichmann and Susan Todd.

Our research on Paleozoic epiboles has been funded by grants from the Petroleum Research Fund, American Chemical Society, and by *NSF EAR*-8816856, and *EAR* 9219807.

References

Arkell, W. J. 1933. *The Jurassic System in Great Britain.* Oxford: Clarendon Press.

Baird, G. C., C. E. Brett, and R. W. Frey. 1989. "Hitchhiking" epizoans on orthoconic cephalopods: Preliminary review of the evidence and its implications. *Senckenbergiana Lethaea* 69:439–65.

Banerjee, I. and S. M. Kidwell 1991. Significance of molluscan shell beds in sequence stratigraphy: An example from the Lower Cretaceous Mannville Group of Canada, *Sedimentology* 38:913–34.

Boucot, A. J. 1983. Does evolution take place in an ecological vacuum? II, *Journal of Paleontology* 57:1–30.

———. 1986. Ecostratigraphic criteria for evaluating the magnitude, character and duration of bioevents: In O. H. Walliser, ed., *Global Bio-events: A Critical Approach. Lecture Notes in Earth Sciences,* vol. 8, pp. 25–45. Berlin: Springer-Verlag.

———. 1988. *Devonian Biostratigraphy: An Update.* Second Symposium on the Devonian System, *Canadian Society of Petroleum Geologists Memoir* No. 14, pp. 211–77.

———. 1990a. Phanerozoic extinctions: How similar are they to each other? In E. G. Kauffman and O. H. Walliser, eds., *Extinction Events in Earth History. Lecture Notes in Earth Sciences,* pp. 5–30. Berlin: Springer-Verlag.

———. 1990b. Silurian and pre-Upper Devonian bio-events. In E. G. Kauffman and O. H. Walliser, eds., *Extinction Events in Earth History. Lecture Notes in Earth Sciences,* vol. 30, pp. 125–32. Berlin: Springer-Verlag.

Brett, C. E. 1988. Paleoecology of hard substrate communities: An overview. *Palaios* 3:374–78.

Brett, C. E. and G. C. Baird. 1992. Coordinated stasis and evolutionary ecology of Silurian-Devonian marine biotas in the Appalachian basin, *Geological Society of America, Abstracts with Programs* 24(7):139.

———. 1995. Coordinated stasis and evolutionary ecology of Silurian to Middle Devonian faunas in the Appalachian basin. In D. Erwin and R. Anstey, eds., *New Approaches to Speciation in the Fossil Record.* New York: Columbia University Press, pp. 285–315.

Brett, C. E., G. C. Baird, and S. E. Speyer. This volume. Fossil Lagerstätten: Stratigraphic record of paleontological and taphonomic events. In C. E. Brett and G. C. Baird, eds. *Paleontological Events: Stratigraphic, Ecologic, and Evolutionary Implications.* New York: Columbia University Press.

Brett, C. E. and J. F. Cottrell. 1982. Substrate selectivity in the Devonian tabulate coral *Pleurodictyum americanum* (Hall). *Lethaia* 25:248–63.

Brett, C. E. and G. K. Kloc. 1988. Faunas, stratigraphy, and depositional environments of the Cherry Valley Limestone and associated carbonates. In E. Landing, ed., *The Canadian Paleontology and Biostratigraphy Seminar, New York State Museum Bulletin* 462:122–25.

Brett, C. E. and W. D. Liddell. 1978. Preservation and paleoecology of a Middle Ordovician hardground community. *Paleobiology* 4:329–48.

Brett, C. E., K. B. Miller, and G. C. Baird. 1990. A temporal hierarchy of paleoecological processes in a Middle Devonian epeiric sea. In W. Miller, III, ed., *Paleocommunity Temporal Dynamics: The Long-Term Development of Multispecies Assemblages. Paleontological Society Special Publication,* No. 5, pp. 178–209.

Brett, C. E. and A. Seilacher. 1991. Fossil Lagerstätten: A taphonomic consequence of event sedimentation, pp. 293–97. In G. Einsele, W. Ricken, and A. Seilacher, eds. *Cycles and Events in Stratigraphy.* New York: Springer-Verlag.

Brett, C. E. and W. L. Taylor. This volume. The *Homocrinus* beds: Silurian crinoid *Lagerstätten* of western New York and southern Ontario. In C. E. Brett and G. C. Baird,

eds., *Paleontologic Events: Stratigraphic, Ecologic and Evolutionary Implications*. New York: Columbia University Press.
Buckman, S. S. 1893. The Bøjocian of the Sherborne district. *Quarterly Journal of the Geological Society* 49:479–522.
———. 1898. On the grouping of some divisions of so-called Jurassic time. *Quarterly Journal of the Geological Society* 54:442–62.
———. 1902–1903. The term Hemera. *Geology Magazine* [4], 9:95–96.
Chlupác, I. and Z. Kukal. 1986. Reflection of possible global Devonian events in the Barrandian area C.S.S.R. In O. H. Walliser, ed., *Global Bioevents. Lecture Notes in Earth Sciences*, vol. 8, pp. 169–79. Berlin: Springer-Verlag.
Conway Morris, S. 1986. Community structure of the Middle Cambrian phyllopod bed (Burgess Shale). *Palaeontology* 29:423–58.
Copper, P. 1986. Frasnian/Famennian extinction and cold-water oceans. *Geology* 14:835–39.
Cuvier, G. 1825. Discours sur les révolutions de la surface du globe et sur les changements qu'elles ont produites dans le règne animal. Paris: Dufour et d'Ocagne.
Donovan, S. K., ed. 1989. *Mass Extinctions*. New York: Columbia University Press.
Dunbar, C. O. and J. Rogers 1957. *Principles of Stratigraphy*. New York: John Wiley.
Eagan, K. E. and W. D. Liddell. This volume. Columnar stromatolites: A bioevent horizon in the Middle Cambrian Ute Formation of northern Utah and southern Idaho. In C. E. Brett and G. C. Baird, eds., *Paleontological Event Strata: Stratigraphic, Ecologic, and Evolutionary Implications*. New York: Columbia University Press.
Eicher, D. L. and R. Diner 1985. Foraminifera as indicators of water mass in the Cretaceous Greenhorn sea, Western Interior. In L. M. Pratt, E. G. Kauffmann, and F. G. Zelt, eds., *Fine-Grained Deposits and Biofacies of the Cretaceous Western Interior Seaway: Evidence of Cyclic Sedimentary Processes, Society of Economic Paleontologists and Mineralogists, Fieldtrip Guidebook*, vol. 4, pp. 60–71.
Eldredge, N. and S. J. Gould 1972. Punctuated equilibria: An alternative to phyletic gradualism. In T. J. M. Schopf, ed., *Models in Paleontology*, pp. 82–115. San Francisco: Freeman Cooper.
Elliot, D. K., ed. 1986. *Dynamics of Extinction*. New York. Wiley.
Ernst, G. and E. Seibertz 1977. Concepts and methods of echinoid biostratigraphy. In E. G. Kauffman and J. E. Hazel, eds., *Concepts and Methods of Biostratigraphy*, pp. 541–63. Stroudsburg, Pa.: Dowdon, Hutchinson, and Ross.
Follmi, K. B. and K. A. Grimm 1990. Doomed pioneers: Gravity-flow deposition and bioturbation in marine oxygen deficient environments. *Geology* 18:1069–72.
Frey, R. C. This volume. The utility of epiboles in the regional correlation of Paleozoic epeiric sea strata: An example from the upper Ordovician of Ohio and Indiana. In C. E. Brett and G. C. Baird, eds., *Paleontologic Events: Stratigraphic, Ecologic and Evolutionary Implications*. New York: Columbia University Press.
Goodwin, P. W. and E. J. Anderson 1986. Punctuated aggradational cycles: A general hypothesis of episodic stratigraphic accumulation. *Journal of Geology* 93:515–33.
Grabau, A. W. 1898. Geology and paleontology of Eighteen-Mile Creek and the lakeshore sections of Erie County, New York. *Buffalo Society of Natural Sciences Bulletin* 6:1–403.
Griffing, D. H. and C. A. Ver Straeten 1991. Stratigraphy and depositional environments of the lower part of the Marcellus Formation (Middle Devonian) in eastern New York State, in J. R. Ebert, ed., *New York State Geological Association, 63d Annual Meeting Guidebook, Oneonta*, pp. 205–49.
Grotzinger, J. P. 1986. Cyclicity and paleoenvironmental dynamics, Rocknest Platform, northwest Canada. *Bulletin of the Geological Society of America* 97:1208–31.
Hagdorn, H. 1982. The "Bank der Kleinen Terebrateln" (upper Muschelkalk, Triassic)

near Schwäbisch Hall-Germany: A tempestite condensation horizon. In G. Einsele and A. Seilacher, eds., *Cycles and Event Stratification*, pp. 263–87. Berlin: Springer-Verlag.

Hansen, T. A. 1980. The influence of larval dispersal and geographic distribution on species longevity in neogastropods. *Paleobiology* 6:193–207.

Havlicek, K. V. and P. Storch 1990. Silurian brachiopods and benthic communities in the Prague Basin in (Czechoslovakia). *Rozpravy Ustredniho Ustavu Geologickeho, Praha* 48:1–275.

Hedberg, H. D. 1972. Summary of an international guide to stratigraphic classification, terminology, and usage. *Lethaia* 5:297–323.

Holland, S. M. This volume. Using time-environment analysis to recognize faunal events in the upper Ordovician of the Cincinnati Arch. In C. E. Brett and G. C. Baird, eds., *Paleontologic Events: Stratigraphic, Ecologic and Evolutionary Implications*. New York: Columbia University Press.

House, M. R. 1962. Observations on the ammonoid succession of the North American Devonian. *Journal of Paleontology* 36:247–84.

———. 1985. Correlation of mid-Paleozoic ammonoid evolutionary events with global sedimentary perturbations. *Nature* 313:17–22.

Jablonski, D. 1980. Apparent versus real biotic effects of transgression and regression. *Paleobiology* 6:398–407.

———. 1985. Marine regressions and mass extinctions: A test using the modern biota. In J. Valentine, ed., *Phanerozoic Diversity Patterns*, pp. 335–54. Princeton, N.J.: Princeton University Press.

Jablonski, D. 1986. Causes and consequences of mass extinctions: A comparative approach. In D. K. Elliot, ed., *Dynamics of Extinction*, pp. 183–229. New York: Wiley.

Jeppsson, L. 1987. Lithological and conodont distributional evidence for episodes of anomalous oceanic conditions during the Silurian. In R. J. Aldridge, ed., *Palaeobiology of Conodonts*, pp. 129–45. Chichester, Eng.: Ellis Horwood.

———. This volume. The anatomy of the mid-Early Silurian Ireviken event and a scenario for events. In C. E. Brett and G. C. Baird, eds., *Paleontologic Events: Stratigraphic, Ecologic and Evolutionary Implications*. New York: Columbia University Press.

Johnson, M. E. 1988. Why are ancient rocky shores so uncommon? *Journal of Geology* 96:469–86.

———. This volume. Silurian event horizons associated with the evolution and ecology of pentamerid brachiopods, In C. E. Brett and G. C. Baird, eds. *Paleontologic Events: Stratigraphic, Ecologic, and Evolutionary Implications*. New York: Columbia University Press.

Kauffman, E. G. 1986. High-resolution event stratigraphy: Regional and global Cretaceous bio-events. In O. Walliser, ed., *Global Bioevents. Lecture Notes in the Earth Sciences*, vol. 8, pp. 279–335. Berlin: Springer-Verlag.

———. 1988. Concepts and methods of high-resolution stratigraphy. *Annual Review of Earth and Planetary Sciences* 16:605–54.

Kidwell, S. M. 1991a. The stratigraphy of shell concentrations. In P. A. Allison and D. E. G. Briggs, eds., *Taphonomy: Releasing the Data Locked in the Fossil Record*, pp. 211–90. New York: Plenum Press.

———. 1991b. Taphonomic feedback (live/dead interactions) in the genesis of bioclastic beds: Keys to reconstructing sedimentary dynamics. In G. Einsele, W. Ricken, and A. Seilacher, eds., *Cycles and Events in Stratigraphy*, pp. 268–82. Berlin: Springer-Verlag.

Kidwell, S. M. and D. Jablonski. 1983. Taphonomic feedback: Ecological consequences of shell accumulation. In M. J. S. Tevesz and P. L. McCall, eds., *Biotic Interactions in Recent and Fossil Benthic Communities*, pp. 195–248. New York: Plenum Press.

Kirchgasser, W. T. and M. R. House. 1981. Upper Devonian goniatite biostratigraphy.

In W. A. Oliver, Jr., and G. Klapper, eds., *Devonian Biostratigraphy of New York*, part 1, International Union of Geological Sciences, Subcommission on Devonian Stratigraphy, Washington, D.C., pp. 39–55.
Kloc, G. J. 1986. Distribution of goniatitic ammonoids in the Middle Devonian (Givetian) Hamilton Group of New York. *Geological Society of America Abstracts with Programs* 18:27.
Koch, W. 1981. Brachiopod community paleoecology, paleobiogeography, and depositional topography of the Devonian Onondaga Limestone and correlative strata in eastern North America. *Lethaia* 14:83–103.
Křiž, J. 1992. Silurian field excursions: Prague Basin (Barrandian), Bohemia In M. G. Bassett, ed., *National Museum of Wales Geological Series*, vol. 13, Cardiff, Wales.
Lang, W. D. 1924. The Blue Lias of the Devon and Dorset coasts. *Proceedings of the Geological Association* 35:169–85.
LoDuca, S. T. and C. E. Brett. This volume. The *Medusaegraptus* epibole and lower Ludlovian Konservat-Lagerstätten of eastern North America. In C. E. Brett and G. C. Baird, eds., *Paleontologic Event Stratigraphy: Stratigraphic, Ecologic and Evolutionary Implications.* New York: Columbia University Press.
Lottmann, J. 1990. The middle Givetian pumilio-events: A tool for high time resolution and event-stratigraphical correlation. In E. G. Kauffman and O. H. Walliser, eds., *Extinction Events in Earth History. Lecture Notes in Earth Sciences*, vol. 30, pp. 145–49. Berlin: Springer-Verlag.
McGhee, G. R., Jr. 1990. The Frasnian-Famennian mass extinction record in the eastern United States. In E. G. Kauffman and D. H. Walliser, eds., *Extinction Events in Earth History. Lecture Notes in Earth Sciences*, vol. 30, pp. 161–68. Berlin: Springer-Verlag.
———. This volume. Late Devonian bioevents in the Appalachian Sea: Immigration, extinction, and species replacements, In G. E. Brett and G. C. Baird, eds. *Paleontological Events: Stratigraphic, Ecologic and Evolutionary Implications.* New York: Columbia University Press.
Morris, P. J., L. C. Ivany, and K. M. Schopf. 1992. Paleoecological stasis in evolutionary theory. *Geological Society of America Abstracts with Programs* 24(7):313.
Mortimore, R. N. and R. Pomerol. 1991. Stratigraphy and eustatic implications of trace fossil events in the Upper Cretaceous Chalk of northern Europe. *Palaios* 6:216–32.
Murphy, J. L. 1973. *Protosalvinia (Foerstia)* zone in the Upper Devonian sequence of eastern Ohio, northwestern Pennsylvania, and western New York. *Geological Society of America Bulletin* 84:3405–10.
Myro, P. M. and E. Landing. 1992. Mixed siliciclastic-carbonate deposition in an Early Cambrian oxygen-stratified basin, Chapel Island Formation, southeastern Newfoundland. *Journal of Sedimentary Petrology* 62:455–74.
Nitecki, M., ed., 1984. *Extinctions.* Chicago: University of Chicago Press.
d'Orbigny, A. 1849–1852. *Cours élémentaire de paléontologie et de géologie stratigraphique*, pp. 299, 382–841. Paris: Masson.
———. 1850–1852. *Prodrome de Paléontologie*, pp. 99, 190, 197, 394, 427. Paris: Masson.
Pemberton, S. G. This volume. The ichnological signature of storm deposits: The use of trace fossils in event stratigraphy. In C. E. Brett and G. C. Baird, eds., *Paleontologic Events: Stratigraphic, Ecological and Evolutionary Implications.* New York: Columbia University Press.
Rosenkranz, D. 1971. Zur Sedimentologie and ökologie von Echinodermen-Lagerstätten. *Neues Jahrbuch für Geologie und Paläontologie Abhandlungen* 138:221–58.
Sageman, B. B. and C. C. Johnson 1985. Stratigraphy and paleontology of the Lincoln Limestone Member, Greenhorn Limestone, Rock Canyon Anticline, Colorado. In L. M. Pratt, E. G. Kauffman, and F. B. Zelt, eds., *Fine- Grained Deposits and Biofacies of the Cretaceous Western Interior Seaway: Evidence of Cyclic Sedimentary Processes*, pp. 100–

109. *Society for Economic Paleontologists and Mineralogists, Fieldtrip Guidebook* 4. Golden, Colorado.

Sageman, B. B., E. G. Kauffman, P. J. Harries, and W. P. Elder. This volume. Cenomanian-Turonian bioevents and ecostratigraphy in the western interior basin: Contrasting scales of local, regional and global events. In C. E. Brett and G. C. Baird, eds., *Paleontological Events: Stratigraphic, Ecologic, and Evolutionary Implications*. New York: Columbia University Press.

Sageman, B. B., P. B. Wignall, and E. G. Kauffman 1991. Biofacies model for oxygen-deficient facies in epicontinental seas: Tool for paleonenvironmental analysis. In G. Einsele, W. Ricken, and A. Seilacher, eds. *Cycles and Events in Stratigraphy*, pp. 542–64. Berlin: Springer-Verlag.

Sarg, J. F. 1988. Carbonate sequence stratigraphy. In C. Wilgus, C. Ross, H. Posamentier, J. Van Wagoner, and C. St. C. Kendall, eds., *Sea-Level Changes: An Integrated Approach. Society of Economic Paleontologists and Mineralogists Special Publication* No.43, pp. 155–81.

Schindler, E. 1990. The Late Frasnian (Upper Devonian) Kellwasser crisis. In E. G. Kauffman and O. H. Walliser, eds., *Extinction Events in Earth History. Lecture Notes in Earth Sciences*, vol. 30, pp. 151–59. Berlin: Springer-Verlag.

Schopf, J. M. and Schwietering, J. F. 1976. The *Foerstia* Zone of the Ohio and Chattanooga Shales, *U.S. Geological Survey Bulletin*, 1294.

Scrutton, C. T. and Zhang-Qui Deng. 1994. Corals. In C. H. Holland, ed., *Testing Precision in Biostratigraphy: An Experiment on the Silurian of the British Isles and China*. Harlow, U.K.: Longman.

Shaw, A. B. 1964. *Time in Stratigraphy*. New York: McGraw-Hill.

Sheehan, P. M. 1985. Reefs are not so different—they follow the evolutionary pattern of level-bottom communities. *Geology* 13:46–49.

———. 1992. Patterns of synecology during the Phanerozoic. In E. C. Dudley, ed., *The Unity of Evolutionary Biology* 1:103–118.

Trueman, A. E. 1923. Some theoretical aspects of correlation. *Proceedings of the Geological Association* 36:11–25.

Tryols-Massoni, M., R. Montesinos, J. L. Garcia-Alcaldes, and F. Leyva. 1990. The Kacak-*Otomari* event and its characterization in Palentine domain (Cantabrian Zone NW Spain). In E. G. Kauffman and O. H. Walliser, eds., *Extinction Events in Earth History. Lecture Notes in Earth Sciences*, vol. 30, pp. 131–143. Berlin: Springer-Verlag.

Tutcher, J. W. and A. E. Trueman 1925. The Liassic rocks of the Radstock district, Somerset. *Quarterly Journal of the Geological Society* 81:595–666.

Vail, P. R., F. Audemard, S. A. Bowman, P. N. Eisner, and C. Perez-Cruz. 1991. The stratigraphic signatures of tectonics, eustasy and sedimentology—an overview. In G. Einsele, W. Ricken, and A. Seilacher, eds., *Cycles and Events in Stratigraphy*, pp. 617–59. Berlin: Springer-Verlag.

Vossler, S. M. and S. G. Pemberton. 1988. Superabundant *Chondrites:* A response to storm-buried organic material? *Lethaia* 21:94.

Walker, K. R. and L. P. Alberstadt. 1975. Ecological succession as an aspect of structure in fossil communities. *Paleobiology* 1:238–57.

Walliser, O. H. 1986a. *Global Bio-events. Lecture Notes in the Earth Sciences*, vol. 8. Berlin: Springer-Verlag.

———. 1986b. Towards a more critical approach to bio-events. In O. H. Walliser., ed., *Global Bio-events: A Critical Approach. Lecture Notes in Earth Sciences*, vol. 8. Berlin: Springer-Verlag.

———. 1990. How to define "global bio- events." In E. G. Kauffman and O. H. Walliser, eds., *Extinction Events in Earth History. Lecture Notes in the Earth Sciences*, vol. 30, pp. 1–3. Berlin: Springer-Verlag.

Wendt, J. and T. Aigner 1985. Facies patterns and depositional environments of Paleozoic cephalopod limestone. *Sedimentary Geology* 44:263–300.

West, R. R., H. R. Feldman, and C. G. Maples. This volume. Some Upper Carboniferous (Pennsylvanian) event beds (epiboles). In C. E. Brett and G. C. Baird, eds., *Paleontologic Events: Stratigraphic, Ecologic and Evolutionary Implications.* New York: Columbia University Press.

Wignall, P. B. and A. Hallam. 1991. Biofacies, stratigraphic distribution and depositional models of British onshore Jurassic black shales. In R. V. Tyson and T. H. Pearson, eds., *Modern and Ancient Continental Shelf Anoxia. Geological Society of America Special Publication* N0. 58, pp. 291–310.

Wilgus, K., H. Posamentier, J. Van Wagoner, C. A. Ross, and C. G. St. C. Kendall, eds. 1988. *Sea-Level Changes: An Integrated Approach. Society of Economic Paleontologists and Mineralogists Special Publication* No. 42.

Williams, H. S. 1901. The discrimination of time-values in geology, *Journal of Geology* 9:570–85.

Wilson, M. A. and T. J. Palmer. 1992. Hardgrounds and hardground faunas. *University of Wales, Aberystwyth, Institute of Earth Studies Publications* No. 9.

Wolosz, T. H. This volume. Thicketing events—a key to understanding the ecology of the Edgecliff Reefs (Middle Devonian Onondaga Formation of New York). In C. E. Brett and G. C. Baird, eds., *Paleontologic Events: Stratigraphic, Ecologic and Evolutionary Implications.* New York: Columbia University Press.

11

Stromatolite Biostromes as Bioevent Horizons: An Example from the Middle Cambrian Ute Formation of the Eastern Great Basin

Keith E. Eagan and W. David Liddell

ABSTRACT

The Middle Cambrian Ute Formation includes some 200 m of cyclically alternating silty shales, silty lime mudstones, and oolitic packstones to grainstones. These are arranged in eight to nine meter-scale, shallowing-upward packages, representing deposition under predominantly subtidal conditions. One cycle includes a stromatolite biostrome that is distributed across more than 2,000 km^2 in northern Utah and southern Idaho. These stromatolites were apparently established just basinward of an ooid shoal and under normal marine conditions. Stromatolites may be moundlike, club shaped, or columnar and reach up to 2^+ m tall. Analysis of preserved laminae suggests that the biostromal structure developed over a relatively short time interval (10^2–10^3 years), despite the stromatolites' large sizes. Both the biostrome's position in the sequence of cycles and changes in the stromatolites' morphology across the depositional dip indicate that it may be essentially isochronous across its outcrop area and thus may be viewed as a bioevent horizon.

The stromatolites also contribute to a better understanding of the paleogeography of the study area during the Middle Cambrian by providing information on relative energy levels and flow directions.

The Middle Cambrian Ute Formation comprises some 200 m of cyclically deposited and predominantly subtidal carbonates and silty shales. The Formation is widespread, with outcrops scattered over 14,000 km^2 in the northern Utah and

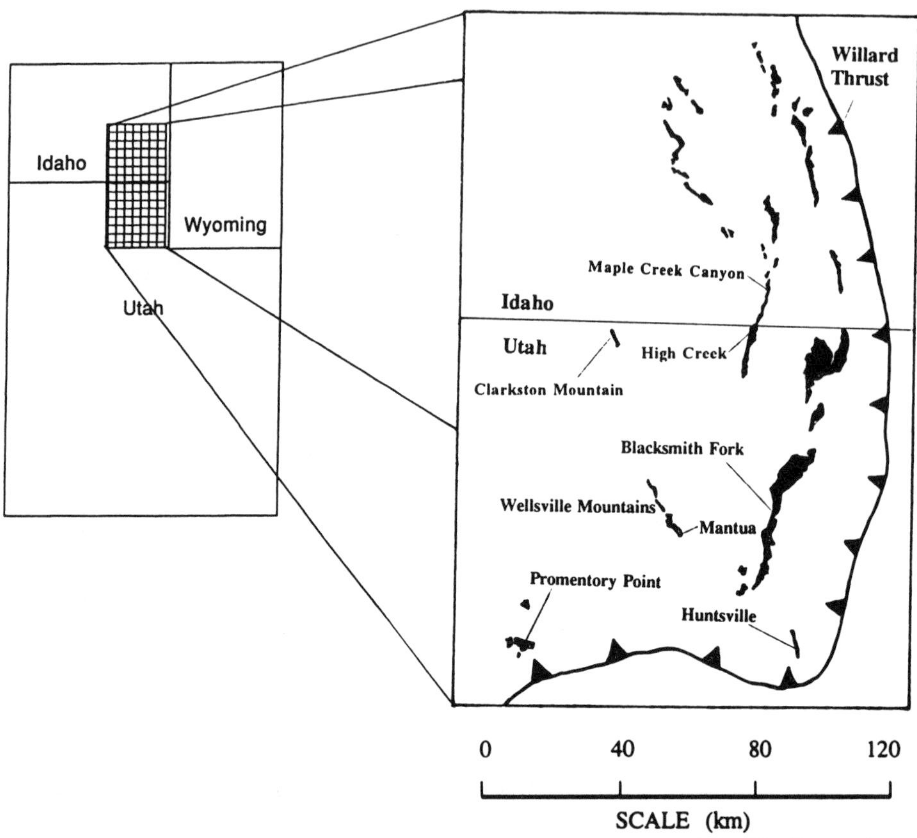

FIGURE 11.1. Map of northern Utah showing the outcrop area of the Ute Formation and study localities. The location of the Ute outcrops is based on Williams (1948), Stokes (1963), Oriel and Platt (1980), and others; the trace of the Willard thrust is from Miller (1990).

southern Idaho portions of the eastern Great Basin (Williams 1948; Maxey 1958; Stokes 1963; Oriel and Platt 1980; fig. 11.1).

The occurrence of widespread, large (2^+ m height) columnar stromatolites in the Ute Formation provides an opportunity to evaluate the utility of these stromatolites in sedimentological and stratigraphic interpretations. For example, analyses of stromatolite laminae may provide information about the horizon's duration and the Ute Formation's sedimentation rates. Also, the wide extent and distinctive nature of the stromatolite horizon may allow it to be used as a biostratigraphic marker (bioevent) horizon. Furthermore, stromatolite morphologies may reveal the paleogeography during the time that the formation was deposited. Finally, this occurrence allows the testing of environmental models for columnar stromatolites.

Stratigraphic and Structural Setting

The Cambrian formations of the western Cordillera were deposited on a passive margin that formed in response to the breakup of a Late Proterozoic supercontinent between 625 and 555 Ma (Bond et al. 1984). During the Middle to Late Cambrian, 2 km of postrift sediments—largely carbonates and fine-grained terrigenous materials—were deposited on the margin (Stewart and Poole 1974; Levy and Christie-Blick 1989). Palmer (1971, 1974), Kepper (1972), and Palmer and Halley (1979) developed a model of a north/south trending carbonate belt that separated the eastern (inner) and western (outer) detrital belts. The eastern Great Basin area has been placed at approximately 10°N latitude during the Cambrian, with the equator bisecting North America in a north/south orientation relative to modern directions (Scotese et al. 1979; Ziegler et al. 1979).

The Middle Cambrian (*Glossopleura* and *Ehmaniella* Zones; Maxey 1958) Ute Formation comprises cyclically alternating carbonate and terrigenous rocks. The lithologies present include green, silty shales, silty limestones (in which orange-weathering, silty laminations alternate with thin, gray lime mudstones to wackestones), and oolitic and oncolitic packstones to grainstones. The contact between the Ute and underlying Langston Formation is sharp, with the Ute's transgressive shales overlying cryptalgal-laminated dolostones of the Upper Limestone Member of the Langston. Overall, the Ute Formation exhibits an upward-shallowing trend from the thick shales and fine-grained carbonates of the Spence Shale Member of the underlying Langston Formation and contains abundant oncolitic and oolitic limestones, commonly displaying well-developed cross-stratification (fig. 11.2). The Ute is conformably overlain by the Blacksmith Formation. This consists of extensive, typically dolomitic, peritidal deposits, including cross-stratified ooid sands and cryptalgal-laminated deposits, which indicate further shallowing after the Ute's deposition. Stratigraphic relationships of the Middle Cambrian units in northern Utah and southern Idaho are discussed by Deiss (1938, 1941), Williams and Maxey (1941), Williams (1948), Maxey (1958), and Oriel and Armstrong (1971). Previous studies of depositional environments within the Ute Formation include those by Maxey (1958), Deputy (1984), and Rogers (1987).

The Ute Formation is largely restricted to the hanging wall of the Willard Thrust (Crittenden 1972). These rocks are thought to have experienced some 50 km of eastward translation during the Sevier orogenic event (Levy and Christie-Blick 1989). The outcrops are discontinuous owing to the effects of Sevier compression and Tertiary extension (Miller 1990; fig. 11.1). The exposures of the Ute Formation in the Wellsville Mountains are generally superior to those at other localities and also contain the best-developed stromatolite horizons. The principal study localities in the Wellsville Mountains were (from north to south) Donation Canyon, Miners Hollow, Cataract Canyon, Antimony Canyon, and Mantua, all in Box Elder County, Utah (Brigham City and Mount Pisgah 7.5′ Quadrangles).

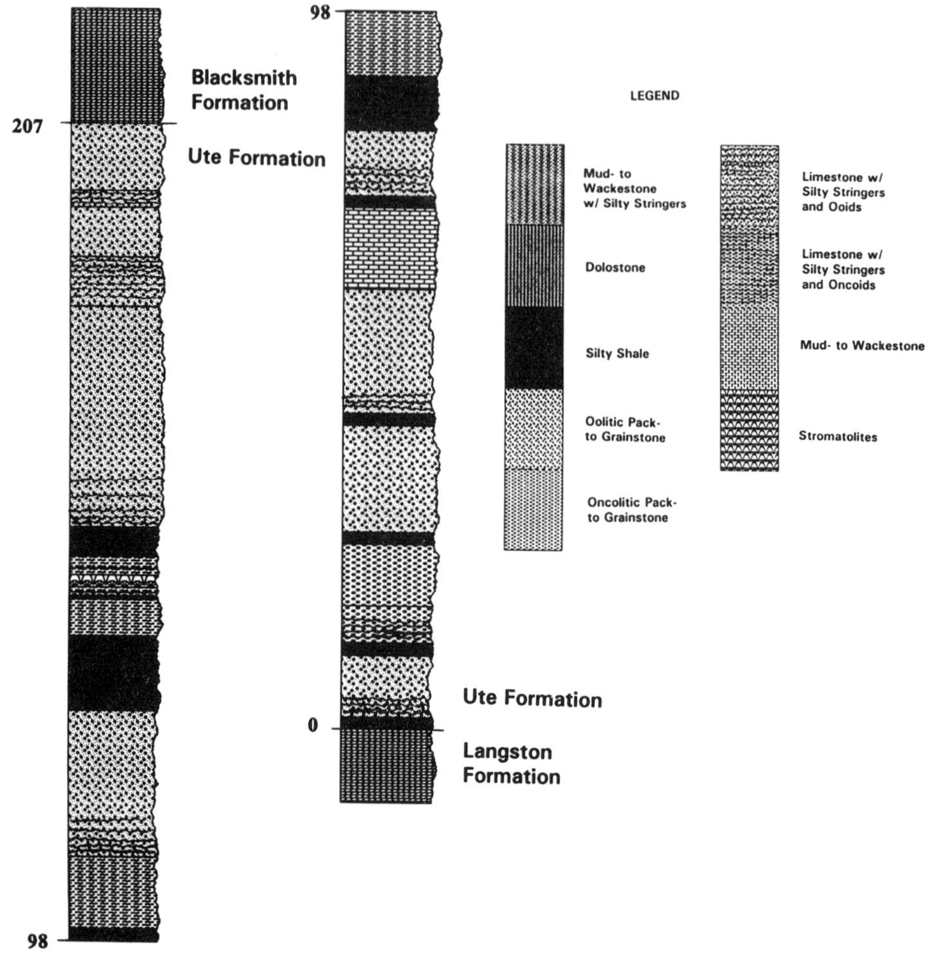

FIGURE 11.2. Generalized stratigraphic section of the Ute Formation in the Wellsville Mountains. Fewer ooid beds occur in the eastern (Bear River Range) sections.

We also found stromatolites in the Ute Formation at Gardner Canyon, Clarkston Mountain, Box Elder County, Utah (Portage 7.5′ Quadrangle) and in the Bear River Range at Maple Creek Canyon, Franklin County, Idaho (Mapleton 7.5′ Quadrangle) and High Creek Canyon and Blacksmith Fork Canyon, Cache County, Utah (Naomi Peak and Porcupine Reservoir 7.5′ Quadrangles), with well-developed horizons occurring at Maple Creek and High Creek Canyons. Another Ute Formation locality that we examined but that apparently lacked large stromatolites is the southern end of the Promontory Mountains, Box Elder County (Promontory Point 7.5′ Quadrangle) (fig. 11.1).

Lithologic Descriptions

The principal lithologies of the Ute Formation are oolitic and oncolitic limestones (packstones to grainstones), peloidal lime mudstones (mudstones to wackestones), silty limestones, and quartzose, silty shales. Conspicuously absent are any features (laminar stromatolites, mudcracks, evaporite casts, bird's-eye structures) that might indicate supralittoral or restricted hypersaline environments. Such features are common in the overlying Blacksmith Dolomite, which represents a further shallowing after the Ute's deposition.

Silty Limestones

The most distinctive rocks of the Ute Formation are the silty limestones. These consist of alternating lime mudstones to wackestones and thin, wavy, silty (quartz) laminae, which may show cross-stratification and often are dolomitic. The limestones are typically tabular and 1–2 cm or more in thickness, while the silty laminae are 1–2 mm thick. The limestones are finely crystalline and largely peloidal; siliceous sponge spicules are also present in places. Both layers are dark to medium gray when fresh. The limestone layers are medium to light gray when weathered, while the silty layers weather in relief and are bright orange. As they weather, the silty layers stain the surrounding rock, making the silty laminae appear thicker than they actually are. An analysis of the insoluble residues of these rocks indicates average insoluble amounts of 22% by weight (High Creek Canyon and Miners Hollow sections; Deputy 1984), with no clear geographic trends in content. These rocks are thought to represent deposition near but still below the fair-weather wave base and basinward of any carbonate shoals. The alternating lithologies may represent the effects of storms.

Oolitic and Oncolitic Limestones

The oolitic and oncolitic limestones range from packstones to grainstones. In addition to ooids, peloids may be present. Cross-stratification, small internal channels, and a mottled appearance are noticeable. The cross-strata are typically inclined to the northwest, although southeastern orientations may also occur. The colors on fresh surfaces are dark to medium gray, but the weathered surfaces are medium to light gray. Occasionally, ooids have been stained by hematite and so are reddish in color. Ooids range from 0.1 to 2.0 mm. Silty laminae are only rarely present in this lithotype. An insoluble-residue analysis of these rocks indicates 3% average insoluble amounts by weight (Miners Hollow; Deputy 1984). The oncoid-bearing limestones (*Girvanella*; Maxey 1958) are volumetrically less important than the oolitic rocks. Oolitic limestones are most abundant in the western and northern (e.g., Donation Canyon) exposures of the Ute, and they represent deposition under turbulent, shallow subtidal conditions. Only rarely do we find evidence for exposure in this lithotype.

LIME MUDSTONES

Limestones vary from mudstones to wackestones. They are generally similar to the silty limestones just described, but without extensive silty laminae. These limestones are much less common than the other lithotypes in the Ute Formation. Because of the general lack of terrigenous materials and the fine grain size, some of these limestones may have been deposited in the peritidal platform behind an ooid shoal.

SHALES

The shales are usually green (olive brown to olive gray), although on fresh surfaces, they are sometimes dark gray. When weathered, they often are orange to rusty red. The shales are silty and frequently micaceous. The green and dark gray colors of the shales, generally well-preserved laminae, and relatively sparse biota (at least when compared with the shales of the underlying Langston Formation) may indicate occasionally dysaerobic conditions (Osleger and Read 1991). These shales were probably deposited in a basinal location below or near storm wave base.

CONTACTS

The silty shales sharply overlie oolitic and/or oncolitic limestones. The shales then grade upward into silty limestones which, in turn, grade upward into oolitic limestones. When oolitic limestones overlie shales, the contacts are sharp. When present, the lime mudstones are usually associated with the oolitic limestones, not with the other lithotypes.

Depositional Setting

The Ute Formation was deposited on a broad, passive margin that existed in northern Utah during the Middle Cambrian (Bond et al. 1984). By the Middle Cambrian Period, the initial ramp had developed into a platform with a carbonate belt separating an intrashelf basin in the east from the slope environments to the west (inner and outer detrital belts; see Palmer 1971, 1974). The intrashelf basin was dominated by siliciclastics to the east (cratonward) and merged with the carbonate belt to the west.

The inferred paleoenvironments of the Ute Formation are illustrated in figure 11.3, which is based on lithologies and vertical facies associations. The silty shales were deposited in basinal settings and below the storm wave base. The

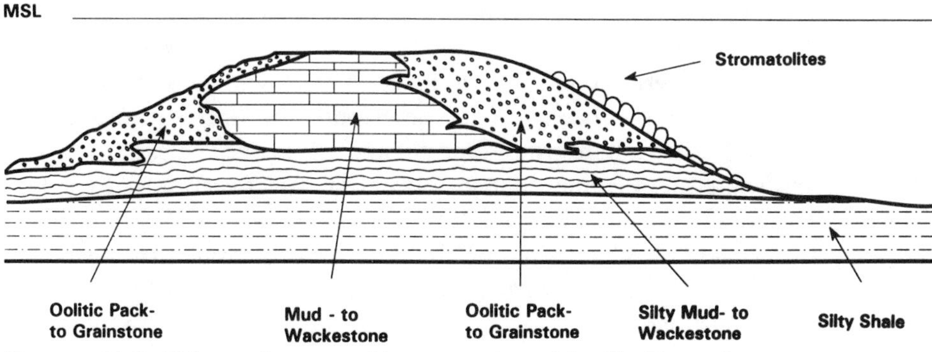

FIGURE 11.3. Paleoenvironmental interpretation of the Ute Formation.

depths of the basin were probably moderate, several tens of meters to perhaps 100 m maximum. A general lack of submarine slump features indicates that the slopes also were moderate. One widespread (several tens of kilometers) horizon in a silty limestone unit approximately 40 m above the base of the Formation does exhibit soft sediment deformation structures indicative of slumping. The oolitic grainstones were deposited in shallow water and under turbulent conditions, perhaps forming shoals. The silty limestones occupied an intermediate position between the previous two lithotypes and were probably subject to the effects of storms, as suggested by the alternation of terrigenous silts and carbonates. Some of the clean carbonate mudstones may have been deposited in a back shoal setting in the interior of the carbonate platform complex, as shown by the lack of terrigenous material and the fine grain size. This depositional model is similar to that developed for the Lower to Middle Cambrian Carrara Formation southwest of the study area (Palmer and Halley 1979).

The craton has generally been considered to be the source of the terrigenous materials in the northern Utah Cambrian section (Deiss 1941; Williams 1948; Maxey 1958), although some terrigenous sediments may have also been derived from the west or northwest (Maxey 1958; Gardiner 1974; Wakeley 1975). The increase in the amount of silt and fine sands in the High Creek section of the Ute Formation (fig. 11.1) and the paleocurrent data (cross-strata and stromatolite orientations) suggest an east to west transport of terrigenous materials.

Sequence Stratigraphy

The Ute Formation exhibits several orders of sedimentary cyclicity. In addition to being part of the Cambrian Sauk Sequence (Sloss 1963), the Middle Cambrian units of northern Utah—including the Ute—exhibit third-order "grand cycles" (Aitken 1966) consisting of formation-scale shallowing-upward sequences that are broadly correlative across the western Cordillera (Bond et al. 1989). The effects of relatively short-term events also are displayed in numerous (eight to nine) meter-scale shallowing-upward packages occurring in the Ute. According to Algeo and Wilkinson (1988), the expected duration of such meter-scale cycles is around 20–225 ky.

An idealized short-term cycle is about 10–20 m thick and consists of a basal silty shale that sharply overlies an oolitic limestone. The upper surface of the limestone may resemble a hardground, and it and the lower few centimeters of the overlying shale often contain dense concentrations of well-sorted trilobite sclerites. This shale is overlain by a silty limestone (composed of lime mudstones with terrigenous silty stringers) that is, in turn, capped by an oolitic packstone or grainstone. Upward transitions between the shale, silty limestone and the oolitic limestone are generally gradational, although of course the cycles vary considerably.

These cycles are thought to be the result of marine flooding, followed by shallowing and progradation of the carbonates when the sea level fell, and then another flooding event. Accordingly, the contact between the oolitic limestone and the overlying silty shale represents a flooding surface. The cycles are predominantly subtidal, rarely shallowing into intertidal depths or developing exposure surfaces. These cycles have been correlated some 60 km along depositional strike (Maple Creek Canyon, Bear River Range to Mantua, Wellsville Mountains) and some 40 km across the approximate depositional dip (Maple Creek Canyon, Bear River Range to Clarkston Mountain, fig. 11.1).

Whether or not these short-term cycles reflect autocyclic (e.g., Ginsburg 1971; Wilkinson 1982; Cloyd et al. 1990; Hardie et al. 1991) or allocyclic (orbitally forced?) control is beyond the scope of this essay. However, the correlatability of these cycles over several tens of kilometers argues against autocyclic control (Osleger and Read 1991). The similarities of these cycles (various scales) and those displayed by Cambrian rocks throughout the Rocky Mountains and the Appalachians (Aitken 1966; Kepper 1972; Palmer and Halley 1979; Moshier 1986; Bond et al. 1988, 1989; Read 1989; Osleger and Read 1991; and others) are striking.

Stromatolites as Paleoenvironmental Indicators

The abundant cyanobacterial structures (stromatolites) of the Proterozoic Eon suffered a drastic decline near the end of the Eon, presumably because of the rise of metazoan grazers and burrowers (Garrett 1970; Awramik 1971; Walter and Heys 1985; and others) and/or competition with eucaryotic macrophytes (Fischer 1965; Gebelein 1969). As a result of this and the fact that the setting of the famous Shark Bay, Australia locality (Logan 1961; Hoffman 1976; Playford and Cockbain 1976), was one of the few available modern occurrences of columnar stromatolites, post-Proterozoic columnar stromatolites have often been assumed to indicate intertidal and restricted, hypersaline environments (although Playford and Cockbain [1976] noted that columnar stromatolites occurred to depths of at least 3.5 m in Hamelin Pool, Shark Bay). Although this interpreta-

FIGURE 11.4. Modern stromatolites associated with ooid sand dunes and occurring at a depth of 6 m off Lee Stocking Island, Exuma Chain, Bahamas.

tion may be correct for many post-Proterozoic occurrences of columnar stromatolites, the recent discovery of columnar stromatolites occurring in nonrestricted subtidal environments of the Bahamas (fig. 11.4; Dill et al. 1986; Aalto and Shapiro 1991; Riding et al. 1991; Shapiro 1991) challenges the universal application of the restricted intertidal environment model to all post-Proterozoic fossil columnar stromatolites.

Even though most of the invertebrate macrofossils associated with the stromatolites are highly comminuted, the fragments in the carbonate units are chiefly trilobites and echinoderms, suggesting high-energy, normal marine conditions. These rocks alternate with quiet-water deposits, such as silty lime mudstones and quartzose siltstones and shales. The shales occasionally contain well-preserved trilobites.

The Ute Formation was deposited in basinal and peritidal platform settings. The columnar stromatolites are thought to have developed just basinward of an ooid shoal (fig. 11.3), similar to the position that Moshier (1986) determined for stromatolites from the Upper Cambrian Richland Formation of Pennsylvania. The subtidal and normal marine setting for the Ute stromatolites and their association with oolites are somewhat analogous to that of the modern Bahamian columnar stromatolites described by Dravis (1983), Dill et al. (1986), and others. This and other ancient occurrences (e.g., Aitken 1967; Playford et al. 1976; Griffin 1987; Southgate 1989; Dromart 1992; Sami and Jones 1993; and others) of Phanerozoic columnar stromatolites in subtidal (even deep-water) and/or nonrestricted settings suggest that paleoenvironmental interpretations based on the occurrence of columnar stromatolites in post-Proterozoic rocks must be carefully evaluated.

Description and Distribution of Stromatolites

Stromatolites have been used in biostratigraphy for many years. The initial attempts involved Proterozoic rocks in the former Soviet Union that were zoned according to the stromatolite assemblages (Raaben 1969). Subsequently, distinctive stromatolite assemblages from other continents were noted (Walter 1972; Preiss 1976; Liang et al. 1985; see Hofmann 1987 for an overview). In addition, because of the often widespread nature of stromatolite horizons, they have been used as marker beds for intrabasinal correlation (e.g., Fenton and Fenton 1937; Rezak 1957; Donaldson 1963; Hoffman 1967, 1968; Bertrand-Sarfati and Trompette 1976; Serebryakov 1976; see Hofmann 1973 for a review).

The largest stromatolites in the Ute Formation occur at a single horizon located approximately 115–135 m above the base of the Formation, where they form a laterally extensive biostrome (figs. 11.2, 11.5, 11.6). The following description applies to the Wellsville Mountains localities, although a similar section can be found at other localities as well. The stromatolite horizon was initially established on an eroded oolitic and oncolitic grainstone. Directly attached to the unconformity surface are small, 15–20 cm tall columnar stromatolites, which typically are surrounded by silty limestones and capped by oolitic or oncolitic limestone. A second horizon of small stromatolites frequently occurs 1 m farther upsection. These small stromatolites are separated from the overlying larger specimens by approximately 1–2 m of silty limestone. The upper stromatolites are 15–30 cm in diameter, 100–200 cm in height, and columnar to club shaped.

We have refrained from referring to these forms as giant stromatolites, owing to their contrast with truly giant forms (e.g., domes 15–20 m in diameter) occurring in Proterozoic rocks (e.g., Hoffman 1974; Southgate 1989). The stromatolite columns are densely packed, branch only rarely, and are round to oval in transverse section (fig. 11.6). They frequently show both inclination and flexure to the northeast (fig. 11.6). Despite the large size of the stromatolites in the Wellsville Mountains, the actual area exposed above the sediment/water interface at any one time—the synoptic profile—was probably only 10–20 cm. When seen in a longitudinal section, the laminae are wavy and moderately convex. These upper stromatolites are typically surrounded by and covered with silty lime mudstones (fig. 11.6), although at one locality (Dry Canyon, Wellsville Mountains), they are surrounded by and capped with oolitic grainstones. At another nearby locality (Antimony Canyon, Wellsville Mountains), prominent channels some 1 m in depth and 2–4 m wide are cut into the stromatolite horizon.

FIGURE 11.5. Field sketch of a stromatolite sequence in the Ute Formation at Cataract Canyon, Wellsville Mountains. The vertical extent shown is approximately 8 m. The lithologic symbols are similar to those of figure 11.2.

Stromatolites from the eastern portion of the study area (Bear River Range) differ in morphology from those to the west (Wellsville and Clarkston Mountains). The Bear River Range stromatolites tend to be less columnar, more isolated, and much more moundlike, ranging from 20 to 75 cm in diameter and from 25 to 100 cm in height. The fact that the stromatolites from the western (Wellsville and Clarkston Mountains) portion of the study area differ morpho-

FIGURE 11.6. Columnar stromatolites in the Ute Formation. Refer to the hammer for scale. (A) Upper stromatolite horizon (see fig. 11.5), Cataract Canyon, Wellsville Mountains, Utah. Note the cap of limestone with silty laminae. (B) Upper stromatolite horizon, Cataract Canyon, Wellsville Mountains. Note the convex laminae, flexure of stro-

matolites, and approximately synchronous termination of stromatolites. (C) Upper stromatolite horizon, Donation Canyon, Wellsville Mountains. Note the close packing of stromatolite columns. (D) Lower stromatolite horizon, Donation Canyon, Wellsville Mountains. Note the limestone with silty laminae filling the gaps between stromatolites.

logically from those to the east (Bear River Range) suggests that the horizon may be isochronous across at least 40 km of depositional dip. The differences in morphology (taller, less synoptic relief?, more contiguous, and more columnar to the west; shorter, more isolated, and more moundlike to the east) can be related to inferred energy regimes, with the energy levels higher to the west. This is also supported by the greater numbers and thicknesses of ooid beds in the western sections. This morphological interpretation is consistent with the distribution of modern, Shark Bay stromatolite morphotypes and energy levels (Hoffman 1976) and changes in stromatolite morphology in ancient shallowing-upward sequences (Southgate 1989; Sami and James 1993). Isochroneity of the stromatolite horizons in the Wellsville Mountains and Bear River Range is also suggested by their position within the cyclic sequences (occurrence within the same, upper cycle).

The highly columnar, western stromatolites are inclined to the east. This is consistent with paleocurrent data from ripple laminations in the silty limestones and cross-stratified ooid beds, which also indicate the flow coming predominantly from the east. The modern Shark Bay (Hofmann 1973) and Bahamian (Shapiro 1991) stromatolites also are inclined into the prevailing currents, and several examples of currentward inclination are known from Proterozoic stromatolite occurrences (Hoffman 1967; Cecile and Campbell 1978; Grotzinger 1986; Sami and James 1993).

The stromatolite horizon has been correlated over a distance of 60 km in a north/south direction along the approximate depositional strike and 40 km in an east/west direction along the approximate dip, for a total areal distribution of 2,400 km^2. It should be noted that the outcrops are not continuous. Further fieldwork may expand the apparent distribution of the stromatolites.

The columnar stromatolites occur unconformably on the top of a shallowing-upward cyclic sequence and mark the initiation (flooding event) of another shallowing-upward cycle. It is unclear why the stromatolites are restricted to this one narrow stratigraphic interval, despite the occurrence of eight similar depositional cycles within the Ute Formation. One difference between this and other cycles is the absence of a basal shale. The stromatolites were established directly on an oolitic and oncolitic grainstone and developed concurrently with the deposition of silty and oolitic limestones, perhaps suggesting a low-magnitude sea-level rise associated with this particular cycle.

The association of columnar stromatolites with oolitic and oncolitic beds or lime conglomerates is pervasive (e.g., Aitken 1967; Hoffman 1967; Serebryakov 1976; Haslett 1976; Griffin 1987; Southgate 1989; Sami and James 1993; and others). Southgate (1989) described the occurrence of mound and columnar stromatolites in Proterozoic-age, shallowing-upward sequences. The stromatolites were initially established on hardgrounds developed in oolitic limestones and associated with rapid deepening at the beginning of a cycle. Finally, microbial biostromes, including small columnar stromatolites, have been implicated as flooding markers in deep-water carbonate sequences (Dromart 1992).

In addition to the columnar and moundlike stromatolites, other microbial structures can be found in the Ute Formation. Oncolitic and oolitic packstones to grainstones occur both above and below the primary layer (fig. 11.2). Approximately 90 m below the primary stromatolite layer, internally structureless mounds or thrombolites (Aitken 1967), up to 25–30 cm in height, are equally persistent and are present in most sections. These were established on an oolitic limestone with an irregular upper surface and are surrounded by and capped with silty limestones grading upward into oolitic limestones. Thus the thrombolites formed in a shallowing-upward sequence similar to that of the large stromatolites.

Duration of the Stromatolite Horizon

Determination of the growth rate of ancient marine organisms such as stromatoporoids and corals may provide information on topics as diverse as sedimentation rates and the earth's past rotation (Wells 1963; Scrutton 1964; Meyer 1981). Difficulties associated with such endeavors include the identification of discrete and unambiguous growth bands (Pannella 1975) and the determination of the forcing factors involved: That is, do the growth bands reflect daily, seasonal, tidal, or some other cycles (Vanyo and Awramik 1985)?

Stromatolites are not skeletal remains; rather, "they are organosedimentary structures that are produced by sediment trapping, binding and/or precipitation as a result of the growth and metabolic activity of microorganisms, principally cyanophytes" (Awramik and Margulis, cited in Walter 1976:1). The characteristic laminae of stromatolites are the result of breaks in accretion, which may be caused by variations in the sediment supply (e.g., influenced by tides or storms) or the growth of the microorganisms (e.g., seasonal or day/night cycles). "The recording of astronomical–geophysical data by stromatolites depends on how well the lamina formation process by microbes can respond to controlling external stimuli and how little the first-order laminae have been altered by secondary processes" (Vanyo and Awramik 1985:126).

Attempts to interpret the rhythmicity of laminae in stromatolites are fraught with hazards. Hofmann (1973:356) provides numerous caveats to be considered when attempting to extract growth-rate information from stromatolites and concludes, "As yet, the significance of the layering and its period have not been ascertained satisfactorily for fossil specimens." For example, the individual laminae may reflect twice daily, daily, seasonal, or yearly rhythmicities, or they may be due to storm events or sun spot cycles.

Our goal was to constrain the duration of the stromatolite horizon within certain probable boundaries in order to evaluate the utility of stromatolite biostromes as stratigraphic markers. To do this, columnar stromatolites from the Wellsville Mountains localities were vertically slabbed, ground, and coated with clear acrylic. We measured the laminae thicknesses under a binocular micro-

TABLE 11.1 *Measurements of Stromatolite Laminae from the Middle Cambrian Ute Formation (means with 95% confidence intervals)*

Location stromatolite ID no.	Number of laminae measured	Average thickness (mm) all laminae	Average thickness (mm) dark laminae	Average thickness (mm) light laminae
Cataract Canyon C-106	280	0.97 ± 0.09	0.83 ± 0.14	1.15 ± 0.18
Dry Canyon D-100	556	0.51 ± 0.04	0.22 ± 0.02	0.64 ± 0.04
Donation Canyon DO-10	413	0.73 ± 0.08	0.69 ± 0.11	0.77 ± 0.10

scope with ocular micrometer, and we used plots of lamina thickness versus lamina number or sequence to identify cyclic patterns in accumulation (Mohr 1975; Pannella 1976; Williams 1989). Our attempts to use acetate peels and photographs of slabbed surfaces proved to be less effective than was direct observation under the microscope. Many of the stromatolites are heavily bioeroded, and this, in addition to diagenetic phenomena, such as stylolitization, made the majority of specimens unsuitable for analyzing the periodicity of the lamination. That the majority of the laminae are not well defined has been frequently observed by other workers (e.g., Pannella 1976; Vanyo and Awramik 1985).

The stromatolites exhibit an alternating type of lamination in which thicker (0.6–1.2 mm), lighter-colored laminae alternate with thinner (0.2–0.8 mm), darker-colored laminae, possibly reflecting day/night variation in growth and sediment trapping (Monty 1976). The average thickness of a lamina (regardless of type) ranges from 0.5–1.0 mm, depending on the specimen and/or locality (table 11.1). These laminae exhibit periodicity in their rate of accumulation, with an average period of 10 light/dark couplets (between 9 and 12) occurring between major "spikes" in lamina thickness. Note that the record contains gaps; that is, one or more of the thinner, darker laminae are frequently absent from a given second-order set (fig. 11.7). In turn, these second-order sets may represent lunar-influenced (tidal) cycles (Pannella 1975).

Our attempt to constrain the duration of the stromatolite horizon required several assumptions: (1) The first-order laminae sets (alternating light and dark laminae) reflect daily (e.g., Monty 1965; Gebelein 1969; Vanyo and Awramik 1985), not twice-daily (Gebelein and Hoffman 1968), seasonal (McGugan 1967), or yearly (Fischer 1964) accumulation. (2) The second-order laminae sets reflect a periodic external forcing function, presumably tides, and not some other, longer-period function. (3) The Cambrian year lasted for about 410–430 days (McGugan 1967; Scrutton 1978; Vanyo and Awramik 1985). (4) The Cambrian year contained approximately 13 synodic months, as determined for the Late Proterozoic (Scrutton 1978; Williams 1989).

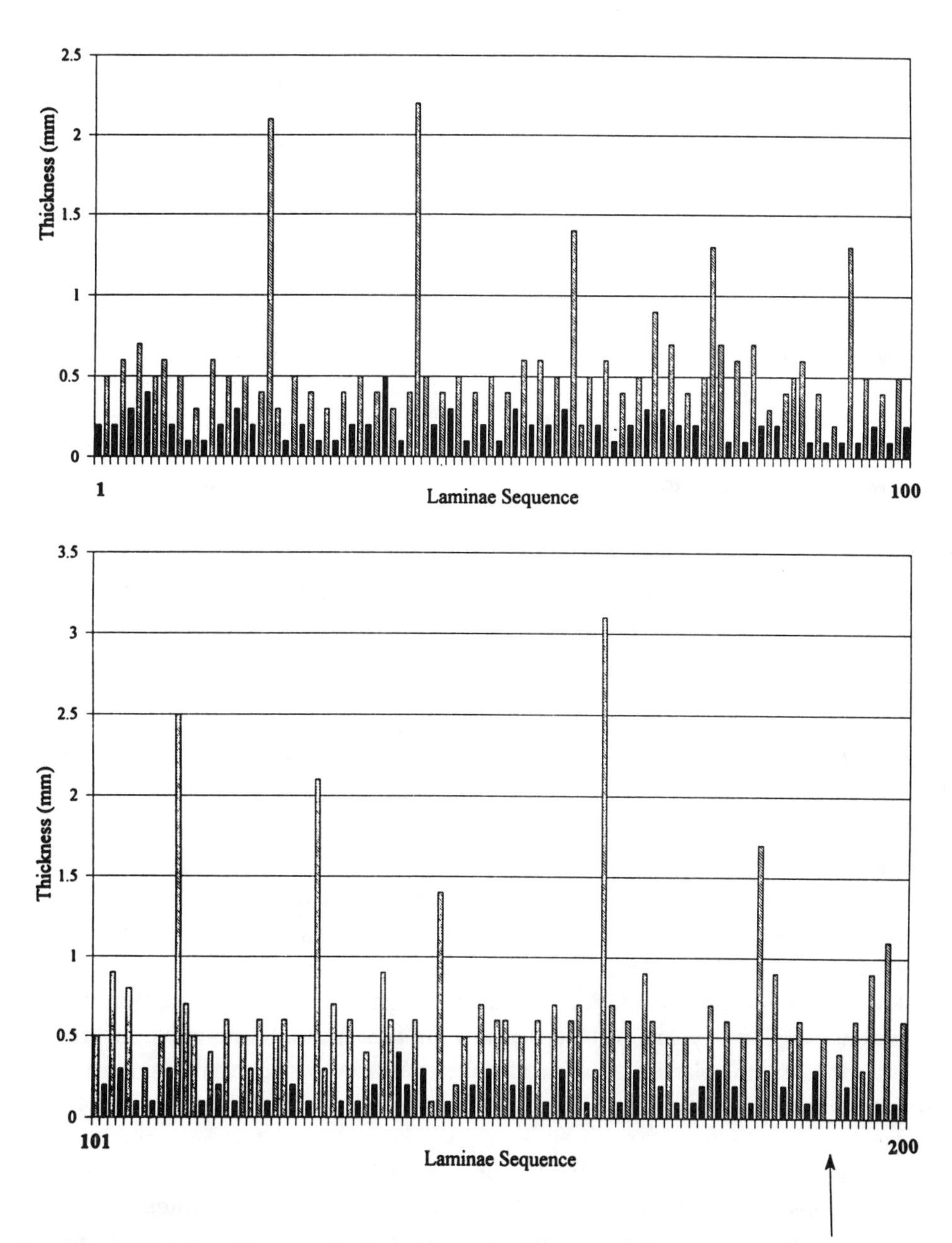

FIGURE 11.7. Plot of lamina thickness versus lamina number (sequence), showing cyclicity in accumulation. Note the first-order sets of thin, dark-colored, and thick, light-colored, laminae, as well as the second-order sets with an average period of 10 first-order sets. The arrow indicates a discontinuity in the record. This is a portion of the record for stromatolite D-100, collected from Dry Canyon, Wellsville Mountains.

Based on average stromatolite laminae thicknesses (1.43 mm for a light/dark couplet) and the estimated duration of the Middle Cambrian year (approximately 410 days), the stromatolite accumulation rates would average approximately 60 cm per year. An alternative value can be derived by noting the thicknesses of second-order (bimonthly?) cycles (14.3 mm; fig. 11.7) and assuming 13 synodic months in the Middle Cambrian year, which yielded accumulation rates of 37 cm per year.

Using the accumulation values of 37–60 cm per year determined by these two methods, the main (2 m thick) stromatolite horizon appears to have lasted for a minimum of only a few years. Accepting this duration for the horizon assumes, of course, that no hiatal surfaces occur within the stromatolites, but this is not really the case, as small gaps frequently occur within the stromatolite records, resulting in the occasional loss of the periodic signal (fig. 11.7). In addition, calculations based solely on the number of first-order cycles within the second-order sets underestimate by about 30% the number of days in the Cambrian year (410–430; cf. Scrutton 1978; Vanyo and Awramik 1985) or days in the Cambrian month (31–32; cf. Pannella et al. 1968). Our data are therefore *minimum* times for the duration of the deposit. If one assumes that breaks in accumulation were frequent, the duration of the horizon is likely to be on the order of decades to centuries, with a maximum duration of 10^3 years. Inasmuch as the stromatolites were forming while the sediment was being deposited around them (as evidenced by the incorporation of certain prominent silty stringers in both the stromatolites and the surrounding sediment), their growth rates can be used to estimate *maximum* possible sedimentation rates. The values so obtained indicate high rates of carbonate sediment accumulation, which may have been a consequence of current baffling and sediment trapping by the stromatolite biostrome.

The stromatolite accumulation rates that we calculated in our study are consistent with those of previous workers. Vanyo and Awramik (1985) documented rates of 8.7 cm per year for the stromatolite, *Anabaria juvensis*, from the 850-Ma Bitter Springs Formation of Australia. Their value was based on counts of laminae (435 per year) and the thickness of an annual sine wave in growth orientation due to solar tracking during the cyanophytes' growth. An additional specimen of the stromatolite, *Inzeria initia*, yielded accumulation rates of 45 cm per year.

Gebelein (1969) determined rates of 1.0 mm per day for modern subtidal stromatolites from Bermuda, and Monty (1965) documented rates of lamina production of up to 0.6 mm during a 24-hour period for stromatolites from the Bahamas, although he noted that the accumulation was discontinuous. Aalto and Shapiro (1991) obtained ^{14}C dates of 3,710 + 90 y.b.p. from the base of a 33-cm-tall, club-shaped stromatolite occurring at 6.7 m depth off Lee Stocking Island, Bahamas. This stromatolite was initiated on a Holocene hardground, and the date is consistent with the establishment of the stromatolite in very shallow water as the Bahamian platform was flooded during the Holocene transgression. This date suggests that other, larger (e.g., 2 m tall) specimens occurring at this locality required only a few thousand years to form, even though these stromatolites have been frequently buried by shifting oolitic sands and were probably quiescent for at least part of that time.

On the other hand, the famous Shark Bay modern stromatolites appear to have reached a state of equilibrium, with growth balanced by erosion (Playford and Cockbain 1976). Also, Playford et al. (1976) concluded, based on conodont biozones, that deep-water (>100 m) columnar stromatolites from the Devonian of the Canning Basin grew at rates of 0.002 mm/year, thus requiring several hundred thousand years to achieve a height of 1 m.

Although the interpretation of the significance of higher-order sets is controversial (Hofmann 1973; Pannella 1976), it is interesting that McGugan (1967) found third-order sets to consist of 400–420 first-order and 9–12 second-order sets in stromatolites from the Upper Cambrian in the Canadian Rocky Mountains. Moreover, Pannella et al. (1968) derived a value of 31.65 + 3.15 days for the synodic month in the Late Cambrian, based on an analysis of stromatolite laminae from Maryland specimens.

The Middle Cambrian Ute Formation contains some 200 m of cyclically alternating silty shales, silty lime mudstones to wackestones and oolitic packstones to grainstones. These are arranged in eight to nine meter-scale, shallowing-upward packages. The cycles are predominantly subtidal, rarely shallowing into intertidal depths or developing exposure surfaces. An idealized cycle is about 10–20 m thick and consists of a basal silty shale that sharply overlies an oolitic limestone. The upper surface of the limestone may resemble a hardground, and it and the lower few centimeters of the overlying shale may contain dense concentrations of well-sorted trilobite cranidia or other skeletal elements. This shale is overlain by a silty limestone (composed of lime mudstones with terrigenous silty stringers), which in turn is capped by oolitic grainstones. Upward transitions between the shale, silty limestone and the oolitic limestone are generally gradational. The cycles are thought to be the result of marine flooding, followed by shallowing and progradation of the carbonates during the drop in sea level. These cycles have been correlated some 60 km along the depositional strike and some 40 km across the approximate depositional dip.

One such cycle includes a stromatolite biostrome that is widely distributed in northern Utah and southern Idaho. The stromatolites were apparently established on a flooding surface located just basinward of an ooid shoal and under normal marine conditions. Stromatolites range from moundlike to club shaped to columnar and extend to 2^{+} m in height. The stromatolites show flexure to the northeast, which is consistent with other paleocurrent data. Furthermore, the stromatolites in the eastern outcrops are smaller and more moundlike, whereas those in the western outcrops are larger and more columnar, suggesting decreasing energy levels to the east. This is consistent with the distributions of ooid beds across the study area. Thus stromatolites contribute to our understanding of the paleogeography of the study area during the Middle Cambrian, by providing information about relative energy levels and flow directions.

The preserved stromatolite laminae exhibit first-order patterns of alternating light and dark bands grouped into larger, second-order sets consisting of 9–12 first-order pairings. Although interpretations of stromatolite laminae are controversial, attributing the first- and second-order laminae sets to daily and bimonthly (lunar) periodicities enables estimations of growth rates that are consistent with other ancient occurrences and, also, modern Bahamian forms. This suggests that the biostromal structure was developed in a relatively short time (10^2–10^3 yrs). The biostrome's position in the sequence of cycles and the changes in stromatolite morphology from eastern to western exposures mean that it may be essentially isochronous across its outcrop area and thus may be considered a bioevent horizon. The widespread nature of this and other Proterozoic and Cambrian stromatolite biostromes indicates considerable utility for stratigraphic correlation.

ACKNOWLEDGMENTS

This study was funded in part by a J. Stewart Williams Graduate Fellowship at Utah State University awarded to Keith E. Eagan. Stephen K. Boss initially called our attention to the stromatolite occurrences in the Ute Formation, gave us his unpublished stratigraphic data, and contributed valuable comments on the manuscript. Carlton E. Brett, Peter T. Kolesar, Dave Lehmann, and Robert Q. Oaks also provided thoughtful comments on the manuscript.

REFERENCES

Aalto, K. R. and R. S. Shapiro. 1991. Petrology of a modern subtidal stromatolite, Lee Stocking Island, Bahamas. In R. J. Bain, ed., *Proceedings of the Fifth Symposium on the Geology of the Bahamas,* pp. 1–9. San Salvador: Bahamian Field Station.

Aitken, J. D. 1966. Middle Cambrian to Middle Ordovician cyclic sedimentation, Southern Canadian Rocky Mountains of Alberta. *Bulletin of Canadian Petroleum Geology* 14:405–41.

———. 1967. Classification and environmental significance of cryptalgal limestones and dolomites, with illustrations from the Cambrian and Ordovician of southwestern Alberta. *Journal of Sedimentary Petrology* 37:1163–78.

Algeo, T. J. and B. H. Wilkinson. 1988. Periodicity of mesoscale Phanerozoic sedimentary cycles and the role of Milankovitch orbital modulation. *Journal of Geology* 96:313–22.

Awramik, S. M. 1971. Precambrian columnar stromatolite diversity: Reflection of metazoan appearance. *Science* 174:825–27.

Bertrand-Sarfati, J. and R. Trompette. 1976. Use of stromatolites for intrabasinal correlation: Example from the Late Proterozoic of the northwestern margin of the Taoudenni Basin. In M. R. Walter, ed., *Stromatolites: Developments in Sedimentology,* vol. 20, pp. 517–22. New York: Elsevier.

Bond, G. C., M. A. Kominz, and J. P. Grotzinger. 1988. Cambro–Ordovician eustasy: Evidence from geophysical modeling of subsidence in Cordilleran and Appalachian

passive margins. In C. Paola and K. Kleinspehn, eds., *New Perspectives in Basin Analysis*, pp. 129–61. New York: Springer-Verlag.
Bond, G. C., M. A. Kominz, M. S. Steckler, and J. P. Grotzinger. 1989. Role of thermal subsidence, flexure, and eustasy in the evolution of early Paleozoic passive-margin carbonate platforms. In P. D. Crevello, J. L. Wilson, J. F. Sarg, and J. F. Read, eds., *Controls on Carbonate Platform and Basin Development*, pp. 39–61. Tulsa: *Society of Economic Paleontologists and Mineralogists, Special Publication* No. 44.
Bond, G. C., P. A. Nickeson, and M. A. Kominz. 1984. Breakup of a supercontinent between 625 Ma and 555 Ma: New evidence and implications for continental histories. *Earth and Planetary Science Letters* 70:325–45.
Cecile, M. P. and F. H. A. Campbell. 1978. Regressive stromatolite reefs and associated facies, middle Goulburn Group (lower Proterozoic), in Kilohigok Basin, N.W.T.: An example of environmental control of stromatolite form. *Bulletin of Canadian Petroleum Geologists* 26:237–67.
Cloyd, K. C., R. V. Demicco, and R. J. Spencer. 1990. Tidal channel, levee, and crevasse-splay deposits from a Cambrian tidal channel system: A new mechanism to produce shallowing-upward sequences. *Journal of Sedimentary Petrology* 60:73–83.
Crittenden, M. D. Jr. 1972. Willard Thrust and the Cache Allochthon, Utah. *Geological Society of America Bulletin* 83:2871–80.
Deiss, C. F. 1938. Cambrian formations and sections in part of the Cordilleran Trough. *Geological Society of America Bulletin* 49:1067–1168.
———. 1941. Cambrian geography and sedimentation in the central Cordilleran region. *Geological Society of America Bulletin* 52:1085–1116.
Deputy, E. J. 1984. Petrology of the Middle Cambrian Ute Formation, north-central Utah and southeastern Idaho. Master's thesis, Utah State University.
Dill, R. F., E. A. Shinn, A. T. Jones, K. Kelley, and R. P. Stienen. 1986. Giant subtidal stromatolites forming in normal salinity waters. *Nature* 234:55–60.
Donaldson, J. A. 1963. Stromatolites in the Denault Formation, Marion Lake, coast of Labrador, Newfoundland. *Geological Survey of Canada Bulletin* 102:1–33.
Dravis, J. F. 1983. Hardened subtidal stromatolites, Bahamas. *Science* 219:385–86.
Dromart, G. 1992. Jurassic deep-water microbial biostromes as flooding markers in carbonate sequence stratigraphy. *Palaeogeography, Palaeoclimatology, Palaeoecology* 91:219–28.
Fenton, C. L. and M. A. Fenton. 1937. Belt Series of the north: Stratigraphy, sedimentation, paleontology. *Geological Society of America Bulletin* 48:1873–1969.
Fischer, A. G. 1964. The Lofer cyclothems of the Alpine Triassic. *Kansas Geological Survey Bulletin* 169:107–49.
———. 1965. Fossils, early life and atmospheric history. *Proceedings of the National Academy of Science* 53:1205–15.
Gardiner, L. L. 1974. Environmental analysis of the Upper Cambrian Nounan Formation, Bear River Range and Wellsville Mountain, north-central Utah. Master's thesis, Utah State University.
Garrett, P. 1970. Phanerozoic stromatolites: Noncompetitive ecologic restriction by grazing and burrowing animals. *Science* 169:171–73.
Gebelein, G. D. 1969. Distribution, morphology, and accretion rate of Recent subtidal algal stromatolites, Bermuda. *Journal of Sedimentary Petrology* 39:49–69.
Gebelein, G. D. and P. Hoffman. 1968. Intertidal stromatolites from Cape Sable, Florida. *Geological Society of America, Abstracts for 1968*, Special Paper No. 121, p. 129.
Ginsburg, R. N. 1971. Landward movement of carbonate mud: New model for regressive cycles in carbonates. Abstract. *American Association of Petroleum Geologists Bulletin* 287:340.
Griffin, K. M. 1987. A comparison of the depositional environments of U. Cambrian

thrombolites and stromatolites of the Nopah Fm., Death Valley region, California and modern stromatolites of the Bahamas. *Geological Society of America, Abstracts with Programs* 19:684.

Grotzinger, J. P. 1986. Evolution of Early Proterozoic passive-margin carbonate platform, Rocknest Formation, Wopmay orogen, Northwest Territories, Canada. *Journal of Sedimentary Petrology* 56:831–47.

Hardie, L. A., P. A. Dunn, and R. K. Goldhammer. 1991. Field and modelling studies of Cambrian carbonate cycles, Virginia Appalachians–Discussion. *Journal of Sedimentary Petrology* 61:636–46.

Haslett, P. G. 1976. Lower Cambrian stromatolites from open and sheltered intertidal environments, Wirrealpa, South Australia. In M. R. Walter, ed., *Stromatolites: Developments in Sedimentology*, vol. 20, pp. 565–84. New York: Elsevier.

Hoffman, P. F. 1967. Algal stromatolites: Use in stratigraphic correlation and paleocurrent determination. *Science* 157:1043–45.

———. 1968. Stratigraphy of the Lower Proterozoic (Aphebian), Great Slave Supergroup, East Arm of Great Slave Lake, District of Mackenzie. *Geological Survey of Canada Paper* 68:93.

———. 1974. Shallow and deep-water stromatolites in lower Proterozoic platform-to-basin facies change, Great Slave Lake, Canada. *American Association of Petroleum Geologists Bulletin* 58:856–67.

———. 1976. Stromatolite morphogenesis in Shark Bay, Western Australia. In M. R. Walter, ed., *Stromatolites: Developments in Sedimentology*, vol. 20, pp. 261–71. New York: Elsevier.

Hofmann, H. J. 1973. Stromatolites: Characteristics and utility. *Earth-Science Reviews* 9:339–73.

———. 1987. Paleoscene #7. Precambrian biostratigraphy. *Geoscience Canada* 14:135–54.

Kepper, J. C. 1972. Paleoenvironmental patterns in Middle to Lower Upper Cambrian interval in eastern Great Basin. *American Association of Petroleum Geologists Bulletin* 56:503–27.

Levy, M. and N. Christie-Blick. 1989. Pre-Mesozoic palinspastic reconstruction of the eastern Great Basin (western United States). *Science* 245:1454–62.

Liang, Y., Sh. Zhu, L. Zhang, R. Cao, Zh. Gao, and D. Bu. 1985. Stromatolite assemblages of the Late Precambrian in China. *Precambrian Research* 29:15–32.

Logan, B. W. 1961. *Cryptozoon* and associated stromatolites from the Recent, Shark Bay, Western Australia. *Journal of Geology* 69:517–33.

Maxey, G. B. 1958. Lower and Middle Cambrian stratigraphy in northern Utah and southern Idaho. *Geological Society of America Bulletin* 69:647–88.

McGugan, A. 1967. Possible use of algal stromatolite rhythms in geochronology. *Geological Society of American, Abstracts for 1967*, Special Paper No. 115, p. 145.

Meyer, F. O. 1981. Stromatoporoid growth rhythm and rates. *Science* 213:894–95.

Miller, D. M. 1990. Mesozoic and Cenozoic tectonic evolution of the northeastern Great Basin. In D. R. Shaddrick, J. A. Kizis Jr., and E. L. Hunsaker III, eds., *Geology and Ore Deposits of the Northeastern Great Basin, Field Trip No. 5*, pp. 43–73, Reno: Geological Society of Nevada.

Mohr, R. E. 1975. Measured periodicities of the Biwabik (Precambrian) stromatolites and their geophysical significance. In G. D. Rosenberg and S. K. Runcorn, eds., *Growth Rhythms and the History of the Earth's Rotation*, pp. 43–56. New York: Wiley Interscience.

Monty, C. L. V. 1965. Recent algal stromatolites in the Windward Lagoon, Andros Island, Bahamas. *Annales Société Géologie Belgique* 94:269–276.

———. 1976. The origin and development of cryptalgal fabrics. In M. R. Walter, ed., *Stromatolites: Developments in Sedimentology*, vol. 20, pp. 193–259. New York: Elsevier.

Moshier, S. O. 1986. Carbonate platform sedimentology, Upper Cambrian Richland Formation, Lebanon Valley, Pennsylvania. *Journal of Sedimentary Petrology* 56:204–16.

Oriel, S. S. and G. C. Armstrong. 1971. Uppermost Precambrian and lowermost Cambrian rocks in southeastern Idaho. *U.S. Geological Survey Professional Paper* No. 394.

Oriel, S. S. and L. B. Platt. 1980. Geologic map of the Preston 1° × 2° Quadrangle, southeastern Idaho and western Wyoming. Washington, D.C.: U.S. Geological Survey Miscellaneous Inventory, Map I-1127.

Osleger, D. and J. F. Read. 1991. Relation of eustasy to stacking patterns of meter-scale carbonate cycles, Late Cambrian, U.S.A. *Journal of Sedimentary Petrology* 61:1225–52.

Palmer, A. R. 1971. The Cambrian of the Great Basin and adjacent areas, western United States. In C. H. Holland, ed., *Cambrian of the New World*, pp. 1–78. London: Wiley Interscience.

———. 1974. Search for the Cambrian world. *American Scientist* 62:216–24.

Palmer, A. R. and R. B. Halley. 1979. Physical stratigraphy and trilobite biostratigraphy of the Carrara Formation (Lower and Middle Cambrian) in the southern Great Basin. *U.S. Geological Survey Professional Paper* No. 1047.

Pannella, G. 1975. Paleontological clocks and the history of the Earth's rotation. In G. D. Rosenberg and S. K. Runcorn, eds., *Growth Rhythms and the History of the Earth's Rotation*, pp. 253–83. New York: Wiley Interscience.

———. 1976. Geophysical inferences from stromatolite lamination. In M. R. Walter, ed., *Stromatolites: Developments in Sedimentology*, vol. 20, pp. 673–85. New York: Elsevier.

Pannella, G., C. MacClintock, and M. N. Thompson. 1968. Paleontological evidence of variations in length of synodic month since the Late Cambrian. *Science* 162:792–96.

Playford, P. E. and A. E. Cockbain. 1976. Modern algal stromatolites at Hamelin Pool, a hypersaline barred basin in Shark Bay, western Australia. In M. R. Walter, ed., *Stromatolites: Developments in Sedimentology*, vol. 20, pp. 389–411. New York: Elsevier.

Playford, P. E., A. E. Cockbain, E. C. Druce, and J. L. Wray. 1976. Devonian stromatolites from the Canning Basin, western Australia. In M. R. Walter, ed., *Stromatolites: Developments in Sedimentology*, vol. 20, pp. 543–63. New York: Elsevier.

Preiss, W. V. 1976. Intercontinental correlations. In M. R. Walter, ed., *Stromatolites: Developments in Sedimentology*, vol. 20, pp. 359–70. New York: Elsevier.

Raaben, M. E. 1969. Columnar stromatolites and Late Precambrian stratigraphy. *American Journal of Science* 267:1–18.

Read, J. F. 1989. Controls on evolution of Cambrian–Ordovician passive margin, U.S. Appalachians. In P. D. Crevello, J. L. Wilson, J. F. Sarg, and J. F. Read, eds., *Controls on Carbonate Platform and Basin Development*, pp. 147–65. Tulsa: Society of Economic Paleontologists and Mineralogists *Special Publication* No. 44.

Rezak, R. 1957. Stromatolites of the Belt Series in Glacier National Park and vicinity, Montana. *U.S. Geological Survey Professional Paper* No. 294–D.

Riding, R., S. M. Awramik, B. M. Winsborough, K. M. Griffin, and R. F. Dill. 1991. Bahamian giant stromatolites: Microbial composition of surface mats. *Geological Magazine* 128:227–34.

Rogers, D. T. 1987. Petrology of the Middle Cambrian Langston and Ute Formations in southeast Idaho. Master's thesis, Utah State University.

Sami, T. T. and N. P. James. 1993. Evolution of an early Paleozoic foreland basin carbonate platform, lower Pethei Group, Great Slave Lake, north-west Canada. *Sedimentology* 40:403–30.

Scotese, C. R., R. K. Bambach, C. Barton, R. Van Der Voo, and A. M. Ziegler. 1979. Paleozoic base maps. *Journal of Geology* 87:217–77.

Scrutton, C. T. 1964. Periodicity in Devonian coral growth. *Palaeontology* 7:552–58.
———. 1978. Periodic growth features in fossil organisms and the length of the day and month. In P. Brosche and J. Sundermann, eds., *Tidal Friction and the Earth's Rotation*, pp. 154–96. New York: Springer-Verlag.
Serebryakov, S. N. 1976. Distribution of stromatolites in Riphean deposits of the Uchur–Maya Region of Siberia. In M. R. Walter, ed., *Stromatolites: Developments in Sedimentology*, vol. 20, pp. 613–33. New York: Elsevier.
Shapiro, R. S. 1991. Morphological variations within a modern stromatolite field: Lee Stocking Island, Exuma Cays, Bahamas. In R. J. Bain, ed., *Proceedings of the Fifth Symposium on the Geology of the Bahamas*, pp. 209–19. San Salvador: Bahamian Field Station.
Sloss, S. L. 1963. Sequences in the cratonic interior of North America. *Geological Society of America Bulletin* 74:93–111.
Southgate, P. N. 1989. Relationships between cyclicity and stromatolite form in the Late Proterozoic Bitter Springs Formation, Australia. *Sedimentology* 36:323–39.
Stewart, J. H. and F. G. Poole. 1974. Lower Paleozoic and uppermost Precambrian Cordilleran miogeocline, Great Basin, western United States. In W. R. Dickenson, ed., *Tectonics and Sedimentation*, pp. 28–58. Tulsa: Society of Economic Paleontologists and Mineralogists *Special Publication* No. 22.
Stokes, W. L. 1963. *Geologic Map of Northwestern Utah*. Salt Lake City: College of Mines and Mineral Industries, University of Utah.
Vanyo, J. P. and S. M. Awramik. 1985. Stromatolites and earth–sun–moon dynamics. *Precambrian Research* 29:121–42.
Wakeley, L. D. 1975. Petrology of the Upper Nounan–Worm Creek Sequence, Upper Cambrian Nounan and St. Charles Formations, Southeast Idaho. Master's thesis, Utah State University.
Walter, M. R. 1972. Stromatolites and the biostratigraphy of the Australian Precambrian and Cambrian. *Palaeontological Association Special Publication* No. 11.
———. 1976. Introduction. In M. R. Walter, ed., *Stromatolites: Developments in Sedimentology*, vol. 20, pp. 1–3. New York: Elsevier.
Walter, M. R. and G. R. Heys. 1985. Links between the rise of the metazoa and the decline of stromatolites. *Precambrian Research* 29:149–74.
Wells, J. W. 1963. Coral growth and geochronometry. *Nature* 197:948–50.
Wilkinson, B. H. 1982. Cyclic cratonic carbonates and Phanerozoic calcite seas. *Journal of Geology Education* 30:180–203.
Williams, G. E. 1989. Precambrian tidal sedimentary cycles and Earth's rotation. *EOS*, January, 17:33–41.
Williams, J. S. 1948. Geology of the Paleozoic rocks, Logan Quadrangle, Utah. *Geological Society of America Bulletin* 59:1121–64.
Williams, J. S. and G. B. Maxey. 1941. The Cambrian section in the Logan Quadrangle, Utah and vicinity. *American Journal of Science* 239:276–85.
Ziegler, A. M., C. R. Scotese, W. S. Mickerrow, M. E. Johnson, and R. K. Bambach. 1979. Paleozoic paleogeography. *Annual Review of Earth and Planetary Sciences* 7:473–502.

12

Using Time/Environment Analysis to Recognize Faunal Events in the Upper Ordovician of the Cincinnati Arch

Steven M. Holland

ABSTRACT

Time/environment analysis, with the time units defined by sequence stratigraphy and the spatial units defined by depositional environments, is an effective tool for recognizing and characterizing global and regional bioevents such as migrations and extinctions. Events of this scale are more properly examined over longer spans of time and wider geographic areas than a few meters spanning an event in a single section. A major faunal immigration in the lower Richmondian, in the Upper Ordovician of the Cincinnati Arch, was timed with a transgression at the base of a stratigraphic sequence. Although the invasion appears to be gradual in a single section, the time/environment approach shows that the invasion was rapid. Although most of the brachiopod genera studied tracked environment and displayed environmental stasis, several did not. The subfamily Sowerbyellinae moved onshore, whereas *Retrorsirostra* moved offshore, and *Rhynchotrema* oscillated between shoreface and offshore habitats. The lack of consistent offshore movements at the generic level suggests that higher taxa move offshore through the production of new taxa in progressively deeper water environments.

Time/environment analysis has been increasingly used in diversity studies to document patterns of origination, extinction, and evolutionary migration on a continental to global scale (Droser and Bottjer 1989; Jablonski and Bottjer 1991; Miller 1990; Sepkoski 1991; Westrop and Ludvigsen 1987). A similar approach can be used in single sedimentary basins to identify basinwide faunal events. In this study the time units are defined by sequence stratigraphy, and the spatial units, by depositional environments. These units automatically impose the temporal and spatial resolution at which an event can be studied.

There is a tendency to study basinwide or even global biotic events on the scale of a single outcrop (e.g., Surlyk and Johansen 1984; Ward et al. 1986). Because environments are continually changing whether or not this change is reflected in lithofacies, the appearance or disappearance of taxa in a section usually cannot be read simply as a temporal phenomenon; rather, the possibility of spatial phenomena such as local migration or local nonpreservation always must be considered.

Time/environment analysis is one approach to this problem. By sacrificing some temporal resolution, the time/environment approach forces the problem to be examined regionally across a range of environments. I believe that many biotic events are more effectively studied in the context of longer spans of time and wider reaches of space than they are by focusing on a short time interval in only a few sections. As an example of this approach, this study examines a major immigration event and the shifting habitat preferences and potential epiboles in the Upper Ordovician of the Cincinnati Arch.

Cincinnatian Sequences

Because sequence stratigraphy is not yet widely used in paleontological studies, sequences of the Cincinnati Series are briefly described here. More complete descriptions can be found in Holland (1990, 1993) and Holland et al. (1993). The Upper Ordovician of the Cincinnati Arch is composed of six sequences, varying in duration from 1 to 6 million years and in thickness from 10 to 115 m (fig. 12.1). The bulk of each sequence is a shallowing-upward cycle, many of which were recognized by previous workers (Anstey and Fowler 1969; Hay et al. 1981; Tobin 1982). Each shallowing-upward cycle corresponds to a highstand systems tract and was formed by the progradation of five facies belts across a northward-sloping, mixed carbonate/clastic ramp (table 12.1). The five lithofacies record deposition in offshore, transition-zone, nearshore, intertidal, and supratidal environments. The supratidal settings are indicated by abundant desiccation cracks and widespread parallel laminites, with only minor burrows. The intertidal settings are characterized by pervasive bioturbation and rare desiccation cracks. Nearshore environments contain abundant proximal storm deposits and robust brachiopod and bryozoan morphologies. The transition-zone settings contain a greater percentage of shale, a greater proportion of medial and distal storm deposits, and morphologically and taxonomically diverse fauna. Offshore settings are dominated by shale, medial to distal storm deposits, and thin and fragile fossil morphologies.

In any given sequence the highstand systems tract contains only a few of the five lithofacies in a particular area, and the upramp sections contain a shallower

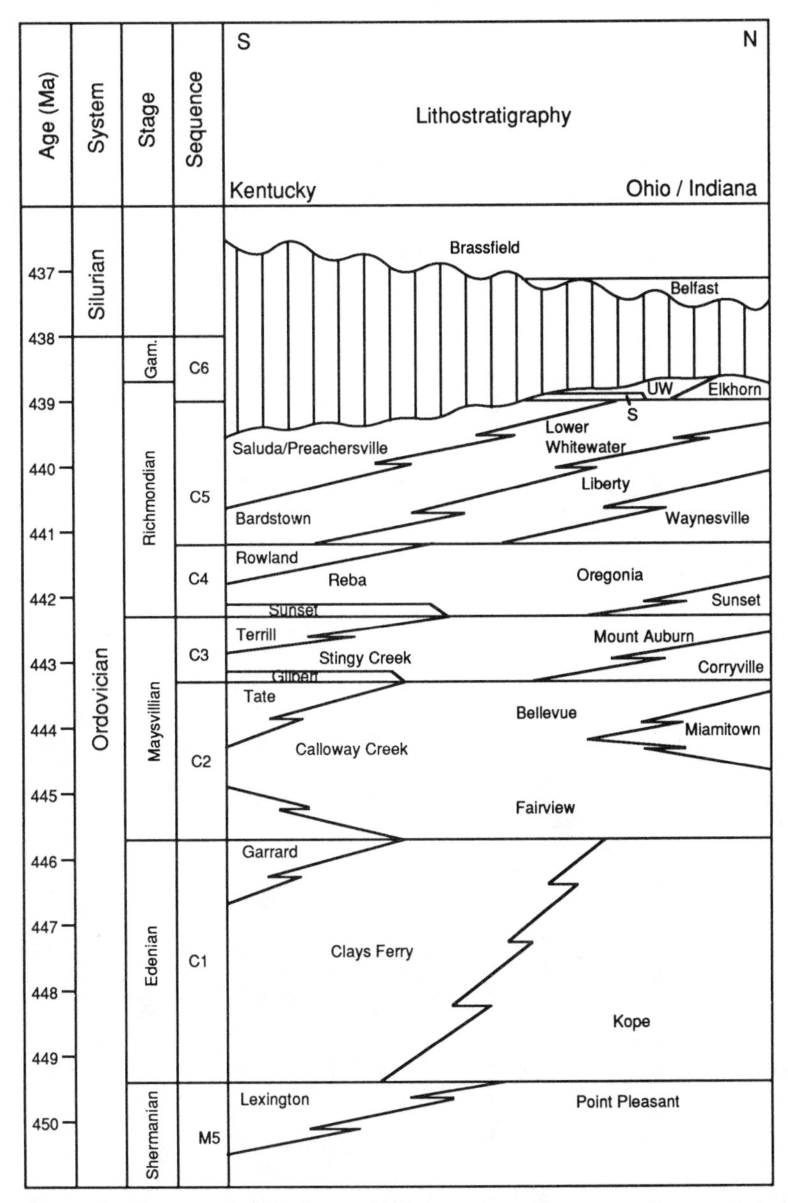

FIGURE 12.1. Sequence stratigraphy and lithostratigraphic nomenclature of the Cincinnati Series in Kentucky, Indiana, and Ohio. The Cincinnatian sequences are numbered upsection from C1 through C6; the underlying Point Pleasant and Lexington Limestones comprise one sequence (M6). Although the transgressive systems tracts are thick enough to be shown only in upramp areas (e.g., Gilbert and Sunset), they are present as cross-bedded calcarenites and bioclastic packstone intervals in downramp areas. Except at the sequence boundaries and maximum flooding surfaces, most of the formational contacts are time-transgressive, gradational-facies contacts and are arbitrarily placed in outcrop (e.g., Waynesville–Liberty contact). Note that many equivalent formations are known by different names in different states (e.g., Stingy Creek and Mount Auburn; Clays Ferry and Fairview). The time scale is based on Sweet (1984) and Sloan (1987a).

TABLE 12.1 *Highstand Facies*

Environment	Lithofacies	Lithologies	Sedimentary structures	Fossil assemblages	Lithostratigraphic units
Supratidal	Laminated carbonate mudstone	Planar laminated calcareous to dolomitic mudstone	Ubiquitous planar lamination, wave-ripple lamination, abundant desiccation cracks	Body fossils absent, scarce vertical burrows	Parts of Saluda, Rowland, Terrill, Tate, and Preachersville
Intertidal to shallow subtidal	Bioturbated carbonate mudstone	Irregularly bedded to bioturbated calcareous to dolomitic mudstone	Rare desiccation cracks, wave-ripple lamination	Colonial rugosan–tabulate–stromatoporoid–ramose trepostome assemblage thoroughly bioturbated	Parts of Saluda, Rowland, Terrill, Tate, and Preachersville
Intertidal	Variegated mudstone	Bioturbated mudstone (95%), siltstone or dolomitized wackestones/packstones (5%)	Ripple-lamination	Elements of *Hebertella* assemblage: *Hebertella*, ramose trepostome bryozoans, cyclostome bryozoans; *Skolithos* ichnofacies	Part of Preachersville
Shoreface	Wavy/nodular limestone	Nodular bedded fossiliferous packstones (70%), fossiliferous mudstone (30%), minor wackestones, grainstones, and calcisiltites	Proximal storm beds, wave-ripple lamination, gutter casts	*Hebertella* assemblage; *Skolithos* ichnofacies	Whitewater, Bardstown, Reba, Oregonia, Stingy Creek, Mount Auburn, Straight Creek, Bellevue, Calloway Creek (in part)
Transition zone	Mixed packstone shale	Medium bedded fossiliferous packstones (45%), mudstone (45%), wackestones, grainstones, and calcisilites (10% total)	Storm beds, planar laminations, wave-ripple lamination, small-scale hummocky cross-lamination	Mixed brachiopodbryozoan assemblage; mixed *Cruziana Skolithos* ichnofacies	Liberty, Sunset (see Tobin 1986, not Weir et al. 1984), Corryville, Fairview, Clays Ferry
Offshore	Shale dominated	Thickly bedded mudstone (70%), thin to medium bedded fossiliferous wackestone and packstone (20%), crinoidal grainstone (5%), calcisiltite (5%)	Storm beds, small-scale hummocky and trough cross-lamination, megaripples, gutter casts, groove casts, runzelmarken	*Onniella* assemblage, ichnofacies	Waynesville, Kope
Deep basin	Black shale	Laminated mudstone (90%), wackestones and calcisiltites (10%)	Planar laminae	Orbiculoid brachiopods, graptolites, trilobites (*Cryptolithus*, *Triarthrus*)	"Utica Shale" (see Mitchell and Bergström 1991)

set of facies than do the downramp areas. The highstand facies are always arrayed in sharp-based, shallowing-up hemicycles. The contacts between the highstand lithofacies are gradational to intertonguing and are often difficult to define in the field.

Each highstand systems tract is capped by a sharp contact marking the sequence boundary. The contact is often noteworthy only for the sharp juxtaposition of dissimilar facies; the sequence boundaries in the Cincinnati Series are not marked by bioturbated horizons or glauconite. Unusual lithofacies immediately overlie the sequence boundaries and include gastropod coquinas, cross-bedded calcarenites, unusually shale-poor micrite-rich wackestones, and anomalous bands of packstones with a taphonomically degraded fauna (table 12.2). The complete transgressive systems tract in any given area may comprise only one or two of these peculiar facies. The transgressive systems tract is typically thin, relative to the highstand systems tract, and is usually only several meters thick, although in some sequences, such as the C1, the transgressive systems tract may be tens of meters thick.

Some lithostratigraphic nomenclature can be easily translated into sequence terms (fig. 12.1, tables 12.1 and 12.2). In Kentucky, the members recognized during the USGS/KGS mapping program (Weir et al. 1984) generally correspond to the lithofacies described in this study. In Ohio and Indiana the formations and members of the traditional lithostratigraphy (Caster et al. 1955; Tobin 1986) correspond to the lithofacies in this study. However, the workers in Ohio and Indiana generally did not separate the thin transgressive systems tracts into separately named units; instead they were usually considered part of the underlying member. In addition, several of the lithostratigraphic units proposed in the past 25 years (e.g., Bull Fork, Dillsboro, Grant Lake, Tanners Creek) comprise multiple lithofacies, straddle sequence boundaries, and offer little precision in either environmental or chronostratigraphic terms.

Sequence boundaries and maximum flooding surfaces have chronostratigraphic significance; that is, all overlying rocks are always younger than all underlying rocks (Haq et al. 1987; Van Wagoner et al. 1988). Accordingly, the sequence boundaries define the chronostratigraphic units C1–C6 that I used for the time/environment analyses in this study. Highstand lithofacies define the time/environment units because they dominate the thickness of the local sections and are units comparable to the environmental units of Droser and Bottjer (1989), Jablonski and Bottjer (1991), Miller (1990), and Sepkoski (1991).

Fossil Assemblages

Of the five depositional environments only the three subtidal environments are abundantly and consistently fossiliferous. Each of these environments contains a characteristic assemblage of genera that is present in each of the shoaling-upward sequences. These assemblages are named for and principally reflect

TABLE 12.2 *Transgressive Facies*

Environment	Lithofacies	Lithologies	Sedimentary structures	Fossil assemblages	Lithostratigraphic units
Lagoonal	Wavy-bedded wackestone	Wavy-bedded wackestone to micstone (80%), mudstone (20%)	Graded bedding, parallel lamination, burrow mottling	Tabulate stromatoporoid assemblage grading upward to *Hebertella* assemblage, *Skolithos* ichnofacies	Sunset (see Weir et al, 1984, not Tobin 1986), Gilbert, Saluda (in part), Marble Hill Bed (upper part locally)
Sand shoals/bars	Cross-bedded calcarenite	Crinoidal to fossil fragmental calcarenite (90%), fossiliferous mudsone (10%)	Trough cross-lamination, megaripples	Highly abraded *Hebertella* assemblage	Included by previous workers in Calloway Creek, Stingy Creek, and Reba
Starved shelf	Bioclastic packstone	Fossiliferous wackestone and packstone (80%), mudstone (20%)	Planar and trough cross-lamination, megaripples	*Hebertella* assemblage grading upward to mixed brachiopod/bryozoan or *Onniella* assemblage	Included by previous workers in Bellevue, Mount Auburn, Straight Creek, and Oregonia
Tidal channels and bar	Gastropod coquina	Crinoidal calcarenite (50%), gastropod calcirudite (50%)	Cross-laminated, sharp-based fining-upward beds	Almost entirely the gastropod *Loxoplocus*, also elements of *Hebertella* assemblage	Marble Hill Bed

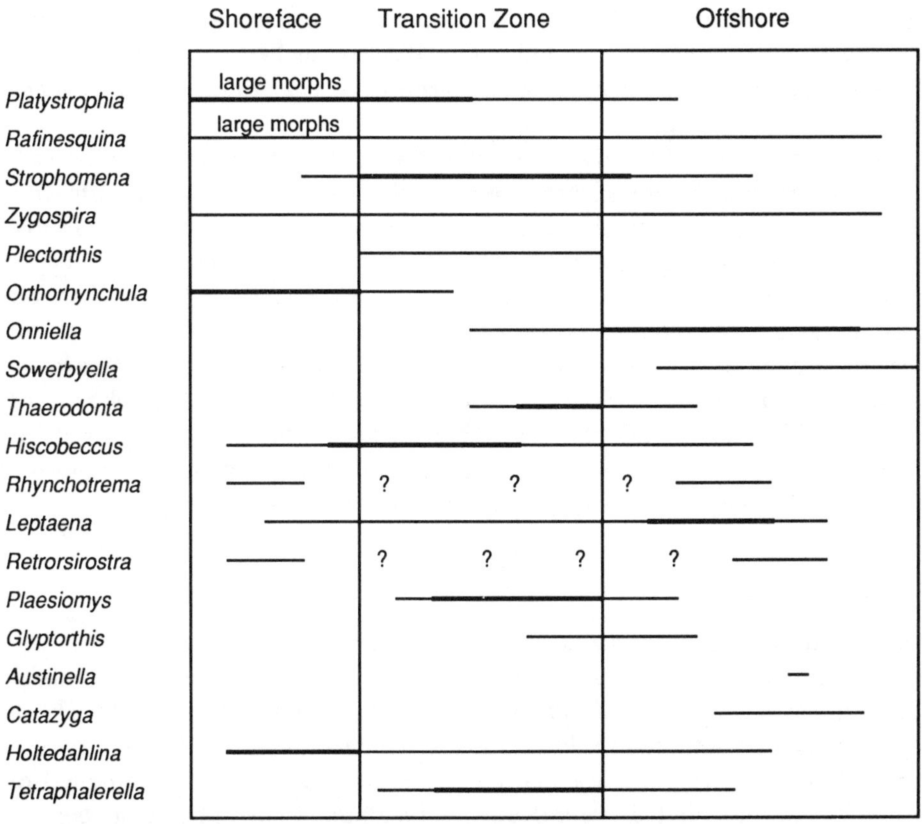

Shoreface Transition Zone Offshore

large morphs
Platystrophia
large morphs
Rafinesquina
Strophomena
Zygospira
Plectorthis
Orthorhynchula
Onniella
Sowerbyella
Thaerodonta
Hiscobeccus
Rhynchotrema ? ? ?
Leptaena
Retrorsirostra ? ? ? ?
Plaesiomys
Glyptorthis
Austinella
Catazyga
Holtedahlina
Tetraphalerella

FIGURE 12.2. Environmental distribution of brachiopod genera in the Cincinnati Series. The thin lines represent the presence of a genus, and the thick lines represent parts of the range where the genus is most abundant. Compiled from Alberstadt (1979), Austin (1927), Caster et al. (1955), Dalvé (1948), Hay (1981), Hay et al. (1981), Holland (1990), Howe (1972), Howe (1979), and Walker (1982).

brachiopod distributions. The brachiopods are common and conspicuous faunal elements, are well preserved, and are easily identified in the field. Although these assemblages are described as discrete units, they grade into one another, just as the lithofacies do (fig. 12.2).

Hebertella Assemblage

The *Hebertella* assemblage is dominated by large and thick-shelled brachiopods and massive and ramose trepostome bryozoans. The brachiopods include globose and coarsely ribbed forms (*Platystrophia, Orthorhynchula*) that have thickened lowermost valves as a stability adaptation in energetic environments (Rudwick 1970). Other brachiopods include inflated and thickly shelled forms (*Hebertella*). The forms with planar shells (*Rafinesquina*) are unusually large and thick-valved compared with similar species in other assemblages, a feature that is a stability adaptation for agitated waters (Alexander 1975). Although most of the brachiopods in the *Hebertella* assemblage also occur in other assemblages, they are unusually large and robust here.

Bryozoan colony morphologies are dominated by ramose forms, but bifoliate sheets, encrusting colonies, and massive colonies are much more numerous than they are in other assemblages. Although bivalves (*Ambonychia, Modiolopsis,* and *Ischyrodonta*) and gastropods (*Cyclonema, Loxoplocus*) can be found, they are only abundant locally. Crinoids and trilobites (*Flexicalymene*) are present, but their multielement skeletons are rarely articulated. Single taxa commonly dominate the *Hebertella* assemblage locally. The *Hebertella* assemblage occurs in the shoreface environment in the wavy/nodular limestone facies. The robust morphologies, the stability adaptations in brachiopods, and the scarcity of articulated multielement skeletons further confirm that the wavy/nodular limestone facies was deposited in an energetic setting such as the shoreface.

Mixed Brachiopod/Bryozoan Assemblage

The mixed brachiopod/bryozoan assemblage contains a wide diversity of taxa and morphologies. The brachiopod morphologies include the nearly planar forms (*Leptaena, Rafinesquina, Strophomena, Trigrammaria, Tetraphalerella, Thaerodonta*), which are adapted for life on soft substrates (Thayer 1975), the coarsely ribbed biconvex forms (*Hiscobeccus, Platystrophia*), which are better adapted for firm bottoms and agitated water (Rudwick 1970), the finely ribbed planoconvex-to-biconvex forms (*Hebertella, Glyptorthis, Plaesiomys, Plectorthis*), and the tiny biconvex form (*Zygospira*). In general, the brachiopods are less massively built than are those in the *Hebertella* assemblage, suggesting somewhat lower daily wave energies.

The bryozoans are abundant and include a wide range of morphologies: ramose, encrusting, bifoliate sheets, massive, and eschariform. Bivalves (*Ambonychia, Modiolopsis, Caritodens*), gastropods (*Cyclonema, Loxoplocus*), and orthoconic nautiloids are common in some horizons. Crinoids (*Iocrinus, Pycnocrinus, Dendrocrinus, Gaurocrinus*), edrioasteroids, and trilobites (*Flexicalymene, Isotelus*) locally occur in beds with numerous, completely articulated specimens (Caster et al. 1955; Meyer 1990; Schumacher and Ausich 1983). Unlike the *Hebertella* and *Dalmanella* assemblages, the mixed brachiopod/bryozoan assemblage is rarely dominated by a single taxon. The mixed brachiopod/bryozoan assemblage occurs in the transition-zone environment, in the mixed packstone/shale facies. The range of morphologies, with adaptations for both soft and firm substrates and both quiet and agitated water, and the increased preservation of multielement skeletons suggests an environment with a variety of bottom and energy conditions. The lack of massively armored shells indicates an environment less energetic than that in the *Hebertella* assemblage.

Dalmanella Assemblage

The *Dalmanella* assemblage is characterized by a small, delicate fauna. Small, nearly planar (*Dalmanella*) or concavo-convex brachiopods (*Sowerbyella*) dominate the assemblage; both their small size and nearly planar forms are adaptations for soft substrates, implying quiet water conditions (Thayer 1975). Small, flattened forms of *Rafinesquina* are present in lesser numbers, and their small size and flattened shape are typical of quiet water conditions (Alexander 1975).

The bryozoans are dominated by thin ramose forms, although bifoliate sheets and encrusting forms are also present. Bivalves and gastropods are locally present and commonly dominate in shales (Miller 1988). Graptolites are conspicuous as well (Bergström and Mitchell 1991; Mitchell and Bergström 1991). Crinoids (*Ectenocrinus, Cincinnaticrinus*) and trilobites (*Cryptolithus, Flexicalymene, Isotelus*) are commonly articulated (Frey 1987b; Schumacher and Ausich 1983). The *Dalmanella* assemblage occurs in the offshore environment, in the shale-dominated facies. The consistently delicate morphologies, small body sizes, and nearly planar brachiopod forms support the interpretation of the shale-dominated facies as an offshore quiet-water environment.

Richmondian Invasion

Pattern of Invasion

This gradient of fossil assemblages changes little within the first three sequences (table 12.3). For example, *Dalmanella* is the principal brachiopod in the offshore; a diverse mixture of orthid, strophomenid, and atrypid brachiopods occurs in the transition zone; and large *Platystrophia* and *Hebertella* dominate in the shoreface. Through the first three sequences, most of the differences among equivalent assemblages occur at the species level, with only minor additions or losses of genera (table 12.3).

During the C5 sequence, an enormous number of new forms appeared throughout this gradient during the Richmondian Invasion (table 12.3, fig. 12.3). The tabulate corals and several genera of nautiloids, trilobites, and brachiopods first appear in the C4 sequence but are more abundant during the C5. The Richmondian invasion is marked by a faunal diversification at the species through class level. Most of these taxa lack precursors in Edenian, Maysvillian, and, in some cases, Franklinian strata of the Cincinnati Arch (Howe 1988; see Sweet 1988 for a definition of Franklinian). For example, rugosan corals appear in the Cincinnati area for the first time since the Franklinian (Browne 1964; Browne 1965; Cressman 1973; Elias 1983). More than 20 genera of nautiloids that were absent in the Edenian and Maysvillian appear in the Richmondian (fig. 12.3; Flower 1946). Likewise, the new genera of trilobites, crinoids, gastropods, pelecypods, and brachiopods were absent in the Edenian and Maysvillian

TABLE 12.3 *Occurrence of Brachiopod Species in the Upper Ordovician of the Cincinnati Arch Region*

	C1	C2		C3		C4		C5			C6
	O	T	S	T	S	T	S	O	T	S	
Anazyga recurvirostris (Hall)	x										
Austinella scovillei (Miller)								x			
Catazyga schuchertana Ulrich								x			
Cyclocoelia crassiplicata (Foerste)		x									
Cyclocoelia sectostriata (Ulrich)		x									
Cyclocoelia sordida (Hall)		x									
Dalmanella emacerata (Hall	x										
Dalmanella fultonensis (Foerste)	x										
Dalmanella meeki (Miller)						x	x	x			
Dalmanella multisecta (Meek)	x	x									
Eridorthis nicklesi Foerste	x										
Glyptorthis insculpta (Hall)								x			
Hebertella alveata Foerste									x	x	
Hebertella occidentalis Hall	x	x	x	x	x	x	x	x	x	x	x
Heterorthina fairmountensis (Foerste)			x								
Hiscobeccus capax (Conrad)						x	x	x	x	x	x
Holtedahlina sulcata (Verneuil)								x	x	x	
Lepidocyclus perlamellosus (Whitfield)								x	x	x	x
Leptaena gibbosa (James)	x										
Leptaena kentuckyensis Pope							x				
Leptaena richmondensis Foerste						x	x	x	x		x
Megamyonia sp.											x
Orthorhynchula linneyi James		x	x								
Pionodema bellula (Meek)		x									
Plaesiomys subquadrata (Hall)										x	
Platystrophia acuminata (James)					x						
Platystrophia acutilirata (Conrad)										x	
Platystrophia annieana Foerste								x	x	x	
Platystrophia clarksvillensis Foerste								x	x		
Platystrophia corryvillensis McEwan				x							
Platystrophia crassa (James)		x	x	x							
Platystrophia cumingsi McEwan									x		
Platystrophia cypha (James)			x		x	x	x	x	x	x	
Platystrophia foerstei McEwan										x	
Platystrophia hopensis Foerste		x									
Platystrophia juvensis McEwan		x			x						
Platystrophia laticosta (Meek)		x	x	x	x			x			
Platystrophia moritura (Cumings)											x
Platystrophia morrowensis (James)				x							
Platystrophia pauciplicata (Cumings)		x									
Platystrophia ponderosa Foerste		x	x	x	x	x	x				
Platystrophia profundosulcata (Meek)		x									
Platystrophia sublaticosta McEwan				x							
Platystrophia wallowayi Foerste						x	x				
Plectorthis aequivalvis (Hall)		x									
Plectorthis fissicosta (Hall)		x									
Plectorthis jamesi (Hall)				x							
Plectorthis neglecta (James)	x	x									
Plectorthis plicatella (Hall)		x									
Rafinesquina alternata (Conrad)	x	x	x	x	x	x	x	x	x	x	x
Rafinesquina squamula (James)	x	x									
Rafinesquina ulrichi (James)	x										
Retrorsirostra carleyi (Hall)							x	x			
Rhynchotrema dentatum (Hall)							x	x	x	x	x
Sowerbyella plicatellus (Ulrich)	x										
Sowerbyella rugosus (Meek)	x										

	C1	C2		C3		C4		C5			C6
	O	T	S	T	S	T	S	O	T	S	
Strophomena concordensis Foerste						x	x				
Strophomena hallie (Miller)	x										
Strophomena maysvillensis Foerste		x									
Strophomena millionensis Foerste	x										
Strophomena nutans Meek								x			
Strophomena planumbona (Hall)						x	x	x	x		
Strophomena sinuata James		x									
Strophomena vetusta (James)								x	x	x	
Tetraphalerella neglecta (James)								x		x	
Thaerodonta recedens (Sardeson)								x			
Trigrammaria planoconvexa (Hall)		x									
Zygospira cincinnatiensis Meek		x									
Zygospira kentuckiensis James								x			
Zygospira modesta Hall	x	x	x	x	x	x	x	x	x	x	x

C1 through C6 refer to Cincinnatian sequences defining the limit of chronostratigraphic resolution. Listed within each sequence are environments: O indicates offshore, T indicates transition zone, and S indicates shoreface. Because facies migrate laterally, environments in each sequence are not chronostratigraphic units. C6 data are preliminary and have not yet been sorted according to environment. Subspecies are included under their parent species; in most cases, this does not affect either the range or the environmental occurrence.

Data sources in reference list; primarily from Dalve 1948; Holland 1990; and Weir et al. 1984.

stages. Some diversification occurred at the species level and possibly originated in species present on the Cincinnati Arch during Edenian and Maysvillian time (Alberstadt 1979). For example, the brachiopods *Platystrophia* and *Strophomena* were present during the C2–C4 sequences but underwent species-level diversification during the C5 sequence (Alberstadt 1979; Dalvé 1948; McEwan 1919).

The appearance of new taxa did not drive preexisting taxa to extinction. For example, *Dalmanella, Rafinesquina, Zygospira, Strophomena, Platystrophia,* and *Hebertella* are widespread and conspicuous elements in C1–C4 strata and are equally widespread and abundant in C5 and C6 strata. The extinction of several C1 and C2 brachiopod genera preceded the C5 invasion. *Orthorhynchula* is present in C1 and C2 shoreface environments of Kentucky (Howe 1979) but is absent in C3 shoreface environments. *Plectorthis* is present in C1, C2, and C3 transition-zone environments but is gone by C4 (Dalvé 1948). Similarly, *Anazyga, Cyclocoelia, Eridorthis, Heterorthina, Pionodema,* and *Trigrammaria* were present in the C1 and C2 sequences but were regionally extinct by the C3 sequence (table 12.3). The new C5 taxa did not replace preexisting taxa but increased the overall diversity.

The new taxa are not limited to a single facies; rather, all facies contain new taxa (table 12.3; Austin 1927; Dalvé 1948). When they are considered individually, many or most of the new taxa seem to be facies-specific because their peak abundance occurs within particular facies. The new taxa are not limited to a single trophic group, either. Some invading taxa were predators (nautiloids), scavengers and/or deposit feeders (trilobites), deposit feeders (some of the bivalves; see Frey 1987a; Pojeta 1971), and suspension feeders (brachiopods,

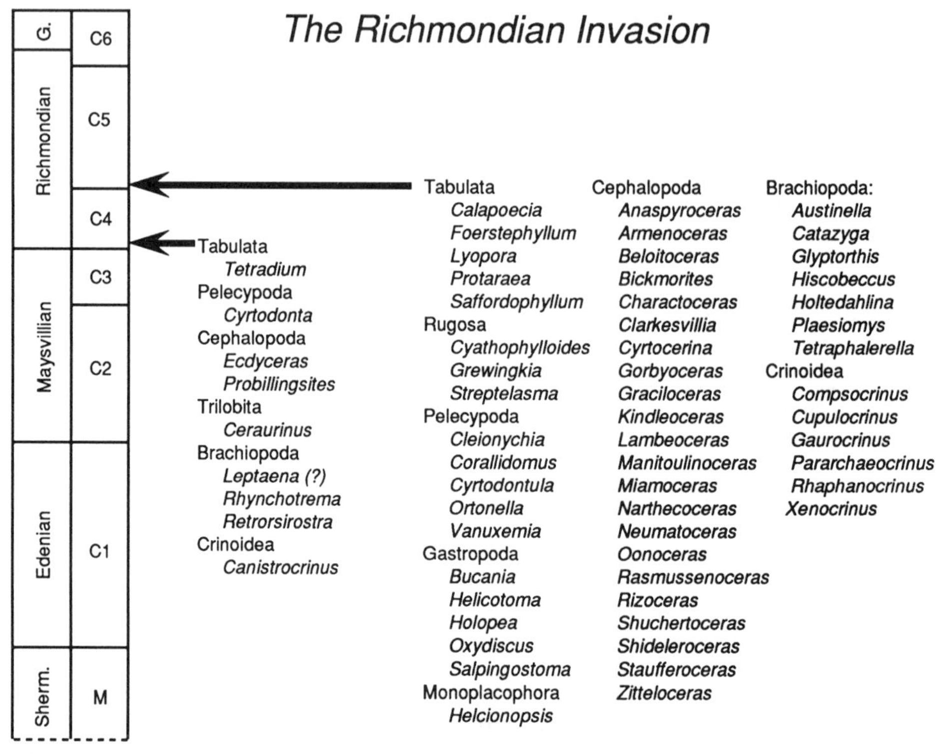

FIGURE 12.3. The timing and new genera of the Richmondian Invasion. Several groups have not been included, awaiting future study: algae, sponges, bryozoans, graptolites, conodonts, noncrinoid echinoderms, and nontrilobite arthropods.

rugosans, crinoids). The new taxa included nektonic (nautiloids), epifaunal benthic (brachiopods, crinoids), and infaunal benthic life-habit groups (some bivalves; Frey 1987a; Pojeta 1971). The invasion was not limited to particular facies, trophic groups, or life-habit groups; rather, the Richmondian Invasion was a major ecological revolution affecting all aspects of the Cincinnatian seas.

ORIGIN OF INVASION

The origin of these new taxa has long been questioned and has resulted in two major hypotheses: a major immigration episode (e.g., Frey 1985; Elias 1985; Mitchell and Sweet 1989) or an evolutionary burst (e.g., Howe 1988). The view in this study favors large-scale immigration for the introduction of most genera and higher-level taxa. The expansion of genera beyond their Maysvillian ranges may have led to speciation events at the edges of their range, producing the strong endemicity of the Cincinnati Arch fauna.

Generic-level immigration is favored because many of the genera belong to the Red River fauna or the so-called Arctic fauna characteristic of the Middle and Upper Ordovician rocks of northern Canada and found in Wyoming, Colorado, South Dakota, Texas, Iowa, and Minnesota (Flower 1946; Macomber 1970; Mitchell and Sweet 1989; Sloan 1987b; Sweet and Miller 1958). Despite its name, the fauna is probably tropical (Flower 1946), in part because these areas would have been at 10°S to 20°S during the Ordovician (Scotese and McKerrow 1991). Many of the "Arctic fauna" brachiopod genera (e.g., *Leptaena, Glyptorthis,* and *Plaesiomys*) are present in the Blackriveran of the southern Appalachians (Cooper 1956). The appearance, disappearance, and subsequent reappearance of these taxa in the eastern United States have confounded correlations with Canadian and western United States Ordovician strata, because the western strata seem to contain a mixture of Trentonian and Richmondian forms (see Sweet and Miller 1958).

The temporary disappearance of Arctic fauna from the eastern United States has been attributed to the influx of clastics from the rising Taconic mountains at the end of the Blackriveran (Sweet and Miller 1958). However, the Blackriveran/Trentonian boundary also marks a major transgression across which carbonates in the eastern United States switch from a tropical character to a temperate character (Holland 1990; Ludvigsen 1978). For example, the Blackriveran Oregon and Tyrone limestones of Kentucky contain abundant micrite, peloids, intraclasts, hardgrounds, and cryptalgalaminites (Cressman and Noger 1976), which collectively indicate tropical carbonate deposition (Nelson 1988). Many of these features are rarer in the Franklinian-age Lexington Limestone and in the Cincinnati Series. In addition, these post-Blackriveran limestones are coarse-grained, seldom contain ooids and aggregates, and contain bioclasts as the principal allochems, which together suggest temperate carbonate deposition.

Similarly, post-Blackriveran carbonates from Ontario have also been interpreted as temperate carbonates (Brookfield 1988). The initiation of a deep foreland basin at the beginning of the Franklinian (Lash 1987) may have allowed cooler waters to spread onto eastern Laurentia and also introduced clastics to the region (Holland 1990). The cool waters may also have altered carbonate deposition and forced the retreat of the tropical Arctic fauna to stable platform and passive-margin areas in western North America and Canada.

The Richmondian reappearance of the Arctic fauna broadly coincides with the major C5 transgression. The transgression may have broken barriers that kept the Arctic fauna away from the Cincinnati Arch, with the Canadian Shield and deep-water Maquoketa belt postulated as barriers (Elias 1983). As the Taconic Foreland Basin was progressively filled with sediment and cooler oceanic waters were progressively excluded from the continental interior, the C5 transgression may have broken down any remaining barriers to invasion. If the cool-water hypothesis is correct, the barriers may have been a thermal oceanographic barrier and not necessarily a topographic one. Such thermal barriers are common in modern oceans (Vermeij 1978). In an alternative explanation, Frey (1985) attributed the invasion to Laurentia's equatorward movement during the latest Ordovician, coupled with a decrease in clastics from the Taconic highlands.

RAPID OR PROLONGED INVASION?

In any given C5 outcrop on the Cincinnati Arch, the appearances of new taxa are staggered through the section, so that the immigration appears to have been prolonged. But is this pattern real, or is it an artifact? I tried to answer this question by examining the occurrences of brachiopods and rugosans. Many of the new genera are limited by facies (Fox 1962, 1968). For example, Fox's (1962) *Strophomena planumbona* assemblage, containing *S. planumbona, Plaesiomys subquadrata, Hiscobeccus capax, Grewingkia canadensis,* and *Thaerodonta recedens,* occurs in the mixed packstone/shale facies of the transition zone throughout the northern outcrop area in Ohio and Indiana (fig. 12.4). In these downramp areas, this assemblage appears approximately midway through the C5 sequence. Most elements of this assemblage occur within the upper part of the shale-dominated facies but do not extend to the base of the C5 sequence in these northern outcrops.

Because facies prograded northward across the Cincinnati Arch ramp (Holland 1990; Weir et al. 1984), any given facies was deposited first in the upramp areas (central Kentucky) and progressively later in the downramp areas (Ohio and Indiana). If the *S. planumbona* assemblage appeared on the Cincinnati Arch midway in time through the C5 sequence—as taking the downramp sections at face value would imply—the mixed packstone/shale facies in the upramp areas of central Kentucky should be devoid of this assemblage. But if the assemblage invaded the transition-zone environments at the onset of the transgression, the assemblage should be present in C5 mixed packstone/shale facies in all areas on the Cincinnati Arch, including central Kentucky. And indeed, the assemblage is present in all C5 mixed packstone/shale facies of the Cincinnati Arch, even where the facies immediately overlies the sequence boundary, suggesting that these taxa entered the region at the beginning of the sequence (fig. 12.5).

Similarly, colonial rugosan and tabulate coral assemblages can be found in the bioturbated carbonate mudstone facies near the top of the C5 sequence in the downramp areas (fig. 12.4; Browne 1964). Although the upramp section at Frederickstown (fig. 12.5) is somewhat condensed, this assemblage again occurs in the bioturbated carbonate mudstone facies. This facies at Frederickstown is situated much closer to the C5 sequence boundary than it is in the northern sections, indicating that the coral assemblage arrived on the Cincinnati Arch earlier than its stratigraphic position in northern sections implies. I suspect that if this facies could be found where it immediately overlies the C5 sequence boundary, it would contain this coral assemblage.

The brachiopods *Rhynchotrema dentatum* and *Leptaena richmondensis* are precursors to the C5 invasion and first occur sparsely in mixed packstone/shale

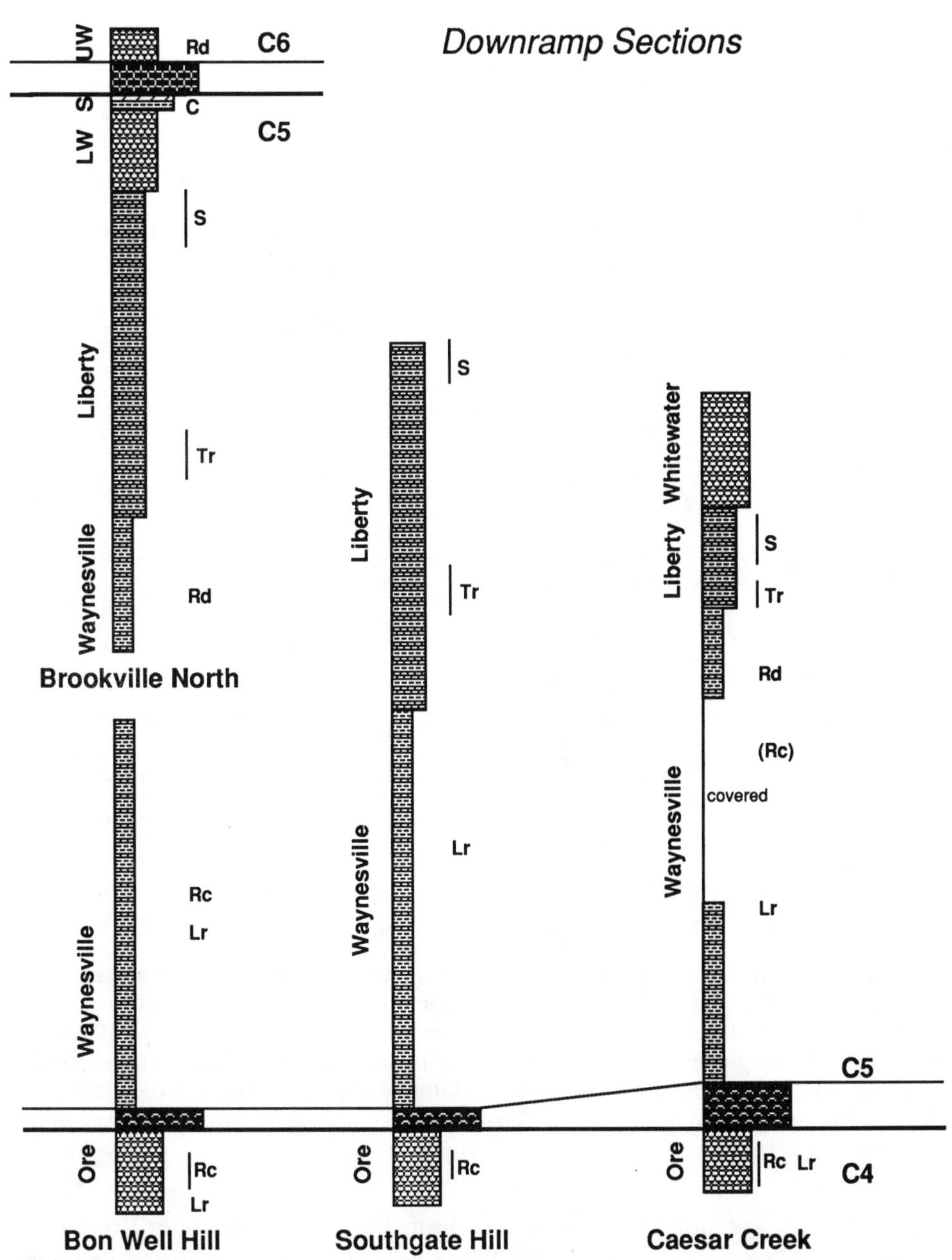

FIGURE 12.4. Four representative downramp sections with the occurrence of several taxa. The paleoenvironments, scale, and taxonomic abbreviations are given in fig. 12.5. Sp indicates peak development of the *Strophomena planumbona* association. Lr, Rd, Rc, and C indicate horizons with *Leptaena richmondensis*, *Rhynchotrema dentatum*, *Retrorsirostra carleyi*, and the colonial rugosan/tabulate association, respectively. Lithostratigraphic abbreviations: Ore (Oregonia), LW (Lower Whitewater), and S (Saluda).

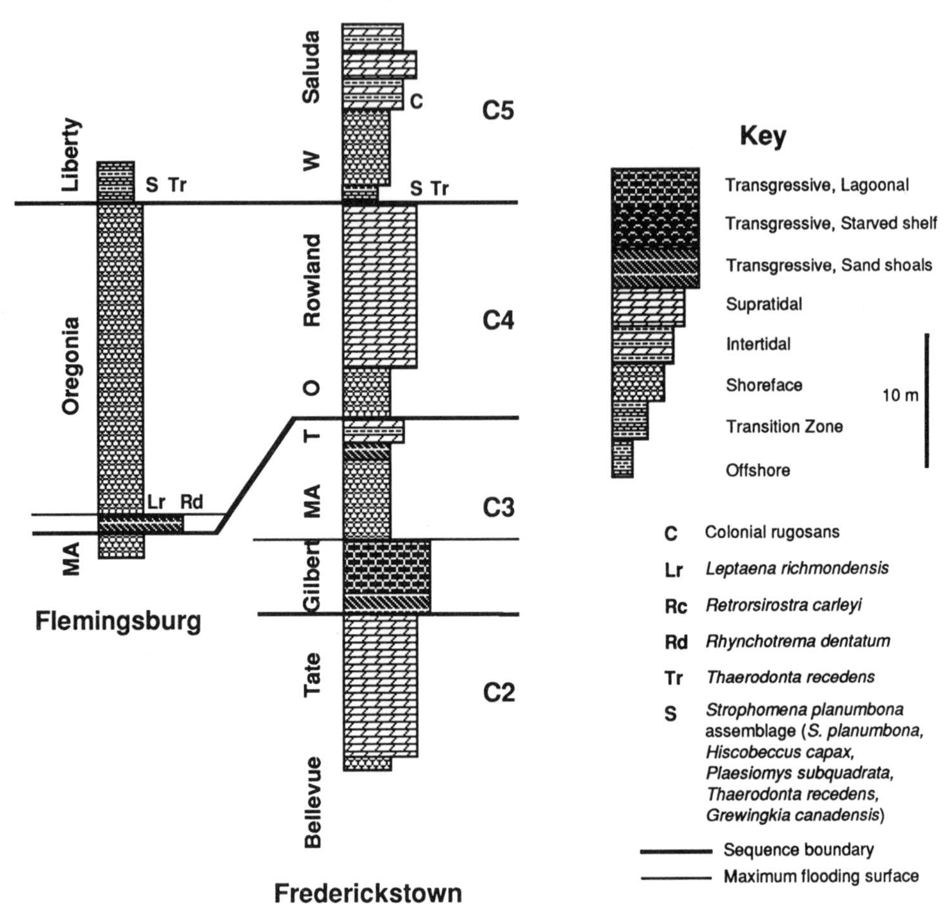

FIGURE 12.5. Two representative upramp sections with the occurrence of several taxa, as illustrated in fig. 12.4. Note that in the Frederickstown section, the more upramp of these two sections, the C4 sequence is thinner and contains a shallower set of lithofacies than in the Flemingsburg section. Note also that the maximum flooding surface and sequence boundary may merge in some places. Lithostratigraphic abbreviations: MA (Mount Auburn [= Stingy Creek]), T (Terrill), O (Oregonia [= Reba]), L (Liberty), and W (Whitewater [= Bardstown]).

facies of the C4 sequence in Ohio and Indiana. At Flemingsburg, in central Kentucky, these two occur immediately above the transgressive ravinement deposits at the base of the C4 sequence, indicating that their occurrence in Ohio and Indiana above the base of the C4 sequence is an artifact of facies control; their true first occurrence is at the beginning of the C4 sequence (fig. 12.5). Although I could not verify this for this study, these two species are reported from C3 strata in a single outcrop in Kentucky (Weir et al. 1984), within a few meters of the inferred position of the sequence boundary.

These three examples are representative of nearly all common taxa. Although the first occurrences of the rarer taxa need to be investigated, the pattern indicates that the invasion came at about the same time as the transgression, instead of following it. However, the timing of the invasion cannot be inferred from a single outcrop or even several paleogeographically similar outcrops, especially the downramp outcrops. The reason is that facies control and a sequence architecture delay such first occurrences, which then are gradually spaced throughout any given section. Therefore the timing of the Richmondian Invasion as a single episode of faunal introductions rather than as a prolonged succession of introductions is apparent only by examining a widely spaced series of outcrops perpendicular to the paleogeographic strike. This time/environment approach confirms that this is indeed a faunal event, despite its gradual nature in single outcrops. This approach also shows the limits to what can actually be inferred from single outcrops in regard to global or regional bioevents such as migration events and mass extinctions (Surlyk and Johansen 1984; Ward et al. 1986). Facies control of taxa and sequence architecture contribute to both the Lazarus effect, in which taxa disappear near a boundary and reappear further upsection (Jablonski 1986), and the Signor/Lipps effect, in which a sudden bioevent appears gradual because of incomplete sampling (Signor and Lipps 1982).

Shifting Habitat Preferences

In most cases in this study, the genera and higher taxa faithfully tracked the environment (table 12.3). Indeed, they did this so consistently that characteristic assemblages could be named and used to interpret paleoenvironments. Furthermore, these assemblages can often be recognized in the same environments away from the Cincinnati Arch. For example, shoreface environments are dominated by a mixture of *Platystrophia, Hebertella,* and *Orthorhynchula* not only in the Cincinnati Series but also in the Reedsville Formation of the Appalachians (Bretsky 1970; Holland 1990; Springer and Bambach 1985). *Dalmanella* and related forms are consistently the hallmark of offshore facies in the Cincinnati Series and the Martinsburg/Reedsville/Trenton complex of the central and southern Appalachians (Bretsky 1970; Holland 1990; Springer and Bambach 1985) and in the Utica Shale of New York (Cisne and Rabe 1978). The pervasiveness of habitat stasis in time and space is reflected in depth-parallel communities (Walker and Laporte 1970), benthic assemblages (Boucot 1981), and dual biostratigraphy (Ludvigsen et al. 1986).

In contrast, some of the taxa in this study do not track environments; their occurrences are facies independent. Three brachiopod taxa, the genus *Rhynchotrema,* the genus *Retrorsirostra,* and the the subfamily Sowerbyellinae all exhibit this lack of environmental fidelity.

Retrorsirostra carleyi first appears in a thin zone within the shoreface facies at the top of the C4 sequence (fig. 12.4; Austin 1927; Caster et al. 1955; Hay et al. 1981). It appears sparingly (as the subspecies *R. carleyi insolens*) a second time in a thin zone within the offshore facies of the C5 sequence (fig. 12.4; Austin 1927). Because it is comparatively rare and limited to thin stratigraphic intervals (especially in the offshore occurrence), it is difficult to prove its absence in correlative strata on the Cincinnati Arch. Nonetheless, *Retrorsirostra carleyi* appears to occur first in a shoreface environment and later in an offshore environment (fig. 12.6).

Rhynchotrema dentatum can be found in three stratigraphically separated zones within the C4, C5, and C6 sequences (figs. 12.4 and 12.5). It occurs abundantly but locally in the C4 shoreface facies of Kentucky (fig. 12.5; Foerste 1905), and it occurs sparsely within a thin zone of the C5 offshore facies (fig. 12.4; Austin 1927) and more abundantly within the C6 shoreface facies (fig. 12.4; Hay et al. 1981). *Rhynchotrema dentatum* appears to occur first within a shoreface setting, later in an offshore setting, and still later in a shoreface setting (fig. 12.6).

Cincinnatian Sowerbyellinae consist of two closely related genera: *Sowerbyella* and *Thaerodonta*. *Sowerbyella* occurs only in the offshore environment of the C1 sequence (Howe 1972), usually co-occurring with *Dalmanella*. It is not present in the Cincinnati area during the C2, C3, and C4 sequences, presumably because the offshore facies of these sequences is not exposed. In the C5 sequence, *Thaerodonta* appears for the first time, but *Sowerbyella* is absent (Howe 1972). *Thaerodonta* is most numerous in a thin zone within the transition-zone environment—not the offshore environment in which *Sowerbyella* occurs and stratigraphically above the acme zone of *Dalmanella* (fig. 12.4; Fox 1962). These closely related taxa appear first within an offshore setting and later in a transition-zone setting (fig. 12.6).

There could be two reasons for the the changing habitat preferences of these three taxa (fig. 12.7). First, these patterns may represent true shifts in habitat preferences, similar to those expressed by numerous higher taxa over longer periods of time (Droser and Bottjer 1989; Jablonski and Bottjer 1991; Miller 1990; Sepkoski 1991). Higher taxa tend to originate in nearshore environments, then spread offshore, and eventually retreat to offshore or deeper settings (Jablonski and Bottjer 1990).

Alternatively, these apparent habitat shifts may be artificially produced by taxa that recur throughout the section in epiboles that are not facies limited (fig. 12.7). Brett et al. (1990) distinguished three types of epiboles: *Ecologic epiboles* are faunal bursts in response to unusual environmental conditions. *Taphonomic epiboles* form under extraordinary burial or diagenetic conditions. *Incursion epiboles* are associated with the first occurrence of a taxon in a region, often in tandem with the transgression and opening of barriers. If they represent epiboles, *Retrorsirostra* and *Rhynchotrema* would be ecological epiboles because peculiar preservation would not be required for these thick-shelled brachiopods, and their occurrences are not directly associated with transgressions. If these taxa are responding to some unpreserved environmental fluctuation, they

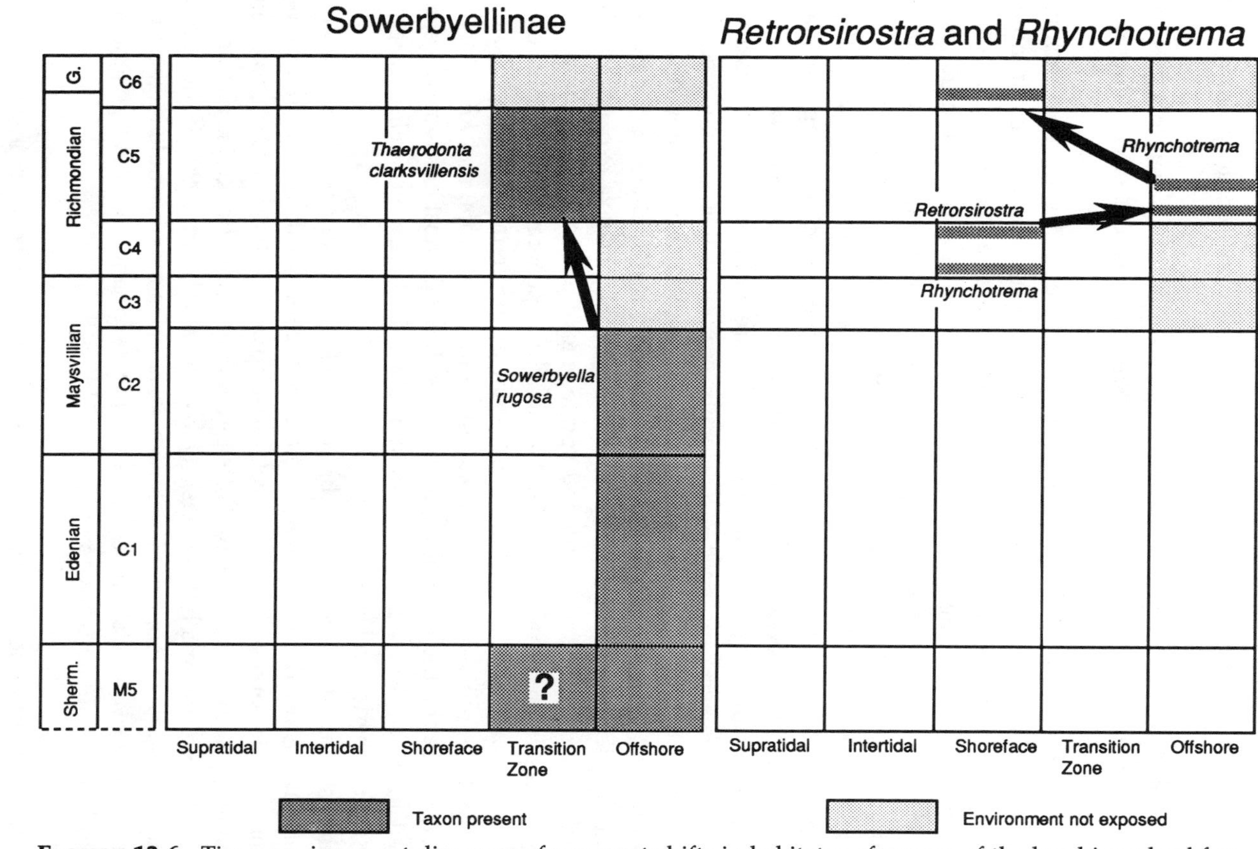

FIGURE 12.6. Time–environment diagrams of apparent shifts in habitat preferences of the brachiopod subfamily Sowerbyellinae and genera *Retrorsirostra* and *Rhynchotrema*.

might appear to occur several times in a section and in different facies. And if they are relatively rare, they might be missed during sampling, thereby further distorting their environmental distribution. However, none of these brachiopods were found in correlative strata of other environments during this study, and none was reported in the USGS/KGS mapping in Kentucky (Weir et al. 1984).

These brachiopods may still represent epiboles, but they were not environmentally widespread during any single epibole. Their record of changing habitat preferences thus is probably not an artifact.

The Sowerbyellinae do not occur as ecological epiboles in the Cincinnati Series. Rather, they are abundant whenever they occur, and their environmental distribution has been well established. In the Cincinnati Series, the Sowerbyellinae appear to shift their habitat preferences from offshore to the transition zone. However, other studies suggest that the Sowerbyellinae have a broad environmental distribution, ranging from shoreface to offshore and deeper (Cisne and Rabe 1978; Mitchell and Sweet 1989; Titus 1986; Patzkowsky 1995). If this is true, the pattern of habitat shifts in the Sowerbyellinae might be more complex, possibly involving reversing shifts, as seen in *Rhynchotrema*.

Although higher taxa generally show offshore movements (Jablonski and Bottjer 1990), generic and species-level shifts are not as well documented. The driving force behind offshore movements in higher taxa are not clear (see Sepkoski 1991): Are offshore expansions driven by the origination of new genera and species in offshore environments, or do higher taxa expand offshore because their component genera and species move offshore? Even though the sample sizes in my study were far too small for generalization, I still could detect no clear trend of offshore movement at lower taxonomic levels. The Sowerbyellinae move onshore; *Retrorsirostra* moves offshore; and *Rhynchotrema* alternates between the shoreface and offshore. Indeed, most brachiopod genera of the Cincinnati Series show environmental stasis, hence our ability to recognize stable generic-level assemblages.

Other studies at the generic level have also failed to find consistent offshore movements. After the Cambrian biomere extinctions, trilobites tend to move onshore at the genus and species level (Westrop and Ludvigsen 1987). Salenioid echinoderms show no preferential offshore movement at the genus and species level, and tend towards habitat stasis (Jablonski and Smith 1990). If the current scarcity of generic and species level offshore shifts is real, it would suggest that higher taxa moved offshore because of the origination of new genera and species in more offshore environments.

Sequence stratigraphic studies provide a necessary framework for paleocommunity studies by establishing a correlation network that is largely independent of biozones and by interpreting paleoenvironments. A sequence stratigraphic study creates the basic time/environment matrix needed for evolutionary studies at the basin level. This matrix is an effective tool for recognizing and

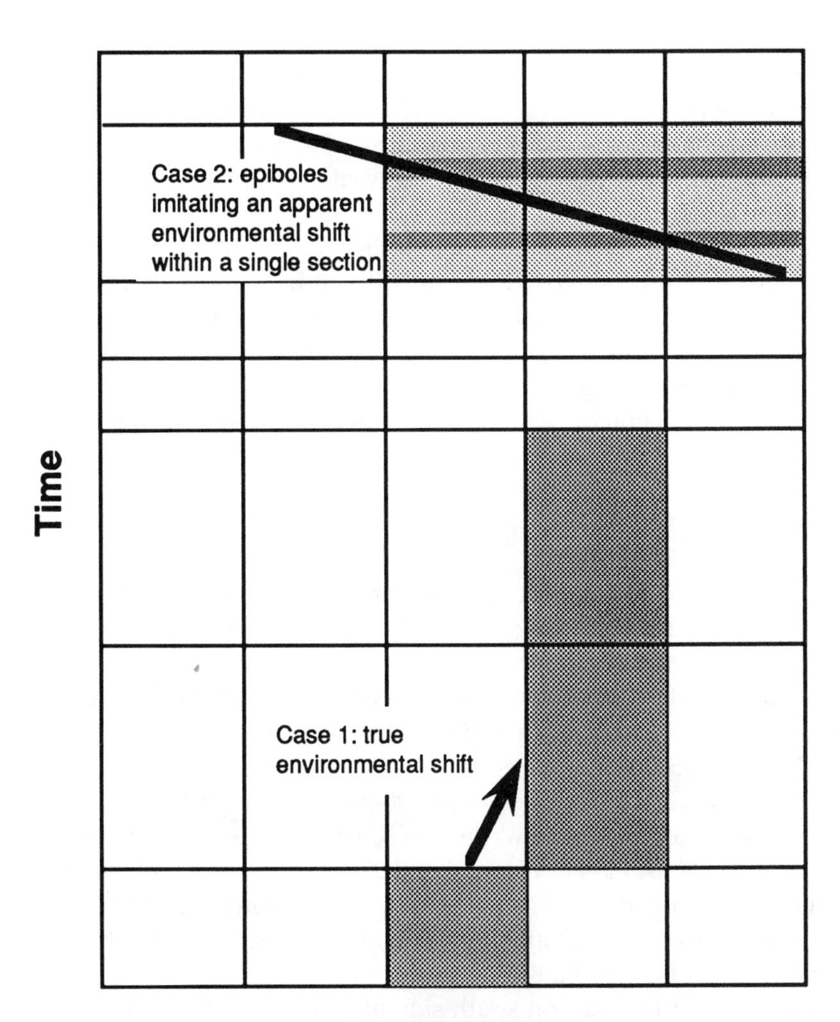

FIGURE 12.7. Two possible origins of apparent habitat shifts. In case 1, the environmental preferences of a taxon change over evolutionary time. In case 2, an incomplete sampling of a rare but widespread taxon may give the illusion of a habitat shift if the taxon becomes locally abundant several times. The heavy diagonal line indicates the path of a single outcrop through "time–environment space." In this case, a shallowing-up section is preserved. The lighter pattern indicates a very low abundance of the taxon, and the darker patterns are times of greater abundance. If a taxon is always difficult to find, even during times of its "greater abundance," the taxon might be encountered several times in a single section. If those horizons cannot be traced to other outcrops, the taxon might appear to have shifted its habitat preference when in fact its appearances mark low-abundance epiboles.

characterizing evolutionary patterns, including faunal migrations, environmental shifts of taxa, and epiboles.

The Richmondian Invasion in the Cincinnati Series was a large-scale faunal immigration episode that occurred at the same time as initial transgression. The invasion can appear gradual in individual outcrops because of facies control of the fauna, but regional studies perpendicular to the paleogeographic strike demonstrate that the invasion was probably more of a limited episode than a prolonged migration.

Time/environment analysis documents the movements of taxa in time and space, and it enables the recognition of shifts in habitat preferences. Three of the brachiopod taxa in this study do not show habitat stasis, in contrast to the majority of genera that I examined. No clear pattern of offshore shifts was present at the generic level, suggesting that higher taxa moved offshore by means of the production of genera and species in offshore environments.

Appendix: Locality Information

Bon Well Hill: Roadcut at the intersection of Indiana State Route 101 and Brookville Dam Road, 1.0 mile north of Brookville, Indiana. Whitcomb, IN quadrangle. 39 26′ 13″ N, 84 58′ 41″W.

Brookville North: Two roadcuts on opposite sides of valley along Indiana State Route 101, 5.0 miles north of Brookville, Indiana. Whitcomb, IN quad. 39 29′ 15″ N, 85 56′ 55″ W.

Caesar Creek: Exposures in emergency spillway along Clarksville Road and hillside cuts immediately below dam of Caesar Creek Lake. Oregonia, OH quad. 39 29′ 04″ N, 84 02′ 49″ W (dam) and 39 48′ 26″ N, 84 03′ 27″ W (spillway).

Flemingsburg: Roadcuts along Kentucky State Route 32, 1.0 to 1.2 miles south of Kentucky State Route 11 west of Flemingsburg, Kentucky. Flemingsburg, KY quad. 38 15′ N, 83 30′ W.

Frederickstown: Roadcut on south side of U.S. Route 150, 0.5 miles west of Nelson/Washington County line. Maud, KY quad. 37 46′ 09″ N, 85 21′ 25″ W.

South Gate Hill: Roadcut on Indiana State Route 1, 1.9 miles south of intersection with U.S. Route 52 at Cedar Grove, Indiana. 39 21′ 30″ N, 84 57′ 17″ W.

ACKNOWLEDGMENTS

I thank Carlton Brett for inviting me to the symposium and for including my essay in this volume. I thank Mark Patzkowsky and David Jablonski for their discussions of several aspects of time/environment studies. Carlton Brett, Mark Patzkowsky, David Lehman, and an anonymous reviewer offered many helpful comments on the manuscript.

This work was supported by a National Science Foundation Graduate Fellowship and by grants from the Geological Society of America Grant and Sigma Xi.

References

Alberstadt, L. P. 1979. The brachiopod genus *Platystrophia*. *U.S. Geological Survey Professional Paper* No. 1066–B.
Alexander, R. R. 1975. Phenotypic lability of the brachiopod *Rafinesquina alternata* (Ordovician) and its correlation with the sedimentologic regime. *Journal of Paleontology* 49:607–18.
Anstey, R. L. and M. L. Fowler. 1969. Lithostratigraphy and depositional environment of the Eden Shale (Ordovician) in the tri-state area of Indiana, Kentucky, and Ohio. *Journal of Geology* 77:668–82.
Austin, G. M. 1927. Richmond faunal zones in Warren and Clinton counties, Ohio. *Proceedings of the U.S. National Museum* 70:3–18.
Bergström, S. M. and C. E. Mitchell. 1991. Trans-Pacific graptolite faunal relations: The biostratigraphic position of the base of the Cincinnatian Series (Upper Ordovician) in the standard Australian graptolite zone succession. *Journal of Paleontology* 64:992–97.
Boucot, A. J. 1981. *Principles of Benthic Marine Paleoecology*. New York: Academic Press.
Bretsky, P. W. 1970. *Late Ordovician Benthic Marine Communities in North-Central New York*. Albany: State University of New York, State Education Department.
Brett, C. E., K. B. Miller, and G. C. Baird. 1990. A temporal hierarchy of paleoecologic processes within a middle Devonian epeiric sea. In W. M. Miller III, ed., *Paleocommunity Temporal Dynamics: The Long Term Development of Multispecies Assemblies*, pp. 178–209. Knoxville, Tenn.: Paleontological Society Special Publication No. 5.
Brookfield, M. E. 1988. A mid-Ordovician temperate carbonate shelf—The Black River and Trenton Limestone Groups of southern Ontario, Canada. *Sedimentary Geology* 60:137–54.
Browne, R. G. 1964. The coral horizons and stratigraphy of the Upper Richmond group in Kentucky west of the Cincinnati Arch. *Journal of Paleontology* 38:385–92.
———. 1965. Some Upper Cincinnatian (Ordovician) colonial corals of north-central Kentucky. *Journal of Paleontology* 39:1177–91.
Caster, K. E., E. A. Dalvé, and J. K. Pope. 1955. *Elementary Guide to the Fossils and Strata of the Ordovician in the Vicinity of Cincinnati Ohio*. Cincinnati: Cincinnati Museum of Natural History.
Cisne, J. L. and B. D. Rabe. 1978. Coenocorrelation: Gradient analysis of fossil communities and its applications stratigraphy. *Lethaia* 11:341–64.
Cooper, G. A. 1956. *Chazyan and Related Brachiopods*. Washington, DC: Smithsonian Institution.
Cressman, E. R. 1973. Lithostratigraphy and depositional environments of the Lexington Limestone (Ordovician) of central Kentucky. U.S. Geological Survey Professional Paper No. 768.
Cressman, E. R. and M. C. Noger. 1976. Tidal-flat carbonate environments in the High Bridge Group (Middle Ordovician) of central Kentucky. *Kentucky Geological Survey Report of Investigations* 18:1–15.
Dalvé, E. 1948. *The Fossil Fauna of the Ordovician in the Cincinnati Region*. Cincinnati: University Museum, Department of Geology and Geography.
Droser, M. L. and D. J. Bottjer. 1989. Ordovician increase in extent and depth of bioturbation: Implications for understanding early Paleozoic eco-space utilization. *Geology* 17:850–52.
Elias, R. J. 1983. Middle and Late Ordovician solitary rugose corals of the Cincinnati Arch region. *U.S. Geological Survey Professional Paper* No. 1066–N.
———. 1985. Solitary rugose corals of the Upper Ordovician Montoya Grolup, southern New Mexico and westernmost Texas. *Paleontological Society Memoir* No. 16.

Flower, R. H. 1946. Ordovician cephalopods of the Cincinnati region, part 1. *Bulletins of American Paleontology* 29:1–656.
Foerste, A. F. 1905. Note on the distribution of brachiopoda in the Arnheim and Waynesville beds. *American Geologist* 36:244–50.
Fox, W. T. 1962. Stratigraphy and paleoecology of the Richmond Group in southeastern Indiana. *Geological Society of America Bulletin* 73:621–42.
———. 1968. Quantitative paleoecologic analysis of fossil communities in the Richmond Group. *Journal of Geology* 76:613–41.
Frey, R. C. 1985. A well preserved specimen of *Schuchertoceras* (Cephalopoda, Ascocerida) from the Upper Ordovician (basal Richmondian) of southwest Ohio. *Journal of Paleontology* 59:1506–11.
———. 1987a. The occurrence of pelecypods in Early Paleozoic epeiric-sea environments, Late Ordovician of the Cincinnati, Ohio area. *Palaios* 2:3–23.
———. 1987b. The paleoecology of a Late Ordovician shale unit from southwest Ohio and southeast Indiana. *Journal of Paleontology* 61:242–67.
Haq, B. U., J. Hardenbol, and P. R. Vail. 1987. Chronology of fluctuating sea levels since the Triassic. *Science* 235:1156–67.
Hay, H. B. 1981. Lithofacies and formations of the Cincinnatian Series (Upper Ordovician), southeastern Indiana and southwestern Ohio. Ph.D. diss., Miami University.
Hay, H. B., J. K. Pope, and R. C. Frey. 1981. Lithostratigraphy, cyclic sedimentation, and paleoecology of the Cincinnatian Series in southwestern Ohio and southeastern Indiana. In T. G. Roberts, ed., *GSA Cincinnati '81 Field Trip Guidebooks. Vol. 1: Stratigraphy, Sedimentology*, pp. 73–86. Falls Church, Va. American Geological Institute.
Holland, S. M. 1990. Distinguishing eustasy and tectonics in foreland basin sequences: The Upper Ordovician of the Cincinnati Arch and Appalachian Basin. Ph.D. diss., University of Chicago.
———. 1993. Sequence stratigraphy of a carbonate–clastic ramp: The Cincinnati Series (Upper Ordovician) in its type area. *Geological Society of America Bulletin* 105:306–22.
Holland, S. M., B. F. Dattilo, A. I. Miller, D. L. Meyer, and S. C. Diekmeyer. 1993. Anatomy of a mixed carbonate–clastic depositional sequence. *Geological Society of America Abstracts with Programs* 25:A338.
Howe, H. J. 1972. Morphology of the brachiopod genus *Thaerodonta*. *Journal of Paleontology* 46:440–46.
———. 1979. Middle and late Ordovician plectambonitacean, rhynchonellacean, syntrophiacean, trimerellacean, and atrypacean brachiopods. *U.S. Geological Survey Professional Paper* No. 1066–C.
———. 1988. Articulate brachiopods from the Richmondian of Tennessee. *Journal of Paleontology* 62:204–18.
Jablonski, D. 1986. Causes and consequences of mass extinctions: A comparative approach. In D. K. Elliot, ed., *Dynamics of Extinction*, pp. 183–229. New York: Wiley.
Jablonski, D. and D. J. Bottjer. 1990. The origin and diversification of major groups: Environmental patterns and macroevolutionary lags. In P. D. Taylor and G. P. Larwood, eds., *Major Evolutionary Radiations*, pp. 17–57. Oxford: Clarendon Press, Systematics Association Special Volume No. 42.
———. 1991. Onshore–offshore trends in marine invertebrate evolution. In R. M. Ross and W. D. Allmon, eds., *Causes of Evolution: A Paleontologic Perspective*, pp. 21–75. Chicago: University of Chicago Press.
Jablonski, D. and A. B. Smith. 1990. Ecology and phylogeny: Environmental patterns in the evolution of the echinoid order Salenioidia. *Geological Society of America Abstracts with Program* 22:A266.
Lash, G. G. 1987. Geodynamic evolution of the Lower Paleozoic central Appalachian foreland basin. In C. Beaumont and A. J. Tankard, eds., *Sedimentary Basins and Basin-*

Forming Mechanisms, pp. 413–23. Calgary, Alberta: *Canadian Society of Petroleum Geologists Memoir* No. 12.
Ludvigsen, R. 1978. Middle Ordovician trilobite biofacies, southern Mackenzie Mountains. In C. R. Stelck and B. D. E. Chatterton, eds., *Western and Canadian Biostratigraphy*, pp. 1–33. *Geological Association of Canada Special Paper* No. 18.
Ludvigsen, R., S. R. Westrop, B. R. Pratt, P. A. Tuffnell, and G. A. Young. 1986. Dual biostratigraphy: Zones and biofacies. *Geoscience Canada* 13:139–54.
Macomber, R. W. 1970. Articulate brachiopods from the Upper Bighorn Formation (Late Ordovician) of Wyoming. *Journal of Paleontology* 44:416–50.
McEwan, E. D. 1919. A study of the brachiopod genus *Platystrophia*. *U.S. National Museum Proceedings* 56:383–448.
Meyer, D. L. 1990. Population paleoecology and comparative taphonomy of two edrioasteroid (echinodermata) pavements: Upper Ordovician of Kentucky and Ohio. *Historical Biology* 4:155–78.
Miller, A. I. 1988. Spatio-temporal transitions in Paleozoic bivalvia: An analysis of North American fossil assemblages. *Historical Biology* 1:251–73.
———. 1990. The relationship between global diversification and spatio-temporal transitions in Paleozoic bivalvia. In W. Miller III, ed., *Paleocommunity Temporal Dynamics: The Long-Term Development of Multispecies Assemblages*, pp. 85–98. Knoxville, Tenn.: *Paleontological Society Special Publication* No. 5.
Mitchell, C. E. and S. M. Bergström. 1991. New graptolite and lithostratigraphic evidence from the Cincinnati region, U.S.A., for the definition and correlation of the base of the Cincinnati Series (Upper Ordovician). *Geological Survey of Canada Paper* No. 90–9.
Mitchell, C. E. and W. C. Sweet. 1989. Upper Ordovician conodonts, brachiopods, and chronostratigraphy of the Whittaker Formation, southwestern District of Mackenzie, N.W.T., Canada. *Canadian Journal of Earth Sciences* 26:74–87.
Nelson, C. S. 1988. An introductory perspective on non-tropical shelf carbonates. *Sedimentary Geology* 60:3–14.
Patzkowsky, M. E. 1995. Gradient analysis of Middle Ordovician brachiopod biofacies: Biostratigraphic, biogeographic and macroevolutionary implications. *Palaios* 10:154–79.
Pojeta Jr., J. 1971. Review of Ordovician pelecypods. *U.S. Geological Survey Professional Paper* No. 695.
Rudwick, M. J. S. 1970. *Living and Fossil Brachiopods*. London: Hutchinson University Library.
Schumacher, G. A. and W. I. Ausich. 1983. New Upper Ordovician echinoderm site: Bull Fork Formation, Caesar Creek Reservoir (Warren County, Ohio). *Ohio Journal of Science* 83:60–64.
Scotese, C. R. and W. S. McKerrow. 1991. Ordovician plate tectonic reconstructions. *Geological Survey of Canada Paper* No. 90–9.
Sepkoski, J. J. 1991. A model of onshore–offshore change in faunal diversity. *Paleobiology* 17:58–77.
Signor, P. W. and J. H. Lipps. 1982. Sampling bias, gradual extinction patterns, and catastrophes in the fossil record. *Geological Society of America Special Paper* No. 190.
Sloan, R. E. 1987a. Black River/Trenton extinction, paleooceanography and chronology of the Middle and Late Ordovician of the Upper Mississippi Valley. *Geological Society of America Abstracts with Program* 19:246.
———. 1987b. Introduction to the Middle and Late Ordovician field trips. In N. H. Balaban, ed., *Field Trip Guidebook for the Upper Mississippi Valley, Minnesota, Iowa, and Wisconsin*, pp. 45–52. *Minnesota Geological Survey Guidebook Series* No. 15.
Springer, D. A. and R. K. Bambach. 1985. Gradient versus cluster analysis of fossil as-

semblages: A comparison from the Ordovician of southwestern Virginia. *Lethaia* 18:181–98.

Surlyk, F. and M. B. Johansen. 1984. End-Cretaceous brachiopod extinctions in the chalk of Denmark. *Science* 223:1174–77.

Sweet, W. C. 1984. Graphic correlation of upper Middle and Upper Ordovician rocks, North American Midcontinent Province, U.S.A. In D. L. Bruton, ed., *Aspects of the Ordovician System*, pp. 23–35. Oslo: University of Oslo Press.

———. 1988. Mohawkian and Cincinnatian chronostratigraphy. *New York State Museum Bulletin* 462: 84–90.

Sweet, W. C. and A. K. Miller. 1958. Ordovician cephalopods from Cornwallis and Little Cornwallis Islands, District of Franklin, Northwest Territories. *Geological Survey of Canada Bulletin* 38:1–86.

Thayer, C. W. 1975. Morphologic adaptations of benthic invertebrates to soft substrata. *Journal of Marine Research* 33:177–89.

Titus, R. 1986. Fossil communities of the Upper Trenton Group (Ordovician) of New York State. *Journal of Paleontology* 60:805–24.

Tobin, R. C. 1982. A model for cyclic deposition in the Cincinnatian Series of southwestern Ohio, northern Kentucky and southeastern Indiana. Ph.D. diss., University of Cincinnati.

———. 1986. An assessment of the lithostratigraphic and interpretive value of the traditional "biostratigraphy" of the type Upper Ordovician of North America. *American Journal of Science* 286:673–701.

Van Wagoner, J. C., H. W. Posamentier, R. M. Mitchum, P. R. Vail, J. F. Sarg, T. S. Loutit, and J. Hardenbol. 1988. An overview of the fundamentals of sequence stratigraphy and key definitions. *Society of Economic Paleontologists and Mineralogists Special Publication* No. 42.

Vermeij, G. 1978. *Biogeography and Adaptation.* Cambridge, MA: Harvard University Press.

Walker, K. R. and L. F. Laporte. 1970. Congruent fossils communities from Ordovician and Devonian carbonates of New York. *Journal of Paleontology* 44:928–44.

Walker, L. G. 1982. The brachiopod genera *Hebertella, Dalmanella,* and *Heterorthina* from the Ordovician of Kentucky. U.S. Geological Survey Professional Paper No. 1066–M.

Ward, P., J. Wiedmann and J. F. Mount. 1986. Maastrichtian molluscan biostratigraphy and extinction patterns in a Cretaceous/Tertiary boundary section exposed at Zumaya, Spain. *Geology* 14:899–903.

Weir, G. W., W. L. Peterson, W. C. Swadley, and J. Pojeta Jr. 1984. Lithostratigraphy of Upper Ordovician strata exposed in Kentucky. *U.S. Geological Survey Professional Paper* No. 1151–E.

Westrop, S. R. and R. Ludvigsen. 1987. Biogeographic control of trilobite mass extinction at an Upper Cambrian "biomere" boundary. *Paleobiology* 13:84–99.

13

The Utility of Epiboles in the Regional Correlation of Paleozoic Epeiric Sea Strata: An Example from the Upper Ordovician of Ohio and Indiana

Robert C. Frey

ABSTRACT

The *Onniella meeki* epibole consists of 1.5–2.5 m of interbedded shale and thin planar packstone beds composed almost entirely of the valves of the small orthid brachiopod *Onniella meeki* (Miller). These storm-deposited limestone beds can be traced on outcrop from Clinton County, Ohio, in the east, 135 km west into western Franklin County, Indiana. The thickness varies from one outcrop to another, but the zone retains its distinctive lithologic, biostratinomic, and faunal identity throughout the outcrop belt.

The *O. meeki* epibole is a significant marker horizon within the Upper Ordovician section exposed in the area. There is a major influx of new taxa into the Cincinnatian region within 10 m of the base of the epibole. This event marks a gradual change from primarily mudstone to increasingly more calcareous substrates across the Cincinnati Shelf in early Richmondian time. The faunas in these more calcareous strata are dominated by a diverse set of articulate brachiopods, trepostome bryozoans, and the coral *Grewingkia*. This is the classic "Richmond fauna" derived from older Red River and Maquoketa faunas characteristic of carbonate facies to the north and west of the Ohio valley.

Causal factors leading to the development of the epibole center on the development of unique and widespread environmental conditions across the Cincinnati Shelf, conditions resulting from the waning of a period of clastic influx into the Cincinnati Shelf that was associated with a regional transgressive event. In these environments sedimentation rates were too high and substrates were too muddy for most other epifaunal filter feeders, but they were optimal for *Onniella* and, to a lesser extent, the plectambonitid *Thaerodonta*. These conditions resulted in the explosive proliferation of *O. meeki* populations. With the increase in skeletal substrates and the influx of a new, diverse benthic fauna adapted to these substrates, there is a gradual but marked decline in the abundance of *O. meeki*, with the subsequent extinction of the species below the top of the Waynesville Formation.

Other studies of the Ordovician of the Cincinnati Arch region show a similar development of dalmanellid-dominated facies, marking transgressive maxima in the late Middle Ordovician (Kirkfieldian), the early Late Ordovician (Edenian), and the middle Late Ordovician (early Richmondian). This suggests that the development of such beds can be used to mark these transgressive events regionally across the Cincinnati Arch outcrop belt and possibly elsewhere in Middle to Late Ordovician strata across eastern North America.

The *Onniella meeki* epibole differs in some respects from proposed epibole models that describe epiboles as consisting of an exceptional abundance of normally rare taxa that often crosscut facies. *O. meeki* is a common species outside the epibole, and it appears that the *O. meeki* epibole is distributed along depositional strike, restricted to the Waynesville facies in southeastern Indiana and southwestern Ohio. This indicates there are distinct geographic limits to the utility of this zone and similar event beds in regional correlation. We need further study of such epiboles in the Ordovician of the Cincinnati Arch region and in other similar cratonic shelf sequences to define more precisely the various types of epiboles, their limitations, and their utility in detailed regional stratigraphic correlation.

Epiboles, peak zones, and *acme zones* are all terms that have been used to describe biostratigraphic units marked by the exceptional abundance of a particular fossil taxon. The epibole was the first special type of biostratigraphic zone to be described (Trueman 1923), exclusive of the concept of a range zone, as proposed by Oppel (1856). Arkell (1933) elaborated further on this type of zone, stating that each species during the course of its evolutionary history reaches an "acme" in its development, marked by the species' maximum abundance and geographic distribution. This acme defined an interval of time, the *hemera*—represented by a thickness of strata—the epibole, and it provided the rationale for using epiboles as time-significant biostratigraphic zones. This concept of an epibole interfaced well with the prevailing "layer-cake" concept of stratigraphy promoted by E. O. Ulrich and other stratigraphers working in cratonic and Appalachian North America in the early twentieth century.

With the advance of the idea of facies change espoused by A. W. Grabau (1920–21) and the application of these Waltherian principles by Chadwick (1924) and Cooper (1930, 1933) to the interpretation of the classic Devonian strata exposed in New York, this layer-cake approach to stratigraphy was effectively discredited, case by case, by the middle third of this century. With the demise of this principle and the recognition that facies and the physical environments they represent control the distribution and abundance of fossil taxa, epiboles lost favor as time-significant biostratigraphic units. Dunbar and Rogers described the "acme of a form (as being) so largely a matter of ecology that it is of little use in correlation or chronology" (1957:300). Shaw (1964) agreed, noting that the concept of an acme and hemera for each species was not a proven biological fact and that it was difficult to picture any species that would be equally abundant over its entire geographic range simultaneously. Instead the numerical maxima defining these "zones" could represent the diachronous migration of favorable environmental conditions across a geographic region, unconnected local populations, or a mechanical accumulation of dead shells.

Hedberg (1972) described an acme zone as the body of strata representing the acme or maximum development of a given taxon but not representing its total range, as the zone boundaries are rather subjective. Ernst and Seibertz (1977) described peak zones as the periods of a guide fossil's maximum abundance. They claimed that such peak zones might be better for regional correlation than range zones but pointed out that the zones' usefulness in biostratigraphy depends on the population densities of a species increasing simultaneously over a wide geographic area so that they become areally extensive and can be traced out as laterally continuous.

The Code of Stratigraphic Nomenclature (1983) defines an abundance zone as a biozone characterized by the quantitatively distinctive maxima of a relative abundance of one or more taxa. The code notes that the distribution of these zones may reflect strong local ecological control and that such zones are informal.

In the past decade the bulk of the stratigraphic record has been recognized as the sum of periods of nondeposition punctuated by short-term, episodic, sedimentological events (Dott 1983). Ager emphasized the importance of "rare" cataclysmic events in interpreting this record, and he described the rock record as "a lot of holes tied together with sediment," with the associated geologic history consisting of "long periods of boredom and brief periods of terror" (1981:35,107).

Stratigraphic studies by the Tubingen group (Einsele and Seilacher 1982; Aigner 1985) and Brett et al. (1986) have demonstrated that these events can be traced throughout areally extensive Paleozoic and Mesozoic epeiric sea sequences, enabling a previously unrealized and detailed regional "microstratigraphic" correlation of these strata. Brett and Baird (1989 and essay 10 of this volume) found that widespread, stratigraphically restricted paleontological events, including epiboles, are likewise characteristic of Phanerozoic epeiric sea strata, with fossil horizons proving to be exceptionally useful for detailed stratigraphic correlation.

Brett and Baird (1989) described epiboles as thin intervals of strata that display an unusual abundance of normally rare taxa that often crosscut facies within depositional basins. They recognized a number of different types of epiboles, including beds reflecting unusual preservation conditions (taphonomic epiboles), peculiar environments (ecological epiboles), episodic immigration of foreign taxa (incursions), and the abrupt evolutionary appearance and spread of new taxa (saltation epiboles).

This essay reports on a study of the *Onniella meeki* epibole, an "ecological epibole" from the Late Ordovician Waynesville Formation in the Indiana/Ohio area, and discusses its stratigraphic significance.

Stratigraphy

The Upper Ordovician section exposed in southeastern Indiana and southwestern Ohio consists of cyclic packets of fine-grained siliclastic and skeletal carbonate facies (fig. 13.1), 213–305 m thick and usually described as representing alternating "transgressive" and "regressive" conditions across a northward-ramping shelf (Fox 1962; Hay 1981; Tobin 1986; Holland 1993). The shalier units, including the Waynesville Formation, are generally inferred to represent deeper water, lower-energy, mud-bottom environments, with the carbonates indicative of shallower water, higher-energy shoaling conditions. These clastic/carbonate packets may also represent sedimentary as well as eustatic fluctuations, with the clastic units representing pulses of sediment influx into the Cincinnati Shelf from Taconic source areas to the east. The carbonate facies in the area may indicate the development of an autochthonous accumulation of biogenic skeletal debris, undiluted by the influx of these terrigenous sediments.

Studies by Hay (1981) and Tobin (1986) recognized a minimum of 12 distinct lithofacies types arranged into three major shallowing-upward cycles. The basal cycle includes the Kope Formation (70% shale) and the overlying Fairview Formation (50% shale and 50% planar packstone), capped by the Bellevue Member of the Grant Lake Formation (less than 20% shale, nodular rudstone/grainstone). The second cycle is represented by a thinner sequence, beginning with the Corryville Member of the Grant Lake Formation (mostly shale with planar siltstone and packstone), the Mount Auburn Member of the Grant Lake Formation (skeletal wackestone and shale), and the Arnheim Formation (complex facies of basal shale and thin planar limestone capped by nodular limestone). The Richmondian cycle forms the third transgressive/regressive sequence in the area in the Late Ordovician, consisting of the Waynesville Formation (70% shale and 30% thin planar packstone) and the Liberty Formation (50% shale and 50% planar packstone) and culminating in a complex of shallow-water facies. These include the Saluda Formation (dolomicrites), the Whitewater Formation (rubbly argillaceous limestone), and the Drakes Formation (planar

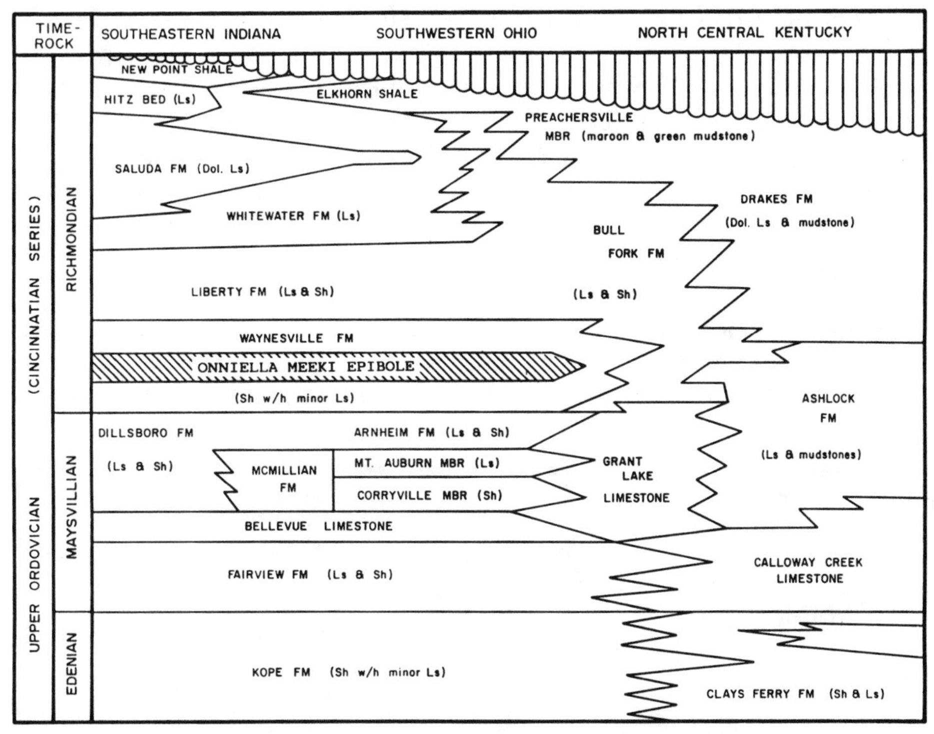

FIGURE 13.1. Lithostratigraphic units within the Upper Ordovician Cincinnatian Series exposed in the Cincinnati, Ohio, area, showing their lateral and vertical facies relationships. The relative thicknesses of these units do not express their true thicknesses. Modified from J. Pojeta, in Weir et al. 1984:fig. 50.

packstone and maroon and green mudstone). This complex is truncated locally by an erosional unconformity that marks the Ordovician/Silurian boundary.

The *Onniella meeki* epibole consists of 1.5 m of interbedded thin shale and thin packstone situated approximately 14–15 m above the base of the Waynesville Formation, in the deeper-water phase of this last transgressive/regressive cycle. The Waynesville Formation can be traced on the outcrop from southern Clinton County, Ohio, west to westernmost Franklin County, Indiana, a distance of 135 km (fig. 13.2). Throughout this outcrop belt, the formation, 32–36 m thick, consists of 70% or more blue clay shale or claystone and 30% or less thin-bedded (less than 10 cm thick), planar, very fossiliferous, skeletal wackestone or packstone. Foerste (1909) subdivided the Waynesville Formation into three divisions, later renamed *members* by Wolford (1930) and Caster, Dalvé, and Pope (1955).

The lower 12-m-thick Fort Ancient member consists of blue fissile shale and claystone (70% to 80%), with subordinate thin, planar packstone and wackestone (less than 30%) that overlies the rubbly, irregularly bedded phosphatic fossiliferous wackestone of the Arnheim Formation (fig. 13.1). This member is capped by the *Treptoceras duseri* or Trilobite Shale unit, 1.5 m of massive-bedded blue claystone. Immediately overlying this claystone unit is the Clarksville member, consisting of 8–9 m of blue shale, with an increase in the number and thickness of thin limestone beds (60% shale and 40% limestone). The basal 1.5 m of this member comprises the *O. meeki* epibole, with the valves of this small orthid brachiopod comprising 80% to 96% of the larger allochems (more than 5 mm) in the thin packstone beds. The top of this member is marked by a 7–10-cm-thick cluster of closely spaced, thin (1–3-cm-thick) brachiopod packstones with a diverse fauna of orthids and atrypaceans.

The upper Blanchester Member, 12 m thick, is somewhat more shaly (65% to 70% shale) than the Clarksville Member but has a greater number of thicker packstone beds compared with the Fort Ancient Member. The base of this

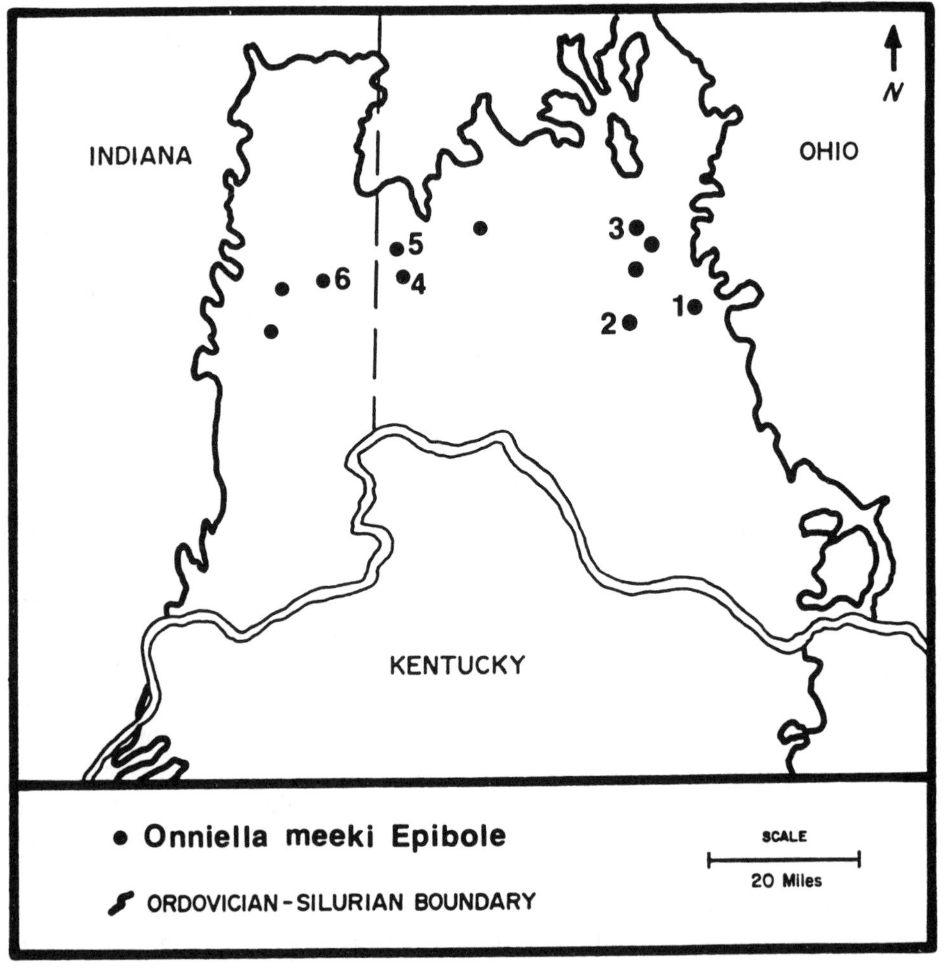

FIGURE 13.2. Map of Indiana/Ohio area showing the locations of exposures of the *Onniella meeki* epibole that were studied. Numbered localities correspond to those localities more intensively studied and sampled for this essay.

member is marked by about 1 m of compact blue claystone. The succeeding strata consist of alternating thin (less than 30 cm thick) blue shale and thin packstone beds. The upper contact with the overlying Liberty Formation is gradational. Shrake et al. (1988) distinguished the contact between the Waynesville and Liberty Formations in Warren County, Ohio, by the increased limestone content of the Liberty Formation (43% compared with 26%) and by the increase in the average thickness of individual limestone beds (10 cm versus less than 5 cm thick).

The Paleogeographic and Paleoenvironmental Setting

The Ordovician strata exposed in the Indiana/Ohio region were deposited on the southeastern portion of a cratonic platform that comprised much of the Laurentian continent in the early Paleozoic. Recent paleogeographic reconstructions of the Ordovician (Scotese and McKerrow 1990) place the Indiana/Ohio area in tropical or subtropical latitudes, 20–30 degrees south of the equator, in the "carbonate belt." For much of this period this region was inundated by a shallow epicontinental seaway, with lithofacies, fossils, and sedimentary structures all indicating intertidal to subtidal environments with maximum water depths of 25–30 m.

To the south the central Kentucky region was the site of nearly continuous, very-shallow-water, carbonate platform deposition throughout most of the Ordovician (Cressman 1973; Weir et al. 1984). Adjacent portions of the northern Kentucky/Indiana/Ohio area formed the shallow northward-ramping margin of this "Bluegrass Platform," gently sloping toward deeper-water basins to the north and west (Cressman 1973). During the Ordovician, several volcanic island-arc systems were present off the eastern margin of Laurentia. These progressively collided with Laurentia throughout the period, culminating in the Middle Ordovician Taconic orogeny, with the formation of a clastic source area along the eastern margin of the continent (Bird and Dewey 1970; Ettensohn 1991).

The vast quantities of detrital sediments eroded off this rising landmass led to the development of the Queenston Deltaic Complex, which spread these clastics north and west into adjacent epeiric sea-shelf areas (Dennison 1976). These clastics, mostly finer silts and clays, began to filter into the Kentucky/Indiana/Ohio area in the later Middle Ordovician (Kirkfieldian), with an increasing influx of these clastics into the area throughout the succeeding Late Ordovician.

As mentioned earlier, these clastic wedges are associated with transgressive maxima and mark the deepest-water facies of these sedimentary packets. Holland (1993) argued that these transgressions were the result of eustatic rise rather than regional subsidence. These eustatic sea-level rises—possibly resulting from the ebb of continental glaciation in North Africa (Denison 1976;

Sheehan 1988)—carried these clastic sediments far out onto the shelf and were responsible for the pulses of sediment influx represented by the shaly Kope, Corryville, and Waynesville facies in the Cincinnatian region (fig. 13.1).

During Late Ordovician Richmondian times, the Cincinnati Shelf area was a gently sloping, muddy epeiric seafloor that graded into somewhat deeper areas

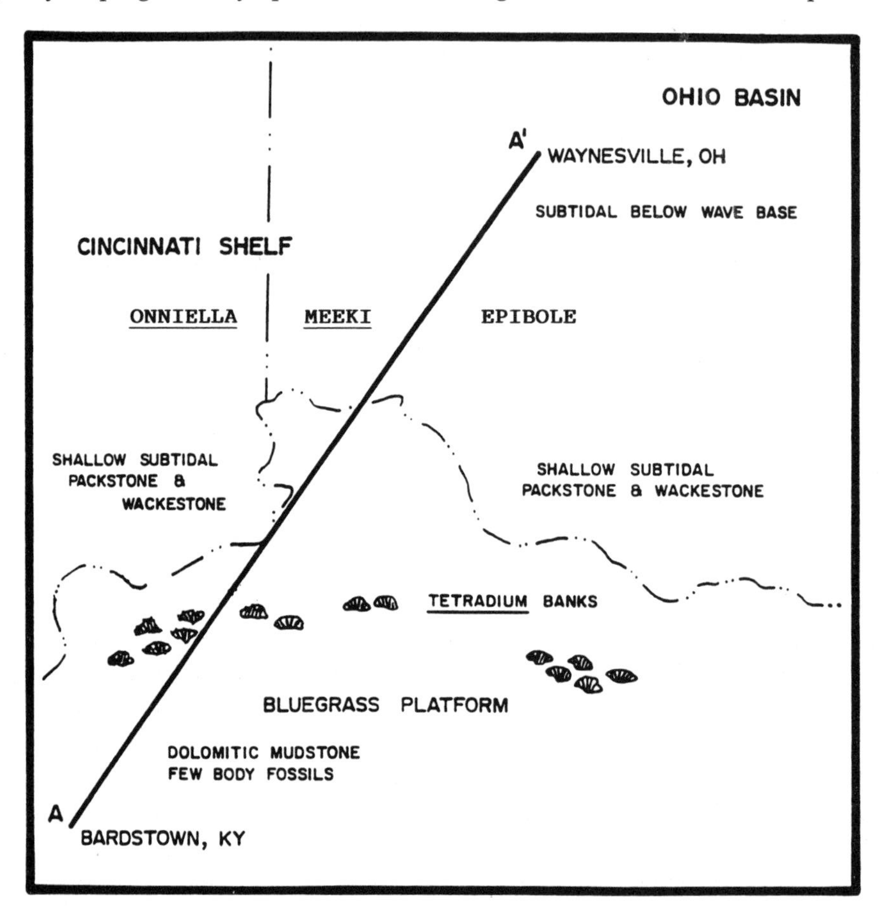

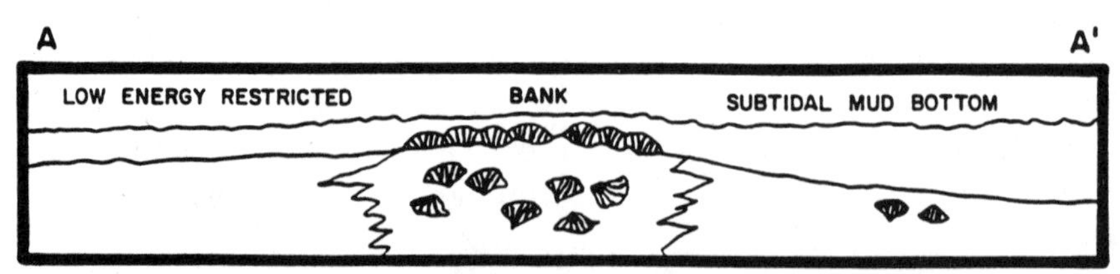

FIGURE 13.3. A reconstruction of local paleoenvironmental conditions in the Indiana/Ohio/Kentucky area during the time the *Onniella meeki* epibole was deposited in the Indiana/Ohio portions of the Cincinnati Shelf, north of the Bluegrass Platform. Cross section A-A′ shows the reconstructed environments from the platform in the south to the gently sloping muddy shelf area to the north. From Hay (1981) and field studies by the writer.)

to the north. This mud-bottom shelf graded southward into shallower-water environments represented by low-energy shale/wackestone facies exposed along the Ohio River in Indiana, Ohio, and northern Kentucky (fig. 13.3). The depositional strike for these facies was from west to east, roughly aligned parallel to the present-day course of the Ohio River.

Weir et al. (1984), citing observations by Bucher (1917) and Anstey and Fowler (1969) of the Kope Formation, described their lithofacies Group 1-D as being deposited in water depths of 20–25 m. The Waynesville Formation in Indiana and Ohio is lithologically similar to the Kope and so would be classified as part of the same lithofacies group. This suggests that the water depths during the deposition of the *Onniella meeki* epibole were similar, 20–25 m in depth, below wave base but within the zone of storm-current reworking. The underlying *T. duseri* Shale indicates periodically high sedimentation rates, with the increase in thin packstone beds in the overlying *O. meeki* epibole reflecting a slowing of sedimentation and stabilization of the substrate, leading to the increased colonization of these bottoms by shelly, filter-feeding epifaunal benthos.

The Occurrence of Onniella Meeki

Onniella meeki is a small (less than 20 mm wide), strophic, ventribiconvex, costellate orthid brachiopod with a weakly developed pedicle fold and brachial sulcus. Its shell microstructure is punctate. The adult of the species was equipped with a functional pedicle, as indicated by its open pedicle notch. Specimens preserved in place in limestone beds show that this species lived with its beak down and the commissure perpendicular to the substrate surface (Richards 1970). *O. meeki* has a deltidiodont hinge.

Alexander (1990) suggested that this comparatively weak type of hingement, coupled with this species' erect life position, contributed to the occurrence of *O. meeki* in fossil assemblages as primarily disarticulated specimens. His study of a suite of specimens from the Waynesville Formation at Weisburg, Indiana, revealed that nearly 85% of the 350 specimens that he examined were disarticulated. Foerste (1909) and Hall (1962) noted that *O. meeki* was restricted in its occurrence to the Arnheim and Waynesville Formations (Late Ordovician, early Richmondian) in southeastern Indiana and southwestern Ohio, with the species most abundant in the Fort Ancient and Clarksville members of the Waynesville Formation. Wolford (1930) and Foerste (1909) recorded the earliest occurrence of the species from just below the *Retrorsirostra carleyi* Bed in the Oregonia Member of the Arnheim Formation, with the species ranging through into the lower 3 m of the Blanchester Member of the Waynesville Formation. Austin (1927) did not record the presence of *O. meeki* in the Arnheim Formation in Warren County, Ohio, but lists the species as being abundant throughout the Fort Ancient and Clarksville divisions of the Waynesville Formation. He also did not record the species as occurring above the basal 1.5 m of the Blanchester division of the same formation.

My own studies in Indiana and Ohio indicate that *O. meeki* is rarely present in the upper 5 m of the Arnheim Formation in southwestern Ohio and is somewhat more common in equivalent strata in Franklin County, Indiana. The species is found frequently in the thin packstone beds throughout the Fort Ancient Member of the Waynesville Formation in both Indiana and Ohio (Harris and Martin 1979) and reaches its greatest abundance in the basal 1.5 m of the Clarksville Member (*O. meeki* epibole), occurring in tremendous numbers nearly to the exclusion of all other benthic organisms. The species continues to be abundant throughout the remainder of the Clarksville member but does not exhibit the extraordinary numbers or the high degree of dominance characteristic of the basal 1–2 m of the member.

Like these other researchers, I did not observe *O. meeki* in strata above the basal 2–3 m of the Blanchester Member in the Waynesville Formation. The genus *Onniella* does not recur in the Upper Ordovician section exposed in the Indiana/Ohio area, except for the presence of a finely costellate, unidentified species (cf. *O. quadrata* Wang) in the peculiar blue-green, friable limestones of the Newpoint Member of the Brainard Shale (latest Richmondian) in portions of southeastern Indiana (Frey 1976; Fluegeman and Pope 1982).

The Onniella Meeki Epibole

STRATIGRAPHIC CHARACTER

The *Onniella meeki* epibole in the Waynesville Formation consists of 1.5–2.5 m of interbedded blue clay shale and thin (0.5–13 cm thick) brachiopod packstone distinguished by (1) an exceptional abundance (densities up to 6,500 individuals per square meter) of the disarticulated valves of the small orthid *Onniella meeki* (Miller), typically comprising 80% to 96% of the larger skeletal allochems (greater than 5 mm in size); and (2) an association of *O. meeki* with the plectambonitid brachiopod *Thaerodonta clarksvillensis* (Foerste). The base of the *O. meeki* epibole in the Indiana/Ohio region is marked by the first packstone bed containing *O. meeki* that is 5 cm or more in thickness, everywhere capping a 1.5-m-thick blue claystone unit that has a locally abundant fauna of trilobites, large pelecypods, nautiloids, the monoplacophoran *Sinuites*, and inarticulate brachiopods (= *Treptoceras duseri* or Trilobite Shale).

At all the localities studied as part of this investigation, the *O. meeki* epibole is overlain by 8–9 m of interbedded blue clay shale (1–40 cm thick), thin, planar packstone beds (2–20 cm thick), and minor blue-gray, rubbly mudstone (3–10 cm thick) (fig. 13.4). Lithologically, this section exhibits little to distinguish it from the *O. meeki* epibole, except for a slight increase in the number and thickness of the limestone beds. It contains a distinct fauna, which provides evidence of an increasingly diverse benthic biota. The top of the epibole is placed at the first

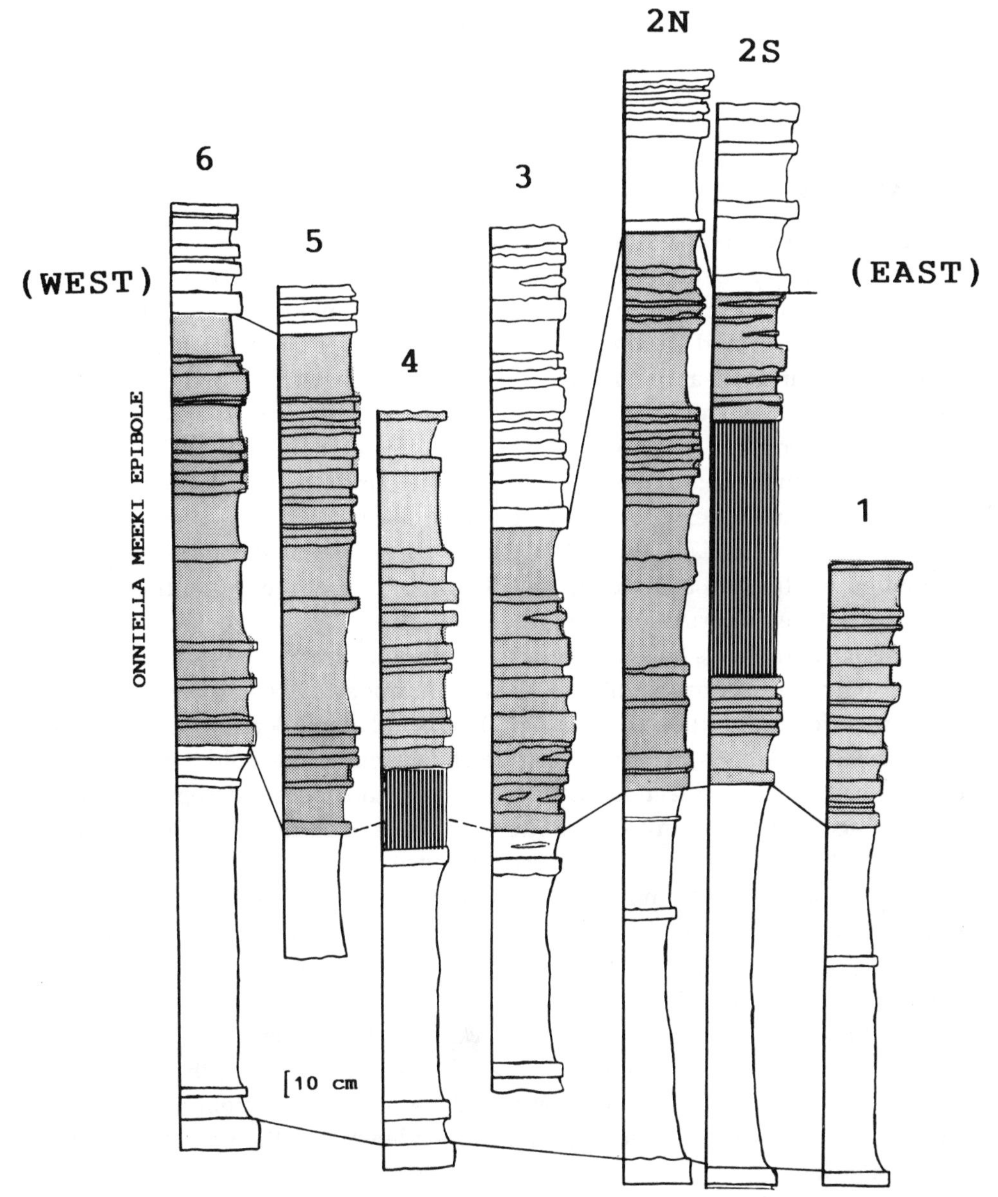

FIGURE 13.4. Correlation of the measured stratigraphic sections in the Waynesville Formation studied in detail as part of this investigation. Shaded portions correspond to the thickness of the *Onniella meeki* epibole observed at each locality. Vertically striped areas represent slumped or covered sections at these outcrops. Note the 1– to 2–m-thick basal claystone unit common to all these localities, as well as the alternating thin packstone/shale couplets characteristic of the overlying *O. meeki* epibole. The upper boundary of the epibole is less well defined, based on a gradual increase in the diversity of the enclosed fossil fauna, with no obvious changes in the lithology other than an increase in the number and thickness of the limestone beds.

limestone in which *O. meeki* makes up less than 75% of the larger skeletal allochems. This also coincides with the first occurrence of the geniculate strophomenid brachiopod *Leptaena richmondensis* and an increase in the abundance of the brachiopods *Platystrophia laticosta* and *Strophomena planumbona*.

LOCALITIES

The *Onniella meeki* epibole has been traced on outcrop from the Clarksville area in Clinton County, Ohio, west across southern Greene, Warren, Montgomery, Butler, and Preble counties in Ohio, into western Franklin County, Indiana, as far as Oldenburg, and southward to St. Leon, in Dearborn County, Indiana. This is a distance of approximately 135 km. My previous work (1976, 1983) established the occurrence of the *O. meeki* epibole at numerous localities in this east/west outcrop belt. I studied in detail the *O. meeki* epibole at six localities in the Indiana/Ohio area: one locality in Clinton County, two in Warren County, and one each in Butler and Preble counties, all in Ohio; and one locality in Franklin County, Indiana (fig. 13.2). These areas represent the best-exposed, continuous sections in the outcrop belt. The appendix at the end of this chapter provides detailed information about these localities.

Onniella meeki also is locally prolific in equivalent strata exposed to the south at Weisburg, Dearborn County, Indiana (Alexander 1990); in the vicinity of Canaan, Jefferson County, Indiana (Fox 1968); and along Beasleys Creek, north of Manchester, Adams County, Ohio (which I examined). I could not, however, directly correlate these sections with those in the northern outcrop belt and so did not include them in my study. The occurrence of abundant *O. meeki* associated with *T. clarksvillensis* suggests the presence of the *O. meeki* epibole in the lower part of Fox's Assemblage Zone C (1968) in the Canaan area. Fox also indicates in a note that the solitary rugose coral *Streptelasma* (= *Grewingkia canadense*) occurs within 1 m of this zone's base. Its presence corresponds to the similar appearance of this coral in the northern outcrop belt.

The known distribution of the *O. meeki* epibole parallels the trend of the regional depositional strike for the Upper Ordovician. As a result, the zone is restricted to the Waynesville Formation, along a broad arc from southeastern Indiana to southwestern Ohio. There is no evidence that the epibole crosscuts regional facies.

LITHOLOGY

Limestone Beds A majority of the limestone beds associated with the *O. meeki* epibole are grain-supported, muddy packstones. The amount of mud present in the matrix varies, with the bed sampled at locality 4 (see fig. 13.2 and the appendix) having the most mud and the sample from locality 3 (10% micrite/clay) having the least mud.

TABLE 13.1 *Biostratinomic Features of the Sampled Packstone Specimens from the* Onniella meeki *Epibole*

Feature/Locality	1(T)	(B)	2N(T)	(B)	2S(T)	(B)	3(T)	(B)	4(T)	(B)	5(T)	(B)	6(T)	(B)
Sample size (*N*)	298	194	195	87	228	102	126	213	131	152	307	300	134	207
Valves articulated (%)	1	—	1	—	1	—	6	1	11	5	10	10	5	1
Brachial valves (%)	44	51	51	34	42	39	57	55	41	49	48	42	35	39
Pedicle valves (%)	56	49	49	66	58	61	43	45	59	51	52	58	65	61
Worn or abraded valves (%)	15	15	24	13	11	11	22	14	11	6	12	9	1	—
Valves convex-up (%)	83	75	91	81	79	54	73	47	68	52	86	74	77	75
Onniella meeki (%)	91	92	92	86	88	80	92	93	87	93	85	83	93	96
Rafinesquina loxorhytis (%)	—	1	—	2	1	3	—	2	—	—	2	3	3	2
Thaerodonta clarksvillensis (%)	8	7	5	6	8	14	3	—	3	—	—	—	—	—
Zygospira modesta (%)	1	—	1	2	1	—	5	2	10	7	10	11	4	2

Note: T- Top B-Bottom Numbers correspond to localities in figure 13.2.

The larger allochems (more than 5 mm) are predominantly (85% to 100%) articulate brachiopod shells, with fewer pelecypod shells, spar-replaced gastropod shells, disarticulated crinoid columnals, fragmentary trepostome bryozoan zooaria, trilobite fragments, and small, calcareous conical shells of uncertain affinity. All the brachiopod specimens are typically disarticulated (89% to 100% of the specimens counted), with nearly equal numbers of brachial and pedicle valves. The sampled limestone beds from epibole contain, on average, 45% brachial valves and 55% pedicle valves (see table 13.1). The valves are primarily whole, complete specimens that are fresh and unabraded. Worn, abraded, or fragmentary valves comprised an average of 11.7% of the specimens I studied. These large allochems were not extensively encrusted by epizoans, and less than 3% of these shells were encrusted by bryozoans or craniid brachiopods or pierced by borings. Brachiopod valves accounted for 30% to 60% of these packstones, averaging 35% of the volume of these rocks. Indeterminant finer, sand-sized shell fragments, ostracode valves, and other shelly detritus made up the bulk of these packstones, comprising 45% to 65% of these rocks, averaging 45%. The matrix in these packstone beds consists primarily of carbonate mud (micrite), clays, and minor coarse sparry calcite, precipitated interstitially in shelter voids beneath brachiopod valves. Muds and spar together made up 10% to 40% of these packstones and averaged 20% by volume.

Individual limestone beds range from 0.5 to 13 cm thick and are typically planar, even bedded, with sharply defined bases and upper surfaces grading into overlying clay shales. Thicker packstone beds (more than 2 cm thick) are generally laterally continuous across the outcrop for distances of up to 100 m, but most exhibit variable amounts of pinch and swell. Thinner beds tend to be more discontinuous, often forming bundles of lenslike beds separated by thin shale partings. The upper third of the individual limestone beds is typically bioturbated to some extent. One skeletal wackestone bed from Collins Run (locality 4) is extensively bioturbated, resulting in the destruction of much of the rock's original fabric. Identifiable ichnofossils include *Chondrites* type A (Osgood 1970), a small-diameter (1–2 mm) branching network of burrows.

Brachiopod valves on both basal and upper bedding plane surfaces of individual packstone beds are typically oriented parallel to the bedding, with these valves comprising 87% to 100% of the valves present. The number of valves inclined or perpendicular to the bedding is slightly higher for the upper bedding plane surfaces than for the bases (4.5% versus 3.75%). Most of the slabbed packstone specimens from the *O. meeki* epibole reveal a three-part internal structure, consisting of (1) a thick basal sequence, with the brachiopod valves parallel to the bedding; (2) an equally thick medial section, with the valves exhibiting a more random orientation relative to the bedding; and (3) a thin upper section, in which the valves are typically parallel to the bedding (fig. 13.5B). An average of 72.5% of these brachiopod valves are oriented in a hydrodynamically stable convex-up position. Typically, more valves are in this orientation on the upper surfaces of the packstone beds than on the basal surfaces (79.5% versus 65.4%). In addition, there is a higher percentage of worn and abraded valves on these upper bedding plane surfaces (13.7% compared with 9.7%). These worn valves are distinct in color and texture from the fresh valves, black and dull rather than the silver-gray, more glossy appearance of the fresh valves. The convex-up brachiopod valves provided a number of shelter-void features, with some of these areas filled with lime mud and others with sparry calcite. Perched, micrograded mud and shelly detritus also can be found between the brachiopod valves in these packstones.

Shale Beds Interbedded with these thin brachiopod packstones in the *O. meeki* epibole are 1–30-cm-thick blue clay shale layers. The texture of these mudstones ranges from fissile to massive and blocky. In composition, these beds are primarily clay (Illite = 70% to 80%), silt (20% to 24%), and carbonate (less than 8%), although the carbonate content increases greatly next to shell beds.

The shales typically contain no preserved body fossils or even trace fossils. The lack of ichnofossils may be the result of an absence of a contrasting sediment type, because clay-filled burrows are often distinct in the upper surfaces of the immediately underlying packstone bed. These shale beds comprise 65% to 80% of the *O. meeki* epibole, exceptions being the sections exposed at localities 3 and 4 in Warren and Butler counties in Ohio. The *O. meeki* epibole at

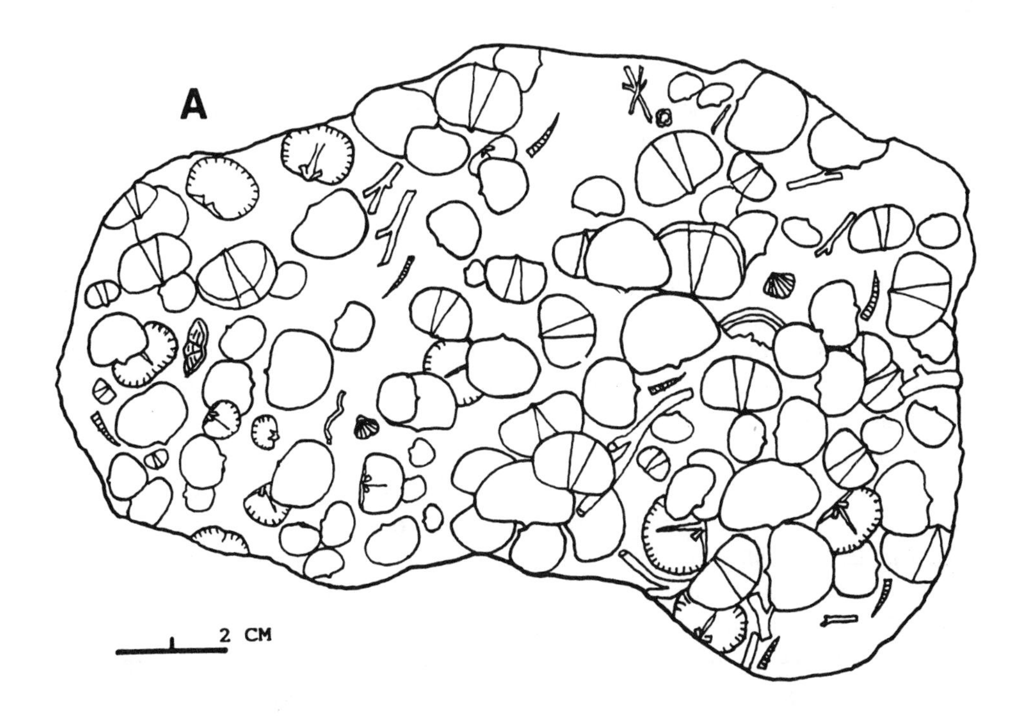

FIGURE 13.5. Diagrammatic drawings illustrating taphonomic features characteristic of the thin packstone beds within the *O. meeki* epibole. (A) The upper surface of a typical packstone bed. Note the abundance of *O. meeki* valves, their predominant convex-up orientation, and the lack of a preferred azimuthal alignment; from locality 5, Four-Mile Creek, Preble County, Ohio. (B) A section through a packstone slab from the north fork of Harpers Run (locality 2N), Warren County, Ohio. Note the rough development of a three-part internal fabric in this "tempestite" bed.

locality 3 is composed of 44% limestone and 56% shale as a result of the average greater thickness of limestone beds there (mean thickness = 6.5 cm). The limestone beds at this locality are also mud-poor, with a greater incidence of spar in the matrix. The section at locality 4 is composed of 37% packstone and 63% shale, with the increase in limestone percentage also resulting from a combination of a greater number of thicker limestone beds (mean thickness = 5.0 cm).

Biostratinomic, Lithologic, and Faunal Trends

Packstone beds from the *Onniella meeki* epibole were sampled across the study area (fig. 13.2). The sample interval was the same for each locality, consisting of the basal packstone bed overlying the claystone of the *Treptoceras duseri*/Trilobite Shale. An exception was the Collins Run locality (locality 4), where this stratigraphic interval is not exposed on outcrop. Instead the bed sampled was the first limestone bed observed with large numbers of *O. meeki* associated with *Thaerodonta clarksvillensis,* which was a limestone bed estimated to be 65 cm above the base of the epibole at this locality. The sampled packstone at the Four-Mile Creek locality (locality 5) overlies a massive claystone unit of indeterminate thickness (at least 25 cm thick, with the remainder below stream level) but lacks *T. clarksvillensis*. Based on its stratigraphic position, it is thought to approximate the basal bed of the *O. meeki* epibole. The stratigraphic sections for each locality that I studied are illustrated in figure 13.4.

The individual packstone slabs that I examined ranged from 2 to 8.5 cm in thickness, averaging 5 cm thick, with an average studied surface area of 600 cm^2. I scanned both the basal and upper surfaces of each slab, using a binocular scope. All allochems on both surfaces more than 3 mm were counted, identified, and measured, and the orientation and condition of each shell were noted. The slabs were then sectioned, polished, and etched with dilute HCL for further study of these limestones' internal fabric and texture. I collected bulk samples at six localities, with one sample each from localities 1, 3, 4, 5, and 6 and two samples from locality 2, representing exposures on both the north and south forks of Harpers Run (see fig. 13.2 and the appendix for the location of these exposures).

BIOSTRATINOMY

Little variation in the biostratinomy of these packstone beds could be discerned, as their taphonomic features were quite consistent throughout the study area (table 13.1). The larger allochems are unaltered articulate brachiopod valves, with the original shell microstructure preserved intact. Specimens are dominated by the shells of *O. meeki* (80% to 96%), typically disarticulated (89% to 100%), with nearly equal numbers of pedicle and brachial valves (55% pedicle

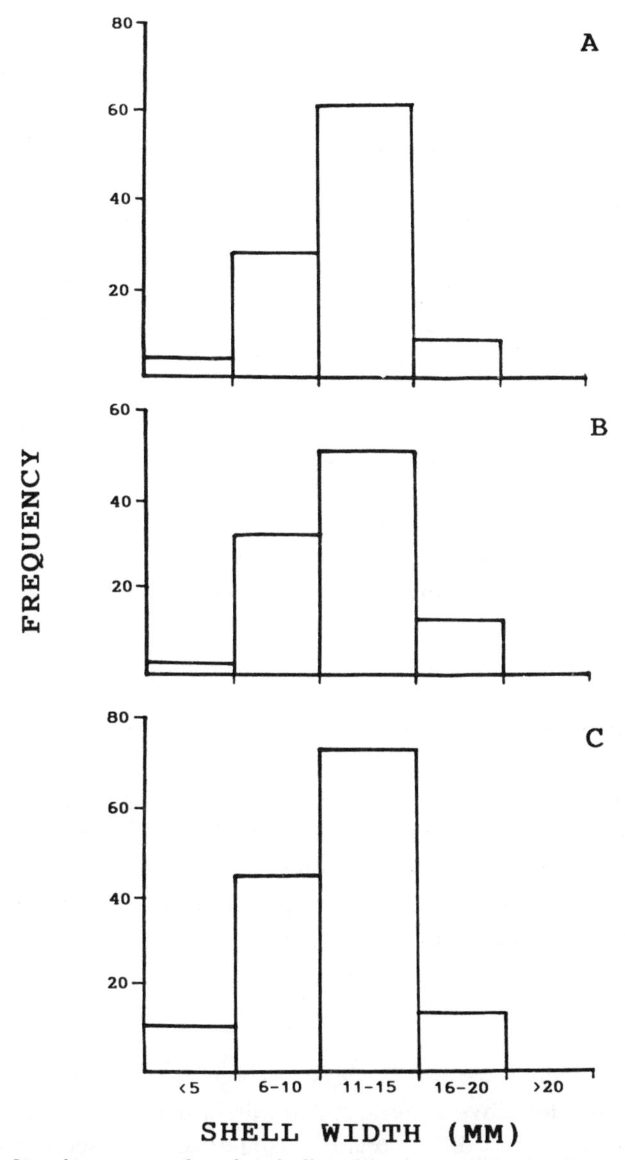

FIGURE 13.6. Size frequency plots for shell widths for censused collections of *Onniella meeki* from the studied localities in the *O. meeki* epibole in Indiana and Ohio. Note the predominance of specimens in the 11–15–mm range and the low numbers of observed juveniles less than 5 mm in width. (A) Censused valves from the upper surface of the studied packstone bed at locality 6, Bon Well Hill, Franklin County, Indiana. (B) Censused valves from the upper surface of the limestone bed, collected from locality 3, Route 42 roadcut, Warren County, Ohio. (C) Censused valves from the upper surface of the sampled limestone bed from locality 2N, north fork of Harpers Run, Warren County, Ohio.

versus 45% brachial), few worn or abraded valves (up to 24% of specimens counted, average = 11.7%), and with most valves oriented in a convex-up position parallel to the substrate surface (an average of 72.5% of the specimens).

No geographic trends in shell bed biostratinomy could be identified. Pedicle valves were consistently slightly more abundant than brachial valves across the entire study area. The exception was sample 3 (locality 3, Warren County, Ohio), where both the top and basal surfaces of the limestone bed had greater numbers of the thinner, less convex brachial valves (table 13.1). The size frequency of *O. meeki* valves was likewise uniform in these beds across the study area. *O. meeki* valves, which ranged in width from 2.5 to 21 mm in the studied suites of specimens, exhibited the highest frequencies in the 11–15-mm size class in all the slabs studied (fig. 13.6). Sample 1T (locality 1, Clinton County, Ohio) and 4T (locality 4, Butler County, Ohio) also contained significant numbers of specimens in the 6–10-mm size range. The percentage of worn, abraded, or fragmental valves remained uniformly low across the outcrop belt, with most (24% for sample 2N and 22% for sample 3 valves worn and/or abraded) on the top surfaces of samples 2N (locality 2, north branch of Harpers Run) and 3 (locality 3, Route 42 roadcut), both in Warren County, Ohio. Similarly, the percentage of encrusted or bored brachiopod valves was low across the region, less than 3% of the valves counted, with no geographic trends identified.

The convex-up attitude parallel to the bedding planes was the dominant orientation for the clear majority (72.5%) of brachiopod valves on both the upper and lower surfaces of the studied packstone beds at all the outcrops. In sample 3 (locality 3, Warren County, Ohio), however, the basal surface of the sampled limestone bed had a slightly greater number (53%) of valves in a convex-down orientation. None of the studied limestone slabs exhibited any indications of the development of a preferred azimuthal alignment of shells on these bedding plane surfaces. Long axes of the brachiopod valves were distributed randomly over the bedding plane surfaces (fig. 13.5A).

LITHOLOGY

The measured sections for the *O. meeki* epibole at the sampled localities (fig. 13.4) revealed no definitive geographic trends in regard to the lithology of the entire epibole. The ratio of carbonate to shale for each section was a higher percentage of limestone (more than 35%) at localities 1 (41%) and 3 (44%) in Clinton and Warren counties in Ohio and locality 4 (37%) in Butler County, Ohio. The remainder of the studied sections contained a higher percentage of shale relative to limestone, with the percentage of limestone being less than 30%. The lowest amount of limestone (22%) was found at locality 5 in Preble County, Ohio. The analysis of the limestone/shale ratios for these sections was complicated by missing covered intervals at both locality 2S (south fork of Harpers Run) and locality 4 (Butler County, Ohio), both of which are thought to represent less resistant shaly intervals.

The limestone beds were consistent throughout the study area in terms of general petrography, bedding type, and the taphonomic features of the larger allochems. The sample at locality 3 was somewhat thicker (8.5 cm thick) than the samples from the same interval elsewhere (average thickness = 5 cm). Again, most of the limestone beds at this locality tended to be thicker than was typical of the *O. meeki* epibole elsewhere. The composition of the matrix in these limestone beds ranged from being entirely insoluble clays at locality 6 (Franklin County, Indiana) in the west, to being almost entirely calcareous (micrite/spar) at locality 3 to the the east in Warren County, Ohio. This suggests a trend toward more calcareous strata from the west to the east in the study area.

FAUNA

As is the case with shell biostratinomy and bedrock lithology, the shelly macrofauna associated with the sampled packstone beds from the *Onniella meeki* epibole is distributed more or less uniformly and consistently across the studied outcrop belt. The small orthid brachiopod *O. meeki* (Miller) comprises an average of 89% of the larger allochems in these packstones, ranging from a low of 80% in a limestone sample from the south fork of Harpers Run in Warren County, Ohio (locality 2S), to a high of 96% in the sample from the Bon Well Hill roadcut in Franklin County, Indiana (locality 6) (see table 13.1). Specimens of *O. meeki* were equally abundant on the upper and basal surfaces of the limestone slabs I studied.

The small (less than 25 mm wide), concavo-convex plectambonitid brachiopod *Thaerodonta clarksvillensis* (Foerste) is a locally common element associated with *O. meeki* in these packstone beds, comprising up to 14% of the counted fauna. This species was most abundant in the southeastern portion of the study area, at localities 1 and 2 in Clinton County and adjacent portions of Warren County, Ohio. The sampled packstone beds indicated that *T. clarksvillensis* comprised 7% of the fauna at locality 1, 9% to 14% of the fauna at locality 2S, and 6% of the fauna at locality 2N. The species was absent from censused packstone slabs in the northwestern portion of the study area: localities 5 (Four-Mile Creek, Preble County, Ohio) and 6 (Bon Well Hill, Franklin County, Indiana). Although no specimens of the species were observed in the samples from locality 5, a number of rare specimens of *T. clarksvillensis* were observed in slabs other than the censused specimen at locality 6. These occurrences indicate that this species increases in abundance toward the southeast within the *O. meeki* epibole exposed in Indiana and Ohio. Data presented by Fox (1968) indicated some abundance of *T. clarksvillensis* (= *Sowerbyella rugosus*) in association with *O. meeki* in the lower several meters of his Zone C in Jefferson County, Indiana. Their presence suggests that there may be a north/south rather than a northwest/southeast trend in the abundance of *T. clarksvillensis* in the *O. meeki* epibole.

Exhibiting a somewhat reverse relationship with *T. clarksvillensis* in these strata is the small (2.5–7-mm-wide) plicate atrypacean *Zygospira modesta* (Say). In contrast to the other brachiopods in these packstone beds, specimens of *Z. modesta* are typically articulated (98% to 100% of specimens censused). This finding also supports Alexander's contention (1990) that hingement type partly controls the ratio of disarticulated to articulated specimens in fossil assemblages. *Z. modesta* has a stronger cyrtomatodont hinge compared with the weaker deltidiodont hinge of *Onniella, Rafinesquina,* and *Thaerodonta,* all of which occur as largely disarticulated valves in these strata. *Z. modesta* is most abundant in the three northwestern localities in the study area, making up 10% of the fauna at localities 4 (Collins Run, Butler County) and 5 (Four-Mile Creek, Preble County) in Ohio and 4% of the fauna at locality 6 (Bon Well Hill, Franklin County) in Indiana (see fig. 13.2 and table 13.1).

Other minor elements of the preserved fauna in these packstone beds include the large strophomenid brachiopod *Rafinesquina loxorhytis* (Meek), the ramose trepostome bryozoan *Parvohallopora onealli* (James), the calymenid trilobite *Flexicalymene meeki* (Foerste), and the problematic, small calcareous tubes assigned to *Ancienta ohioensis* Ross. Throughout the study area, all these taxa maintained consistently low numbers in terms of comparative abundance, and no geographic trends in the abundance of these species could be identified from the sampled limestone beds.

Depositional Environment

EVIDENCE OF STORM-EVENT REWORKING

The packstone/shale couplets that make up the *O. meeki* epibole exhibit all the basic features that have been ascribed to storm-event deposits or "tempestites" (Brenner and Davies 1973; Kreisa 1981; Kreisa and Bambach 1982; Aigner 1985). These include poorly sorted packstones with sharp, erosional bases; a rough fining-up, graded texture; shells disarticulated but fresh in appearance, not wave worn or abraded, and typically aligned parallel to the bedding in the hydrodynamically stable convex-up position; and the development of a number of distinctive shelter-void and sediment infiltration fabrics within packstone beds. These latter features include mud-filled or spar-filled shelter voids beneath convex-up brachiopod valves, mud-coated allochems, and perched, micrograded sediment. Also suggestive of storm-event deposition is the close physical association of basal shelly packstones overlain by barren, fissile clay shales, with the packstones representing storm-peak deposition and the shale beds the waning phase of the storm event, as the finer sediment dropped rapidly from suspension (Kreisa and Bambach 1982). These tempestite features are well represented in all the packstone beds sampled from the *O. meeki* epibole.

This storm-event fabric may be disrupted by postdepositional events, including low to intensive bioturbation of the upper one-fourth to one-half the individual packstone beds. These events often lead to the development of a basal, undisturbed tempestite layer with aligned, convex-up shells, followed by a disturbed layer of oblique and perpendicular shells and mud-filled burrows. The common development of a thin (less than 1 cm thick), upper-winnowed, condensed layer that contains a somewhat higher percentage of dull black, worn, or abraded shells may also point to postdepositional reworking of the upper layers of these tempestite beds by weaker currents, which removed the finer sediment and left the resultant shell pavement exposed as a lag deposit.

On closer examination, some of the thicker packstone beds represent more complex, composite beds deposited by closely spaced storm and winnowing events. Samples 2S and 5 both feature such composite units, consisting of packstone beds separated by thin (1–4 mm thick) clay partings. Other thick limestone beds, however, show no evidence of more than one depositional event. Such is the case with sample 3 from locality 3, Warren County, Ohio. This 8.5-cm-thick packstone exhibits no apparent breaks in sedimentation and consists of variably aligned *O. meeki* valves in a matrix that grades up from fine shelly detritus to micrite and spar. The internal fabric of the 7-cm-thick wackestone bed sampled at locality 4 in Butler County, Ohio, was destroyed by extensive bioturbation, leading to the development of a more homogeneous texture of mud and allochems.

Numerous authors (Meyer et al. 1981; Kreisa et al. 1981; Tobin 1986; Frey 1987) have suggested a storm-current origin for many of the Upper Ordovician shallow/marine shelf deposits exposed in the Cincinnati, Ohio, region. Marsaglia and Klein's paleogeographic studies (1983) place this area into a predicted hurricane-dominated zone in the Ordovician. They and Kreisa (1981) concluded that such storms originated in subtropical highs developing in the ocean basin to the east and subsequently swept southwestward across the southeastern portion of Laurentia, situated between 10 and 25 degrees south of the Ordovician equator. Packstone/shale couplets indicating a similar storm-event deposition may be found throughout the entire Waynesville Formation in Indiana and Ohio, with even the thicker, seemingly homogeneous claystone intervals revealing evidence of a storm reworking of these epeiric sea mud bottoms (Frey 1987, 1988).

The Environmental Setting

The environment of deposition for the bulk of the *O. meeki* epibole was a low-energy, muddy-bottom marine shelf with water depths of 20–25 m, below the fair-weather wave base but within the zone of storm-current reworking. This uniformly deep, muddy shelf area extended west to east across what is now southeastern Indiana and south Ohio, offshore and north of the shallower waters of the Blue Grass Platform situated to the south in northern and central Kentucky (fig. 13.3). The lack of associated epifaunal filter-feeding organisms and the presence of the underlying 1.5-m-thick, massive-bedded claystone unit suggests that the colonization of the bottom by these nearly monospecific brachiopod populations was associated with the waning phase of a period of heavy sediment influx into the Cincinnati Shelf, resulting from a eustatic rise and marking the last major transgressive/regressive cycle documented for the Ordovician of the Kentucky/Indiana/Ohio area. In this environment, the sedimentation rates were still too high and the bottoms too muddy for most other epifaunal filter feeders.

Onniella meeki and *Thaerodonta clarksvillensis* were species apparently well adapted to living on these mud bottoms, which they were able to colonize rapidly, over broad areas. Both species have small (less than 25 mm wide), rather thin shells, have a minimal biomass, and evidently were more tolerant of turbid bottom conditions than were other contemporary brachiopod species (Richards 1972). Richards (1970) described *O. meeki* as having rapid growth rates, probably early sexual maturity, and a prodigious reproductive capacity, all characteristics of r-selected species. Such features allowed this species to cope successfully with the high juvenile mortality rates associated with these unstable substrates (Richards 1970). Both these brachiopod species seem to match the criteria used to distinguish "opportunistic species" as defined by Levinton (1970).

The associated brachiopods *Rafinesquina loxorhytis* and *Zygospira modesta* were environmental generalists, being ubiquitous over the entire Cincinnatian section exposed in the area and occurring in a variety of lithofacies, especially the muddier facies. *Rafinesquina* used a snowshoe strategy to live on mud substrates, possessing a large, broad, very thin, concavo-convex shell that essentially "floated" on these sediments, with the geniculate shell margins elevating the shell commissure above the sediment/water interface (Alexander 1975; Pope 1976). *Zygospira* used its small size and gregarious, peduculate nature to colonize isolated shells or other hard stable substrates on these mud bottoms, forming large clusters of tightly packed individuals completely covering surfaces on these hard substrates (Richards 1972; my own observations).

Fox remarked on the occurrence of "many bedding planes within [his *Resserella-Zygospira* and *Resserella-Sowerbyella* zones] entirely covered by the small brachiopod *Resserella meeki* (= *Onniella meeki*)" (1968:620), both in the lower half of the Waynesville Formation in southeastern Indiana. He described these beds as representing "population explosions," with *O. meeki* moving into the area and reproducing rapidly, covering the seafloor with its shells. Fox calculated population densities as high as 6,500 individuals per square meter, with the species forcing out almost all other organisms from that part of the seafloor. Similar colonization events resulting in dense accumulations of individuals of a single species have been described by Levinton (1970).

Richards (1970, 1972) also observed that *O. meeki* occurred in large numbers in muddy facies, in contrast to most other orthid brachiopods found in the Upper Ordovician of the Cincinnati area. Noting that the dense populations in the Waynesville Formation were disarticulated and not in a life position, Richards labeled these shell beds *reworked lag deposits of dead shells*. He calculated the shell densities of living populations to be much lower than Fox's estimates, on the order of 20–200 per square meter, pointing out that Fox's populations were time-averaged samples consisting of many generations.

As Richards stated (1970), these *Onniella* shell beds are storm-reworked populations of individuals that reflect the time averaging of more than one generation of shells. Most of the shells' fresh, pristine appearance, a lack of shell corrosion, the absence of extensive encrustation by epibionts, the nearly equal numbers of pedicle and brachial valves, and the range of size groups (fig. 13.6) all imply, however, a lack of prolonged exposure of these shells on the seafloor. After the storm events, most of the shells remained buried beneath the sediment surface.

The packstone beds that I studied did indicate some limited reworking of the uppermost shell layers by weak currents, which had winnowed away some of the finer sediment and left the upper surfaces of these shelly accumulations exposed. Other packstone beds exhibited obvious multicycle storm-event reworking, which created thicker packstone beds with complex fabrics and taphonomic histories. The packstone beds in the *O. meeki* epibole run the gamut from typically thin (less than 2 cm thick), mud-rich packstones that appear to represent a single storm event to thicker beds (up to 13 cm thick) that indicate multicycle reworking of the shells, mixing shells of several successive generations. The original population densities of *O. meeki* represented by these packstone beds would be almost impossible to estimate in the thicker, multicycle beds and would have to be carefully calculated in the thinner shell layers that lack evidence of these multiple reworking events.

The *O. meeki* beds appear to represent locally dense populations of *O. meeki* that were vulnerable to dislodgement by waves and currents as a result of the species' upright position and its probably weak attachment to the muddy substrates that it colonized. This vulnerability limited *O. meeki* to normally low-energy, quiet-bottom environments below wave base. The effects of episodic storm events would have been devastating to these populations, uprooting enormous numbers of individuals and forming widespread shell pavements of disarticulated, convex-up valves. Rapid growth rates and early sexual maturity may have been behaviors used by the species to quickly repopulate these storm-disrupted mud-bottom environments.

The Rise and Fall of the Onniella meeki *Epibole*

Onniella meeki is a common, but not dominant, species in thin packstone beds throughout much of the Fort Ancient member of the Waynesville Formation across the study area. It occurs in these strata with a rather low-diversity fauna that includes the brachiopods *Rafinesquina loxorhytis* and *Zygospira modesta* and a variety of byssate mussels, including *Ambonychia alata* (Meek), *A. suberecta* (Ulrich), and *Corallidomus versaillesensis* (Miller). Locally, within a 3-m-thick section exposed at the Bon Well Hill locality (locality 6), *O. meeki* becomes abundant, forming shell pavements with densities approaching those of the overlying *O. meeki* epibole (Harris and Martin 1979). With the deposition of the massive claystone of the *Treptoceras duseri*/Trilobite Shale unit, however, bottom conditions became untenable for *O. meeki*, as well as other brachiopod species, and they disappeared from these strata across the study area in Indiana and Ohio.

The lower meter of this shale unit consists of a monotonous sequence of blocky claystone beds indicative of mud-bottom environments with episodically high rates of sedimentation and water depths below the fair-weather wave base. These claystones support a distinctive fauna dominated by mobile forms that could avoid being smothered by these muds by adjusting their position relative to the sediment/water interface. The fauna includes epifaunal vagrant deposit-feeding calymenid and asaphid trilobites and the monoplacophoran *Sinuites*; a number of semi-infaunal filter feeders, including linguloid brachiopods and large byssate ambonychiid and modiomorphid mussels; and a diverse assemblage of nekto-benthic nautiloids (Frey 1987). Sessile epifaunal filter feeders are restricted to isolated, large, mushroom-shaped stromatoporoid sponges and similarly shaped *Tetradium* corals.

Within the upper 50 cm of this shale unit are thin (less than 2 cm thick) packstone beds, which increase in both abundance and thickness toward the top of the unit. These initially consist of 1-cm-thick packstones that mark the return of articulate brachiopods to these bottom environments, primarily *Rafinesquina loxorhytis*, with fewer numbers of *O. meeki*, although the fauna is still dominated by mollusks and trilobites. In the upper 10 cm of the shale, the packstone beds are thicker (more than 2.5 cm thick), and the fixed epifaunal filter feeders, including articulate brachiopods (especially *O. meeki*), ascend to a position of dominance. Mollusks—in particular, large filter-feeding clams and nautiloids—become rare, and vagrant epifaunal deposit-feeders are uncommon. Ostracodes, ramose trepostome bryozoans, and crinoids may be locally abundant. The shift from a predominance of mobile benthos to fixed, epifaunal filter feeders, coupled with the increase in the abundance and thickness of skeletal packstones upsection, indicates a waning of the sediment influx into the Cincinnati Shelf at this time, associated with the initiation of a shallowing-up sequence following a transgressive event. These conditions enabled the development of locally diverse, patchy communities of shelly epifaunal organisms on these mud bottoms (Frey 1987). The abundance of *O. meeki* parallels this trend across the study area, with a gradual increase in the abundance and dominance of individuals upsection.

This trend culminates in the deposition of the brachiopod packstone/barren shale couplets of the *O. meeki* epibole forming the basal 1.5–2.5 m of the Clarksville member. The interval is marked by the formation of more continuous, thicker limestone beds (2–12 cm thick), in which the disarticulated valves of *O. meeki* make up 80% to 96% of the preserved fauna. *O. meeki* completely dominates this low-diversity fauna, with estimates of population densities ranging from 20 to 200 individuals per square meter (Richards 1970). Episodic storm events repeatedly disrupted these populations, dislodging, smothering, and burying these shells essentially in place on the seafloor. Less intense storms periodically reworked the upper layers of these shell beds, washing away muds and leaving the subsequent shell pavements exposed on the bottom for brief periods of time. The formation of shell beds probably inhibited the activities of burrowing infauna, mobile deposit-feeders, and semi-infaunal filter feeders (Kidwell 1986), thereby contributing to the low numbers of these types of organisms occurring with *O. meeki* in these strata (fig. 13.7).

As the clastic influx into the Cincinnati Shelf area from the east continued to slow and depths in this shelf area shallowed, bottom conditions in the Waynesville sea changed to more stable, skeletal-influenced substrates, aided by the development of the shell pavements generated by successive generations of these "pioneer" brachiopod populations. The consequent stable, firmer, more current-affected bottom environments promoted colonization of this shelf area by a new benthic fauna. This was not simply the result of the northward migration of faunas from the shallow shelf areas in Kentucky to the south. Rather, most of these taxa were new to the Cincinnati region and are not found in older Late Ordovician shallow-water facies in Kentucky. They appear to be derived from faunas originating in the midcontinent portions of Laurentia,

FIGURE 13.7. A reconstruction of the *Onniella meeki* community that colonized the muddy bottom environments across the Cincinnati Shelf area during the Late Ordovician (early Richmondian) time. (A) The orthid brachiopod *Onniella meeki* (Miller); (B) the plectambonitid brachiopod *Thaerodonta clarkvillensis* (Foerste); (C) the atrypacean brachiopod *Zygospira modesta* (Say); (D) the ramose trepostome *Parvohallopora onealli* (James); (E) the calymenid trilobite *Flexicalymene meeki* (Foerste); and (F) the problematic calcareous tubes of *Ancienta ohioensis* Ross.

north and west of the Cincinnati Shelf (Foerste 1909, 1912; Macomber 1970; Frey 1981; Elias 1982). A lesser number of taxa, including a number of mollusks and the orthids *Hebertella* and *Platystrophia*, apparently did migrate north from already established shallow-water environments in Kentucky (Foerste 1909; Howe 1979; Shideler 1991) at this time.

This "Richmond" benthic fauna is associated with the local development of extensive thickets of branching trepostome bryozoans that provided a habitat for a diverse attached fauna, in particular, nestling brachiopods—including the orthids *Hebertella, Glyptorthis,* and *Platystrophia,* and the strophomenid *Holtedahlina.* Other elements of the fauna are the solitary rugose coral *Grewingkia canadense* and the reclining strophomenids *Leptaena, Strophomena,* and *Tetraphalerella.* This gradual changeover from a mud-bottom to a shelly-bottom fauna took place in the upper part of the Clarksville member, concurring with the increase in thin packstone beds in these strata throughout the study area. Influxes of clastic sediment, possibly related to minor eustatic rises, are represented by 1-m-thick, blocky "trilobite shale" beds that periodically disrupt the trend toward more skeletal substrates. At such times, conditions in the Cincinnati Shelf would temporarily return to mud bottoms (Austin 1927; Wolford 1930; Frey 1987; Schumacher and Shrake, essay 6 of this volume; Wolford 1930). These clastic events covered the entire shelf and can be traced throughout the study area in the overlying Blanchester member of the Waynesville Formation. This trend toward the establishment of shallower-water, more skeletal carbonate substrates, however, continued throughout the remainder of the Ordovician in the Cincinnati region (fig. 13.1).

These more current-affected, more grain-rich substrates were not optimal environments for *O. meeki.* With the gradual decrease in shale and the corresponding increase in limestones up through the Clarksville member, there is a corresponding decrease in the abundance of *O. meeki.* The common occurrence of this species seems to be linked closely to the development of a particular bottom environment, specifically low-energy, muddy bottoms below the fair-weather wave base, but with only moderate to low sedimentation rates. These conditions resulted in lithologies composed of 60% to 70% shale and 30% to 35% skeletal packstone. With higher rates of sedimentation, these mud bottoms were too unstable for *O. meeki,* and a mobile mollusk/trilobite fauna became prevalent. As indicated above, *O. meeki* became extinct within 3 m of the base of the Blanchester member of the Waynesville Formation (early middle Richmondian). *Onniella* does not reappear in the remainder of the Ordovician section in the area, even though similar optimal conditions returned for brief intervals of time.

In contrast, *Thaerodonta clarksvillensis,* which disappeared with *O. meeki* at the base of the Blanchester Member, briefly reappears 2–3 m above the base of the overlying Liberty Formation, where it covers bedding planes of thin (2–5 cm thick) planar packstone beds capping a 1-m-thick trilobite-bearing claystone. This 1-m-thick, thin packstone/shale interval was widespread, occurring throughout the study area in Indiana and Ohio (Cumings 1908; Austin 1927; Wolford 1930; Shideler 1991). This species then disappears, to return again at the top of the Ordovician section in the *Megamyonia* Bed (Frey 1976) and at the base of the Elkhorn Shale Member of the Whitewater Formation, both in southeastern Indiana.

Similar to the development of the *O. meeki* epibole in the Waynesville Formation is the occurrence in the Cincinnati, Ohio, area of horizons at the top of the shaly Kope Formation (Late Ordovician, Edenian) and the top of the overlying Mount Hope member of the Fairview Formation (early Maysvillian) of thin packstones filled with the valves of *Onniella multisecta*, nearly to the exclusion of all other shelly benthos. The lithologies and inferred environments of deposition are similar to those for the Waynesville Formation. Interestingly, *O. multisecta* is associated in these packstone beds with the small plectambonitid *Sowerbyella rugosus*. This latter species is thought to be a possible evolutionary precursor of *Thaerodonta clarksvillensis* (Howe 1972, 1979). These occurrences further support the idea that throughout the Late Ordovician in the Cincinnati area, both *Onniella* and these small plectambonitids were well adapted to a very specific environmental setting. This was the waning phase of a period of clastic sediment influx into this shelf area, typically marking the beginning of a shallowing-up cycle.

Holland (1993) surmised that this facies marked the deepest-water environment in "highstand" sequences following transgressive events within the Cincinnatian Series. Such an environment was too muddy for most fixed epifaunal filter feeders but was ideal for these small brachiopod species. Austin (1927) recognized this trend early in this century, noting a correlation between the occurrences of these monospecific brachiopod beds and the thick claystone layers that invariably underlay these beds in the Cincinnatian section. He suggested that species like *O. meeki* and *T. clarksvillensis* were well adapted to the "unsettled" conditions that followed the deposition of these thick clay beds. Austin decided that the formation of these brachiopod shell beds resulted from the colonization of the mud bottoms by great numbers of these two species, with no competition from other epifaunal filter-feeding organisms.

Middle Ordovician Dalmanellid Epiboles

The older Middle Ordovician (Kirkfieldian/Shermanian) strata in the same general Kentucky/Indiana/Ohio area contain similar dalmanellid-defined epiboles, apparently deposited under somewhat similar environmental conditions as these Late Ordovician examples. Cressman (1973) described the "Brachiopod Coquina" in the Logana Member of the Lexington Limestone (Middle Ordovician, Kirkfieldian) in central Kentucky as consisting of tightly packed masses of valves of the small orthid *Dalmanella sulcata* Cooper, oriented parallel to the bedding, forming wavy beds up to 15 cm thick. The rock was described as poorly washed biosparrudite with brachiopod valves disarticulated but not broken or abraded. The Logana Member was seen as representing the basal transgressive stage in the deposition of the Lexington Limestone—deposited under deeper-water conditions than those of either the underlying or overlying strata. Cressman inferred that the interbedded "coquinoid limestone" and silty shales were deposited below the wave base and represented the height of the initial Lexington transgression, deposited at water depths of 25–30 m, under quiet-water conditions. The "Brachiopod Coquina," up to 4 m thick, can be traced on outcrop from Owen and Nicholas counties, Kentucky, in the north, 50 miles south into Jessamine County, Kentucky. The outcrop belt is 80 miles wide along strike, with the unit trending northeast to southwest, thinning to the southeast and thickening to the north.

Cressman (1973) also described the "Macedonia Bed" as a recurrence of these conditions in the overlying Grier Limestone Member of the same formation (Middle Ordovician, Shermanian). *Dalmanella sulcata,* however, is replaced by the dalmanellid orthid *Heterorthina macfarlani* Neuman in these younger strata. This unit consists of even-bedded coquinoid limestone beds 2–4 cm thick, interbedded with 10% to 40% shale.

Equivalent to these Kentucky strata is the "*Dalmanella* Coquina" that Wilson described (1962) from the Hermitage Formation (Middle Ordovician, Kirkfieldan) in central Tennessee. This unit has been traced along a narrow north/south trending belt 15 miles wide and extending some 130 miles through Giles, Maury, Williamson, and Davidson counties in Tennessee. Wilson (1962) described this "coquina" as consisting of up to 30 feet of massive-bedded gray limestone with silt and laminated shale, containing a profusion of unfragmented *Dalmanella* shells, tightly packed in both vertical and horizontal positions. He inferred that these deposits were the result of storms disrupting a normally low-energy, deep-water, muddy-bottom environment. These bottoms were "covered with countless millions of *Dalmanella,* living closely packed together" (1962:497), with nearly no other forms of benthic life. This unit also grades up into increasingly shallow-water facies of the Bigsby/Cannon Limestone (Shermanian), paralleling the same trend in the equivalent Lexington Limestone in Kentucky.

From these examples it is evident that in the Ordovician, dalmanellid brachiopods like *Onniella* and *Heterorthina* proliferated in low-energy, mud-bottom, comparatively deep water (20–30 m deep) cratonic sea environments. These environments are preserved as dense accumulations of storm-reworked dalmanellid shells interbedded with shale or mudstones. Epiboles defined by these shelly accumulations of dalmanellid brachiopod valves often mark the waning phases of transgressive events, succeeded by shallowing-up cycles within these cratonic facies. Such dalmanellid epiboles, coupled with sedimentological data, may prove to be of value in identifying similar transgressive events in these cratonic strata.

The *Onniella meeki* epibole consists of 1.5–2.5 m of interbedded shale and thin (2–12 cm thick) packstone beds composed almost entirely of the valves of the small orthid brachiopod *O. meeki* (Miller). These storm-deposited limestone beds can be traced on outcrop from Clinton County, Ohio, in the east, 135 km west into western Franklin County, Indiana. The thickness varies from one outcrop to another, but the zone retains its distinctive lithologic, biostratinomic, and faunal identity throughout the outcrop belt. This identity includes a lithology consisting of thin, planar, packstones with a rough graded texture; biostratinomic features, including shell disarticulation, with nearly equal numbers of brachial and pedicle valves oriented primarily parallel to bedding in a convex-up position; distinctive shelter-void and infiltration fabrics; and preserved fauna, 80% to 96% of which are the shells of *O. meeki.* Throughout the study area in Indiana and Ohio, the *O. meeki* epibole caps a massive claystone bed 1–2 m thick and is overlain by a series of increasingly more calcareous strata consisting of interbedded, thin planar packstones and thin clay shales.

The causal factors leading to the development of the epibole center on the rapid development of a unique set of widespread environmental conditions across the Cincinnati Shelf. These conditions are associated with the waning phase of a period of clastic influx into the Cincinnati Shelf that marked the maximum extent of a transgressive event and the initiation of a shallowing-up cycle. These conditions led to the explosive proliferation of *O. meeki.* In such an environment, the sedimentation rates were too high and the substrates too muddy for most other epifaunal filter feeders, but they were optimal for *Onniella* and, to a lesser extent, the plectambonitid *Thaerodonta.* With the increase in skeletal substrates and the influx of a new, diverse fauna adapted to these substrates, there was a gradual but noticeable decline in the abundance of *O. meeki.* The species subsequently became extinct locally just below the top of the Waynesville Formation.

Other studies of the Ordovician of the Cincinnati Arch region indicate a similar development of dalmanellid-dominated facies marking transgressive maxima in the late Middle Ordovician (Kirkfieldian) and the early Late Ordovician (Edenian). This suggests that the development of these beds can be used to mark these transgressive events regionally across the Cincinnati Arch outcrop belt and possibly elsewhere in Middle and Late Ordovician strata across eastern North America.

The *O. meeki* epibole in the Waynesville Formation differs in several respects from Brett and Baird's definition (1989) of an epibole. They described epiboles as typically being defined by the unusual abundance of a normally rare species. Although *O. meeki* does not occur in the prolific numbers characteristic of the shell beds described here, it is a relatively common species in the underlying strata throughout the study area. It similarly occurs in low-density populations in the overlying 1–3 m of strata before finally becoming extinct below the top of the Waynesville Formation. Brett and Baird also stated that epiboles often crosscut facies in depositional basins. However, the *O. meeki* epibole does not crosscut facies but occurs along the depositional strike, thereby limiting the development of these beds to the Waynesville facies in southeastern Indiana and adjacent portions of southwestern Ohio. A fuller documentation of the extent of this epibole north and south of the studied outcrop belt is complicated by a lack of outcrops to the north, coupled with a scarcity of subsurface cores available for detailed lithologic and paleontological study. To the south there are no continuous exposures between the studied sections to the north and equivalent strata to the south along the Ohio River (i.e., the Madison, Indiana, and the Manchester, Ohio, sections).

The known distribution of the *O. meeki* epibole suggests geographic limits to the utility of this epibole and other similar types of paleontological event beds in regional correlation. More studies of these fossil accumulations are needed to define more accurately the various types of epiboles and determine their usefulness and limitations with regard to detailed stratigraphic correlation.

Future candidates for such studies are other similar brachiopod epiboles in the Upper Ordovician of the Cincinnati area that are equally widespread across Indiana and Ohio. These include other monospecific shelly packstones dominated by *Onniella* and/or *Sowerbyella* or *Thaerodonta,* typically capping thick claystone beds. In addition to these, I have traced, across my study area in Indiana and Ohio, other short-lived, regionally widespread faunal zones identified by previous authors (Foerste 1909, 1912; Austin 1927; Wolford 1930) within the Arnheim, Waynesville, and Liberty Formations. These primarily brachiopod-defined marker horizons include the *Retrorsirostra carleyi* Zone in the Arnheim Formation, the first and second *Glyptorthis insculpta* Zones and the *Strophomena nutans/Tetraphalerella neglecto* Zone in the Waynesville Formation, and the *Plaesiomys subquadrata/Hiscobeccus capax* Zone in the Liberty Formation. In addition, Meyer (1990) traced a distinctive edrioasteroid bed stratigraphically across Late Ordovician strata exposed in the Greater Cincinnati area of Kentucky and Ohio, and various coralliferous zones have been used to correlate younger Richmondian strata in northern Kentucky, southeastern Indiana, and adjacent portions of western Ohio (Foerste 1909; Shideler 1914; Hattin 1961). The occurrence and distribution of these various epiboles in the Ordovician of the Cincinnati region indicate the potential utility of these paleontological event beds for detailed stratigraphic correlation within similar cratonic shelf sequences.

Appendix: Studied Localities of Onniella meeki Epibole

1. Stream-bank exposures along Stony Hollow, just north of Todds Fork, 1 km north of Clarksville, Clinton County, Ohio. Clarksville, Ohio, 7.5′ Quadrangle, long. 83°59′ W, lat. 39°24′30″ N.
2. Stream-bank exposures along the north (2N) and south (2S) forks of Harpers Run, south of Strout Road and west of Middleboro Road, at Camp Whip-Poor-

Will G.S.A. Outdoor Center, Warren County, Ohio. Oregonia, Ohio, 7.5′ Quadrangle, long. 84°05′ W, lat. 39°23′ N.
3. Roadcut along U.S. Route 42, west side of road, 3 km northeast of Waynesville, Warren County, Ohio. Waynesville, Ohio, 7.5 Quadrangle, NE1/4 of the SE1/4 sec. 26, R 4, T 4.
4. Stream-bank exposures along Collins Run, west of Pfieffer Park, just south of Miami University campus, Oxford, Butler County, Ohio. Millville, Ohio, 7.5′ Quadrangle, center of SE1/4 sec. 27, R 1 E, T 5 N.
5. Stream-bank exposure where a small stream enters Four-Mile Creek from the east, 0.5 km south of Acton Lake Spillway at Huston Woods State Park, Preble County, Ohio. Oxford, Ohio, 7.5′ Quadrangle, NE1/4 of NW1/4 sec. 11, R 1 E, T 5 N.
6. Roadcut along Route 101, along north side of road at Bon Well Hill, 3 km east of Brookville, Franklin County, Indiana. Mount Carmel, Indiana, 7.5′ Quadrangle, SE1/4 sec. 16, R 2 W, T 9 N.

ACKNOWLEDGMENTS

I would like to thank Carlton Brett and Gordon Baird for inviting me to contribute to this volume, J. K. Pope and J. Marek for giving me access to preparatory and lab equipment at Miami University, and W. A. Ausich and S. M. Bergstrom at The Ohio State University for providing me with office space and access to the facilities of the Department of Geological Sciences. G. Baird, C. Brett, D. Meyer, and an anonymous reviewer critically read an earlier draft of this paper and offered many helpful suggestions to improve its quality.

References

Ager, D. V. 1981. *The Nature of the Stratigraphic Record*, 2d ed. New York: Wiley.
Aigner, T. 1985. *Storm Depositional Systems*. Berlin: Springer-Verlag.
Alexander, R. R. 1975. Phenotypic lability of the brachiopod *Rafinesquina alternata* (Ordovician) and its correlation with the sedimentologic regime. *Journal of Paleontology* 49:607–18.
———. 1990. Disarticulated shells of Late Ordovician brachiopods: Inferences on strength of hinge and valve architecture. *Journal of Paleontology* 64:524–32.
Anstey, R. L. and M. L. Fowler. 1969. Lithostratigraphy and depositional environment of the Eden Shale (Ordovician) in the tri-state area of Indiana, Kentucky, and Ohio. *Journal of Geology* 77:668–82.
Arkell, W. J. 1933. *The Jurassic System in Great Britain*. Oxford: Oxford University Press.
Austin, G. M. 1927. Richmond faunal zones in Warren and Clinton counties, Ohio. *Proceedings of the U.S. National Museum* 70:1–18.
Bird, J. M. and J. F. Dewey. 1970. Lithosphere plate-continental margin tectonics and the evolution of the Appalachian orogen. *Geological Society of America Bulletin* 81:1031–60.
Brenner, R. L. and D. K. Davies. 1973. Storm-generated coquinoid sandstone: Genesis of high-energy marine sediments from Upper Jurassic of Wyoming and Montana. *Geological Society of America Bulletin* 84:1685–98.
Brett, C. E., ed. 1986. Dynamic stratigraphy and depositional environments of the Hamilton Group (Middle Devonian) in New York. *New York State Museum Bulletin* 457:156.

Brett, C. E. and G. C. Baird. 1989. Paleontological event stratigraphy: Fossil beds, epiboles, and extinction horizons in the Paleozoic of the Appalachian Basin. *Geological Society of America Abstracts with Programs* 21:6.

Bucher, W. H. 1917. Large current-ripples as indicators of paleogeography. *Proceedings of the U.S. National Academy of Science* 3:285–91.

Caster, K. E., E. A. Dalvé, and J. K. Pope. 1955. Elementary guide to the fossils and strata of the Ordovician in the vicinity of Cincinnati, Ohio. *Cincinnati Museum of Natural History Popular Publication Series* No. 2.

Chadwick, G. H. 1924. The stratigraphy of the Chemung Group in western New York. *New York State Museum Bulletin* 251:149–57.

Cooper, G. A. 1930. Stratigraphy of the Hamilton Group of New York. *American Journal of Science* 19:116–34, 214–36.

———. 1933. Stratigraphy of the Hamilton Group of eastern New York. *American Journal of Science* 26:537–51; 27:1–12.

Cressman, E. R. 1973. Lithostratigraphy and depositional environments of the Lexington Limestone (Ordovician) of central Kentucky. *U.S. Geological Survey Professional Paper* No. 768.

Cumings, E. R. 1908. The stratigraphy and paleontology of the Cincinnatian Series in Indiana. *Indiana Department of Geology and Natural Resources 32d Annual Report*, pp. 607–1188.

Dennison, J. M. 1976. Appalachian Queenston Delta related to eustatic sea level drop accompanying Late Ordovician glaciation centered in Africa. In M. G. Bassett, ed., *The Ordovician System*, pp. 107–20. Cardiff: University of Wales Press and National Museum of Wales.

Dott, R. H. 1983. Episodic sedimentation: How normal is average? How rare is rare? Does it matter? *Journal of Sedimentary Petrology* 53:5–23.

Dunbar, C. O. and J. Rodgers. 1957. *Principles of Stratigraphy*. New York: Wiley.

Einsele, G. and A. Seilacher, eds. 1982. *Cyclic and Event Stratification*. Berlin: Springer-Verlag.

Elias, R. J. 1982. Latest Ordovician solitary rugose corals of eastern North America. *Bulletins of American Paleontology* 81(314):1–116.

Ernst, G. and E. Seibertz. 1977. Concepts and methods of echinoid biostratigraphy. In E. G. Kauffman and J. E. Hazel, eds., *Concepts and Methods of Biostratigraphy*, pp. 541–63. Stroudsburg, Pa: Dowden, Hutchinson, and Ross.

Ettensohn, F. R. 1991. Flexural interpretation of relationships between Ordovician tectonism and stratigraphic sequences, central and southern Appalachians, U.S.A. In C. R. Barnes and S. W. Williams, eds., *Advances in Ordovician Geology*, pp. 213–24. *Geological Survey of Canada Paper* No. 90–9.

Fluegeman, R. and J. K. Pope. 1982. Brainard Shale outliers (Upper Ordovician), Maquoketa Group in southeastern Indiana. *Geological Society of America Abstracts with Programs* 15:574.

Foerste, A. F. 1909. Preliminary notes on Cincinnatian and Lexington fossils. *Bulletin of the Science Laboratories, Denison University* 14:289–324.

———. 1912. Strophomena and other fossils from Cincinnatian and Mohawkian horizons, chiefly in Ohio, Indiana, and Kentucky. *Bulletin of the Science Laboratories, Denison University* 17:17–174.

Fox, W. T. 1962. Stratigraphy and paleoecology of the Richmond Group in southeastern Indiana. *Geological Society of America Bulletin* 73:621–42.

———. 1968. Quantitative paleoecologic analysis of fossil communities in the Richmond Group. *Journal of Geology* 76: 613–40.

Frey, R. C. 1976. The biostratigraphy of the Richmond Group (Upper Ordovician) in Franklin County, Indiana. Master's thesis, Miami University.

———. 1981. Narthecoceras (Cephalopoda) from the Upper Ordovician (Richmondian) of southwestern Ohio. *Journal of Paleontology* 55:1217–24.
———. 1983. The paleontology and paleoecology of the *Treptoceras duseri* Shale unit (Late Ordovician, Richmondian) of southwest Ohio. Ph.D. disseration, Miami University.
———. 1987. The paleoecology of a Late Ordovician shale unit from southwest Ohio and southeastern Indiana. *Journal of Paleontology* 61:242–67.
———. 1988. The paleoecology of *Treptoceras duseri* from the Upper Ordovician of southwest Ohio. In D. L. Wolberg, ed., *Contributions to the Paleozoic Paleontology and Stratigraphy in Honor of Rousseau H. Flower*, pp. 79–101. *New Mexico Bureau of Mines and Mineral Resources Memoir* No. 44.
Grabau, A. W. 1920–21. *A Textbook of Geology*, 2 vols. Boston: Heath.
Hall, D. D. 1962. Dalmanellidae of the Cincinnatian. *Palaeontographica Americana* 4:131–63.
Harris, F. W. and W. D. Martin. 1979. Benthic community development in limestone beds of the Waynesville Formation (Cincinnatian Series, Upper Ordovician) of southeastern Indiana. *Journal of Sedimentary Petrology* 49:1295–1305.
Hattin, D. E. 1961. Notes on Richmond stratigraphy in southeastern Indiana. *Guidebook for Field Trips, Cincinnati Meeting of the Geological Society of America*, pp. 328–31.
Hay, H. B. 1981. Lithofacies and formations of the Cincinnatian Series (Upper Ordovician) in southeastern Indiana and southwestern Ohio. Ph.D. dissertation, Miami University.
Hedberg, H. D. 1972. Summary of an international guide to stratigraphic classification, terminology, and usage. *Lethaia* 5:297–323.
Holland, S. M. 1993. Sequence stratigraphy of a carbonate-clastic ramp: The Cincinnatian Series (Upper Ordovician) in its type area. *Geological Society of America Bulletin* 105:306–22.
Howe, H. J. 1972. Morphology of the genus *Thaerodonta*. *Journal of Paleontology* 46:440–46.
———. 1979. Middle and Late Ordovician plectambonitacean, rhynchonellacean, syntrophiacean, trimerellacean, and atrypacean brachiopods. *U.S. Geological Survey Professional Paper* No. 1066–C.
Kidwell, S. M. 1986. Taphonomic feedback in Miocene assemblages: Testing the role of dead hardparts in benthic communities. *Palaios* 1:239–55.
Kreisa, R. D. 1981. Storm-generated sedimentary structures in subtidal marine facies with examples from the Middle and Upper Ordovician of southwestern Virginia. *Journal of Sedimentary Petrology* 51:823–48.
Kreisa, R. D. and R. K. Bambach. 1982. The role of storm processes in generating shell beds in Paleozoic shelf environments. In G. Einsele and A. Seilacher, eds., *Cyclic and Event Stratigraphy*, pp. 200–207. Berlin: Springer-Verlag.
Kreisa, R. D., S. L. Dorobek, P. J. Accorti, and E. P. Ginger. 1981. Recognition of storm-generated deposits in the Cincinnatian Series, Ohio. *Geological Society of America Abstracts with Programs* 13:285.
Levinton, J. S. 1970. The paleoecological significance of opportunistic species. *Lethaia* 3:69–78.
Macomber, R. W. 1970. Articulate brachiopods from the Upper Bighorn Formation (Late Ordovician) of Wyoming. *Journal of Paleontology* 44:416–50.
Marsaglia, K. M. and G. D. Klein. 1983. The paleogeography of Paleozoic and Mesozoic storm depositional systems. *Journal of Geology* 91:117–42.
Meyer, D. L. 1990. Population paleoecology and comparative taphonomy of Two edrioasteroid pavements: Upper Ordovician of Kentucky and Ohio. *Historical Biology* 4:155–78.

Meyer, D. L., R. C. Tobin, W. A. Pryor, W. B. Harrison, and R. G. Osgood. 1981. Stratigraphy, sedimentology, and paleoecology of the Cincinnatian Series (Upper Ordovician) in the vicinity of Cincinnati, Ohio. In T. G. Roberts, ed., *Field Trip Guidebooks, 1981 Annual Meeting of the Geological Society of America,* pp. 31–71. Falls Church, Va.: American Geological Institute.

North American Commission on Stratigraphic Nomenclature. 1983. North American Stratigraphic Code. *American Association of Petroleum Geologists Bulletin* 67:841–75.

Oppel, A. 1856. Die Juraformation Englands, Frankreich, und des sudwestlichen Deutchlands, nach ihren einzelnen gliedren eingetheit und verglichen. *Abdruck der Wurttemburg naturwissenenschaft Jahreshefte* 12–14:1–438. Stuttgart: von Ebner & Seubert.

Osgood, R. G. 1970. Trace fossils of the Cincinnati area. *Palaeontographica Americana* 6:281–444.

Pope, J. K. 1976. Comparative morphology and shell histology of the Ordovician Strophomenacea (brachiopoda). *Palaeontographica Americana* 4:166–216.

Richards, R. P. 1970. Paleoecology of the brachiopod species of the Richmond Group (Late Ordovician) of Indiana and Ohio. Ph.D. diss., University of Chicago.

———. 1972. Autecology of Richmondian brachiopods (Late Ordovician of Indiana and Ohio). *Journal of Paleontology* 46:385–405.

Schumacher, G. A. and D. L. Shrake. 1989. The Isotelus gigas bed: Revisited. *Geological Society of America Abstracts with Programs* 21:65.

Scotese, C. R. and W. S. McKerrow. 1990. Revised world maps and introduction. In W. S. McKerrow and C. R. Scotese, eds., *Palaeozoic Palaeogeography and Biogeography: Geological Society Memoir No. 12,* pp. 1–21. London: Geological Society.

Shaw, A. B. 1964. *Time in Stratigraphy.* New York: McGraw-Hill.

Sheehan, P. W. 1988. Late Ordovician extinction events and the terminal Ordovician extinction. In D. L. Wolberg, ed., *Contributions to Paleozoic Paleontology and Stratigraphy in Honor of Rousseau H. Flower.* New Mexico Bureau of Mines & Mineral Resources Memoir No. 44, pp. 405–15.

Shideler, W. H. 1914. The Upper Richmond beds of the Cincinnati Group. *The Ohio Naturalist* 14 (3):229–35.

———. 1991. The Richmond Group of the Cincinnati province. In J. H. Marek, ed., unpublished manuscript. Oxford, OH: Limper Museum of Geology, Miami University.

Shrake, D. L., G. A. Schumacher, and E. M. Swinford. 1988. *Field Guidebook to the Stratigraphy, Sedimentology, and Paleontology of the Upper Ordovician Cincinnati Group of Southwest Ohio.* SEPM 5th Midyear Meeting. Columbus: Ohio Department of Natural Resources, Division of Geological Survey.

Tobin, R. C. 1986. An assessment of the lithostratigraphic and interpretive value of the traditional "biostratigraphy" of the type Upper Ordovician of North America. *American Journal of Science* 286:673–701.

Trueman, A. E. 1923. Some theoretical aspects of correlation. *Geological Society of London, Proceedings* 34:193–206.

Weir, G. W., W. L. Peterson, and W. C. Swadley. 1984. Lithostratigraphy of the Upper Ordovician strata exposed in Kentucky, with a section on biostratigraphy by John Pojeta. *U.S. Geological Survey Professional Paper* No. 1151–E.

Wilson, C. W. 1962. Stratigraphy and geologic history of Middle Ordovician rocks of central Tennessee. *Geological Society of America Bulletin* 73: 481–504.

Wolford, J. J. 1930. The stratigraphy of the Oregonia–Fort Ancient region, southwestern Ohio. *Ohio Journal of Science* 30:301–8.

14

The Medusaegraptus *Epibole and Lower Ludlovian Konservat-Lagerstätten of Eastern North America*

Steven T. LoDuca and Carlton E. Brett

ABSTRACT

In 1925 Rudolf Ruedemann described a remarkable soft-bodied fossil biota from the "Gasport Channel," an inferred interreef deposit in the Silurian Lockport Group of western New York State. New stratigraphic evidence reveals a laterally extensive (nearly 100 km), sheetlike geometry for this Konservat-Lagerstätte, referred to herein as the *Medusaegraptus* epibole.

The unusual biota of the *Medusaegraptus* epibole is dominated by thallophytic algae and dendroid graptolites; well-preserved annelid worms (*Protoscolex*) also occur. However, typical Silurian shelly fossils, such as brachiopods, bryozoa, trilobites, and crinoids, are rare or absent.

Other, quite similar Konservat-Lagerstätten from North America (lower Mississinewa Shale, *Lecthaylus* Shale) apparently were deposited concomitant with the *Medusaegraptus* epibole. Comparative study of the sedimentology, taphonomy, and biotic composition of these laterally extensive units reveals that sediment composition, oxygen availability, and relation to storm and fair-weather wave bases controlled exceptional preservation, whereas other factors, such as water depth and the thickness of the initial burial layer, evidently were insignificant in this regard.

In 1925 Rudolf Ruedemann described a remarkable Silurian soft-bodied fossil biota recovered by amateur paleontologist E. Reinhard from the Wickwire Steel Company Quarry at Gasport, New York (the bulk of Reinhard's collection was sold to the New York State Museum in 1923). Ruedemann (1925) noted that the unusual, dendroid-graptolite-dominated biota occurred in a laterally restricted, lens-shaped interval of argillaceous Lockport Group strata situated between two large bioherms in the south wall of the quarry at Gasport, and he referred to the deposit as the "Gasport Channel." Although the biota of the Gasport Channel eventually became widely cited as an example of a Silurian dendroid graptolite association, this occurrence remained largely unstudied, as the Gasport exposure was rendered inaccessible by quarrying shortly after Ruedemann's report was published.

During this study, several recently discovered exposures of this important fossil occurrence, including a fresh bedding plane surface in the still-active quarry at Gasport (currently operated by Frontier Dolomite, Inc.), were examined along the Silurian outcrop belt in western New York. These localities reveal the fossil-bearing unit as a laterally extensive interval of argillaceous dolomite that *overlies* the associated bioherms. Thus to convey the widespread nature of this thin deposit, and the abundance of the noncalcified dasyclad alga *Medusaegraptus mirabilis* within it, this interval is referred to herein as the *Medusaegraptus* epibole (LoDuca 1990a). (An *epibole* is a biostratigraphic zone characterized by an unusual abundance of the namesake fossil species; the term is synonymous with *acme zone*.)

Because the *Medusaegraptus* epibole contains fossils of soft-bodied organisms, including annelid worms and thallophytic algae, it represents a Konservat-Lagerstätte (see Seilacher 1970). In this study, descriptions of the stratigraphy, lithology, sedimentology, biota, taphonomy, and paleoenvironment of the *Medusaegraptus* epibole Lagerstätte are provided, several similar and apparently age-equivalent Lagerstätten are reviewed, and the physical and environmental factors that controlled exceptional preservation in these Lagerstätten are considered.

Study Area and Methods

The study area spanned the Silurian outcrop belt in western New York State from Niagara Falls in the west to Rochester in the east. The principal localities investigated were the Niagara Gorge, the Niagara Stone Quarry at Niagara Falls, the Frontier Dolomite quarries at Lockport and Gasport, the Genesee-LeRoy Stone Quarry at Clarendon, and the Iroquois Stone Quarry at Brockport (fig. 14.1). Deposits similar to the *Medusaegraptus* epibole were investigated in northeastern Indiana (Mississinewa Shale), northeastern Illinois (*Lecthaylus* Shale), and southern Ontario (the Ancaster Member of the Goat Island Formation) (fig. 14.2).

At the Frontier Dolomite Quarry in Gasport, a 70- by 100-m bedding plane exposure of the *Medusaegraptus* epibole was mapped with a plane table and alidade. The area was subdivided into 10-m^2 plots, and a 1-m^2 sample was removed from the center of each plot in an effort to determine the spatial distribution of the biota.

In the lab, insoluble residue analyses were conducted and representative thin

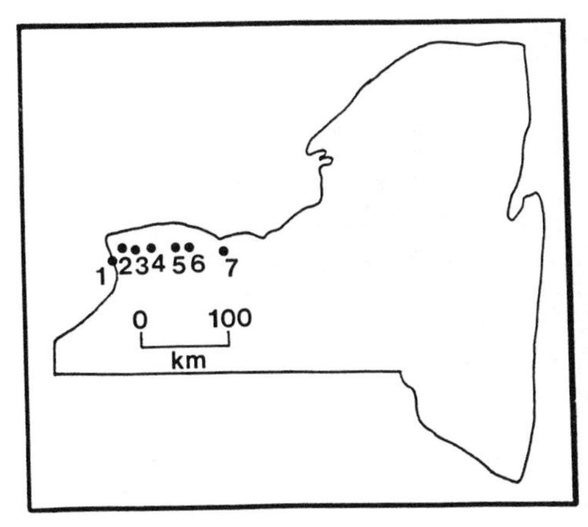

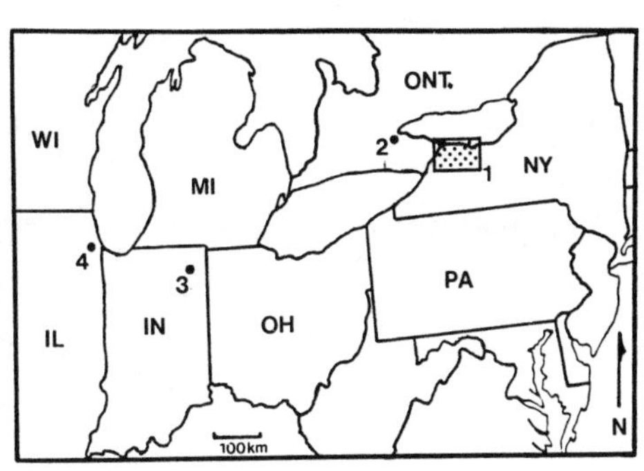

FIGURE 14.1. The primary localities investigated in western New York State: (1) The Niagara Gorge; (2) Niagara Stone Quarry; (3) Frontier Dolomite Quarry, Lockport; (4) Frontier Dolomite Quarry, Gasport; (5) Genesee–LeRoy Stone Quarry, Clarendon; (6) Iroquois Stone Quarry, Brockport; and (7) Iroquois Stone Quarry, Penfield.

FIGURE 14.2. Geographic distribution of the deposits considered in this study. Numbered localities are as follows: (1) the principal study area in western New York; (2) the Niagara escarpment at Hamilton, Ontario; (3) Huntington, Indiana; and (4) Blue Island, Illinois.

sections and polished slabs were prepared. X-ray diffraction was employed to determine the calcite/dolomite ratio of the matrix.

The Medusaegraptus *Epibole*

GEOLOGIC SETTING

The Lockport Group in New York State was deposited during the early Late Silurian in the northern end of the Appalachian Foreland Basin, a broad, northeast-trending trough approximately 300 km wide developed during the Middle Ordovician Taconic Orogeny. Along the southeastern margin, the basin was bordered by a linear belt of uplifted Middle to Upper Ordovician shales and sandstones and, beyond, by the Taconic orogenic belt. The northwestern margin of the Appalachian Basin was bordered by the northern extension of the Findlay Arch, locally referred to as the Algonquin Arch (fig. 14.3). During deposition of the Lockport Group, the western part of the basin was a broad carbonate platform (Zenger 1965).

Throughout the Silurian the northern Appalachian Basin was located approximately 20–25 degrees south of the equator (Van der Voo 1988). Given this position, the climate was likely tropical to subtropical.

STRATIGRAPHY

The Lockport Group in western New York consists of 50–70 m of massive to thin-bedded, pure to argillaceous limestones and dolomites subdivided into

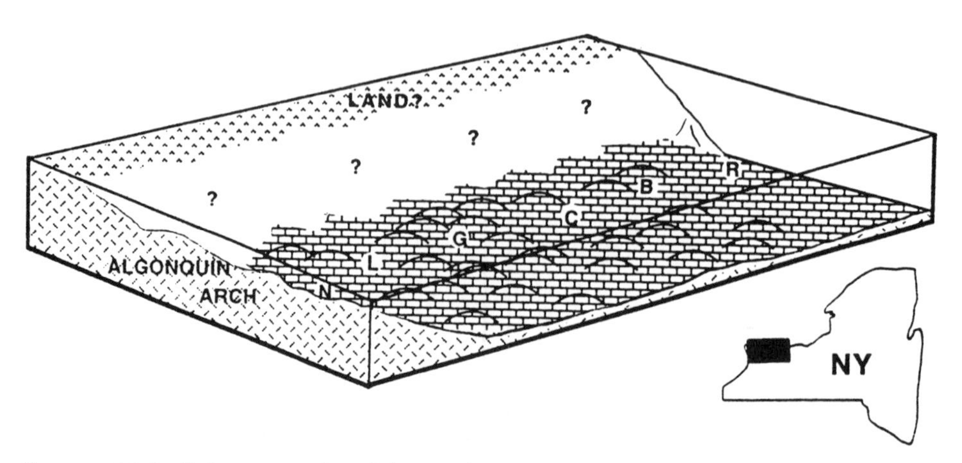

Figure 14.3. Paleogeography of the northwestern Appalachian Basin during the early Ludlovian. Abbreviations are as follows: (N) Niagara Gorge, (L) Lockport, (G) Gasport, (C) Clarendon, (B) Brockport, and (R) Rochester.

four formations primarily on the basis of bedding thickness, grain size, and siliciclastic content. These formations are, in ascending order, the Gasport, Goat Island, Eramosa, and Guelph (stratigraphic nomenclature follows Brett et al. 1995) (fig. 14.4).

The Goat Island Formation is further subdivided into a lower, gray to buff, medium- to thick-bedded, biohermal, dolomitic grainstone termed the Niagara Falls Member; a middle, buff, thin-bedded, fine-grained, cherty dolomite termed the Ancaster Member; and an upper, very fine-grained, largely non-cherty dolomite termed the Vinemount Member. At localities between the Niagara Gorge and Rochester, the Niagara Falls Member is itself characterized by tripartite subdivision into a lower, 0.3–1.3-m-thick interval of fine- to coarse-grained dolomitic crinoidal packstone and grainstone with *Cladopora*/stromatoporoid bioherms; a middle, 0.1–1.0-m-thick interval of thin-bedded, very fine-grained argillaceous dolomite; and an upper, 0.3–3.0-m-thick interval of massive, fine-grained dolomite with a profusion of stromatoporoid heads.

The *Medusaegraptus* epibole is the argillaceous dolomite at the middle of the Niagara Falls Member, and is the source of the fossils described by Ruedemann in 1925, and in this report. This unit overlies the smaller bioherms in the lower part of the Niagara Falls Member and butts sharply against the flanks of the larger bioherms without any evidence of interfingering (fig. 14.5). Thus, the *Medusaegraptus* epibole represents a depositional event that postdates the growth of the associated bioherms. Furthermore, this unit is not restricted to a single "channel" between bioherms, as Ruedemann (1925) maintained, but is laterally extensive. Indeed, this unit spans the entire Frontier Dolomite Quarry exposure at Gasport and has been located in every quarry exposure between Niagara Falls and Brockport, a distance of nearly 100 km (fig. 14.6).

East of Brockport, the sub-Guelph portion of the Lockport Group becomes

BOLTON, 1957	ZENGER, 1965	RICKARD, 1975	THIS REPORT
ALBEMARLE GROUP	LOCKPORT FORMATION	LOCKPORT GROUP	LOCKPORT GROUP
GUELPH FORMATION	OAK ORCHARD MEMBER	OAK ORCHARD FORMATION	GUELPH FORMATION
LOCKPORT FORMATION: ERAMOSA MEMBER			ERAMOSA FM.
	ERAMOSA MBR.	ERAMOSA FM.	GOAT ISLAND FM.: VINEMOUNT MBR.
GOAT ISLAND MEMBER	GOAT ISLAND MEMBER	GOAT ISLAND FORMATION	ANCASTER MBR.
			NIAGARA FALLS MEMBER
GASPORT MEMBER	GASPORT MEMBER	GASPORT FORMATION	GASPORT FM.: PEKIN MBR.
			GOTHIC HILL MEMBER

FIGURE 14.4. Stratigraphic nomenclature of the Lockport Group.

quite sandy. This sandy facies persists eastward to Sodus, beyond which the Lockport Group is composed primarily of stromatolitic carbonates and black shales (Sconondoa and Ilion formations). To date, no vestige of the *Medusaegraptus* epibole has been observed in these eastern facies of the Lockport Group.

In the Niagara Gorge and localities to the west, the *Medusaegraptus* epibole appears to be missing as the result of erosional truncation beneath the upper part of the Niagara Falls Member. In the north end of the Niagara Gorge, erosional events evidently also removed the lower part of the Niagara Falls Member and the underlying Pekin Member of the Gasport Formation (Brett et al. 1995). Stratigraphic unconformity in this area may reflect episodic uplift along the Algonquin Arch.

The lower contact of the *Medusaegraptus* epibole is abrupt and well-defined in the eastern and western parts of the study area, but becomes somewhat cryptic in areas of the Frontier Dolomite Quarry at Gasport where the epibole overlies interreef strata of the lower part of the Niagara Falls Member. Where bioherms are present in the lower Niagara Falls Member at Gasport, however, the contact is sharply delineated. Dark-stained debris surrounding these bioherms, and pitted and corroded bioherm surfaces at this and other localities, including the Frontier Dolomite Quarry at Lockport and the Iroquois Stone Quarry at Brockport, indicate subaerial exposure of these structures prior to deposition of the epibole sediments (Crowley 1973).

At the Niagara Stone Quarry in Niagara Falls and the Frontier Dolomite Quarry in Lockport, the *Medusaegraptus* epibole is approximately 10 cm thick. The interval is considerably thicker at the Frontier Dolomite Quarry in Gasport, where a maximum of 100 cm is exposed, and even this may not represent the full thickness because the top is unconstrained. At the Genesee-LeRoy Stone Quarry in Clarendon, the unit is 15 cm thick, and at the Iroquois Stone Quarry

Figure 14.5. West face of a small quarry exposure located on the north side of Upper Mountain Road, across from the main Frontier Dolomite pit at Gasport. Small bioherm at **(a)** is developed in the lower part of the Niagara Falls Member of the Goat Island Formation. The *Medusaegraptus* epibole extends over the top of the structure. The bioherm at **(a)** is situated near the apex of a larger bioherm (b) in the underlying Pekin Member of the Gasport Formation but is separated from the underlying structure by a thin zone (30 cm) of nonbiohermal lower Niagara Falls grainstone. (Note stadia rod, in feet, for scale.)

in Brockport 10 cm have been observed, although at Brockport, as at Gasport, the unit represents top of bedrock and therefore may originally have been thicker.

Age

Conodont evidence indicates that the *Medusaegraptus* epibole is located near the top of the range of *Ozarkodina sagitta sagitta*, a position that suggests an earliest Ludlovian age for this unit (M. Kleffner, personal communication). The stratigraphic distribution of *M. mirabilis* is also suggestive of an earliest Ludlovian age, as this species is otherwise known only from the lower part of the well-dated, earliest Ludlovian Mississinewa Shale of northern Indiana (Shrock 1928; Erdtmann and Prezbindowski 1974; Wood 1975). Fragmentary graptolites yielded by the *Medusaegraptus* epibole were recently described as the early Ludlovian taxa *Saetograptus chimaera*, *S. colonus?*, and *Spinograptus spinosus* (LoDuca and Brett 1990, 1991). However, the *Spinograptus* specimens evidently represent fragmentary dendroids, and the identity of the other species remains open to question because of poor preservation (D. Loydell, personal communication). Acritarch specimens from this interval also are too poorly preserved to provide conclusive age information in this regard (M. Miller, personal communication). Thus available evidence suggests, and is consistent with, an earliest

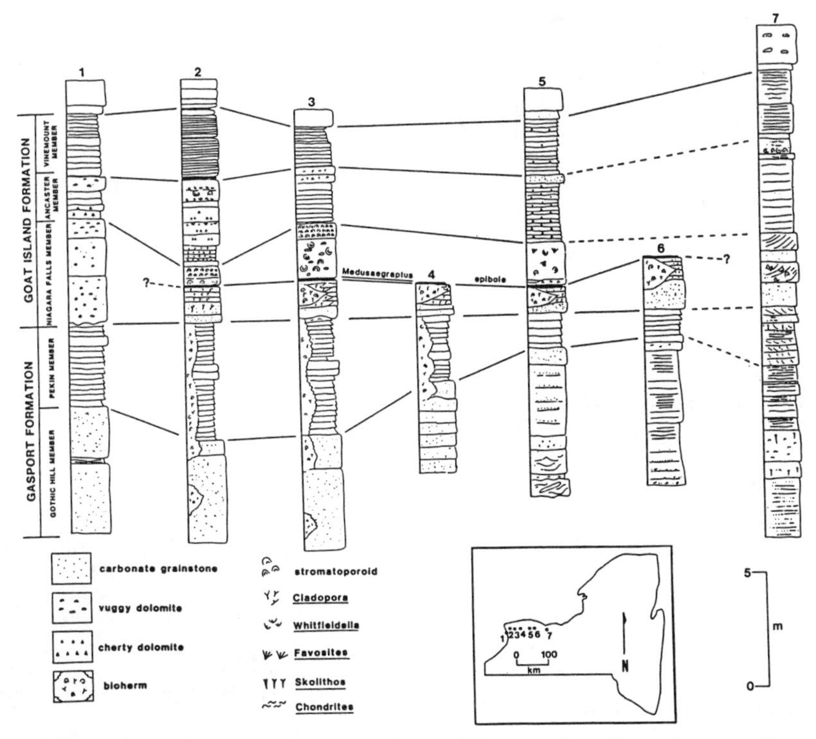

FIGURE 14.6. Correlated section across the western New York study area, showing the position and lateral extent of the *Medusaegraptus* epibole. The numbered localities are the same as in fig. 14.1.

Ludlovian age for the *Medusaegraptus* epibole (fig. 14.7), but, pending the recovery of well-preserved graptoloid graptolite material, this assignment must be regarded as tentative.

LITHOLOGY AND SEDIMENTOLOGY

The *Medusaegraptus* epibole consists predominantly of fine- to medium-grained, thin-bedded dolomite. Weathered exposures are buff to ocher; fresh surfaces are dark brown to gray. Where the epibole is particularly fine-grained, as at Gasport, concave-up conchoidal fractures 0.3 m in diameter are common on weathered bedding plane surfaces.

The fine-grained component of the sediment is composed primarily of euhedral dolomite rhombs approximately 20 μm in length. Insoluble residue analyses revealed a consistent siliciclastic content of approximately 15% across the study area, with most of this material being very fine quartz silt (fig. 14.8). Locally, muscovite is present, usually as flakes less than 1 mm across. The persistence of equable siliciclastic values away from the primary siliciclastic source to the east and perpendicular to basin strike suggests the influence of a sediment source, and shoreline, to the north. Zenger (1965) also noted a similar

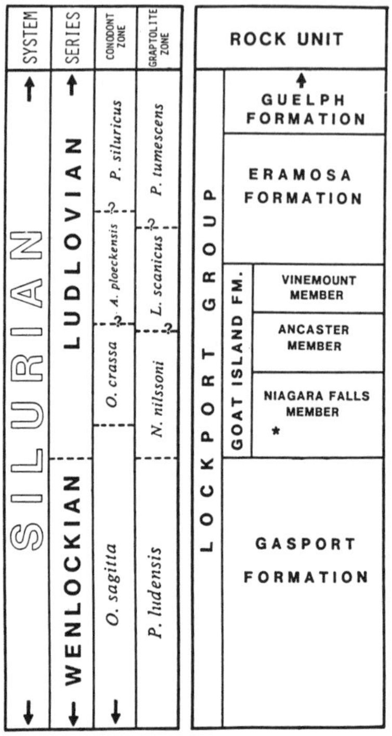

SYSTEM
SERIES
CONODONT ZONE
GRAPTOLITE ZONE
ROCK UNIT
SILURIAN
LUDLOVIAN
WENLOCKIAN
P. siluricus
A. ploeckensis
O. crassa
O. sagitta
P. tumescens
L. scanicus
N. nilssoni
P. ludensis
?
LOCKPORT GROUP
GOAT ISLAND FM.
GUELPH FORMATION
ERAMOSA FORMATION
VINEMOUNT MEMBER
ANCASTER MEMBER
NIAGARA FALLS MEMBER
*
GASPORT FORMATION

PERCENT INSOLUBLE
0
10
20
NIAGARA
LOCKPORT
GASPORT
BROCKPORT

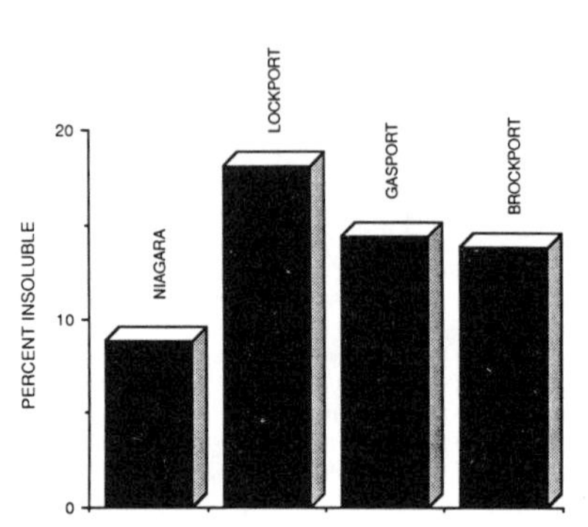

Figure 14.7. Chronostratigraphic placement of the *Medusaegraptus* epibole (indicated by *).

Figure 14.8. Insoluble-residue content of the *Medusaegraptus* epibole.

trend in the insoluble content of other units within the Lockport Group and likewise postulated the presence of a nearby northern shoreline. The total organic content of the sediment, at 0.1%, is quite low.

The origin of the dolomite in the *Medusaegraptus* epibole is somewhat problematic, as undistorted and detailed soft-bodied fossil preservation appears to be antithetic to a replacement origin. Erdtmann and Prezbindowski (1974) were similarly puzzled by the presence of detailed soft-bodied fossils in the dolomites of the Mississinewa Member at Huntington, Indiana, and suggested a detrital origin for an originally dolomitic sediment. However, decalcified calcium carbonate shells and large vugs up to 10 cm in diameter occur in both the *Medusaegraptus* epibole and Mississinewa deposits and are consistent with secondary dolomitization, and dolomitization of lithographic limestones evidently can occur without appreciable changes in grain size (Van Tuyl 1916). Thus the dolomite of the epibole is believed to be of secondary origin.

The *Medusaegraptus* epibole is composed of a series of 3–8-cm thick, normally graded beds with sharp, irregular bases (fig. 14.9). The lower, 1–5-cm-thick, "lag" parts of the beds are composed primarily of crinoid debris and are cross-bedded locally. These lags are relatively coarse-grained in the eastern and western parts of the study area (1.5–2.5-mm clasts) (fig. 14.9B, C, D) and become finer-grained toward Gasport (0.5–1.5-mm clasts) (fig. 14.9A). Laterally discontinuous laminations approximately 0.1 mm thick are well developed within the upper, fine-grained, dolomicrite "caps" of the graded beds, and these occasionally fill small, concave-up depressions 2–4 cm wide and 4–8 mm deep. These small-scale scour-and-fill structures are analogous to the "crescent structures" described by Sunderman and Mathews (1975) from the Mississinewa Shale of northeastern Indiana. Burrows are rare and other sedimentary structures, including ripples, are absent.

The sharp-based, graded beds of the epibole are regarded as tempestites. The scour-and-fill structures in the dolomicrite caps were probably developed during the waning stages of storm activity, and wave energy on the bottom between storm events likely was minimal.

Biota

The unusual biota of the *Medusaegraptus* epibole is dominated by thallophytic algae ("seaweeds") and dendroid graptolites. Typical Silurian shelly fossils, such as brachiopods, bryozoa, bivalves, trilobites, and crinoids, are rare or absent. Overall, the unit is poorly fossiliferous, although thallophytic algae and dendroid graptolites are abundant locally.

Fossils of soft-bodied organisms are most abundant and diverse at Gasport. Both east and west of this locality, the diversity of the soft-bodied biota decreases significantly, although a few dendroid graptolite taxa and the alga *Medusaegraptus* continue to be abundant, and calcified taxa, including the corals *Cladopora* and *Favosites* and stromatoporoids, become important constituents.

A summary of the various elements of the biota is provided below and in table 14.1. Much of the formal taxonomic treatment of the biota was adequately and ably addressed by Ruedemann (1925). However, he originally described *Medusaegraptus mirabilis* and *Diplospirograptus goldringae* as dendroid graptolites; these forms have since been redescribed as thallophytic algae (LoDuca 1990a, b).

Porifera Small stromatoporoids, up to 10 cm in diameter, occur infrequently in the epibole at Gasport and are common in the epibole at Lockport and Brockport. Other sponge fossils, even isolated spicules, are absent.

Brachiopoda Inarticulate brachiopods are common in the epibole and occur at all localities. Inarticulate representatives include *Lingula lamellata, Schizotreta tenuilamellata,* and *Trematis spinosa* (fig. 14.10d, g-i). Specimens of *T. spinosa* are of particular interest because the numerous, long, hairlike spines have been preserved. None of the inarticulates were cemented to hard substrates.

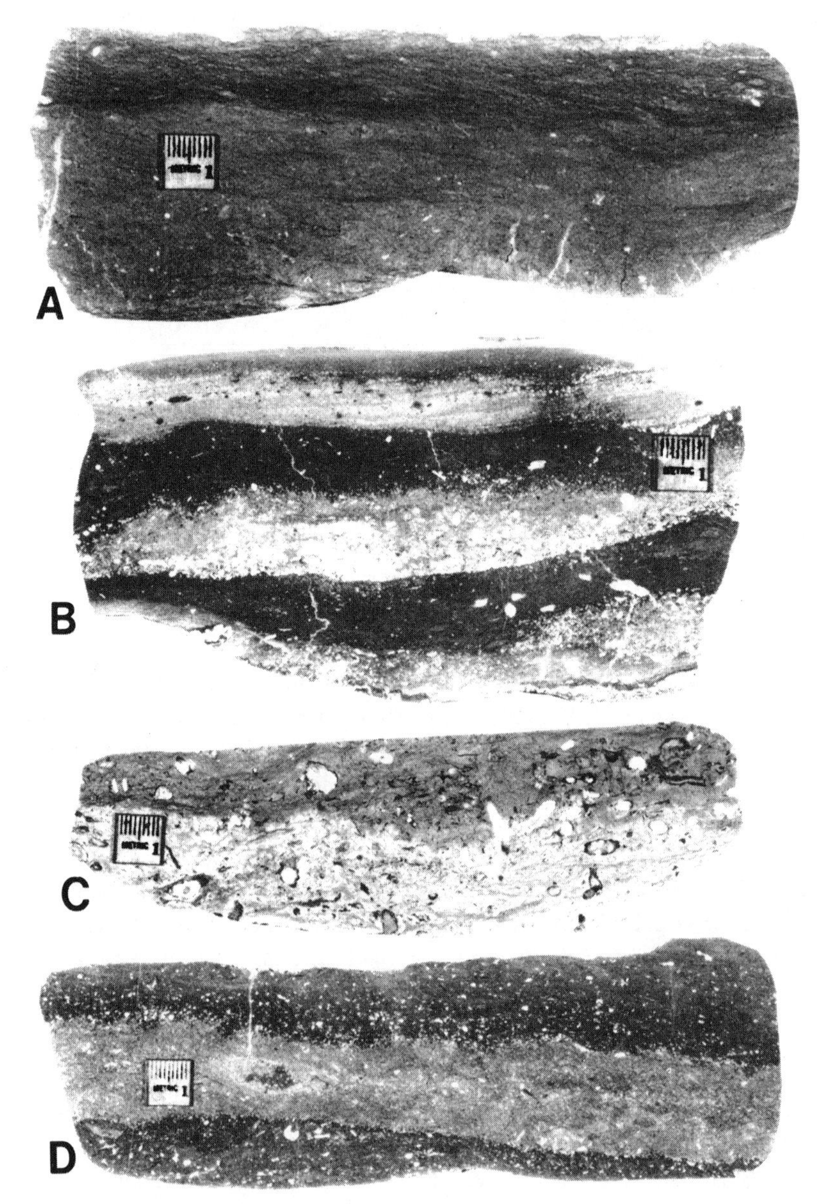

FIGURE 14.9. Polished slabs of the *Medusaegraptus* epibole from the following localities: (A) Frontier Dolomite Quarry, Gasport, a single tempestite. Note the lower coarse lag bed and the upper fine-grained cap. (B) Frontier Dolomite Quarry, Lockport. (C) Iroquois Stone Quarry, Brockport. Note large fragments of the coral *Cladopora* at the base. (D) Niagara Stone Quarry, Niagara.

TABLE 14.1 *Summary of the biota from various exposures of the* Medusaegraptus *epibole in western New York State.*

	N	*L*	*G*	*C*	*B*
DENDROID GRAPTOLITES					
Family Dendrograptidae					
Dendrograptus praegracilis Spencer*		R	R		
Callograptus niagarensis Spencer			C		
Calyptograptus cyathiformis Spencer		C	R		
Dictyonema crassibasale Gurley		C	A	C	R
D. infundibuliforme Ruedemann			A		
D. expansum Spencer			C		
D. filiramus Gurley		C	A	A	C
D. retiforme Hall			R		
D. tenellum Spencer			C		
Desmograptus micronematodes (Spencer)		A	A	A	A
Family Acanthograptidae					
Inocaulis ramulosus Spencer			R		
I. plumulosus Hall		A	A	C	A
Acanthograptus walkeri (Spencer)			A		
A. granti Spencer			C		
A. pulcher Spencer*			C		
GRAPTOLOID GRAPTOLITES					
Saetograptus sp.			?		
ALGAE					
Medusaegraptus mirabilis Ruedemann	R	C	A	C	A
Diplospirograptus goldringae Ruedemann			A		
"Chondrites" versus Ruedemann			C		
Powysia sp.*			C		C
PORIFERA					
Stromatoporoid*	C	C	R	R	C
COELENTERATA					
Cladopora sp.*	A	A	C	A	A
Favosites sp.*		C	R	R	C
Enterolasma sp.*					R
BRACHIOPODA					
Inarticulata					
Lingula lamellata Hall	R	C	A	C	A
Trematis spinosa Ruedemann			A		R
Schizotreta tenuilamellata (Hall)*			C		
Articulata					
Eospirifer sp.*			?		
Coolinia striata (Hall)*			R		R
MOLLUSCA					
Cephalopoda					
Dawsonoceras annulatum (Sowerby)*		R	C		C
Kionoceras cancellatum (Hall)*		R	C		C
Orthoceras simulator (Hall)*			R		
Cyrtiactinoceras sp.*			C		C
Gastropoda					
Eotomaria sp*.			A		C
ANNELIDA					
Protoscolex batheri Ruedemann				R	
Serpulites sp.*			C	R	
ARTHROPODA					
Trilobita					
Bumastus niagarensis (Whitfield)*			R		
Dalmanites sp.			R		
Phyllocarida					
Ceratiocaris sp.*			R		

Table 14.1 *Summary of the biota from various exposures of the* Medusaegraptus *epibole in western New York State. (Continued)*

	N	*L*	*G*	*C*	*B*
CONULARIA					
Conularia niagarensis Hall			C		C
C. tenuicosta Ruedemann			C	R	C
TRACE FOSSILS					
Lockportia conspicua (Ruedemann)			C	R	

Site abbreviations: (N) Niagara Stone Quarry, Niagara Falls; (L) Frontier Dolomite Quarry, Lockport; (G) Frontier Dolomite Quarry, Gasport; (C) Genesee—LeRoy Stone Quarry, Clarendon; (B) Iroquois Stone Quarry, Brockport. Asterisks indicate species not reported by Ruedemann (1925). R = rare, 1–5 specimens; C = common, 5–15 specimens; A = abundant, more than 15 specimens.

Articulate brachiopods are extremely rare in the epibole. Three specimens of *Coolinia* sp. (fig. 14.10l) and a single, poorly preserved spiriferid were recovered from the epibole at Gasport. Articulate brachiopods were not observed at any of the other localities.

Conularia Conulariids occur in abundance in the epibole at several localities, most notably Gasport, and are represented by two species, *Conularia niagarensis* and *C. tenuicosta* (fig. 14.10a, b). No conulariid specimens have been observed attached to hard substrates, although Ruedemann (1925, plate 22, fig. 3) figured two specimens of *C. tenuicosta* joined at the appices. Unfortunately, conulariids have not been recovered with preserved soft parts.

Coelenterata Corals are rather rare in the epibole at Gasport, although thickets of *Cladopora* represent a significant part of the biota at Brockport, Lockport, and Niagara (see fig. 14.9c). Other coelenterate representatives include the tabulate *Favosites* (fig. 14.10c) and the rugosan *Enterolasma*.

Mollusca The gastropod *Eotomaria* occurs in the epibole at most localities and is most abundant at Gasport (fig. 14.10f). Cephalopods representing the genera *Dawsonoceras, Kionoceras,* and *Cyrtiactinoceras* are common. Most occur as fragments (fig. 14.10k), although occasionally complete phragmocones are found. (The original aragonite is often completely dissolved, and all that remains is a thin organic film that overlies an internal mold.) Bivalves are absent.

Annelida Although worms are very well preserved in the epibole, they are rare. Two complete and several incomplete specimens of the oligochaete(?) *Protoscolex batheri* have been recovered from the epibole at Gasport (fig. 14.11a–e). One partial specimen apparently displays details of the musculature (fig. 14.11f). Specimens of *Protoscolex* have not been recovered from the other localities examined, although this may simply reflect the fact that the surface area

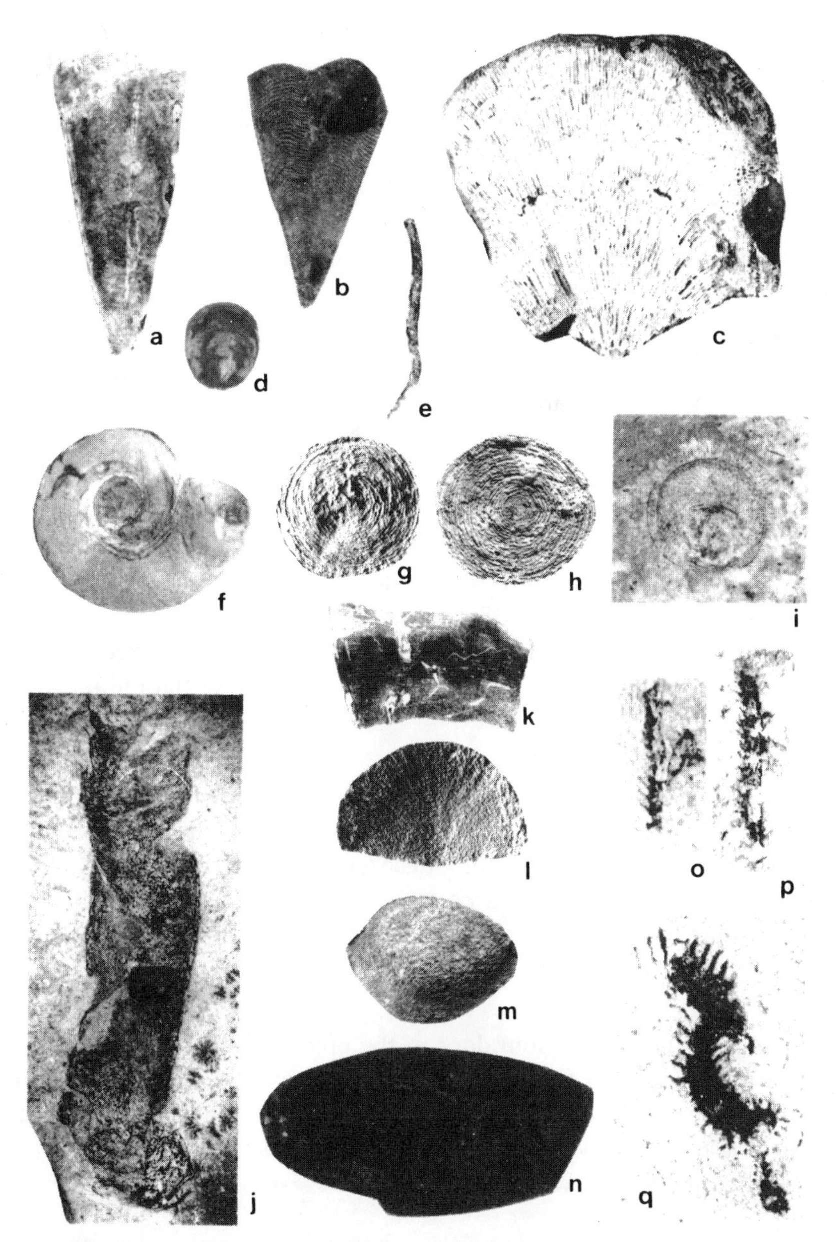

FIGURE 14.10. Biota of the *Medusaegraptus* epibole (all specimens were collected by LoDuca unless otherwise noted). (a) *Conularia tenuicosta* Ruedemann, Gasport, × 1; (b) *Conularia niagarensis* Hall, Gasport, × 3/4; (c) *Favosites* sp., Gasport, × 1/3; (d) *Lingula lamellata* Hall, Gasport, × 1; (e) *Serpulites* sp., Gasport, × 3/4; (f) *Eotomaria* sp., Gasport, × 3/4; (g) *Schizotreta tenuilamellata* (Hall), pedicule valve, Gasport, × 1; (h) *Schizotreta tenuilamellata* (Hall), brachial valve, Gasport, × 1; (i) *Trematis spinosa* Ruedemann, brachial valve, Gasport, × 1; (j) *Lockportia conspicua* (Ruedemann), an organic-lined trace fossil, Gasport, × 3/4; (k) *Dawsonoceras* sp., portion of a phragmocone, Gasport, × 1; (l) *Coolinia* sp. Gasport, × 3/4; (m) *Bumastus* sp. pygidium, Gasport, × 1 1/4; (n) *Ceratiocaris* sp., carapace, Gasport, × 3/4; (o) *Saetograptus?* sp., Gasport, × 2.25, NYSM 15067; (p) *Saetograptus?* sp., Gasport, × 2.25, NYSM 15066; (q) unidentified graptolite, previously misidentified as *Spinograptus spinosus*, Clarendon, × 2.25, NYSM 15068. (NYSM = New York State Museum)

available for investigation at these sites is considerably smaller than at the Gasport locality.

Scolecodonts are abundant in the epibole at all localities and are represented by *Arabellites* sp. and several as yet unidentified forms. To date, no soft parts of the scolecodont-bearing animal have been recovered. The problematic phosphatic worm (?) tube *Serpulites* has been recovered from the epibole at Gasport and Clarendon (fig. 14.10e).

Arthropoda Arthropods are extremely rare within the epibole. A phyllocarid referrable to the genus *Ceratiocaris* occurs in the epibole at Gasport (fig. 14.10n). Trilobites are represented by a bumastid (a single pygidium) (fig. 14.10m) and a dalmanitid (one complete specimen), both from Gasport.

Dendroid Graptolites Dendroid graptolites are the most diverse taxon within the epibole. At Gasport, 15 species were recovered (fig. 14.12a–h). The abundance and diversity of the dendroid graptolite fauna decrease both east and west of Gasport, although *Dictyonema* remains an important faunal element at all localities.

Graptoloid Graptolites Graptoloid graptolites possibly representing the genus *Saetograptus* were recovered during this study from the Frontier Dolomite Quarry at Gasport (fig. 14.10o–p). These specimens are the first graptoloid graptolites reported from the Lockport Group, although their identity remains open to question because of poor preservation (D. Loydell, personal communication).

Acritarchs Acritarchs are abundant in the epibole, although in general they are too poorly preserved to permit species-level identification. The following genera were recovered from the epibole at Gasport (M. Miller, personal communication): *Leiosphaeridia, Multiplicisphaeridium, Diexallophasis, Veryhachium, Leiofusa, Tunisphaeridium,* and *Micrhystridium.*

Algae Thallophytic algae, primarily the large noncalcified dasyclad *Medusaegraptus mirabilis* (fig. 14.13b–d) (LoDuca 1990b), are the most abundant fossils within the *Medusaegraptus* epibole at Gasport, where *M. mirabilis* accounted for 57% of the fossils recovered in a systematic survey. This corresponds to an average density of two individuals per square meter, although the distribution actually was uneven and patchy. A second, smaller alga, *Diplospirograptus goldringae* (fig. 14.13f) also occurs at Gasport, but is not nearly as abundant. Other likely green algae recovered from the epibole are *Powysia* sp. (fig. 14.13e), "*Chondrites*" *verus* (perhaps related to the extant green alga *Codium*) (fig. 14.13a), and an as yet undescribed species of *Chaetocladus* (fig. 14.13g–i).

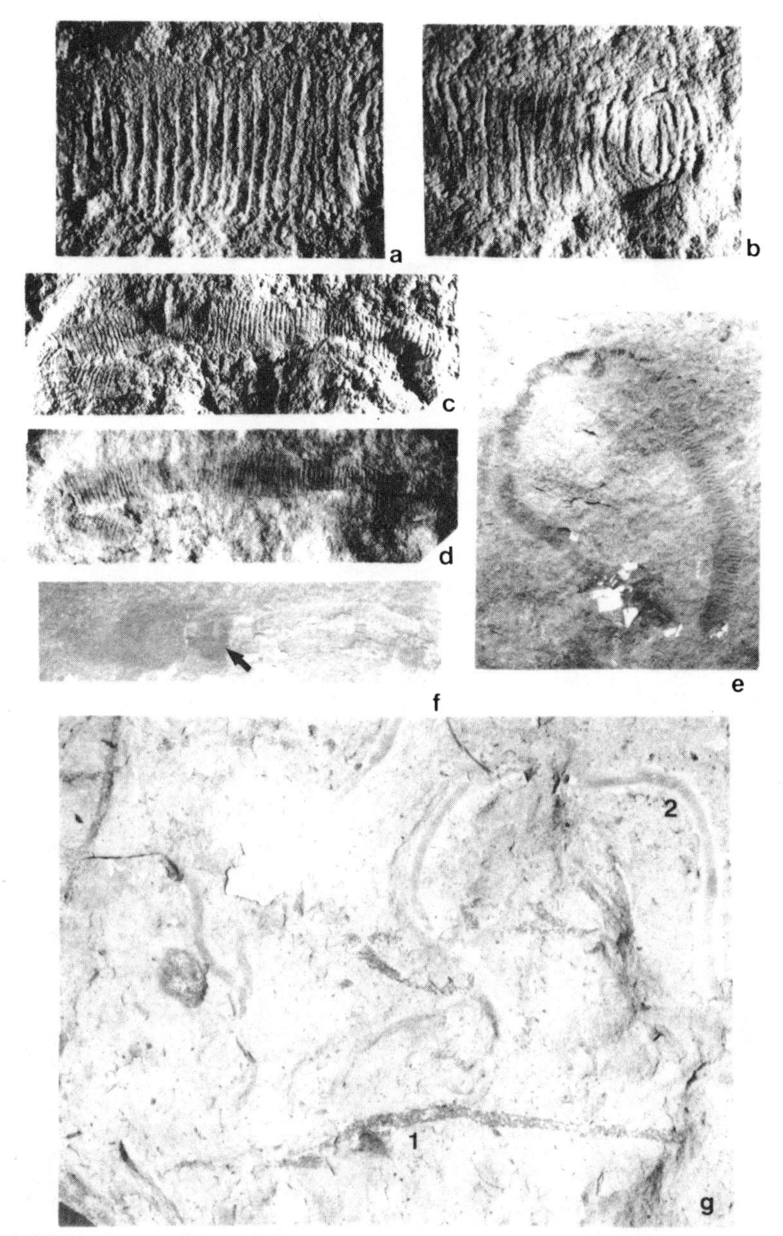

FIGURE 14.11. Biota of the *Medusaegraptus* epibole. (a) *Protoscolex batheri* Ruedemann, Gasport, × 4, close-up of the midarea of the specimen shown in (c); (b) close-up of the proximal end of the specimen shown in (c), × 4; (c) Complete specimen of *P. batheri*, × 1; (d) specimen in (c) photographed under water. Note the presence of a gut trace, × 1; (e) type specimen, NYSM 7848, × 1; (f) partial specimen that appears to retain traces of longitudinal musculature (arrow), × 1; (g) trace fossils from the *Medusaegraptus* epibole at Gasport, 1 = Type 3 trace, 2 = Type 2 trace, × 3/4.

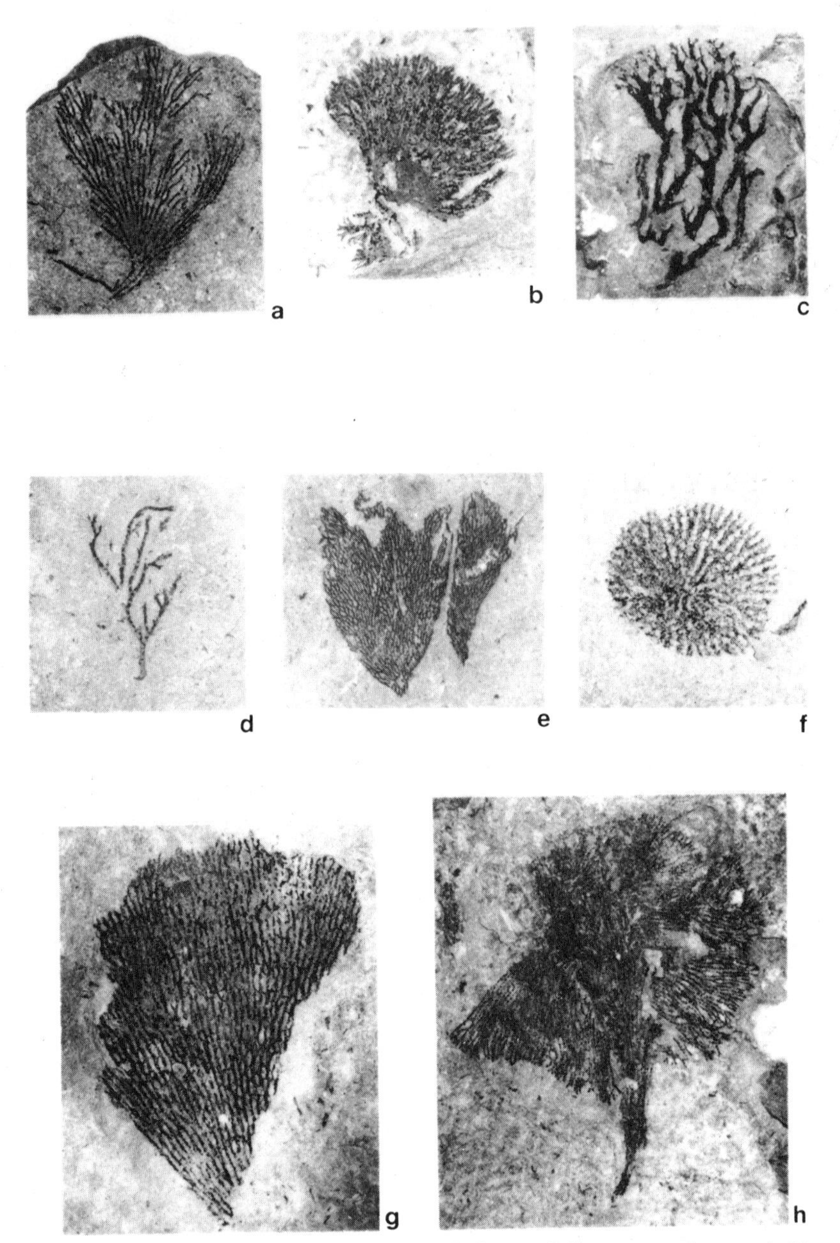

FIGURE 14.12. Biota of the *Medusaegraptus* epibole. (a) *Callograptus niagarensis* (Spencer), Gasport, × 1; (b) *Acanthograptus pulcher* Spencer, Gasport, × 3/4; (c) *Inocaulis ramulosus* Spencer, Gasport, × 3/4; (d) *Dendrograptus praegracilis* Spencer, Gasport × 3/4; (e) *Desmograptus micronematodes* (Spencer), Clarendon, × 1; (f) *Dictyonema crassibasale* Gurley, juvenile, Gasport, × 2.25; (g) the same as (f), adult, Gasport, × 2/3; (h) *Dictyonema tenellum* (Spencer), Gasport, × 3/4.

The abundance and diversity of thallophytic algae within the epibole decrease both east and west of Gasport. *Medusaegraptus mirabilis,* however, has been recovered from every locality where the epibole is exposed, although specimens are significantly better preserved at Gasport than elsewhere.

Trace Fossils Trace fossils are rare within the *Medusaegraptus* epibole and are represented only by shallow, horizontal forms. Three types of traces are recognized:

Type 1: Blind-ended, unbranched, organic-lined burrows up to 3 cm wide and 30 cm long, termed *Lockportia conspicua* (Ruedemann) (probably dwelling traces) (fig. 14.10j).
Type 2: Long, sinuous, smooth-walled, unbranched burrows up to 5 mm wide (probably made by *Protoscolex batheri*) (fig. 14.11g).
Type 3: Long, sinuous, unbranched burrows up to 7 mm wide, with a distinctive stippled pattern (possibly made by gastropods) (fig. 14.11g).

All burrows, with the exception of localized (30 cm^2) clusters of Type 2 burrows, are widely scattered.

Taphonomy

All of the soft-bodied fossils in the *Medusaegraptus* epibole occur in the fine-grained caps of the graded beds. Thallophytic algae and dendroid graptolites are preserved as rather sharply delineated carbonized compressions. Peels have failed to reveal cellular-level details of the algae. Annelid worms are preserved as detailed impressions with a coating of light blue phosphate, and one specimen (fig. 14.11f) apparently reveals portions of the musculature. The phosphatic skeletons of conulariids and inarticulate brachiopods are essentially unaltered. Calcium carbonate shell material is decalcified and is often only represented by molds, as has been noted in other marine Lagerstätten, including the Silurian (Llandoverian) Brandon Bridge Lagerstätte of southeastern Wisconsin (Mikulic et al. 1985) and the Jurassic Osteno Limestone Lagerstätte of northern Italy (Pinna 1985).

Soft-bodied fossil preservation is best at Gasport. Here, complete and *in situ* examples of *Medusaegraptus* and dendroid graptolite specimens are common, and transport of the biota prior to burial appears to have been minimal. At all other localities, however, most of the algae and graptolites are damaged or fragmentary and *in situ* specimens are absent, evidently indicating that in these areas preburial transport was more substantial.

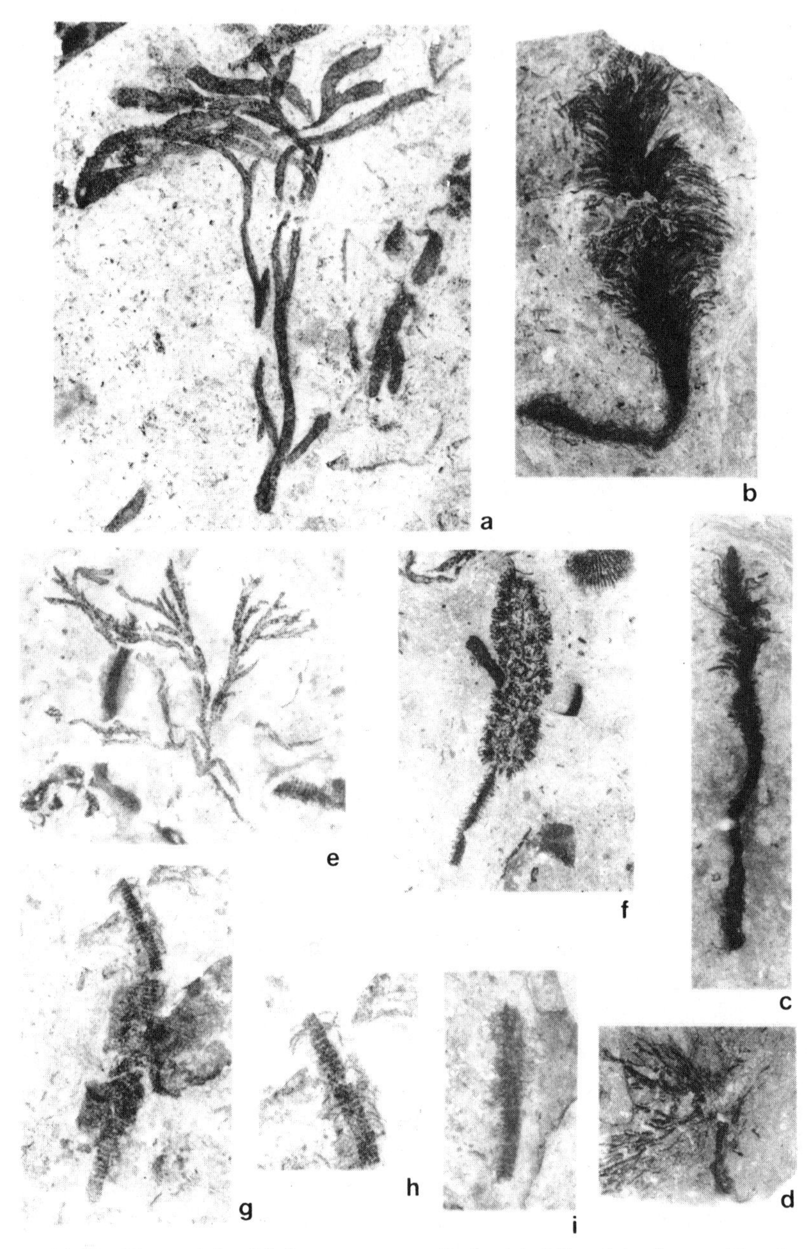

FIGURE 14.13. Biota of the *Medusaegraptus* epibole. (a) "*Chondrites*" *verus* Ruedemann, type specimen (NYSM 6491), Gasport, × 2/3; (b) *Medusaegraptus mirabilis* Ruedemann, Gasport, × 3/4; (c) *M. mirabilis*, Brockport, × 3/4; (d) *M. mirabilis*, Niagara, × 3/4; (e) *Powysia* sp., Gasport, × 3/4; (f) *Diplospirograptus goldringae* Ruedemann, Gasport, × 3/4; (g) *Chaetocladus* sp., Gasport, × 3/4; (h) enlargement of (g), note the euspondyl ramifications, × 1.5; (i) *Chaetocladus* sp., Gasport, × 3/4.

Paleosynecology

The biota of the *Medusaegraptus* epibole, in the strictest sense, does not represent an interreef community, because the bioherms that were present were relict structures. However, the relief provided by the underlying bioherms likely had an important role in providing the sedimentological and environmental conditions necessary for the establishment and subsequent preservation of this biota.

Through detailed mapping and sampling of a large bedding plane surface exposed at the Frontier Dolomite Quarry at Gasport, it became apparent that the distribution of species in the epibole was neither random nor uniform (fig. 14.14). Two distinct associations are recognized: the dendroid graptolite association and the *Medusaegraptus mirabilis* association.

The dendroid graptolite association at Gasport is characterized by an abundant and diverse dendroid graptolite fauna in association with conulariids; inarticulate brachiopods; annelid worms; thallophytic algae, including *"Chondrites" verus* and the dasyclad *Medusaegraptus mirabilis;* and occasional, small, stromatoporoids and corals (*Cladopora*) (fig. 14.15). No fewer than 10 dendroid graptolite species are represented. This diverse association—which yielded most of the taxa recovered from Gasport—is not representative of the epibole at Gasport as a whole, however, and is restricted in distribution to two small areas, both located along topographic highs provided by the underlying bioherms (fig. 14.14).

The *Medusaegraptus mirabilis* association dominates the Gasport site and is characterized by an abundance of the dasyclad alga *Medusaegraptus mirabilis* and the near absence of any associated taxa (fig. 14.16). The thallophytic alga *Diplospirograptus goldringae* and the gastropod *Eotomaria* sp. are the only other commonly associated species. Conulariids, corals, and stromatoporoids have never been observed. Dendroid graptolites are rare in this association and are represented principally by widely scattered specimens of *Dictyonema crassibasale.* Distribution of the *Medusaegraptus mirabilis* association at the Gasport site was patchy, not uniform; many of the densest patches of *M. mirabilis* occurred in the topographically lowest portions of the site (fig. 14.14).

The distribution of the two associations recognized at Gasport appears to have been controlled by the topography afforded by the underlying bioherms. The dendroid graptolite association is restricted to the topographically highest surfaces at Gasport, areas corresponding to the shallowest, and perhaps best oxygenated water during deposition. In the slightly deeper and perhaps more stagnant areas away from these highs, the biota was dominated by *Medusaegraptus mirabilis.*

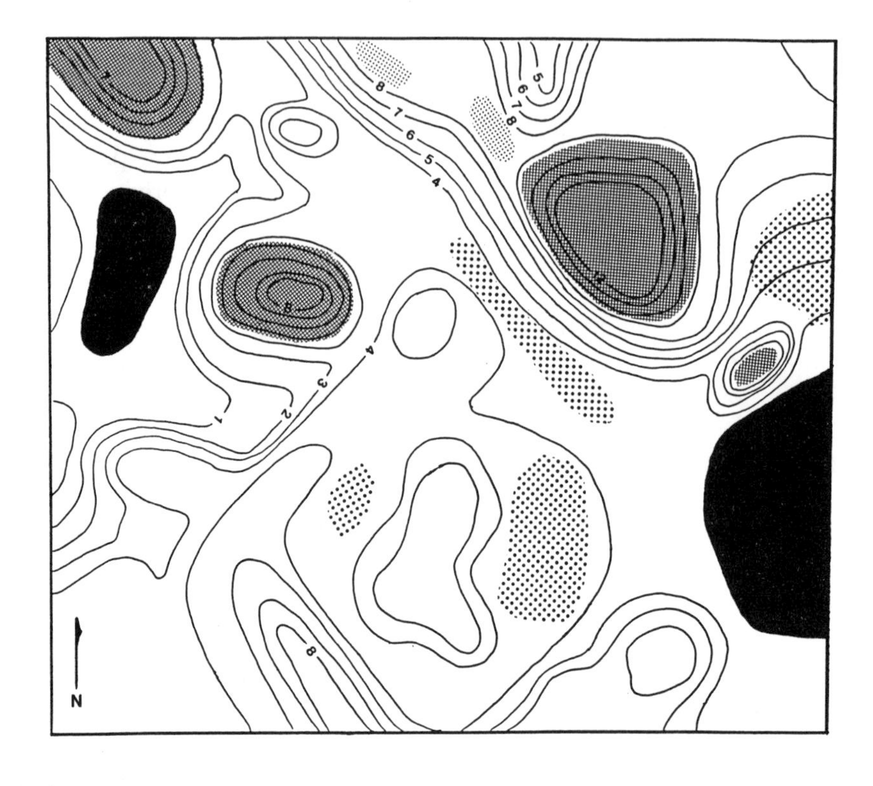

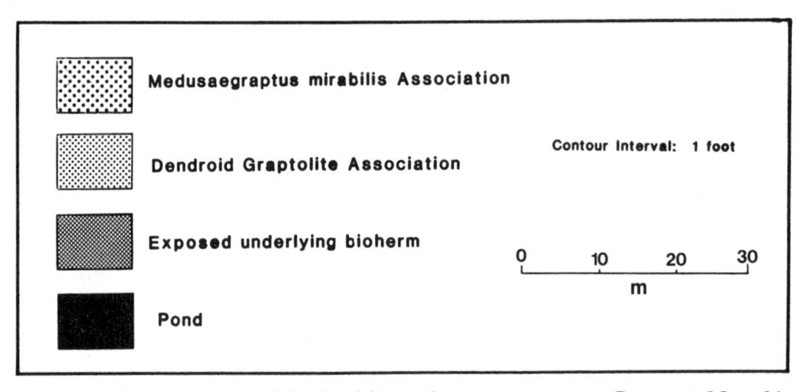

Figure 14.14. Contour map of the bedding plane exposure at Gasport, New York, showing the distribution of the dendroid graptolite association and the *Medusaegraptus mirabilis* association.

Comparison with Modern Analogues

With the exception of dendroid graptolites, which have been extinct since the Carboniferous, alga-dominated communities remarkably similar to the biota represented in the *Medusaegraptus* epibole occur in modern, shallow marine settings. For instance, Upchurch (1970) described a biota dominated by patches of green algae (80% of the total biota), worms, and gastropods from a dysaero-

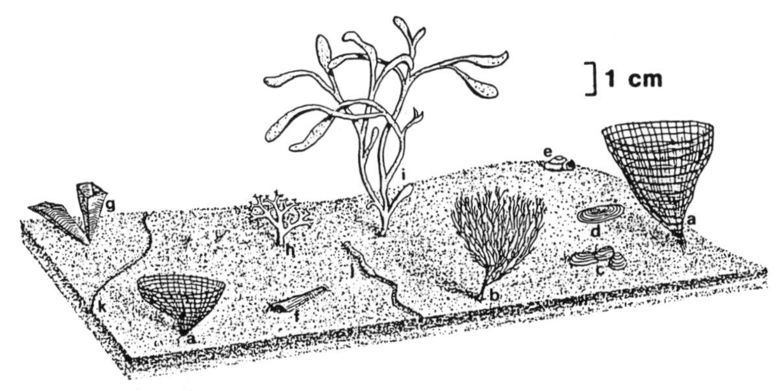

FIGURE 14.15. Reconstruction of the dendroid graptolite association. The biota is as follows: (a) *Dictyonema crassibasale,* (b) *Callograptus niagarensis,* (c) *Lingula lamellata,* (d) *Schizotreta tenuilamellata,* (e) *Eotomaria* sp., (f) fragment of *Dawsonoceras* sp., (g) *Conularia niagarensis,* (h) *Powysia* sp., (i) *"Chondrites" versus,* (j) *Protoscolex batheri,* (k) Type 2 trace.

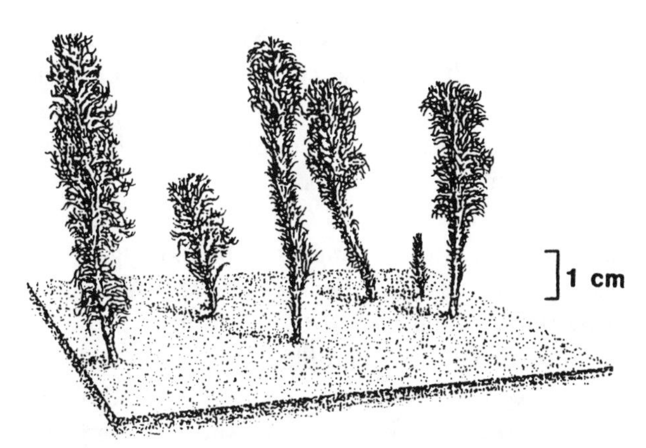

FIGURE 14.16. Reconstruction of the *Medusaegraptus mirabilis* association (a monospecific "patch" is depicted).

bic lagoon in Bermuda. Similar alga-dominated associations have been reported from lagoons and inlets surrounding the Caribbean island of Curaçao (Van den Hoek et al. 1972), and in general are typical of restricted, shallow, tropical marine environments (Taylor 1960).

PALEOENVIRONMENT

Depth The abundance of the dasyclad alga *Medusaegraptus mirabilis* in the *Medusaegraptus* epibole sets strict limitations on water depth during deposition. Modern dasyclads occur in abundance only in waters shallower than 10 m (Wray 1977; Rezak 1990, personal communication). Elliott (1968), based on a study of Permian to Paleocene dasyclad algae, concluded that fossil taxa were similarly restricted with regard to depth. Thus it is unlikely that the epibole was deposited in water deeper than 10–15 m (fig. 14.17). In addition, the presence of the coral *Cladopora* in the epibole at Brockport and Lockport is indicative of a Benthic Assemblage 3 association (see Boucot 1975). According to Eckert and Brett (1989), the BA-3 association was typically developed in waters 10–20 m deep, an estimate consistent with the depth suggested by algal evidence.

Through the use of Airy wave theory, a minimum depth of deposition for the *Medusaegraptus* epibole can be calculated based upon the relationship between wave-generated current velocity and sediment deposition from suspension (see LoDuca 1990a for a complete discussion). At 25 degrees south latitude, the Appalachian Basin would have been subjected to average fair-weather wind speeds of approximately 5 knots. Under these conditions, a minimum water depth of 5 m is required before wave energy is sufficiently reduced to allow fine-grained sediment, such as that of the *Medusaegraptus* epibole, to settle from suspension and accumulate on the bottom. Note, however, that this figure is based upon open-ocean conditions; the current-baffling effect imposed by the relict bioherms present during deposition of the *Medusaegraptus* epibole would have significantly reduced the minimum depth at which the accumulation of these fine-grained sediments could have occurred.

Oxygen During deposition *Medusaegraptus* epibole, oxygen was likely entirely absent below, and perhaps for some distance above, the sediment/water interface, as indicated by the unusual and restricted nature of the biota, the lack of bioturbation, and the retardation of aerobic decay (figs. 14.17–14.20). Current-baffling imposed by the relict bioherms and decay of the abundant benthic algae likely contributed to the development of anoxia. The fact that deposition of this unit apparently coincided with the initiation of a eustatic, earliest Ludlovian transgression (see Hurst 1975; McKerrow 1979) may also be significant (fig. 14.21). Kauffman (1986) noted that regional anoxia of the Cretaceous Western Interior Seaway often occurred during eustatic rise as the result either of the onset of density stratification in deeply embayed areas with high freshwater runoff or of an incursion of the oceanic oxygen minimum zone onto the craton. Each of these scenarios could account for the development of widespread anoxia during accumulation of the epibole sediments. The latter is particularly intriguing, as Wilde and Berry (1984) predicted—based on their model of climatically induced oceanic turnover—an episode of near-surface ocean anoxia during the latest Wenlockian–earliest Ludlovian. Thus the *Medusaegraptus* epibole may record a widespread anoxic event during the early Late Silurian associated with the onset of transgression, in which case similar deposits of equivalent age might be expected to occur on other parts of the North American craton. As will be discussed later, this appears to be true.

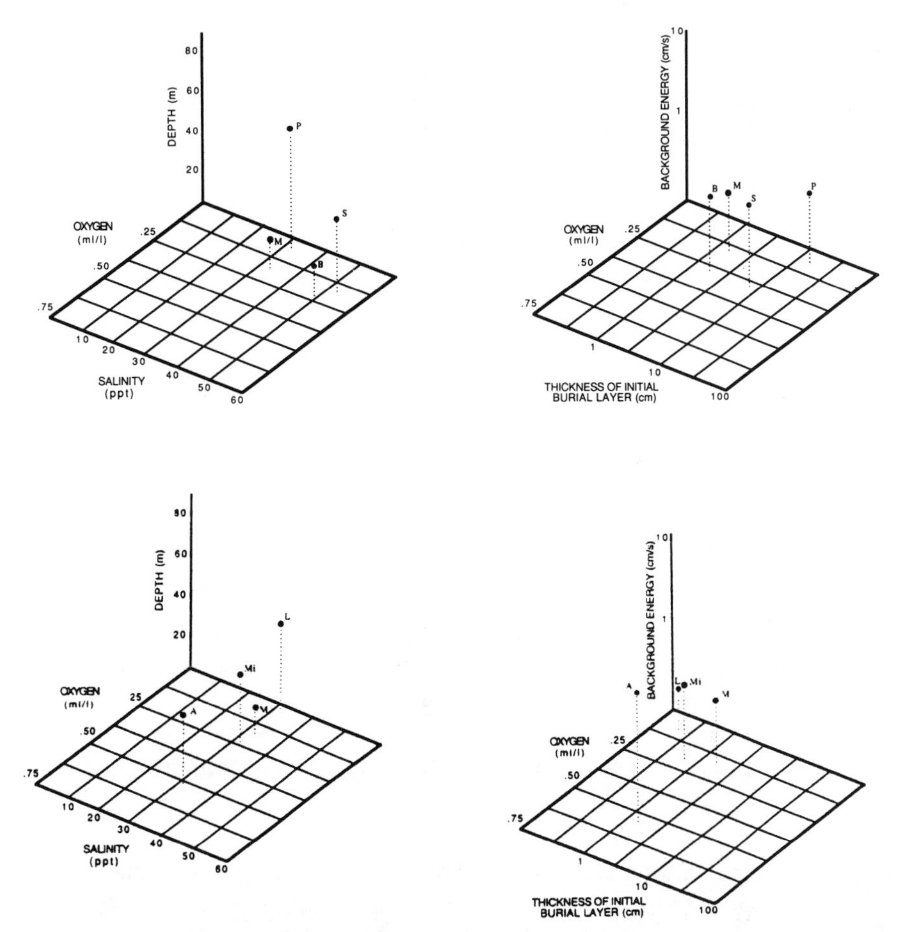

FIGURE 14.17. Three-dimensional diagram illustrating the environmental parameters during deposition of the *Medusaegraptus* epibole. Other classic Konservat-Lagerstätten are plotted for comparison. Abbreviations are as follows: (M) *Medusaegraptus* epibole; (P) Posidonia Shale, Jurassic of Germany; (S) Solnhofen Limestone, Jurassic of Germany; (B) Bertie Group (Williamsville Formation), Upper Silurian of western New York. Paleoenvironmental data for the Solnhofen Limestone are from Barthel et al. 1990; Posidonia Shale from Seilacher and Westphal 1971; Bertie Group from LoDuca, in preparation.

FIGURE 14.18. Three-dimensional diagram illustrating the depositional regime of the *Medusaegraptus* epibole. Other classic Konservat-Lagerstätten are plotted for comparison. Abbreviations and sources are the same as for fig. 14.17.

FIGURE 14.19. Three-dimensional diagram illustrating the environmental conditions during deposition of the *Medusaegraptus* epibole and the similar Ludlovian deposits considered in this report. Abbreviations are as follows: (A) Ancaster Member, (M) *Medusaegraptus* epibole, (Mi) Mississinewa Shale, (L) *Lecthaylus* Shale.

FIGURE 14.20. Three-dimensional diagram illustrating the depositional regimes of the *Medusaegraptus* epibole and the similar Ludlovian deposits considered in this report. Abbreviations as for fig. 14.19.

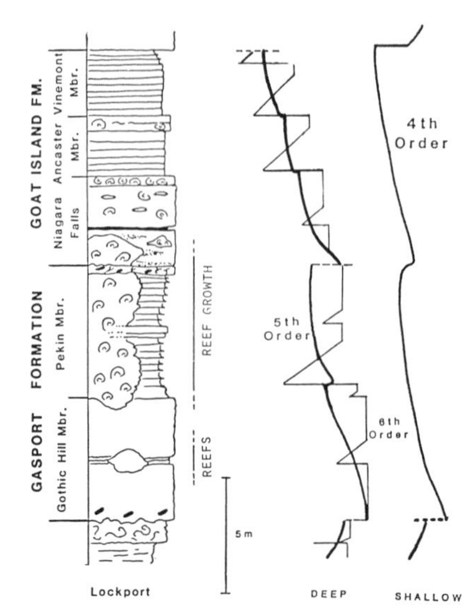

Figure 14.21. Sea-level curve for the lower part of the Lockport Group. The *Medusaegraptus* epibole is the dark band at the middle of the Niagara Falls Member.

Salinity Based on the presence of stenohaline forms in the epibole, including articulate brachiopods and coelenterates, this unit likely was deposited in waters with near-normal marine salinity (fig. 14.17, 14.19). Evidence for either hyper- or hyposaline conditions, such as mud cracks or evaporite minerals, is entirely lacking, as are faunal elements that might reflect abnormal salinities, such as eurypterids.

Energy The depositional environment of the *Medusaegraptus* epibole evidently was rather stagnant, as indicated by the near absence of current-generated sedimentary structures, the amount of fine-grained sediment present, the maintenance of anoxia, and the presence of delicate thallophytic algae and dendroid graptolites (fig. 14.18, 14.20). The sharp-based and graded nature of the beds, however, indicates that this otherwise tranquil environment was at least occasionally disturbed by storms.

A Depositional Model

The depositional environment of the *Medusaegraptus* epibole appears to have been that of a shallow, stagnant, anoxic body of marine water established at one end of a broad carbonate platform during the early Ludlovian. Relict bioherms created an irregular sea floor topography and restricted circulation. A similar depositional environment, including relict bioherms (in this case composed of siliceous sponges and cyanobacteria), was proposed for the Solnhofen Limestone by Barthel et al. (1990), although hypersaline bottom waters evidently were established during deposition of this Jurassic Lagerstätte.

Deposition of the *Medusaegraptus* epibole sediment commenced with the onset of a minor, but apparently eustatic, transgression. As the sea crept across the formerly exposed, udulating relict bioherm surface, a stagnant shallow marine habitat was established. Anoxia may have developed in this setting solely in response to restricted circulation or the decay of benthic algae, but a regional "anoxic event," reflecting either widespread salinity stratification or perhaps an incursion of the oceanic oxygen minimum zone onto the craton during eustatic sea level rise, may be implicated as well.

Most marine organisms found the resulting environment uninhabitable; instead, a low diversity composed of thallophytic algae, dendroid graptolites, inarticulate brachiopods, and annelid worms colonized the area. Storms occasionally disturbed this otherwise tranquil setting, and these events resulted in the suspension, grading, and perhaps delivery of the epibole sediment, in addition to the obrution-style burial of the benthos. Following burial, the delicate biotic remains were largely undisturbed by infaunal scavengers and bacterial decomposers, and the enveloping carbonate muds lithified quickly.

Similar Ludlovian Deposits

Several Ludlovian deposits have been compared to, and correlated with, the "Gasport Channel" (= *Medusaegraptus* epibole) (Shrock 1928; Roy and Croneis 1935; Ross 1962; Erdtmann and Prezbindowski 1974). An overview of each of these occurrences is provided below, and the biotas of each are summarized in table 14.2.

The Mississinewa Shale of Northeastern Indiana

The Mississinewa Shale of the Wabash Formation in northeastern Indiana consists of 25–75 m of gray to buff, fine-grained, thin- to medium-bedded, argillaceous and silty dolomite. The unit interfingers with large (up to 70 m high) bioherms that comprise the "Huntington Facies" of the Wabash Formation (Cumings and Shrock 1928; Shaver 1961; Sunderman and Mathews 1975).

In 1928 Shrock described a diverse dendroid graptolite fauna recovered near the base of the Mississinewa Shale at Yorktown, Indiana. Because many of the taxa, especially the dendroid graptolite species, were identical to those reported by Ruedemann (1925) from the "Gasport Channel" occurrence (= *Medusaegraptus* epibole), he considered the two deposits correlative. Of particular interest is the presence of *Medusaegraptus mirabilis* among the Yorktown biota (Shrock 1928). This is the only documented occurrence of this species outside the *Medusaegraptus* epibole.

Unfortunately, the Yorktown site has long been inaccessible. In 1974, however, Erdtmann and Prezbindowski reported a nearly identical biota from the lower part of the Mississinewa Shale at Huntington, Indiana, some 50 km north

TABLE 14.2 *Summary of the biota from the Silurian deposits considered in the text*

	ME	*MI*	*A*	*L*
DENDROID GRAPTOLITES				
Family Dendrograptidae				
Dendrograptus spinosus Spencer			X	
D. parallelus Shrock		X		
D. praegracilis Spencer	X	X	X	
Callograptus niagarensis Spencer	X	X	X	
C. pulchellus Shrock		X		
Calyptograptus cyathiformis Spencer	X		X	
C. radiatus Spencer			X	
Dictyonema crassibasale Gurley	X	X	X	X
D. expansum Spencer	X	X	X	
D. filiramum Gurley	X	X	X	
D. lyriforme Shrock		X		
D. retiforme Hall	X			X
D. tenellum Spencer	X	X	X	X
Reticulograptus polymorphus (Gurley)		X	X	X
Desmograptus micronematodes (Spencer)	X	X	X	X
D. cumingsi Shrock		X		
Family Acanthograptidae				
Inocaulis ramulosus Spencer	X	X		
I. plumulosus Hall	X	X	X	
Acanthograptus walkeri (Spencer)	X	X		
A. granti Spencer	X	X		
A. pulcher Spencer	X	X	X	
GRAPTOLOID GRAPTOLITES				
Diplograptus sp. Shrock		X		
Pristiograptus jaegeri		X		
M. vomerinus (Nicholson)				X
M. dubius (Suess)				X
Saetograptus colonus (Barrande)	?			X
ALGAE				
Medusaegraptus mirabilis Ruedemann	X	X		
Diplospirograptus goldringae Ruedemann	X			
"Chondrites" versus Ruedemann	X			
Powysia sp.	X			
Receptaculites sp.		X		
PORIFERA				
Astylospongia sp.		X	X	
Stromatoporoid		X		
COELENTERATA				
Cladopora ap.	X		X	
Favosites sp.	X		X	
Enterolasma sp.	X		X	
Heliolites elegans Hall			X	
BRACHIOPODA				
Inarticulata				
Lingula lamellata Hall	X	X	X	X
Trematis spinosa Ruedemann	X			
Schizotreta tenuilamellata (Hall)	X	X	X	X
Articulata				
Platystrophia biforatus (Schlotheim)			X	
Rynchotreta americana (Hall)			X	
Stegerhynchus neglectum (Hall)			X	
Amphistrophia striata (Hall)		X	X	
Leptaena rhomboidalis (Wilckens)		X	X	

	ME	*MI*	*A*	*L*
Anastrophia sp.		X		
Gypidula sp.		X		
Dolerorthis sp.		X	X	
Plectatrypa nodostriata (Hall)			X	
Atrypa reticularis (Linnaeus)			X	
Eospirifer niagarensis (Conrad)	?		X	
Eoplectodonta transversalis (Wahlenberg)			X	
Protomegastophia profunds (Hall)			X	
Coolinia sp.	X	X	X	
MOLLUSCA				
Cephalopoda				
Dawsonoceras annulatum (Sowerby)	X	X	X	
Kionoceras cancellatum (Hall)	X	X	X	
Orthoceras simulator (Hall)	X	X		
Cyrtiactinoceras sp.	X	X		
Bivalvia				
Necklania sp.		X		
Cypricardinia sp.		X		
Pterinia sp.		X	X	
Gastropoda				
Coelocaulus sp.			X	
Naticonema niagarense (Hall)			X	
Eotomaria sp.	X			
Hormotoma sp.		X		
Holopea sp.		X	X	
ANNELIDA				
Protoscolex batheri Ruedemann	X	X		
P. ruedemanni Roy and Croneis				X
Lecthaylus gregarious Weller				X
Serpulites sp. A	X			
Serpulites sp. B		X		
ARTHROPODS				
Trilobita				
Dalmanites limulurus (Green)			X	
Dalmanites sp.	X	X		
Encrinurus indianensis Kindle & Breger		X		
Odontopleura ortoni (Foerste)		X		
Bumastus niagarensis (Whitfield)	X	X	X	
Phyllocarida				
Ceratiocaris sp.	X			
Ceratiocaris markhami Roy				X
CONULARIA				
Conularia niagarensis Hall	X		X	
C. tenuicosta Ruedemann	X			
Metaconularia manni Roy				X
TRACE FOSSILS				
Lockportia conspicua (Ruedemann)	X	X		X

Locality abbreviations: (ME) *Medusaegraptus* epibole, western New York State; (MI) lower Mississinewa Shale, northeastern Indiana; (A) Ancaster Member, Hamilton, Ontario; (L) *Lecthaylus* Shale, northeastern Illinois.

of Yorktown; this site remains available for study. Specimens of the graptoloid graptolite *Pristiograptus jaegeri* and acritarchs of the *Duenffia eisenackii* subfacies of Cramer and Deiz (1974) indicate an earliest Ludlovian age for this interval (Erdtmann and Prezbindowski 1974; Wood 1975). Although graptoloid graptolites and acritarchs are unknown from the Yorktown site, the similar stratigraphic position of the Yorktown and Huntington deposits (lower Mississinewa) indicates that they are likely equivalent.

The dendroid graptolite–dominated biota at Huntington occurs in an interval of fine-grained argillaceous dolomite approximately 30 cm thick composed of a series of thin (2–4 cm), normally graded beds with sharp bases (Erdtmann and Prezbindowski 1974). Insoluble content is approximately 52%, with most of this material being illite clay and fine quartz silt. Bioturbation is minimal, and trace fossils are represented only by rare, horizontal, unbranched burrows approximately 0.75 cm in diameter.

The biota of the dendroid-graptolite-bearing interval at Huntington, as that of the *Medusaegraptus* epibole, is quite unlike that of a "typical" Silurian shelly association: Bryozoa, coelenterates, and stromatoporoids are absent, and shelly organisms overall are poorly represented (table 14.2). In contrast, dendroid graptolites, often buried *in situ*, are abundant and diverse. Bladelike algae, scolecodonts, often in pairs but without associated soft parts, and inarticulate brachiopods, represented principally by *Lingula* and *Schizotreta*, are common. Articulate brachiopods occur much less frequently; specimens of *Gypidula, Leptaena, Strophonella,* and *Dolerorthis* are indicative of the Benthic Assemblage 3–4 boundary (see Boucot 1975). Molluscs are represented by gastropods, ortho- and cyrtoconic nautiloids, and bivalves. Only the cephalopods, the nektonic molluscan component, are common. Sponges (*Astylospongia*), annelid worms (*Protoscolex*), trilobites (typically articulated and mainly represented by a bumastid, a dalmanitid, and an encrinurid), and crinoids (typically as complete crowns and represented by at least two camerate species) are rare.

Evidently, this graptolite-dominated association is not characteristic of the Mississinewa Shale as a whole but is restricted to a thin interval near the base of this unit: a careful search of stratigraphically higher intervals within the Mississinewa failed to reveal similar associations. Although the bulk of the Mississinewa is composed of strata essentially identical in composition to the Huntington and Yorktown deposits, a significant difference is evident when trace fossils are taken into consideration. Burrows within the Huntington and Yorktown deposits are rare and only occur as shallow, horizontal types, whereas the remainder of the Mississinewa is heavily bioturbated, with both horizontal and deeply penetrating vertical burrows present in abundance. Thus it appears that during deposition of the lower Mississinewa Shale, unusual anoxic conditions were briefly established below, and for some distance above, the sediment/water interface, thereby prohibiting bioturbation and restricting colonization of the bottom to organisms either tolerant of low-oxygen conditions—such as graptolites, algae, and inarticulate brachiopods—or able to elevate themselves some distance off the seafloor into less stagnant water (e.g., cephalopods by swimming). A similar depositional environment was envisioned by Erdtmann and Prezbindowski (1974), although they suggested that this association represented a recurrent pioneer community in the sequence of reef succession. However, as this diverse graptolite assemblage evidently is restricted to a thin interval near the base of the unit, such a relationship is considered unlikely.

Overall, the lower Mississinewa biota is remarkably similar to that of the *Medusaegraptus* epibole. In both cases diverse and suggested dendroid graptolite/algal associations are preserved in sections otherwise characterized by typical Silurian shelly biotas. Sedimentologically, the units are identical; both are composed of a series of graded tempestites. The lower, lag portions of the Mississinewa beds are finer-grained than those of the *Medusaegraptus* epibole, but this may simply reflect the lack of a proximal source (e.g., bioherm) for the coarse material.

While the lower Mississinewa and *Medusaegraptus* epibole deposits are very similar, there is one potentially significant difference: the composition of the algal flora. The flora of the *Medusaegraptus* epibole is dominated by dasyclad green algae, particularly *Medusaegraptus mirabilis*, whereas with the exception of a few fragmentary specimens of *M. mirabilis* reported by Shrock (1928) from the Yorktown occurrence, none of these species occur in the lower Mississinewa. Instead this interval yields bladelike algae referred (although questionably, because of insufficient detail) by Erdtmann and Prezbindowski (1974) to the Rhodophyta. If these algae are indeed affiliated with the Rhodophyta, the discrepancy between the two algal floras may reflect a difference in the depth at which these units were deposited, as red algae flourish at greater depths than do dasyclads. Brachiopod and coral evidence, although meager, appears to support this interpretation.

The *Lecthaylus* Shale of Northeastern Illinois

The *Lecthaylus* Shale is a thin (45–60-cm) interval of argillaceous, dark gray, thinly laminated dolomite that occurs in the otherwise massive dolomite of the Racine Formation in the vicinity of Blue Island, Illinois (south Chicago) (Weller 1925; Roy and Croneis 1935; Roy 1935a,b; Lowenstam 1948). Insoluble content is high, averaging about 50%. This interval was briefly accessible during construction of the Calumet feeder of the Chicago Drainage Canal, although upon completion of the canal this exposure was submerged. A second exposure of this unit existed for a time in the Midlothian Quarry, although this too has long been covered over. Currently, no exposures of the *Lecthaylus* Shale exist, but blocks of the unit remain scattered along the banks of the canal and are available for study.

Graptoloid graptolites recovered from the *Lecthaylus* Shale by Ross (1962), including *Saetograptus colonus* and *Monograptus dubius*, indicate an earliest Ludlovian age for this unit. Associated acritarchs (Cramer and Deiz 1974) are identical to those reported from the lower Mississinewa Shale (Wood 1975) and belong to the narrowly defined early Ludlovian *Duenffia eisenackii* subfacies.

Fossils are rare within the *Lecthaylus* Shale and include dendroid and graptoloid graptolites, inarticulate brachiopods, worms *(Protoscolex ruedemanni* and *Lecthaylus gregarius),* conulariids, and phyllocarids. The planktonic chlorophycean alga *Tasmanites* is locally abundant. Trace fossils are rare and are represented only by small (0.5 cm diameter) horizontal burrows. Corals, bryozoa, stromatoporoids, sponges, trilobites, echinoderms, and molluscs are absent. The benthic biota appears to be autochthonous (Kluessendorf and Mikulic 1991).

The laminated nature of the *Lecthaylus* Shale suggests that this unit was deposited in calm waters below fair-weather wave base. *Dicoelosia* brachiopods in the immediately overlying limestones represent a BA-4 association, but the absence of benthic algae in the *Lecthaylus* Shale, despite apparently favorable taphonomic conditions, may reflect deposition in deeper waters below the photic zone, at depths more typically associated with BA-5 associations. Anoxic conditions during deposition of this unit are indicated by the unusual composition of the biota, the near absence of bioturbation, and the preservation of soft tissues. Similar depositional conditions were suggested by Kluessendorf and Mikulic (1989, 1991).

The Ancaster Member of Hamilton, Ontario

The dendroid graptolite fauna recovered from the "Niagaran Formation" exposed at Hamilton, Ontario, is an often cited example of a Silurian dendroid graptolite-dominated association. Most of the specimens recovered from this area were collected in the late nineteenth century by an amateur named C. C. Grant, when large quarries, such as the Corporation Quarry, operated along the Niagara escarpment. Taxonomic descriptions of the graptolite fauna were provided by Spencer (1884), Gurley (1898), and Bassler (1909). Although the quarry exposures that produced these specimens have long been filled, Grant (1892, 1893, 1894, 1897, 1899) published detailed accounts of his collecting forays, and these can provide valuable insights. Indeed, it is evident from these accounts, and those of Spencer (1883, 1884), that the graptolite specimens were not recovered from a single interval, but from four separate stratigraphic units:

1. The lower part of the Irondequoit Formation (early Wenlockian; one species).
2. The Stoney Creek Member of the Rochester Shale, especially the upper, more calcareous part, which was referred to locally as the "Blue Building Beds" (late Wenlockian; 16 species).
3. The Ancaster Member of the Goat Island Formation, locally referred to as the "Chert Beds" (early Ludlovian; 12 species).

4. The basal part of the Eramosa Formation, locally referred to as the "Barton Beds" (middle Ludlovian, two species, probably not graptolites).

Thus to consider the entire graptolite fauna as a contemporary association representing a stratigraphically restricted interval, as has been done in the past, is a considerable oversimplification. In this report, only the dendroid graptolite fauna of the Ancaster Member will be considered.

The Ancaster Member of the Goat Island Formation in the Hamilton, Ontario, area consists of 5–7 m of buff, silty to argillaceous (15% insoluble) fine-grained dolomite composed of 3–6-cm-thick, normally graded beds with sharp bases. Like the other examples discussed, these beds are interpreted as tempestites. Chert is locally abundant and is often concentrated in horizontal bands. Both horizontal and vertical burrows are abundant.

The Ancaster Member immediately overlies the Niagara Falls Member. Therefore, although the Ancaster graptolite occurrence has been correlated with the "Gasport Channel"(= *Medusaegraptus* epibole) (Ruedemann 1925; Shrock 1928), it is actually slightly younger.

Within the Ancaster Member in the Hamilton region, dendroid graptolites are not confined to a specific bed or level but occur scattered throughout. Overall, they are not particularly common. The graptolites occur in direct association with a diverse shelly fauna composed of tabulate and rugose corals, bryozoa, trilobites, brachiopods, and sponges; the latter are locally quite numerous. The presence of the brachiopod *Eoplectodonta* in this unit is indicative of a Benthic Assemblage 4 association. Neither thallophytic algae nor soft-bodied animals have been recovered from this interval.

Based on the association of dendroid graptolites with an abundant shelly fauna, the amount of bioturbation present, and the tempestite nature of the beds, the Ancaster Member is considered to have been deposited in marine waters with near normal oxygen content and at depths within storm wave base but below fair weather wave base.

Depositional Criteria of Ludlovian Konservat-Lagerstätten

Of the deposits considered above, only the *Medusaegraptus* epibole, lower Mississinewa Shale, and *Lecthaylus* Shale rank as bona fide Konservat-Lagerstätten, as the Ancaster Member lacks fossils of soft-bodied organisms. A comparison of these deposits suggests that certain environmental parameters were essential for the preservation of soft-bodied biotas, whereas other factors apparently were insignificant.

Age

The *Medusaegraptus* epibole, lower Mississinewa Shale, and *Lecthaylus* Shale apparently are equivalent; the Ancaster Member is slightly younger. The fact that the Konservat-Lagerstätten were deposited during roughly the same time frame suggests that age may be significant. Why this might be so is explored under the heading of Anoxia.

Lithology

All the soft-bodied biotas considered occur within fine-grained sediments and, although the insoluble content varies (fig. 14.22), all are preserved within carbonates. Thus both sediment grain size and sediment composition evidently controlled exceptional preservation (fig. 14.23). Carbonate sediment composition likely facilitated early diagenesis (Allison 1988), whereas fine-grained sediment provided an appropriate medium for recording the impressions of the soft-bodied organisms.

Depth

Depth apparently played little direct role in controlling the preservation of the biotas considered. The *Medusaegraptus* epibole was deposited in rather shallow water (< 15 m), the lower Mississinewa and Ancaster at intermediate depths (20–30 m), and the *Lecthaylus* Shale in relatively deep water (40 m), based on brachiopod and/or algal evidence (figs. 14.19, 14.24).

Salinity

All of the Ludlovian Lagerstätten considered in this report yield crinoids and/or coelenterates. Thus, these units apparently accumulated in waters with near normal marine salinities (figs. 14.19, 14.24). It should be noted, however, that

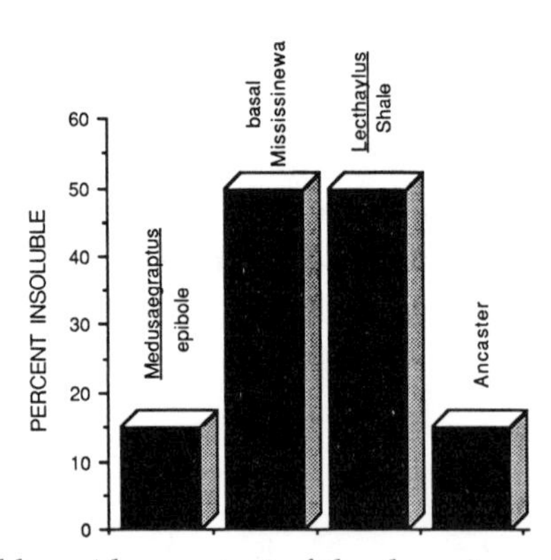

Figure 14.22. Insoluble-residue content of the deposits considered.

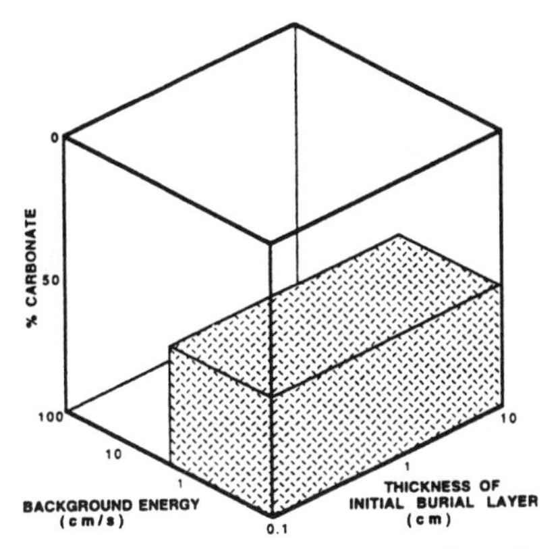

FIGURE 14.23. Depositional regime (patterned field) of Ludlovian Konservat-Lagerstätten considered in this report expressed as a three-dimensional hyperspace with respect to carbonate content, background energy, and thickness of the initial burial layer.

other Silurian Lagerstätten, including a newly discovered, somewhat younger occurrence from the Ludlovian Eramosa Formation in the southern Bruce Peninsula (Tetreault 1995) and the eurypterid-bearing intervals of the Pridolian Williamsville Formation in western New York and southern Ontario (S. Ciurca, personal communication), are associated with salt hoppers, indicating hypersaline conditions existed either during or shortly after deposition of these units. In these cases, elevated salinity almost certainly acted to retard scavenging and decay. Nevertheless, based on study of the Ludlovian Lagerstätten considered in this report, it appears that abnormal salinity was not a prerequisite for exceptional preservation.

Anoxia

Anoxia appears to have been essential for the preservation of all the soft-bodied biotas considered (figs. 14.19, 14.20 and 14.24). In the *Medusaegraptus* epibole, lower Mississinewa Shale, and *Lecthaylus* Shale, anoxic conditions are indicated immediately below the sediment/water interface by the near absence of horizontal burrows, the lack of vertical burrows, the absence of scavenging, and the retardation of decay. The restricted and unusual composition of the biotas (dendroid graptolites, algae, and worms) indicates that oxygen-starved conditions extended above the seafloor as well.

Although the depositional setting of each of these deposits was influenced by restricted circulation imposed by bioherms, this alone may not have been responsible for the development of anoxia, as similar, immediately underlying and overlying units (e.g., Ancaster Member) fail to yield exceptionally preserved biotas. The fact that the soft-bodied biotas are roughly the same age and occur abruptly within thin, laterally extensive intervals strongly suggests a widespread anoxic event. As noted earlier, this anoxic event may have been caused by widespread density stratification or an incursion of the oceanic oxygen minimum zone onto the craton associated with the onset of eustatic sea level rise during the early Ludlovian. Current-baffling by bioherms may have acted to prolong and intensify this event.

Energy (Wave and Current)

Sedimentologic, taphonomic, and biotic features indicate that all of the Ludlovian Lagerstätten considered in this report accumulated in stagnant bottom waters episodically agitated by storms (figs. 14.20 and 14.23). In these settings, anoxia persisted during fair-weather intervals in the absence of circulation between bottom and surface waters, and obrution occurred during brief storm events. In modern marine basins, bottom waters characterized by this combination of energy conditions occur at depths ranging from deep (> 50 m) to shallow (< 1 m), depending, in part, on basin size, geometry, bottom topography, and geographic location. Thus Lagerstätten formed under these conditions could have accumulated at rather different depths within and between basins, and such appears to be the case with regard to the Ludlovian examples considered.

Thickness of the Initial Burial Layer

In the Ludlovian Lagerstätten considered, the initial burial layer was quite thin, typically less than 1–2 cm (figs. 14.20 and 14.23). Thus it appears that only a thin veneer of sediment over the top of the biota was required. Indeed, deep burial was likely unnecessary because of the near absence of bioturbation.

Bacterial Sealing

Although cyanobacterial mats have been implicated in the preservation of some Lagerstätten (see Seilacher et al. 1985), evidence to support the presence of such

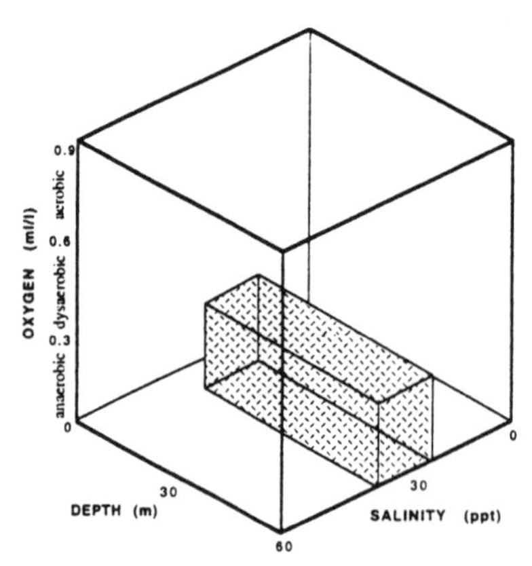

Figure 14.24. Depositional regime (patterned field) of Ludlovian Konservat-Lagerstätten considered in this report expressed as a three-dimensional hyperspace with respect to oxygen, depth, and salinity.

a mat, such as radiating ridges around shells, has not been observed in any of the examples considered.

Summary of Depositional Criteria

Based on comparative study of the sedimentology, taphonomy, and biotic composition of Ludlovian Konservat-Lagerstätten, it appears that the following conditions promoted exceptional preservation: (1) burial in very fine-grained sediment composed of at least 50% calcium carbonate (to preserve fine details as molds and casts and to facilitate early diagenesis, respectively); (2) near or total absence of oxygen below, and for some distance above, the sediment/water interface (to inhibit scavenging and decay, and also to facilitate early diagenesis); and (3) deposition below fair-weather wave base but near the lower limit of storm wave base (to maintain anoxic conditions and to promote rapid, obrution-style burial, respectively). Factors relevant to the establishment of these conditions are proximity to an abundant source of carbonate mud sediment, such as a carbonate platform (condition 1); the presence of irregular seafloor topography, such as that afforded by bioherms (conditions 2 and 3); and deposition during the initial stages of a transgression (condition 2). Factors considered to have little influence over the preservation of these biotas include water depth and the thickness of the initial burial layer.

By identifying stratigraphic intervals within a sedimentary basin that satisfy these criteria, it should be possible to prospect efficiently for soft-bodied biotas similar to those described in this report. Given the environmental controls on exceptional preservation outlined above, however, it appears unlikely that autochthonous "typical" Silurian fossil associations will be found with either the soft parts of the shelly organisms or the entirely soft-bodied faunal and/or floral elements preserved.

Acknowledgments

Bill Goodman, Sam Ciurca, and Elizabeth LoDuca provided valuable field assistance during this study. J. Gray and M. Miller examined and identified organic microfossils. W. B. N. Berry assisted with graptolite identification. Brian Parmal illustrated the community reconstructions, and Ed Landing provided specimens from the New York State Museum. This manuscript was improved by the helpful suggestions of J. Kluessendorf and A. Boucot. LoDuca's fieldwork was supported by grants from the New York State Museum, the Geological Society of America, the Northeast Science Foundation, and the Paleontological Society of America. This research was also supported by a grant from the donors to the Petroleum Research Fund (American Chemical Society) awarded to Brett.

References

Allison, P. 1988. Konservat-Lagerstätten: cause and classification. *Paleobiology* 14:331–34.

Barthel, K. W., N. H. M. Swinburne, and S. Conway Morris. 1990. *Solnhofen*. Cambridge: Cambridge University Press.

Bassler, R. 1909. Dendroid graptolites of the Niagaran dolomites at Hamilton, Ontario. *U.S. National Museum Bulletin* 65:1–65.

Berry, W. B. N. and A. J. Boucot. 1970. Correlation of the North American Silurian rocks. *Geological Society of America Special Paper* No. 102.

Boucot, A. 1975. *Evolution and Extinction Rate Controls*. New York: Elsevier.

Brett, C. E., D. Tepper, W. M. Goodman, S. T. LoDuca, and B. Y. Eckert. 1995. Revised stratigraphy and correlations of the Niagaran Provincial Series (Medina, Clinton, and Lockport Groups) in the type area of western New York. *U.S. Geological Survey Bulletin* No. 2086.

Cramer, F. and M. Deiz. 1974. North American Silurian palynofacies and their spatial arrangement. *Palaeontographica B* 138:107–80.

Crowley, D. 1973. Middle Silurian patch reefs in Gasport Member (Lockport Formation) New York. *American Association of Petroleum Geologists Bulletin* 57:283–300.

Cumings, E. and R. Shrock. 1928. Niagaran coral reefs of Indiana and adjacent states and their stratigraphic relations. *Geological Society of America Bulletin* 39:579–619.

Eckert, B. Y. and C. Brett. 1989. Bathymetry, paleoecology, and depositional environments of Silurian benthic assemblages, late Llandoverian, New York State. *Paleogeography, Paleoclimatology, and Paleoecology* 74:297–326.

Elliott, G. 1968. Permian to Paleocene calcareous algae (Dasycladaceae) of the Middle East. *British Museum of Natural History Bulletin* Supplement 4, pp. 1–111.

Erdtmann, B. D. and D. R. Prezbindowski. 1974. Niagaran (Middle Silurian) interreef fossil burial environments. *Neues Jarhbuch für Geologie und Paläontologie Abhandlungen* 144:342–72.

Grant, C. 1892. Fragments of Silurian sea-floors from Hamilton. *Journal and Proceedings of the Hamilton Scientific Association* 8:149–54.

———. 1893. Geological notes. *Journal and Proceedings of the Hamilton Scientific Association* 9:97–135.

———. 1894. Remarks on the annual excursion. *Journal and Proceedings of the Hamilton Scientific Association* 10:74–78.

———. 1897. List of local fossils not previously reported in the *Journal of Proceedings*. *Journal and Proceedings of the Hamilton Scientific Association* 13:128–36.

———. 1899. Geological notes. *Journal and Proceedings of the Hamilton Scientific Association* 15:48–66.

Gurley, R. 1896. North American graptolites; new species and vertical range. *Journal of Geology* 4:63–102, 291–311.

Hurst, J. 1975. The diachronism of the Wenlock Limestone. *Lethaia* 8:301–14.

Kauffman, E. 1986. High-resolution event stratigraphy: Regional and global Cretaceous bio-events. In O. Walliser, ed., *Global Bio-events*, pp. 279–337. Berlin: Springer-Verlag.

Kluessendorf, J. and D. Mikulic. 1989. Konservat-Lagerstätten from the Silurian of North America. *The Murchison Symposium Programme and Abstracts*, pp. 55–56.

———. 1991. The role of anoxia in the formation of Silurian Konservat-Lagerstätten. *Geological Society of America Abstracts with Programs* 23:54.

LoDuca, S. T. 1990a. The *Medusaegraptus* epibole: Paleontology, stratigraphy, taphonomy, and depositional environment of Silurian (Ludlovian) Konservat-Lagerstätten. Ph.D. diss., University of Rochester.

———. 1990b. *Medusaegraptus mirabilis* Ruedemann as a noncalcified dasyclad alga. *Journal of Paleontology* 64:469–74.

LoDuca, S. T. and C. Brett. 1990. Placement of the Wenlockian/Ludlovian boundary in western New York State. *Geological Society of America Abstracts with Programs* 22:31.

———. 1991. Placement of the Wenlockian/Ludlovian boundary in western New York State. *Lethaia* 24:255–64.

Lowenstam, H. 1948. Biostratigraphic studies of the Niagaran interreef formations in northeastern Illinois. *Illinois State Museum Bulletin* 9:1–146.

McKerrow, W. 1979. Ordovician and Silurian changes in sealevel. *Journal of the Geological Society of London* 136:137–45.

Mikulic, D., J. Kluessendorf, and D. Briggs. 1985. A new exceptionally preserved biota from the Lower Silurian of Wisconsin, U.S.A. *Philosophical Transactions of the Royal Society of London* 311B:75–85.

Pinna, G. 1985. Exceptional preservation in the Jurassic of Osteno. *Philosophical Transactions of the Royal Society of London* 311B:171–81.

Ross, C. A. 1962. Silurian monograptids from Illinois. *Palaeontology* 5:59–72.

Roy, S. 1935a. A new Niagaran Conularia. *Fieldiana* 6:147–54.

———. 1935b. A new Silurian phyllopodous crustacean. *Fieldiana* 6:141–46.

Roy, S. and C. Croneis. 1935. A Silurian worm and associated fauna. *Fieldiana* 4:229–47.

Ruedemann, R. 1925. Some Silurian (Ontarian) faunas of New York. *New York State Museum Bulletin* 265:1–84.

Seilacher, A. 1970. Begriff and bedeuntung der Fossil-Lagerstätten. *Neues Jarhbuch für Geologie und Paläontologie Abhandlungen* 46:34–39.

Seilacher, A., W. Reif, and F. Westphal. 1985. Sedimentological, ecological, and temporal patterns of Fossil-Lagerstätten. *Philosophical Transactions of the Royal Society of London* 311B:5–23.

Shaver, R. 1961. Stratigraphy of the Silurian rocks of northern Indiana. *Indiana Geological Survey Field Conference Guidebook* 10:1–62.

Shrock, R. R. 1928. A new graptolite fauna from the Niagaran of northern Indiana. *American Journal of Science* 16:1–38.

Spencer, J. 1883. Paleozoic geology of the region about the western end of Lake Ontario. *Canadian Naturalist and Quarterly Journal of Science* 10:129–71.

———. 1884. Niagaran fossils. *Transactions of the Academy of Science of St. Louis* 4:555–610.

Sunderman, J. and G. Mathews, eds. 1975. *Silurian Reef and Interreef Environments.* Fort Wayne, Ind.: Society of Economic Mineralogists and Paleontologists *5th Annual Field Conference Guidebook.*

Taylor, W. 1960. *Marine Algae of the Eastern Tropical and Subtropical Coasts of the Americas.* Ann Arbor: University of Michigan Press.

Tetreault, D. K. 1995. An unusual Silurian anthropod/echinoderm dominated soft-bodied fauna from the Eramosa Member (Ludlow) of the Guelph Formation, southern Bruce Peninsula, Ontario, Canada. *Geological Society of America Abstracts with Programs* 27:A-114.

Upchurch, S. 1970. Sedimentation on the Bermuda platform. Department of the Army Corps of Engineers Resource Report No. 2.

Van den Hoek, C., F. Colijn, A. Cortel-Breeman, and J. Wanders. 1972. *Algal Vegetation Types Along the Shores of Inner Bays and Lagoons of Curaçao, and the Lagoon de Lac, Netherlands Antilles.* Amsterdam: North Holland.

Van der Voo, R. 1988. Paleozoic paleogeography of North America, Gondwana, and intervening displaced terranes: Comparisons of paleomagnetism with paleoclimatology and biogeographical patterns. *Geological Society of America Bulletin* 100:311–24.

Van Tuyl, F. 1916. New points on the origin of dolomite. *American Journal of Science* 42:249–60.
Walliser, O. H. 1964. Conodonten des Silurs. *Abhandlungen des Hessichen Landesamtes für Bodenforschung, Wiesbaden* 41:1–106.
Weller, S. 1925. A new type of Silurian worm. *Journal of Geology* 33:540–44.
Wilde, P. and W. B. N. Berry. 1984. Destabilization of the oceanic density structure and its significance to marine "extinction" events. *Paleogeography, Paleoclimatology, and Paleoecology* 48:143–62.
Wood, G. 1975. Acritarchs and trilete spores from the Mississinewa Shale of northern Indiana. In J. Sunderman and G. Mathews, eds., *Silurian Reef and Interreef Environments,* pp. 91–94. Fort Wayne, Ind.: Society of Economic Mineralogists and Paleontologists *5th Annual Field Conference Guidebook.*
Wray, J. 1977. *Calcareous Algae.* New York: Elsevier.
Zenger, D. H. 1965. Stratigraphy of the Lockport Formation (Middle Silurian) in New York State. *New York State Museum Bulletin* 404:1–120.

15

Thicketing Events: A Key to Understanding the Ecology of the Edgecliff Reefs (Middle Devonian Onondaga Formation of New York and Ontario, Canada)

Thomas H. Wolosz

ABSTRACT

Thicketing events in the Edgecliff Member of the Onondaga Formation reflect the widespread colonization of preexisting low, shield-shaped crinoidal sandbanks by phaceloid colonial rugosans that led to the formation of a single rugosan thicket covering the entire bank. No further recruitment of colonial rugosans occurred, and the thicket was buried by crinoidal grainstone/packstone. The cause of the event is thought to be a sea-level fluctuation or some other short-lived ecological event that allowed the colonial rugosans to become established. However, their phaceloid morphology was poorly adapted to the shallow-water environment, leading to the rapid accumulation of sediment in the colonies and the resultant high vulnerability to burial in storm-deposited sediment. The similarity of thicket/bank cycles to mound/bank cycles in other Edgecliff reefs leads to the conclusion that the rugosan/crinoidal sandbank transition in all Edgecliff reefs was controlled by water depth, with extremely shallow water being beyond the tolerance limits of the colonial rugosans.

The importance of event stratigraphy as a reflection of physical phenomena is clear (Einsele et al. 1991; and papers in this volume). What is less well understood is the usefulness of the event bed as a tool for interpreting the ecology of poorly understood rock units. Although it is known that epiboles can be traced to taphonomy or episodic incursions, they may also represent unusual ecological environments (Brett and Baird, essay 10 of this volume) and should therefore be examined in that light.

An unusual environment may be defined as one with conditions normally just beyond the tolerance limits of a given species, which, because of some short-lived environmental change, becomes available for colonization by that species. The Centerfield Horizon coral thickets within the Hamilton group are an example of such an unusual environment. Normally inhospitable to coral growth, an apparent sea-level fluctuation modified the environment sufficiently to allow the formation of extensive coral thickets (Savarese et al. 1986; Wolosz and Wallace 1981). If the environmental fluctuation is of geologically short duration, the result will be an event bed. Furthermore, investigation of the paleoecology of such an event will lead to a fuller understanding of the overall environment.

The Edgecliff Reef Problem

The Eifelian Edgecliff Member of the Onondaga Formation contains abundant reefs in New York State and Ontario, Canada. These reefs consist of dense accumulations of phaceloid colonial rugosans surrounded by, and in some cases interbedded with, bedded crinoidal grainstones and packstones containing a fauna of large sheetlike-to-domal favositids. Although they generally are small (the largest known surface exposure is roughly 20 m thick, but most are much smaller; subsurface "pinnacle reefs," however, reach about 60 m), these structures are unusual because they lack the major reef-building paleocommunity of the Devonian, namely, stromatoporoids and calcareous algae (Oliver 1956a; Wolosz 1992a).

The Edgecliff member is itself unusual, consisting primarily of crinoidal packstones and grainstones but lacking a well-defined basal peritidal facies in western New York and Ontario, Canada, where it is transgressive over a disconformable surface. (See Oliver [1954, 1956b, 1976], for a descriptive stratigraphy. Rickard [1975] and Cassa and Kissling [1982] illustrate the magnitude of the disconformity.)

The absence of both the typical Devonian reef builders and the associated shallow-water facies has led to differing interpretations of these reefs' paleoecology. Crowley and Poore (1974), Williams (1980), Wolosz (1985, 1992a), Lindemann (1988), and Wolosz and Paquette (1988) all favor growth in shallow water, whereas Cassa and Kissling (1982) and Kissling (1987) regard the absence of stromatoporoids, algae, and the associated peritidal deposits as indicative of reef growth in deep water, possibly below the photic zone. Wolosz (1990a, 1991) and Wolosz and Paquette (1995) recently argued for reef growth in shallow but temperate waters, a hypothesis that has received some support from studies of both brachiopod (Koch and Boucot 1982) and gastropod (Blodgett et al. 1988) distributions. An analysis of the ecology of "thicketing events" in the Edgecliff will add to our understanding of the overall ecology of these unusual structures.

Stratigraphy

Our understanding of the stratigraphy of the Onondaga Formation is based mainly on the work by Oliver (1954, 1956b, 1976), who used gross lithology and fauna to subdivide the Onondaga into the Edgecliff, Nedrow, Moorehouse, and Seneca members; and on that by Lindholm (1967), who studied the Onondaga microfacies (for a detailed discussion and extensive reference list, see Oliver 1976).

In central and western New York and in Ontario, Canada, there is a major unconformity at the base of the Onondaga Formation, with the basal Edgecliff Member a massive, biostromal, very coarsely crystalline grainstone to packstone that marks a major westward transgression. Eastern New York offers no evidence for a major unconformity, although the contact with the underlying clastics of the Tristates Group may be considered as disconformable (Chadwick 1944). East of Utica, New York, the basal Edgecliff is a gray calcisiltite (C1 unit of Oliver [1956b]) up to 2 m thick. The C1 unit is absent in the northeasternmost part of the strike belt in the vicinity of Albany, New York, and beneath the North Coxsackie reef (fig. 15.1) but is present beneath Roberts Hill to the south and the Thompson Lake and Mount Tom reefs to the west. The C1 facies thickens in a downramp direction (southeastern facies of Oliver [1956b]). Overlying the C1, the Edgecliff is a coarse-grained crinoidal grainstone/packstone that is overlain by the shaley Nedrow member. From Albany south to Roberts Hill (fig. 15.1), the entire Onondaga Formation above the C1 is a coarse crinoidal packstone that can be divided into members only on paleontological grounds (Oliver 1956b).

In contrast to the purely transgressive nature of the Edgecliff in the central and western portion of New York, Wolosz (1984:264) and Wolosz and Lindemann (1986) suggested that the eastern Edgecliff represents a shallowing from the Tristates through the C1 Edgecliff into the upper Edgecliff packstones and grainstones, followed by a deepening into the overlying Nedrow member.

Reef Paleocommunities

Wolosz (1992a, 1992b) interpreted the growth patterns in the Edgecliff reefs as the result of an environmentally controlled interplay between two distinct paleocommunities, one dominated by phaceloid colonial rugosans and the other a diverse, biostromal, favositid/crinoidal sand paleocommunity.

The dense growth of rugosan colonies in the low-diversity phaceloid colonial rugosan paleocommunity resulted in the construction of mounds and the

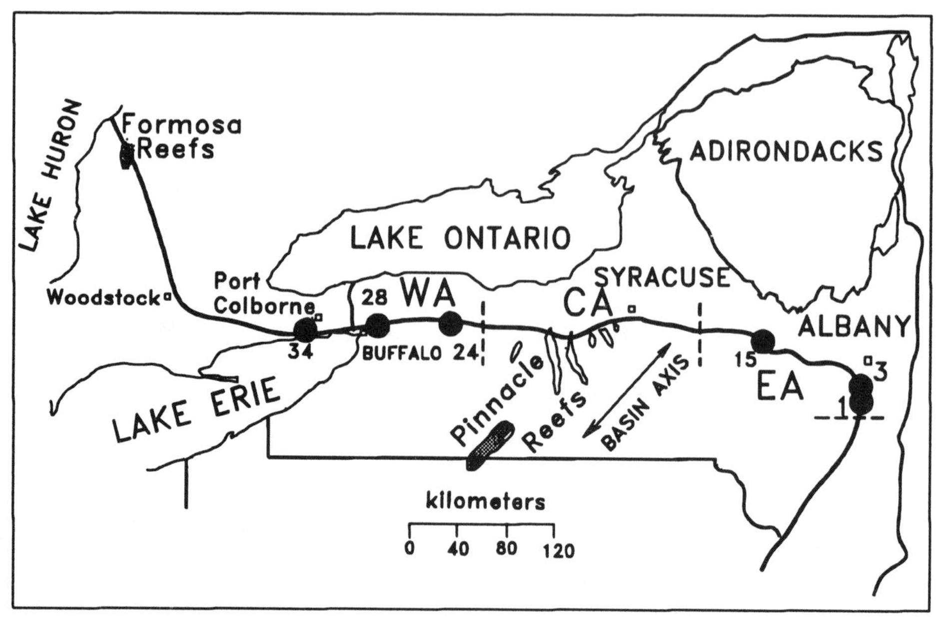

FIGURE 15.1. Distribution of known Edgecliff reefs. Numbered dots are surface exposures referred to in text: (1) Roberts Hill Reef, (3) North Coxsackie Reef, (15) Mount Tom, (24) LeRoy Bioherm, (28) Country Club of Buffalo Reef, and (34) small bioherms west of Port Colborne. Pinnacle reefs are subsurface, and Formosa reefs are considered to be equivalent to Edgecliff reefs. WA = Western Area, CA = Central Area, EA = Eastern Area. (Redrafted with modification from Oliver 1976.)

"thicketing events" that I describe in this report. Although the thickets are biostromal, according to Cumings's definition (1932:334), they are differentiated from the average biostrome through their domination by, and the dense growth of, phaceloid colonial rugosans, to the nearly total exclusion of other organisms. Common colonial rugosan genera include *Acinophyllum, Cylindrophyllum,* and *Cyathocylindrium,* with *Eridophyllum, Synaptophyllum,* and, less commonly, phaceloid colonies of *Heliophyllum* as accessories. Favositids (both domal and branching) are small and rare; brachiopods are uncommon; and bryozoans are mainly fragmentary encrusters.

The favositid/crinoidal sand paleocommunity is characteristic of bedded Edgecliff facies and those beds that onlap and interfinger with rugosan mounds to form large reef structures (referred to as *flank beds* by Oliver [1956a]). This biostromal paleocommunity displays a much higher diversity than does the phaceloid rugosan paleocommunity. Large sheetlike-to-domal favositids are abundant but do not form constructional masses. Crinoids were the greatest contributors to this paleocommunity—ossicles make up the bulk of the rock and indicate the abundant growth of these organisms—but complete calyces are never found. Solitary rugose corals are locally extremely abundant, as are fenestrate bryozoan colonies. Brachiopods and other reef dwellers are common, but single colonies of the mound-building phaceloid rugosans are uncommon to rare. Stromatoporoids and massive colonial rugosans are extremely rare in the Edgecliff reefs but, when found, are part of this paleocommunity.

Wolosz (1992a) classified the Edgecliff reefs based on the relative contribution of these two paleocommunities to the overall growth of the reef. In summary, mounds consisting entirely of the phaceloid colonial rugosan paleocommunity with a distinct developmental succession of rugosan genera are classified as *successional mounds,* and the cyclic intergrowth of rugosan mounds with banks of the favositid/crinoidal sand paleocommunity produced *mound/bank reefs.* Finally, banks of the favositid/crinoidal sand paleocommunity that contain only thin horizons of phaceloid rugosans are classified as *thicket/bank reefs* (fig. 15.2).

Geographic Distribution and Ecology

Edgecliff reef exposures are found from just south of Albany, New York, westward along the Onondaga strike belt to Buffalo and then on into Ontario, Canada (fig. 15.1). Large pinnacle reefs are found in the subsurface in New York and Pennsylvania along a northeast/southwest trend.

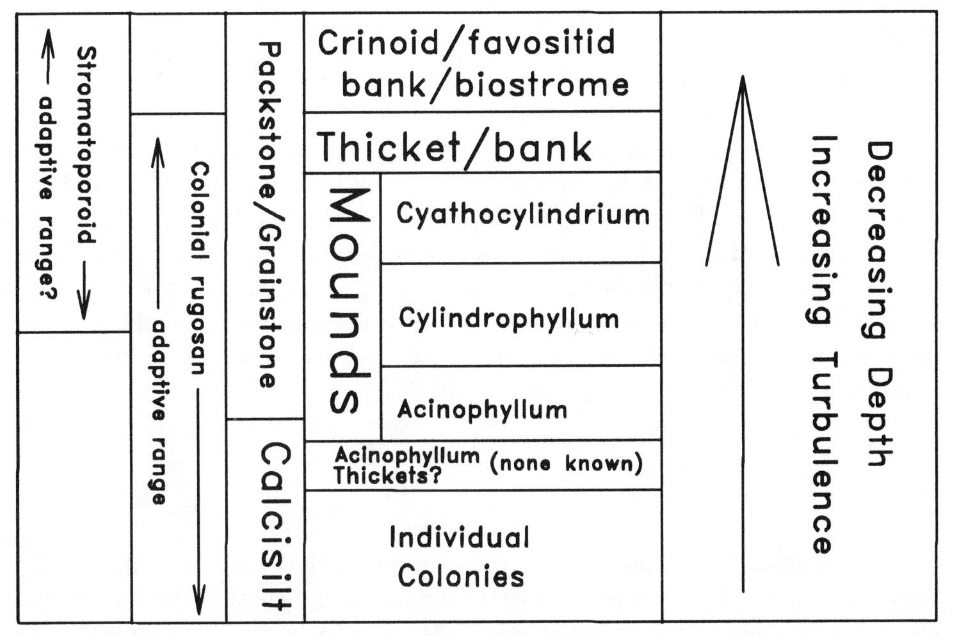

FIGURE 15.2. Interpreted adaptive ranges of phaceloid colonial rugosans and stromatoporoids. Under warm-water conditions, the stromatoporoids would quickly exclude the rugosans. Removal of stromatoporoids by environmental conditions results in the colonial rugosans' extending to the extreme end of their adaptive range–"thicket/bank"–reefs. Calcisilt is generally equivalent to Edgecliff "C1" facies but can also represent quiet shallow- water environments preserved in Ontario, Canada. Packstone/grainstone would be equivalent to normal Edgecliff facies across New York State and Ontario. The rugosan succession is based on Wolosz (1985, 1992b), also see Fig. 15.10.

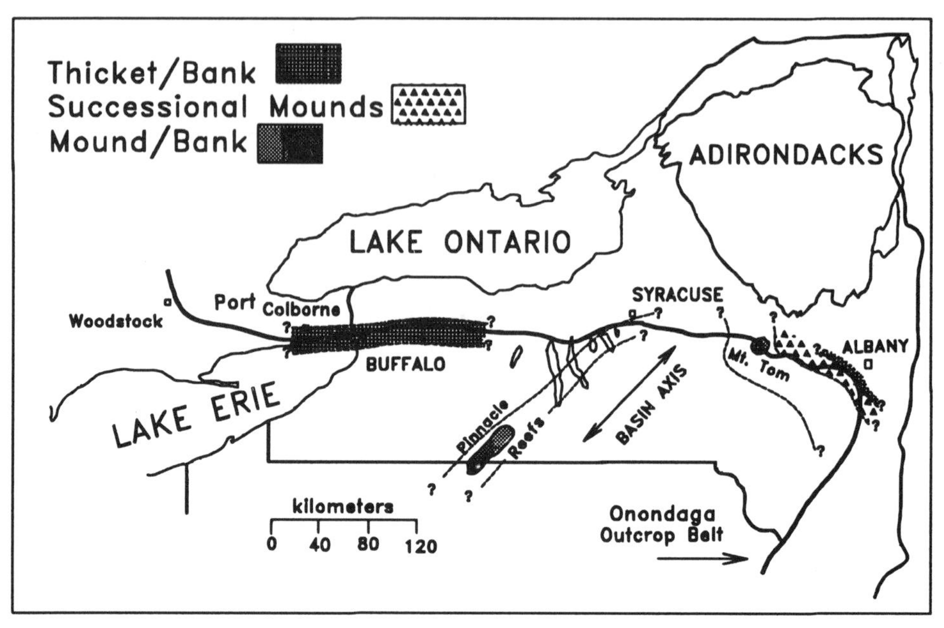

FIGURE 15.3. Interpreted distribution of Edgecliff reefs by type. The dashed lines represent interpreted boundaries between areas of distinct reef types, and the patterning indicates the confirmed presence of reefs in either the outcrop or the subsurface. Question marks show the limits of interpretable area owing to the lack of exposure or subsurface information, or the erosional removal of the Onondaga. (After Wolosz 1992a.)

Classifying the Edgecliff reefs by means of growth pattern (fig. 15.3) reveals a distinct geographic distribution showing control by the water depth and the rate of basinal subsidence (Wolosz 1992a). The large mound/bank structures rim the axis of major basinal subsidence (as located by Lindholm [1967] and Mesolella [1978]), whereas the thicket/bank structures are restricted to an assumed shallow-water area in the northeastern corner of the strike belt and the western transgressive facies of the Edgecliff from the vicinity of Rochester westward into Ontario, Canada.

Thicketing Events

Thickets of phaceloid colonial rugosans were undoubtedly common during the Edgecliff's deposition, but in most cases the continued coral growth led to the formation of rugosan mounds (fig. 15.2). An examination of the exposed base of any rugosan successional mound rooted in the C1 calcisilts of the Edgecliff (e.g., the lower mound at Mount Tom; see Wolosz et al. 1991) leads to the conclusion that these structures began as *Acinophyllum* thickets. The internal bedding in the small lenticular mounds in the vicinity of Port Colborne, Ontario, Canada, indicate little relief above the seafloor during deposition.

"Thicketing events" in the Edgecliff are, however, characteristic of, and restricted to, structures classified as *thicket/bank* reefs, or crinoidal grainstone/packstone banks in which the thickets are the sole contribution of the phaceloid colonial rugosan paleocommunity (Wolosz 1992a). The "events" can be defined as the widespread colonization of preexisting low, shield-shaped crinoidal sandbanks by phaceloid colonial rugosans that leads to the formation of a single rugosan thicket covering the entire bank. The extent of these thickets can be determined by tracing them along outcrops (North Coxsackie Reef) and/or by correlating the thickets between remnant portions of the reef in a quarry (LeRoy Bioherm). The thickets appear to have been controlled entirely by the size of the preexisting crinoidal sandbank.

At the North Coxsackie Reef, two such events cover areas, as exposed, of 18,000 m^2 and 15,000 m^2, but the total thicket coverage is estimated to have been at least twice as extensive as is currently preserved or exposed. Crinoidal packstones lap up onto an eroded biohermal mound at the LeRoy Bioherm (Wolosz and Paquette 1994) and contain a thicket with an exposed area of roughly 16,000 m^2. But again, it is estimated to have covered at least twice that area. Near Port Colborne, Ontario, Canada, small banks averaging 250 m^2 were also covered by phaceloid rugosan thickets (Wolosz 1990b).

Colonies within the thickets often appear recumbent, owing to a combination of lateral corallite growth, compaction, and, in some cases, storm damage. Overall, however, the colonies are robust and fully developed (fig. 15.4). Despite the apparently normal growth of the colonies in the thickets, the phaceloid colonial rugosan paleocommunity did not develop further, and the thickets eventually were buried by crinoidal sands of the favositid/crinoidal sand paleocommunity.

Wolosz (1992a) noted that all these reefs are either restricted to the shallow-water, western transgressive facies of the Edgecliff or, in the case of the North Coxsackie Reef, represent the shallowest of the eastern reefs, because of the absence of the deeper-water C1 facies, as the reef is rooted directly in Edgecliff grainstone/packstone (Wolosz 1992b). In all cases, the thicket is enclosed in packstones of the favositid/crinoidal sand facies.

The packstone favositid fauna varies from well developed at the Port Colborne mounds, the North Coxsackie Reef, and the Country Club of Buffalo Reef, to weak at LeRoy. In all cases, the colonies are often overturned (figs. 15.5 and 15.6), indicating at least intermittent periods of high-energy conditions. The thickets themselves are characterized by those colonial rugosan genera typical of higher-turbulence conditions (fig. 15.2; see Wolosz [1992b] for a discussion of the turbulence control of colonial rugosan succession): mainly *Cylindrophyllum* at Country Club of Buffalo Reef and LeRoy, *Cyathocylindrium* at the North Coxsackie Reef, and *Cyathocylindrium* and *Heliophyllum* in the small Port Colborne mounds. As in the later stages of mound succession, *Acinophyllum* is found as an accessory in the thickets (Wolosz 1985, 1992b).

The most important clue to understanding these thickets is that they are single events in packstone/grainstone facies. As already noted, thickets rooted

FIGURE 15.4. Country Club of Buffalo Reef. Most of structure is bedded favositid/crinoidal sand paleocommunity. Thicket is indicated by arrows.

in the C1 facies continued to develop, and they eventually formed mounds. The small, bedded lenticular structures near Port Colborne, however, are restricted to protected areas (Wolosz 1992a:11). Also important is the observation that well-developed phaceloid colonies are seldom found in the bedded grainstone/packstones of the Edgecliff (Oliver 1976:39; Wolosz, personal observation) but are common in calcisilts above the thin Edgecliff facies along the Erie Peninsula to the west of Buffalo (Clarence member of Ozol [1963]), and in the Onondaga calcisilts exposed along the strike belt for a few kilometers to the south of Roberts Hill (the southernmost known reef). No isolated thickets have been identified from the calcisilts in either area (Wolosz, personal observation).

These patterns reveal that the phaceloid rugosans were better adapted in some way to the calcisilt facies than to the nonreefal grainstone/packstone. Indeed, simply the survival rates of larvae in different sedimentary environments could explain this distribution of colonial rugosans. It is well known from studies of recent coral (Sato 1985; Babcock and Davies 1991) that in quiet water, larvae that find suitable attachment sites are able to survive, whereas in more turbulent waters the higher sedimentation rates on nearly horizontal surfaces quickly kill off newly settled larvae. However, since the favositid/crinoidal sand paleocommunity is characterized by abundant favositid coralla and solitary rugosans, it is doubtful that a simple "high sedimentation rate" argument applies here.

Instead, the initial colonial growth rate and the colonial morphology, in conjunction with the lateral movement of sediment caused by intermittent water turbulence, probably controlled the survival of the phaceloid rugosans. Examination of the bases of favositid colonies from the packstone facies reveals an initial reptant growth pattern with widely spaced tabulae within the corallites (Wolosz, personal observation). The wide spacing of the tabulae is assumed to indicate rapid growth (Stel 1978:81), which would lead to the early stabilization of the colony on the substrate. Slow initial growth among phaceloid rugosans could have been detrimental in the crinoidal sands of the Edgecliff, but the protocorallite and initial offsets in such colonies have only rarely been preserved (Oliver, personal communication); hence no data on initial growth rates in phaceloid rugosans are available.

The open structure of the phaceloid colony would, however, have acted as an excellent sediment baffle, being a highly advantageous morphology in environments of moderate to low sedimentation, in which the sediment filling the intercorallite spaces would help anchor the colony to the substrate. This

FIGURE 15.5. *Cylindrophyllum* thicket at LeRoy Bioherm. Note overturned and eroded favositid colony beneath thicket.

Figure 15.6. Favositid/crinoidal sand paleocommunity at Country Club of Buffalo Reef. Bottom of photo = top of figure 15.4. Note abundant favositids (arrows) and large overturned colony (O).

appears to be the case in the reef mounds. At the Roberts Hill Reef, Wolosz (1984:110–119) noted that calcisilt intercorallite fills are common and are often partially washed out and replaced by bioclastic packstone. However, under conditions of higher sedimentation—or sediment mobility—baffling would become disadvantageous, since the coralla would become sites of rapid sediment accumulation, eventually swamping the colony.

Figure 15.7 shows two possible models of a rugosan thicket in the Edgecliff. In fig. 15.7A, we see what may be considered a standard image of colonies in a thicket, in which only the base of the colony is in contact with the sediment. This is common in both modern shallow-water (James 1983, fig. 24) and deep-water reefs (Wilson 1979, Plate II). However, examination of a colony from the LeRoy thicketing event (fig. 15.8) reveals that the second model (fig. 15.7B), in which only a small part of the colony exists above the sediment, may be closer to the truth.

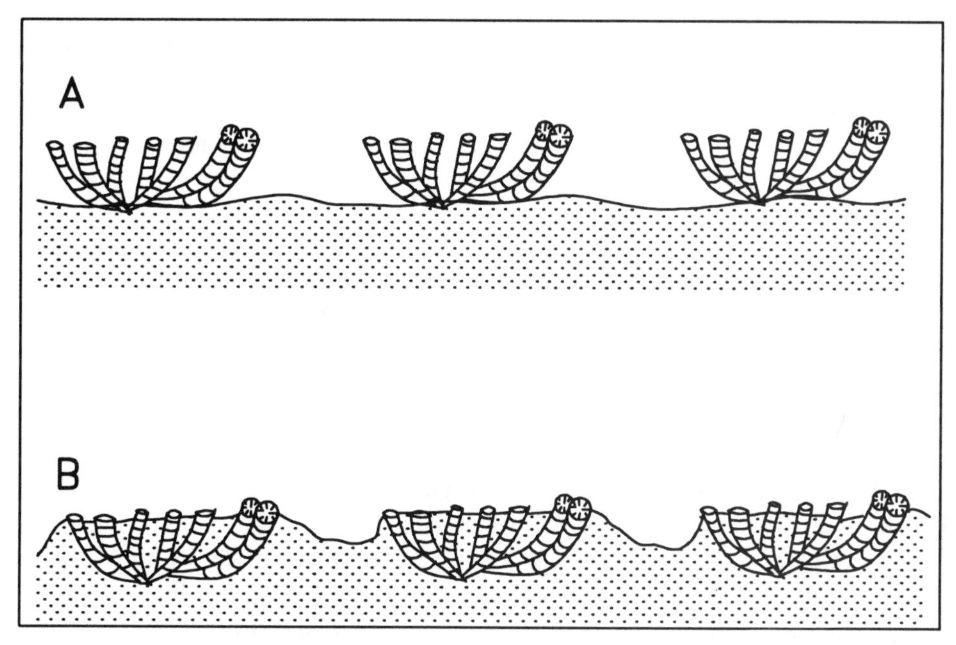

FIGURE 15.7. Two possible models for colonial rugosan thickets during growth. (A) Most of colony exists well above the sediment water interface, as is commonly observed in modern reef settings. (B) Most of intercorallite spaces are filled with sediment because of the baffling effect of phaceloid colonial morphology.

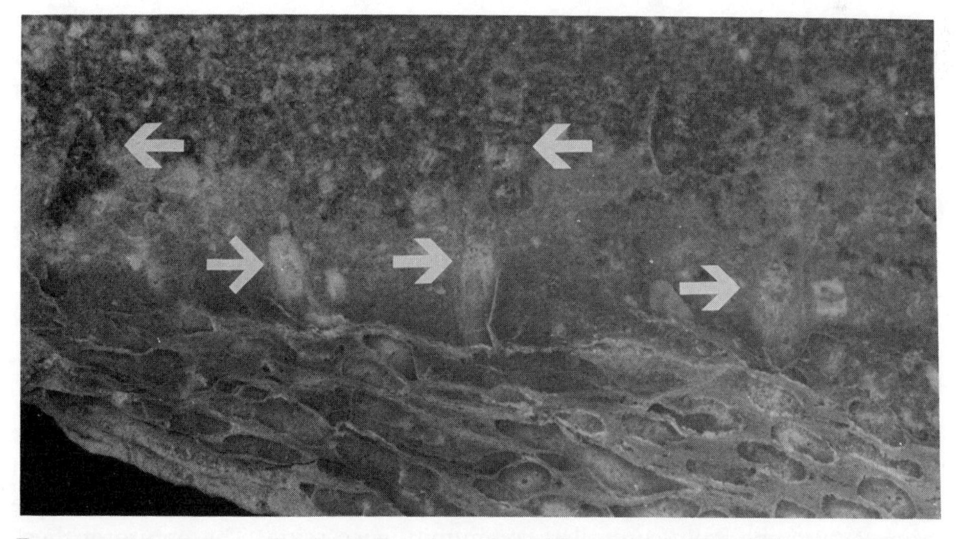

FIGURE 15.8. Colony of *Acinophyllum* from the thicket at the LeRoy Bioherm. Note that the intercorallite lithology is very fine grained, compared with the overlying coarse bioclastic lithology. Only five corallites (arrows) continue across the contact of these two lithologies and are regarded as the few survivors following the swamping of most of the colony by the deposition of coarse sediment (storm deposit?). The dark spots in the uppermost lithology is bitumen staining of small pores. Polished slab.

In fig. 15.8 the colony of *Acinophyllum* has a matrix of bioclastic calcisiltite and green terrigenous clay filling the intercorallite spaces almost to the top of the colony. The clay-rich bioclastic calcisiltite is overlain by grainstone, in which a few corallites are preserved. These corallites represent the few members of the colony that survived a storm that buried most of the polyps, owing to their lack of relief above the substrate (note that burial by storm deposits permits the thickets to be classified as *Obrution Deposits,* according to Brett and Seilacher 1991).

Sediment that accumulated rapidly enough to fill the phaceloid colony while it was growing, making it vulnerable to burial by a storm deposit, would also be the main reason for the inability of new larvae to become established in the thicket. Outside the thicket, juvenile colonies would be either rapidly swamped or overturned and buried. The more rapidly upward-growing solitary rugosans and the rapidly lateral-growing favositids would survive.

These thicketing events therefore represent the appearance of the phaceloid rugosans at the extreme edge of their adaptive range—"fortuitous beach-heads"—which may be attributable to a sea-level fluctuation but may more simply have been due to a few years without a major storm. The rugosans, however, were unable to capitalize on these "fortuitous beachheads" because the environmental conditions were ephemeral.

With the return to conditions of normal Edgecliff deposition—which may just have been very shallow water (in the case of the Port Colborne mounds) or intermittent water turbulence and sediment movement (on the larger thicket/banks)—the colonies in the thicket were able to survive, but their continued development into a mound through the recruitment of spat was prevented, with the result that the thickets became "events." In this sense, the Edgecliff thickets are ecological "events" similar to those of the Centerfield Member in the Hamilton Group—another marginal environment—in which phaceloid coral thickets are common in the siltstone facies yet never develop into mounds (Wolosz and Wallace 1981; Savarese et al. 1986).

Ecology of the Edgecliff Reefs

As illustrated in fig. 15.2, the transition from coral mounds to the favositid/crinoidal sandbanks and biostromes is believed to have occurred as the water depth fell, leading Wolosz (1992b) to argue that the transition from phaceloid colonial rugosa to the favositid/crinoidal sand paleocommunity is an example of an environmentally controlled community succession. The nature of these thicketing events supports both arguments and is the key to our understanding of the Edgecliff reefs. Both mound/bank reefs (see fig. 15.9, Mount Tom) and successional mounds (see fig. 15.10, Roberts Hill) exhibit a pattern similar to that of the thicket (event)/bank reef, but on a much different scale. In each case the phaceloid colonial rugosan paleocommunity is onlapped by the favositid/

crinoidal sand paleocommunity, which is itself then capped by a return to the rugosan community (second mound stage at Mount Tom; recolonization horizon at Roberts Hill; second thicket at any thicket/bank reef). Since the second-stage mound in a large structure like Mount Tom is restricted to the top of the structure (fig. 15.9), it appears that the shifts from one paleocommunity to another were controlled by relative sea-level fluctuations in which the decreasing water depth moved the initial mound beyond the adaptive range of the phaceloid rugosans and led to the formation of a bank stage populated by the favositid/crinoidal sand paleocommunity. Hence the termination of mound growth would have been controlled by the same physical features of the environment that controlled the thicketing. Later deepening resulted in a return of the colonial rugosans to the top of the bank and the formation of a new mound (Wolosz 1985; Wolosz and Paquette 1988; Wolosz et al. 1991).

Unlike the thickets, however, the second-stage mounds in the mound/bank reefs were maintained at sufficient depth to allow continued growth. It is therefore notable that although the thicket/bank structures are located in the assumed shallowest-water areas farthest from the axis of basinal subsidence,

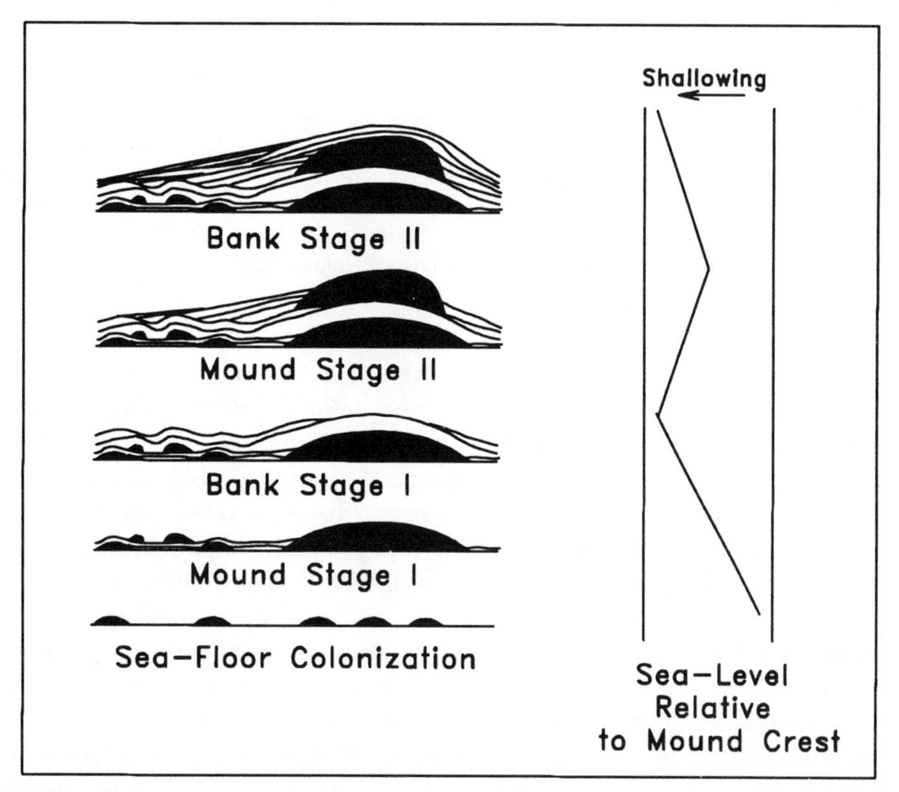

FIGURE 15.9. Sequence of development and interpreted depth curve for reef crest of Mount Tom Reef, an example of a mound/bank structure. Note that second colonial rugosan mound stage does not drape the structure, only caps it. (After Wolosz et al. 1991.)

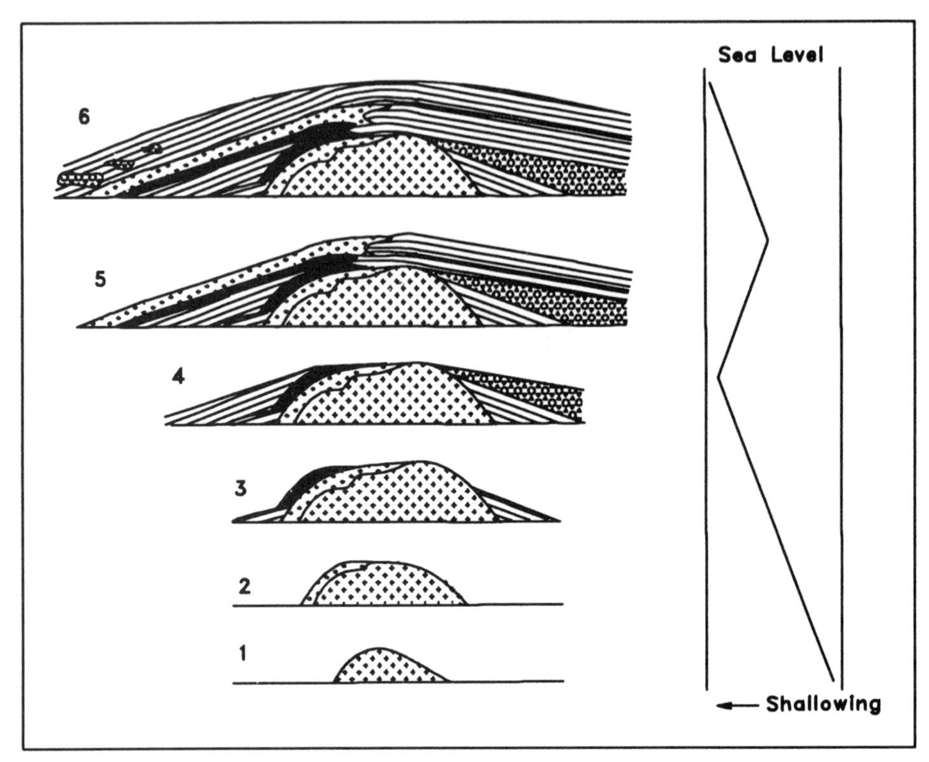

FIGURE 15.10. Sequential development of Roberts Hill reef (an example of a successional mound) with interpreted depth curve for reef crest. Steps 1–3 illustrate the initial succession of mound-building rugosa (see fig. 15.2). Steps 4 and 5 show the swamping of the mound by crinoidal sands, followed by recolonization by phaceloid rugosans. Step 6 gives the final burial by crinoidal sand. (After Wolosz 1992b.)

the mound/bank reefs are the closest to the axis (fig. 15.3), where their crests would undergo the greatest range of relative sea-level fluctuation owing to either incipient anticlinal folding (Willette and Robinson 1979) or simple subsidence versus upward reef growth (Wolosz and Paquette 1988).

At this point the reason for the absence of stromatoporoids becomes important to our understanding of the ecology of these thicketing events and of the Edgecliff in general. Owing to their rapid lateral growth, the Devonian stromatoporoids dominated high-energy reef facies (Meyer 1981). As already noted, the favositids employed the same adaptation—rapid lateral growth—to stabilize themselves on the packstone substrate. The healthy stromatoporoids should also have been able to survive in this environment. Yet a careful microscopic examination of thin sections from the grainstone/packstone flank beds at Roberts Hill reveals that although very small stromatoporoid colonies are moderately common (Wolosz 1990a), they appear to have been quickly swamped or overturned and killed. Some environmental factor must have suppressed the stromatoporoids, because under normal conditions they would have grown rapidly and outcompeted the corals near the initial point of overlap of their adaptive ranges (see Walker and Alberstadt [1975] and many other examples of Middle Paleozoic reefs successions too numerous to mention).

Wolosz (1990a, 1991) and Wolosz and Paquette (1995) contended that stromatoporoids are rare and small in the Edgecliff because of the cool-water temperatures during deposition. Indeed, the growth rate of modern sponges does appear to decrease with decreasing temperature (Storr 1976), and stromatoporoids are generally considered to be restricted to tropical and subtropical environments (Stock 1990; Nestor 1984).

A cool-water Edgecliff is not beyond reason, and accordingly, Heckel and Witzke (1979) place the Appalachian Basin at roughly 30°S latitude during the Middle Devonian, but with waters entering the basin from a cool subtropical oceanic gyre. This paleogeographic model was later revised (Witzke and Heckel 1988), with the Appalachian Basin placed farther south, closer to 40°S latitude.

Therefore if the cool-water model for the Edgecliff deposition is correct, then the phaceloid rugosans would meet Grime's (1977) criteria for stress-resistant species with respect to their ability to survive cool temperatures (stress), but not high-disturbance environments (shallow water). The stromatoporoids were clearly a competitive species, adapted to high-energy environments, but only under warm-water conditions. Removal of the stromatoporoids by adverse environmental conditions would thus have allowed the rugosans to spread to the extreme edge of their environmental range, thereby permitting the formation of the observed thicketing events.

Thicketing Events and Mound/Bank Cycles: Correlative Events?

If mound/bank and thicket/bank cycles are in fact due to sea-level fluctuations within the basin, they should be correlative across the basin. Interpretive depth curves for the reef crests at Mount Tom (fig. 15.9) and Roberts Hill (fig. 15.10) are nearly identical and so support the correlation of mound/bank cycles between reefs. Future testing will determine whether these cycles can be directly correlated to the thicketing events in shallower water.

The thickets that define the thicket/bank reefs in the Edgecliff were ephemeral events. They were caused by some environmental change—a rise in sea level or a more simple change—that allowed the phaceloid colonial rugosans to become established on the banks of the favositid/crinoidal sand paleocommunity. The return to conditions of normal Edgecliff deposition prevented the thickets from developing into mounds and led to their return to a favositid/crinoidal sandbank.

These thicket/bank structures can be used as a model for Edgecliff reefs, since all Edgecliff reefs exhibit—albeit on a much larger scale—similar rugosan mound (thicket)/favositid—crinoid bank cycles. In each case the cycles can be attributed to the rugosan mounds' either being moved into (because of a drop in the sea level) or growing into a high-energy environment beyond the adaptive range of the phaceloid colonial rugosans. A subsequent relative rise in the sea level led to the return of the mound-building colonial rugosans.

It also has been suggested that the presence of the colonial rugosans at the extreme edge of their adaptive range was in large part due to the inhibition of stromatoporoid growth by cool-water conditions in the Edgecliff sea.

ACKNOWLEDGMENTS

I wish to thank Douglas E. Paquette for his assistance in the field and his discussions of the Edgecliff reefs. Reviews by William A. Oliver Jr. and Richard H. Lindemann greatly improved this report. Thomas Mooney, James Edel II, and James Quinn helped with the fieldwork. My research was supported by the U.S. Department of Energy Special Research Grants Program Grant #DE-FG02–87ER13747.A000.

References

Babcock, R. and P. Davies. 1991. Effects of sedimentation on settlement of *Acropora millepora. Coral Reefs* 9:205–208.

Blodgett, R. B., D. M. Rohr, and A. J. Boucot. 1988. Lower Devonian gastropod biogeography of the Western Hemisphere. In N. J. McMillan, A. F. Embry, and D. J. Glass, eds., *Devonian of the World, Proceedings of the Second International Symposium on the Devonian System,* vol. 3. *Canadian Society of Petroleum Geologists Memoir* No. 14, pp. 281–94.

Brett, C. E. and A. Seilacher. 1991. Fossil Lagerstätten: A taphonomic consequence of event sedimentation. In G. Einsele, W. Ricken and A. Seilacher, eds., *Cycles and Events in Stratigraphy,* pp. 283–97. New York: Springer-Verlag.

Cassa, M. R. and D. L. Kissling. 1982. Carbonate facies of the Onondaga and Bois Blanc Formations, Niagara peninsula, Ontario, pp. 65–97. In *New York State Geological Association, 54th Annual Meeting, Field Trip Guidebook.*

Chadwick, G. H. 1944. Geology of the Catskill and Kaaterskill quadrangles, part II: Silurian and Devonian geology. *New York State Museum Bulletin* 336.

Crowley, D. and R. Z. Poore. 1974. Lockport (Middle Silurian) and Onondaga (Middle Devonian) patch reefs in western New York, pp. A1–A41. In *New York State Geological Association, 46th Annual Meeting, Field Trip Guidebook.*

Cumings, E. R. 1932. Reefs or bioherms? *Geological Society of America Bulletin* 43:331–52.

Einsele, G. W. Ricken and A. Seilacher, eds. 1991. *Cycles and Events in Stratigraphy.* New York: Springer-Verlag.

Grime, J. P. 1977. Evidence for the existence of three primary strategies in plants and its relevance to ecological and evolutionary theory. *The American Naturalist* 111:1169–94.

Heckel, P. H. and B. J. Witzke. 1979. Devonian world palaeogeography determined from distribution of carbonates and related lithic palaeoclimatic indicators. In M. R. House, C. T. Scrutton, and M. G. Bassett, eds., *The Devonian System,* pp. 99–124. *Special Papers in Palaeontology* No. 23.

James, N. P. 1983. Reef. In P. A. Scholle, D. G. Bebout, and C. H. Moore, eds., *Carbonate Deposition Environments,* pp. 345–440. *American Association of Petroleum Geologists Memoir* No. 33.

Kissling, D. L. 1987. Middle Devonian Onondaga pinnacle reefs and bioherms, Northern Appalachian Basin. In *Second International Symposium on the Devonian System, Calgary, Alberta, Canada, Program and Abstracts,* p. 131.

Koch, W. F. II and A. J. Boucot. 1982. Temperature fluctuations in the Devonian Eastern Americas Realm. *Journal of Paleontology* 56:240–43.
Lindemann, R. H. 1988. The LeRoy Bioherm, Onondaga Limestone (Middle Devonian), Western New York. In H. H. J. Geldsetzer, N. P. James, and G. E. Tebbutt, eds., *Reefs–Canada and Adjacent Areas*. pp. 487–91. *Canadian Society of Petroleum Geologists Memoir* No. 14.
Lindholm, R. C. 1967. Petrology of the Onondaga Limestone (Middle Devonian). Ph.D. diss., Johns Hopkins University.
Mesolella, K. J. 1978. Paleogeography of some Silurian and Devonian reef trends, Central Appalachian Basin. *American Association of Petroleum Geologists Bulletin* 62:1607–44.
Meyer, F. O. 1981. Stromatoporoid growth rhythms and rates. *Science* 213:894–95.
Nestor, H. 1984. Autecology of stromatoporoids in Silurian cratonic seas, In M. G. Bassett and J. D. Lawson, eds., *Autecology of Silurian Organisms*, pp. 265–80. *Special Papers in Palaeontology* No. 32.
Oliver, W. A. Jr. 1954. Stratigraphy of the Onondaga Limestone (Devonian) in central New York. *Bulletin of the Geological Society of America* 65:621–52.
———. 1956a. Biostromes and bioherms of the Onondaga Limestone (Devonian) in eastern New York. *New York State Museum and Science Service Circular* No. 45.
———. 1956b. Stratigraphy of the Onondaga Limestone in eastern New York. *Bulletin of the Geological Society of America* 67:1441–74.
———. 1976. Noncystimorph colonial rugose corals of the Onesquethaw and Lower Cazenovia stages (Lower and Middle Devonian) in New York and adjacent areas. *United States Geological Survey Professional Paper* No. 869.
Ozol, M. A. 1963. Alkali reactivity of cherts and stratigraphy and petrology of cherts and associated limestones of the Onondaga Formation of central and western New York. Ph.D. diss., Rensselaer Polytechnic Institute.
Rickard, L. V. 1975. Correlation of Silurian and Devonian rocks in New York State. *New York State Museum and Science Service Map and Chart Series* No. 24.
Sato, M. 1985. Mortality and growth of juvenile coral *Pocillopora damicornis* (Linnaeus). *Coral Reefs* 4:27–34.
Savarese, M., L. M. Gray, and C. E. Brett. 1986. Faunal and lithologic cyclicity in the Centerfield Member (Middle Devonian: Hamilton Group) of western New York: A reinterpretation of depositional history. In C. E. Brett, ed., *Dynamic Stratigraphy and Depositional Environments of the Hamilton Group (Middle Devonian) in New York State*, Part 1, pp. 32–56. *New York State Museum Bulletin* No. 457.
Stel, J. H. 1978. *Studies on the Palaeobiology of the Favositids*. Groningen: Stabo/All-Round B.V.
Stock, C. W. 1990. Biogeography of Devonian stromatoporoids. In W. S. McKerrow and C. R. Scotese, eds., *Palaeozoic Palaeogeography and Biogeography*, pp. 257–66. *Geological Society Memoir* No. 12.
Storr, J.F. 1976. Ecological factors controlling sponge distribution in the Gulf of Mexico and the resulting zonation. In F. W. Harrison and R. R. Cowden, eds., *Aspects of Sponge Biology*, pp. 261–76. New York: Academic Press.
Walker, K. R. and L. P. Alberstadt. 1975. Ecological succession as an aspect of structure in fossil communities. *Paleobiology* 1:238–57.
Willette, P. D. and J. E. Robinson. 1979. Structural control of the Onondaga Reefs (Devonian) in south-central New York State. *American Association of Petroleum Geologists Bulletin* 63:1591.
Williams, L. A. 1980. Community succession in a Devonian patch reef (Onondaga Formation, New York)–Physical and Biotic Controls. *Journal of Sedimentary Petrology* 50:1169–86.

Wilson, J. B. 1979. "Patch" development of the deep-water coral *Lophelia pertusa* (L.) on Rockall Bank. *Journal of the Marine Biological Association of the United Kingdom* 59:165–77.

Witzke, B. J. and P. H. Heckel. 1988. Paleoclimatic indicators and inferred Devonian paleolatitudes of Euramerica. In N.J. McMillan, A. F. Embry, and D. J. Glass, eds., *Devonian of the World, Proceedings of the Second International Symposium on the Devonian System,* vol. 1, pp. 49–66. *Canadian Society of Petroleum Geologists Memoir* No. 14.

Wolosz, T. H. 1984. Paleoecology, sedimentology and massive favositid fauna of Roberts Hill and Albrights reefs. Ph.D. diss., State University of New York at Stony Brook.

———. 1985. Roberts Hill and Albrights Reefs: Faunal and sedimentary evidence for an eastern Onondaga sea-level fluctuation, pp. 169–85. In *New York State Geological Association, 57th Annual Meeting, Field Trip Guidebook.*

———. 1990a. Edgecliff reefs of New York and Ontario, Canada–Middle Devonian Temperate water bioherms. *Geological Society of America, Abstracts with Program* 22(7):A220.

———. 1990b. Shallow water reefs of the Middle Devonian Edgecliff Member of the Onondaga Limestone, Port Colborne, Ontario, Canada, pp. E1–E17. In *New York State Geological Association, 62d Annual Meeting, Field Trip Guidebook.*

———. 1991. Edgecliff Reefs–Devonian temperate water carbonate deposition. *American Association of Petroleum Geologists Bulletin* 75:696.

———. 1992a. Patterns of reef growth in the Middle Devonian Edgecliff Member of the Onondaga Formation of New York and Ontario, Canada and their ecological significance. *Journal of Paleontology* 66(1):8–15.

———. 1992b. Turbulence controlled succession in the Middle Devonian Edgecliff reefs of eastern New York State. *Lethaia* 25:283–90.

Wolosz, T. H., H. R. Feldman, R. H. Lindemann, and D. E. Paquette. 1991. Understanding the east central Onondaga Formation (Middle Devonian): An examination of the facies and brachiopod communities of the Cherry Valley section, and Mt. Tom, a small pinnacle reef, pp. 373–412. In *New York State Geological Association, 63d Annual Meeting, Field Trip Guidebook.*

Wolosz, T. H. and R. H. Lindemann. 1986. Evidence of sea level fluctuations within the Onondaga Formation in eastern New York. *Geological Society of America, Abstracts with Program* 18(1):77.

Wolosz, T. H. and D. E. Paquette. 1988. Middle Devonian reefs of the Edgecliff Member of the Onondaga Formation of New York. In N. J. McMillan, A. F. Embry, and D. J. Glass, eds., *Devonian of the World, Proceedings of the Second International Symposium on the Devonian System,* vol. 2, pp. 531–39. *Canadian Society of Petroleum Geologists Memoir* No. 14.

Wolosz, T. H. and D. E. Paquette. 1994. The LeRoy Bioherm revisited: Evidence of a complex developmental history, pp. 445–56. In *New York State Geological Association, 66th Annual Meeting, Field Trip Guidebook.*

Wolosz, T. H. and D. E. Paquette. 1995. Middle Devonian temperate water bioherms of eastern New York State (Edgecliff Member, Onondaga Formation), pp. 227–50. In *New York State Geological Association, 66th Annual Meeting, Field Trip Guidebook.*

Wolosz, T. H. and R. J. Wallace. 1981. Coral population variations in a colonizing community (Devonian). In J. Gray, A. J. Boucot, and W. B. N. Berry, eds., *Communities of the Past,* pp. 223–42. Stroudsburg, Pa.: Hutchinson Ross.

16

Some Upper Carboniferous (Pennsylvanian) Event Beds (Epiboles)

Ronald R. West, Howard R. Feldman,
and Christopher G. Maples

ABSTRACT

Taxa from three different phyla (Porifera, Coelenterata, and Mollusca) are recorded in eight separate event beds (epiboles) in the Upper Carboniferous (Pennsylvanian) of the Midcontinent. Two different types of epiboles are recognized, composed of (1) organisms preserved in, or nearly in, their life position and (2) predominantly disarticulated and transported fossils. Coralline (chaetetid-grade) demosponge and syringoporoid corals make up the former, and rugosoid corals and myalinid pelecypods typify the latter. The occurrence of all these epiboles was controlled mainly by the organisms' ecological requirements, but the widespread distribution of the rugosoid corals and myalinid pelecypods was in part due to within-habitat reworking by waves, currents, and tides during and between storms.

Available evidence suggests that the sponges, corals, and pelecypods were smothered by rather sudden influxes of siliciclastic and carbonate debris. Potential causes of such influxes are eustatic, climatic, tectonic, or a combination of these three. Although these epiboles are recognized over areas covering hundreds to thousands of square kilometers, even more extensive, detailed stratigraphic studies over larger geographic areas are needed to evaluate their temporal significances.

Different types of event beds have been, and can be, recognized in the rock record. One such event bed is an epibole. An *epibole,* as originally defined by Trueman (1923), is "a stratigraphical unit to cover deposits accumulated during a hemera." This definition of epibole is essentially the same as that given by Bates and Jackson (1987:218). As used in this definition of epibole, a *hemera* is a chronological unit that marks the acme of one or more species (Arkell 1933; see also Bates and Jackson 1987:303). Brett et al. added that "epiboles are typically thin intervals, rarely more than a few meters thick, that are geographically very widespread (10's to 100's of kilometers) and characterized by an unusual abundance of fossil taxa that are normally rare or absent from a given facies or region" (1990:193). If used as a component of event stratigraphy (Seilacher 1981), event stratinomy (Seilacher 1984b), or event communities/species (Kauffman and Sageman 1990), an event bed is synonymous with an epibole (we use *bed* rather than *horizon* here because horizon implies a two-dimensional feature and epibole clearly implies three dimensionality, including some degree of time averaging), except that epiboles are generally limited to biotic or biotically influenced events.

Brett et al. also stated that epiboles *may* crosscut litho- and biofacies and that *some* are useful as marker beds for high-resolution correlation (1990:193–94). We emphasize the *may* and *some* because not all epiboles are isochronous—something we all know but do not always remember. Ideally, an epibole (in a strict sense) is isochronous, and some epiboles approach this ideal more closely than others do. Unfortunately, independent tests of isochroneity rarely exist, and so in this paper, *epibole* is used in a broader sense.

How does an *abundance* zone (= *acme zone*, ISSC 1976:59) differ from an epibole? Our understanding of epibole, and the way in which we use it here, is that all epiboles are abundance zones, but relatively few abundance zones have the temporal acuity necessary to be epiboles (see North American Stratigraphic Code 1983, Article 48c, p. 862, for an example). Our assumption here is that epiboles, although not necessarily perfectly isochronous, are nearly so, especially when compared with standard biostratigraphic time scales (units).

When studying epiboles, it is therefore necessary to define, and limit, the processes responsible for the events, in order to (1) understand the autecology of the organism(s) involved, (2) document and interpret the lithology surrounding the organism(s), (3) determine the areal extent of the event bed(s), and (4) calculate the relative isochroneity of the event bed(s). This last factor is especially critical for extensive epiboles, because the greater the geographic extent of the event bed(s) is, the greater the potential for heterochroneity will be.

Autecology of the organism(s), associated taphonomic signatures, and characteristics of the entombing rock(s) provide clues to the conditions that produce epiboles and their relative isochroneity. But they alone may not adequately explain the geographic distribution of these event beds. Distinctive palaeontological event beds that occur at only one outcrop probably were produced by short-lived local processes, such as colonization of the seafloor following deposition of a layer of mud. Epiboles that extend for tens or hundreds of kilometers are more likely to result from regional or global processes, such as the lower sedimentation rate associated with changes in sea level, tectonics, or climate. The geographic distribution of an epibole (event bed) may thus be the most crucial factor in identifying the processes of its genesis.

Studies, by us and others, of Pennsylvanian exposures in the northern Midcontinent, primarily Kansas, have revealed several deposits that we judge to be epiboles. Although there surely are others that will be revealed by further study, we focus in this essay on eight epiboles based on the accumulation of three different major taxa: coralline demosponges with a chaetetid-grade skeleton, corals (syringoporoid tabulates and the rugosoid *Pseudozaphentoides*), and pelecypods (*Orthomyalina*). Stratigraphically, these occur in the Desmoinesian (chaetetids), Missourian (syringoporoids), and Virgilian (*Pseudozaphentoides*, syringoporoids, and *Orthomyalina*). Our current understanding of the timing of these epiboles is general, although we do know that they are associated with small-scale events. Future studies of the temporal extent and lateral distribution of these epiboles will improve our understanding of their cause, importance, and timing.

Sponge Epiboles

Coralline demosponges with a chaetetid-grade skeleton range from the Silurian to the Triassic (West 1992), and the extant demosponges *Spirastrella* (*Acanthochaetetes*) and *Ceratoporella* are probably living relatives and extend the range to the Recent. Chaetetids are most conspicuous in rocks of the Lower and Middle Pennsylvanian (Morrowan, Atokan, and Desmoinesian) and occur in a variety of growth forms (Kershaw and West 1991), ranging from thin laminar bodies to columns several meters tall (West 1988). Although some occurrences of chaetetids are limited to one or a few outcrops, that is, have only a local distribution, chaetetid-rich beds in Morrowan rocks of Oklahoma and Arkansas and Atokan rocks of the Great Basin have been used biostratigraphically (West 1988, 1992). Because we have no detailed firsthand knowledge of these chaetetid beds, we will discuss only two examples from the Desmoinesian of the Midcontinent, which we have studied in some detail.

Our two examples of chaetetid epiboles occur in two different limestones of the Marmaton Group (Desmoinesian) of southeastern Kansas and northeastern Oklahoma. The two limestones are the Houx-Higginsville Limestone Member and the Blackjack Creek Limestone Member of the Fort Scott Limestone, the basal formation of the Marmaton Group (fig. 16.1). Suchy (1987) studied in detail the Houx-Higginsville limestone in southeastern Kansas, and Roth (1991) provided a similar detailed study of the Blackjack Creek limestone in southeastern Kansas and northeastern Oklahoma. In both of these epiboles, chaetetids occur in their life position with minor disruption and breakage.

Blackjack Creek Limestone Member

Careful field and laboratory study by Roth (1991) of 12 outcrops and three cores from Bourbon County, Kansas, to Rogers County, Oklahoma, revealed two (0.5–3.0 m and 2.5–5.0 m) small-scale units (sixth-order cycles or PACs) within

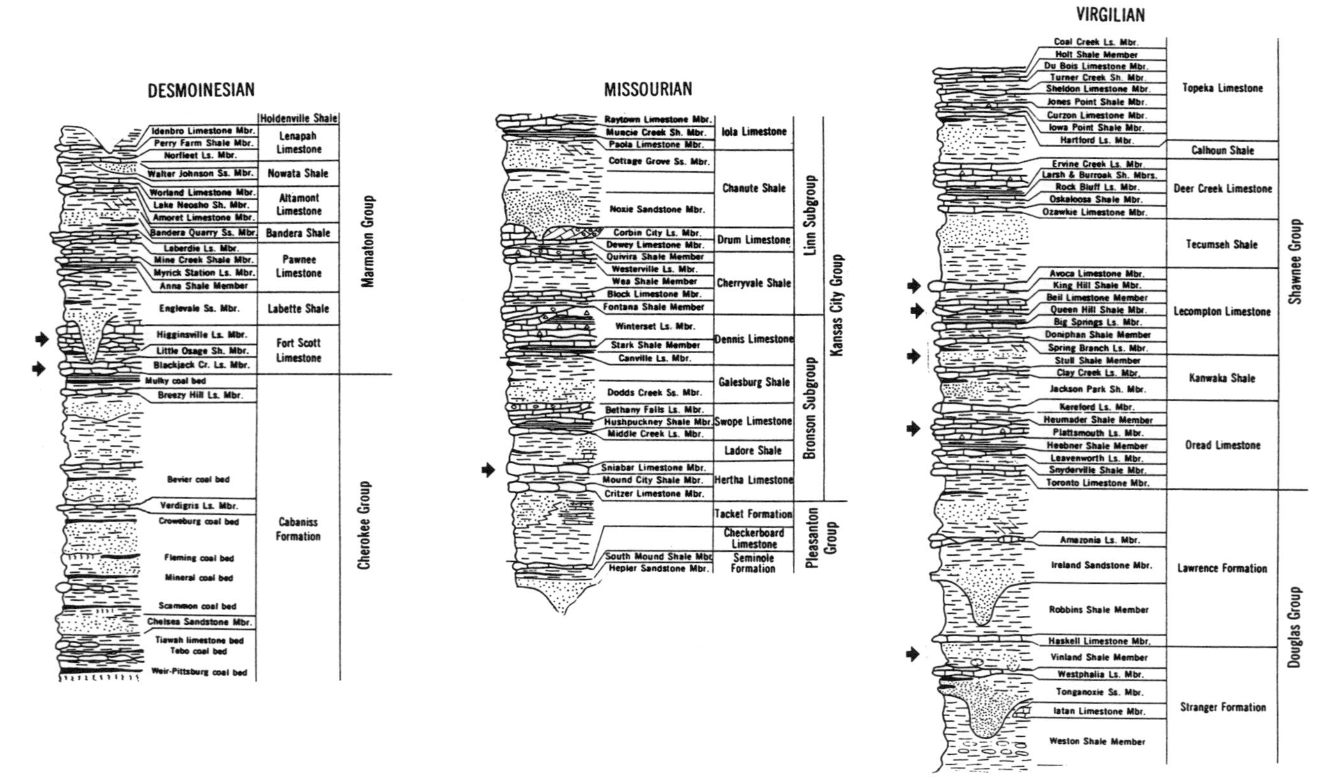

FIGURE 16.1. General stratigraphic chart for parts of the Virgilian, Missourian, and Desmoinesian rocks of Kansas. Arrows mark the stratigraphic position of the epiboles discussed in this essay. (Based on Zeller 1968.)

the Blackjack Creek Limestone Member that contained chaetetids (PACs 2 and 3 of fig. 16.2A). In outcrops and cores from northern Crawford County into central Labette County (all in Kansas) (fig. 16.3), the top of PAC 2 is characterized by thin laminar chaetetids in a matrix dominated by silt- and clay-sized siliciclastic grains (fig. 16.2A). The area of this epibole is nearly 2,500 km^2 (113 by 22 km). Although additional data (cores) are currently lacking, given the

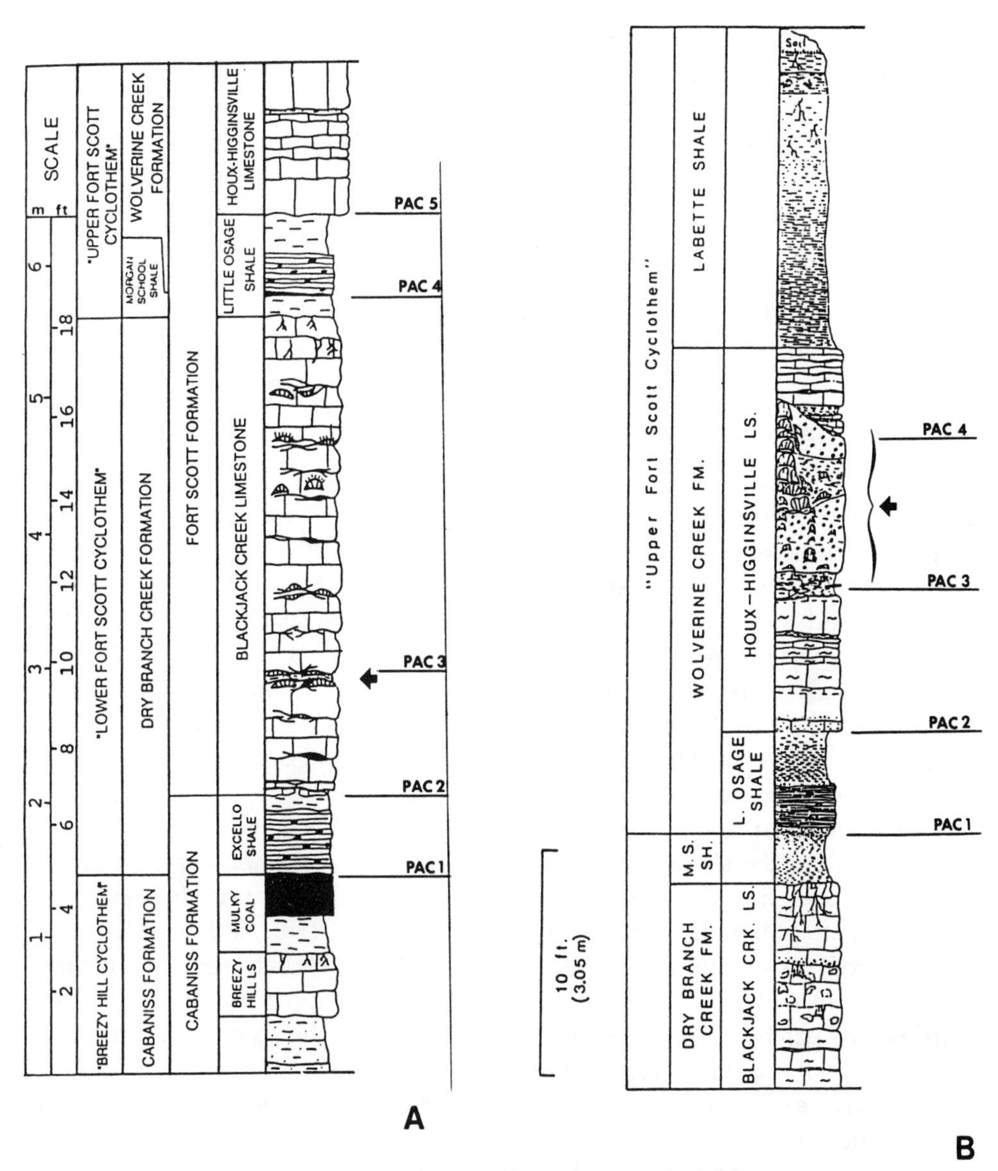

FIGURE 16.2. Generalized stratigraphic section showing the position (arrow) of the chaetetid epibole: (A) in the Blackjack Creek Limestone Member (Roth 1991), and (B) in the Houx-Higginsville Limestone Member (Suchy and West 1991).

regional geology (fig. 16.4), this epibole probably extends farther west and south.

Chaetetid growth-form in this interval (epibole) is primarily laminar, but at some localities low-domical forms are associated with the laminar forms. The low-domical forms have ragged margins, suggesting episodic sedimentation that interrupted growth (Kershaw and West 1991), and indeed the low-domical forms are nothing more than a succession of laminar forms separated by very thin (0.5–2.0 mm) argillaceous partings. Ecological studies of chaetetids (Connolly et al. 1989; West and Kershaw 1991) suggested that they grew in shallow water, often above normal wave base and possibly intertidally (*Spirastrella* [*Acanthochaetetes*] is reported from intertidal habitats in the south Pacific; see Saunders and Thayer 1987). Laminar and ragged, low-domical growth of chaetetids appears to be associated with rather shallow, turbid water, compared with the "cleaner" water habitats of domical and columnar forms (West and Roth 1991). The dominant siliciclastic matrix that encases the laminar and ragged, low-domical chaetetids of this epibole at the top of a shallowing-upward sequence (Roth 1991) suggests a shallow, turbid-water environment.

Houx-Higginsville Limestone Member

Suchy (1987) examined nine exposures of the Houx-Higginsville Limestone Member in southeastern Kansas and western Missouri. He supplemented the data from these nine with data from the literature on 20 additional exposures, which extended the area of study from north-central Missouri across western Missouri and southeastern Kansas into northeastern Oklahoma. Within the Houx-Higginsville Limestone Member in this area, two (1.0–3.0 m and 2.0–5.0 m) small-scale units (sixth-order cycles or PACs) were identified. The uppermost (thinner of the two) unit of this limestone contains abundant chaetetids (fig. 16.2B) at five localities, extending from western Missouri (Butler County) into Kansas (Bourbon County), with the southernmost occurrence in southwestern Crawford County, Kansas (note that two of these five localities are the same as two of the six localities where we recognize a chaetetid epibole in the Blackjack Creek limestone; see fig. 16.3). Chaetetids also occur in the same unit in west-central Missouri (Suchy 1987; Suchy and West 1991), although they are absent from this sixth-order cycle at several localities between west-central Missouri and southeastern Kansas. In addition, the west-central Missouri occurrences are "patchy." Thus we will limit this essay to the southeastern Kansas occurrences of the Houx-Higginsville chaetetid epibole.

The areal extent of the Houx-Higginsville chaetetid epibole is more than 1,250 square kilometers (114 by 11 km), about half the area of the Blackjack Creek chaetetid epibole. Although we lack subsurface data, the regional geological setting (fig. 16.4) indicates that this epibole also extends farther south and west. Indeed, with more careful field study, it might also be extended north and east, tying together the "patchy" occurrences in west-central Missouri.

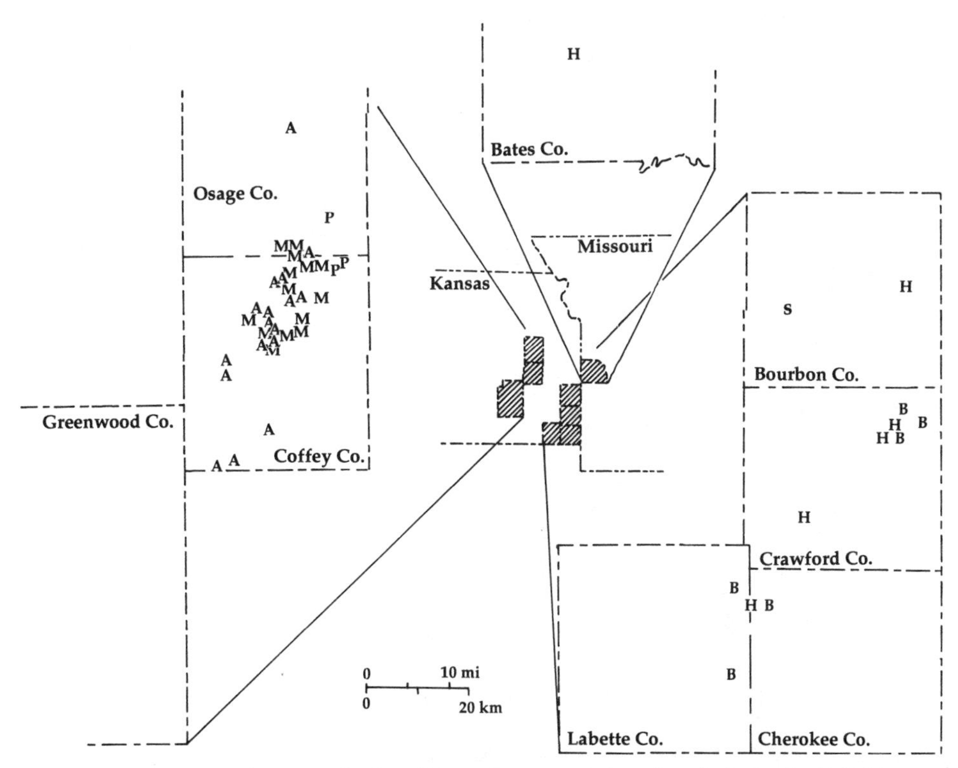

FIGURE 16.3. Map showing the localities where most (coral epibole in the Beil Limestone Member is not shown) epiboles discussed in this essay occur. A = coral epibole in the Avoca Limestone Member, B = chaetetid epibole in the Blackjack Creek Limestone Member, H = chaetetid epibole in the Houx-Higginsville Limestone Member, M = myalinid epibole in the Stull Shale Member, P = coral epibole in the Plattsmouth Limestone Member, S = coral epibole in the Sniabar Limestone Member.

The Houx-Higginsville chaetetid epibole is different from the one in the Blackjack Creek limestone. The growth of the chaetetids in the Houx-Higginsville chaetetid epibole is ragged domical to columnar, and the enclosing matrix is most often a fusulinid packstone to grainstone. Growth interruptions occur, but they are less often marked by the thin siliciclastic partings common in the ragged, low-domical chaetetids of the Blackjack Creek chaetetid epibole. Instead, such interruptions are (1) a line marked by a slightly darker color of the skeleton, (2) a slight change in the growth direction of the skeleton, or (3) a thin layer of the tabulate coral *Multithecopora*. The ragged margins of the growth form and the growth interruptions indicate episodic sedimentation, and the columnar forms appear to be composed of a "stacking" of low- to high-domical forms. Some silt- to clay-sized siliciclastic debris is present, but there appears to be less of it than in occurrences where laminar to low-domical growth dominates (Roth 1991; West and Roth 1991). This chaetetid epibole thus developed in a shallow-water environment marked by higher energy and less turbidity than was typical of the Blackjack Creek chaetetid epibole. Both these chaetetid

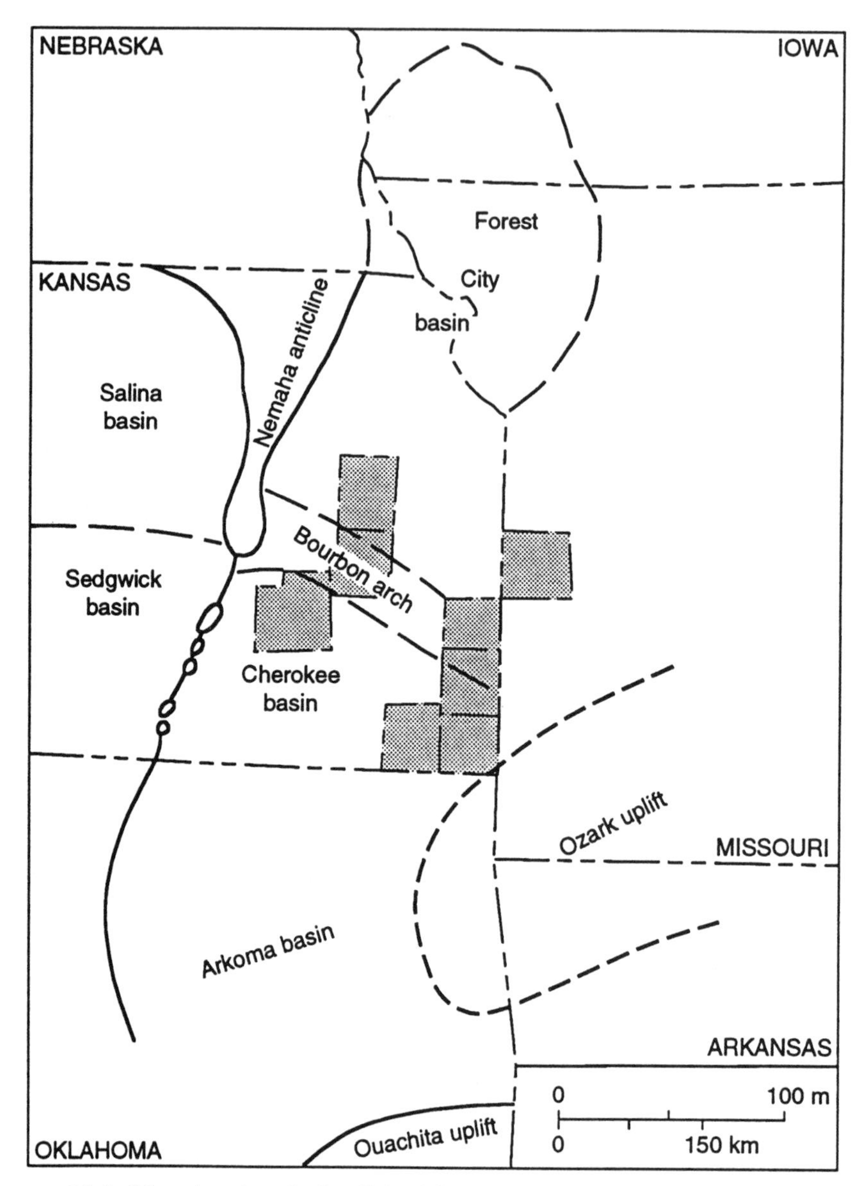

FIGURE 16.4. Map showing the localities (shaded counties) relative to the major structural features of the northern Midcontinent. The localities in each shaded county are shown in fig. 16.3.

epiboles occur in a shallowing-upward sequence (regressive phase), at a time in earth history (late Desmoinesian) during which the climate in North America was fluctuating between wetter and drier conditions (Parrish and Barron 1986; Cecil 1990; Archer and West 1992).

In this context, one possible explanation is that the Houx-Higginsville chaetetid epibole accumulated during a drier climatic period. During a drier climate there would be less freshwater runoff available to transport terrigenous detritus onto the shallow marine shelf and thus less siliciclastic detritus in the Houx-Higginsville chaetetid epibole.

Coral Epiboles

TABULATE CORALS

Syringoporoid corals are common constituents of some Pennsylvanian limestones in Kansas, but they do not occur as scattered colonies throughout the entire section. Most commonly they are present in distinct layers that can be traced for tens of meters to tens of kilometers. Thus there is a spectrum of expression of coral-rich layers, from local single-outcrop occurrences to laterally extensive layers that fit the concept of epiboles. Unlike many other fossils, syringoporoid corals usually are preserved in growth position and so may be used to interpret conditions at the sediment surface.

Syringoporoid corals are typically preserved in shallow-water limestones with a low siliciclastic content. They apparently grew best in shallow, clear water of low turbidity, because they were highly susceptible to smothering. If the rate of deposition exceeded the coral's growth rate, the coral died.

Growth histories of the species examined for this study all are similar and probably typify syringoporoid corals in general. Early stages of corallum growth record rapid lateral expansion of the colony by repeated lateral budding of corallites. Small colonies are typically thin (less than 5 mm) throughout, with little additional height in the central, older part of the colony. Individual corallites are oriented tangential to the substrate away from the founding corallite within the lowest 2–3 mm of the colony and then are directed vertically with subsequent growth. (As an aside, these young colonies of syringoporoids are commonly referred to the form genus *Aulopora*, irrespective of their "real" taxonomy.) The lower surface of each colony thus records the topography of the substrate during early astogeny. In many cases, the bedding planes, defined by colony bases, are otherwise not easily discerned.

Most of the colonies encountered in this study are surrounded by wackestone. Micrite-dominated carbonate rocks have been interpreted as accumulating slowly in low-energy environments. Recent reinterpretations, however, are that many Palaeozoic siliciclastic mudrocks, also formerly thought to accumulate slowly and continuously, were deposited in two distinct modes: slow background deposition and rapid episodic deposition (see Brett and Baird 1986; Brett et al. 1986). Background deposition occurs during normal conditions as mud settles slowly from suspension. Rapid depositional events typically represent redeposition of mud below the storm wave base; the mud was originally scoured from shallow-water settings during storms. Syringoporoid colonies from the three rock units that we discuss next show both rapid and slow depositional modes.

Laterally Restricted Coral Layers A syringoporoid, similar to *Neosyringopora,* is common in the lower 1 m (3.3 ft) of a phylloid-algal mound in the Sniabar Limestone Member (Hertha Limestone) at Uniontown, Kansas (figs. 16.1, 16.3, and 16.5A). The following is a summary of Feldman and Maples's (1989) discussion of this locality. The colonies occur in a fossiliferous phylloid-algal wackestone with brachiopods and gastropods. Syringoporoid colonies

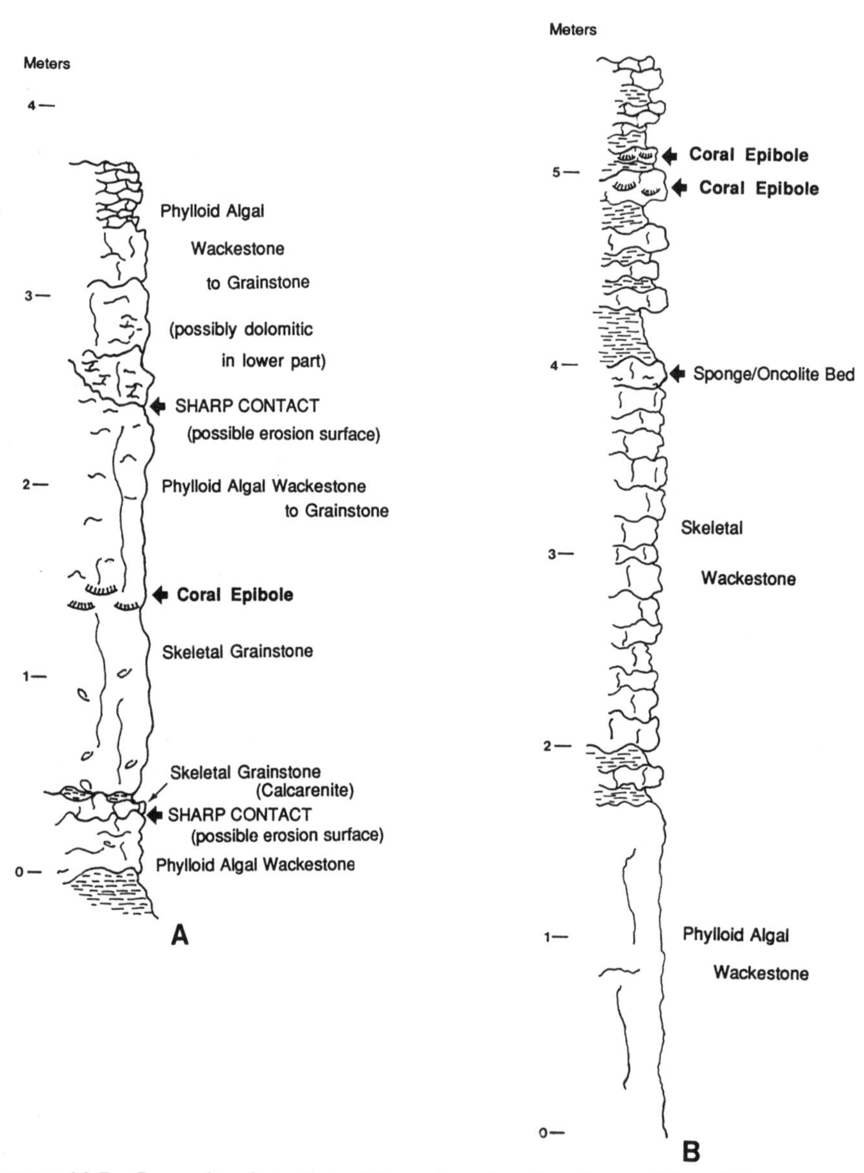

Figure 16.5. Generalized stratigraphic section showing the position of the syringoporoid coral epibole: (A) in the Sniabar Limestone Member at Uniontown, Kansas, and (B) in the Plattsmouth Limestone Member near Waverly, Kansas.

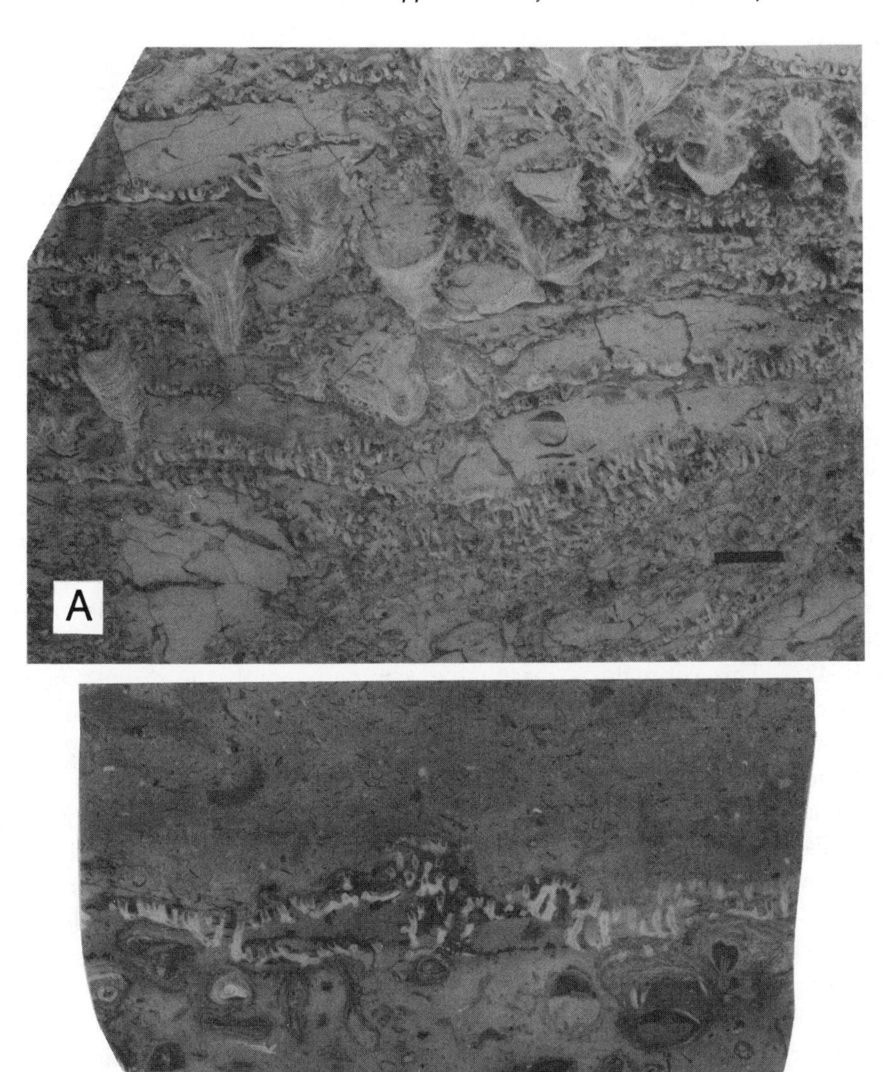

FIGURE 16.6. (A) Syringoporoid coral layers intergrown with an unidentified rugose coral from the Sniabar Limestone Member at Uniontown, Kansas. (B) Syringoporoid corals on an oncolite layer in the Plattsmouth Limestone Member near Waverly, Kansas. Scale bars are 2 cm.

occur in thin sheets (approximately 5 mm) to thick (up to tens of centimeters) hemispherical domes. Rugose corals are intimately associated with the colonies where the layers are thicker than about 2 cm (fig. 16.6A).

The thin sheets of syringoporoids are commonly stacked, separated by layers of lithified sediment. Each coral layer records the early expansional and later vertical growth of the colony. The stacking of coral layers does not result from a random occurrence of new colonies over old colonies, because few colonies lack multiple layers and the area occupied by colonies is much smaller than the area with no colony growth. We interpret each coral layer as growing during periods of slow, background deposition. Corallum growth likely took place in clear water with little or no deposition of carbonate mud (at least the amount of deposition was less than the thickness of the coral layers, which is commonly about 5 mm). The intervening layers represent rapidly deposited sediment that smothered most of or all the underlying coral layer. Usually, a few tall corallites from a lower layer initiate the succeeding layer, which then repeats the colony's early astogenetic history by expanding laterally across the substrate. Tall corallites represent those parts of each colony that survived smothering and established a new coral layer.

The thickness of each coral layer is directly related to the length of time between rapid depositional events. This time can be estimated if growth rates are known, although growth rates of extinct organisms are difficult to estimate. Favositid corals commonly display growth banding (interpreted as annual), which indicates growth rates of up to about 2 cm (0.8 in)/yr (Philcox 1971; Hill 1981:F428). This estimate is reasonable, compared with the growth rates of living corals, and is probably a reasonable estimate for syringoporoid corals. The thinnest layers of syringoporoids in the Sniabar limestone are 2–3 mm thick, suggesting that successive, rapid depositional events may have occurred within less than a year of one another (assuming that the coral began growing immediately after each rapid event). Most of the corals observed at the Uniontown locality have layers about 5 mm thick. The thickest coral layer observed is a hemispherical colony over 20 cm (8 in) thick. This colony has several restrictions around its edges, indicating that the central part of the colony survived several depositional events essentially unharmed.

Corallites still exposed after a rapid depositional event were able to initiate a new layer. Thus the distance between coral layers records the thickness of each rapid depositional event. The thickness between corallum layers ranges from 4 to 90 mm and averages 25 mm ($n = 11$). The thickness between coral layers can vary considerably, although coral layers in the upper parts of the colonies generally have the same surface features as the lower layers or exhibit subdued relief, suggesting that mud was deposited as a drape over the surface of the corallum, with somewhat thicker accumulations in depressions.

From these data it is possible to estimate the ratio of sediment accumulation during normal versus rapid events. Coral growth during periods of slow, background deposition must have at least kept pace with carbonate-mud deposition, or the colony would have been smothered and died. In fact, as noted earlier, corallum growth had to be faster than background deposition, so that the thickness of a coral layer is a maximum estimate of sedimentation between rapid events. Applying this reasoning to all colonies observed at Uniontown, the minimum ratio of rapidly versus slowly deposited sediment is about three to one. This is equivalent to at least 75% of the sediment being deposited during short-lived events and at most only 25% representing normal day-to-day conditions. Given that coral growth took up to one year and assuming that the blanket of smothering sediment was deposited episodically (i.e., hours or days), then, fully three-fourths of the depositional package represents less than 1% of the time encompassed by the entire unit.

Tabulate Epiboles Other coral occurrences in Kansas are more widespread. A syringoporoid-rich interval near the top of the Plattsmouth Limestone Member (Oread Limestone, fig. 16.1) can be correlated for about 10 km (fig. 16.3). The Plattsmouth is a phylloid-algal-rich limestone generally more than 6 m thick in the area studied. Most of the Plattsmouth Limestone Member in the study area consists of fossiliferous wackestone (fig. 16.5B), but the upper part becomes oncolitic with thick interbedded fossiliferous mudrocks. These mudrocks contain abundant encrusting organisms, primarily bryozoans and sponges. Nearly every fossil (e.g., brachiopods, crinoid fragments) in the upper part of the Plattmouth is encrusted, suggesting a slow deposition during which the organisms were exposed on the seafloor for long periods of time.

The syringoporoid colonies are developed on two thin beds of oncolitic limestone (fig. 16.6B). The seafloor, covered with oncolites, must have been stabilized for long enough periods of time to allow colonization by syringoporoids. Algal coatings on many oncolites are uneven, indicating that lengthy periods of stabilization (low water turbulence) were common. Once established, the corals would have anchored the substrate and inhibited further reworking of the oncolites. Syringoporoids in the Plattsmouth Limestone Member are less than 2 cm tall, and some small colonies are overturned, suggesting occasional movement of the oncolites on the seafloor even after syringoporoid colonization. Such movement could have been caused by turbulent water (e.g., storms), bioturbation by fish (Dullo and Hecht 1990), or both. Beyond the area of this syringoporoid layer, well-developed beds of oncolites are absent; thus the original extent of this layer may have been greater than it is now.

The most extensive syringoporoid epibole we encountered is in the upper part of the Avoca Limestone Member (Lecompton Limestone, fig. 16.1). In the study area the Avoca consists of two micritic limestones, each 30–50 cm thick separated by a 1- to several-meters-thick mudrock bed. The syringoporoid epibole is near the top of the upper limestone bed. This epibole can be correlated from central Greenwood County, Kansas, to central Osage County, Kansas, a distance of 80 km (fig. 16.3). Expression of the epibole ranges from small (10–20 cm in diameter) colonies to thick (more than 20 cm) coral-rich intervals that can be traced up to about 200 m within a single outcrop. Where coral-rich intervals are traceable across outcrops, individual coral layers can be up to 6 cm thick (fig. 16.7), so there must not have been a smothering event in this area for several years. Thus, the thickest known, tabular layers of corals are within the most extensive syringoporoid epibole.

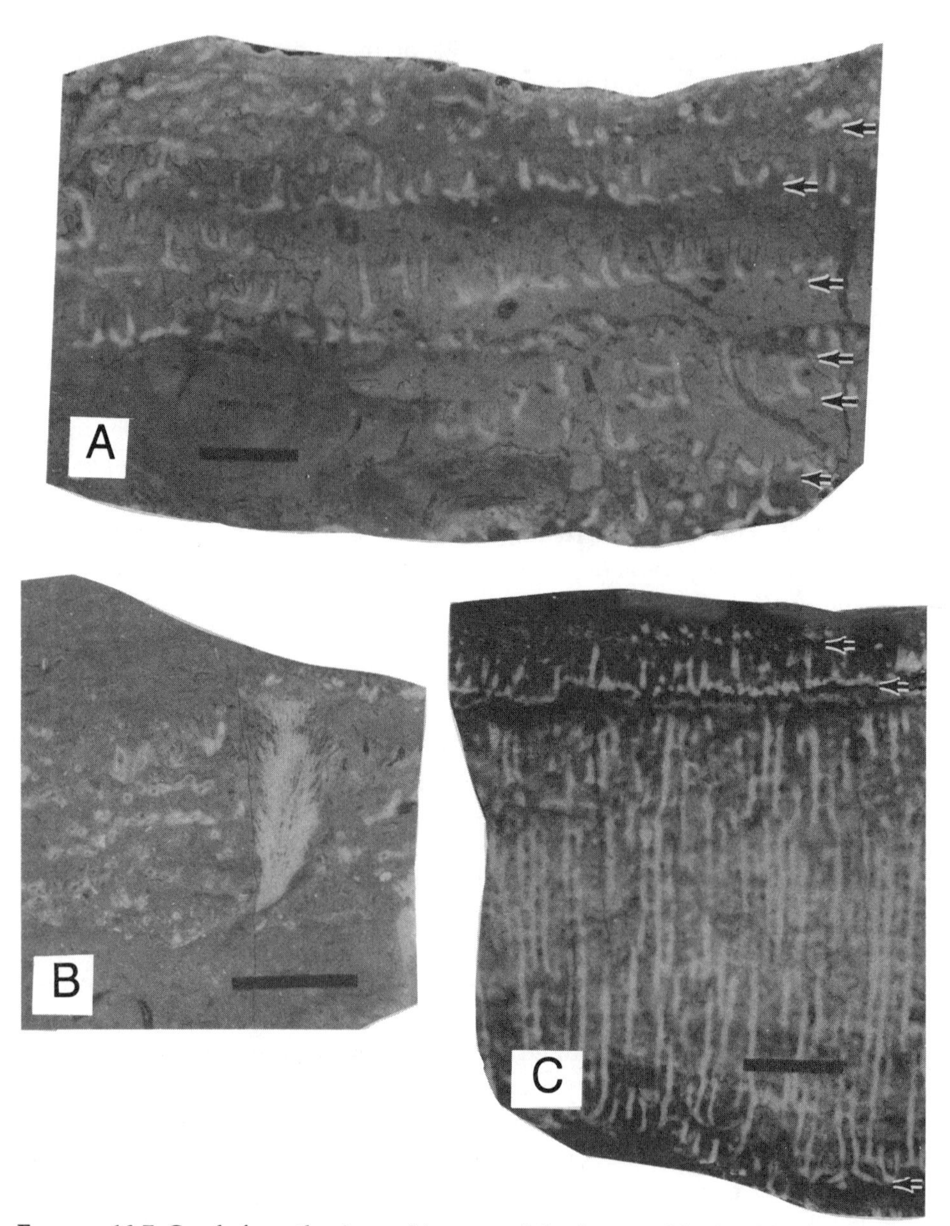

Figure 16.7 Corals from the Avoca Limestone Member at midpoint, North edge, NW 1/4, Sec. 34, T. 20S., R. 15E., Coffey County, Kansas. (A) Multiple coral layers. (B) Intergrown syringoporoid and rugose corals. (C) An unusually thick coral layer. Small arrows point to the base of the coral layers. Scale bars are 2 cm.

The Avoca Limestone Member is well known by field geologists as a limestone characterized by the abundance of syringoporoid corals, and this feature continues to be useful in geologic mapping. It is not surprising, therefore, that almost all outcrops of the upper Avoca we encountered contained syringoporoid corals. The epibole is developed within a bedded wackestone, and the associated fossils include brachiopods, bryozoans, gastropods, and echinoderm debris. Rugose corals are again associated with the thick syringoporoid layers, as was noted for syringoporoids in the Sniabar Limestone Member. Bedding in the upper Avoca is not apparent except where syringoporoids are present.

The syringoporoid colonies are always in growth position, and most have nearly flat bases. Where the colonies are less conspicuous, they tend to be small, mound-shaped, and composed of one or a few coral layers, and the bedding surfaces on which the colonies grew cannot be traced beyond the colony's edges. These small mound-shaped colonies have frequently colonized a single brachiopod valve, small bryozoan colony, or aggregate of skeletal debris (similar occurrences were noted in the Beil Limestone Member near Leavenworth, Kansas, by G. Baird, personal communication). Where the colonies are abundant, they tend to have flat bases and multiple layers. Localities with fewer colonies and those where colonies are conspicuous are interspersed throughout the geographic range of the epibole.

Rugose Corals

Rugose corals also are common in Pennsylvanian rocks in Kansas. The most extensive rugose coral epibole in Kansas, known to us, is that of *Pseudozaphentoides torquia* (Owen), in the Beil Limestone Member (Lecompton Limestone, fig. 16.1), as documented by Baird (1971) and Maerz (1978). The Beil is a 2–3-m-thick limestone with an abundant and diverse assemblage of marine fossils. *Pseudozaphentoides torquia* (formerly referred to as *Caninia torquia*) is common to profuse in several facies throughout the 280-km-long outcrop of the Beil Limestone Member in Kansas and, like the syringoporoid occurrence in the Avoca, has been known and used by field geologists for many years (Moore 1935, 1949; Zeller 1968).

Unlike the syringoporoid epiboles, rugose corals are not typically in growth position and do not occur in thin beds that can be traced with absolute certainty between outcrops, yet the abundance of this coral makes the Beil distinct. Maerz (1978) determined that *P. torquia* is most abundant in facies representing the Beil's maximum water depth, but he did not suggest the cause of this unusual concentration of rugose corals. Syringoporoid corals commonly are associated with *Pseudozaphentoides torquia,,* but generally are much less abundant and apparently are absent, or nearly so, from exposures in Nebraska (G. Baird, personal communication).

Pelecypod Epiboles

Mollusk-bearing mudrocks and limestones are common components of Pennsylvanian (Virgilian) cyclothems in the Midcontinent (Moore 1935). Most often these mollusk-bearing lithologies are immediately above or a few centimeters to meters below coal beds (Moore 1935:24–25). Newell commented that on these beds

> at many horizons, particularly in the Pennsylvanian and Early Permian rocks of the Mid-Continent region, there are thin beds of limestone or shale, up to a foot or so in thickness, filled with the massive valves of pelecypods. Scores of beds of this kind are traceable for tens of miles, and constitute excellent datum horizons for field mapping. Certain "*Myalina* beds," such as occur in the Meadow limestone (Lansing), Vinland shale (Douglas), Stull shale (Shawnee), Caneyville formation (Wabaunsee), and Moran formation, to mention only a few, are incredibly persistent, and represent old shell "banks" comparable with the extensive *Mytilus* and *Ostrea* beds of the present day seas. (1942:11)

These *Myalina* (*Orthomyalina*) beds in the Vinland and Stull shales are the focus of this section of our essay. Using Kidwell and Holland's criteria (1991), these would be densely to loosely packed and well sorted to bimodal bioclastic deposits.

The dominant myalinid in these beds is *Myalina* (*Orthomyalina*), which Rollins et al. (1979) and West et al. (1990) have suggested, following Newell (1942), occupied an ecological "niche" more or less analogous to that of the extant oyster *Crassostrea virginica* along tidal channels in marginal-marine environments, that is, oyster banks or oyster biostromes. This does not mean that the myalinid had the same mode of life (i.e., cemented) as the oyster, just that it inhabited a similar environment. As Newell pointed out (1942), *Mytilus* also is a possible modern analogue, as is the ribbed marsh mussel *Geukensia demissa*. Both *Mytilus* and *Geukensia*, however, are byssally attached to hard/firm surfaces (shell fragments, plant roots, etc.). Although *Geukensia* is the dominant pelecypod of low-marsh (intertidal) dense grass beds, *Mytilus* is more commonly associated with open, normal-marine intertidal areas.

The Pennsylvanian myalinids, although possessing a byssal notch, are regarded as a semi-infaunal edgewise recliner that lived in soft mud (Seilacher 1984a). Articulated in situ specimens of myalinids in the units discussed in this essay (see Seilacher 1984a, and Maples and West 1991) support this interpretation, and indeed if these forms had been byssally attached, their siphons would have been fouled by the soft muddy substrate. The semi-infaunal *Geukensia* might be a reasonable extant analogue except that its shells are thin and when they do accumulate the assemblage has a low diversity (one or two taxa) of eurytopic forms. On the other hand, extant oysters and Pennsylvanian myalinids have more durable shells and dominate assemblages that are more diverse, with more stenotopic ("normal" marine) taxa.

Using accumulations of either *Crassostrea* or *Geukensia* as a reasonable analogy, living and dead individuals would be subject to distribution by water movement, waves, currents, and tides. Accordingly, our myalinid epiboles would be classed as event concentrations and would fall somewhere along the line between "behavior of shell producers" and "within-habitat reworking," as illustrated by Kidwell (1991, fig. 6).

Given the density of myalinid valves—as noted by Newell (1942) and observed by us—these pelecypods were prolific producers of very durable shells. In this respect, myalinids are more similar to extant oysters than to *Geukensia*, for, as Kidwell (1990:322) pointed out, the rate of egg production and benthic carbonate production is highest among extant oysters. Thus the myalinid epiboles can be interpreted, using Kidwell (1990:25), as the results of an advantageous combination of fecundity and skeletal durability that overcame vulnerable shell characteristics or aggressive taphonomic environments to produce significant bodies of bioclastic debris.

Stranger Formation

An epibole of the pelecypod genus *Myalina* occurs below the base of the Haskell Limestone Member (basal member of the Lawrence Formation) at the top of the Vinland Shale Member (the uppermost member of the Stranger Formation) (figs. 16.1 and 16.8A). Although several people have noted the abundance of *Myalina* (*Orthomyalina*) on measured sections, including Newell (1942), S. M. Ball (1964) documented the geographic extent and stratigraphic usefulness of this particular occurrence. Later, D. S. Ball stated, "This highly fossiliferous zone is a key marker unit where the Haskell Limestone is discontinuous in southern Kansas and northeastern Oklahoma exposures" (Ball 1985:36).

Extending from Missouri, across Kansas, and into Oklahoma, this bed is dominated by myalinid clams that are "invariably . . . single valves aligned parallel to stratification in excavated exposures" (Ball 1964:115) (fig. 16.8). These pelecypods, and associated fossils, are partially or entirely covered by a calcareous coating, which in most cases is algal (Ball 1964:115) (fig. 16.8B and C), but bryozoan encrustations (fig. 16.8D) also can be found. Their occurrence is particularly useful for sorting out stratigraphic problems, because the myalinids and associated fossils occur "in very fine-grained quartzose sandstones . . . rather than the usual calcareous shales or mudstone-argillaceous limestone facies" (Ball 1964:115).

Referring to the upper Vinland epibole, Ball noted that gastropods, bryozoans, echinoderms, brachiopods, and corals, in addition to myalinids, are present (1985:36), which agrees with our observations of this interval. A map (Ball 1985:37, fig. 9) showed this upper Vinland epibole extending from Buchanan County, Missouri, southwestward across southeastern Nebraska to Chautauqua County, Kansas, a distance of more than 350 km. Interestingly, this epibole is not present in most of the exposures of the upper Vinland in Anderson County, Kansas. In addition to the myalinids, brachiopods are abundant in the exposures of northwestern Missouri and are the dominant fossil in exposures in southeastern Nebraska where myalinids are sparse (Ball 1985:36).

The genesis of this epibole, according to Ball (1985:85), is that the myalinid shells were concentrated in shore-zone environments by one or more storms. Deposition of single valves in a hydrodynamically stable position and further

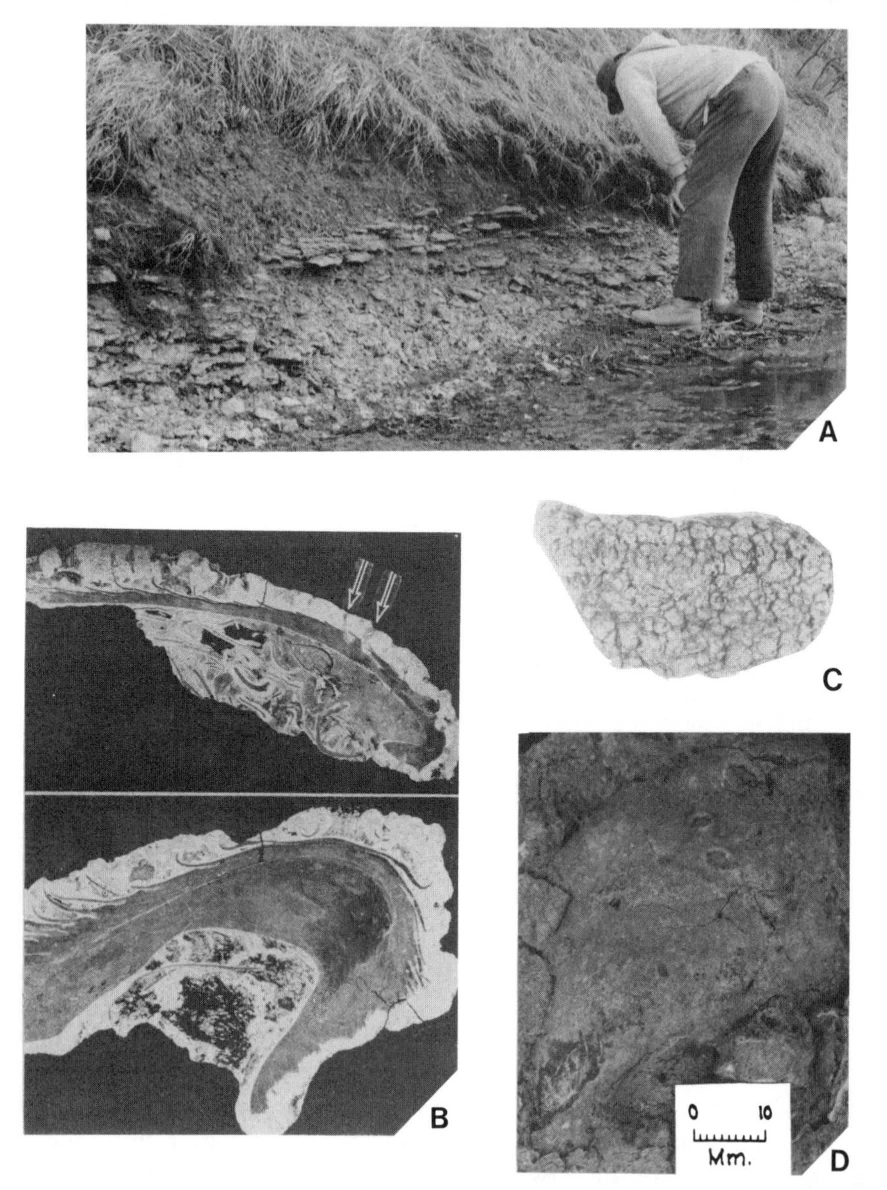

FIGURE 16.8. *Myalina* (*Orthomyalina*) epibole at the top of the Vinland Shale Member (Stranger Formation). (A) Outcrop of epibole along the Neosho River, near center of South line, SW 1/4, Sec. 28, T. 22S., R. 16E., Coffey County, Kansas. Haskell Limestone Member (Lawrence Formation) is directly above this epibole. (B) Acetate peel prints of algal coated and bored (arrows) valves of *Myalina* (*Orthomyalina*) (from Ball 1964, Plate 22). (C) Algal-coated single right valve of *Myalina* (*Orthomyalina*) (from Ball 1964, Plate 23). (D) Single left valve of *Myalina* (*Orthomyalina*) encrusted by a fistuliporoid bryozoan.

exposure before their final burial allowed them to become bored and encrusted by algal/foraminiferal osagid encrustation and bryozoans. Some boring took place before the algal encrustation, and some came after it (fig. 16.8B). Because of the lack of articulated in situ myalinids, the hydrodynamically stable orientation of single valves, the degree of postmortem alteration (borings and encrustations), and the relatively high diversity (including a number of "normal" marine taxa), this epibole, in the Kidwell (1991, fig. 6) scheme, would be an event concentration nearer the "within-habitat reworking" apex rather than the "behavior of shell producers" apex. Obviously, the "behavior of shell producers" (high fecundity and high skeletal durability) was a prerequisite, but in this case "within-habitat reworking" (postmortem alteration and mixing) largely obscures the former.

Although we do not disagree with the role of storms, we do not think that storms are the only relevant process. Tides, currents, and waves can produce such bioclastic deposits along strand lines and in shallow subtidal environments in and along the mouths and edges of estuaries and more open marine coastlines.

Kanwaka Shale

The second *Myalina* (*Orthomyalina*) epibole is at the top of the Stull Shale Member (uppermost member of the Kanwaka Shale), just below the Spring Branch Limestone Member (basal member of the Lecompton Limestone, figs. 16.1, 16.3, and 16.9A). This epibole was first studied by Roth et al. (1989) and the Kansas State University Paleoecology Seminar (1989). At that time, it was thought that the epibole was at the top of the Doniphan Shale Member (fig. 16.1), but subsequent field mapping demonstrated that it is at the top of the Stull shale.

More than 30 localities (fig. 16.3) of the upper Stull Shale Member in Coffey, Osage, Franklin, Woodson, and Greenwood counties in Kansas contain an abundance of myalinid pelecypods. According to the currently available data, this epibole extends over an area of more than 2,000 km^2 and has been very useful in geologic mapping (see also Newell 1942). In the inferred setting recorded by the upper Stull Shale Member (Doniphan shale of Kansas State University Paleoecology Seminar 1989), the dispersion of the valves would have been dominated by tidal currents, with the area of geographic distribution increased by the occurrence of storm surges. The processes responsible for the distribution of this epibole, therefore, are not unlike those envisioned for the myalinid epibole in the upper Vinland shale.

As Maples and West noted (1991), the inferred mode of life of *Myalina* (*Orthomyalina*) is supported by localities in this upper Stull Shale Member epibole where articulated specimens are oriented with their dorsal margins down (fig. 16.9B). Such in situ specimens are associated with hydrodynamically stable single valves (fig. 16.9C) that frequently show evidence of bioerosion and/or

FIGURE 16.9. *Myalina* (*Orthomyalina*) epibole in the Stull Shale Member near the SW corner, NW 1/4, SW 1/4, Sec. 17, T. 20S., R. 15E., Coffey County, Kansas. (A) Weathered surface of epibole; (B) freshly exposed surface of epibole; (C) vertical profile of epibole, with arrows indicating myalinid valves.

algal encrustation, as seen in the myalinid epibole at the top of the Vinland Shale Member. In general, the Stull epibole is less diverse overall. In some localities the many myalinids and small gastropods are the only fossils, but articulated in situ myalinids indicate little postmortem alteration and mixing. Therefore, at some localities, this epibole (event concentration) would fall close to the "behavior of shell producers" apex and, at others, closer to the "within-habitat reworking" apex (Kidwell 1991, fig. 6). Additional details on the encrusting and boring assemblages associated with the Stull myalinids are available in Baker (1995).

Table 16.1 summarizes what we know about the eight epiboles described in this essay. Essentially there are two different types. One type is composed of the organisms preserved in, or nearly in, their life position, namely, the sponge and syringoporoid/coral epiboles. The other type, pelecypod and rugose-coral epiboles, is composed predominantly of disarticulated and transported fossils, although some localities contain in situ specimens. The occurrence of both was controlled by the organisms' ecological requirements, but the organisms' widespread distribution is partly due to the within-habitat reworking by waves, currents, and tides during both storm and nonstorm periods.

The sponge- and coral-rich epiboles represent times when conditions were favorable for the colonization and growth of these organisms, that is, a period of "quality" ecological conditions. A primary environmental factor was the probably low overall rates of sedimentation. Localized concentrations of sponges and corals can be attributed to local factors such as changing currents, seasonal fluctuations of sediment influx, and evolving seafloor topography. These factors probably were important for initial colonization by the sponges and corals. The size of these organisms may indicate the relative amount of time that ecological conditions were favorable for their growth: larger sizes represent a longer time; smaller sizes, less time. Death ensued when conditions changed radically, and in most cases it was recorded by a major change in the lithology, the organism, or both.

Extensive, detailed stratigraphic and paleogeologic studies covering larger areas are needed to determine a general pattern, if one exists (e.g., Baker 1995). However, based on what we know now, the most obvious changes are recorded by lithologic changes, such as the periodic influxes of siliciclastic or carbonate debris that smothered the sponges or corals. These periodic episodes of sedimentation resulted in nearly instantaneous smothering of the sponges and corals that had spread over a stable substrate during times of low, or no, sedimentation. In some cases these encrusting organisms were only partially smothered by the sediment. Asexual reproduction, from the exposed parts of the organisms, resulted in formation of the next layer of chaetetids, or the coral colony, which spread over the smothered parts of the preexisting layer during the subsequent episode of little or no sedimentation.

Eustatic, tectonic, and climatic events, or a combination of these three, could change the sea level, water clarity, nutrient supply, water turbulence, and other environmental conditions, thereby altering the depositional regimes that could have produced such results over a large geographic area. This relationship (during the Carboniferous) between climatic perturbations and these variables is reasonable, given the studies by Parrish and Barron (1986), Cecil (1990), and Archer and West (1992).

Table 16.1 *The Five Basic Characteristics of the Eight Epiboles*

Dominant taxa	Known autecology of dominant taxa	Lithology of unit containing dominant	Inferred paleoenvironmental	Areal extent of epibole
Chaetetid-grade demosponges	Epifaunal suspension feeders; firm to hard substrate	Carbonate wackestone to grainstone/packstone	Shallow water, turbid to clear turbulent marine waters	Blackjack Creek = $2500km^2$ Houx–Higginsville Limestone = $1250km^2$
Syringoporiod Corals: (?) *neosyringopora*	Epifaunal suspension feeders; firm to hard substrate	"Clean" carbonate wackestone/packstone	Shallow, "clear" turbulent marine waters	Sniabar Limestone = local Plattsmouth Limestone = local(?) Avoca Limestone = 400 km^2
Rugose Corals: *Pseudozaphrentoides torquia*	Epifaunal suspension feeders; loose to firm mud substrate	"Dirty" carbonate; clean carbonate algal wackestone to calcareous mudrock	Shallow, clean to turbid subtidal to low intertidal	Beil Limestone = >600 km^2
Myalina (Orthromyalina) sp. Pelecypoda	Semi-infaunal (edgewise recliner) suspension feeders; loose mud substrate	Mudrock	Shallow subtidal to low intertidal, along tidal creeks and estuaries	Vinland Shale = ± 1000 km^2 Stull Shale = >200 km^2

Conditions favorable to sponge and coral colonization might indicate a flooding event. Indeed, this appears to be the case for the chaetetid epibole in the Houx-Higginsville limestone, in that it begins at the base of a traceable, 1–3-m-thick (sixth-order cycle, PAC), shallowing-upward sequence (Suchy 1987; Suchy and West 1991). In addition, some interruption partings associated with both the chaetetids and the tabulate corals could be reflecting still smaller-scale events, as suggested by Gyllenhaal and Kidwell (1989) for interruption partings in subspherical bryozoan colonies.

The *Myalina* (*Orthomyalina*) epiboles are very suggestive of the modern-day oyster biostromes so conspicuous along coastal Georgia and elsewhere. The valves of these pelecypods are disarticulate easily after death and are transported over very large areas of the coastal marshes by high and spring tides, as well as by storm tides and storm surges. The mode of life, semi-infaunal edgewise recliner, and durability of the shells of *Myalina* (*Orthomyalina*) probably made it rather easy to transport the valves over hundreds of square kilometers. These widespread shell beds could be buried by something as simple as seasonal rainfall, which would increase the terrigenous input into coastal marshes and tidal flats, thereby making conditions unsuitable for settling spat and smothering the adult sessile inhabitants.

Many questions remain unanswered. What factors controlled the generally low diversity of these epiboles? Each chaetetid, syringoporoid, and *Myalina* bed apparently is dominated by one or two species, with small amounts of associated organisms. Why did only one or two species of corals find the conditions favorable; why not a taxonomically diverse coral epibole? What is the ecological and/or evolutionary relationship between (1) chaetetids and *Multithecopora* in the Houx-Higginsville limestone and (2) the syringoporoid corals and the intergrown rugose corals?

Finally, one critical factor needed to evaluate possible processes is the relative synchroneity of each layer over its entire range. If the most extensive epiboles are synchronous, then basinwide processes must be involved; if the epibole beds are synchronous with other facies, then a dynamic seafloor with migration of biofacies would be apparent and hence important, with smaller-scale processes responsible for the smaller patches. Although as yet we have no method for determining the timing of these epiboles, the recognition and evaluation of small-scale events and their association with event "packages" should move us in the right direction.

ACKNOWLEDGMENTS

We express our sincere thanks to Carl Brett for inviting us to contribute this essay and to Carl Brett, Gordon Baird, and Susan Kidwell for extremely helpful and constructive reviews.

Our work was supported in part by funds from the U.S. Geological Survey Cooperative Geologic Mapping Program (COGEOMAP), the Kansas Geological Survey, and the National Science Foundation (grant no. EAR-8816678 to West). The views and conclusions in this document are ours and do not necessarily represent the official policies, either expressed or implied, of the U.S. government.

REFERENCES

Archer, A. W., and R. R. West. 1992. Climato-stratigraphic analysis of Permo-Carboniferous section of Kansas. *Geological Society of America, Abstracts with Programs* 24:2.

Arkell, W. J. 1933. *The Jurassic System in Great Britain.* Oxford: Clarendon Press.

Baird, G. C. 1971. Paleoecology of the Beil Limestone (Upper Pennsylvanian) in the northern Midcontinent region. Master's thesis, University of Nebraska.

Baker, J. A. 1995. Quantitative assessment of bioerosion and encrustation in *Orthomyalina* from shell beds of the Stull Shale Member (Kanwaka Shale, Upper Pennsylvanian, Virgilian) of eastern Kansas. Master's thesis, University of Kansas.

Ball, D. S. 1985. The Pennsylvanian Haskell—Cass Section, a perspective on controls of Midcontinent cyclothem deposition. Master's thesis, University of Kansas.

Ball, S. M. 1964. Stratigraphy of the Douglas Group. (Pennsylvanian, Virgilian) in the northern Midcontinent region. Ph.D. diss., University of Kansas.

Bates, R. L. and J. A. Jackson, eds. 1987. *Glossary of Geology,* 3d ed. Alexandria, Va.: American Geological Institute.

Brett, C. E. and G. C. Baird. 1986. Comparative taphonomy—A key to paleoenvironmental interpretation based on fossil preservation. *Palaios* 1:207–27.

Brett, C. E., K. B. Miller, and G. C. Baird. 1990. A temporal hierarchy of paleoecologic processes within a Middle Devonian eperic sea. In W. Miller III, ed., *Paleocommunity Temporal Dynamics: The Long-Term Development of Multispecies Assemblies,* pp. 178–209. *Paleontological Society Special Publication* No. 5.

Brett, C. E., S. E. Speyer, and G. C. Baird. 1986. Storm-generated sedimentary units: Tempestite proximality and event stratification in the Middle Devonian Hamilton Group of New York. In C. E. Brett, ed., *Dynamic Stratigraphy and Depositional Environments of the Hamilton Group (Middle Devonian) in New York State, Part 1,* pp. 129–56. *New York State Museum Bulletin* No. 457.

Cecil, C. B. 1990. Paleoclimate controls on stratigraphic repetition of chemical and siliciclastic rocks. *Geology* 18:533–36.

Connolly, M. W., L. L. Lambert, and R. J. Stanton Jr. 1989. Paleoecology of Lower and Middle Pennsylvanian (Middle Carboniferous) chaetetes in North America. *Facies* 20:139–68.

Dullo, W-C. and C. Hecht. 1990. Corallith growth on submarine alluvial fans. *Senckenbergiana Maritima* 21:77–86.

Feldman, H. R. and C. G. Maples. 1989. Sedimentologic implications of encrusting organisms from the phylloid-algal mound in the Sniabar limestone near Uniontown, Kansas. In W. L. Watney, J. A. French, and E. K. Franseen, eds., *Sequence Stratigraphic Interpretations and Modeling of Cyclothems in the Upper Pennsylvanian (Missourian), Lansing and Kansas City Groups in Eastern Kansas,* pp. 173–78. *Guidebook for Field Conference, Kansas Geological Society 41st Annual Field Trip,* Kansas Geological Survey.

Gyllenhaal, E. D. and S. M. Kidwell. 1989. Growth histories of subspherical bryozoan colonies: Ordovician and Pliocene evidence for paleocurrent regimes and commensalism. *Geological Society of America, Abstracts with Programs* 21:A112.

Hill, D. 1981. *Coelenterata: Anthozoa, Subclass Rugosa, Tabulata: Treatise on Invertebrate Paleontology, Part F, Supplement 1.* Lawrence: University of Kansas and Geological Society of America.

Holland, C. H. 1989. Trueman's epibole. *Proceedings of the Geological Association* 100:457–60.

International Subcommission on Stratigraphic Classification (ISSC). 1976. *International Stratigraphic Code.* New York: Wiley.

Kansas State University Paleoecology Seminar (R. R. West, V. Voegeli, S. Roth, C. G. Maples, K. Leonard, H. R. Feldman, and C. Cunningham). 1989. Stop 8: Waverly, Kansas, trace fossil locality. In R. K. Pabian and R. F. Diffendal Jr., comps., *Late Pennsylvanian and Early Permian Cyclic Sedimentation, Paleogeography, Paleoecology, and Biostratigraphy in Kansas and Nebraska,* pp. 35–39. *Nebraska Geological Survey Guidebook.*

Kauffman, E. G., and B. B. Sageman. 1990. Biological sensing of benthic environments in dark shales and related oxygen-restricted facies. In R. N. Ginsburg and B. Beaudoin, eds., *Cretaceous Resources, Events, and Rhythms: Background and Plans for Research,* pp. 121–38. Dordrecht, Netherlands: Kluwer. NATO ASI Series, *Series C, Mathematical and Physical Sciences* No. 304.

Kershaw, S. and R. R. West. 1991. Chaetetid growth form and its controlling factors. *Lethaia* 24:333–46.

Kidwell, S. M. 1990. Phanerozoic evolution of macroinvertebrate shell accumulations: Preliminary data from the Jurassic of Britain. In W. Miller III, ed., *Paleocommunity Temporal Dynamics: The Long-Term Development of Multispecies Assemblies,* pp. 309–27. *Paleontological Society Special Publication* No. 5.

———. 1991. The stratigraphy of shell concentrations. In P. A. Allison and D. E. G. Briggs, eds., *Taphonomy: Releasing Data Locked in the Fossil Record,* pp. 211–90. Vol. 9 of *Topics in Geobiology.* New York: Plenum Press.

Kidwell, S. M. and S. M. Holland. 1991. Field description of coarse bioclastic fabrics. *Palaios* 6:426–34.

Maerz, R. H. Jr. 1978. Paleoautecology of *Caninia torquia* (Owen) from the Beil Limestone Member (Pennsylvanian, Virgilian), Kansas. *University of Kansas Paleontological Contributions Paper* No. 92.

Maples, C. G. and R. R. West. 1991. Dependent and independent data in paleontology: Tools for the sedimentary modeler. In E. K. Franseen, W. L. Watney, C. G. Kendall, and W. Ross, eds., *Sedimentary Modeling: Computer Simulations and Methods for Improved Parameter Definition,* pp. 177–84. *Kansas Geological Survey Bulletin* No. 233.

Moore, R. C. 1935. Stratigraphic classification of the Pennsylvanian rocks of Kansas. *State Geological Survey of Kansas Bulletin* No. 22.

———. 1949. Division of the Pennsylvanian System in Kansas. *State Geological Survey of Kansas Bulletin* No. 83.

Newell, N. D. 1942. *Late Paleozoic Pelecypods: Mytilacea.* Vol. 10, part 2 of *State Geological Survey of Kansas.*

North American Commission on Stratigraphic Nomenclature. 1983. North American stratigraphic code. *American Association of Petroleum Geologists Bulletin* 67:841–75.

Parrish, J. T. and E. J. Barron. 1986. Paleoclimates and Economic Geology. *SEPM Short Course* No. 18.

Philcox, M. E. 1971. Growth, form, and role of colonial coelenterates in reefs of the Gower Formation (Silurian), Iowa. *Journal of Paleontology* 45:338–46.

Rollins, H. B., M. Carothers, and J. Donahue. 1979. Transgression, regression, and fossil community succession. *Lethaia* 12:89–104.

Roth, S. M. 1991. Regional stratigraphic analysis of the Blackjack Creek Limestone (Desmoinesian, Middle Pennsylvanian) in southeast Kansas and northeast Oklahoma. Master's thesis, Kansas State University.

Roth, S. M., C. Cunningham, K. Leonard, and V. Voegeli. 1989. The paleoecology of a myalinid shell bed. *Geological Society of America, Abstracts with Programs* 21(1):39.

Saunders, W. B., and C. W. Thayer. 1987. A cryptic intertidal brachiopod/sclerosponge community in Palau, West Carolina Islands. *Geological Society of America, Abstracts with Programs* 19:829.

Seilacher, A. 1981. Towards an evolutionary stratigraphy. In J. Martinell, ed., *Concepts and Methods in Paleontology*, pp. 39–44. *Acta Geological Hispanica* No. 16.

———. 1984a. Constructional morphology of bivalves: Evolutionary pathways in primary versus secondary soft-bottom dwellers. *Palaeontology* 27:207–37.

———. 1984b. Storm beds: Their significance in event stratigraphy, pp. 49–54. *American Association of Petroleum Geologists, Studies in Geology* No. 16.

Suchy, D. R. 1987. Regional stratigraphic setting and paleoecology of a chaetetid "reef" in the Houx-Higginsville Limestone (Pennsylvanian) of Southeast Kansas. Master's thesis, Kansas State University.

Suchy, D. R., and R. R. West. 1991. Genetic stratigraphy of the Fort Scott Limestone (Pennsylvanian, Desmoinesian), southeastern Kansas. In E. K. Franseen, W. L. Watney, C. G. Kendall, and W. Ross, eds., *Sedimentary Modeling: Computer Simulations and Methods for Improved Parameter Definition*, pp. 195–206. *Kansas Geological Survey Bulletin* No. 233.

Trueman, A. E. 1923. Some theoretical aspects of correlation. *Proceedings of the Geological Association* 34:193–206.

West, R. R. 1988. Temporal changes in Carboniferous reef mound communities. *Palaios* 3:152–69.

———. 1992. Chaetetes (Demospongiae): Its occurrence and biostratigraphic utility. In P. K. Sutherland and W. L. Manger, eds., *Recent Advances in Middle Carboniferous Biostratigraphy—A Symposium*, pp. 163–69. *Oklahoma Geological Survey Circular* No. 94.

West, R. R. and S. Kershaw. 1991. Chaetetid habitats. In J. Reitner and H. Keupp, eds., *Fossil and Recent Sponges*, pp. 445–55. Berlin: Springer-Verlag.

West, R. R., H. B. Rollins, and R. M. Busch. 1990. Taphonomy and an intertidal palimpsest surface: Implications for the fossil record. In W. Miller, ed., *Paleocommunity Temporal Dynamics: Long-Term Development of Multispecies Assemblages*, pp. 351–69. *Paleontological Society Special Publication* No. 5.

West, R. R. and S. M. Roth. 1991. Siliciclastic content of chaetetid habitats: Preliminary results. *Geological Society of America, Abstracts with Programs* 23:A343.

Zeller, D. E. 1968. *The Stratigraphic Succession in Kansas*. *State Geological Survey of Kansas Bulletin* No. 189.

17

The Anatomy of the Mid–Early Silurian Ireviken Event and a Scenario for P-S Events

Lennart Jeppsson

ABSTRACT

This essay derives a theoretical scenario for stepwise extinction events from Jeppsson's earlier model by incorporating the effects of Milankovitch cyclicity. This scenario relates the datum points (extinction steps) to a decrease in the supply of nutrients for primary organisms that consumers use as food, directly or indirectly. Four phases of such events can be distinguished in the fossil record by differences in the spacing of the datum points. The scenario also predicts those differences caused by Milankovitch cyclicity and a pattern in the relative effects of the datum points (i.e., extinction steps). The effects and combination of other characters vary for each event because of the differences in the interference pattern of the Milankovitch cycles. Identification of these cycles provides a chronometric time scale for the event.

This scenario is used to analyze the important extinction event near the Llandovery-Wenlock boundary, which is termed the Ireviken Event (Jeppsson 1993). This event affected many taxonomic groups. For example, of the global conodont record of at least 60 conodont species, only 12 have a continuous record through the event, and more than 50% of trilobite species disappeared on Gotland. The global conodont record reveals eight datum points, numbered individually, whose spacing is described from the Ireviken 3 locality on Gotland. Their relative effects were found by analyzing the global conodont record of extinctions for each point. Phases 1, 2, and 4 were identified. The spacing of and differences among the datum points' relative effects are exactly as expected for an event that started with a Milankovitch pattern in which every second amelioration caused by the obliquity cycle (known today as the 41,000-year cycle) was reinforced by one caused by the precession cycle.

The Ireviken Event, an important extinction and replacement event straddling or just after the Llandovery-Wenlock boundary, has only recently been recognized as such and so has not been included in discussions of the cause and effects of such events.

Nonetheless, changes in the conodont faunas during these approximately 200,000 years had more far-reaching consequences than did all subsequent Silurian events. Furthermore, many other taxonomic groups were affected as well. For example, trilobites suffered a major extinction during this event. Ramsköld (1985:4) noted that at least half of the number of trilobite species on Gotland (Sweden) disappeared at a level that now can be identified as at or close to Datum 2 (Jeppsson 1993; Aldridge et al. 1993; Jeppsson and Männik 1993) of the Ireviken Event. It is at this same level that the greatest effects are also seen in the conodonts.

This event was named the Ireviken Event after the locality on Gotland where the most detailed records were available (Jeppsson 1993). Although there also is a fairly good record in North America, Europe, and Australia, this information has not been readily available, since it is scattered throughout descriptions of local and regional faunas.

The faunal changes affecting conodonts near the end of the *Pterospathodus amorphognathoides* Zone already were evident in 1964 when this zone was named and when Walliser described some of its conodonts, but Aldridge (1976) seems to have been the first to discuss the possibility of a global crisis for the conodonts during the Wenlock. Subsequent descriptions of conodont faunas from many parts of the world have substantially reinforced this scenario (for references, see Jeppsson 1987). When combined, these results form a very good base for describing the global extent of the event, together with some of its effects and the sequence of changes. In fact, we now know more about this event than about some of the frequently quoted, Treatise-based, end-stage "events." (The latter approach would have extended the ranges of many or all of the affected taxa to the end of the Wenlock, thereby creating an "end-Wenlock event" instead of, or in addition to, a more correctly dated "end-Llandovery event.")

The exact age of this event must be related to the type locality for the Llandovery-Wenlock boundary. Mabillard and Aldridge (1985) made a very detailed study of the conodonts' ranges there. In the series of collections described, the middle part of the event can readily be identified in the very earliest Wenlock (Aldridge et al. 1993). But the details of the earliest part of the Ireviken Event cannot be recognized at the type locality, although changes in the conodont faunas at this level were much more profound (Aldridge et al. 1993). It follows that the present basal-Wenlock type locality does not offer enough precision to tell if the event started slightly before the beginning of the Wenlock or if the event was entirely within the Wenlock. Accordingly, I have here dated it as *mid–Early Silurian.* The Llandovery-Wenlock Boundary would have been more explicit, but the event itself could well have been entirely in the earliest Wenlock. The new standard conodont zonation (Jeppsson in manuscript) is more precise than what can be established from the type locality, so the date of the beginning of the event must wait for a more precise definition of the boundary.

The Ireviken Event occurs within a cyclicity in which a typical cycle consisted of two distinct episodes, *secundo* and *primo,* followed by an *event* (Jeppsson 1990a, b; Aldridge et al. 1993). The average duration of such a cycle during the Silurian can be calculated as 3 Ma or less, and thus there were also several such other primo-secundo events during the Silurian. Of these, the Ireviken Event is the best known and had the most far-reaching effects.

Before the detailed conodont stratigraphy now possible for this interval (Jeppsson in manuscriptc), macrofossil-based global biostratigraphy rarely reached a precision of better than a few million years, and often even less. Such a lack of precision precluded the differentiation of even the individual episodes of the cycles. Although graptolite stratigraphy has better resolution than other macrofossil, it is applicable mainly in shale sequences. Furthermore, the analysis of the Ireviken Event has been hampered by the thinness of the strata in those regions where most conodonts have been studied. The combined thickness of the strata deposited during the preceding primo episode (about one order of magnitude longer) and the Ireviken Event itself is less than 1 m—and often much less—in large parts of the most extensively studied areas of Silurian strata (Jeppsson 1987). This situation is analogous to the current study of the end-Mesozoic, in which the necessary centimeter-by-centimeter stratigraphy began only after the iridium anomaly was discovered. Both for the said interval and for the Ireviken Event much data on taxonomic ranges is available and can be used once the importance of such detailed studies is realized.

To analyze the sequence of changes during the Ireviken Event, we need a time scale with a resolution good enough to separate its different components. The very great effects on the conodont faunas make them suitable for this purpose. In this essay I bring together a large amount of conodont data from three continents that are relevant to the changes during this event. In most sections, range ends are now known with a precision in the order of 10% to 25% of the event, but in some instances a still better precision has been achieved. One such example concerns the conodont faunal changes found in the Silurian sequence on the island of Gotland (Sweden), where for most of the Ireviken Event a precision of better than 0.5% of the duration of the event has been achieved.

On Gotland the Ireviken Event is represented by argillaceous limestones and calcareous mudstones similar to those of the previous primo episode. The lower strata recording the event are referred to as the uppermost Lower Visby Beds. (Lower Visby is the oldest formation exposed on Gotland.) This sequence is 4.5 m thick at Ireviken 3 on northern Gotland but is 2.5 to 4 times as thick (in the southernmost outcrops), 36 and 40 km, respectively, south southwest of Ireviken. All the succeeding Upper Visby Beds—around 8–10 m—were formed during this event. The succeeding, postevent Högklint Beds consist largely of reefs and reef-related pure limestones. The thickness of the strata of the Ireviken Event can be compared with the total thickness of all the Wenlock stratigraphic units on Gotland, which is at least 250 m (Hede 1960), but probably considerably more.

In addition to describing the sequence of changes during the Ireviken Event, I will consider how the faunal changes may be interpreted—their timing and their different magnitude. These are envisaged not as a series of unrelated events, but as a sequence of changes, of which the timing and characters have a single clock and a single cause. Ultimately, we seek to identify what that cause has been. To do so, I have derived a scenario for events of this kind based on the recently developed oceanic model, which enables the nature of the various episodes to be understood (Jeppsson 1990a, b). This model can now be derived independently, as is further discussed in the next section of this paper. The last two sections compare this scenario with the record as revealed by the conodont faunal changes. The final chapter discusses some of the additional knowledge about the sequence of changes that can be gained by such a comparison. This discussion has been continued to the point where it consists largely not only of explanations of known details, but of testable predictions. Future research should both increase empirical knowledge considerably and result in important improvements in the scenario. A deeper understanding of other stepwise extinction events may derive from the present analysis and equally provide an independent test for it.

There are three other kinds of events that can now be predicted from my model. The only Silurian example of the former, a secundo-secundo event, that has been identified with certainty so far started with a near extinction of the graptoloids before the end of the *C. lundgreni* Zone (Jeppsson 1993; Jeppsson et al. 1995; Jeppsson, in manuscript). Empirically (i.e., this event specifically) and theoretically (events of this kind), such episodes start with a major extinction, after which conditions improve either gradually or in a stepwise manner. In contrast, during the Ireviken and other primo-secundo events, conditions deteriorated stepwise through time.

There are several other Silurian events of the same kind as the Ireviken Event. Furthermore, faunal changes of such a kind are now known throughout the whole time range of the conodonts (for a summary and references see Jeppsson 1990a). Each event can be expected to be unique in its severity and details, because of different initial conditions, although the basic principles remain the same. The latter indicates that the scenario outlined herein may have a very wide applicability, from the Cambrian onward.

In regard to the nomenclature of Silurian episodes and events, a *datum* is a point in time, and in stratigraphy it is represented by a datum plane (Aldridge et al. 1993). The intervals between the datum planes may be described as zones and subzones, and the datum points may be named (cf. Cooper 1980). Both methods have advantages, with the second being better suited for studying the cause and effects of an event. To make the order of the datum points easier to remember, they are numbered and not named individually.

A Theoretical Model of Cycles in the Oceanic Carbon Dioxide Storage Capacity

Carbon is present in the atmosphere, the ocean, and the sediments and other rocks. These may in short be treated as the gas, dissolved, and solid phases, respectively. The concentration of carbon in the gas phase regulates the earth's climate—the greenhouse effect. The possible maximum concentration of carbon in the dissolved phase is inversely related to temperature, as it decreases with any increase in temperature. The range of temperatures of the dissolved phase is discontinuous. Except for its thin surface layer, the ocean consists of either cold-dense water (as it does today) or much warmer, saline-dense water. Because the carbon concentration of oceanic water tends to be close to saturation, any change in the ocean's characteristics produces either a surplus or a shortage of carbon. This imbalance is relatively rapidly regulated by the discharge of carbon to, or its extraction from, the gas phase. The processes that transfer carbon from one phase to another are largely independent of one another, but variations in the rate of transfer between the three phases can, under certain circumstances, result in cyclic changes. Models of these changes in turn enable a better understanding of global changes in faunas and sediments, which may be termed *interpretative* stratigraphy. In this essay I summarize one such model, develop it further, and apply it to the study of the Ireviken Event.

Many years of work on the cause of correlated, cyclic changes in particular faunas and the characteristic lithologies in which they occur have resulted in an empirical model (Jeppsson 1990a, b). During this time, the number of initial assumptions were gradually reduced to a few, while the number of identified changes caused by each cycle grew. The next stage of that work, developing a theoretical explanation for these correlated global cycles, is now in progress (Jeppsson 1990a, b). Most of the changes during these cycles can be derived as consequences of this preliminary theoretical model. (For a fuller account and references see Jeppsson 1990a.) In addition, my most recent approach opens up some new dimensions. (See also the supplementary notes at the end of this essay.)

The well-known effect of changes in the concentration of atmospheric carbon dioxide is that the average global temperature increases with increasing CO_2 concentration ("the greenhouse effect"). Because there is about 58 times as much dissolved and ionized carbon dioxide in the ocean as there is in the atmosphere, even small changes in the oceanic CO_2 balance have major effects. Furthermore, the solubility of carbon dioxide in water decreases with an increase in temperature, as is well known to anybody who has opened a heated container of a carbonated beverage. Except for that of a thin surface layer, the present-day oceanic water has a temperature near zero, because the principal source of the deep water is cold-dense high-latitude waters, since it is the densest surface water today. The alternative source of deep ocean water at other times in earth's history is saline-dense mid-latitude water. For instance, if the high-latitude temperature increases to about 5°C, the saline-dense water will become the densest water and thus will take over as the deep-water source.

Replacing the cold high-latitude water by the mid-latitude water, which is about 12° to 15°C warmer, results in a very rapid release of oceanic carbon dioxide (fig. 17.1). The maximum time it takes to complete such a release is the oceanic mixing time for an ocean with saline-dense bottom water. This has been calculated at 10,000 years (Bralower and Thierstein 1984), about 10 times the mixing time for the present ocean with cold-dense deep water. However, the CO_2 release time may be much shorter because rapid transport of heat down to the deep ocean water would lead to more rapid heating of the existing cold-dense deep water than was the case earlier. Thus a release time even shorter than that for the present ocean, around 1,000 years, might be expected. Furthermore, if water saturated with carbon dioxide is heated, gas bubbles will start forming, thereby lowering the density of the mixture of water and gas bubbles. As the water rises to depths where the hydrostatic pressure is lower, more gas is released, and the water is forced further upward. The result is similar to the mechanism of geysers, though in this case new water is constantly sucked in from below and lifted up until it too starts rising on its own, resulting in a continuous transport. At the surface the CO_2-rich water would spread out; the gas would be released into the atmosphere; and the water would gain density and start sinking. Such gas-driven upwelling could shorten the release time for the carbon dioxide still further. In this way the concentration of atmospheric carbon dioxide can rise very quickly. Aside from whether the time scale is a matter of tens or thousands of years, the result will be a very marked increase in the concentration of atmospheric carbon dioxide.

Another greenhouse gas is methane. Large quantities of methane hydrates (clathrate) are stored in the deep-shelf and deep-sea sediments. Clathrates are highly sensitive to increases in temperature, and so the rise in the sea-bottom temperature following a change to saline deep water would release methane, some of which would reach the atmosphere where it would gradually be broken down. Before that, however, its greenhouse effects would add to those of the carbon dioxide. The result would be a rise in the average global temperature far beyond that of the climatic amelioration that started it. Indeed, the high-latitude temperatures would be far too high to cause a return to the former oceanic state, even during a cold spell.

If, however, for some reason the atmospheric carbon dioxide concentration could be lowered sufficiently, an unusually cold period could lower the temperature of the high-latitude water so much that its density would be high enough to form oceanic deep water. Such water could be more or less saturated with

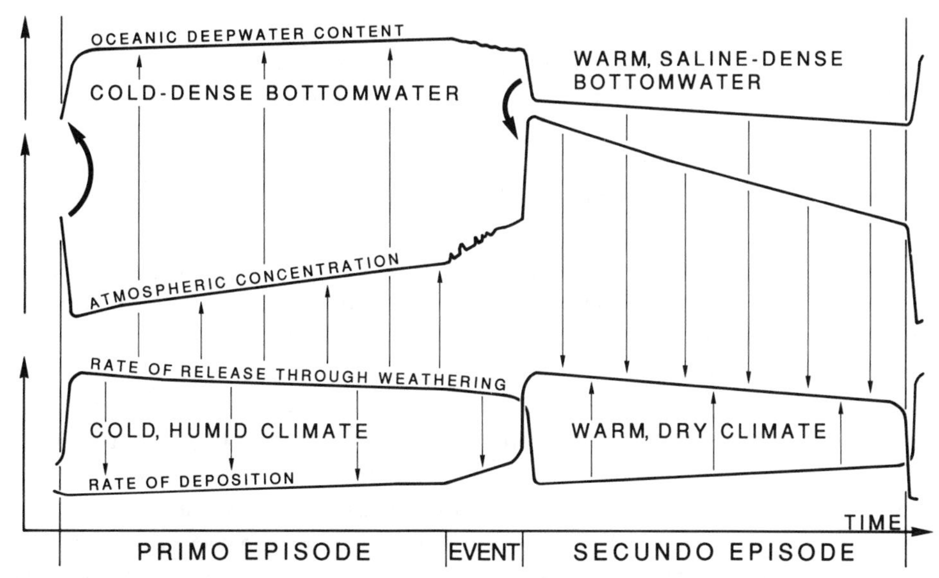

FIGURE 17.1. Changes in carbon distribution through a primo episode, a primo-secundo event, a secundo episode, and, in the right margin, the initial part of a succeeding primo episode. Arrows indicate where the net surplus goes. For example, during the cold, humid climate of a primo episode, the carbon released through weathering is only partly deposited again as carbonates. The surplus is distributed between the atmosphere and the oceanic deep water. The thick arrows mark the effects of the two rapid changes in the deep-water storage capacity connected with a cycle. The curves are not drawn to scale; they are intended only to indicate the direction of change. Furthermore, the relative magnitude of different changes is not yet known. The cycle illustrated is an ideal one; that is, at the end, conditions were exactly the same as they were at the beginning.

carbon dioxide when it started to sink, and it would draw down much more carbon dioxide than would an equal volume of saline-dense water, with an immediate strengthening feedback effect on the climate. Furthermore, as it sank, the increased hydrostatic pressure would increase the storage capacity. This would be filled by the dissolution of carbonate and by the oxidation of the organic matter in particles raining down from the surface waters.

The warm saline-dense water, gradually forced up by the new bottom water, would contain less carbon dioxide per volume than the water that sank. The net effect would be a drawdown of carbon dioxide, since the atmospheric carbon dioxide would dissolve into the surface water, thereby maintaining a balance between atmospheric and oceanic surface-water concentrations. A balance would be reached only when all the saline-dense deep water was replaced and the upwelling water consisted of cold-dense water, which released about as much carbon dioxide as was brought down. The time scale is probably similar to that of the circulation in the present ocean, around 1,000 years. After such a period of time, so much carbon dioxide would have been drawn down from the atmosphere and stored in the deep oceanic waters (fig. 17.1) and the global average temperature would have dropped so far that not even a strong climatic amelioration would cause a return to an oceanic circulation driven by saline-dense water.

It follows that both oceanic circulation states were so stable initially that no known external forcing can cause either of them to flip over to the other. The mechanism that can destabilize either state is in the interactions among the gas phase (the atmosphere), the dissolved phase (the ocean), and the solid phase (carbonates and other minerals). The rate of exchange with the third reservoir, the solid phase, is affected by temperature and humidity.

In the Wenlock and Ludlow sediments (the interval that I know best) there is abundant evidence that temperature and humidity were inversely correlated (see the last paragraph of this section for a comparison with Quaternary conditions). Thus the intervals with cold-dense oceanic deep water—named primo episodes by Aldridge et al. (1993)—not only were colder but also were more humid than the intervals with saline-dense deep water—the secundo episodes.

During primo episodes the more humid climate resulted in a faster rate of carbon release through weathering. At the same time, the net carbonate deposition was suppressed by the less favorable conditions on the carbonate ramps and platforms because of the cooler, more or less freshened water, rich in nutrients. The result was that more carbon dioxide was lost from the solid phase than was added to it, and so its concentration in the ocean and in the atmosphere increased gradually from its very low initial level.

In contrast, the warm, dry climate during secundo episodes was characterized by less weathering and by widespread formation of carbonate rocks. The result was that more carbon dioxide was transferred to the solid phase than was released from it, and so its concentration in the ocean and atmosphere gradually fell from its initial high value (fig. 17.1).

The result was that both the initially highly stable oceanic states gradually became less and less stable. In the ideal model this process would continue until the system flipped over to the alternative state (fig. 17.1). When the effects of external forcing (e.g., by Milankovitch cycles) were added, minor climatic fluctuations triggered the change to the alternative oceanic state once the stability had been lowered enough. The group of fossils that I know best, the conodonts, strongly reacted to the cyclic change between primo and secundo episodes. Such changes have now been documented to the end of the "conodontozoic," that is, the end of the Triassic (Clark 1987).

The transport of carbon dioxide to the oceanic deep water by the rain of fecal pellets and the like is limited by the amount of oxygen available (equal to that drawn down initially). Once it is used up, the rest of the organic material will be buried, and the deep water will be anoxic. During the Early Palaeozoic the ocean was predominantly anoxic.

The transport of CO_2 as carbonate is much more efficient, since carbonate does not require oxygen to dissolve. Therefore the transport of carbon dioxide down to the deep water became much more efficient during the post-Triassic, the "calcaroplanktobiotic," when the "carbonate compensation depth" could develop. This may have increased removal of atmospheric carbon dioxide during primo episodes to such an extent that the corresponding oceanic circulation state took over.

During the conodontozoic there were some intervals when pelagic carbonate-secreting, filter-feeding metazoa may have been abundant enough temporarily to create conditions comparable to those of the "calcaroplanktobiotic." Thus the effects of, for example, *Scyphocrinites* and the dacryoconarid tentaculites (nowakiids) may be worth studying. During parts of the Pridoli, a carbonate compensation depth has been identified (Jaeger 1976:279) when *Scyphocrinites* was widespread. Furthermore, the evolution of forest and peat-forming land plants already during the Devonian affected carbon dioxide storage in such a way that the primo episodes may have begun to lengthen. These changes may have been one cause of the gradual change from the predominantly anoxic Early Palaeozoic deep-water sediments to the predominantly oxic younger sediments.

Similarly, during the "calcaroplanktobiotic," there were brief intervals when the planktic carbonate production was lowered or inhibited. During these the oceanic conditions may have resembled those of the conodontozoic.

There is also a brief third oceanic state characterized by inhibited vertical circulation. The source of oceanic deep water may briefly stop production of deep water for one of two reasons. First, the density of an alternative source may increase enough to replace the original source. If this happens, the rate of vertical circulation will change, but the circulation will continue uninterrupted. Second, the original source may stop producing water dense enough to maintain circulation, perhaps because of a rapid change in climate. As long as no source is producing water dense enough to force away the existing deep water, the vertical circulation will be inhibited. However, the density of the surface water in the alternative source will fluctuate with the climatic changes; for example, the Milankovitch cycles will turn the deep-water source on and off. Such intervals can be distinguished and named as events (Jeppsson 1990, 1993; Aldridge et al. 1993; Jeppsson et al. 1995).

In the "undisturbed model" (Jeppsson 1990), such events are found only between primo and secundo episodes and are called *primo-secundo events (P-S events)* (Aldridge et al. 1993; Jeppsson 1993; Jeppsson et al. 1995). This sort of event ends when the density of the deep water has been lowered so much that it is in balance with that of the saline-dense source. When this has happened, the latter can start producing new bottom water. The climatic effects during such a time include an injection of carbon dioxide into the atmosphere, causing a change to a hotter, dryer climate (fig 1, Jeppsson 1990).

In the undisturbed model the saline-dense source would continue to produce deep water until the cold-dense source had become dense enough to replace the original source directly. Thus it would not be an extended event, only a single datum at the end of a secundo episode. When the effects of external forcing are taken into account—three other kinds of events, primo-primo, secundo-primo, and secundo-secundo—can be outlined (Jeppsson 1990b).

Nine cycles have now been identified during the Silurian (Jeppsson 1990, 1993; Aldridge et al. 1993; Jeppsson et al. 1995). That is to say, the average duration for a cycle during that period was on the order of 3 Ma (or, expressed differently, the frequency's order of magnitude was 10^1 fHz).

Before analyzing the theoretical model for constructing a scenario for primo-secundo events, I will outline a few of the cyclicity's effects. These are important, since they can be observed in the stratigraphic and biological record, whereas the past large-scale cyclicity in oceanic circulation is no longer directly observable.

Of the many consequences of the cyclicity, those in two fields are the best known. One is sedimentation in the marginal carbonate areas and slightly more distal areas, where more argillaceous sediments were formed. During primo episodes the net carbonate deposition nearly ceased in the marginal carbonate areas, owing to both a lowered rate of deposition resulting from the less favorable conditions and an increase in the biological destruction of carbonate caused by the higher amounts of nutrients brought out from the land. An increased nutrient concentration benefits those organisms that break down carbonates (Hallock 1988). At the same time, large volumes of terrigenous clastics were formed on land, owing to the humid weather, and were transported farther out. The result was the deposition of argillaceous, condensed sediments (or even a hiatus) in the carbonate areas and thick strata in areas where siliciclastics were deposited.

During secundo episodes, the warm, clear, salty coastal waters favored the deposition of thick units of pure limestone in the carbonate areas, whereas little terrigenous material was available for other environments, in which deposits would be thin or absent.

This "seesaw sedimentation" has added to the perennial problems of correlating carbonate and shale sequences more closely. During each interval of time, well-developed sediments were formed in either of these two environments, but the lack, or poor development of, contemporaneous sediments in the other area adds the differences between the primo and the secundo episode taxa to the facies differences. It follows that those strata formed during secundo episodes in the carbonate areas often must be compared with primo episode strata in areas with terrigenous clastics.

The second field, in which some of the effects of the cyclicity can be studied, concerns changes in the biota. Planktic taxa would, of course, thrive during primo episode conditions, because of the increased amount of nutrients available. Furthermore, the rich supply and diversity of food enabled the evolution of taxa specialized for very narrow trophic niches. In addition, many benthic taxa and communities are parts of food chains based on the planktic primary production, as larvae, as adults, or both.

During secundo episodes, reef formation was widespread, since the coastal waters were warmer, clearer, and less freshened. In addition, other taxa, which were parts of food chains based on the benthic primary production, flourished during secundo episodes. Typical planktic taxa during secundo episodes were those specialized for scarce food resources.

Many conodont lineages, which specialized in either primo or secundo conditions, have now been traced through several cycles. They became widespread when conditions were favorable for them and became restricted to refugia again when the conditions changed. These intermittent records agree well with the cycles deduced from the lithologic changes. I should stress that this is not a case of ordinary facies dependence. When the right kind of conditions prevailed, these taxa can be found in both pure carbonates and argillaceous sediments, and they were absent from the same lithologies when adverse oceanic conditions (for them) prevailed. For most of the Silurian, we can now identify the sequence of oceanic episodes and events, name them, and pinpoint their boundaries in the stratigraphic columns, with an exactness limited only by precision of sampling (Aldridge et al. 1993; Jeppsson and Männik 1993; Jeppsson et al. 1995, in manuscript). In the following text I will analyze what the theoretical model predicts regarding a primo-secundo event and then compare that scenario with the record of changes during the best-known primo-secundo event.

Many other causal models for cycles have been published. This is not the place to compare these in detail with that given in my 1990 study, although a few distinguishing characteristics should be stressed.

1. At least during the Silurian the average duration of the cycles was fairly short, only a few million years. Most other models describe cycles of longer (e.g., 25–35 Ma) or shorter (11,000 to some 100,000 years) duration. Herein I explore the power of the model for that time interval of cycles, during which Milankovitch cyclicity is important. (Feed-back mechanisms and quantification etc. need to be considered for a full description of the changes, but before doing so, it may be interesting to see how much of the stratigraphic and palaeontologic record can be understood using only the present simple model. Furthermore, quantification is very difficult in the Early Palaeozoic, where a very large part of the sediments that once were deposited are now destroyed or have not yet been studied in enough detail.)
2. My model (Jeppsson 1990, and discussion herein) predicts very rapid changes. Thus, the boundaries between the different parts of each cycle can be pinpointed with an exactness limited only, in most sections, by the effects of bioturbation. This character is shared by the biomeres (Palmer 1984). The origin of these boundaries is probably to be found in changes described by this model (Jeppsson 1990). Comparisons with other published models (e.g., those for sea-level cycles) are hampered by the lack of precision in most such cycles described.
3. Although I based my model on detailed studies of the Silurian record, the model can now be derived from both theoretical and experimental data and

well-documented recent and ancient oceanographic relationships. Accordingly, my model may more properly be called a theory rather than a model and thus differs in this respect from most other models for cycles longer than 1 Ma.

4. The Quaternary ice age cycles and the models developed to understand them differ from those cycles documented in the Paleozoic and described by my model, in that temperature and humidity are correlated. The empirical evidence for both the Silurian and the Quaternary relationships between these kinds of cycles is good, and neither can easily be shown to be wrong. The cause of these differences is unknown, but they may be one or more of the many fundamental differences between the two intervals. Earlier I noted the fundamental differences in the principal entries in the carbon budget. Another difference is that no changes in the oceanic regime of the kind discussed earlier accompanied the ice age cycles. The latter cycles also were accompanied by strong albedo changes. Such albedo changes would be expected during Silurian cycles too, but supposedly changes in the ice-covered area would cause them to be much stronger.

A Scenario for Primo-Secundo Events

The theoretical model, outlined in the previous section permits the prediction of a detailed scenario of the changes during a primo-secundo event.

The primo episode was characterized by a slow, long-term increase in the content of carbon dioxide in the gas and liquid reservoirs. Milankovitch cycles and other, still shorter cycles probably caused minor fluctuations in this trend. At the beginning of the primo-secundo event the pause in vertical circulation stopped both the drawing down of carbon dioxide and its release from the water that returned to the surface. The latter contained both the carbon dioxide that had been drawn down at the time when the deep water was formed and the carbon dioxide that had been added by the oxidization of organic matter and the dissolution of hemipelagic carbonate particles. The pause in vertical circulation transformed the deep water into a "black hole," with regard to not only carbon but also nutrients lost from surface waters. Thus planktic productivity dropped considerably, thereby also reducing the rate of transfer of carbon to the deep-water reservoir. The net transfer of carbon from the solid phase to the gas and dissolved phases continued, however, but most of this surplus stayed in the atmosphere and in the surface waters. As a result, the rate of addition of CO_2 to the atmosphere increased, as did the average global temperatures. The deep water gradually lost density through heating. Thus once the amelioration caused by the extreme phase of a Milankovitch cycle ended, high-latitude temperatures did not need to be equally cold, as before, in order to resume the formation of new deep water. A new balance was reached between the rate of carbon dioxide brought down and that brought up. If the new deep water was warmer than earlier, then its storage capacity was smaller, and a surplus amount of carbon dioxide developed and was released into the atmosphere. The next amelioration caused by Milankovitch cycles repeated this pattern. This amelioration was stronger, even if the external disturbance was as strong as the previous one, owing to the greater concentration of atmospheric carbon dioxide caused by the previous pause in vertical circulation. Compared with that of the primo episodes, the long-term average rate of carbon transported down was lower, because the "net drop" was lower on average, and the atmospheric concentration and the average high-latitude temperatures rose. As the event progressed, weaker and weaker ameliorations and thereby smaller and smaller climatic cycles could stop the vertical circulation, with the result being more frequent pauses. However, after a critical time, the normal state was a continuous intermission, with only the strongest cycles able to lower temperatures enough to start the vertical circulation again. Finally, toward the end of the event, no cycle was strong enough. Thus the deep water continued to lose density unchecked. The total increase of the deep-water temperatures needed to be only a few degrees before much warmer saline-dense water started sinking and forming new deep water. The very rapid changes caused thereby have been described earlier (Jeppsson 1990a, b).

Empirical data show that the long-term increase of the global temperature during an event led to a decrease in humidity (Jeppsson 1990a, b). The result was a drop in the rate of release of nutrients through weathering and transport to the ocean, and the long-term trend in the primary planktic production was a decrease. Strong fluctuations, caused by the Milankovitch cycles, produced a complicated pattern, but on average, each time the recycling of nutrients from the deep oceanic waters ceased, the effect on the primary planktic productivity was stronger than during previous pauses, because of the lowered rate of supply from the land.

If the response in the primary planktic production had been the same during each pause, then all of them would have caused temporary faunal changes, but only the first one would have caused any extinctions. Those taxa that survived the first pause would do so also during succeeding pauses. (The only exception would be those taxa that survived the first pause only "by chance.") However, when each pause on average meant a new record low productivity, then the survival of one pause did not guarantee survival of the next. The result would thus be stepwise extinction. Pauses that broke the average trend and were less severe than the previous ones were not marked by extinctions. Furthermore, the rapid loss in diversity, caused by previous extinctions, may have drained the pool of sensitive taxa to a level at which their number was too low to mark every pause.

The expected biological pattern would be a series of extinctions, each at the beginning of a pause, forming distinct datum points in time, identifiable as datum planes in the sediments. The extinctions might be complete—wherein every population of the affected taxon became extinct—but could just as likely have been incomplete. In the latter case there could have been a range of effects, from only a regional extinction of the taxon to the survival of only a small population in a small area. The changes at the datum points would also include the reappearance of taxa adapted to conditions of low planktic productivity—that is, oceanic deserts—and taxa utilizing the benthic production.

Once the pause ended, the productivity of the plankton would rise again. If the increase were strong and long enough, then those taxa confined to refugia at a previous datum might spread widely and appear as Lazarus taxa. The time of return of Lazarus taxa would be influenced not only by the time when world conditions had improved enough but also by the taxa's rate of dispersal. For those taxa with very widespread uniform populations, the delay caused by dispersal would probably be negligible if conditions improved very rapidly. If, however, improvement was slow, areas with lingering unfavorable conditions might have prevented the spread of some taxa for a long time. Frequency changes in taxa, which never disappeared, may thus be a better guide to register the end of the pauses in vertical circulation.

Two alternatives regarding the resumption of vertical circulation should be considered. First, the resumption may be gradual, and the rate of circulation low. Alternatively, the oceanic state may flip over rapidly, and full-scale vertical circulation may start at once. The latter would require a strong feedback mechanism like that strengthening the high-latitude climatic deterioration.

The most obvious change in the carbon storage pattern, however, would have the opposite effect. The new bottom water would draw down less carbon dioxide per volume than that given back by the water that returned to the surface. The latter would be saturated with carbon dioxide at the high pressure of the deep ocean, whereas new deep water that sank down would be in equilibrium with the atmosphere, at the atmospheric pressure. The surplus in the rising water would be the result of the dissolution of calcium carbonate and the oxidization of organic carbon in particles that had rained down from the surface during the whole time of the pause in vertical circulation. If the planktic primary production increased quickly, the downward transport of carbon would quickly have led to an equilibrium. To achieve that, however, the plankton would have needed plenty of nutrients, which they could get only through a vigorous vertical circulation, preferably combined with a high rate of transport from the land. Even in that case, the concentration of atmospheric carbon dioxide would probably have increased during the first 1,000 years—the time it today takes to replace the deep waters. This would act as a negative feedback. After the first pause, when the atmospheric concentration was still rather close to the threshold level for the start of an event, the likelihood for resuming the previous rate of vertical circulation would be the greatest, especially if the pause had been short. However, the likelihood would decrease with each pause, since both the pause and the end of it would inject extra carbon dioxide into the atmosphere.

This discussion is summarized in table 17.1 as a number of specific predic-

TABLE 17.1 *Predictions of observable changes based on the scenario for a primo-secundo event*

1. The datum points are distinguished by extinctions and other faunal changes.
2. The first few datum points are separated by relatively long intervals of time.
3. These intervals may be of equal duration, corresponding to the period of the Milankovitch cycle with the strongest high-latitude effect.
4. During the second phase of the event, the intervals between the datum points are shorter.
5. During the third phase of the event, the intervals between the datum points again lengthen.
6. The fourth and final phase of the event may have no datum points, except that one marking the end of the event.
7. In addition to these punctuations, there was a decrease in the average planktic productivity during the event. Successive intervals with similar oceanic conditions show a gradual change in addition to the extinctions, and so on, found at the datum points.
8. Taxa that had disappeared but survived in refugia reappeared as Lazarus taxa.
9. These reappearances are unlikely to be clustered as well in time as are those marking the datum points, which are caused by the pause in recycling.
10. The decrease in planktic productivity leads to less turbid waters and thereby better conditions for benthic productivity.
11. The major changes in benthic productivity, however, are found at the end of the event when the humidity dropped and the coastal waters became clear, un-freshened, and poor in nutrients.

tions, which can be compared with the fossil record. Here I will limit the comparisons to the best-known primo-secundo event, the Ireviken Event.

In the next three sections of this paper the empirical data for that event are assembled.

Datum Points During the Ireviken Event

The faunal changes caused by the Ireviken Event are known everywhere where conodonts of this age have been studied, and the order and the clustering of the changes are similar. At least eight datum points marked by extinctions may be distinguished during the Ireviken Event (fig. 17.2). The distance between these datum points may be found in the levels of the datum planes in the well-known section Ireviken 3 on Gotland (for locality data, see Laufeld 1974). This is the section where the ends of the ranges have been identified with the highest precision. At Ireviken 3 the exposed, richly fossiliferous sequence includes the Lower Visby Beds (10.5 m exposed above sea level), Upper Visby Beds (8.5 m), and Högklint Beds. The first two formations consist of thin beds of argillaceous limestone alternating with calcareous mudrocks. The Högklint Beds consist of reefal and lagoonal limestones. The weight of the reefs has compacted the sediments below them so strongly that the thickness and dip of the underlying strata have been affected.

There are two distinct bentonites in the Lower Visby Beds, about 2.46 m apart, where I first collected at Ireviken 3, but this distance increases away from the reef. The base of the lower bentonite is used as the reference level (Laufeld 1974), although the level of each sample has been measured to the nearest

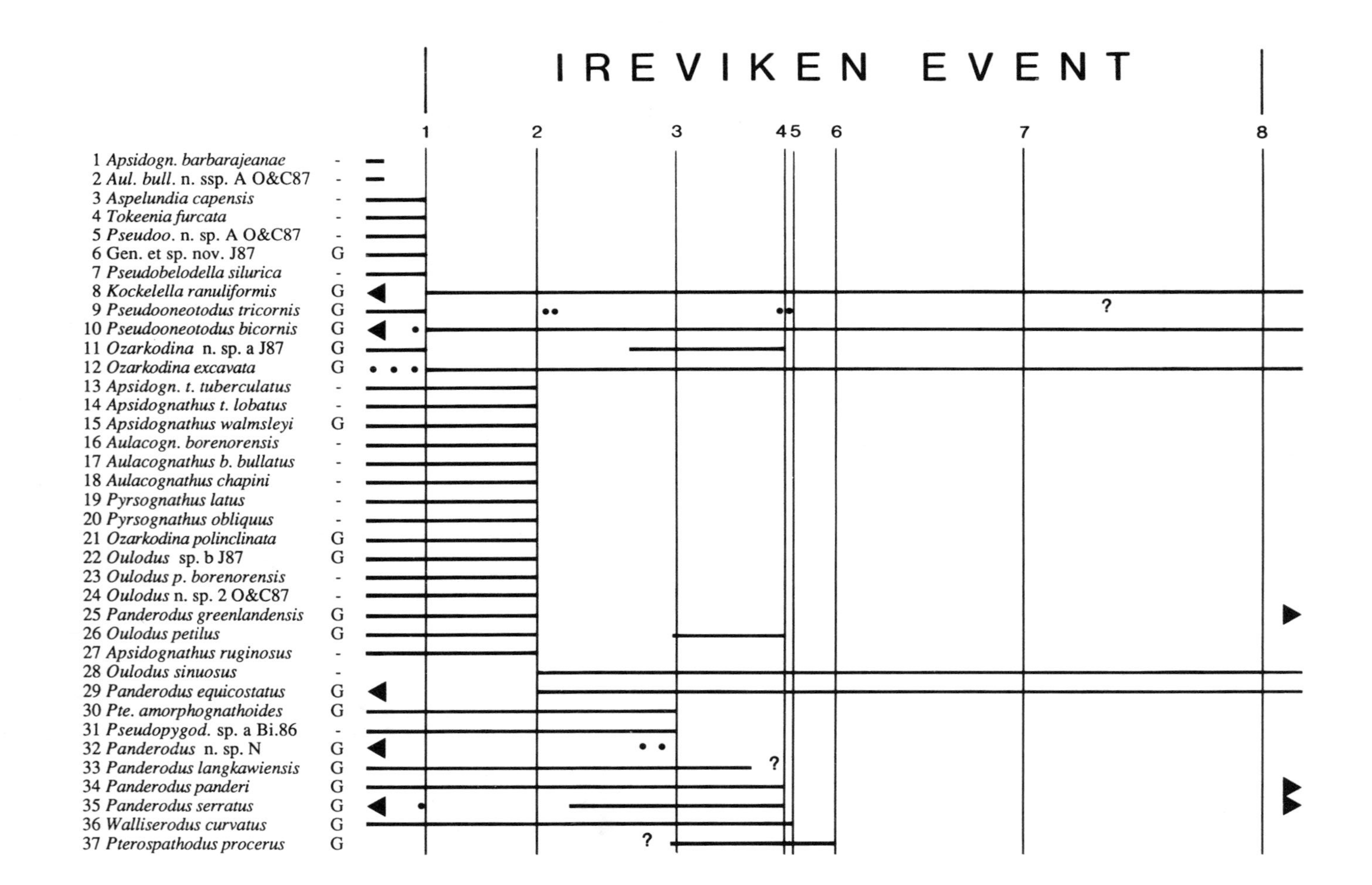
IREVIKEN EVENT
1
2
3
4
5
6
7
8
1 Apsidogn. barbarajeanae -
2 Aul. bull. n. ssp. A O&C87 -
3 Aspelundia capensis -
4 Tokeenia furcata -
5 Pseudoo. n. sp. A O&C87 -
6 Gen. et sp. nov. J87 G
7 Pseudobelodella silurica -
8 Kockelella ranuliformis G
9 Pseudooneotodus tricornis G
10 Pseudooneotodus bicornis G
11 Ozarkodina n. sp. a J87 G
12 Ozarkodina excavata G
13 Apsidogn. t. tuberculatus -
14 Apsidognathus t. lobatus -
15 Apsidognathus walmsleyi G
16 Aulacogn. borenorensis -
17 Aulacognathus b. bullatus -
18 Aulacognathus chapini -
19 Pyrsognathus latus -
20 Pyrsognathus obliquus -
21 Ozarkodina polinclinata G
22 Oulodus sp. b J87 G
23 Oulodus p. borenorensis -
24 Oulodus n. sp. 2 O&C87 -
25 Panderodus greenlandensis G
26 Oulodus petilus G
27 Apsidognathus ruginosus -
28 Oulodus sinuosus -
29 Panderodus equicostatus G
30 Pte. amorphognathoides G
31 Pseudopygod. sp. a Bi.86 -
32 Panderodus n. sp. N G
33 Panderodus langkawiensis G
34 Panderodus panderi G
35 Panderodus serratus G
36 Walliserodus curvatus G
37 Pterospathodus procerus G
?
?
?

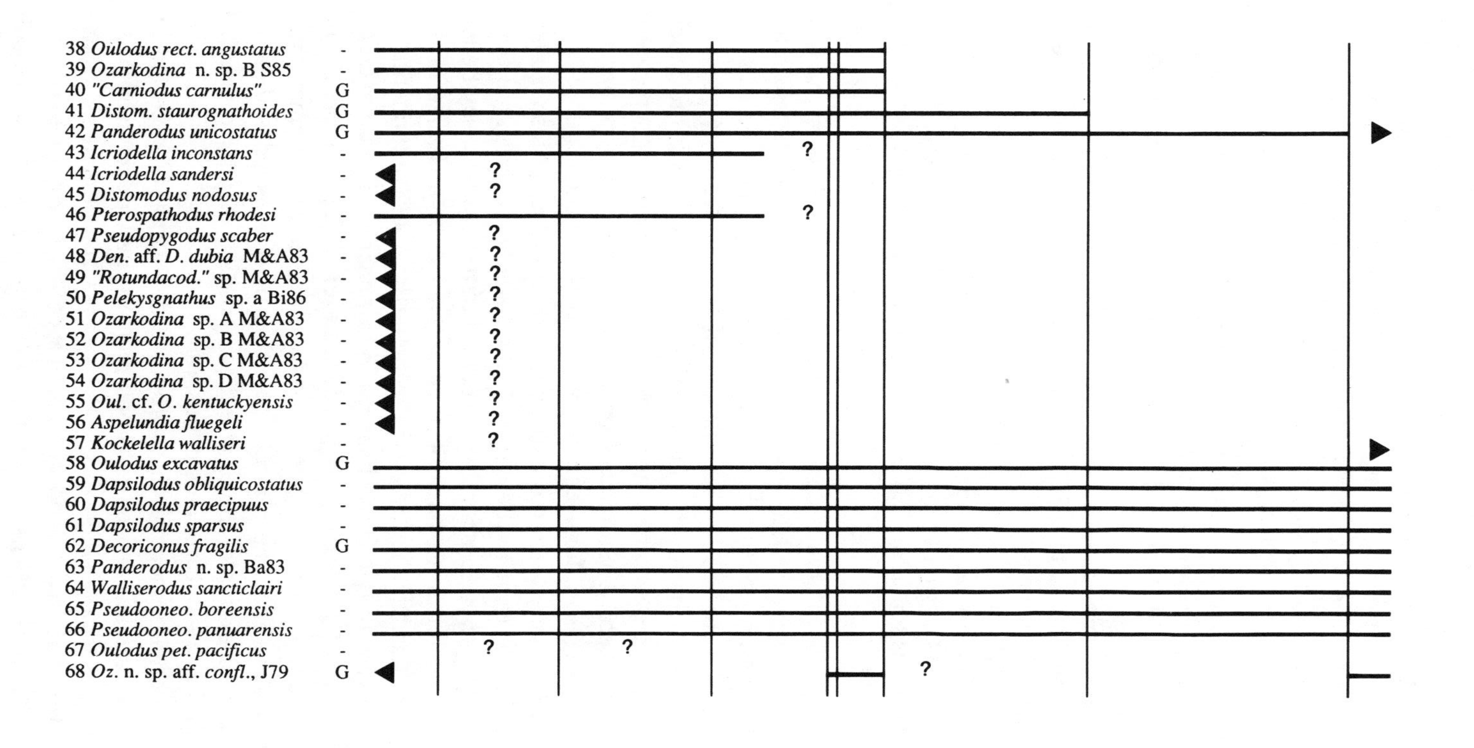

FIGURE 17.2. The range of the taxa discussed in the text. The letter G indicates that the range is based on the detailed record from Gotland. Many of these end points have been used to identify the datum points elsewhere. Spacing of the datum points is the same as in fig. 17.3.

bentonite and recalculated, using the value 2.46 as a standard, to minimize the error. The top of the Lower Visby Beds is at about +5.50 m and that of the Upper Visby Beds is at about +14 m. The section has been collected to +9.3 m. Repeated collecting has narrowed the uncertainty of the levels of extinctions to the values given below. The location of the two highest datum levels was calculated partly using data from other sections. The datum levels are as follows (the range ends marking these datum points are labeled G in fig. 17.2):

1. The First Datum is found at +1.01 ± 0.02 m above Laufeld's (1974) reference level, the base of the lower bentonite.
2. The Second Datum is found at +2.40 ± 0.03 m, that is, 0.06 m below the upper, major bentonite.
3. The Third Datum is found at +4.17 ± 0.05 m.
4. The Fourth Datum is found at +5.50 ± 0.06 m, that is, at or very near the base of the Upper Visby Beds.
5. The Fifth Datum is found at +5.62 ± 0.03 m.
6. The Sixth Datum is found at +6.15 ± 0.20 m.
7. The Seventh Datum is found at about $9.3^{+2.0}_{-0.0}$ m. (See also the addenda.)
8. The Eighth Datum is found at 14.0 ± 1.0 m, that is, at or near the top of the Upper Visby Beds.

The length of the time interval, represented between each two of the first four datum points, has been calculated to about 30,000 years using average rates of sedimentation (Jeppsson 1987b). This duration agrees with the Milankovitch obliquity term, which I cited as the likely cause of the extinctions (Jeppsson 1990). The strata between the fourth, fifth, and sixth datum levels are much thinner, whereas the sixth, seventh, and eighth are more widely separated. Before comparing this record with the scenario, I will discuss whether the difference is due to changes in, for example, the rate of sedimentation.

The Upper Visby Beds are much more calcareous (Sanford and Mosher 1985; note that they mistakenly refer to the Upper Visby Beds as the Högklint Beds and use the name *Upper Visby Beds* for the upper part of the Lower Visby Beds). This change in composition must have been accompanied by a change in the rate of sedimentation of either the carbonate or the terrigenous components, or both. Only in such a case in which the increase in carbonate net sedimentation was exactly compensated by a decrease in the rate of sedimentation of the terrigenous component would the rate of sediment formation remain unchanged. It thus follows that the rate of formation of compacted sediments most likely changed; for example, the different composition may have resulted from an increase in the rate of carbonate deposition.

A more favorable environment for net carbonate deposition is indicated by the bigger, more widespread, and more varied reefs in the Upper Visby Beds. Alternatively, the rate of deposition of the terrigenous component may have decreased before the carbonate deposition increased. This would have produced a thinner pile of sediments deposited between the Fourth and Sixth

Datum points, but a considerable decrease in the deposition of both components must be assumed in order to interpret the Fourth and Fifth Datum points as separated by the same length of time as the older ones. Such an explanation is not likely.

A second alternative is that all the taxa that survived four datum points may have been hardy enough to survive some more datum points. Since each of the Fifth to Eighth Datum planes identified is marked by the extinction of only one taxon on Gotland, it is evident that the number of survivors after the Fourth Datum is too low to mark more datum planes than have been identified. The presence of datum planes not marked by an extinction or a temporary disappearance of at least one taxon may be revealed by future research on abundance fluctuations, studies of other faunal groups, or sedimentological and geochemical work. Also, this second alternative fails to explain the close spacing of the Fourth to Sixth Datum planes, but it could explain the distance between the Sixth and Seventh Datum planes.

Once changes in the rate of sedimentation have been ruled out as the explanation for the uneven distribution of the Fourth to Eighth Datum points, we can compare them with the predicted scenario.

The Database for Worldwide Recognition of the Ireviken Event

Many studies of the conodont faunas of the Ireviken Event are so detailed that individual datum points can be identified. In turn, the conodont faunal changes can be compared globally, and the severity of each datum can be judged, since it may be based on the total effects on the conodont faunas. A list of some of these studies may be useful for those not familiar with Silurian conodonts. I note the number of collections representing the Ireviken Event. However, the basis for referring these to that event is the identification of at least two of its datum points; in other words, it is equally important to have collections older and younger than the Ireviken Event. Furthermore, many of these studies clearly show the contrast between the brief Ireviken Event and the long uneventful interval before it.

The number of specimens is quoted here from the original description, except when the study covered a very long interval. In such cases the figure refers to the event interval only. Inevitably, however, this figure is too low, since the immediately preceding and succeeding faunas are equally important to establishing the first and the last datum levels identified.

The datum planes identified are noted. The numbers in brackets refer to the bed number, sample number, and so on, used by the original author to identify the samples adjacent to the datum planes. To avoid lengthy repetitions, the faunal changes in each section or area are summarized in the next section. In a few cases the collections were so small that the exact level of a datum plane is uncertain. However, when the order of the faunal changes is consistent with those identified elsewhere, I have included them here.

For other considerations of reliability, see the next section, where the last record of each species is identified.

Cape Island, SE Alaska Thirty-three samples from an 8-m-thick limestone sequence, a total of 875 kg of rock, produced more than 1,500 coniform elements (not yet described) and 2,688 other conodont elements (Savage 1985). The sequence starts in the *P. amorphognathoides* Zone, and seven samples represent the Ireviken Event. Datum levels 1 (24/25), 2 (25/26), 6 (29/30), and 7 (31/32) can be identified.

Southern Mackenzie Mountains, Northwest Territories, Canada The sequence studied ranges from the base of the Silurian up into the *K. patula* Zone (Over and Chatterton 1987). That zone has been correlated with the *C. rigidus* Zone. At least four samples, 3 and 2.5 m apart, respectively, from two limestone sections represent the Ireviken Event. More than 25 kg of rock produced 384 conodont elements. Datum levels 2 (AV 2 157/160 m; AV 4 5/12 m), 6 (AV 2 above 160 m; AV 4 12/14.5 m), and perhaps 1 (AV 2 156/157 m; AV 4 1/5 m) may be identified in both sections.

Southeastern California The dolomite sequence studied (364 m) ranges from the *P. celloni* Zone to the Devonian (Miller 1978). At least two samples, 5 m apart, represent the Ireviken Event. About 2.2 kg of rock produced more than 300 conodont elements. Datum levels 3 (HV3 85/93 m) and 6 (HV3 98/110 m) can be identified.

Eastern North America More than 30 publications are available for the area ranging from Texas to Ontario, New York State, and Virginia (for references, see Jeppsson 1987a). In a large part of this area, the whole *P. amorphognathoides* Zone is represented by only a few decimeters of shale. In consequence, the Ireviken Event, which represents a much shorter interval of time spanning the end of that zone, would not be expected to be identifiable. However, by combining numerous large collections from many sections, Barrick (Barrick and Klapper 1976; Barrick 1977, 1983) has compiled range lists that do permit identifying some of the datum levels.

The Estill Shale in Ohio is an exception because of the very rapid sedimentation. Kleffner (1987) studied several sections. Thirty-two samples from the 42-m-thick Jacksonville Estill section produced about 1,000 conodont elements. Probably 30 of these samples represent the Ireviken Event. Datum 1 is either older or present between 1 and 2 m above the base, Datum 2 between 26 and 28 m, and Datum 3 between 40 and 42 m.

From West Virginia and nearby, Helfrich (1980) described the faunas from 33 samples from five sections based on approximately 1,000 conodont elements.

The interval ranges from the *P. celloni* Zone to and probably somewhat beyond the extinction of *P. amorphognathoides*. Most samples are from strata older than the Ireviken Event, but the extinction of *P. amorphognathoides* indicates Datum 3.

North Greenland Armstrong's monograph (1990) includes the most recent revisions of many of the important taxa, although it deals mainly with faunas and strata older than the Ireviken Event.

Southwestern Wales in Great Britain Nearly 7,000 conodont elements were identified from 21 samples of scattered limestone beds from about 60 m of section with terrigenous clastics (Mabillard and Aldridge 1983). Most samples are older than the Ireviken Event. Datum 3 might be found between the two youngest samples.

The Type Locality for the Base of the Wenlock in the Welsh Borderland More than 18,000 conodont elements have been obtained in 20 collections from slightly more than 2 m of strata, starting in the *P. celloni* Zone (Mabillard and Aldridge 1985). Datum planes 3 (38/39), 4 (39/40), and 6 (40/41) can be identified (Aldridge et al. 1993).

Gotland, Sweden The strata of calcareous mudstone and argillaceous limestone, formed during the Ireviken Event, are less than 15 m thick in the northern part of the outcrop belt but are probably much thicker in the south (where the oldest parts are not available) (Jeppsson 1979, 1983, 1987a, 1993; Jeppsson and Männik 1993). About 200 samples from several sections and some spot samples have produced more than 200,000 conodont elements from the upper part of the *P. amorphognathoides* Zone and the strata just above it. More than 150 of these samples containing in excess of 150,000 conodont elements represent the Ireviken Event. Repeated collecting has aimed at pinpointing the range ends for all conodont taxa. All eight datum levels have been identified. (See also the supplementary notes at the end of this essay.)

New South Wales, Australia The interval studied ranges through a large part of the Silurian (Bischoff 1986). Eleven samples from the Boree section, in which the Ireviken Event is represented by the best set of collections, produced more than 5,000 conodont elements (plus coniform taxa, which have not yet been described, with the exception of *Pseudooneotodus*) from an interval of about 9 m of limestone. Datum levels 1 (35/36), 2 (38/39), 3 (39/40), 6 (45/46), and 7 (46/47) can be identified.

Summary

As is evident from these summaries, the number of datum levels that can be identified is related to the number of samples and specimens. In many cases, additional samples would reveal more details, but the cost per analyzed sample is high and none of these studies, except the study of the sequence of Gotland, were aimed at the datum levels. Even so, the fact that so many studies record these datum levels is evidence for the high stratigraphic precision of routine conodont fieldwork.

In many cases, only one sample between each pair of datum levels had been collected. If considered separately, it could only be said that the faunal changes took place in a certain order, and not that there was any clustering at particular datum levels. However, if all these sections are combined, the clustering of the faunal changes becomes evident. (Since so many samples are available, any artificial clustering could easily be detected, independently of whether they were taken in a single section or derived from different sections. In the latter case, the sampled levels would be spread statistically over the interval, thereby being comparable to a more detailed sampling in a single section.)

Global Conodont Faunal Changes During the Ireviken Event

All the reports considered in this essay were accepted at face value, unless there was an obvious anomaly. For example, if a particular species was recorded in a section up to, but not beyond, the datum when it is known to have become extinct elsewhere, then it was registered as extinct at that datum, even though most of the database is statistically insufficient for this conclusion, when based on a single section.

Similarly, the occurrence of a single specimen, especially outside the expected range, should be considered with regard to contamination, bioturbation, reworking, and identification. This statement is not a criticism of those who have not done this, as it is only now that the distinct datum planes have been found to be significant and that such a discussion is relevant. I have judged my own stray specimens as follows.

1. Contamination can be excluded if the discovery is repeated (i.e., if the taxon is found in two samples from that general level or in a new sample from the same level). Contamination is unlikely if both the involved taxon is rare in all potential source samples (all those containing the taxon) and the taxa, which are much more abundant in the potential source samples, are absent from the sample with the stray specimen.

2. Bioturbation is unlikely to have moved a specimen downward more than 0.5 m. Reworking is expected to have led to the specimen becoming at least slightly darker or more stained than the indigenous specimens. Both kinds of smearing at the range ends are also assumed to affect the abundant taxa in addition to the rare ones, in the same way as was discussed with regard to contamination. On Gotland, the range ends are so distinct that blurring seems to have had much less impact than did the

thickness of the sampled interval. Often the thickness is about 0.1 m, except where samples have been taken to identify more precisely the levels of the datum planes. These sampling intervals were mostly limited to a single layer of limestone or nodules; that is, the sampled interval is a few centimeters to about 5 cm thick.

3. In the sample with the last (or first) specimens of a taxon, the range end was judged on its frequency in the collection. If that were the same as in the previous (or succeeding) samples, then the taxon was assumed to range through the sampled interval and halfway to the lower (upper) boundary of the next sample. If the frequency was considerably lower than in the previous samples, then the range end was assumed to be within the interval represented by the sample.

4. Identification of a specimen can be based on only that specimen or partly on co-occurring specimens. That is, when all specimens in a collection are identified, a proportion of these identifications must be of the following kind: "This specimen agrees, in all details that can be studied, with species X, the presence of which is already established in the sample (area, Zone)." Ramiform and coniform elements are especially difficult to identify, and many taxa have ramiform elements that I am not confident about identifying unless I use more than one element or even two or more specific elements. A rigorous identification of a taxon's range end requires that all such supporting data be from the range-end sample only. A rigorous evaluation of single specimens should be done by somebody familiar with the collections. I have therefore done this only for my own collections and otherwise have accepted the published identifications. For each species I have accepted the judgment of the latest author, who has commented on the taxonomy and nomenclature, for all the taxa with which I am not familiar myself. This may have led to some taxa being considered under more than one designation. On the other hand, a large proportion taxa not identified by a species name has been left out. But these omissions probably more than compensate for any inflation of the diversity due to undetected synonymies.

Finally, my own data from Gotland indicate that in addition to the changes in taxa present, there are large changes in the frequency of the taxa. In most cases, however, I shall not discuss these here. Similarly, brief gaps in the presence of species and brief reappearances of "Lazarus taxa" are not expected to be evident in the published data. A few have recently been found on Gotland.

With these reservations, I offer the following description of the faunal changes. The ranges are illustrated in fig. 17.2, and the record is summarized at the end of this essay.

THE FIRST DATUM

Apsidognathus barbarajeanae (Savage 1985) was very rare in both southeast Alaska (Savage 1985) and the Canadian Northwest Territories (Over and Chatterton 1987). In both areas the last record is well before the First Datum. The record of *Aulacognathus bullatus* n. ssp. A of Over and Chatterton (1987) also ended before the First Datum.

Gen. et sp. nov. of Jeppsson (1987a) from Gotland became extinct. *Pseudobelodella silurica* Armstrong 1990, known from California (Miller 1978), the Canadian Northwest Territories (Over and Chatterton 1987), and Greenland (Armstrong 1990), probably became extinct at the First Datum. *Aspelundia capensis* Savage 1985 and *Tokeenia furcata* Savage 1985 were not found after the First Datum in Alaska, and *Pseudooneotodus* n. sp. A of Over and Chatterton 1987 has a similar distribution in the Canadian Northwest Territories.

Kockelella ranuliformis (Walliser 1964) is known also from strata much older than the *P. amorphognathoides* Zone (Aldridge 1985). It is widespread in the interval considered here and in younger strata. However, nearly everywhere it (re-?)appears at the First Datum and is one of the best indicators for that datum. On Gotland a single specimen was found in a very large collection (perhaps more than 10,000 specimens) from 0.15 m below the datum; thus bioturbation cannot be excluded. At Leasows (Mabillard and Aldridge 1985), where the *P. amorphognathoides* Zone is only about 0.9 m thick, there is a gap of about 0.2 m in the record of this species. The return of the taxon may mark the First Datum. At Marloes (Mabillard and Aldridge 1983), where none of the datum levels have been identified, the species ranges through most collected horizons. If this sequence extends to near below the interval considered here, the presence of *K. ranuliformis* there may not have been interrupted.

Pseudooneotodus tricornis Drygant 1974 and *Ps. bicornis* Drygant 1974 occur together in many collections, but usually their ranges are separated. The most likely interpretation is that these two taxa competed for the same niche and normally excluded each other and that small changes in some parameter (e.g., food resources) affected the one that had the upper hand in the competition. Their co-occurrence in collections may be due to seasonal changes or less rapid oscillations (although rapid enough not to be separable by the sampling). In many areas one effect of the First Datum was a stable change from *Ps. tricornis* to *Ps. bicornis* faunas, but *Ps. tricornis* lived on elsewhere. For example, on Gotland it was replaced by *Ps. bicornis* at the First Datum in the surface sections, as expected, but it reappeared locally, at a very low frequency in younger strata, and survived at least to the Fifth Datum. In New South Wales it seems to have survived beyond the Seventh Datum (Bischoff 1986). On Gotland *Ozarkodina* n. sp. A disappeared, only to reappear as a Lazarus taxon about halfway between the second and Third Datum points.

Ozarkodina excavata (Branson and Mehl 1933) was present but very rare before the First Datum on Gotland, and it occurred in Britain from low in the *Ps. celloni*

Zone and thereafter continuously onward, although it increased considerably in frequency at about the horizon where *K. ranuliformis* returned (Mabillard and Aldridge 1985), that is, at the level that may be the First Datum. It also was promoted by the First Datum elsewhere and appeared widely in North America at that level. This appearance may reflect only a considerable increase in frequency, since Barrick (1983), who based his conclusion on very large collections, recorded it as present through the main part of the *P. amorphognathoides* Zone and onward.

The Second Datum

A large number of taxa became extinct at the Second Datum. This group includes *Apsidognathus t. tuberculatus* Walliser 1964 (widespread), *Ap. t. lobatus* Bischoff 1986 (New South Wales), *Ap. walmsleyi* Aldridge 1974 (Norway and Gotland), *Aulacognathus borenorensis* (New South Wales), *Au. bullatus bullatus* (Nicoll and Rexroad 1969) (ranged this high at least in the Canadian Northwest Territories; Over and Chatterton 1987), *Au. chapini* (Savage 1985) (southeast Alaska, Canadian Northwest Territories; Over and Chatterton 1987), *Pyrsognathus latus* Bischoff 1986, and *Py. obliquus* Bischoff 1986 (both from New South Wales), all with well-developed platform elements. *Ozarkodina polinclinata* (Nicoll and Rexroad 1969) (rather widespread), *Oulodus* sp. b Jeppsson 1987a (Alaska; Savage 1985; Gotland), *Ou. planus borenorensis* Bischoff 1986 (New South Wales), and *Oulodus* n. sp. 2 of Over and Chatterton 1987 (Canadian Northwest Territories) have ramiform elements. *Panderodus greenlandensis* Armstrong 1990 also may have disappeared at this time. On Gotland it is seldom found in samples from the youngest layers and is not found in the sample immediately below the datum plane.

P. panderi (= *P. recurvatus*) is represented by distinct populations before and after the Second Datum, at least on Gotland. Like the other taxa affected by this datum, both populations are present immediately below and above the datum, respectively. The differences between the two taxa are so pronounced that it could be a case of competition between two distinct subspecies, one being kept away until the dominant one became extinct. In addition, *P. langkawiensis* changes before becoming extinct, but the differences are minor and the exact level has not yet been identified. Planned studies may reveal more such taxonomic pairs.

It seems likely that e.g. *Ap. ruginosus* Mabillard and Aldridge 1983) (Wales), for which there is yet no record of the level of extinction, similarly became extinct not later than the Second Datum, since it would otherwise be expected to invade niches left vacant in other areas by the extinction of the platform-equipped taxa listed earlier, once the "Indian summer" indicated by the presence of Lazarus taxa was established. On Gotland, *Oulodus petilus* disappeared at this datum but reappeared as a Lazarus taxon, at or just before the Third Datum.

Ou. sinuosus Bischoff 1986 (New South Wales) appeared or reappeared, and *Panderodus equicostatus* Rhodes 1953 (sensu Jeppsson 1983) reappeared at the Second Datum.

More than half of the number of trilobite species on Gotland disappeared at or very close to the Second Datum (Ramsköld 1985:4).

THE THIRD DATUM

The victims at the Third Datum included the very widespread species *Pterospathodus amorphognathoides* Walliser 1964. It probably disappeared as early as during the First Datum in the Canadian Northwest Territories (Over and Chatterton 1987) and was rare or absent from some other places in northern North America (Nowlan 1983; Savage 1985).

Pseudopygodus sp. a. Bischoff 1986 (New South Wales) and *Panderodus* n. sp. N. (Gotland) may also be among those that became extinct at this datum.

THE FOURTH DATUM

With regard to the number of taxa that disappeared on Gotland, the Fourth Datum constitutes the second major extinction level. The victims include *Oulodus p. petilus* Nicoll and Rexroad 1969 (widespread), *Ozarkodina* n. sp. A (as yet only known from Gotland), *Panderodus langkawiensis* Igo and Koike 1967 (widespread), *P. panderi* Stauffer 1940 (widespread), and *P. "serratus"* Rexroad 1967. The first three lineages became extinct, but the last two reappeared in younger strata. The first two are Lazarus taxa in that they disappeared temporarily after the Second Datum and the First Datum, respectively. Similarly, *P. langkawiensis* and *P. panderi* were exceedingly rare after the Second Datum and about halfway toward the Third Datum. Elsewhere the reappearance of *Oulodus p. petilus* is not recorded, possibly because of its rarity and the fairly brief "Indian summer" with Lazarus taxa.

P. langkawiensis has not been found up to the Fourth Datum and may have succumbed before this datum.

Ozarkodina n. sp. aff. *confluens* sensu Jeppsson (1979) appeared on Gotland at this datum (Jeppsson 1979) only to become rare or disappear (again) soon afterward, probably at or near the Sixth Datum.

THE FIFTH DATUM

The Fifth Datum is hitherto recorded only on Gotland and in Estonia (Jeppsson and Männik 1993), where *Walliserodus curvatus* (Branson and Mehl 1933) disappeared. At Leasows, in Britain, *W.* cf. *curvatus* continued about 0.4 m higher than *Pt. procerus*. It is difficult to determine whether the order of disappearance there was real or an artifact related to the lithology, from which it is not easy to extract conodonts and from which only rather small collections have been isolated. The Fourth and Sixth Datum points were separated by dense sampling at Leasows by Mabillard and Aldridge (1985).

The Sixth Datum

The Sixth Datum is widely characterized by the extinction of *Pt. procerus* (Walliser 1964). In New South Wales *Oulodus rectangulus angustatus* Bischoff 1986 and in southeast Alaska *Ozarkodina* n. sp. B Savage 1985 disappeared at the same time. *"Carniodus carnulus"* probably ranged to the Sixth Datum on Gotland but disappeared earlier in other areas.

The Seventh Datum

Our knowledge about faunal changes after the Sixth Datum is based on much smaller collections than those used in establishing the record of the older intervals. However, it is well documented that *Distomodus staurognathoides* (Walliser 1964) survived beyond the Sixth Datum in most areas. It became gradually less frequent, and the exact level of extinction is difficult to determine.

The Eighth Datum

The Eighth Datum has not yet been studied in detail anywhere. On Gotland *Panderodus unicostatus* Branson and Mehl 1933 disappeared temporarily at this level, and a few species appeared (Jeppsson 1979).

Other Extinctions and Appearances

In addition to these species, a large number of conodont taxa became extinct (or disappeared) during the Ireviken Event, but as yet their ranges are so poorly known that the time of extinction cannot be pinpointed. Such taxa include *Icriodella inconstans* Aldridge 1972 (later than the Third Datum; Aldridge 1985), *I. sandersi* Mabillard and Aldridge 1983, *Distomodus nodosus* Helfrich 1980, *Pt. rhodesi* Savage 1985 (after the Third Datum), *Pseudopygodus scaber* (Drygant 1984), *Dentacodina* aff. *D. dubia* Mabillard and Aldridge (1983), *"Rotundacodina"* sp. Mabillard and Aldridge 1983, *Pelekysgnathus* sp. a. Bischoff 1986, and *Ozarkodina* sp. A, *O.* sp. B, *O.* sp. C, *O.* sp. D, *Oulodus* cf. *O. kentuckyensis* all from Mabillard and Aldridge 1983, and *Aspelundia fluegeli* (Walliser 1964).

Similarly, *Kockelella walliseri* (Helfrich 1975) probably appeared and/or disappeared somewhere during this interval but at an unknown level. It was reported together with *Pterospathodus amorphognathoides* from the Canadian Northwest Territories by Over and Chatterton (1987:15) and from West Virginia together with *Apsidognathus tuberculatus* by Helfrich (1980). We may conclude that it was present before the Second Datum, although the records may well be much older. Its closest relative was *K. ranuliformis*. These taxa possibly competed for the same niche and tended to exclude one another in the same way as discussed earlier regarding *Pseudooneotodus tricornis* and *P. bicornis*.

OTHER CONODONT TAXA PRESENT DURING THE IREVIKEN EVENT

A small group of taxa passed through the Ireviken Event without any known gaps in their presence (brief gaps, embracing just a few zones, would not have been detected). Similarly, considerable changes in absolute or relative frequencies or areas of distribution would also mostly escape detection, owing to the paucity of studies of taxa with coniform elements (many of the most detailed studies of the conodonts from this interval were limited to the ramiform- and/or platform-equipped taxa, although in some cases the genus *Pseudooneotodus* was also covered). This group of apparently unaffected taxa includes *Oulodus excavatus* (Branson and Mehl 1933) and many taxa with coniform elements, such as *Dapsilodus obliquicostatus* (Branson and Mehl 1933), *D. praecipuus* Barrick 1977, *D. sparsus* Barrick 1977, *Decoriconus fragilis* (Branson and Mehl 1933), *Panderodus* n. sp. Barrick 1983, *W. sancticlairi* Cooper 1976, *Pseudooneotodus boreensis* Bischoff 1986, and *Ps. panuarensis* Bischoff 1986.

Finally, the record of a few taxa does not show any consistent pattern. There are many possible explanations, ranging from a local response to oceanic changes, because the datum affected them differently in different areas, to a need for taxonomic revision. This group includes *Oulodus petilus pacificus* (Savage 1985) and *Ozarkodina hadra* (Nicoll and Rexroad 1969).

SUMMARY OF GLOBAL CONODONT FAUNAL CHANGES DURING THE IREVIKEN EVENT

The result of the changes during the Ireviken Event was that a fauna of about 60 conodont taxa was reduced to around 12. To these numbers must be added at least four that reappeared during the event, plus many that reappeared during the succeeding secundo episode. These latter have no known ancestors, either during the Ireviken Event or during the previous primo episode. Such ancestors must be sought in sediments representing the Llandovery secundo episodes. The total number of such absent taxa is thought to be at least between 10 and 15.

The Second Datum had the largest effect. Probably more than a third of the taxa present before that datum disappeared during an interval of time that may have been less than 2,000 years. The "Indian summer" before the Fourth Datum caused a minor temporary increase in diversity, indicating that those taxa that failed to return during this "Indian summer" had already become extinct. It is thus possible, for instance, that still more of the taxa—which disappeared widely in connection with the faunal reorganization at the First Datum—will turn out to have survived somewhere, at least to the Second Datum. If so, it would further emphasize this, the Second, as the main datum. The closely spaced Fourth, Fifth, and Sixth Datum points delivered the coup de grâce for most of the "extinction-bound" survivors, and so began for most surviving stenotopic taxa a long-term confinement to refugia that lasted until the next primo episode.

Characteristics of the Datum Points

The eight datum points differ markedly in the magnitude of their faunal effects and accordingly in how readily they can be identified. Even the two that involve the largest number of species are very different. On Gotland, all the taxa affected by the changes connected with the Fourth Datum are subordinate faunal components in adjacent strata, with either coniform or ramiform elements. Therefore the Fourth Datum is difficult to recognize, even though so many taxa are affected, sometimes even when large collections have been studied. The reason is the low frequency of the affected taxa, the more difficult identification of these kinds of conodonts, and the lack of data on the distribution of taxa with coniform elements in many of the published studies. In contrast, the Second Datum affected widespread, well-known, platform-equipped taxa, as well as other taxa that had formed a numerically important part of previous faunas. The Second Datum can be identified nearly everywhere where conodont faunas have been studied.

The least pronounced datum is the Fifth. Its only effect documented so far is the regional disappearance of *Walliserodus curvatus.* It is possible that the species became extinct at this level and that *W. curvatus* persisted to it in many areas, but at present, little is known about the distribution of taxa with coniform elements.

A Comparison of the Scenario and the Record

The predictions forming the scenario are summarized in table 17.1, and the changes that form the record as it is now known are summarized at the end of the previous section. This record agrees well with most of the scenario's predictions. There are several distinct datum points during the event, most of which can be traced worldwide. The spacing between the first four datum points is relatively long, as expected. Furthermore, the intervals between them match what would be expected, 30,797 years, if the obliquity term were responsible for the cycles. (This and succeeding numbers were calculated by interpolation between the data points of Berger et al. [1989], using the radiometric dating of the Ireviken Event, 430.5 Ma, by Odin et al. [1986].) Berger et al.'s numbers were not rounded off; abbreviations are discussed in the final section of this essay.)

The Fourth, Fifth, and Sixth Datum points are very closely spaced, and as pointed out above, this cannot be related to the changes in the rate of sedimentation predicted by the model, whereas the prediction in the scenario that cycles of weaker effects would be able to cause changes in the oceanic circulation during the second phase of the event fits very well. Possible candidates are the two major precession cycles, which today have periods of about 19,000 and 21,000 years. During the time of the Ireviken Event their periods were about 16,470 and 19,394 years, respectively (interpolated from data in Berger et al. [1989] and Odin et al. [1986]).

The spacing between the Sixth and succeeding datum points was, as predicted, much longer. Identification of a distinct third phase of the event is slightly uncertain in that only one datum is recorded and the exact level of this datum remains uncertain owing to the rarity of the taxon whose extinction marks this datum. The Fourth phase of the event is represented by a thick pile of strata.

I have not yet discussed quantitative faunal changes, and much work is needed before this can be done on a worldwide basis. However, on Gotland, the changes in faunal composition during the event fit very well with the seventh prediction in the table.

Similarly, identifying the Lazarus taxa requires a more detailed record than is available for most areas, but such taxa are known from Gotland. Much less work has been done on pinpointing the exact levels of their reappearances than that done on the datum planes. (The likelihood of that phenomenon during this event was discovered only recently, through analysis of the consequences of the empirical model.) No clear-cut reappearance datum is evident; instead the taxa seem to return one after another, as would be expected.

The changes in benthic productivity have not been studied, apart from the fact that reefs started growing during the later part of the event, at least on Gotland.

The strong conodont faunal changes at the end of the event have been described elsewhere (Jeppsson 1979).

Further Interpretations of the Ireviken Event Based on the Model

The model and the scenario also permit conclusions regarding recorded changes that may be specific to the Ireviken Event, because during such a short interval of time the Milankovitch interference pattern was unlikely to be repeated exactly during another such brief interval.

The weak effect of the First Datum may well indicate that it caused a very brief disturbance in the oceans' circulation. This conclusion is based on the following argument: The strongest high-latitude effect of the Milankovitch cycles is due to the obliquity term. It follows that as late as about 31,000 years before the First Datum there was a similar amelioration, although its effects were not strong enough to start the event. During these 31,000 years, the gradual changes in the level of atmospheric carbon dioxide, typical for a primo episode, would change the concentration only slightly, and so only during the very warmest part of the amelioration that caused the First Datum did new bottom water stop forming. Once that warmest time was over, the high-latitude temperatures again dropped enough and recycling of nutrients started again. The humidity during this pause was largely unchanged, and nutrients transported from land could support a primary planktic production that was high enough, over wide areas, to permit nearly all taxa to survive, at least somewhere. The length of this pause has not yet been quantified, although it may have been a few thousand years or perhaps less.

The catastrophic effect of the Second Datum indicates that the drop in planktic productivity was considerable during the second inhibition of deep-water formation. The extreme phase of the obliquity cycle may have been reinforced by one or more of the other Milankovitch cycles; under certain conditions a pattern in this reinforcement may develop.

The relationship of the various cycles' lengths may have been such that the pattern of interference was similar through several cycles, either reinforcing or dampening the effects of one another. The precession terms, which are 19,000 and 23,000 years today, can be calculated as 16,470 and 19,394 years in duration 430 Ma ago, based on Berger et al. (1989). These values are close to half and two-thirds of that of the obliquity term, 30,797 years. It follows that if the shorter cycle happened to be in such a phase that at one point in time, it reinforced the obliquity cycle, then it would also have reinforced the nearest obliquity beats. Similarly, the 19,394-year cycle would have reinforced the second previous and successive obliquity beats. Alternatively, for example, if the 19,394-year cycle were halfway out of phase at one point in time, then there would have been an interval during which every second obliquity beat was dampened.

The considerably stronger effects of the Second and Fourth Datum points, compared with the weak ones of the First and Third Datum points, fit well with either of these alternatives. If such interference did cause the differences between the datum points, it would probably also explain the characters of the intervals between the datum points (interdatum), since the interference would also have caused each strong datum to be preceded or succeeded by unusually favorable conditions. That is, it might explain both the nearly normal first interdatum and the "Indian summer" during the third interdatum, when several taxa could return.

An additional point regarding reinforcement is that the relative length of the Milankovitch cycles changed through time. For example, the 23,000-year precessional cycle is today about 57% of the 41,000-year cycle, but it was about 63% 430 Ma ago, when calculated on the values given by Berger et al. (1989). The uncertainty limit, indicated by Berger et al., is on the other side of the two-thirds value. The nearer to two-thirds it was, the longer sequences would have occurred with a regular pattern of reinforcing or dampening of every second cycle. The value 63% gives at the most a sequence of three reinforced obliquity beats, separated by weaker ones. Thus the regular pattern of reinforcement—evident in every second of the four first obliquity cycles during the Ireviken Event—was not present in the previous few cycles. This agrees with the fact that the obliquity beat 30,800 years before the Ireviken Event was not strong enough to start the event. In fig. 17.3 the ameliorations caused by the obliquity cycle during the event are numbered. If the 19,394-year precession cycle were behind the strong effects of the Second Datum, it must have been the second amelioration caused by that cycle during the event (the first would have culminated around 10,000 years after the event started). The fifth 19,394-year amelioration reinforced the third obliquity cycle amelioration, causing the severe effects of the Fourth Datum (fig. 17.3).

The end of the event would have come when the density of the deep water had decreased enough to permit saline-dense water to replace it. The density of the latter must also have varied with the average global temperature. It follows that the end of the event probably ended when the average global climate was warmest, or slightly before, when it had reached the critical threshold. Earlier (1990) I calculated the maximum duration of an undisturbed event as 90,000 years, which is also the maximum duration of each interdatum, including the fourth phase of the event. However, this calculation pertains to the probable conditions at the beginning of the event. It is likely that the deep-water temperature was much higher after the several earlier pauses in vertical circulation. We thus may assume that the maximum duration of the intervals between each two datum points of the third and fourth phases was considerably shorter than 90,000 years. If not, then the effects of the eighth obliquity cycle (fig. 17.3) would have ended the event. This means that the average rate of sedimentation

FIGURE 17.3. Changes during the Ireviken Event, as discussed in the text. The distances between the datum planes at Ireviken 3 have been used as a relative time-scale, although the places of the Seventh and the Eighth Datum points have been adjusted (see discussion in the text). The curves for planktic productivity and atmospheric carbon dioxide concentration have the same limitations as do those in fig. 17.1, although I have tried to indicate the relative magnitude at adjacent datum points, based on the degree of conodont faunal changes. The curve for external forcing is drawn as a regular curve, which was modified by a second cycle, so that every second cycle became less extreme. The only exception is found during the third phase of the event, where the curve is modified to reflect the registered datum points, assuming that each of them was due to external forcing. The real curve would differ in several aspects, as is partly discussed in the text. Even so, the curve fits astonishingly well with the record, in both the relative magnitude of the extinction at different datum points and the degree of recovery in between, during both the first and the third phase of the event. Those parts of the curve that are not reflected in the record are dashed (as explained in the text, these parts are not expected to be reflected in the record). The two bentonites (B), the boundaries between the Lower and Upper Visby Beds, and the Upper Visby and Högklint Beds (LVB, UVB, and HB), and subdivisions of the former two, b, c, d, e, and a, b, c, d, respectively, are marked.

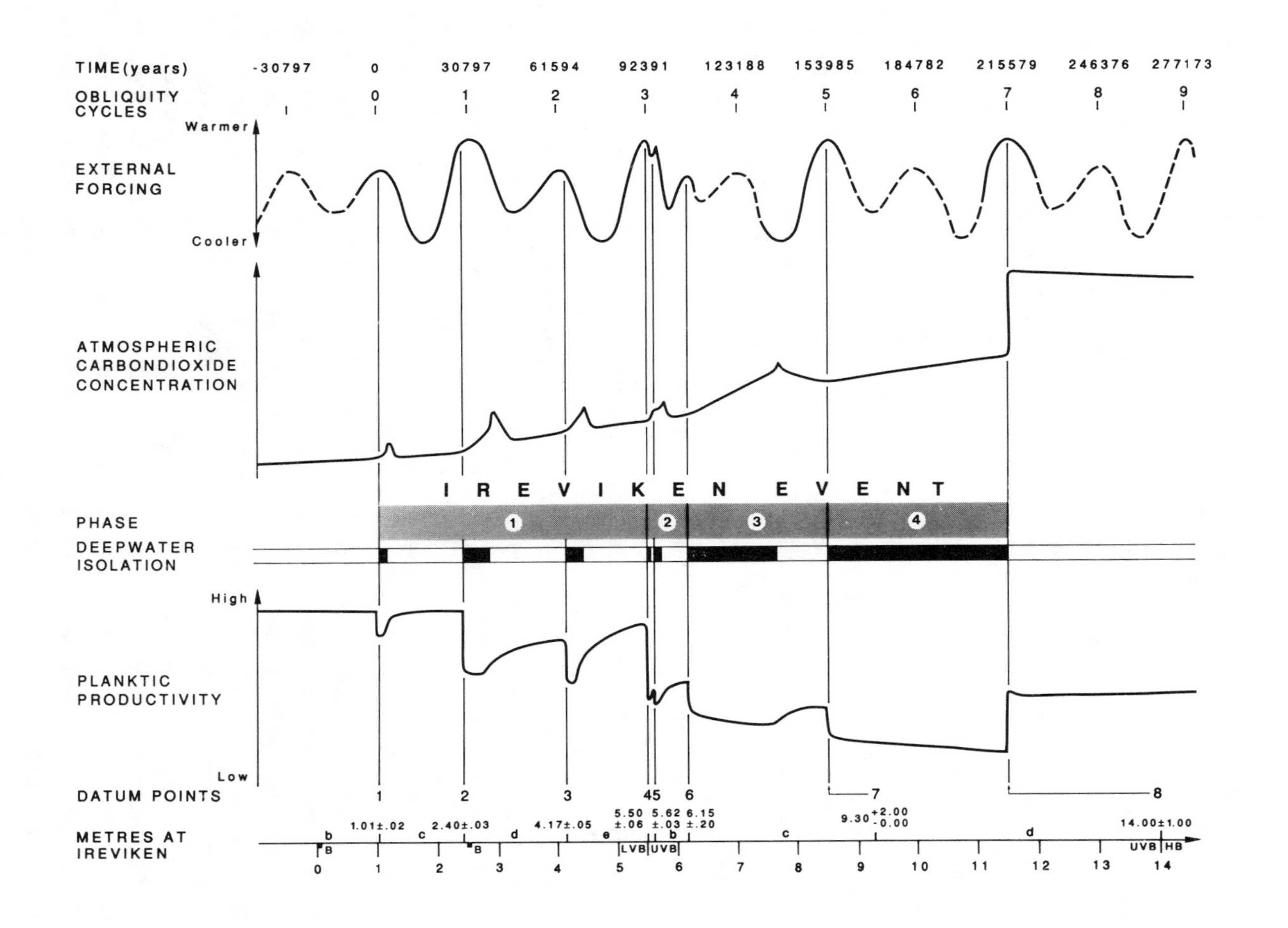

TIME(years)
-30797
0
30797
61594
92391
123188
153985
184782
215579
246376
277173
OBLIQUITY CYCLES
Warmer
EXTERNAL FORCING
Cooler
ATMOSPHERIC CARBONDIOXIDE CONCENTRATION
IREVIKEN EVENT
PHASE
DEEPWATER ISOLATION
High
PLANKTIC PRODUCTIVITY
Low
DATUM POINTS
METRES AT IREVIKEN
1.01±.02
2.40±.03
4.17±.05
5.50 ±.06
5.62 ±.03
6.15 ±.20
9.30 +2.00 -0.00
14.00±1.00
LVB
UVB
HB

during the third and fourth phases was 1.2 times that of the first phase at Ireviken 3. A still greater difference in the rate of sedimentation seems likely, since reefs were formed during the deposition of the Upper Visby Beds. The alternative numbers to consider in this case are that the duration of the fourth phase was about 60,000 years, that the seventh obliquity cycle ended the event, and that the average rate of sedimentation was 1.5 times. This means that enough deep water during the fourth cycle was formed to reset oceanic conditions to only 30% away from the conditions at the beginning of the event.

Considering the likely changes in the concentration of atmospheric carbon dioxide during the previous phase of the event, such a strong reset seems to be the maximum possible. We can therefore conclude that more probably the seventh rather than the eighth obliquity cycle ended the event. Indeed, the event may well have ended earlier, since it seems more likely that the new deep water, formed during the last interval with deep-water formation, would be very close to the maximum possible high-latitude temperature that permits the formation of cold-dense bottom water. If so, a still shorter duration of the third and fourth phases would be expected, but that would also mean that the rate of sedimentation had increased to 2.1 times the rate during the first phase. This seems quite possible, although it would probably mean that the Seventh Datum was not related to an obliquity beat.

The best suggestion regarding the duration of the Ireviken event is that it consisted of seven obliquity cycles. An interpolation between Berger et al.'s (1989) two nearest data points results in a duration of 215,579 years for seven cycles. If we consider, first, that there is some latitude in their calculations and, second, the likelihood that the end of the event came slightly before the time of maximum forcing, a value of around 210,000 seems likely, or if the uncertainty in the calculations is still greater, a value of around 200,000 years. As I noted, I do not think that it could have been longer, but it cannot be ruled out that it was the Sixth obliquity cycle that ended the event; this would give a duration of about 180,000 years.

The pattern of dampening and reinforcing would not be repeated for hundreds of thousands of years, even if only two Milankovitch cycles are taken into account. This is longer than the duration of a primo-secundo event. Therefore the characters of different events were influenced by whatever phase of such reinforcement-dampening cycles happened to affect the climate, especially at the beginning of the event. An event with strong reinforcement would have had much more severe faunal effects than would one with weak reinforcement. In the latter case the dampening of the climatic cycles would have produced longer pauses in vertical circulation. Furthermore, the less cold climates between the pauses would have caused much less cooling of the deep ocean water. Both effects would have helped shorten the total duration of the event, thereby further reducing the effects on the fauna. The duration and effects of primo-secundo events thus varied with the strength of the Milankovitch-induced climatic effects during the event. The rebound effect also contributed to the weaker effects of most primo-secundo events, for only when the diversity has been restored could an extinction event be as devastating as the Ireviken Event.

Addenda

Since this essay was submitted for publication on December 15, 1990, many new data have come to light, and so this discussion should be extended.

Supplementary Note (April 28, 1993)

The Viki core from Saaremaa (Estonia) has been studied in detail comparable to that for the Ireviken 3 section (Männik, in press; Jeppsson and Männik 1993). The study of the Lusklint 1 section on Gotland is now even more detailed than the Ireviken 3 section. The fieldwork and treatment of samples were more uniform and the collections are much larger and in some intervals much more closely spaced. Most of the new data only reinforce the picture, widening the number of published sequences and taxa. In addition, some uncertainties about the ranges have been removed. (Most of those new data pertaining to the Ireviken Event can be found in Jeppsson and Männik [1993].)

The new results include two more datum points. Their designation and position in figs. 17.2 and 17.3 are as follows: Datum 3.3—the extinction of *Panderodus langkawiensis*—is at 4.8 m, and Datum 6.2—the temporary disappearance of *P. unicostus*—is at 6.9 ± 0.4 m. Furthermore, the larger collections show that *P. procerus* survived slightly longer, moving Datum 6 to about 6.6 m.

Datum 3.3 is exactly where the fourth amelioration caused by the longer precession cycle would occur if it reinforced the Second and Fourth Datum points. The event apparently had progressed to Phase 2 already at Datum 3.3, about 70,000 years after the event started. Considering the duration of this cycle, the conditions characterizing the second phase may have been established even somewhat earlier, perhaps at Datum 3.

An unchanged rate of sedimentation would relate Datum 6 to the sixth amelioration caused by the longer precession cycle, and Datum 6.2 to the fourth obliquity amelioration (see fig. 17.3). Only if the rate of sedimentation increased, and had done so already at Datum 4, could the fifth precession amelioration have caused Datum 6.2. If reef debris and the like were the main cause for the increase in the rate of sedimentation, this increase would have taken place later than Datum 4, since the reef did not begin to grow until somewhat higher up in the Upper Visby (K. Mitchell, personal communication, and my own observations).

Other details include the fact that Datum 4 coincides at the highest precision possible (probably better than ±1 cm) with the extinction of the button coral, *Palaeocyclus porpita*, and thereby with the top of the Lower Visby Beds, as defined by Hede (1921). At Lusklint, the boundary bed consists of a 0.5- to 2-cm-thick limestone bed, in the middle of which is a horizon of pyrite crystals. If this horizon can be found elsewhere too, it may indicate that the pause in vertical circulation resulted in oxygen-depleted (or possibly even anoxic) bottom waters and anoxic conditions immediately below the sea bottom, even in as shallow settings as those of Benthic Assemblage 4, found at depths of around 70 to 175 m (data in Gray et al. 1974).

I found a distinct marker-bed with *Phaulactis* immediately above the Fourth Datum Plane at two localities south of Visby and, in 1992, also at Lusklint. This *Phaulactis* Layer can be found at every outcrop studied as far to the north as the contact is exposed (K. Mitchell, personal communication), 50 km north northeast of the southernmost locality studied. *Phaulactis* must have benefited in some way from the changes connected with or following Datum 4.

Distomodus staurognathoides turned out to have survived until near the end of the event, although it is represented by less than (sometimes far less) one element per kilogram. Identification of the change from phase 3 to 4 is therefore not possible with the available data. The lack of any conodont extinction may be due only to the depleted supply of sensitive conodonts. Frequency data or information from other taxonomic groups or other scientific methods may help identify one or more datum points in the silent interval after Datum 6.2. If none is found, then the heating time for the oceanic deep water will limit the maximum possible duration of this part of the event to less than 90,000 years.

Very rare, probably juvenile specimens of *Walliserodus* sp. are found from about 11 m and onward, and a few have also been found deeper down. A distinct taxon of *Panderodus*, *P.* n. sp. J, is regularly present from about 9.2 m to Datum 8. *P. boreensis* can be removed from the small group of taxa ranging through the event without known effects.

Boucot (1991; cf. Hede 1921) noted that also the brachiopods suffered extinctions during the Ireviken Event, but which specific datum point(s) terminated these lineages is (are) yet unknown. Boucot also quoted Chatterton (probably as yet unpublished) about the termination of many trilobite lineages, and Brett and Eckert (1989) about a significant Late Llandovery pelmatozoan extinction event.

The most recent discussion of the effects of carbonate deposition for the concentration of atmospheric carbon dioxide and thereby for the greenhouse effect (Gauffin and Fujiita 1992; Sundquist 1993) concluded that the effects should be the opposite of those included in my model. If this is correct, some changes are needed in that model. The empirical model (Jeppsson 1990:663–71) would not be affected. (A remarkably similar model—illustrations, arguments, and their sources—by Tucker was published in 1992, with reference to the Precambrian-Cambrian interval. The lack of a reference to Jeppsson 1990 implies that it was derived independently, from a part of the stratigraphic record quite different from that on which I reported in 1990. If really so, then this apparently independent derivation of this empirical model may be seen as support for these cycles in the stratigraphic and biotic record and as support for the empirical model, including the temperature-humidity relationship.)

Most parts of my theoretical model remain unchanged, for example, the difference in the oceanic carbon dioxide storage capacity between the two oceanic states, the abrupt changes connected with the alterations in oceanic state, and thereby the initial strong reinforcement of the triggering climatic change. The main change in the theoretical model is that another "engine" for the cyclicity must be found. Since both oceanic states would have been very stable initially, because of the effects of oceanic change, something that gradually destabilized the prevailing atmosphere-ocean system must be sought. That or those "engine(s)" would need to have had such profound effects that it (they) also compensated for those caused by the changes in carbonate deposition (documented in the stratigraphic record). However, before searching for such alternative engines, it may be worth studying the extent to which modern discussions of the geochemical carbon cycle cover oceanic cycles of the duration and nature that my model describes. I can offer here only a few preliminary comments.

Modern discussions of the geochemical carbon cycle concern its effects on a "multimillion-year time scale" of the "Arrhenius-Urey reactions" (e.g., Berner 1990; although Urey [1952] suggested a mechanism that would help regulate the concentration of atmospheric carbon dioxide in the prebiotic earth). This cycle involves the uptake of atmospheric carbon dioxide in fresh water during the weathering of, for example, carbonates and precipitation of HCO_{3-} as marine $CaCO_3$ (e.g., Berner 1990). Today about 90% of the dissolved carbon dioxide consists of bicarbonate ions (HCO_{3-}), whereas around 5% consists of carbonate ions (CO_3^{2-}) and about 2% of dissolved carbon dioxide (e.g., Broecker 1983). My discussion was based on marine carbonate formation from carbonate ions and standard chemical equilibrium reactions (the decrease in carbonate ion concentration is partly restored through such equilibrium reactions from atmospheric carbon dioxide via dissolved carbon dioxide and bicarbonate ions). These reactions would affect pH, but this is regulated through different processes (e.g., modern results indicate the importance of the circulation through fresh oceanic basalt for regulating the oceans' composition).

Bicarbonate ions are the dominant negative ions in fresh water and thus in weathering. This fact alone does not conflict with my model. Rather, the question is how much of this dissolved carbon is returned to the atmosphere in the transformation of riverine water to oceanic water (This transformation includes, for example, a rise in pH, which within years might contribute to the return of both the CO_2 borrowed from the atmosphere during weathering and some of that resulting from carbonate weathering.) This question is not included in discussions of the geochemical carbon cycle.

Berner (1990) noted that little is known about weathering before 350 Ma ago.

(My empirical model was derived from studies of the interval from 430 to about 410 Ma ago, although I also discussed evidence for similar cycles during a much longer interval.)

In sum, a much more thorough discussion is needed than merely a reference to standard geochemical textbooks (Tucker 1992) to find out to what extent my theoretical model is incompatible with the present theoretical models of the geochemical carbon cycle and their basis in calcaroplanktobiotic conditions.

SUPPLEMENTARY NOTE (MAY 15, 1995)

I have recently (in manuscript on a new Wenlock conodont zonation) correlated the intervals of the Ireviken Event with the *Cyrtograptus insectus* to the *C. bohemicus* zones. This interval forms the first acme of cyrtograptids (Richards et al. 1977:76). Graptolites of this kind were adapted to nutrient-poor waters (see Underwood 1993 for a summary and references). This is in very good agreement with the conclusion herein that the Ireviken Event caused reduced primary planktic production.

Furthermore, considerable extinctions took place in the graptolites during this interval, especially in the *C. insectus* Zone and at the end of the *C. murchisoni* Zone (Melchin 1994; Štorch 1994). Correlations are not yet good enough to identify the specific datum points, but the extinctions in the former zone were probably due to one or more of the first four datum points, whereas those at the end of the *C. murchisoni* Zone are at least very close to—if not at—Datum 8. Taken together, the effects of the Ireviken Event on the graptolites were about as severe as those on the conodonts and the trilobites. Similar high-resolution data for the extinction of other macrofossil lineages will be very important for judging the effects of this event on different ecosystems.

D. staurognathoides had a Lazarus gap between datum points 4 and 5, and "*Carniodus*" is present between these datum points. These new data are derived from the section Lusklint 1, where the sampled intervals are 1—5 cm, and the sampling was continuous near most datum planes. These samples have yielded more than 1,000 specimens per centimeter of section (i.e., on average more than five specimens per year) at some datum planes, giving a much more detailed picture of the changes. Further, these samples confirm that the range ends as registered above are real and not due to only a slight lowering in frequency below that detectable with the Ireviken 3 samples.

The collections from Lusklint 1 also permit recognition of reworking of conodonts upwards. Reworked material can be recognized by being darker, and including only mature specimens of the most wear-resistant elements in a frequency far below that in samples older than the true range end. The largest distance found is at Datum Plane 3 (at 4.25 m), where *P. amorphognathoides* is found in all three younger samples up to that from 4.35–4.39 m. A similar distance is expected at Ireviken 3. The last record of *O. polinclinata* there was based on two mature specimens (but only very slightly darkened) from the last

10 cm below the bentonite. It seems now probable that Datum 2 is at or a few centimeters below 2.36 m at Ireviken 3.

The pyrite layer at Datum 4 has now also been found at Ireviken 3 (by Peep Männik), Nygårdsbäcksprofilen 1, and Buske 1; the distance between the first and the last is 35 km. Less well-developed pyrite layers are found between this layer and Datum 6 at some localities.

The specimens of *Phaulactis* in the *Phaulactis* Layer are big—usually 10 to more than 20 cm tall and up to 7 cm in diameter. All specimens seen have fallen and are lying horizontally. Dr. Männik pointed out that their "upper" side is absent at Ireviken. I have observed this destruction elsewhere too. It is not the result of mechanical erosion—the cut is very uneven, and the specimens lie in soft clay. It looks more like etching. The local conditions seem to have included brief intervals of inhospitable conditions, where the *Phaulactis* layer may record a dysaerobic interval ended by still worse conditions. Literature studies have provided some indications that these conditions were not local but more widespread at this time.

My theoretical model requires that carbonate deposition is a net sink for carbon dioxide, whereas the Arrhenius-Urey equations state that carbonate production releases carbon dioxide. The debate regarding this part of the Arrhenius-Urey equations has been intensive during the last years. Direct measurements on a coral reef have now shown that reefs do form a net sink for carbon dioxide (Kayanne et al. 1995).

ACKNOWLEDGMENTS

The content of this essay was influenced by discussions with Richard J. Aldridge, Leicester; Maurits Lindström, Stockholm; and Peep Männik, Tallinn. Carlton Brett invited me to contribute to this volume. Linguistic and other errors in the manuscript were corrected by him, Euan Clarkson, Doris Fredholm, Ann-Sofi Jeppsson, and Ron Harris. Comments from the reviewers—one anonymous, Harold B. Rollins, and Ronald R. West, and Alfred G. Fischer—led to further improvements. Amy Gayle, Erna Hansson, and Margaretha Kihlblom typed the manuscript, and Britt Nyberg drafted the illustrations. My research, done at the Department of Historical Geology and Palaeontology in Lund, was financed by the Swedish National Science Research Council. To everyone, my sincere thanks.

REFERENCES

Aldridge, R. J. 1972. Llandovery conodonts from the Welsh Borderland. *Bulletin of the British Museum (Natural History) Geology* 22:125–231.

———. 1974. An *amorphognathoides* Zone conodont fauna from the Silurian of the Ringerike area, south Norway. *Norsk Geologisk Tidskrift* 54:295–303.

———. 1976. Comparison of macrofossil communities and conodont distribution in the British Silurian. In C. R. Barnes, ed., *Conodont Paleoecology. The Geological Association of Canada Special Papers* 15:91–104.

———. 1985. Conodonts of the Silurian system from the British Isles. In A. C. Higgins and R. L. Austin, eds., *A Stratigraphical Index of Conodonts*, pp. 68–92. Chichester: Ellis Horwood.

Aldridge, R. J., L. Jeppsson, and K. J. Dorning. 1993. Early Silurian oceanic episodes and events. *Journal of the Geological Society of London* 150:501–13.

Armstrong, H. A. 1990. Conodonts from the Upper Ordovician–Lower Silurian carbonate platform of North Greenland. *Grøndlands Geologiske Undersøgelse Bulletin* 159:1–151.

Barrick, J. E. 1977. Multielement simple-cone conodonts from the Clarita Formation (Silurian), Arbuckle Mountains, Oklahoma. *Geologica et Palaeontologica* 11:47–68.

———. 1983. Wenlockian (Silurian) conodont biostratigraphy, biofacies, and carbonate lithofacies, Wayne Formation, central Tennessee. *Journal of Paleontology* 57: 208–39.

Barrick, J. E. and G. Klapper. 1976. Multielement Silurian (late Llandoverian-Wenlockian) conodonts of the Clarita Formation, Arbuckle Mountains, Oklahoma, and phylogeny of *Kockelella*. *Geologica et Palaeontologica* 10:59–99.

Berger, A., M. F. Loutre, and V. Dehant. 1989. Influence of the changing lunar orbit on the astronomical frequencies of Pre-Quaternary insolation patterns. *Paleoceanography* 4:555–64.

Berner, R. A. 1990. Atmospheric carbon dioxide levels over Phanerozoic time. *Science* 249:1382–86.

Bischoff, G. C. O. 1986. Early and Middle Silurian conodonts from midwestern New South Wales. *Courier Forschungsinstitut Senckenberg* 89:1–337.

Boucot, A. J. 1991. Developments in Silurian studies since 1839. In M. G. Bassett, P. D. Lane, and D. Edwards, eds., *The Murchison Symposium*. Proceedings of an International Conference on the Silurian System. *Special Papers in Palaeontology* 44:91–107.

Bralower, T. J. and H. R. Thierstein. 1984. Low productivity and slow deep-water circulation in mid-Cretaceous oceans. *Geology* 12:614–18.

Branson, E. B. and M. G. Mehl. 1933. Conodont studies number 1. *University of Missouri Studies* 8:1–72.

Brett, C. E. and J. D. Eckert. 1989. Paleoecology and evolution of pelmatozoan echinoderm associations during the Silurian-Early Devonian interval in eastern North America. *Abstracts of the 29th International Geological Congress* 1:1-197–1-198.

Broecker, W. C. 1983. The ocean. *Scientific American* 249:100–12.

Clark, D. L. 1987. Conodonts: The final fifty million years. In R. J. Aldridge, ed., *Paleobiology of Conodonts*, pp. 165–74. Chichester: Ellis Horwood.

Cooper, B. J. 1976. Multielement conodonts from the St. Clair Limestone (Silurian) of southern Illinois. *Journal of Paleontology* 50:205–17.

———. 1980. Towards an improved Silurian conodont biostratigraphy. *Lethaia* 13:209–27.

Drygant, D. M. 1974. Simple conodonts of the Silurian and lowermost Devonian of the Volyno-Podolian. *Paleontologicheskij Sbornik* 10:64–70. In Russian with an English summary.

———. 1984. Korreljacia i Konodonty Silurijskich-Nizhnedevonskich Otlochenij Volyno—Podolii. *Akademia Nauk Ukrainskoj SSR Institut Geologii i Geochemii Gorinich*. Kiev: Akademia Nauk Ukrainskoj SSR.

Gauffin, S. R. and M. F. Fujiita. 1992. Comments on "The Significance of Coral Reefs as Global Carbon Sinks—Response to Greenhouse" by D. W. Kinsey and D. Hopley. *Palaeogeography, Palaeoclimatology, Palaeoecology* (Global and Planetary Change Section) 97:365–67.

Gray, J., S. Laufeld and A. J. Boucot. 1974. Silurian trilete spores and spore tetrads from Gotland: Their implications for land plant evolution. *Science* 185:260–63.

Hallock, P. 1988. The role of nutrient availability in bioerosion: Consequences to carbonate buildups. *Palaeogeography, Palaeoclimatology, Palaeoecology* 63:275–91.

Hede, J. E. 1921. Gottlands silurstratigrafi. *Sveriges Geologiska Undersökning* C 305:1–100.

———. 1960. The Silurian of Gotland. In G. Regnéll and J. E. Hede, eds., *The Lower Paleozoic of Scania: The Silurian of Gotland.* International Geological Congress, 21st session, *Norden 1960 Guidebook Sweden d.* The Geological Survey of Sweden, Stockholm. Lund 91:44–89.
Helfrich, C. T. 1975. Silurian conodonts from Wills Mountain Anticline, Virginia, West Virginia, and Maryland. *Geological Society of America Special Paper* No. 161.
———. 1980. Late Llandovery-Early Wenlock conodonts from the upper part of the Rose Hill and the basal part of the Mifflintown Formations, Virginia, West Virginia, and Maryland. *Journal of Paleontology* 54: 557–69.
Igo, H. and T. Koike. 1967. Ordovician and Silurian conodonts from the Langkawi Islands, Malaya, part 1. *Geology and Palaeontology of Southeast Asia* 3:1–29.
Jaeger, H. 1976. Das Silur und Unterdevon vom thüringischen Typ in Sardinien und seine regionalgeologische Bedeutung. *Nova Acta Leopoldina neue Folge* 45 (224): 263–99.
Jeppsson, L. 1979. Conodonts. In V. Jaanusson, S. Laufeld, and R. Skoglund, eds., *Lower Wenlock Faunal and Floral Dynamics—Vattenfallet Section, Gotland. Sveriges Geologiska Undersökning* C 762:225–48.
———. 1983. Silurian conodont faunas from Gotland. *Fossils and Strata* 15:121–44.
———. 1987a. Lithological and conodont distributional evidence for episodes of anomalous oceanic conditions during the Silurian. In R. J. Aldridge, ed., *Palaeobiology of Conodonts,* pp. 129–45. Chichester: Ellis Horwood.
———. 1987b. Some thoughts about future improvements in conodont extraction methods. In R. L. Austin, ed., *Conodonts: Investigative Techniques and Applications,* pp. 35, 45–53. Chichester: Ellis Horwood.
———. 1990a. An oceanic model for lithological and faunal changes tested on the Silurian record. *Journal of the Geological Society of London* 147:663–74.
———. 1990b. A climatic and oceanic model for events. *Global Biological Events, Precambrian–Cambrian Event Stratigraphy,* September 25–27, 1990, Oxford. Also in *International Geological Correlation Programme Project 303, Late Precambrian and Cambrian Event Stratigraphy* 3:11H.
———. 1993. Silurian events: The theory and the conodonts. *Proceedings of the Estonian Academy of Sciences* 42:23–27.
———. Manuscript a. The early Silurian Lower and Upper Visby Beds on Gotland and their conodont faunas.
———. Manuscript b. The first identified secundo-primo event?
———. Manuscript c. A new early and midde Wenlock standard conodont zonation.
Jeppsson, L., R. J. Aldridge, and K. J. Dorning. 1995. Wenlock (Silurian) oceanic episodes and events. *Journal of the Geological Society, London* 152:487–98.
Jeppsson, L. and P. Männik. 1993. High resolution correlations between Gotland and Estonia near the base of the Wenlock. *Terra Nova* 5:348–58.
Jeppsson, L., V. Viira, and P. Männik. 1994. Silurian conodont-based correlations between Gotland (Sweden) and Saaremaa (Estonia). *Geological Magazine* 131(2): 201–18.
Kayanne, H., S. Atsushi, and H. Saito. 1995. Diurnal changes in the partial pressure of carbon dioxide in coral reed water. *Science* 267:214–16.
Kleffner, M. A. 1987. Conodont of the Estill Shale and Bisher Formation (Silurian, southern Ohio): Biostratigraphy and distribution. *Ohio Journal of Science* 87 (3):78–89.
Laufeld, S. 1974. Reference localities for paleontology and geology in the Silurian of Gotland. *Sveriges Geologiska Undersökning C* 705:1–172.
Mabillard, J. E. and R. J. Aldridge. 1983. Conodonts from the Coralliferous Group (Silurian) of Marloes Bay, south-west Dyfed, Wales. *Geologica et Palaeontologica* 17:29–43.
———. 1985. Microfossil distribution across the base of the Wenlock Series in the type area. *Palaeontology* 28:89–100.
Männik, P. In press. Evolution of the conodont genus *Pterospathodus* and the *celloni-amor-*

phognathoides Zones in the Silurian of Estonia. The aspects of the phylogeny in palaeontology. *Contributions to the 35th Sessions of the All-union Palaeontological Society,* St. Petersburg. In Russian.

Melchin, M. J. 1994. Graptolite extinction at the Llandovery-Wenlock boundary. *Lethaia* 27:273–362.

Miller, R. H. 1978. Early Silurian to Early Devonian conodont biostratigraphy and depositional environments of the Hidden Valley Dolomite, southeastern California. *Journal of Paleontology* 52:323–44.

Nicoll, R. S. and C. B. Rexroad. 1969. Stratigraphy and conodont palaeontology of the Salamonie Dolomite and Lee Creek Member of the Brassfield Limestone (Silurian) in southeastern Indiana and adjacent Kentucky. Indiana *Geological Survey Bulletin* 40:1–73.

Nowlan, G. S. 1983. Early Silurian conodonts of eastern Canada. *Fossils and Strata* 15:95–110.

Odin, G. S., J. C. Hunziker, L. Jeppsson, and N. Spjeldnaes. 1986. Ages radiométriques K-Ar de biotites pyroclastiques sedimentées dans le Wenlock de Gotland (Suède). In G. S. Odin, guest-ed., *Calibration of the Phanerozoic Time Scale. Chemical Geology (Isotope Geoscience Section)* 59:117–25.

Over, D. J. and D. E. Chatterton. 1987. Silurian conodonts from the southern Mackenzie Mountains, Northwest Territories, Canada. *Geologica et Palaeontologica* 21:1–49.

Palmer, A. T. 1984. The biomere problem: evolution of an idea. *Journal of Paleontology* 58:599–611.

Ramsköld, L. 1985. Studies on Silurian trilobites from Gotland, Sweden. Stockholm. Department of Geology, University of Stockholm, and Department of Palaeozoology, Swedish Museum of Natural History.

Rexroad, C. B. 1967. Stratigraphy and conodont paleontology of the Brassfield (Silurian) in the Cincinnati Arch Area. *Indiana Geological Survey Bulletin* 36:1–71.

Rhodes, F. H. T. 1955. Some British Lower Palaeozoic conodont faunas. *Philosophical Transactions of the Royal Society of London B* 237:261–334, pls. 21–23.

Rickards, R. B., J. E. Hutt, and W. B. N. Berry. 1977. The evolution of the Silurian and Devonian graptoloids. *Bulletin of the British Museum (Natural History), Geology* 28:1–120.

Sandford, J. T. and R. E. Mosher. 1985. Insoluble residues and geochemistry of some Llandoverian and Wenlockian rocks from Gotland. *Sveriges Geologiska Undersökning* C811:1–31.

Savage, N. M. 1985. Silurian (Llandovery-Wenlock) conodonts from the base of the Heceta Limestone, southeastern Alaska. *Canadian Journal of Earth Sciences* 22:711–27.

Stauffer, C. R. 1940. Conodonts from the Devonian and associated clays of Minnesota. *Journal of Paleontology* 14:417–35.

Štorch, P. 1994. Graptolite biostratigraphy of the Lower Silurian (Llandovery and Wenlock) of Bohemia. *Geological Journal* 29:137–165.

Sundquist, E. T. 1993. The global carbon dioxide budget. *Science* 259:934–41.

Tucker, M. E. 1992. The Precambrian-Cambrian boundary: Seawater chemistry, ocean circulation and nutrient supply in metazoan evolution, extinction and biomineralisation. *Journal of the Geological Society,* London 149:655–68.

Underwood, C. J. 1993. The position of graptolites within Lower Palaeozoic planktic ecosystems. *Lethaia* 26:189–205.

Walliser, O. H. 1964. Conodonten des Silurs. *Abhandlungen des Hessischen Landesamtes für Bodenforschung* 41:1–106.

18

Late Devonian Bioevents in the Appalachian Sea: Immigration, Extinction, and Species Replacements

George R. McGhee, Jr.

ABSTRACT

A "Great Interchange" of marine species occurs in the Middle and Late Devonian world, comparable to the Pleistocene interchange of terrestrial species between North and South America. The interchange is also of unequal directionality in the Devonian: In most cases Old World Realm species successfully immigrate into North America, while few species of the Eastern Americas Realm successfully migrate to the west. Particularly affected are the dominant spiriferacean brachiopods of North America, which become extinct and are replaced by Old World immigrants. Reanalysis of this presumed example of competitive replacement of ecologically equivalent species reveals a more complex pattern of interactions than the simple model of extermination of local endemic faunas by more cosmopolitan immigrants.

Introduction: The Great Devonian Interchange

The Early Devonian world is characterized by extensive biological provinciality and localized endemic faunas. Early Devonian marine faunas are separated into three major biogeographic realms—the Old World Realm, the Eastern Americas Realm, and the cooler southern Malvinokaffric Realm (Boucot et al. 1968; Oliver 1977; Boucot 1988). Within these realms exist localized smaller provincial biogeographic regions of endemic species (Oliver 1990).

The Early Devonian world is in stark contrast to the Late Devonian world, where a cosmopolitan fauna spans the globe. The spread of cosmopolitan marine faunas coincides with the progressive breaching of land barriers with rising sea levels from the Early to Late Devonian. The breakdown of the Eastern Americas Realm and its provinces is seen as due to the breaching (perhaps in several phases) and eventual elimination of the Transcontinental Arch, which previously was a barrier to migration between western and eastern marine areas in North America (Johnson 1970; Oliver 1977, 1990).

Of considerable interest is the unequal directionality of the species interchange that follows the late Givetian to early Frasnian migration events in North America. In the great majority of cases species of the Old World Realm in western North America successfully immigrate into and proliferate in seas in eastern North America, while very few species of the Eastern Americas Realm successfully migrate to the west (Oliver 1977, 1990; Oliver and Pedder 1989). Instead, many of the endemic Eastern Americas Realm species become extinct and are replaced by immigrant species of the Old World Realm (Boucot 1988).

In a previous study of the comparative ecology of benthic marine communities in the Appalachian Basin during this critical interval of time it was concluded that community ecological structures remained relatively constant during the Frasnian but that major changes occurred in community species compositions (McGhee 1981). The most important of these changes is the total replacement of the dominant spiriferacean brachiopods of the early Frasnian by their ecological equivalents in the late Frasnian, within a framework of persistent or stable community structures.

At that time it was considered that a series of species ecological relays was exhibited by the replacement of the Givetian species *Mucrospirifer mucronatus* by its ecologically equivalent Frasnian colonist species *Tylothyris mesacostalis* (fig. 18.1A) and by the similar replacement of *Orthospirifer mesastrialis* by *Cyrtospirifer chemungensis* (fig. 18.1B). The demise of *M. mucronatus* and *O. mesastrialis* was viewed as another possible example of the decline of endemic species of the Eastern Americas Realm following the incursion of Old World Realm immigrant species, although the evolutionary impact of immigration was not the focus of the previous study and the possibility of purely passive replacement was seen as an alternative explanation (McGhee 1981:280).

Further data now indicate that the observed Frasnian replacement in dominant spiriferacean species is not quite as simple as the scenario of endemic species extinctions following the immigration of more cosmopolitan colonist species. Interactions between the species affected appear to have been more complex than a series of one-on-one relays and are the subject of the present study.

The Present Study

The previous study was based upon only three data sets: the studies of the New York Genesee Group by Thayer (1974), the Sonyea Group by Sutton et al.

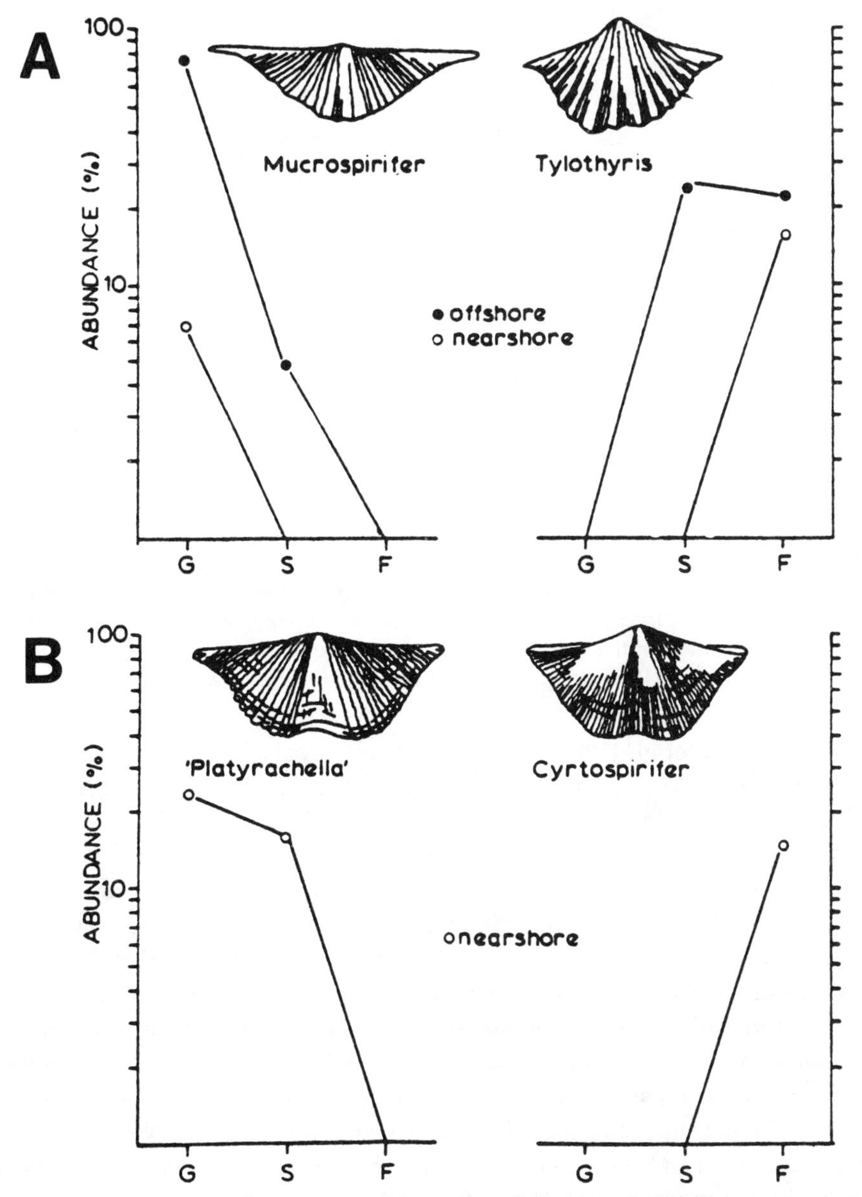

FIGURE 18.1. Patterns of extinction of Givetian spiriferacean holdover species, and the reciprocal expansion of morphologically similar spiriferacean colonist species, in the Frasnian of the eastern United States. **(A)** Givetian *Mucrospirifer mucronatus* versus Frasnian *Tylothyris mesacostalis,* and **(B)** Givetian *Orthospirifer (= "Platyrachella") mesastrialis* versus Frasnian *Cyrtospirifer chemungensis.* Abbreviations: G = Genesee Group, S = Sonyea Group, F = Foreknobs Formation. Modified from McGhee (1981).

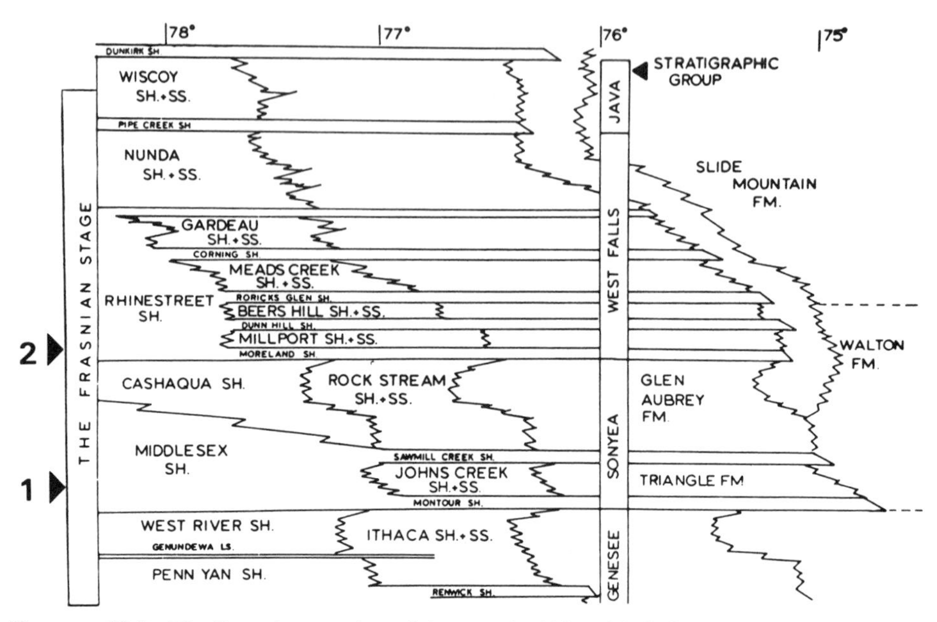

FIGURE 18.2. The Frasnian stratigraphic record of New York State. First occurrences of the colonist species *Tylothyris mesacostalis* marked by the arrow headed "1," and of the colonist species *Cyrtospirifer chemungensis* by the arrow headed "2." See text for discussion of the delimitation of the stratigraphic boundaries of the Frasnian in New York. Modified from Rickard (1975) and Sutton and McGhee (1985).

(1970) and Bowen et al. (1974), and the central Appalachian Foreknobs Formation by McGhee (1976). In the intervening decade many more studies have been conducted that allow more complete coverage of the Frasnian interval of time, particularly in New York.

The Frasnian stratigraphic nomenclature of New York State is summarized in figure 18.2. Traditional New York State usage considered the Genesee Group to be entirely within the Frasnian (Rickard 1975). The subsequent decision by the ***IUGS*** Subcommission on Devonian Stratigraphy (Ziegler and Klapper 1985) to place the formal base of the Frasnian at the base of the Lower *asymmetricus* conodont Zone results in the placement of only the upper part of the Genesee Group within the Frasnian. The detailed work of Kirchgasser et al. (1988) has narrowed the placement of the Givetian/Frasnian boundary in New York to be within the Penn Yan Shale Member of the Genesee Group, below the Renwick Shale but above the Lodi Limestone (fig. 18.2).

The base of the Famennian has traditionally been placed in New York State at the base of the Dunkirk Black Shale (Rickard 1975). The ***IUGS*** Subcommission on Devonian Stratigraphy's decision to place the base of the Famennian at the base of the Lower *triangularis* conodont Zone moves the Frasnian/Famennian boundary below the Dunkirk Black Shale horizon (McGhee 1990) to within the Java Group of Fisher et al. (1970). The exact placement of the Java Group (fig. 18.2) awaits the further work of biostratigraphic specialists.

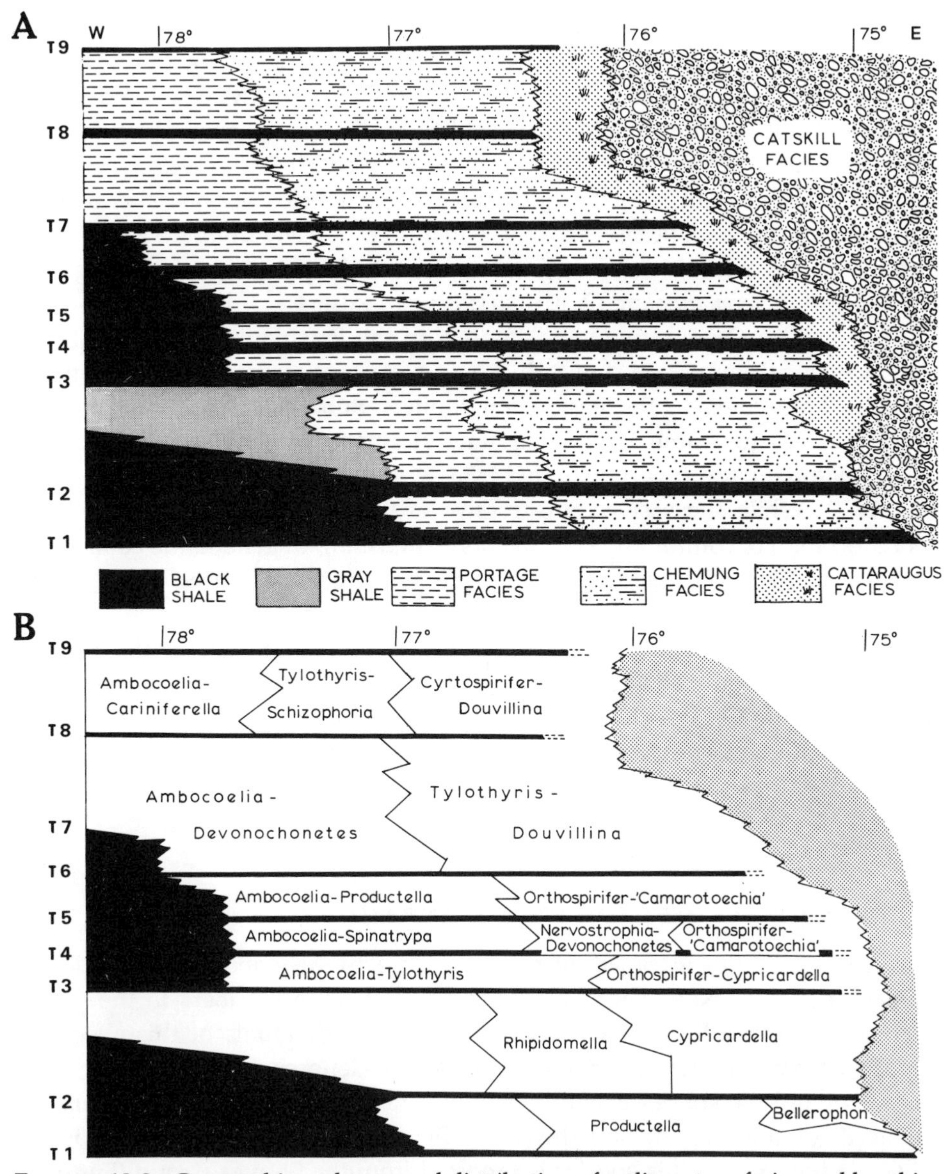

FIGURE 18.3. Geographic and temporal distribution of sedimentary facies and benthic marine communities in the Sonyea, West Falls, and Java Groups (see fig. 18.2 for the names of stratigraphic intervals). The very lowest Frasnian (the upper Genesee Group, see fig. 18.2) is here not included as faunal data were not available for this study. **(A)** Black shale facies represent deepest water conditions, gray shale, Portage, and Chemung facies represent progressively shallower water conditions, Cattaraugus facies are shore sediments, and Catskill facies are nonmarine terrestrial. The continuous progradation of facies to the west was punctuated by several transgressive events (marked "T" in the left margin). **(B)** Communities of benthic shellfish species during the Frasnian. See text for discussion. From McGhee et al. (1990).

Paleoecological analyses of marine fauna and environments have now been completed for the Genesee Group (Thayer 1974), the Sonyea Group (Sutton et al. 1970; Bowen et al. 1974; MacDonald 1975), the lower West Falls Group (McGhee and Sutton 1985), the upper West Falls Group (McGhee and Sutton 1983), and the Java Group (McGhee and Sutton 1981). A general synopsis of Frasnian marine paleoecology is given in Sutton and McGhee (1985).

The first appearances of the Frasnian colonist species *Tylothyris mesacostalis* and *Cyrtospirifer chemungensis* are marked stratigraphically in figure 18.2. By the end of the Frasnian these two species were the dominant spiriferaceans, and the older *M. mucronatus* and *O. mesastrialis* were extinct.

The overall paleoenvironmental context of the Frasnian Appalachian Sea is one of regional regression due to influx of terrestrial sediments shed from the Acadian Mountains to the east, even though global sea level was probably rising during this same interval of time (Johnson et al. 1985). This overall progradational pattern is punctuated by a series of smaller-scale transgressions (fig. 18.3). While successive transgressive events do appear to produce some reorganization of community compositions (fig. 18.3), in general the community structures of benthic species assemblages remain static during the Frasnian (Sutton and McGhee 1985).

Within the preceding outlined paleoenvironmental context for the Appalachian Sea is introduced the additional biological factor of new species immigrants from the Old World Realm, and the disappearance of the last remnant holdovers of the Eastern Americas Realm endemic species.

The Givetian Holdovers

Mucrospirifer mucronatus (Conrad)

The brachiopod *Mucrospirifer mucronatus* is an extremely abundant Middle Devonian species in New York, often occurring in large numbers to the virtual exclusion of other fossils. A count of 8,885,235 individuals of this species is reported from the late Givetian/early Frasnian Genesee Group (Thayer 1974, table 1A), and the species constitutes fully 75% of the number of individuals in offshore communities during this time (fig. 18.1A). An approximate total of 88,000 individuals of this species is reported from the Sonyea Group (Sutton et al. 1970), a drop of two orders of magnitude. While these differences in numbers no doubt also reflect differing sample sizes, proportional numbers indicate that the species has clearly declined from 75% to only 5% of the total number of individuals in the offshore communities (fig. 18.1A).

The stratigraphic and geographic distribution of *M. mucronatus* in New York is given in figure 18.4A. The species is quite widespread in the Triangle Formation of the Sonyea Group (cf. figs. 18.4A and 18.2), occurring in both nearshore and offshore habitats. Its frequency of occurrence drops markedly in the Glen Aubrey Formation, such that it constitutes only 5% of the individuals of the

Rhipidomella community (cf. figs. 18.4A and 18.3B), as reported by Bowen et al. (1974). Its stratigraphically highest occurrences are in the nearshore *Orthospirifer-"Camarotoechia"* community, and the species is no longer present above the Corning Shale.

Orthospirifer mesastrialis (Hall)

The species *Orthospirifer mesastrialis* (formerly *"Platyrachella" mesastrialis*, see Pitrat 1975) is also a characteristic late Givetian species in central and eastern North America (Pitrat 1975). A count of 3,510,925 individuals of this species is reported from the late Givetian/early Frasnian Genesee Group, where the species is the third most abundant fossil taxon (Thayer 1974, table 1A). Only 22,000 individuals of the species are estimated to occur in the Sonyea Group (Sutton et al. 1970), and it declines from 23% to 16% of the number of individuals in nearshore communities during this interval of time (fig. 18.1B). By the end of the Frasnian it is extinct.

Further data reveal a somewhat more complex stratigraphic and geographic distribution of *O. mesastrialis* than previously thought (fig. 18.4B). Rather than simply dwindling away with time, the species actually expands its geographic range from the Sonyea Group into the West Falls Group (cf. figs. 18.4B and 18.2) and constitutes a dominant element in the nearshore *Orthospirifer-Cypricardella* and *Orthospirifer-"Camarotoechia"* communities in the lower West Falls Group (cf. figs. 18.2 and 18.3B). Only then does it decline in distribution, and the species does not survive into the Java Group.

The Frasnian Colonists

Cyrtospirifer chemungensis (Conrad)

The Cyrtospiriferidae are extremely abundant worldwide, and the genus *Cyrtospirifer* itself is commonly used as a Frasnian index fossil (Dineley 1984). In Europe the base of the *Assise de Fromelennes* coincides with the "sudden entry of the cyrtospiriferids" (Ziegler 1979). This horizon is now actually within the Givetian, occurring in the upper part of the *varcus* conodont Zone.

Cyrtospirifer cannot be used as a "Frasnian" index fossil in the area of the Eastern Americas Realm, as species of this genus do not migrate into the central Appalachians and New York until the mid-Frasnian (fig. 18.2). *Cyrtospirifer chemungensis* first appears in New York at the base of the West Falls Group (cf. figs. 18.4C and 18.2). The species *C. chemungensis* itself was once thought to have evolved in the Appalachian Basin (Greiner 1957), but it is now known to be an immigrant from the Canadian west (Greiner 1973). *C. chemungensis* is perhaps the very best brachiopod example of the successful proliferation (fig. 18.4C) of an Old World Realm species immigrant in the formerly isolated Appalachian Basin. It is a dominant element in nearshore communities in the latest Frasnian (fig. 18.3B) and survives the Frasnian/Famennian mass extinction (McGhee 1989) to proliferate further in the Famennian.

Tylothyris mesacostalis (Hall)

Tylothyris presents somewhat of an enigma in the Eastern Americas Realm area. The type species of the genus is a Carboniferous form, and the genus is considered cosmopolitan (Pitrat 1965). Occurrences of the genus older than the Frasnian are considered questionable by Pitrat (1965). Johnson (1971), however, lists *Tylothyris* as an Appalachian endemic in the Givetian. *Tylothyris pauliformis* is listed in the Mahantango Formation (Givetian) of the central Appalachians by Ellison (1974), yet Dutro (1981) indicates that *pauliformis* is only questionably assignable to the genus *Tylothyris*.

Outside of the Appalachian Basin, the genus is reported to be present in Iowa in the late Givetian (Day 1988). The oldest occurrences of the species *Tylothyris mesacostalis* (under the names of *Spirifer [Delthyris] mesacostalis*) listed by James Hall (1867), author of the species, are from the "Chemung beds at Ithaca," which are mapped today as the Ithaca Formation (Fisher et al. 1970) of the upper Genesee Group (see fig. 18.2), and hence of early Frasnian age, yet the species is not listed in the Ithaca Formation in the comprehensive study of Thayer (1974).

Tylothyris mesacostalis is the eighth most abundant species reported from the Sonyea Group by Sutton et al. (1970), and Dutro (1981) considers the species so characteristic of the Sonyea horizon that this interval was designated the *Tylothyris mesacostalis* brachiopod Zone.

T. mesacostalis appears first in two sections in the Triangle Formation of the Sonyea Group (cf. figs. 18.4D and 18.2), expands geographically in the Glen Aubrey Formation, and is a dominant element in nearshore and mid-platform communities in the later Frasnian (fig. 18.3B).

The exact origin of the species remains unclear. Did it evolve within the Appalachian Basin in the early Frasnian, perhaps descended from immigrants of older Givetian species located to the west in Iowa? Iowa is still within the Eastern Americas Realm area. Thus is *Tylothyris* a rare example of an originally derived Eastern Americas Realm taxon that successfully migrated *out* to become of worldwide distribution?

Analysis of Species Interactions

Tests of Association

Abundance data for species occurring at each section indicated in figure 18.4 were originally tabulated using a logarithmic scale with an integer format of 1 (1–10 individuals), 2 (10–100), 3 (100–1,000), and 4 (more than 1,000 individuals; Sutton et al. 1970; McGhee and Sutton 1981). Correlation coefficients be-

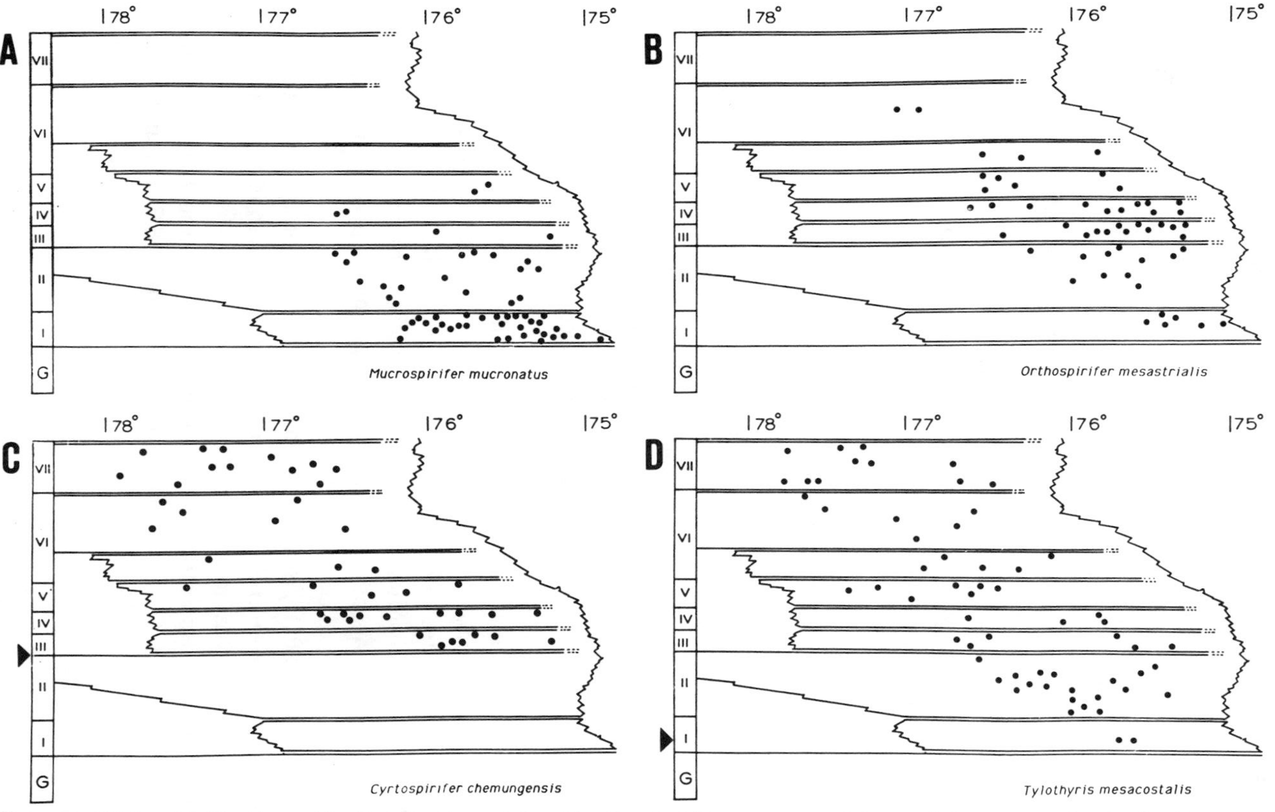

FIGURE 18.4. **(A)** Stratigraphic distribution of the Givetian holdover species *Mucrospirifer mucronatus* in the Sonyea (I, II), West Falls (III -VI), and Java (VII) Groups (see fig. 18.2 for the names of stratigraphic intervals). Section data not available for the upper Genesee Group (G). Stratigraphic distribution of sections from Sutton and McGhee (1985, fig. 4). **(B)** Stratigraphic distribution of the Givetian holdover species *Orthospirifer mesastrialis* in the Sonyea, West Falls, and Java Groups. **(C)** Stratigraphic distribution of the Frasnian colonist species *Cyrtospirifer chemungensis* in the Sonyea, West Falls, and Java Groups. **(D)** Stratigraphic distribution of the Frasnian colonist species *Tylothyris mesacostalis* in the Sonyea, West Falls, and Java Groups.

tween all possible two-species permutations were then calculated (table 18.1) using these data in a log-transformed format.

Quite surprisingly, *Cyrtospirifer chemungensis* positively correlates with all three of the other species (table 18.1). From the previous study (McGhee 1981) it was expected that a negative correlation would exist between the Eastern Americas Realm endemic species *Orthospirifer mesastrialis* and its ecological equivalent, the Old World Realm immigrant species *Cyrtospirifer chemungensis,* which was seen as replacing *O. mesastrialis* with time. Although the term *ecological replacement,* in the strict sense, describes the historical event of one species being succeeded by its ecological equivalent in time and does not imply a causal mechanism such as that implied by *competitive displacement* (McGhee 1981), a negative association was expected. It does not exist. Instead the largest *positive* correlation that *C. chemungensis* exhibits is with its presumed competitor *O. mesastrialis.*

An expected negative correlation (if one assumes competitive displacement) does exist for the other supposed species relay, *Mucrospirifer mucronatus* versus *Tylothyris mesacostalis.* Surprisingly again, however, a negative correlation also exists *between* the Givetian species *M. mucronatus* and *O. mesastrialis* (table 18.1).

Significance of species geographic associations were further explored by non-parametric chi-square techniques. The data were transformed to binary presence/absence format, and contingency tables constructed tabulating the number of sections where the species pairs occur with one another, where one species occurs and the other does not, and where neither occurs. The results of these calculations are given in tables 18.2 and 18.3.

A significant geographic association is exhibited by the species pair *M. mucronatus* and *T. mesacostalis* (table 18.2). This association is also negative (table 18.3), in that sections where one species is present and the other is not are overrepresented (positive residuals), and sections where both species are present are underrepresented (negative residuals), from chance expectations alone.

A weak association is exhibited by the species pair *O. mesastrialis* and *C. chemungensis* ($0.05 > p > 0.01$). Again, however, this association is *positive* (table 18.3). The species occur with one another more frequently than is expected by chance alone (positive residual for co-occurrences, table 18.3).

Interestingly, negative association is again seen here for the two Givetian species *M. mucronatus* and *O. mesastrialis* (tables 18.2 and 18.3), though it is

TABLE 18.1 *Pair-wise comparisons of species correlations*

	Correlation Coefficient			
	A	B	C	D
Mucrospirifer mucronatus	A 1			
Orthospirifer mesastrialis	B −0.166	1		
Tylothyris mesacostalis	C −0.275	−0.040	1	
Cyrtospirifer chemungensis	D +0.041	+0.264	+0.098	1

Table 18.2 *Chi-square tests of species associations*

Species pair	N	Chi-square	Probability
Tylothyris mesacostalis vs *Mucrospirifer mucronatus*	176	26.76	$p<0.001$
Cyrtospirifer chemungensis vs *Orthospirifer mesastrialis*	92	5.53	$p=0.019$
Orthospirifer mesastrialis vs *Mucrospirifer mucronatus*	176	4.17	$p=0.041$
Tylothyris mesacostalis vs *Orthospirifer mesastrialis*	204	0.32	$p=0.572$
Tylothyris mesacostalis vs *Cyrtospirifer chemungensis*	109	0.06	$p=0.802$
Cyrtospirifer chemungensis vs *Mucrospirifer mucronatus*	64	0.00	$p=0.955$

Table 18.3 *Standardized residuals for the first three Chi-square tests given in table 18.2*

	Species occurrences:			
Species pairs	Neither	1 alone	2 alone	Both
1 *Tylothyris mesacostalis* 2 *Mucrospirifer mucronatus*	−1.20	+2.28	+1.71	−3.26
1 *Cyrtospirifer chemungensis* 2 *Orthospirifer mesastrialis*	+0.85	−1.19	−1.08	+1.52
1 *Orthospirifer mesastrialis* 2 *Mucrospirifer mucronatus*	−0.59	+0.98	+0.84	−1.39

statistically "weak" ($0.05 > p > 0.01$). All other possible two-pair permutations of the species concerned are insignificant (table 18.2).

Species Geographic Distributions and Abundances

Relationships between the species utilizing their relative abundances (in addition to geographic occurrences, as earlier) were explored with linear regressions (figs. 18.5, 18.6, and 18.7). Only the three top species pairings given in table 18.2 were treated. The Givetian holdover *M. mucronatus* and the Frasnian colonist *T. mesacostalis* exhibit a near perfect negative association (i.e., a potential slope of −1.0) with one another (fig. 18.5).

The regression function between *O. mesastrialis* and *C. chemungensis* is negative here but not significant (fig. 18.6). The two species frequently co-occur. Moreover, the abundance of individuals of one species does not appear to be a function of the abundance of the other.

Interestingly, a strong negative association is revealed between the two Givetian holdover species during the Frasnian (fig. 18.7). The abundances of the two species do appear to be a function of each other—where *M. mucronatus* is abundant, *O. mesastrialis* is not, and vice versa.

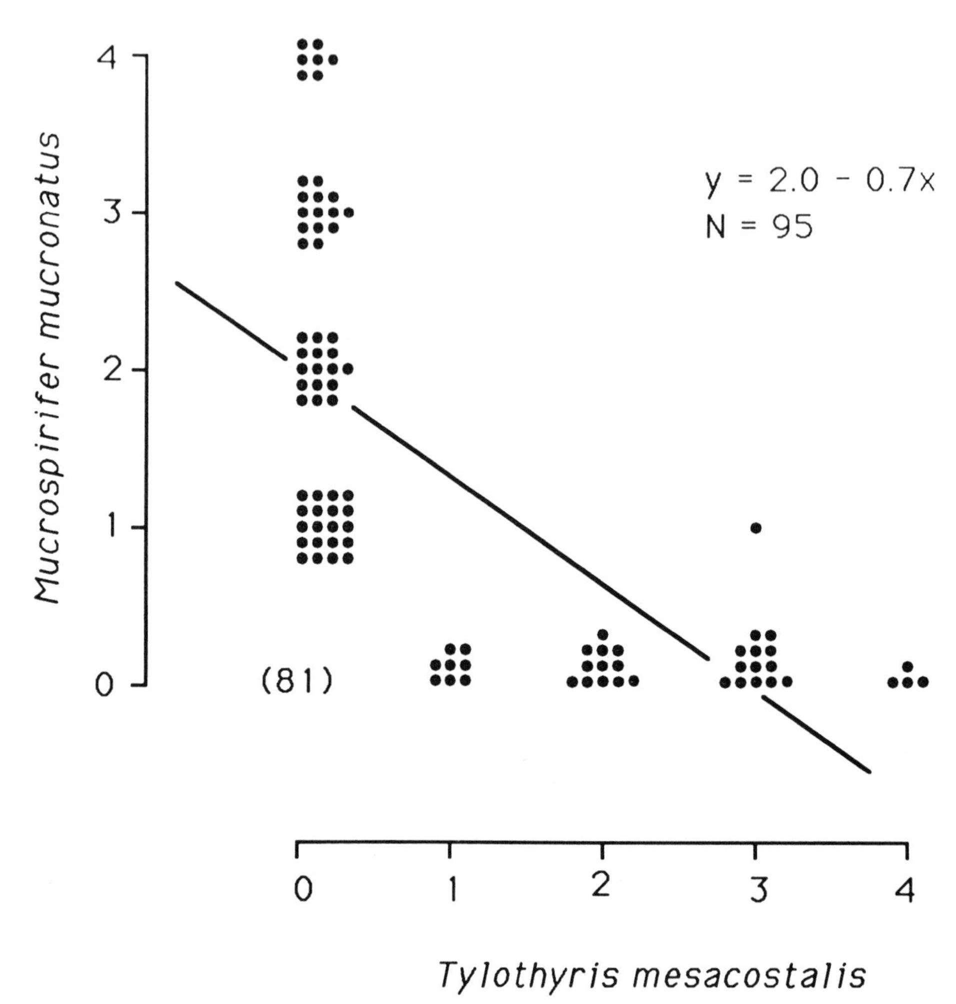

Figure 18.5. Species abundances and co-occurrences of *Mucrospirifer mucronatus* and *Tylothyris mesacostalis*. Each point represents a section, abundances of each species are given in integers on a logarithmic scale (see text for discussion). The number of sections in which neither species is present is given in parentheses at the zero-zero locus. Axes are displaced for purposes of clarity. Regression function for the two species given by the line in the figure and equation in the upper right corner ($F = 89.23$, $p < 0.001$). The two species exhibit a near perfect negative relationship with one another.

What might originally be viewed as a classic brachiopod example of the ecological replacement of Eastern Americas Realm endemic species under the invasion into the Appalachian Sea of new cosmopolitan species immigrants during the passage of the Frasnian appears to be more complex, given further analysis.

If the Old World Realm immigrant *Cyrtospirifer chemungensis* can be said in any sense to have replaced the Eastern Americas Realm endemic species *O. mesastrialis*, it can only be in a passive sense. The two species positively co-occur, indicating they probably preferred similar marine conditions (substrate, water depth, etc.), yet seemingly were not competitors. Though the two species appear to have been ecologically equivalent (morphologically very similar, taxonomically related, identical trophic adaptations, etc.), no evidence of competitive exclusion exists between the two. Whatever eventually caused the demise of *O. mesastrialis* appears to have been unrelated to the appearance of the immigrant *C. chemungensis*.

A strong case can be made for adverse interactions between the Givetian holdover species *Mucrospirifer mucronatus* and the Frasnian colonist species *Tylothyris mesacostalis*. A nearly perfect negative association exists between the two ecologically equivalent species. The two species clearly overlapped in habitat preferences, yet where one species is found the other is generally absent (unlike *O. mesastrialis* and *C. chemungensis*).

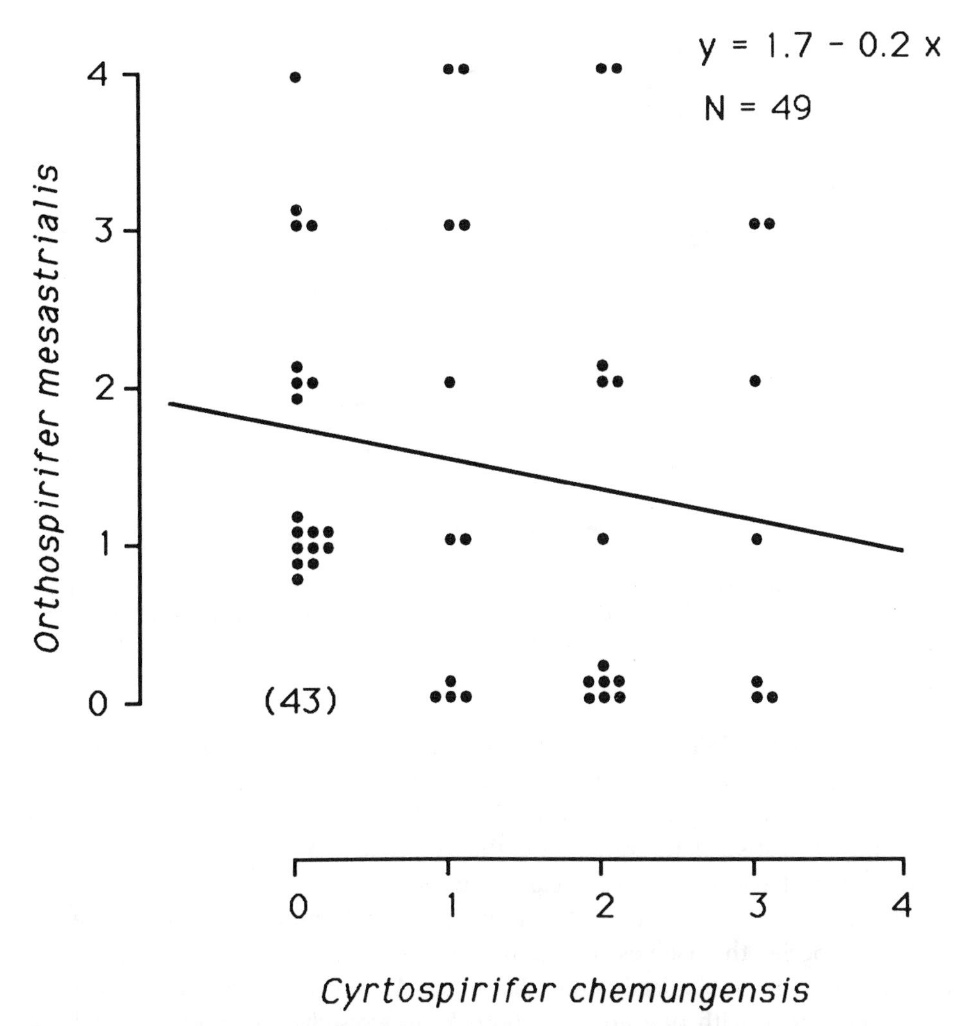

FIGURE 18.6. Species abundances and co-occurrences of *Orthospirifer mesastrialis* and *Cyrtospirifer chemungensis*. Format identical to figure 18.5. The regression function for the two species is negative, but is not significant ($F = 1.30$, $p = 0.26$).

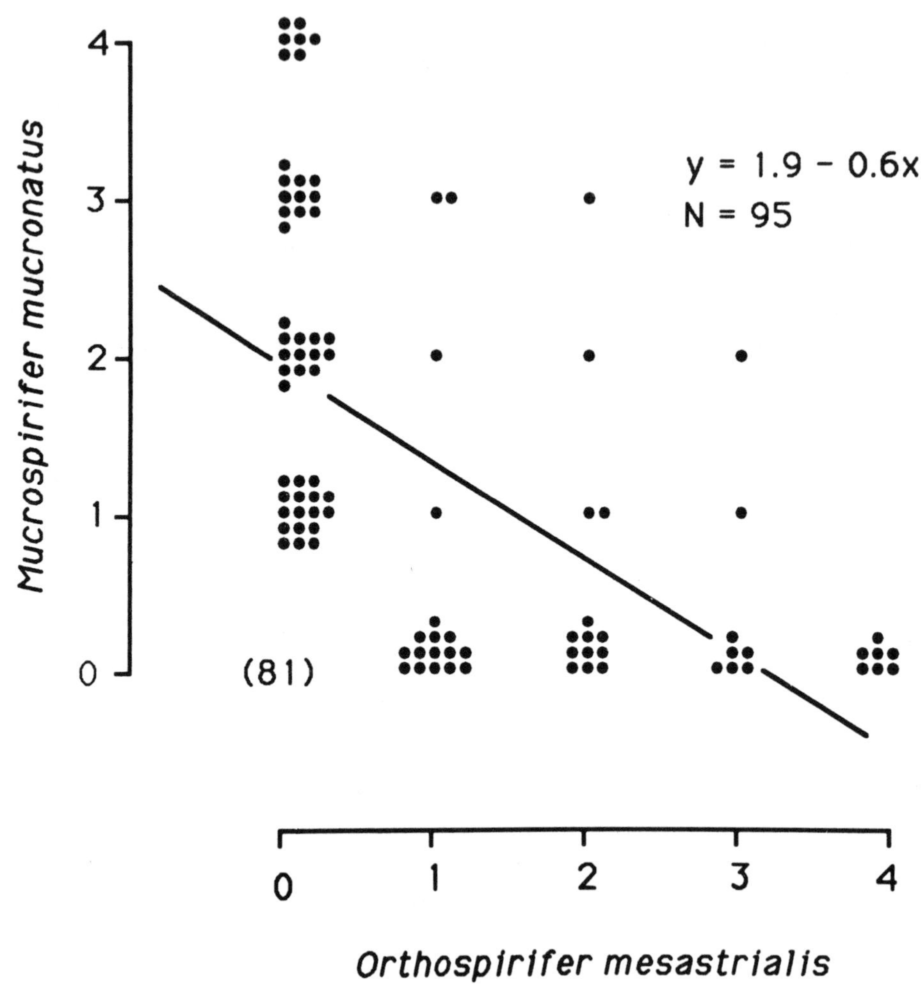

FIGURE 18.7. Species abundances and co-occurrences of *Mucrospirifer mucronatus* and *Orthospirifer mesastrialis*. Format identical to figure 18.5. The two Givetian species exhibit a negative relationship with one another in the regression function ($F = 50.17$, $p < 0.001$).

The enigmatic origin of *T. mesacostalis* does not support the Old World Realm versus Eastern Americas Realm scenario, however. A strong possibility exists that *T. mesacostalis* originated within the Eastern Americas Realm itself, a descendant of eastward migrants into the Appalachian Sea from the western margin of the Eastern Americas Realm (e.g., Iowa).

Finally, a case can be made for adverse interactions between the Givetian holdover species themselves, independent of any impact of new colonist species. Though morphologically quite dissimilar, *M. mucronatus* and *O. mesastrialis* negatively occur with one another. Both *T. mesacostalis*, generally in offshore regions, and *O. mesastrialis*, generally in near-shore regions, have negative relationships with *M. mucronatus*—while these same two species appear to have a totally neutral response to each other.

In general, these conclusions support the thesis of Oliver (1990) that the late Givetian and early Frasnian periods of high extinctions (Bayer and McGhee 1986) were more complex than previously thought, and surely more complex than the simple extermination of local endemic faunas by more cosmopolitan immigrants, triggered by late Givetian marine transgressions.

ACKNOWLEDGMENTS

This is Contribution #92-05 of the Institute of Marine and Coastal Sciences, Rutgers University.

REFERENCES

Bayer, U. and G. R. McGhee. 1986. Cyclic patterns in the Paleozoic and Mesozoic: Implications for time scale calibrations. *Paleoceanography* 1:383–402.

Bowen, Z. P., D. C. Rhoads, and A. L. McAlester. 1974. Marine benthic communities in the Upper Devonian of New York. *Lethaia* 7:93–120.

Boucot, A. J. 1988. Devonian biogeography: An update. In N. J. McMillan, A. G. Embry, and D. J. Glass, eds., *Devonian of the World*, vol. 3, pp. 211–27. Calgary: *Canadian Society of Petroleum Geologists Memoir* No. 14.

Boucot, A. J., J. G. Johnson, and J. A. Talent. 1968. Lower and Middle Devonian faunal provinces based on Brachiopoda. In D. H. Oswald, ed., *International Symposium on the Devonian System*, vol. 2, pp. 1239–54. Calgary: Canadian Society of Petroleum Geologists.

Day, J. 1988. The brachiopod succession of the late Givetian–Frasnian of Iowa. In N. J. McMillan, A. G. Embry, and D. J. Glass, eds., *Devonian of the World*, vol. 3, pp. 303–25. Calgary: *Canadian Society of Petroleum Geologists Memoir* No. 14.

Dineley, D. L. 1984. *Aspects of a Stratigraphic System: The Devonian*. New York: Wiley.

Dutro, J. T. 1981. Devonian brachiopod biostratigraphy of New York State. In W. A. Oliver and G. Klapper, eds., *Devonian Biostratigraphy of New York*, pp. 67–82. Washington, D.C.: International Union of Geological Sciences.

Ellison, R. L. 1974. Stratigraphy and paleontology of the Mahantango Formation in south-central Pennsylvania. *Pennsylvania Geological Survey General Geology Report* 48:1–298.

Fisher, D. W., Y. W. Isachsen, and L. V. Rickard. 1970. *Geologic Map of New York State. New York State Museum and Science Service, Map and Chart Series* No. 15.

Greiner, H. 1957. *"Spirifer disjunctus"*: Its evolution and paleoecology in the Catskill Delta. *Peabody Museum of Natural History Bulletin* 11:1–75.

———. 1973. Upper Devonian *Cyrtospirifer* and related genera of the Canadian West and a provisional comparison with those from the Appalachians. *Palaeogeography, Palaeoclimatology, Palaeoecology* 13:129–41.

Hall, J. 1867. Descriptions and figures of the fossil Brachiopoda of the Upper Helderberg, Hamilton, Portage and Chemung Groups. *New York State Geological Survey, Paleontology* 4(1):1–428.

Johnson, J. G. 1970. Taghanic onlap and the end of North American Devonian provinciality. *Geological Society of America Bulletin* 81:2077–2106.

———. 1971. A quantitative approach to faunal province analysis. *American Journal of Science* 270:257–80.

Johnson, J. G., G. Klapper, and C. A. Sandberg. 1985. Devonian eustatic fluctuations in Euramerica. *Geological Society of America Bulletin* 96:567–87.

Kirchgasser, W. T., G. C. Baird, and C. E. Brett. 1988. Regional placement of Middle/Upper Devonian (Givetian–Frasnian) boundary in western New York State. In N. J. McMillan, A. G. Embry, and D. J. Glass, eds., *Devonian of the World*, vol. 3, pp. 113–17. Calgary: *Canadian Society of Petroleum Geologists Memoir* No. 14.
MacDonald, K. B. 1975. Quantitative community analysis: Recurrent group and cluster techniques applied to the fauna of the Upper Devonian Sonyea Group, New York. *Journal of Geology* 82:473–99.
McGhee, G. R. 1976. Late Devonian benthic marine communities of the central Appalachian Allegheny Front. *Lethaia* 9:111–36.
———. 1981. Evolutionary replacement of ecological equivalents in Late Devonian benthic marine communities. *Palaeogeography, Palaeoclimatology, Palaeoecology* 34:267–83.
———. 1989. The Frasnian-Famennian extinction event. In S. K. Donovan, ed., *Mass Extinctions: Processes and Evidence*, pp. 133–51. New York: Columbia University Press.
———. 1990. The Frasnian-Famennian mass extinction record in the eastern United States. In E. G. Kauffman and O. H. Walliser, eds., *Extinction Events in Earth History*, pp. 161–68. Berlin: Springer-Verlag.
McGhee, G. R., U. Bayer, and A. Seilacher. 1990. Biological and evolutionary responses to transgressive-regressive cycles. In G. Einsele, W. Ricken, and A. Seilacher, eds., *Cycles and Events in Stratigraphy*, pp. 696–708. Berlin: Springer-Verlag.
McGhee, G. R. and R. G. Sutton. 1981. Late Devonian marine ecology and zoogeography of the central Appalachians and New York. *Lethaia* 14:27–43.
———. 1983. Evolution of late Frasnian (Late Devonian) marine environments in New York and the central Appalachians. *Alcheringa* 7:9–21.
———. 1985. Late Devonian marine ecosystems of the lower West Falls Group in New York. *Geological Society of America Special Paper* No. 201, pp. 199–209.
Oliver, W. A. 1977. Biogeography of Late Silurian and Devonian rugose corals. *Palaeogeography, Palaeoclimatology, Palaeoecology* 22:85–135.
———. 1990. Extinctions and migrations of Devonian rugose corals in the Eastern Americas Realm. *Lethaia* 23:167–78.
Oliver, W. A. and A. E. Pedder. 1989. Origins, migrations, and extinctions of Devonian Rugosa on the North American Plate. *Memoirs of the Association of Australasian Palaeontologists* 8:231–37.
Pitrat, C. W. 1965. Spiriferidina. In R. C. Moore, ed., *Treatise on Invertebrate Paleontology, Part H, Brachiopoda*, H667–H728. Lawrence: University of Kansas Press.
———. 1975. *Orthospirifer*: new genus of Devonian spinocyrtiid brachiopods. *Journal of Paleontology* 49:387–94.
Rickard, L. V. 1975. Correlation of the Silurian and Devonian rocks in New York State. *New York State Museum and Science Service, Map and Chart Series* 24.
Sutton, R. G., Z. P. Bowen, and A. L. McAlester. 1970. Marine shelf environments of the Upper Devonian Sonyea Group of New York. *Geological Society of America Bulletin* 81:2975–92.
Sutton, R. G. and G. R. McGhee. 1985. The evolution of Frasnian marine "community-types" of south-central New York. *Geological Society of America Special Paper* No. 201, pp. 211–24.
Thayer, C. W. 1974. Marine paleoecology in the Upper Devonian of New York. *Lethaia* 7:121–55.
Ziegler, W. 1979. Historical subdivisions of the Devonian. *Special Papers in Palaeontology* 23:23–47. London: The Palaeontological Association.
Ziegler, W. and G. Klapper. 1985. Stages of the Devonian System. *Episodes* 8:104–109.

19

Regional Encrinites: A Vanished Lithofacies

William I. Ausich

ABSTRACT

Regional encrinites are a vanished lithofacies defined as crinoidal grainstones and packstones that are composed of more than 50% pelmatozoan debris by volume. Minimum dimensions of a regional encrinite are an average stratigraphic thickness more than 5–10 m and an areal extent of more than 500 km^2. Regional encrinites represent the nearly complete domination of an entire shelf by crinoids (and in some cases other stalked echinoderms) for extensive periods of time. This lithofacies is only known to have occurred from the Ordovician through the Jurassic, and the acme of development of this lithofacies was during the Lower Mississippian. Details of the Burlington-Keokuk limestone are given as an example of the regional encrinite lithofacies.

Pelmatozoan echinoderms, especially crinoids, were important carbonate sediment producers in open marine facies throughout much of the Paleozoic and Mesozoic (Wilkinson 1979). Crinoids in particular thrived in numerous shallow-water carbonate settings from reef flanks (Lane 1971; Brett 1985) to open platform habitats to siliciclastic-dominated settings (Ausich et al. 1979; Brett 1984; Carozzi and Gerber 1985; Ausich and Meyer 1990). In these settings crinoidal limestones developed in relatively restricted microfacies, such as reef flank beds, hydraulically sorted components of tempestites, or crinoidal buildups.

The number of pelmatozoan individuals necessary to account for the skeletal debris forming crinoid reef flank beds is considerable. However, most crinoidal limestones are very small in extent when compared to certain thick and geographically extensive crinoidal limestone epiboles, such as the Lower Mississippian Burlington-Keokuk limestone. The Burlington-Keokuk limestone represents an immense number of crinoid individuals[1] and records vast areas of the seafloor that were dominated by pelmatozoans for a considerable length of time. This contribution proposes that massive sheet accumulations of crinoidal limestone (pelmatozoan epiboles) like the Burlington-Keokuk limestone are a unique lithofacies in Earth history, which should be referred to as *regional encrinites* (Ausich 1990a, 1990b). Similar accumulations of echinodermal biogenic sediments are not being deposited today (Carozzi and Gerber 1985); therefore regional encrinites are a time-restricted, vanished lithofacies.

Definitions

Regional encrinites are defined herein as crinoidal grainstones and packstones composed of more than 50% pelmatozoan debris by volume. Minimum dimensions of a unit qualifying as a regional encrinite are an average stratigraphic thickness of more than 5–10 m and an areal extent of more than 500 km^2 (Ausich 1990a). The adjective *crinoidal* in the term *crinoidal limestone* actually refers to any pelmatozoan (crinozoan or blastozoan) debris. Crinoids are the principal biogenic component of most of these limestones; however, other pelmatozoans were important locally, e.g., rhombiferans or diploporans during the Ordovician and Silurian or blastoids during the Lower Mississippian (fig. 19.1).

Complete crinoids are typically preserved by catastrophic burial (Blyth Cain 1968; Meyer 1971; Liddell 1975; Lewis 1980; Meyer and Meyer 1986). In normal circumstances a crinoid animal will rapidly disarticulate into component ossicles. These isolated ossicles are easily transported, and most crinoidal remains of regional encrinites have undergone transportation from their living site. However, crinoidal remains of regional encrinites are interpreted to be in situ on a facies-wide basis or parautochthonous. Even where tempestites dominated a regional encrinite sequence, redistribution of sediment only within the facies was typically present.

Wilson (1975:65) considered encrinites to be a special microfacies of his standard microfacies 12 (SMF-12) within the shoal environment in agitated water. Regional encrinites are interpreted herein to be from somewhat deeper settings.

Examples

The Lower Mississippian Burlington and Keokuk limestones from Iowa, Missouri, and Illinois are a striking example of a regional encrinite (Macurda and

[1]Calculations of the number of crinoid individuals is highly speculative. However, given estimates of 40–50 m of Burlington Limestone thickness over tens of thousands of km^2 and 15,000 crinoid individuals per m^3, the remains of as many as 1×10^{15} to 5×10^{16} crinoid individuals comprise the Burlington-Keokuk limestones.

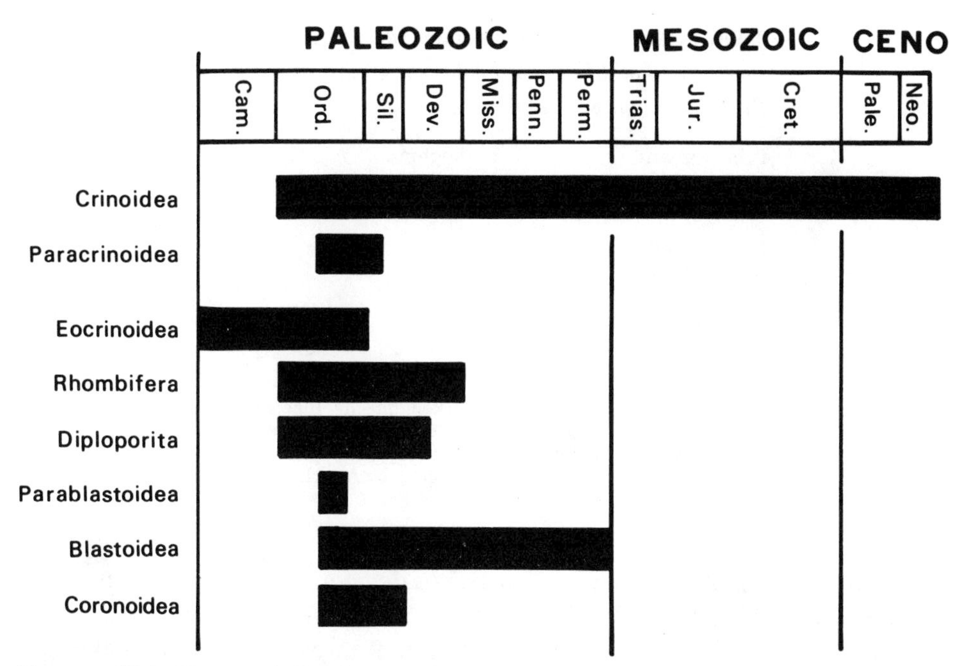

FIGURE 19.1. Temporal distribution of stalked (pelmatozoan) echinoderm classes that may have contributed debris to regional encrinites. Crinoidea and Paracrinoidea belong the subphylum Crinozoa, and the other classes belong to the subphylum Blastozoa. (Adapted from Ausich and Bottjer [1985].)

Meyer 1983). These Osagean units have an outcrop exposure of as much as 74,000 km^2 (28,500 mi^2) (Thompson 1986) with a combined thickness of up to 53.34 m (175 ft) (Lane 1978; Thompson 1986). Although other carbonate lithologies are present in these units, crinoidal packstones and grainstones dominate. The Burlington and Keokuk limestones were deposited on a carbonate ramp bordered by deeper water of the Illinois Basin to the east and the Transcontinental Arch far to the west.

Pelmatozoan echinoderms are well known from the Burlington and Keokuk limestones, although only the Keokuk fauna has received comprehensive modern systematic treatment (e.g., Ausich and Kammer 1990; Kammer and Ausich 1994). Volumetrically, the Burlington Limestone was probably dominated by camerate crinoid debris, although this is very difficult to determine accurately. Based on an uncorrected faunal list (Bassler and Moodey 1943)—undoubtedly oversplit at the species level—59% of the Burlington pelmatozoan species are monobathrid camerates (table 19.1). In the Keokuk Limestone 41% are monobathrid camerate crinoids (table 19.1).

Considered by itself, the Burlington-Keokuk limestone regional encrinite is remarkable. However, these units are a small part of approximately coeval carbonate platform/ramp deposits that stretch from the western margin of the Illinois Basin across the southern margin of the Laurentian continent to Arizona

Table 19.1 *Percent species (and subspecies) composition of different pelmatozoan groups in the Burlington and Keokuk limestones*

Pelmatozoan Group	Burlington Limestone	Keokuk Limestone
Blastoids	11%	5%
Diplobathrids	3%	4%
Monobathrids	59%	41%
Disparids	2%	3%
Primitive Cladids	7%	14%
Advanced Cladids	19%	28%
Flexibles	4%	5%

Burlington Limestone data is taken directly (uncorrected) from Bassler and Moodey (1943:81–88); 457 species considered. Data for the Keokuk Limestone estimated from systematic revisions published, in press, and in progress by Ausich and T. W. Kammer; 74 species considered.

and northward to Alaska. Regional encrinites developed discontinuously along this entire continental margin, stretching for more than 6,400 km (4,000 mi). In addition to the Burlington-Keokuk limestones, examples include the Lake Valley Formation, Hachita Formation, Redwall Limestone, Leadville Limestone, Argus Limestone, Madison Limestone, Livingstone-Mount Head formations of the Rundle Group, and the Kogruk Formation (table 19.2).

The Osagean of the Laurentian continental margin is the most extensive example of regional encrinites, although regional encrinites occurred at other times from the Middle Ordovician to the Jurassic. Brett (1991) considered pelmatozoans as probably the most significant Silurian carbonate grain producers, volumetrically, and Lowenstam (1957) considered pelmatozoans the "*Halimeda* of the Paleozoic." Selected examples include the Ordovician Holston Formation, Silurian Brassfield Formation, Silurian Amabel-Wiarton, Devonian Sadler Ranch Formation, Triassic Lower Muschelkalk, and Jurassic Smolegowa Limestone (table 19.2).

Discussion

The most remarkable aspect of regional encrinites is that they record the occupation of an entire carbonate platform by a vast crinoid garden for considerable lengths of time. Depositional rate is impossible to estimate in these facies, but it seems likely that crinoid populations dominated a single large area for hundreds of thousands to millions of years. A fairly limited range of physical/environmental parameters must have been required for persistent deposition of this lithofacies.

The Burlington-Keokuk regional encrinites and coeval regional encrinites are considered in detail to identify important depositional features of this lithofa-

TABLE 19.2 *Selected Examples of Regional Encrinites*

Formation	Period	Area	References
Holston Formation	Ordovician	Eastern Tennessee	Walker et al. (1989)
Brassfield Limestone	Silurian	Ohio, Kentucky	Ausich (personal obs.)
Irondequoit, Gasport, Wiarton	Silurian	New York, Pennsylvania, Ontario	Bolton (1957) Brett (1985)
Coeymans, Keyser, and New Creek Limestones	Devonian	New York, West Virginia	Smosna (1988)
Sadler Ranch Formation	Devonian	Nevada, California	Kendall et al. (1983)
Edgecliff Member of Onondaga Limestone	Devonian	New York	Cassa and Kissling (1982)
Burlington/Koekuk Limestone	Mississippian	Iowa, Missouri, Illinois	Lane (1978); Lane and DeKeyser (1980)
Tierra Blanca Member of the Lake Valley Formation	Mississippian	New Mexico	Ahr (1989)
Hachita Formation	Mississippian	New Mexico, Arizona	Armstrong et al. (1980)
Redwall Limestone	Mississippian	Arizona	Gutschick et al. (1980); Kent and Rawson (1980)
Leadville Limestone	Mississippian	Colorado	Armstrong and Mamet (1976)
Argus Limestone	Mississippian	California	Saul et al. (1979)
Madison Limestone	Mississippian	Wyoming	Gutschick et al. (1980)
Joanna Limestone	Mississippian	Nevada, Utah	Gutschick et al. (1980)
Livingston/Mount Head Formations of the Rundle Group	Mississippian	Alberta, Canada	Walpole and Carozzi (1961) MacQueen and Bamber (1968)
Kogruk Formation	Mississippian	Alaska	Sable and Dutro (1961)
Lower Muschelkalk	Triassic	Germany	Aigner (1985)
Smolegowa Limestone	Jurassic	Poland	Birkenmajer (1977)

cies. Parameters to be considered are climate, depositional setting, water depth, position in a facies gradient, dominant depositional mode, character of the sediment, and taphonomic feedback.

The Osagean Laurentian regional encrinites were principally situated within 15° of the paleoequator, although the Kogruk Formation was positioned at greater than 20°N paleolatitude (fig. 19.2). Most Osagean regional encrinites were deposited on a carbonate ramp and the adjacent carbonate platform (e.g., Burlington-Keokuk, Lane 1978). In most instances the ramp was marginal to a cratonic basin or was on a passive continental margin. However, in the northwestern United States the Madison Limestone was deposited on a carbonate platform cratonward of the Antler Orogeny, and the Joana Limestone accumulated in the Antler Foreland Basin (Gutschick et al. 1980). Interpretation of depositional environments of regional encrinites varies somewhat among authors. A combination of grainstones and packstones with common cross stratification and graded bedding indicates conditions shallow enough for traction movement of crinoidal debris. Carozzi and Gerber (1985) reconstructed the Burlington Limestone as a bank shoal area developed within normal wave base (fig. 19.3A). Facies reconstructions by others place the regional encrinite lithofacies in water deeper than normal wave base but well within storm wave base (Redwall, Hachita, and Lake Valley [Armstrong et al. 1980]; Redwall [Kent

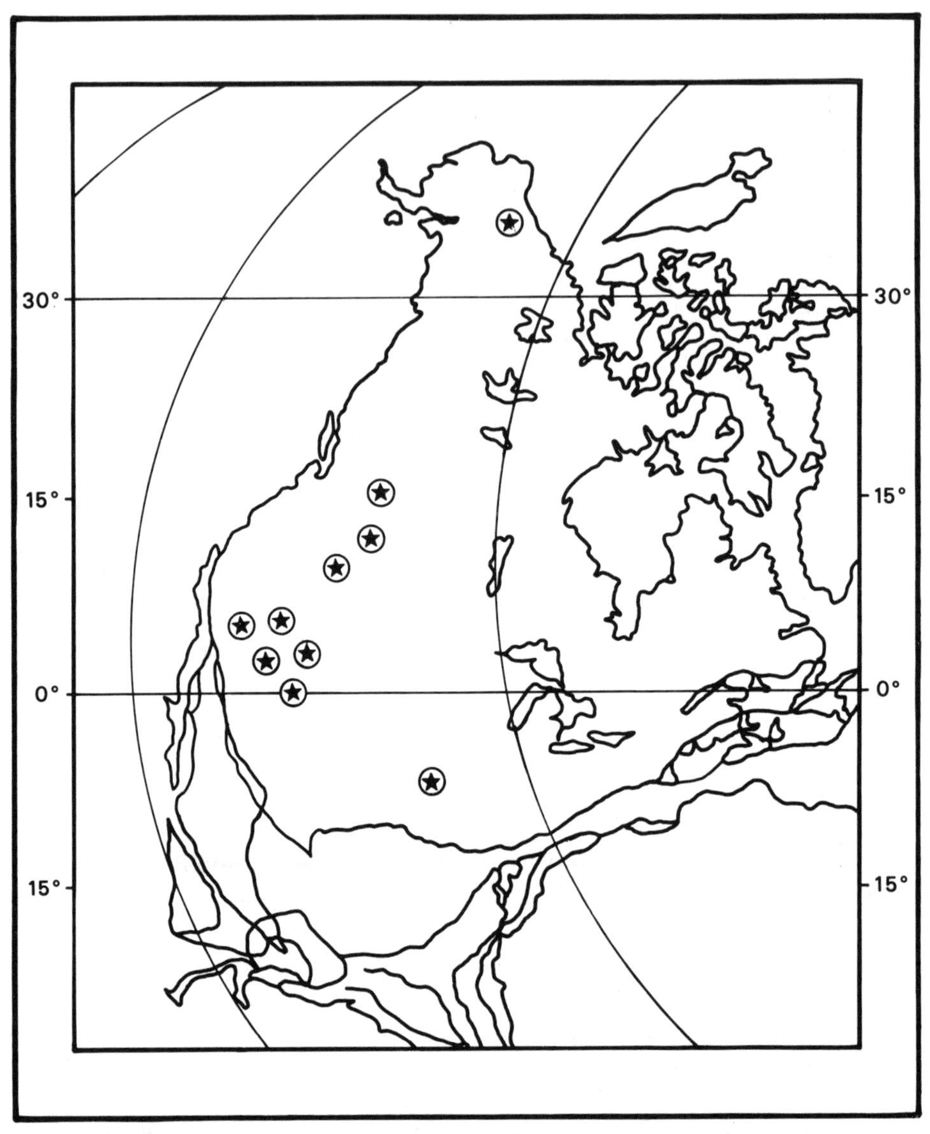

FIGURE 19.2 Approximate locations of Early Mississippian regional encrinites on Laurentia. Paleogeographic map from 347 million years ago reconstruction by Scotese and Denham (1988).

and Rawson 1980, fig. 19.3C; Madison [Gutschick et al. 1980]; and Burlington Limestone, inferred from Lane [1978], fig. 19.3B). In some instances regional encrinites are basinward/oceanward of an oolitic shoal facies belt; therefore in these instances they are clearly in water deeper than normal wave base.

In summary, regional encrinites were typically deposited below normal wave base but within storm wave base. In this general setting and with consideration of lithologies and sedimentary structures, Osagean regional encrinites are inter-

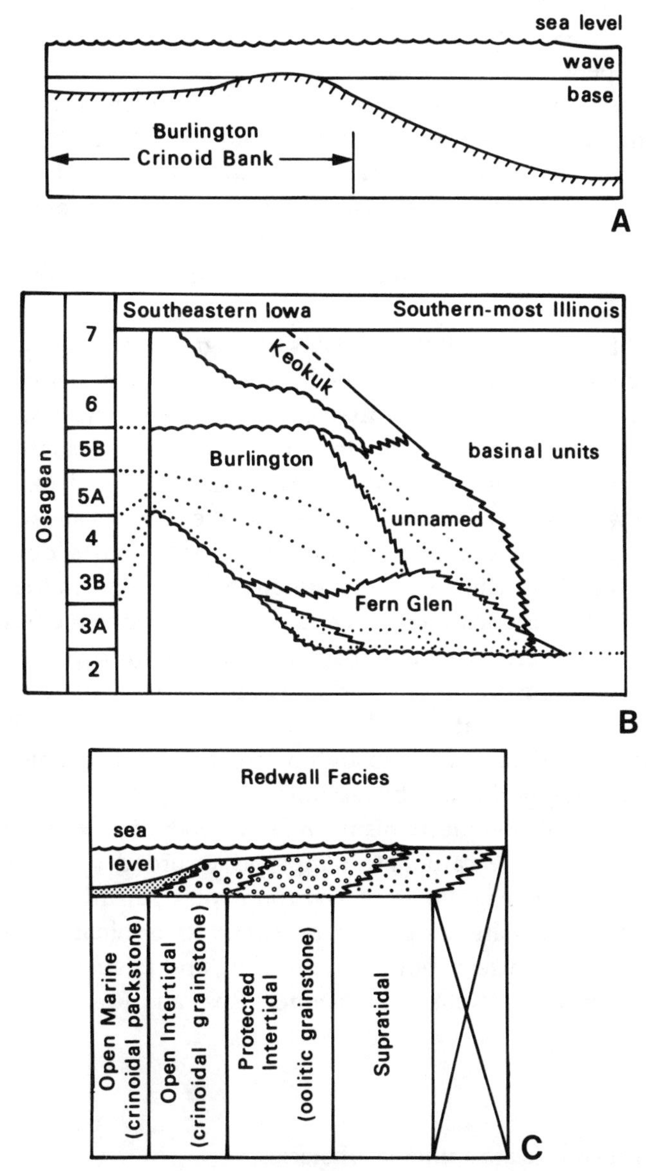

FIGURE 19.3. Environmental reconstruction and cross sections of Lower Mississippian regional encrinites. **(A)** Depositional reconstruction of the Burlington Limestone by Carozzi and Gerber (1985). **(B)** Cross section showing lithostratigraphic and chronostratigraphic relationships between the Burlington Limestone and Keokuk Limestone carbonate ramp deposits and coeval units in the center of the Illinois basin. Dotted lines indicated biostratigraphic zone boundaries. (From Lane [1978].) **(C)** Depositional reconstruction of the Redwall Limestone by Kent and Rawson (1980).

preted as units dominated by tempestites. This interpretation corresponds to that of the Lower Muschelkalk by Aigner (1985).

Because of cross-stratification, graded bedding, and common crinoid holdfasts with cirri penetrating into the sediment, the substratum of Osagean regional encrinites must have been unconsolidated. With tempestites a dominant depositional mode and the sediment unconsolidated, the bottom would have been episodically mobile and unsatisfactory for most benthic invertebrates. Macurda and Meyer (1983:356) estimated that more than 99% of the Burlington Limestone was composed of pelmatozoan remains, and most benthos are unsuited for life on an unconsolidated, episodically mobile, poorly sorted debris substratum.

During the Silurian (Brett 1991) and Triassic (Aigner 1985), regional encrinites apparently accumulated landward of pelmatozoan habitats. Because these pelmatozoans typically had holdfasts that attached to firm substrata, they were concentrated on small, isolated bioherms (Brett 1991) but separated from adjacent areas where skeletal debris accumulated. So during these times, even pelmatozoans were not well adapted to a regional encrinites substratum. Alternatively, during the Lower Mississippian, many crinoids had developed a robust holdfast with cirri, which was well suited for anchorage in this substratum type. In a Lower Mississippian encrinite, a positive taphonomic feedback situation was probably established on regional encrinites (Kidwell and Jablonski 1983), whereby the remains of dead crinoids provided a substratum for subsequent generations. During at least the Lower Mississippian, a well-established crinoid garden would produce substratum conditions that could perpetuate continued occupation of the area by crinoids.

Whatever the stabilizing mechanisms, regional encrinites were stable, persistent lithofacies on certain carbonate ramps, as recorded by the stratigraphic record. The type and magnitude of environmental perturbation, other than elimination of the carbonate ramp/shelf, necessary to terminate the stability of a regional encrinite are unclear, but rapid lowering and raising of sea level separating the Burlington and Keokuk limestones (Ross and Ross 1985) were insufficient.

In summary,

1. Regional encrinites are a unique lithofacies.
2. Regional encrinites are grainstones and packstones composed of more than 50% pelmatozoan debris and that have an average stratigraphic thickness of more than 5–10 m and an areal extent of more than 500 km^2.
3. Regional encrinites occur at low paleolatitudes, were typically deposited below normal wave base but within storm wave base, and had tempestites as a dominant sedimentation mode. Positive taphonomic feedback was an important process in maintenance of regional encrinites.
4. Regional encrinites are a time-restricted (Ordovician to Jurassic), vanished lithofacies.

ACKNOWLEDGMENTS

C. E. Brett provided data and information for this study, and this manuscript was improved by the suggestions of C. E. Brett, S. K. Donovan, and G. C. McIntosh. H. H. Hayes typed the manuscript, and K. Tyler and B. Daye aided with the illustrations. This project was fully supported by the National Science Foundation (EAR-8706430 and EAR-8903486).

REFERENCES

Ahr, W. M. 1989. Sedimentary and tectonic controls on the development of an Early Mississippian carbonate ramp, Sacramento Mountains area, New Mexico. In P. D. Crevello, J. L. Wilson, J. F. Sarg, and J. F. Read, eds., *Controls on Carbonate Platform and Basin Development*. SEPM Special Paper 44, pp. 203–12.

Aigner, T. 1985. *Storm Depositional Systems. Lecture Notes in the Earth Sciences*, vol. 3. New York: Springer-Verlag.

Armstrong, A. K. and B. L. Mamet. 1976. Biostratigraphy and regional relations of the Mississippian Leadville Limestone in the San Juan Mountains, southwestern Colorado. *U.S. Geological Survey Professional Paper* 985.

Armstrong, A. K., B. L. Mamet, and J. E. Repetski. 1980. The Mississippian System of New Mexico and southern Arizona. In T. D. Fouch and E. R. Magathan, eds., *Rocky Mountain Paleogeography Symposium 1, Paleozoic Paleogeography of the West-Central United States*, pp. 82–99.

Ausich, W. I. 1990a. Regional encrinites: A vanished lithofacies. *13th International Sedimentological Congress Abstracts, Papers*, pp. 30–31.

———. 1990b. Regional encrinites: How can 5×10^9 m^3 of crinoidal limestone be accumulated? *Geological Society of America Abstracts with Program* 22:A219.

Ausich, W. I. and D. J. Bottjer. 1985. Echinoderm role in the history of Phanerozoic tiering in suspension-feeding communities. In B. F. Keegan and B. D. S. O'Connor, eds., *Echinodermata*, pp. 3–11. Rotterdam: A. A. Balkema.

Ausich, W. I. and D. L. Meyer. 1990. Origin and composition of carbonate buildups and associated facies in the Fort Payne Formation (Lower Mississippian, south-central Kentucky): An integrated sedimentologic and paleoecologic analysis. *Geological Society of America Bulletin* 102:129–146.

Ausich, W. I. and T. W. Kammer. 1990. Systematics and Phylogeny of the late Osagean and Meramecian crinoids *Platycrinites* and *Eucladocrinus* from the Mississippian stratotype region. *Journal of Paleontology* 64:759–78.

Ausich, W. I., T. W. Kammer, and N. G. Lane. 1979. Fossil communities of the Borden (Mississippian) delta in Indiana and northern Kentucky. *Journal of Paleontology* 53:1182–96.

Bassler, R. S. and M. W. Moodey. 1943. Bibliographic and faunal index of Paleozoic pelmatozoan echinoderms. *Geological Society of America Special Papers* No. 45.

Birkenmajer, K. 1977. Jurassic and Cretaceous lithostratigraphic units of the Pienny Klippen Belt, Carpathians, Poland. *Studia Geologica Polonica* 45:1–158.

Blyth Cain, J. D. 1968. Aspects of the depositional environment and paleoecology of crinoidal limestones. *Scottish Journal of Geology* 4:191–208.

Bolton, T. E. 1957. Silurian stratigraphy and paleontology of the Niagara escarpment in Ontario. *Geological Survey of Canada Memoir* No. 289.

Brett, C. E. 1984. Autecology of Silurian pelmatozoan echinoderms. In M. G. Bassett and J. D. Lawson, eds., *Autecology of Silurian Organisms*, pp. 87–120. *Special Papers in Palaeontology* No. 32.

———. 1985. Pelmatozoan echinoderms on Silurian bioherms in western New York and Ontario. *Journal of Paleontology* 59:820–38.
———. 1991. Organism-sediment relationships in Silurian marine environments. *Special Papers in Palaeontology* No. 44:301–44.
Carozzi, A. V. and M. S. Gerber. 1985. Crinoid arenite banks and crinoid wacke inertia flows: A depositional model for the Burlington Limestone (middle Mississippian), Illinois, Iowa, and Missouri, USA. In E. S. Belt and R. W. MacQueen, eds., Neuvième Congrès International de Stratigraphie et de Géologie du Carbonifère, *Compte Rendu*, vol. 3, pp. 452–60.
Cassa, M. R. and D. L. Kissling. 1982. Carbonate facies of the Onondaga and Boise Blanc Formations Niagara Peninsula, Ontario. In E. J. Buehler and P. E. Calkin, eds., *Guidebook for Field Trips in Western New York, Northern Pennsylvania and Adjacent Southern Ontario. New York State Geological Association 54th Annual Meeting*, pp. 65–97.
Gutschick, R. C., C. A. Sandberg, and W. J. Sando. 1980. Mississippian shelf margin and carbonate platform from Montana to Nevada. In T. D. Fouch and E. R. Magathan, eds., *Rocky Mountain Paleogeography Symposium 1, Paleozoic Paleogeography of the West-Central United States*, pp. 111–28.
Kammer, T. W. and W. I. Ausich. 1994. Advanced cladid crinoids from the middle Mississippian of the East-Central United States: Advanced-grade calyces. *Journal of Paleontology* 66.
Kendall, G. W., J. G. Johnson, J. O. Brown, and G. Klapper. 1983. Stratigraphy and facies across Lower Devonian–Middle Devonian boundary, central Nevada. *American Association of Petroleum Geology Bulletin* 67:2199–2207.
Kent, W. N. and R. R. Rawson. 1980. Depositional environments of the Mississippian Redwall Limestone in northeastern Arizona. In T. D. Fouch and E. R. Magathan, eds., *Rocky Mountain Paleogeography Symposium 1, Paleozoic Paleogeography of the West-Central United States*, pp. 101–109.
Kidwell, S. M. and D. Jablonski. 1983. Taphonomic feedback: Ecological consequences of shell accumulation. In M. J. S. Tevesz and P. L. McCall, eds., *Biotic Interactions in Recent and Fossil Benthic Communities*, pp. 195–248. New York: Plenum Press.
Lane, H. R. 1978. The Burlington shelf (Mississippian, north-central United States. *Geologica et Palaeontologica* 12:165–76.
Lane, H. R. and T. L. De Keyser. 1980. Paleogeography of the late Early Mississippian (Tournaisian) in the central and southwestern United States. In T. D. Fouch and E. R. Magathan, eds., *Rocky Mountain Paleogeography Symposium 1, Paleozoic Paleogeography of the West-Central United States*, pp. 149–162. Denver: The Rocky Mountain Section, Society of Economic Paleontologist and Mineralogists.
Lane, N. G. 1971. Crinoids and reefs. *Proceedings of the North American Paleontological Convention* Part J:1430–1443.
Lewis, R. D. 1980. Taphonomy. In T. W. Broadhead and J. A. Waters, eds., *Echinoderms*, notes for a short course. University of Tennessee Department of Geological Sciences, *Studies in Geology*, vol. 3, pp. 27–39.
Liddell, W. D. 1975. Recent crinoid biostratinomy. *Geological Society of America Abstracts with Programs* 7:1169.
Lowenstam, H. 1957. Niagaran reefs in the Great Lakes area. In H. S. Ladd, ed., Treatise on marine ecology and paleoecology, Part 2, *Geological Society of America Memoir* No. 3, pp. 215–48.
MacQueen, R. W. and E. W. Bamber. 1968. Stratigraphy and facies relationships of the Upper Mississippian Mount Head Formation, Rocky Mountains and foothills, southwestern Alberta. *Bulletin of Canadian Petroleum Geology* 16:225–287.
Macurda, D. B., Jr., and D. L. Meyer. 1983. Sea lilies and feather stars. *American Scientist* 71:354–65.

Meyer, D. L. 1971. Post-mortem disarticulation of Recent crinoids and ophiuroids under natural conditions. *Geological Society of America, Abstracts with Programs* 3:645.
Meyer, D. L. and K. B. Meyer. 1986. Biostratinomy of Recent crinoids (Echinodermata) at Lizard Island, Great Barrier Reef, Australia. *Palaios* 1:294–301.
Ross, C. A. and J. R. P. Ross. 1985. Late Paleozoic depositional sequences are synchronous and worldwide. *Geology* 13:194–97.
Sable, E. G. and J. T. Dutro, Jr. 1961. New Devonian and Mississippian formations in DeLong Mountains northern Alaska. *American Association of Petroleum Geologists Bulletin* 45:585–93.
Saul, R. B., O. E. Bowen, C. H. Stevens, G. C. Dunne, R. G. Randall, R. W. Kistler, W. J. Nokleberg, J. A. D'Allura, E. M. Moores, R. Watkins, E. M. Baldwin, E. H. Gilmour, and W. R. Danner. 1979. The Mississippian and Pennsylvanian (Carboniferous) Systems in the United States—California, Oregon, and Washington. *U.S. Geological Survey Professional Paper 1110-CC.*
Scotese, C. R. and C. R. Denham. 1988. *Terra Moblis™.*
Smosna, R. 1988. Paleogeographic reconstruction of the Lower Devonian Helderberg Group in the Appalachian Basin. In N. J. McMillan, A. F. Embry, and D. J. Glass, eds., *Devonian of the World,* Proceedings of the Second International Symposium on the Devonian of the World, Calgary, Canada, vol. 1, pp. 265–75. Calgary: Canadian Society of Petroleum Geologists.
Thompson, T. L. 1986. Paleozoic Succession in Missouri, Part 4, Mississippian System. *Missouri Department of Natural Resources Division of Geology and Land Survey, Report of Investigations,* 70.
Walker, K. R., J. F. Read, and L. A. Hardie. 1989. Cambro-Ordovician carbonate banks and siliciclastic basins of the United States Appalachians, *28th International Geological Congress Field Trip Guidebook,* T161.
Walpole, R. L. and A. V. Carozzi. 1961. Microfacies study of Rundle Group (Mississippian) of the Front Ranges, central Alberta, Canada. *American Association of Petroleum Geologists Bulletin* 45:1810–46.
Wilkinson, B. H. 1979. Biomineralization, paleoceanography, and the evolution of calcareous marine organisms. *Geology* 7:524–27.
Wilson, J. L. 1975. *Carbonate Facies in Geologic History.* New York: Springer-Verlag.

20

Cenomanian/Turonian Bioevents and Ecostratigraphy in the Western Interior Basin: Contrasting Scales of Local, Regional, and Global Events

Bradley B. Sageman, Erle G. Kauffman,
Peter J. Harries, William P. Elder

ABSTRACT

Cenomanian/Turonian strata of the Western Interior Basin contain a hierarchy of paleontological event horizons/intervals. These features, which range from millimeters-thick bedding plane surfaces (bioevents) to meters-thick intervals (ecozones), can be defined taxonomically as well as in terms of relative increase or decrease in diversity, abundance, and trophic complexity of benthic faunas. Diversification and diversity reduction represent the direct response of biotas to changes in the marine environment, and reflect changes in the physical, chemical, and/or biological factors that controlled development of benthic communities. Basinwide, high-resolution chronostratigraphy in the Western Interior has allowed testing of the local, regional, or global extent of bioevents and ecozones, and provides an opportunity to compare and contrast paleobiological phenomena of wide-ranging spatial and temporal scales. The results suggest that there was a consistent biological response at all levels of the spatial and temporal hierarchy, resulting in widely correlative ecostratigraphic markers. Western Interior paloeocommunitites were dominantly controlled by changes in bottom-water oxygen content and substrate characteristics which, in turn, were strongly influenced by changes in relative sea level and the prevailing climatic regime. The opportunistic nature of low-oxygen-adapted Cenomanian/Turonian community elements was most likely responsible for their sensitivity to paleoenvironmental changes.

Event-driven paleobiological phenomena (bioevents) comprise an important part of the preserved fossil record. Such bioevents include geologically short-term changes in the evolutionary, ecological, and/or biogeographic character of biotas, as well as their extinction, and are commonly expressed as distinct paleontological surfaces or intervals. Bioevents are recognized at the population, species, and community level, and can be classified as diversification events (e.g., short-term colonization, population expansion and/or immigration events, and rapid speciation), which may produce horizons or thin stratigraphic intervals of increased species richness and/or faunal abundance, or as reduction events (e.g., mass mortality, rapid restriction of community or biogeographic range, and extinction), which can result in intervals that are relatively depauperate or barren of fossils. Both types of bioevents provide distinct stratigraphic markers that are commonly correlative over considerable distances within and between sedimentary basins (Kauffman 1986, 1988a; Kauffman et al. 1991). As such, they provide important chronostratigraphic surfaces/intervals for regional and, in special cases, global correlation.

Ecostratigraphy, as described by Boucot (1986), Kauffman (1986, 1988a), and others, represents the integration of paleobiological data in stratigraphic studies. Its twofold application includes (1) the use of unique paleoecological event horizons or thin stratigraphic intervals as regional chronostratigraphic tools, with resolution surpassing that of many traditional biostratigraphic methods, and (2) the opportunity to study the spatial and temporal evolution of ecological interactions, communities, and ecosystems within successive, narrowly constrained time slices. In this study, paleobiological data (e.g., taxa sets and their relative abundance and diversity characteristics) are used to define widespread bioevent horizons and ecostratigraphic intervals. Different types of ecostratigraphic units representing variable temporal (short- vs. long-term) and spatial (local, regional, and global) scales are recognized. By comparing events of different temporal and spatial scale we seek an improved understanding of the forcing mechanisms that control the evolution of populations, species, and paleocommunities.

Cretaceous strata of the Western Interior Basin, which have been studied intensively, provide an ideal opportunity to examine different scales of events within a high-resolution integrated chronostratigraphic and biostratigraphic framework. Numerous examples of short-term pioneer colonization events, local to regional population blooms, and rapid immigration of new species and biotas, as well as intervals of mass mortality, species loss, and mass extinction are preserved in the Cenomanian/Turonian (C/T) sequence of the Greenhorn Limestone. Bioevent surfaces/intervals dominate most of the C/T stratigraphic section in basinal sites. These biotic events represent a dynamic history of paleoenvironmental changes within the basin, but they also reflect phenomena recognized outside the Western Interior, such as the Cenomanian/Turonian "Oceanic Anoxic Event" (OAE) of Schlanger and Jenkyns (1976), and a global mass extinction (i.e., Raup and Sepkoski 1982; Kauffman 1984a, 1988).

The purpose of this report is to present a preliminary synthesis of the ecostratigraphy of the C/T interval in order to interpret better the primary agents of paleobiological change. Analysis of ecostratigraphic trends within a framework of detailed and integrated stratigraphic, sedimentologic, and geochemical data for the Western Interior Basin, and comparison to the published record of biotic change at C/T sections outside the Western Interior, makes possible evaluation of the ecological/environmental forces that control paleocommunities on short and long time frames, and on local, regional, and global scales. The results suggest that relative changes in climate, sea level, and related oceanographic factors played a major role in controlling the development of mid-Cretaceous paleocommunities and paleobiological event horizons.

Ecological Units in Stratigraphy

A great diversity of names and concepts for paleoecological units exists in the stratigraphic literature. Many are poorly defined, contradictory among different definitions, or in some cases not logically defensible as originally defined and/or currently utilized. Because this study deals with several levels of ecological units, we first provide a set of definitions for the terms used herein, and a classification of ecostratigraphic events.

Paleocommunity We follow the definition of Kauffman and Scott (1976) in that a paleocommunity, or fossil association, is the preserved subset of an implicitly more diverse original biological community that is preserved at or very near its original habitat. A paleocommunity reflects a relatively tightly defined set of ancient environmental conditions and is characterized throughout its spatial and temporal range by a specific taxa (species) set that may be fully to partially developed or preserved at different times or localities. Paleocommunities are repetitive in space and time but have more limited ranges than biofacies (see later) because of the greater taxonomic restrictions placed on their definition.

Biofacies From the time of its introduction (Lowman 1945), the term *biofacies* has been much debated (e.g., Kay 1947; Moore 1949; Sloss et al. 1949; Weller 1958). Most authors have defined a biofacies as a distinct assemblage or association (*sensu* Kauffman and Scott 1976) of fossils within a stratigraphic unit(s) that reflects, in its biological and ecological characteristics alone, an explicit set of paleoenvironmental conditions, independent of associated physical or chemical characteristics of the containing strata. Many definitions limit a single biofacies to an explicitly defined stratigraphic unit or interval and are therefore mainly concerned with the spatial expression of a biofacies. Some authors (e.g., Bates and Jackson 1987) suggest that a biofacies represents the fossil record of a biocoenosis, or life assemblage, in which case the term equates with a paleocommunity or fossil association. We find this composite of definitions vague and temporally too limiting when one considers that a lithofacies (e.g., a lower shoreface sand) or chemofacies (e.g., C_{org}-rich black laminated shale) is repetitive in both time and space.

In this paper a biofacies is defined as collectively comprising all fossil associations (e.g., derived from resident paleocommunities) that are closely similar with regard to species composition and/or details of adaptive morphology among component taxa, and that represent a specifically defined set of environmental parameters. Biofacies may be repetitive in both space (e.g., within a laterally persistent stratigraphic unit) and time (e.g., repeated one or more times within a stratigraphic column), whenever the favorable environmental niche is established. Biofacies may have the same basic taxa set throughout their occurrence, but more commonly they are composed of different, commonly related species with similar or nearly identical adaptive form and environmental niche requirements. Thus different species of byssate epibenthic Inoceramidae (Bivalvia) with morphologically similar but taxonomically distinct species of epibiont oysters and pteriid bivalves comprise a single biofacies in the Cenomanian of the Western Interior Basin, even though some of the component species of the biofacies change by immigration or lineage evolution through time. Similarly, a stratigraphically discrete set of forereef communities through time dominated by massive scleractinian head corals with similar colony form comprise a single biofacies, even though species and even genera may change through time.

Ecostratigraphic Zone (Ecozone) Vella (1964) and others defined an ecozone as corresponding to a body of rock characterized by a specific taxonomic/ecologic assemblage, with discrete upper and lower boundaries that mark reversible biotic changes. Vella (1964) believed that the biological reversals from one ecozone to another marked secular lateral shifts of facies belts and that their boundaries were therefore diachronous. In the Western Interior Basin, many regionally correlative stratigraphic intervals are characterized by strong dominance of a single paleocommunity, or a closely related set of paleocommunities. Typically, such intervals are marked by relatively sharp boundaries at which community characteristics (species composition and richness, faunal abundance) show marked shifts, in many cases reflecting relative diversification or diversity reduction events. Within these intervals, community characteristics may change very little, reflecting periods of relative stability in the benthic community, or they may vary on scales that are similar within an interval but different between intervals. Although these ecologically defined intervals reflect secular environmental changes, they are typically forced by allocyclic climatic and/or eustatic events. In fine-grained basinal facies of the Western Interior Basin, these events produce widely correlative ecostratigraphic zones that cross Waltherian facies boundaries and whose boundaries are chronostratigraphically significant (essentially isochronous). We designate these time-based ecological event packages herein as *ecozones* (fig. 20.1). Their duration is typically less than the biostratigraphic ranges of component taxa, and similarly constructed ecozones may occur at several stratigraphic levels, separated by dissimilar ecozones or nonfossiliferous intervals. Because global climatic/eustatic cycles are hierarchical, ecozones are hierarchical: In the Western Interior we recognize at least two levels, which we designate short-term (climate cycle or shorter scale) and long-term (sea-level cycle scale) (fig. 20.1).

Biological Event (Bioevent) Bioevents are short-term (hours or days to kyr) locally, regionally, or interregionally pervasive changes in the ecological, biogeographic, and/or evolutionary character of biotas that are isochronous or nearly so throughout their range. They result from various diversification processes (rapid colonization, short-term population and biomass increase, rapid immigration, punctuated evolution, etc.) or diversity reduction events (population mass mortality, species-level extinction, or major community/ecosystem collapse and mass extinction). Each bioevent is an individual surface or very thin stratigraphic interval potentially useful as a chronostratigraphic horizon. The highest-resolution ecostratigraphic unit, they commonly comprise the boundaries of ecozones (fig. 20.1).

A Classification of Ecostratigraphic Units

Bioevents and ecozones fall into two broad categories: (1) *diversification bioevents/ecozones* (cf. "Constructional Bioevents," Kauffman 1986, 1988a), which reflect expansion of ancient ecosystems through increase in population size and addition of new taxa through immigration and/or evolution; and (2) *reduction bioevents/ecozones* (cf. "Destructional Bioevents," Kauffman 1986, 1988a), which reflect such changes in the ecosystem as lowering of population size and decrease in community diversity through ecological restriction, diminished biogeographic ranges, and/or extinction. Diversification and reduction events are driven by a wide range of physical, chemical, and biological causes, examples of which are discussed later.

Bioevents are related to several different types of stratigraphic units cited in the North American and International Codes of Stratigraphic Nomenclature (ISSC 1976; NACSN 1983). In practice, they are employed as data-based chronostratigraphic units in the development of an integrated event stratigraphic framework for a basin. For example, global to regional origination/extinction bioevents are used to construct composite assemblage biostratigraphic zonations. Other types of bioevents (ecological, biogeographic) provide additional regional to local ecostratigraphic data. In some cases certain physical stratigraphic units at the marker bed or lentil level may also reflect bioevents (e.g., widespread mass mortality, colonization, or productivity events that produce regionally correlative, physically distinct shell/bone/plant beds), providing further chronostratigraphic data. Overall, bioevents are characterized by thin inter-

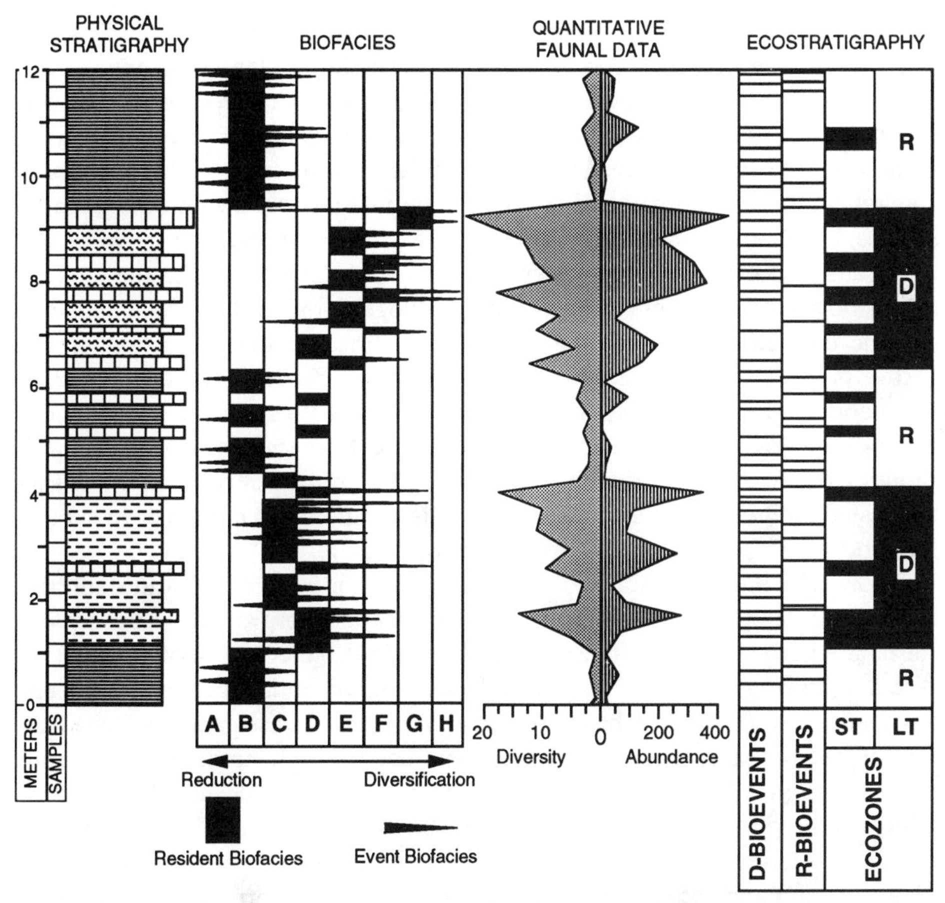

FIGURE 20.1. The system used in this study for recognition of ecostratigraphic units is illustrated for an idealized Western Interior Cretaceous stratigraphic sequence. Diversification (D) and reduction (R) bioevents and ecozones are defined based on relative change in quantitative faunal data (species richness and faunal abundance) and biofacies (following the model of Kauffman and Sageman 1990; Sageman et al. 1991). Biofacies levels A–H represent a gradation of increasing community complexity, as would be observed in a transition from highly dysoxic to fully oxygenated paleoenvironments. Resident biofacies represent continuously colonizing background communities; event biofacies reflect short-term colonization or mortality episodes. Bioevents are defined as the short-term shifts in biofacies (single horizons/bedding planes or cm's—thick units representing hours or days to thousands of years) represented by event communities and/or shifts in resident communities; ecozones correspond to thicker intervals through which resident communities are stable (e.g., short-term ecozones corresponding to limestone/shale hemicycles) or shifting in a consistent direction (long-term ecozones spanning several-meters-thick packages that represent more than 100 kyr).

vals, sharp boundaries, and a geologically short time span (hours/days up to <100 kyr). In mud-dominated depositional systems such as the Western Interior Basin, rock accumulation rates for offshore facies are typically quite low (<1–5 cm/kyr), and condensation is a common phenomenon. Thus many Western Interior bioevents are preserved as discrete bedding plane horizons.

Ecozones may also correspond to biostratigraphic units. In cases where species richness and faunal abundance data reveal diversification ecozones on the scale of one to several meters in Western Interior strata and where such intervals reflect the widespread and short-term population expansion of key "index species," the abundance zone, or "epibole," of traditional usage would be represented. Brett and Baird (paper 10 in this volume) have suggested a redefinition of the epibole concept as a paleoecological rather than as a biostratigraphic unit and have expanded the term to include many features similar to the bioevents and ecozones described herein. Because we developed an independent classification system during analysis of Western Interior paleobiological data (described later), and because the term *epibole* is still primarily thought of as a biostratigraphic unit, we have not applied the new usage suggested by Brett and Baird (paper 10 in this volume), but note instead similarities and differences to their terms where appropriate.

Figure 20.1 illustrates the system used herein for recognition and definition of ecostratigraphic units based on quantitative and interpretative paleoecological data (species richness, faunal abundance, and biofacies). Biofacies are depicted following a model suggested by Kauffman and Sageman (1990) and Sageman et al. (1991) in which each sample interval is assigned resident (taxa persist through sample) and event (taxa confined to single horizons) community or biofacies values based on combined taxonomic, trophic, life habit, abundance, and taphonomic characteristics. When plotted against physical stratigraphy and quantitative faunal data (species richness and faunal abundance), diversification and reduction bioevents and ecozones (short term or long term) can be identified (fig. 20.1).

Diversification Bioevents

Diversification bioevents can be classified into three main categories:

1. *Ecological diversification bioevents* include rapid changes in preexisting biotas within a basin that produce greater biomass and larger, more widely dispersed species populations, or rapid addition of taxa resulting in more complex, more diverse, and/or more highly structured ecological relationships within communities. Rapid population and biomass expansion reflect abrupt changes in environmental parameters that tend to favor a given taxon or group of taxa. These parameters include habitat change or expansion, rapid changes in nutrient and food supply, or abrupt reduction of predator and competitor populations. Examples include colonization

events related to abrupt physical changes in current energy and substrates (e.g., deposition of storm beds or turbidites), rapid chemical changes within the substrate and/or water column (e.g., oxygenation during large storms or advection events), or development of favorable biosubstrates (colonization of shell islands by epibionts). Brett and Baird (paper 10 in this volume) have described similar features as ecological epiboles.

2. *Biogeographic diversification bioevents* mainly involve rapid shifts in paleobiogeographic boundaries that lead to immigration of new taxa and biotas into a basin. Identification of biogeographic diversification bioevents depends on documentation that the immigrant taxa preexist outside the region being studied. On entering a basin, new immigrant taxa commonly integrate with resident communities (e.g., through niche partitioning), which leads to increased diversity and ecological complexity. In some cases, however, they may outcompete and eventually replace portions of preexisting biotas, leading potentially to community disruption (and an ecological reduction bioevent). Biogeographic diversification bioevents take place within geologically short time periods and are regional in scale. They may be related to abrupt changes in water mass characteristics (temperature, salinity, etc.) or oceanic/atmospheric circulation patterns, or to the development of new migratory pathways, which result from plate or regional tectonic modification of continental positions and ocean basins, and/or changes in sea level. Brett and Baird (paper 10 in this volume) have referred to similar features as *incursion epiboles.*

3. *Evolutionary diversification bioevents* are rapid origination events that produce new taxa, new anatomical/morphological structures, and/or new behavior/habitat patterns (as in trace fossils), coupled with rapid dispersal of these new traits or taxa. Mechanisms for such rapid change include punctuated evolution (Eldredge and Gould 1972), allopatric evolution (Mayr 1965), macromutation and other macroevolutionary processes (see Stanley 1979 and references therein). Cited authors present many modern and fossil examples. Rapid spread of these new traits or taxa is normally accomplished by passive dispersal processes (transport of germ cells or larvae by wind and currents) or active juvenile and adult mobility. Identification of evolutionary bioevents within a basin is based on analysis of composite taxonomic range data (first and last occurrences plotted against a standardized section) and documentation that new taxa were not preexisting within or outside the region under study.

Reduction Bioevents

Reduction bioevents can also be classified into three main categories:

1. *Ecological reduction bioevents* include abrupt reductions in population size, community size, structure and diversity, and larger-scale decline in

ecosystem complexity without evidence of species-level extinction (i.e., affected species are through-ranging or show Lazarus distribution). The most common ecological reduction bioevents are catastrophic mass mortalities of species resulting in abrupt decline in population size and biomass. They range from local demise of subpopulations (but not entire species populations) to rapid decimation of regional communities and ecosystems. These diversity reduction events are due to (a) shock from environmental perturbations, such as advection of chemically deleterious water masses, decline in oxygen levels, or abrupt temperature changes at or above the sediment/water interface (SWI), in surface waters or on land; (b) abrupt habitat change or restriction; (c) introduction of disease, unusually high levels of predation, cropping, and competition from new or associated species; or (d) introduction and replacement by lower-diversity communities.

2. *Biogeographic reduction bioevents* result from abrupt restriction in the biogeographic distribution of through-ranging taxa, caused by rapid local to regional climatic and/or oceanographic changes. Such events result from the erection of biogeographic barriers, most commonly due to plate or regional tectonic regulation, changes of sea level, and/or changes in water mass characteristics between different basins. Like immigration events, these may occur within very short time frames, reflecting dynamic shifts in paleobiogeographic provinces and subprovinces during the Phanerozoic (e.g., the Cretaceous; Kauffman 1984a). In addition, the immigration of one or more devastating species into a complex ecosystem (e.g., a virus, parasitic insect, or general predator) may lead to short-term, local to regional depletion of paleobiogeographic units. Recognition of these events requires an interbasinal high-resolution chronostratigraphic framework against which taxonomic range data and biogeographic distributions can be calibrated. In cases where ecological or biogeographic reduction bioevents produce intervals that are highly depleted or devoid of fossils, they are similar to the outages described by Brett and Baird (paper 10 in this volume).

3. *Extinction bioevents* involve the regional to global disappearance of individual taxa during background extinction, or of ecologically and genetically diverse taxa during mass extinction. These may occur within thin stratigraphic intervals and thus comprise important reduction bioevents (Alvarez et al. 1984; Kauffman 1988a). Abrupt, widespread loss of individual species may reflect a variety of causes,including competitive displacement by newly evolved taxa, overcropping by predators that causes severe reduction in population size to nonviable levels, the rapid spread of a new disease (as in the American Chestnut), abrupt loss or change of preferred habitat resulting from internal or external forces, and many other factors (e.g., Hallam 1984; Kauffman 1984b). Mass extinctions reflect larger-scale, more global forcing mechanisms, usually a composite of dramatic changes

in the ocean/climate system that exceed the adaptive ranges of ecologically and genetically diverse taxa. The recognition of extinction bioevents at the species level requires highly refined biostratigraphic and chronostratigraphic frameworks within a basin, knowledge of taxonomic range data in time equivalent strata of other regions, and a set of independent chronostratigraphic surfaces against which ranges can be compared.

The Consistency of Paleobiological Response

Despite differences in scales, there are many inherent similarities between short-term bioevents and longer-term patterns in Cretaceous paleocommunities (e.g., ecozones), and between Cretaceous paleobiological phenomena of local, regional, and global extent. This suggests to us that there was a consistent biotic response in populations, communities, and even ecosystems to similar environmental forcing mechanisms at a wide range of spatial and temporal scales (in the same way that community evolution might be viewed as a more complex and longer-term extension of lineage evolution). This consistent response is recorded as a correlative increase or decrease in community diversity and robustness. It crosses the boundaries between different (time equivalent) populations and communities and may be geologically instantaneous (bioevents) or relatively prolonged (ecozones). This hypothesis has significant implications for understanding paleocommunity dynamics and their relationship to environmental forcing, and will be discussed subsequently in light of the data presented herein.

Cenomanian/Turonian Bioevents and Ecostratigraphy, Western Interior, United States

Geological Setting

The Western Interior Basin of North America (fig. 20.2) developed during Late Jurassic and Cretaceous time as an Andean-style foreland in response to crustal loading in the tectonically active Sevier Orogenic Belt (Price 1973; Jordan 1981; Kauffman 1984a). There are five principal tectonic zones in the basin, including the rapidly subsiding western foredeep basin; an incipient, discontinuous zone of forebulge uplifts; the broad, moderately subsiding central axial basin; an east-central hinge zone; and the stable to slowly subsiding eastern cratonic platform (Kauffman 1984a). As a result of both subsidence and large-scale tectono-eustatic sea-level fluctuations, the basin was repeatedly flooded by marine waters from Albian to Maastrichtian time (Williams and Stelck 1975; Kauffman 1977, 1984a). During the Late Cenomanian to Early Turonian portion of the Greenhorn Marine Cycle, sea level reached peak highstand for the Cretaceous (Kauffman 1977; 1984a; Hancock and Kauffman 1979; Haq et al. 1987). At this

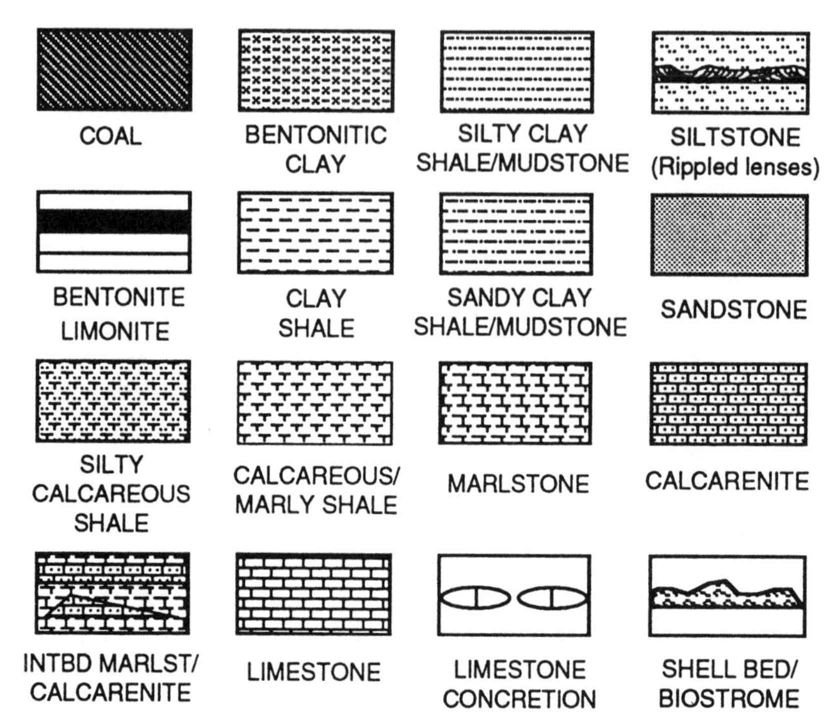

FIGURE 20.2. **(A)** A key for lithologic/lithofacies symbols used in figures 20.2–20.7 is shown. **(B)** Correlation of the study interval across the Western Interior Basin: Map shows distribution of lithofacies and approximate shorelines during the Late Cenomanian/Early Turonian (from Kauffman and Beeson 1992), and the locations of three representative sections of this interval in southern Utah (KPS), south central Colorado (PRS), and central Kansas (BHS). Bentonite marker beds are correlated with dashed lines; letter designations refer to marker beds (M-Beds) of Elder (1985, 1987a, 1988), and the sections are hung on his B-marker bentonite. Carbonate-rich marker beds are indicated by an SL, LL, or LS number corresponding to skeletal limestone, laminated limestone, or pelagic limestone, respectively, and are correlated with solid lines (SL, LL designations from Sageman 1991; LS designations from Elder 1985). Marine flooding surfaces at KPS are indicated with an FS designation. Stable isotope curve reflects $\delta^{13}C$ of organic matter from samples of a core taken near the PRS locality (from Pratt 1985); the positive excursion marks the position of OAE II. Note that meter thickness at PRS is from base of Hartland Shale Member, but only the middle to upper part of the member is shown (see fig. 20.4).

time the basin reached its greatest size and extended some 4,800 km in a north/south direction, connecting the circumpolar "Boreal" Ocean (present-day Arctic Canada and Alaska) with western Tethys (present-day Gulf of Mexico and Caribbean). East to west the seaway had a maximum width of about 1,600 km, with Early Turonian shorelines from New Mexico through Montana on the west and through Minnesota and Iowa on the east (fig. 20.2).

The Sevier Fold and Thrust Belt was the upland source for the bulk of siliciclastic sediments entering the basin, with only minor contributions from the cratonic lowlands to the east (Kauffman 1984a). As a result, the basin has a

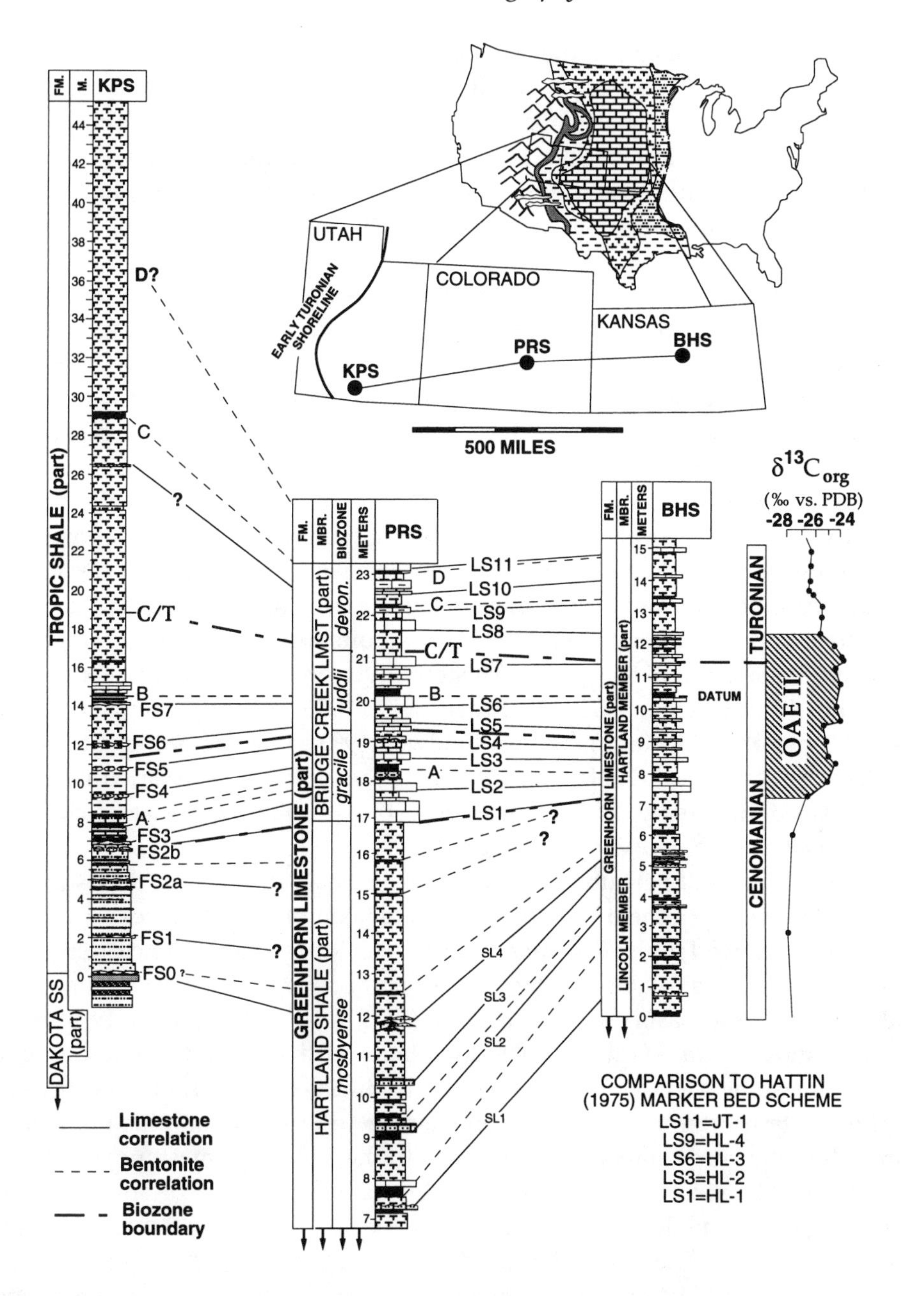

characteristic distribution pattern of marine sediments at maximum flooding of the Greenhorn Cyclothem that generally corresponds to the principal tectonic zones: (1) The western foredeep and forebulge zones (e.g., New Mexico, Arizona, Utah) are characterized by packages of coarse-grained marginal marine and shoreface marine siliciclastics interfingering with finer-grained proximal offshore non- to moderately calcareous mudrocks; (2) the central axial basin and hinge zones (Colorado) are dominated by distal offshore calcareous clay shales and marls with thinner interbedded pelagic limestones; and (3) the eastern stable cratonic shelf (Kansas) is dominated by sequences of proximal to distal offshore carbonate-rich shales, marlstones, chalky limestones, and (shoreward) clay shales with thin interbedded siltstones/sandstones that reflect high carbonate productivity and/or the diminished influence of siliciclastic sediments derived from western source areas and relatively low sediment input from the east.

The distribution of lithofacies and biotas in the basin suggests that the deepest water lay near its central and east-central portions, and extended southward into the Chihuahua Trough of northern Mexico (Frush and Eicher 1975; Kauffman 1984a). Maximum water depth probably did not exceed 200–300 m in the central part of the basin during the Early Turonian second-order highstand (Kauffman 1977, 1984a, 1985), and evidence suggests that smaller-scale relative sea-level fluctuations (10–50 m) may have been common throughout the Greenhorn Cycle. The basin had a favorable configuration for the development of stratified water masses and benthic oxygen deficiency during sea-level rise and around eustatic highstand, as indicated by the prevalence of organic carbon-rich strata and depauperate benthic faunas in offshore areas; various hypotheses have been offered to explain Western Interior oceanographic conditions (e.g., Frush and Eicher 1975; Kauffman 1977, 1986, 1988a; Eicher and Diner 1985; Hay et al. 1993; Kauffman and Caldwell 1993). Biogeographic data, summarized by Kauffman (1984a), indicate that the part of the basin considered herein was dominated by moist warm temperate to subtropical marine climates during the Cenomanian/Turonian interval.

Because of overall net subsidence of the axial basin throughout deposition of the Greenhorn Cyclothem, some of the most complete sequences of marine strata are preserved in the Colorado region. Accordingly, a reference section for the Cretaceous was established by Cobban and Scott (1972) near Pueblo, Colorado (central axial basin). At this locality (PRS in fig. 20.2) the study interval includes most of the Hartland Shale Member and the lower to middle Bridge Creek Limestone Member of the Greenhorn Limestone. Lateral facies changes in the Western Interior Basin result in changes in lithostratigraphic nomenclature. Thus the study interval corresponds to the Lincoln Limestone and Hartland Shale Members of the Greenhorn Limestone in central Kansas (Hattin 1975; Sageman and Johnson 1985) and to the Dakota Formation and Tropic Shale in southern Utah (Peterson and Kirk 1977) (see fig. 20.2). These units have been the subject of considerable study over the years (e.g., Cobban and Scott 1972; Hattin 1975; Kauffman 1977; Pratt et al. 1985), and were recently the focus of detailed basin-wide investigations by the authors (Elder 1985, 1987, 1988, 1989; Elder and Kirkland 1985; Sageman 1985, 1989, 1991; Harries and Kauffman 1990; Harries 1993).

Study Interval and Methodology

The interval chosen for study spans the middle Late Cenomanian through the Early Turonian and contains a mass extinction-recovery interval (Raup and Sepkoski 1982; Kauffman 1984b; Elder 1989; Harries and Kauffman 1990), and several intervals of organic carbon-rich strata, including Oceanic Anoxic Event II of Schlanger and Jenkyns (1976) and Arthur et al. (1990). A detailed west/east transect of the study interval, spanning the major sedimentologic regimes of the Western Interior Basin (described earlier), has been established based on high-resolution event stratigraphic (HIRES) analysis of numerous sections (Elder 1985, 1987, 1989, 1991; Sageman 1985, 1989, 1991; Harries and Kauffman 1990; Harries 1993). This paper presents data from three sections representative of this cross-basin transect (fig. 20.2) to illustrate the history of Cenomanian/Turonian bioevents and ecostratigraphy. These sections include the Kaiparowits Plateau Section (KPS), the Pueblo Reference Section (PRS), and the Bunker Hill Section (BHS) (fig. 20.2).

The detailed chronostratigraphic framework used to correlate between KPS, PRS, and BHS (fig. 20.2) was developed though application of HIRES methods (see Kauffman 1986, 1988a; and Kauffman et al. 1991 for overviews). The methodology depends on four main types of analysis:

1. *Biostratigraphy.* Based on established molluscan biozonations of Kauffman et al. (1978, 1993), Cobban (1984), and Kennedy and Cobban (1991), ammonite and bivalve taxa documented in bulk sample analyses from the study interval confirm a Late Cenomanian through Early Turonian age (Elder 1985, 1987, 1989; Sageman 1985, 1991). Recent advances in biostratigraphic analysis of the Western Interior have resulted in the recognition of up to 12 molluscan assemblage biozones within the study interval (summarized in Kauffman et al. 1993). However, the scheme employed earlier by Cobban (1984), Elder and Kirkland (1985), and Elder (1985, 1991), in which only five major biozones are defined based on total composite range zones of well-known ammonite index taxa, is utilized here (fig. 20.2) because it is most directly comparable to the current European zonation of Kennedy and Cobban (1991). At PRS the Hartland Shale Member corresponds to the Late Cenomanian *Metoicoceras mosbyense* Biozone. The basal beds of the Bridge Creek Limestone Member (e.g., LS1–LS3: fig. 20.2) contain representatives of the *Sciponoceras gracile* Biozone. The latest Cenomanian *Neocardioceras juddii* Biozone occurs from marker beds LS4 to LS7, and the earliest Turonian *Watinoceras* spp. Biozone (including *W. devonense* Wright and Kennedy and *W. coloradoense* Cobban) ranges from LS8 to LS11 (fig. 20.2). Above LS11, *Mammites nodosoides* Biozone taxa have been described from PRS (Cobban 1985; Kennedy and Cobban 1991). The corresponding five biozones of the European Cretaceous (e.g., southern England) include the *Calycoceras guerangeri* Biozone,

the *Metoicoceras geslinianum* Biozone, the *N. juddii* Biozone, the *Watinoceras coloradoense* Biozone, and the *Mammites nodosoides* Biozone, respectively (Cobban 1984; Kennedy and Cobban 1991).

2. *Physical event units.* Following the work of Hattin (1971), the predominant physical event units that are used for regional correlation of the Greenhorn Limestone include bentonite beds, representing altered volcanic ash fall deposits, and carbonate-rich units, such as limestone or skeletal limestone beds, which commonly represent offshore condensed deposits (e.g., Hattin 1975; Elder 1985, 1987, 1988, 1991; Sageman 1985, 1991). Major bentonite beds that were correlated throughout most of the Western Interior Basin using stratigraphic techniques (e.g., beds A and B of Elder 1988, fig. 20.2) have been tested by chemical fingerprinting methods (see Kowallis et al. 1988), providing independent confirmation of the correlations. They represent the most dependable chronostratigraphic markers for distal to proximal offshore facies. In addition, limestone beds have been shown to grade from central basin distal offshore facies westward into concretion horizons of proximal offshore areas, and subsequently into transgressive lags or marine flooding surfaces on top of prograding parasequence deposits of the seaway's western margin (Kauffman et al. 1987; Elder et al. 1993). Five such units (FS3–7, fig. 20.2) have been confidently correlated within the study interval (Elder et al. 1993) and provide a set of high-resolution markers that encompass a complete transect from the basin center to the strandline.

3. *Chemostratigraphy.* Within the study interval there are a series of distinctive geochemically defined horizons marked by shifts or spikes in stable isotopic and trace element data. For example, Scholle and Arthur (1980) first recognized a spike in the $\delta^{13}C$ record of C/T strata that was correlative throughout Europe. Pratt and Threlkeld (1984) and Pratt (1985) later identified this marker in the Western Interior as a 2–4 per mil positive shift in the $\delta^{13}C$ signature of organic matter spanning about 5 m of strata within the study interval at PRS and BHS (fig. 20.2). Kauffman (1988a) suggested that individual peaks within this excursion could be used for even-finer-resolution correlations. The excursion extends from the base of the *S. gracile* Biozone into the *Watinoceras* Biozone and maintains a consistent stratigraphic position relative to biozone boundaries and lithostratigraphic marker beds throughout the west/east transect of the basin. Trace element data presented by Orth et al. (1988) showed similar potential for correlation, with distinct enrichments at several horizons in the lower part of the Bridge Creek Member at PRS (e.g., Mn, Ir). These geochemical markers have since been detected at other Cenomanian/Turonian sections, both within and outside the Western Interior Basin (Orth et al. 1993).

4. *Ecostratigraphy.* Paleoecological analyses are based on continuous collection of uniform-size bulk samples averaging 10–40 cm in thickness through each stratigraphic section. The resulting data set, including both

biostratigraphic range data and values of total species richness and faunal abundance, are statistically comparable within and between sections. Fossil assemblages are also analyzed for basic taphonomic data (i.e., levels of fragmentation, disarticulation, shell abrasion, and bioerosion), as well as trends in shell density, size distribution, and orientation. These data are used to help distinguish long-term resident community components from short-term event communities and allochthonous biotic elements. Each sample is assigned a resident or event biofacies categorization based on its taphonomic condition, faunal association, trophic group composition, and overall diversity and abundance (e.g., Elder 1989, 1991; Kauffman and Sageman 1990; Sageman 1989; Sageman et al. 1991).

The Hartland Shale and Bridge Creek Limestone Members have long been recognized for their distinctive biotic characteristics. The Hartland Shale at PRS is an interval characterized by finely and evenly laminated, dark calcareous shale with high levels of preserved organic carbon and relatively depauperate faunal assemblages (Kauffman 1977, 1984b; Sageman 1985, 1989, 1991). It is virtually devoid of benthic foraminifera (Eicher and Worstell 1970; Eicher and Diner 1985), contains only very rare burrows, and is dominated by resident communities of epifaunal inoceramid bivalves. These features are best developed in the central axial basin but can be recognized at all localities containing offshore marine strata of the *M. mosbyense* Biozone. As a result, the Hartland Shale Member has been interpreted as an interval of widespread and prolonged oxygen depletion within the Western Interior Basin (Frush and Eicher 1975; Kauffman 1984a; Sageman 1985, 1989).

In contrast, the overlying lower Bridge Creek Limestone Member is characterized by a shift to highly burrowed pelagic limestone beds interbedded with laminated dark marlstones (interpreted to represent Milankovitch climatic cycles: Barron et al. 1985; Fischer et al. 1985), and contains the highest composite number of species recorded for any part of the Greenhorn Cyclothem (Koch 1977; Elder 1987). In the beds immediately above the lower part of the member, however, evidence of the C/T mass extinction and of the Bonarelli OAE is represented by increased burial of organic carbon in shale beds, significant loss of species (especially molluscan) in a series of extinction "steps," and a widely correlative decrease in diversity and abundance within benthic marine communities (Elder 1989). These events are followed by an initially punctuated and subsequently gradual recovery of marine communities during the Early Turonian (Harries and Kauffman 1990; Harries 1993).

Analysis of Local Bioevents and Ecostratigraphy

Taxonomic, ecological, and sedimentologic data are presented and analyzed for each of the three sections of the west/east transect (figs. 20.3–20.6). For more extensive presentations of these data the reader is referred to Elder (1987, 1989, 1991, 1985), Sageman (1985, 1989, 1991), Harries and Kauffman (1990), Kauffman and Sageman (1990), Sageman et al. (1991), and Harries (1993). In this paper, plots of species richness and faunal abundance, commonly viewed as proxies for trends in benthic communities, are employed to illustrate bio-events and ecostratigraphic trends (figs. 20.3–20.6). Identified bioevents are categorized following the classification scheme described earlier.

THE KAIPAROWITS PLATEAU SECTION *(KPS)*

The Kaiparowits Plateau Section is located near Wahweap Wash, south-central Utah (fig. 20.2), at the base of the south-facing escarpment of the Kaiparowits Plateau. The study interval includes the Late Cenomanian/Early Turonian parts of the Dakota Formation and lower Tropic Shale, which were deposited in the rapidly subsiding Western Interior foredeep basin (Peterson and Kirk 1977; Gustason 1989). Coal beds occurring at the base of the section represent coastal plain facies of the uppermost part of the middle nonmarine portion of the Dakota Formation (Gustason 1989). They are overlain disconformably by a bioturbated sandy unit and an oyster concretion horizon (FS0, figs. 20.2, 20.3), reflecting the initial marine flooding of the Western Interior sea into southern Utah. The overlying Tropic Shale grades upward from sandy and silty bioturbated mudstones to calcareous shales and contains seven more calcareous concretion or shell lags beds, each of which can be traced westward to transgressive lags overlying upward-coarsening progradational shoreface sequences (Gustason 1989; Elder et al. 1994). These event beds (FS1–7, figs. 20.2, 20.3) represent the marine flooding surfaces of seven parasequences (*sensu* Van Wagoner et al. 1990) that record relatively short-term, high-frequency fluctuations of the strandline with associated migration of benthic environments. Overall, the study interval at KPS represents a period of progressive deepening from paludal/lagoonal to proximal offshore environments.

Based on recognition of key biostratigraphic index taxa, the lower 6.5 m of the Tropic Shale and the underlying Dakota Formation represent the uppermost part of the Late Cenomanian *Metoicoceras mosbyense* Biozone and correspond to the middle to upper Hartland Shale Member as defined at PRS (fig. 20.2). The base of the overlying interval, equivalent to the base of the Bridge Creek Limestone Member at PRS, is marked by a conspicuous concretion/shell bed (FS2b, figs. 20.2, 20.3) and the first appearance of the *Sciponoceras gracile* Biozone fauna (Elder 1991). The top of the *S. gracile* Biozone occurs between 11 and 12 m above the base of the Tropic Shale and is overlain by an interval containing the latest Cenomanian *Neocardioceras juddii* Biozone fauna (Elder 1991). The Cenomanian/Turonian boundary occurs 19 m above the base of the Tropic Shale and is indicated by the first appearance of *Mytiloides hattini* Elder, marking the base of the Early Turonian *Watinoceras* Biozone (Elder 1991).

KPS Bioevents In figure 20.3 trends in macrofaunal species richness and abundance reflect KPS bioevents and ecozones. Most bioevents identifiable within the study interval are ecological and resulted from rapid migration of benthic environments during eight cycles of transgression and regression of the strandline. Each cycle includes an upward-coarsening progradational phase capped by a discrete marine flooding surface (designated FS0–FS7). Using these flooding surface designations as a guide, the ecostratigraphy of the KPS section is as follows.

FS0 Initial marine flooding at KPS is marked by a colonization bioevent that reflects the local establishment of a distinctive brackish to marine community (fig. 20.3). At the base of the KPS section, the uppermost mappable sandstone bed (and first marine unit) of the Dakota Formation is overlain successively by a thin carbonaceous shale and a 10-cm oyster concretion horizon. The concretion layer reflects the development of a biostrome or shell bed of *Flemingostrea prudentia* (White), an epifaunal suspension feeder with a preference for shallow brackish to marine waters and soft substrates (Fürsich and Kirkland 1986). Development of the biostrome indicates a local colonization bioevent (ecological diversification) that resulted from a decrease in coarse siliciclastic sedimentation and formation of a biosubstrate in a nearshore muddy marine area with brackish influence (a brackish to marine bay?). The concretion bed can be traced for miles to the west as a transgressive disconformity, in some cases directly overlying coal beds, and clearly reflects the initial marine transgression into the KPS region (initial marine flooding event: fig. 20.3).

FS1 TO FS2A Increases in species richness and faunal abundance in the lower 4 m of Tropic Shale at KPS reflect colonization and population expansion related to two upward-coarsening parasequences (nearshore parasequences: fig. 20.3). The generalized lithologic description (upsection) for each of these parasequences includes a dark carbonaceous or slightly calcareous laminated shale, overlain by bioturbated silty mudstone with increasing thin, bioturbated to ripple-laminated lower shoreface siltstone/sandstone beds. Flooding surfaces FS1 and FS2a cap the sequences but are not marked by distinct transgressive lag deposits (concretions or shell beds) at the KPS locality. However, continuous updip exposure along the outcrop belt allows regional tracing of parasequences 1 and 2, and correlation of FS1 and FS2a to transgressive lag beds at nearby localities (Gustason 1989; Elder et al. 1994).

The faunal associations in samples spanning the two parasequences reflect a similar succession in each. They are dominated by somewhat brackish (lower bay or estuarine mouth) associations, including trophically diverse infaunal and epifaunal elements (mainly infaunal deposit and suspension-feeding protobranch, corbulid and venerid bivalves, sparse scaphopods, grazing or carnivorous gastropods, and epifaunal suspension-feeding bivalves). Cycles of increas-

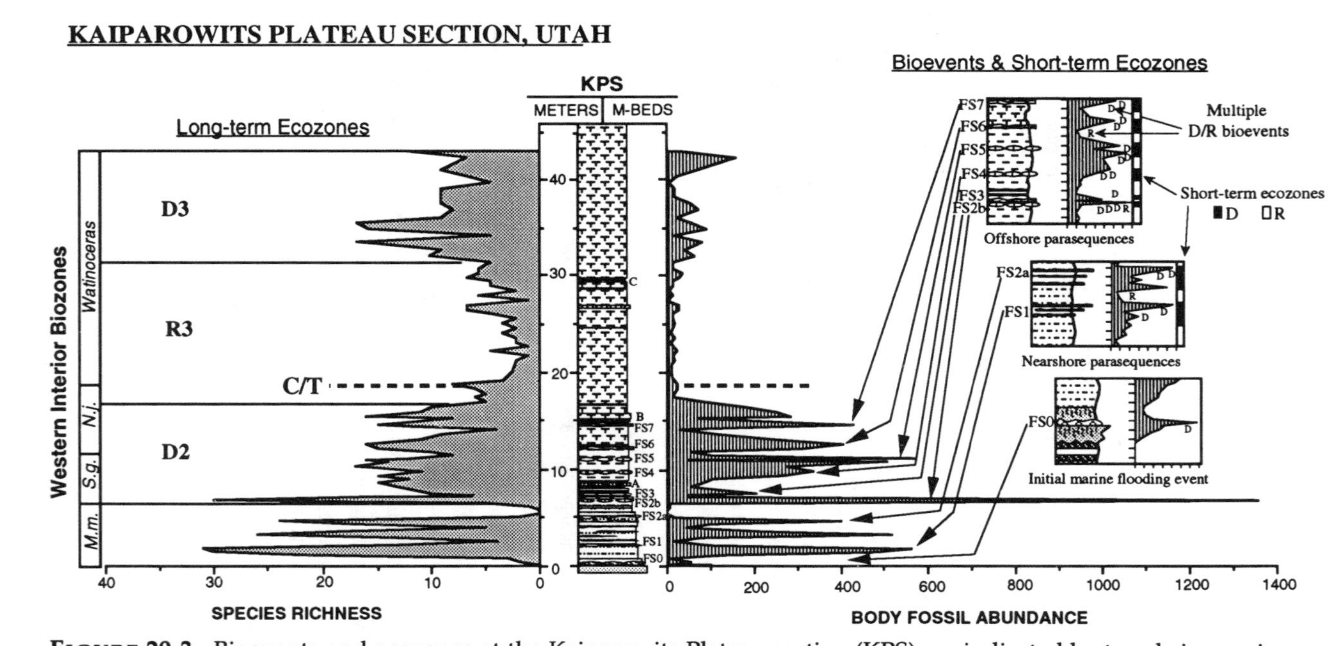

FIGURE 20.3. Bioevents and ecozones at the Kaiparowits Plateau section (KPS) are indicated by trends in species richness and abundance data. Lithologic section and marker beds (M-Beds) are as in figure 20.2. Insets to the right show the relationship of diversification and reduction bioevent horizons (represented by D and R letters) to the time-averaged curve of species richness/abundance data and short-term ecozones. Biostratigraphic zones are indicated at left, with abbreviations as follows: M.m. = *Metoicoceras mosbyense,* S.g. = *Sciponoceras gracile,* N.j. = *Neocadrioceras juddii, Wat.* = *Watinoceras,* and M.n. = *Mammites nodosoides.* Major flooding surfaces are indicated by FS designations. C/T marks the position of the Cenomanian/Turonian boundary.

ing species richness and abundance correlate to trends of increasing shell fragmentation and reach their peaks in the samples just below or at parasequence flooding surfaces FS1 and FS2a. (Note also the faunal peak between the FS1 and FS2a peaks, which is not associated with a widely recognized flooding surface; it may reflect an early pulse of the parasequence 2 progradation.) These cycles reflect trends in benthic colonization and population expansion related to relative shoaling of the parasequences and are interpreted as the result of higher current energy, increased mixing and oxygenation, and greater availability of food resources. Numerous individual colonization bioevents are represented in each cycle, but horizons could not be stratigraphically resolved because of extensive bioturbation. In the samples directly overlying the flooding surfaces, a decline in species richness and abundance, and the occurrence of depauperate faunas (mainly inoceramid, pectinid, pterid, and ostreid bivalves) characteristic of offshore low-oxygen environments and relatively soft substrates reflect reductions in benthic community diversity. These trends, which are much better developed above FS2a than above FS1, are interpreted to reflect rapid deepening and a shift to finer-grained sedimentation, mortality or local emigration of shallow-water brackish forms, and shoreward migration of low-oxygen-tolerant offshore faunas (see "nearshore parasequences" in fig. 20.3).

FS2B TO FS3 The first fully marine, offshore flooding surface occurs at 6.5 m above the base of the KPS section (FS2b, figs. 20.2, 20.3) and correlates eastward to LS1 at the base of the Bridge Creek Limestone Member (fig. 20.2). When traced to the west of KPS, this horizon merges with the underlying flooding surface (FS2a) within 10 km, overlies parasequence 2 for about 70 km to its updip pinchout, and continues 90 km further west as a transgressive surface overlying nonmarine sandstones and coal beds of the Dakota Formation (Gustason 1989; Elder et al. 1994). This surface represents the largest landward shift of the strandline within the study interval. Because there is no preserved evidence of a parasequence deposit between this horizon and the underlying parasequence, and because it coalesces for some distance with the underlying flooding surface, the designations 2a and 2b were applied. At the KPS locality, FS2b is a highly complex condensed unit that includes (upsection) an early diagenetic, shell-rich calcareous concretion horizon, a thin horizon of silty calcareous mudstone, and several dense layers of shell material. The interval is marked by a major peak in abundance of whole fossils, shell fragments and fish debris, and the abrupt appearance of highly diverse subtropical to warm temperate marine taxa belonging to the *S. gracile* Biozone (e.g., numerous molluscs, solitary corals, fodinichnia and domichnia, and abundant planktic foraminifera). These taxa occur mainly as, or associated with, a series of closely spaced or amalgamated colonization surfaces, including horizons with highly abundant surface grazing *Levicerithium* gastropods, lucinid and arciid bivalves, and tiny indeterminate oysters. Whereas the abundance and diversity associated with these colonization events suggests plentiful food resources and favorable benthic conditions, the rapid shifts between distinct associations among successive closely spaced fossil horizons may indicate condensation and/or the presence of disconformities. On some horizons the strong dominance of similarly sized individuals may reflect abrupt mass mortality events (note multiple D/R bioevents, offshore flooding surface 2b, fig. 20.3).

Immediately overlying this fossil-rich zone is a black, organic-rich mudstone unit with dense *Chondrites* burrows followed by a shell-rich layer composed of *Pycnodonte* oysters. Because the *Chondrites* burrow fill consists of sediment from the overlying shell bed, it is possible that burrowing *followed* deposition of the black mudstone and that the mudstone itself represents a time of low benthic oxygen, almost no benthic habitation, and enhanced preservation of organic matter. The shell bed (FS3), however, contains a moderately diverse fauna in addition to the dominant oysters and has been traced by Gustason (1989) westward to a transgressive lag overlying the third upward-coarsening parasequence in the marine part of the Dakota Formation (Elder et al. 1994). It appears that the black mudstone represents the deepest phase of the FS2a transgression, that *Chondrites* represent the pioneer colonization bioevent upon shallowing, and that the oyster colonization bioevent is all that is preserved of the subsequent shoaling/flooding event (FS3, fig. 20.3).

Many oyster shell horizons (usually 5–10 cm thick) associated with flooding surfaces along the basin's western margin consist of a lower biostrome level with shells in life position (representing colonization during the waning stages of progradation). Upon this is superimposed an upper shell lag with disarticulated shells in current stable or chaotic orientations (representing erosion at wave base during development of a ravinement surface). In the case of FS3 the entire unit has the character of a lag deposit and may represent the reworked remains of several colonization events. When this horizon is traced into the basin, it commonly lies directly above the FS2a equivalent, coalesces with the FS2a bed, or is indistinguishable from it. These features support the interpretation of significant condensation and possibly development of disconformities within the FS2a to FS3 interval.

FS4 TO FS7 In the interval overlying FS3 at KPS are four consecutive resistant beds, including fossiliferous concretions, colonization surfaces, and/or shell lags (see offshore parasequences, fig. 20.3). These horizons have been correlated shoreward to four transgressive lags overlying prograding parasequences (Gustason 1989), and basinward to four pelagic limestone beds (Kauffman et al. 1987; Elder 1991; Elder et al. 1994). They are interpreted to represent the offshore expression of the parasequence flooding events. At the KPS locality, each complete cycle includes a 1–2-m-thick unit of variably burrowed claystone or shale representing parasequence progradation, capped by a shelly, carbonate-enriched bed representing the flooding surface.

FS4 is characterized by two distinct short-term colonization bioevents. The lower occurs within an early diagenetic limestone concretion horizon, formed around serpulid worm tubes that colonized a condensed, sediment-starved transgressive surface. The serpulid biostrome is associated with *Gryphaeostrea* and diverse byssally attached bivalves (e.g., *Phelopteria, Lima*), and shows a modest increase in total faunal abundance and species richness from the previous samples (fig. 20.3). In the bioturbated claystone immediately overlying the concretion bed, a colonization surface dominated by two genera of oysters (*Pycnodonte, Rhynchostreon*) creates the final FS4 abundance peak (fig. 20.3). This oyster biostrome represents the second colonization bioevent associated with flooding of parasequence 4.

In the interval between FS4 and FS5 is a fairly diverse and robust resident faunal association characteristic of the *S. gracile* Biozone along the basin's western margin (fig. 20.3). Unlike flooding events 1, 2a, 2b, and 3, the late stage of deepening associated with flooding event 4 produced relatively little reduction overall in species richness and abundance. However, changes in biofacies indicate a distinct shift in benthic conditions between FS4 and FS5: Oysters, which dominate in the lower part of this interval, decline in abundance upward; concurrently, the abundance of shallow infaunal deposit-feeders (*Lucina, Drepanochilus*) increases to a peak just below FS5. These data suggest that infaunal deposit feeders replaced epifaunal oysters because of increasing sedimentation rates and turbidity, and increasing availability of detrital organic matter during progradation of parasequence 5.

FS5 represents an amalgamated diversification bioevent sequence that includes several faunal events. At the base of FS5, a bioturbated limestone concretion contains a diverse assemblage of gastropods and ammonites (e.g., *Sciponoceras, Euomphaloceras, Euspira, Drepanochilus*). Current-aligned *Sciponoceras gracile* (Shumard) within the concretion suggests sediment bypass/starvation and orientation of biotic elements that thrived during the initial shallow, well-mixed phase of parasequence 5 transgression. Sharp burrow walls indicate that following early diagenetic cementation, diverse infauna (including fodinichnia and domichnia) colonized the shell-rich substrate, producing a second diversification bioevent. A *Pycnodonte* oyster biostrome found immediately above the concretion horizon represents the final colonization event associated with FS5 flooding, resulting in a distinctive spike in faunal abundance at FS5. This spike is followed by a marked decrease in species richness/abundance and a shift toward offshore, low-oxygen-tolerant faunas dominated by a byssate epifaunal bivalve association (*Inoceramus, Phelopteria*). This reduction in diversity reflects the deepest stage of the FS5 flooding event and the landward migration of offshore (low-oxygen) benthic environments.

The character of bioevents at FS6 and FS7 reflects progressive deepening at KPS. Lithologically, the flooding surfaces are marked by beds of resistant, burrowed calcareous shale or mudstone (figs. 20.2, 20.3). The offshore character of biofacies in the claystone underlying FS6 does not change appreciably through the flooding surface, but a peak in diversity and abundance directly overlies the resistant bed. This horizon lacks the oyster biostrome typical of underlying (shallower) flooding surfaces but is dominated by an association of *Inoceramus, Phelopteria,* and offshore ammonites (e.g., *Neocardioceras, Kamerunoceras*) that marks the peak flooding stage and maximum offshore condensed interval of the parasequence. Similarly, FS7 is marked by a peak in species richness and abundance dominated by the *Inoceramus, Phelopteria,* and offshore ammonite association. Faunal abundance drops off rapidly in the samples overlying each flooding surface, and biofacies become progressively dominated by low-diversity associations of generalized and infaunal deposit feeders (*Lucina, Drepanochilus*) or epifaunal bivalves (inoceramids and small oysters). These trends reflect diminished benthic oxygen and increased preservation of marine organic matter at KPS. However, unlike the underlying sequences, this pattern of reduced diversity and abundance continues for a considerable distance upsection, including the upper half of the *N. juddii* Biozone and much of the *Watinoceras* spp. Biozone (fig. 20.3). In this interval, inoceramids and small oysters dominate resident biofacies, and there are only rare colonization events by such infaunal taxa as *Lucina* and *Drepanochilus*.

KPS Summary Stratigraphic and faunal data from the KPS locality allow the following conclusions:

1. The section records a late Cenomanian marine incursion across a coastal plain in southern Utah and a series of proximal to medial offshore progradational units and their flooding surfaces (parasequences) that formed as the site progressively deepened toward Early Turonian maximum highstand. A major shift in the character of sediments and faunas occurs midsection (above FS7), where offshore clay-rich facies become dominant and shell-rich flooding surfaces are no longer well developed, suggesting continued deepening to proximal or medial offshore environments that were beyond the perimeter of shallow water shell concentrations.
2. Numerous short-term diversification (D) and diversity reduction (R) bioevents characterize the faunal history of the lower Tropic Shale at KPS. Collectively, they form a pattern of relative D/R cycles that track with a series of seven upward coarsening parasequences in the Dakota Formation. Diversification bioevents are dominantly ecological, and they characterize the shallowest phase of each parasequence cycle (late progradation and early flooding phases), when circulation, benthic oxygenation, nutrient levels, and substrate conditions enhanced opportunities for colonization. Diversification bioevents are commonly complex, amalgamating colonization events that occurred during both relative shallowing and relative deepening periods, and include colonization of transgressive starved and

ravinement surfaces. Diversification of taxa, trophic groups, and adaptive strategies, and increased faunal abundance characterize benthic environments during progressive shoaling and progradation. This effect is enhanced as transgression begins, because of diminished sedimentation rates and initial development of starved flooding surfaces. In many cases benthic communities become dominated by one or two taxa that opportunistically colonize the transgressive ravinement surface, producing distinct abundance peaks. These two types of processes resulting in ecological diversification commonly occur together as a compound event marking KPS flooding surfaces. Reduction bioevents, in contrast, are represented by overall decrease in species richness and abundance because of emigration or localized mortality events, and a shift to offshore biofacies (dominated by low-oxygen-tolerant epifaunal bivalves). They reflect the deepest phases of each parasequence cycle and the furthest landward migration of offshore benthic communities characteristic of oxygen-depleted basinal settings.

3. Plots of species richness and faunal abundance present a time-averaged perspective of faunal change at KPS. Peaks and troughs on the scale of 1–2 m generally correlate with the shallowing and deepening phases of individual parasequences, and these features are bundled in larger relative diversification and diversity reduction zones about 1012 m in thickness (fig. 20.3). Individual peaks and troughs have similar minimum and maximum values within diversification zones (e.g., D2 and D3) and are distinct between diversification and diversity reduction zones (e.g., D3 vs. R3). At each scale the zones are characterized by repetition of similar resident community types and similar multiple bioevents, although taxonomic compositions change through the stratigraphic sequence. Although these short- and long-term ecostratigraphic zones are related to relative sea-level changes and thus appear to reflect normal "faunal tracking" (as described by Brett et al. 1990, and Brett and Baird, paper 10 in this volume), we will demonstrate that they have relatively sharp boundaries that can be correlated at the same stratigraphic level from KPS to BHS. This suggests that the zones reflect more or less synchronous basinwide trends of relative improvement or decline in benthic environments, and thus comprise chronostratigraphically useful ecozones in the Western Interior Basin.

4. The dominant controlling factors for the development of bioevents and ecozones at KPS are interpreted to be (a) levels of benthic oxygenation, reflecting overall water depth, stratification, and mixing intensity; (b) changes in current energy that influenced sediment transport and mixing of the water column; (c) variations in sedimentation rate and resultant changes in substrates, turbidity levels, and transport of land-derived nutrients; and (d) biogenic modification of substrates (e.g., through bioturbation or development of biostromes).

THE PUEBLO REFERENCE SECTION (PRS)

Cobban and Scott (1972) established the standard reference section for basinal Cretaceous facies of the Western Interior Seaway in east-central Colorado (see also papers in Pratt et al. 1985). Predominantly medial to distal offshore facies of the Greenhorn and Niobrara Cyclothems are exposed along escarpments bordering the Arkansas River where it has breached the Rock Canyon Anticline just west of Pueblo, Colorado. The study interval at PRS includes the Hartland Shale and the lower to middle part of the Bridge Creek Limestone Members of the Greenhorn Limestone. The lithostratigraphy and biostratigraphy of these units were introduced earlier under "Geologic Setting" (see also fig. 20.2). Figure 20.4 illustrates the PRS section with species richness and faunal abundance plots and indicates the bioevents and ecozones discussed later.

PRS Bioevents Hartland Shale bioevents are dominantly ecological and reflect rapid changes in benthic oxygen levels and substrate character. Laminated, pyritiferous, olive gray to olive black, C_{org}-rich (3–4 wt. %) Hartland Shale facies are mainly characterized by relatively low-diversity, *Inoceramus*-dominated resident communities. Inoceramids were well adapted to low oxygen levels (Kauffman 1981, 1984a), possibly through chemosymbiosis (Kauffman 1988b; Kauffman and Sageman 1990; McLeod and Hoppe 1992), and acted as the pioneer species for resident Hartland paleocommunities. Bioevents in the Hartland Shale are mainly expressed as moderate to densely colonized bedding plane surfaces (event communities) of epifaunal benthic bivalves (e.g., *Inoceramus, Entolium*) or less commonly, trace fossils of soft-bodied, deposit-feeding infauna (e.g., *Planolites, Chondrites*). Colonization bioevents reflect brief episodes of benthic circulation and oxygenation, and/or changes in substrate character, and are commonly followed by mass mortality bioevents that suggest rapid deoxygenation and H_2S poisoning, or rapid burial.

Bioevents of the Bridge Creek Member at PRS, which is characterized by rhythmic interbedding of highly burrowed micritic limestone beds with laminated, organic-rich shales, include numerous ecological diversification and reduction events. Biofacies of the *Sciponoceras gracile* Biozone represent a significant diversification compared to Hartland Shale faunas, principally through immigration and colonization bioevents. Diversification and reduction bioevents alternate within shale/limestone cycles and reflect regional changes in benthic environments, presumably related to Milankovitch climatic cycles (Barron et al. 1985). In the following sections, PRS bioevents are summarized in ascending stratigraphic order (fig. 20.4).

ENTOLIUM EVENT COMMUNITIES/MASS MORTALITY SURFACES The basal 8 m of the Hartland Shale at PRS are characterized by numerous bedding planes covered by single cohorts of highly abundant, small juvenile (1–2-mm-long), thin-shelled *Entolium gregarium* Kauffman and Powell (Pectenacea), sepa-

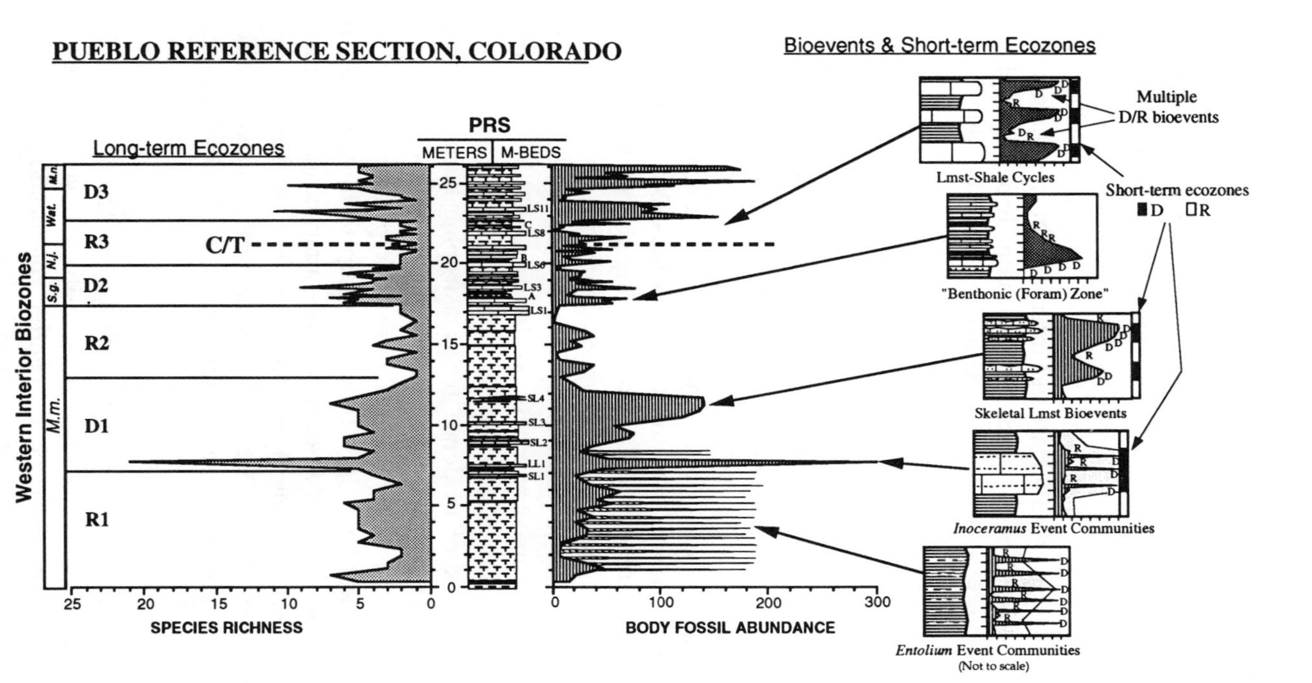

Figure 20.4. Bioevents and ecozones at the Pueblo, Colorado, section (PCS) are indicated by trends in species richness and abundance data. Lithologic section and marker beds (M-Beds) are as in figure 20.2. Insets show relationship of diversification and reduction bioevent horizons (represented by D and R letters) to the time-averaged curve of species richness/abundance data and short-term ecozones. In the case of *Inoceramus* and *Entolium* events, the time-averaged curve is a background stipple, whereas the foreground pattern reflects event community abundance levels. Note that different bioevents may be reflected more strongly by species richness *or* abundance data, as indicated by stipple versus striped patterns. Biostratigraphic zones are abbreviated as in figure 20.3. C/T marks the position of the Cenomanian/Turonian boundary. Diversity trend for "Benthonic (Foram) Zone" inset modified from Eicher and Diner (1985) and Leckie (1985).

rated by thin shale intervals with rare specimens, or lacking *Entolium* (fig. 20.4). These represent opportunistic colonization events or spat falls that occurred during short-term increases in bottom-water oxygenation. Colonization was followed rapidly (weeks–months?) by mass mortality as a result of a rise in the redox boundary to or above the SWI, poisoning by H_2S, and/or smothering by thin mud turbidites. These composite bioevents, modeled in the lower right panel of figure 20.4, have regional expression at the same stratigraphic level in the central part of the Western Interior (Colorado to Oklahoma), but individual colonization/mortality surfaces have not been regionally correlated. Although *Entolium gregarium* ranges throughout the study interval, it occurs in exceptional abundance only in this lower to middle Hartland interval. This *Entolium* abundance zone, which represents the best example of a traditionally defined epibole documented within this study, does not correlate with other benthic faunal indicators, and crosses ecozone boundaries that are traceable well beyond its limits.

INOCERAMUS BIOEVENTS A 10-cm-thick, olive gray laminated limestone, 7 m above the base of the Hartland Shale Member at PRS (marker bed LL1 in fig. 20.4) contains several colonization/mass mortality surfaces representing composite diversification/reduction bioevents related to changes in benthic oxygenation. They are composed of one (predominantly) to two cohorts of *Inoceramus ginterensis* Pergament in or near growth position. Adult-sized *I. ginterensis* (up to 10 cm long) found throughout well-laminated, organic-rich facies of the lower to middle Hartland Shale indicate low-oxygen adaptations for the species. Yet in marker bed LL1 (fig. 20.4), colonization events of *I. ginterensis* were followed within 1 or 2 years by mass mortality, as demonstrated by the strong single size class dominance of juvenile or near-juvenile specimens (0.9 and 1.4 cm long)(Sageman 1989). Lack of evidence for catastrophic burial suggests rapid, extreme oxygen depletion, and/or H_2S poisoning at the SWI. Diversity trends within other samples reflect an additional bioevent type involving *Inoceramus* bivalves in the Hartland Shale. In certain cases, up to three species of *Inoceramus* may be associated with additional byssally attached or cementing taxa (small *Phelopteria, Ostrea, Pycnodonte*) that were unable to survive on soft, oxygen-depleted calcareous muds but that colonized *Inoceramus* bivalves as shell islands. These represent diversification bioevents made possible by biological interactions.

SKELETAL LIMESTONE BIOEVENTS At a number of levels within the middle Hartland Shale (SL2, 3 and 4: fig. 20.4), 1–5-cm-thick, olive gray skeletal limestones (calcarenites) are characterized by benthic colonization events reflecting abrupt changes in physical sedimentation. Ripple lamination and scouring suggest that these are storm-generated, density-flow deposits, which also may have been associated with water mass mixing and short-term benthic oxygenation. The major event communities are composed of trace fossils representing infaunal sediment feeding and dwelling taxa (*Planolites, Thalassinoides,* and *Chondrites* burrows) and scattered surface-crawling organisms (*Crossopodia*). Abundant, fragmented bivalves (mainly *Inoceramus*) comprising these limestones probably represent locally winnowed or transported material, but larger inoceramid fragments and entire valves, and cementing oysters such as *Ostrea* suggest short-term colonization of the coarse-grained substrates. Thus colonization bioevents in skeletal limestones probably reflect a combination of factors, including condensation/deposition (development of a firm substrate) and short-term oxygenation.

BENTHIC FORAMINIFER BIOEVENTS In the uppermost 0.5 m of the Hartland Shale Member, shales become indistinctly burrowed and calcareous benthic foraminifera colonize the central basin for the first time in the Greenhorn Cyclothem (Eicher and Diner 1985; Eicher and Worstell 1970). This initial oxygen-related bioevent of the "Benthonic Zone" (Eicher and Worstell 1970) marks a significant change in substrate and bottom-water conditions in the central Western Interior sea and immediately precedes the beginning of limestone deposition in the central basin ("Benthonic (Foram) Zone," fig. 20.4). Eicher and Worstell (1970) and Eicher and Diner (1985) document the development of a diverse assemblage of calcareous benthic foraminifers in shale samples of the lower Bridge Creek Member in the central Western Interior Basin, with species richness values rising to 30 species (locally reflecting successive colonization events). Species gradually disappear upsection, and benthic foraminiferal diversity and abundance values reach a minimum in association with well-laminated, organic-rich facies of the *N. juddii* Biozone (fig. 20.4). Although species richness and abundance increase again slightly in the lower Turonian *Watinoceras* Biozone (Leckie 1985), calcareous benthic foraminifers remain depauperate throughout the Bridge Creek Member, and disappear completely in the overlying Fairport Chalky Shale Member (Eicher and Diner 1985). Because these benthic taxa range throughout the Greenhorn interval in shallower proximal offshore areas (Eicher and Diner 1985), their presence or absence in the central part of the basin is interpreted to reflect ecological controls on colonization and mortality. (Note that the diversity/abundance acme of the "Benthonic Zone" corresponds to long-term ecozone D2, fig. 20.4).

BIOEVENTS IN SHALE/LIMESTONE CYCLES The Bridge Creek Limestone Member at PRS is characterized by alternating 5–75-cm-thick units of well-laminated, olive gray calcareous shales and marls and 10–50-cm-thick beds of resistant, micritic, gray limestone or marlstone beds (LS1–LS11, fig. 20.2, 20.3). These have been widely interpreted (e.g., Barron et al. 1985; Fischer et al. 1985) to represent changes in sedimentation resulting from alternate phases of Milankovitch climate cycles (characterized predominantly by shifts in evaporation/precipitation balance through changes in monsoonal patterns, or wet/dry cycles). Trends in species richness and abundance, biofacies, and bioevents suggest a strong correlation between changes in benthic communities and these rhythmic depositional cycles, and a general model is shown by the inset in figure 20.4. Immediately overlying highly burrowed limestone beds, well-laminated to microburrowed, organic-rich shales are characterized by dominance of low-diversity associations of broad, flat, resident epifaunal inoceramids. This shift initially reflects a reduction bioevent in response to development of a softer clay-rich substrate and depletion of oxygen in the bottom water, suggesting enhanced stratification. In contrast, limestone/marlstone beds (especially in the basal Bridge Creek Member) are highly bioturbated, with the density, number of generations, tiering, trophic complexity, and diversity of ichnotaxa (*Planolites, Chondrites* and *Thalassinoides* dominant) increasing upward through each carbonate-rich unit (Barron et al. 1985; Elder 1985, 1989, 1991; Kauffman 1988a). A strong correlation between carbonate content and burrow presence, size, and depth of penetration suggests that decrease or cessation of siliciclastic sedimentation and early cementation resulted in firmer carbonate substrates, which in turn encouraged infaunal colonization bioevents. In addition, although higher values of species richness and abundance of aragonitic-shelled ammonites observed in limestone beds may have a slight preservational component, Elder (1989) has correlated these trends into laterally equivalent, nonlimestone facies, suggesting a primary regional ecological signal. Collectively, these events may reflect widespread, short-term changes in community structure in shale/limestone cycles in response to improved benthic oxygen levels, firmer substrate conditions, and greater food availability during times of enhanced mixing.

PRS Summary Observations of lithostratigraphic and faunal trends at the PRS locality allow the following conclusions:

1. The Hartland Shale and Bridge Creek Limestone Members at PRS comprise distal offshore facies deposited prior to peak highstand of the second-order Greenhorn Cyclothem in the central Western Interior Basin. Laminated, organic-rich shales of the Hartland Member indicate a predominantly low-energy, low-oxygen depositional environment at PRS. Shale/limestone bedding couplets of the Bridge Creek Member suggest that, as the basin deepened, the response to Milankovitch climate cycles became pronounced, affecting water mass properties and clay and carbonate depositional processes throughout the basin.
2. Local, short-term D/R bioevents are extremely common in the Cenomanian/Turonian sequence at PRS. Diversification bioevents in the Hartland Shale Member include pioneer colonization of muddy substrates, secondary colonization of inoceramid pioneers by small epibionts, and colonization of the substrate by other free-living epifaunal bivalves or small infaunal burrowers, reflecting a gradient of increasing benthic oxygenation and substrate habitability. Reduction bioevents were mainly due

to oxygen decline, H_2S poisoning, and burial by fine-grained sediment, and in many cases followed colonization events within months to 1 or 2 years to produce single-cohort mass mortality surfaces. Storm events are a likely cause for these short-term changes in Western Interior sedimentation and water mass stratification. Similar ecological bioevents characterize the shale/limestone bedding couplets of the Bridge Creek Member at PRS. Diversity reduction events and low-oxygen-tolerant pioneer or opportunistic taxa dominate laminated, organic-rich shaley intervals, suggesting prevalence of a stratified water column. Diversification bioevents resulting in greater trophic complexity, higher overall species richness and abundance values, and more sustained community growth are best developed in the carbonate-rich (limestone/marlstone) lithofacies, suggesting times of greater water column mixing.

3. Overall patterns of species richness and faunal abundance at PRS are similar to KPS: peaks and troughs at a scale of ± 1 m associated with limestone/shale cycles resemble the biotic patterns observed in KPS parasequences, and suggest regional expression of short-term D/R ecozones. D/R bundles at a scale of 3–7 m (intervals R1–D3, fig. 20.4) are also similar to those of KPS, suggesting long-term ecozones. Cross-basin correlation of these biotic patterns is described in a later section on regional ecostratigraphy.

4. Combined sedimentologic, geochemical, and faunal data suggest that climatically driven, short-term fluctuations in benthic oxygen levels (water column stratification, position and variability of redox boundary) and substrate characteristics (supply of siliciclastic versus biogenic sediment, consistency, porosity) controlled benthic communities in the central Western Interior Basin during Cenomanian/Turonian time.

The Bunker Hill Section (BHS)

The Bunker Hill section is located in roadcut exposures near the town of Bunker Hill, Russel County, Kansas. The study interval is composed of distal offshore carbonate-enriched lithofacies containing numerous disconformities and condensed zones and reflects deposition on the tectonically stable to slowly subsiding, clastic sediment-starved eastern cratonic platform. The high abundance of fecal pellets and planktic foraminifera and the organic richness of laminated facies throughout the study interval (3–6 wt. %) suggest periods of high productivity within the water column and intensified benthic oxygen deficiency. Underlying the study interval, a single thick bed of cross-laminated skeletal limestone contains fossils of the *Calycoceras canitaurinum* Biozone, suggesting high current energy and significant erosion/condensation of the Lincoln Limestone Member (Sageman and Johnson 1985). A typical *Inoceramus*-dominated *M. mosbyense* Biozone fauna occurs in well-laminated, calcareous to marly shales and thin skeletal limestones in the lower 7.5 m of figure 20.5, and reflects low benthic oxygen levels and predominantly soft, muddy substrates. The remainder of the section is comprised of burrowed chalky limestones or marlstones interbedded rhythmically with laminated calcareous to marly shales, suggesting deeper water and/or decrease in siliciclastic sedimentation, and periodic development of firmer substrates and higher benthic oxygenation. These rocks contain the standard upsection biozonal sequence of the *S. gracile, N. juddii, Watinoceras* spp., and *M. nodosoides* biozones (fig. 20.5). Detailed correlation of lithostratigraphic units and biozones from BHS to PRS has been documented by Hattin (1971, 1975), Elder (1988, 1985) and Sageman (1985, 1991) and is represented in figure 20.2. Species richness/abundance plots and models of BHS bioevents are illustrated in figure 20.5 and described later.

BHS Bioevents BHS fossil assemblages are generally similar to those of the PRS locality. Inoceramid bivalves numerically dominate the macrofauna (commonly accounting for 98% to 100% of macrofaunal abundance) and reflect the control of bottom-water oxygen and soft substrate characteristics on colonization and benthic mortality. Yet it is commonly difficult to recognize discrete ecological bioevents in BHS laminated shales because of the highly fragmented nature of many fossils (average values of 70% fragmentation/sample are common). These features appear to reflect high current energy, winnowing, and transport of shell material, especially in the lower part of the section, as well as low sedimentation rates and extensive in-place taphonomic alteration of paleocommunities (through chemical and biological degradation of shell material). As a result, colonization/mass mortality surfaces of epifauna like those described from the PRS locality are poorly preserved. Bioevents involving distinct changes in biofacies, such as shifts to soft-bodied infaunal communities commonly associated with lithologic variations, are more readily identifiable as discrete horizons.

SKELETAL LIMESTONE EVENTS Skeletal limestones in the middle part of the *M. mosbyense* Biozone at BHS are typically about 1 cm thick, wavy to ripple-laminated, and composed of *Inoceramus* prisms, foraminiferal tests, and fish debris (fig. 20.5). The beds are commonly burrowed by *Planolites*, representing brief colonization events following deposition. However, the low diversity and sparse distribution of trace fossils suggest that favorable conditions for infauna were short-lived. Shale samples immediately overlying these skeletal limestones exhibit peak values of faunal abundance and species richness for the *M. mosbyense* Biozone at BHS and are dominated by species of *Inoceramus* and associated shell island taxa (*Phelopteria*, small oysters, and encrusting barnacles), suggesting diversification bioevents due to ecological interactions.

SHALE/LIMESTONE CYCLES Interbedded shale/limestone beds at BHS are lithologically similar to those of the PRS locality and contain many of the same types of bioevents (fig. 20.5). Most limestone beds contain complex cross-

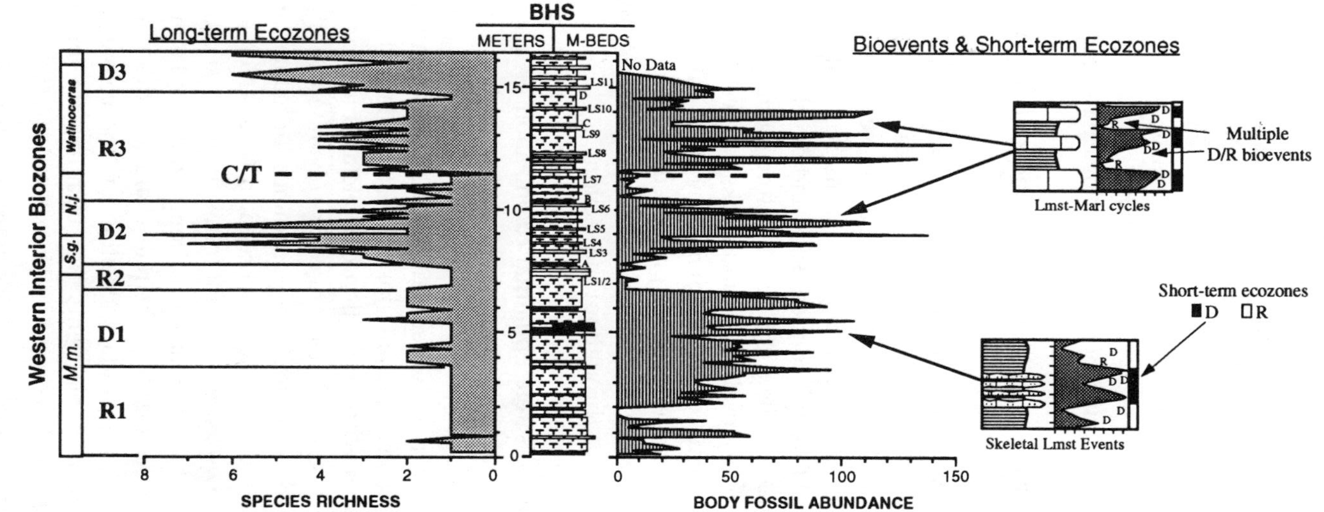

FIGURE 20.5. Bioevents and ecozones at the Bunker Hill section (BHS) are indicated by trends in species richness and abundance data. Lithologic section and marker beds (M-Beds) are as in figure 20.2. Insets show relationship of diversification and reduction bioevent horizons (represented by D and R letters) to the time-averaged curve of species richness/abundance data and short-term ecozones. Note that different bioevents may be reflected more strongly by species richness *or* abundance data, as indicated by stipple versus striped pattern. Biostratigraphic zones are indicated as in figure 20.3. C/T marks the position of the Cenomanian/Turonian boundary.

cutting traces of both deposit-feeding and sediment-dwelling taxa (including *Planolites*, *Chondrites*, and *Thalassinoides*), representing multiple infaunal colonization events. Limestones commonly have higher overall diversity levels than shales, and the shift from infaunal or mixed infaunal/epifaunal biofacies in limestones to strictly epifaunal communities in shales is interpreted to reflect significant changes in benthic conditions, especially the position of the redox boundary. Laminated shale intervals are typically dominated by one to three species of resident epifaunal *Inoceramus* bivalves. In many shale samples the presence of rare shell island taxa, such as *Phelopteria* and small encrusting oysters, suggests limited diversification due to ecological interactions.

BHS Summary Observations of lithostratigraphic and faunal trends at the BHS locality allow the following conclusions:

1. Distal offshore carbonate-enriched lithofacies of the study interval at BHS reflect deposition on the relatively shallow, sediment-starved eastern shelf of the Western Interior Basin. Low rates of sediment supply and subsidence, and/or frequent scouring by storms resulted in significant thinning of the interval through condensation and development of disconformities. Abundant fecal pellets and planktic foraminifera, and high C_{org} levels (4–6 wt. %) in many of the BHS mudrocks suggest high productivity. Intensification of benthic oxygen deficiency may have resulted and would explain above-average levels of preserved organic carbon, below-average levels of species richness, and the dominance of low-oxygen-tolerant *Inoceramus* bivalves throughout the section.
2. As a result of diminished clay input, slow burial and extensive post-mortem alteration, few discrete epifaunal colonization or mass mortality surfaces are preserved at BHS. However, the pervasive occurrence of *Inoceramus* bivalves, shell fragments, and scattered prisms suggests that ecological diversification and subsequent mortality bioevents were common for these low-oxygen-tolerant opportunists. Infaunal colonization bioevents are associated with deposition of skeletal limestone and chalky limestone beds, and suggest that changes in substrate conditions (grain size, firmness) resulted in improved opportunities for habitation by infaunal burrowers. The trace fossil taxa recorded in these events have been described as pioneer low-oxygen-tolerant forms (Bromley and Ekdale 1984; Savrda and Bottjer 1986). Their absence from adjoining laminated shale facies suggests strong control by substrate characteristics, bottom-water oxygen levels, and redox boundary position on the colonization of infauna. Ecological reduction bioevents (for infauna) are inferred at the sharp boundaries between burrowed and laminated strata.
3. Overall patterns of species richness and faunal abundance at BHS are similar to PRS and KPS: peaks and troughs at a scale of ± 1m occur throughout the section but are particularly well developed in association

with limestone/shale cycles, suggesting regional expression of short-term D/R ecozones. D/R bundles at a scale of 3–5 m (intervals R1–D3: fig. 20.5) are also similar to those of PRS and KPS, suggesting long-term ecozones. Cross-basin correlation of these biotic patterns is described later.

4. The dominant controlling factors for benthic communities at BHS are inferred to be bottom-water and interstitial oxygen levels and the position of the redox boundary, and substrate characteristics, such as grain size, porosity, fluidity, firmness, and organic content.

Analysis of Regional Bioevents and Ecostratigraphy

Understanding the relationships between local bioevents and regional to global trends in ecostratigraphy, biogeography, evolution, and extinction depends upon a detailed regional faunal data base. Although community composition and biofacies vary among different Western Interior depositional settings, the basic trends represented in our west/east transect (fig. 20.6) have been confirmed at more than 25 sites extending north to south from Montana to Texas, and west to east from Utah to Kansas (Elder 1987, 1989; Harries and Kauffman 1990; Sageman 1991). The resulting set of paleoecological patterns (figs. 20.6, 20.7) forms the basis of an ecostratigraphic framework for the C/T study interval. Compilation of taxonomic range data from these and other Western Interior localities (recently summarized in Kauffman et al. 1993) further provides a regional biostratigraphic standard against which published data on C/T faunas of the Gulf Coast, Europe, and other parts of the world (e.g., Stephenson 1952; Wright and Kennedy 1981; Kennedy and Cobban 1991; Nishida et al. 1992) can be compared. Thus biogeographic bioevents and origination/extinction bioevents can be identified and analyzed within the regional ecostratigraphic framework. Figure 20.7 summarizes these data.

Short-Term Ecozones Individual peaks in species richness/abundance data on a $\pm$1-m scale occur throughout the west/east transect of the basin and suggest short-term ecozones (fig. 20.6). Many of these features are associated with lithostratigraphic units (e.g., parasequences, shale/limestone bedding couplets), which have been physically correlated using HIRES methods (fig. 20.2). For example, detailed analysis of the D2 interval by Elder et al. (1994) resulted in correlation of six such units from Utah to Kansas (FS2a–FS7 = LS1–LS6: fig. 20.2), suggesting a relationship between nearshore parasequence development and the deposition of shale/limestone cycles in the basin. Simple correlation of associated peaks in species richness and abundance (queried lines in fig. 20.6) is problematic, however, because short-term maxima and minima in species richness/abundance of shelly molluscan faunas are not consistently correlative for every bedding couplet across the transect. Peak abundance values at nearshore to proximal offshore sites (such as KPS) commonly occur just below or at a flooding surface bed and represent the effects of shallowing and community

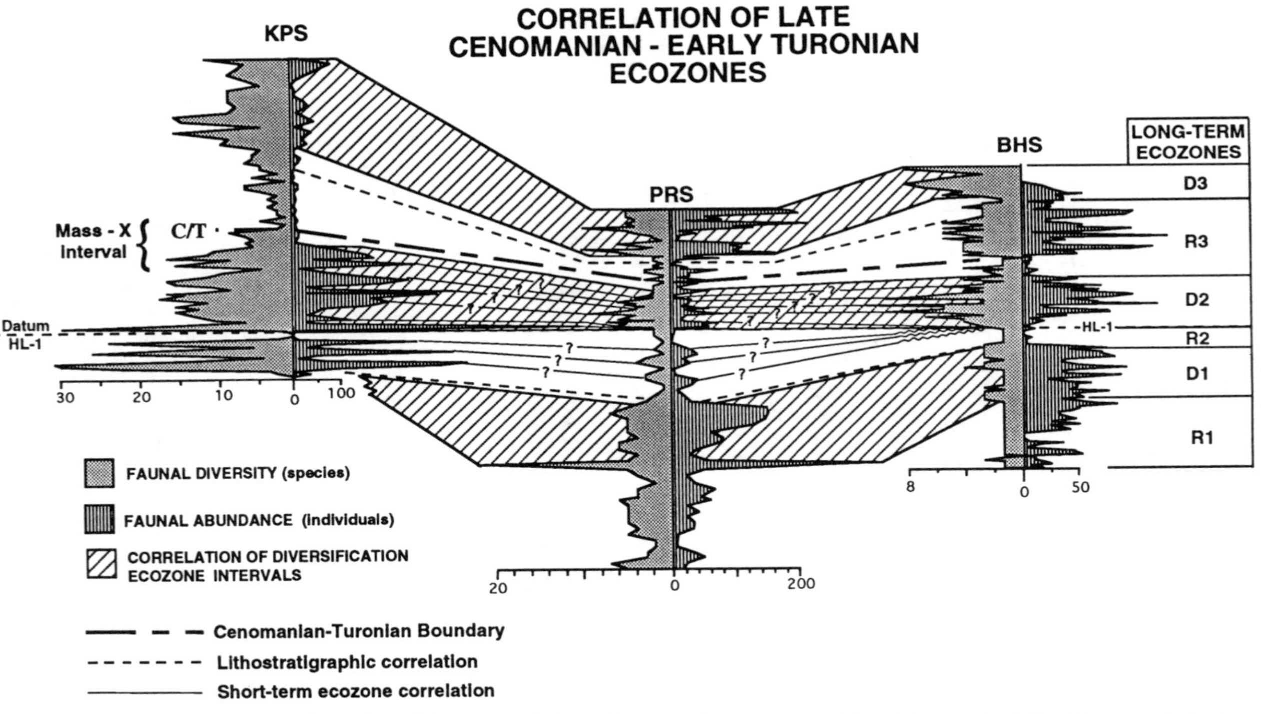

FIGURE 20.6. Correlation of species richness and abundance data across west/east transect of the Western Interior Basin illustrates ecozones. The cross section is hung on the base of the *S. gracile* Biozone, which coincides with the LS1 marker bed. A thick dashed line and C/T label mark the position of the Cenomanian/Turonian boundary, and a bracket shows the mass extinction interval. Selected lithostratigraphic correlations (thin dashed lines) are shown to illustrate the relative isochoneity of ecozone boundaries (compare with fig. 20.1). Long-term ecozones are defined at right, and possible correlations for short-term ecozones are suggested for an interval including the upper Hartland Shale and lower Bridge Creek Limestone Members. Vertical scale varies from west to east (KPS = 43 m, PRS = 26 m, BHS = 16 m).

expansion during progradation combined with sediment cutoff and condensation during the subsequent transgression. In some cases the most closely correlative abundance peaks at offshore sites (such as PRS) occur in shales that are actually stratigraphically higher than the flooding event, whereas the subjacent limestone bed, which is physically traceable to the nearshore flooding surface (Elder et al. 1994), can be relatively low in species richness and abundance of shelly macrofossils. Furthermore, the abundance peak in offshore shale facies is typically dominated by a single faunal group (inoceramid bivalves), and likely reflects opportunistic colonization of muddy substrates under low-oxygen conditions, not improved benthic habitats as indicated by the nearshore trends. Thus short-term faunal abundance peaks that appear closely correlative may reflect disparate local causes and temporally distinct events. Consideration of biofacies data (as represented in fig. 20.1) confirmed that, in fact, relative diversification characterizes the limestones in offshore bedding couplets. Because biofacies integrate body *and* trace fossil taxa, extent of bioturbation, changes in life habit and trophic structure, as well as community diversity and abundance, they normalize for changes in faunal abundance (or species richness) that are restricted to single genera or families of opportunistic taxa. (Such changes commonly reflect intracommunity competition or niche partitioning, rather than widespread environmental fluctuations.) Short-term ecozones are plotted in figure 20.7 based on a combination of species richness/abundance data *and* biofacies interpretations (e.g., fig. 8 in Sageman et al. 1991), and result in the correlation of short-term diversification ecozones to limestones and reduction ecozones to shales in offshore facies.

Shale/limestone cycles in the Western Interior Basin have long been interpreted to reflect Milankovitch climatic cyclicity (e.g., Gilbert 1895), and models for their origin are numerous (e.g., Arthur et al. 1985a; Barron et al. 1985; Eicher and Diner 1985; Fischer et al. 1985; Arthur and Dean 1991; Glancy et al. 1993; Pratt et al. 1993). Short-term ecozones within shale/limestone intervals represent a record of the biotic changes associated with these cycles and provide supporting evidence for a link between climate change, sedimentation, and paleoceanography. In Northern Hemisphere climate-model simulations of a warm, ice-free mid-Cretaceous world, one result of orbitally induced variations in solar insolation was variation in patterns of precipitation, expressed as periodic intensification of seasonal storm tracks across the Western Interior region (Barron et al. 1985; Glancy et al. 1993).

During hypothesized "wet" intervals, high freshwater runoff would enhance sediment delivery, stimulating progradation of the shoreface and deposition of mud in the basin, as well as intensify stratification between surface and deeper water masses by increasing density (mainly salinity) contrasts. Reduction intervals dominated by low-oxygen-tolerant inoceramids support this interpretation by indicating soft, muddy substrates and decreased benthic oxygen levels. During "dry" intervals the effect of decreased sediment delivery to the strandline and continuing second-order tectonoeustatic sea-level rise (Kauffman 1977)

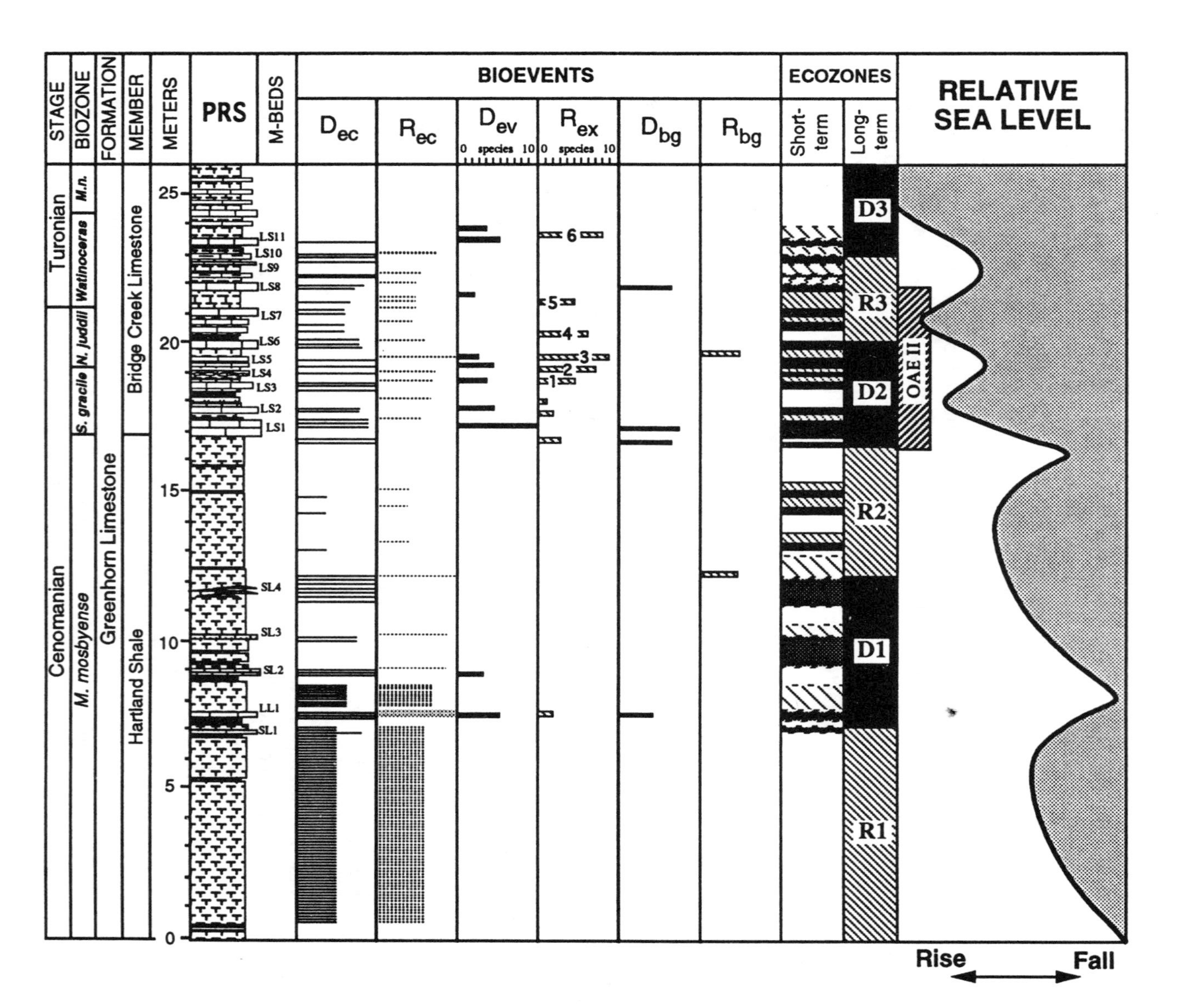
STAGE
BIOZONE
FORMATION
MEMBER
METERS
PRS
M-BEDS
BIOEVENTS
D_{ec}
R_{ec}
D_{ev}
R_{ex}
0 species 10
D_{bg}
R_{bg}
ECOZONES
Short-term
Long-term
RELATIVE SEA LEVEL
Turonian
Cenomanian
M.n.
Watinoceras
N. juddii
S. gracile
M. mosbyense
Greenhorn Limestone
Bridge Creek Limestone
Hartland Shale
25
20
15
10
5
0
LS11
LS10
LS9
LS8
LS7
LS6
LS5
LS4
LS3
LS2
LS1
SL4
SL3
SL2
LL1
SL1
6
5
4
3
2
1
D3
R3
D2
R2
D1
R1
OAE II
Rise
Fall

would be rapid transgression, resulting in a cutoff of siliciclastic sediment to the basin. In addition, normalization of salinity and/or increase in temperature of surface waters could simultaneously enhance primary productivity (see Eicher and Diner 1985 for alternate view); together these factors would result in limestone deposition. Whereas the breakdown of density stratification would allow greater mixing and potential for benthic oxygenation, decrease in clay sedimentation, concentration of bioclastic material, and the early cementation of carbonate-rich sediments would result in firmer substrates; in both cases increased infaunal diversity would result. Diversification intervals, characterized by abundant deposit-feeding and sediment-dwelling ichnofauna, and/or increased diversity and abundance of other taxa support both sedimentologic and oceanographic predictions. As in the bioevent examples described earlier, the dominant controlling factors for short-term ecozones are inferred to be benthic oxygen levels and substrate characteristics.

Long-Term Ecozones In addition to short-term fluctuations in species richness and abundance, "bundles" in species richness/abundance data comprising long-term diversification or reduction zones (long-term ecozones R1–D3: fig. 20.6, 20.7) can be traced across the Western Interior Basin. Relative to biostratigraphic boundaries and bentonite marker beds, the boundaries of these long-term ecozones appear isochronous or nearly so throughout the transect, and thus themselves comprise important bioevent horizons. Yet the characteristics of these intervals vary somewhat from section to section, reflecting the influence of local environments and onshore/offshore environmental gradients. For example, at KPS high species richness and abundance values are found in some intervals (e.g., R2), which are characterized by significantly reduced community development at sites further offshore. The R2 interval coincides with the period of initial transgression into the KPS area, and both lithologic and faunal data

FIGURE 20.7. Using PRS as a Western Interior standard section, a summary of ecostratigraphic data for the Cenomanian/Turonian study interval is shown. Ecological bioevents (D_{ec} and R_{ec}) and short- and long-term diversification (D) or reduction (R) ecozones are replotted from figures 20.3–20.6. (The lengths of D_{ec} and R_{ec} bioevent lines are intended to provide a qualitative reflection of overall species richness/abundance trends and are not scaled.) Short-term ecozones are based on figure 20.6 and data from Elder (1989) and Sageman et al. (1991). In the ecozone columns, solid patterns with letters reflect the ecozones examined in this study; the remaining shaded intervals represent additional short-term ecozones suggested by diversity/abundance data. The stratigraphic positions of biogeographic bioevents are indicated, but D_{bg} and R_{bg} columns are left unscaled as biogeographic analyses of the study interval are still preliminary. (See the text for a discussion of examples.) Origination and extinction data are from Elder (1989), and the position of the OAE II interval is based on the Pratt and Threlkeld (1984) Western Interior $\delta^{13}C$ excursion (fig. 20.1). The relative sea-level curve is based on analysis of transgressive/regressive facies tracts along the western margin of the seaway and is modified from Sageman (1985, 1991) and Elder (1991).

reflect dominance of an estuarine mouth or lower bay community at KPS. The area was too shallow an environment to register the effects of basinal oxygen deficiency. Just below the LS1 datum, however, about a meter of laminated, organic-rich facies is characterized by species richness/abundance values more like those of offshore samples (fig. 20.6), reflecting deepening and the western onlap of the R2 ecozone.

Similarly, species richness and abundance characteristics at BHS reflect regional variations in the ecological response to environmental fluctuations (fig. 20.6). Ecozones R1 and R3 show relatively high abundance values, with maxima similar to those of diversification intervals. Yet consideration of biofacies data reveals that these high values are solely attributable to a single family of taxa, the Inoceramidae, which were extremely opportunistic, low-oxygen-tolerant colonizers that thrived in the carbonate-rich, muddy substrates of the basin's eastern margin. Thus at BHS, diversity trends and knowledge of biofacies provide important constraints on understanding paleoecological data.

The cross-basin correlation in figure 20.6 reflects several regional trends, such as thinning of the entire section from west to east (long-term ecozones grade from 10 to 3 m in thickness), and development of condensed zones/disconformities at certain horizons (truncation of upper R2 ecozone and condensation of basal D2 ecozone at BHS). Overall, however, long-term diversification and diversity reduction ecozones maintain a consistent character throughout the transect. Reduction ecozones (especially R1 and R2) are dominated by well-laminated, organic-rich shales, low faunal diversity and abundance, and inequitable inoceramid-dominated biofacies. Despite the presence of burrowed limestone beds in ecozone R3 (reflecting overprint of short-term carbonate cycles), levels of burrowing are generally lower, and shales have higher percentages of inoceramid bivalves and higher levels of preserved C_{org} as compared with shales of ecozones D2 and D3. These factors suggest that oxygen deficiency, which developed particularly during deposition of the shale beds, was more intense during ecozone R3 than during the adjoining D2 and D3 intervals. In contrast, diversification intervals represent overall improvement in benthic oxygenation and substrate habitability, as indicated by increased species richness and abundance, and appearance of relatively diverse infaunal and mixed infaunal/epifaunal biofacies reflecting greater trophic complexity and community equitability. These changes are commonly associated with deposition of coarser-grained and/or more calcareous facies, reflecting the importance of substrate and sediment/water interface conditions on patterns of colonization, and underscoring the linkage between sedimentologic and paleoceanographic controls on benthic ecology.

The correspondence between long-term ecozones and a relative sea-level curve developed from analysis of transgressive/regressive patterns in nearshore facies of the study interval (fig. 20.7) suggests that sea level fluctuations had a secular influence on benthic communities. Diversification is associated with evidence of shallowing and/or rapid transgression at or near the base of each long-term ecozone (D1, D2, D3). It is suggested that relative shallowing initiates diversification in the basin, primarily through the impact of storms on stratification and the mixing of deep, oxygen-deficient water masses. Diversification persists as relative deepening begins, enhanced by the development of firmer bioclastic substrates in offshore areas, which result from cutoff of siliciclastic sediment and accumulation of bioclastic material during shoreline transgression (e.g., skeletal limestones in the D1 ecozone). As deepening proceeds, however, the eventual reestablishment and/or intensification of stratification results in a return to unfavorable conditions.

Because these fluctuations, which probably represent no more than 10–50 m of relative sea-level rise and fall, occur on a background of progressively rising sea level in the basin, the character of long-term ecozones changes upsection. Whereas the transgressive event associated with the D1 interval resulted in deposition of skeletal limestones in offshore areas, suggesting largely condensation, a similar event at the base of ecozone D2 initiated a significant increase in carbonate production and widespread pelagic limestone deposition, suggesting normal marine circulation and influence of warmer southern (Tethyan) water masses. Apparently, the establishment of open marine circulation and relatively rapid immigration of southern water masses, and/or other oceanographic factors during ecozone D2 were sufficient to dampen the onset of stratification as deepening continued (although the $\delta^{13}C$ record suggests that intensification of oxygen deficiency and burial of C_{org} began in areas outside the Western Interior at this time), and ecozone D2 persists in the central basin through another apparent shallowing cycle (fig. 20.7). During the subsequent deepening pulse, however, the onset of ecozone R3 reflects a return to oxygen-deficient conditions, either through reestablishment or intensification of stratification, and/or expansion and incursion of the oceanic oxygen minimum zone, as suggested by numerous authors (e.g., Arthur et al. 1987; Leckie 1985; Leckie et al. 1991). The next shallowing/deepening cycle corresponds with the cessation of OAE II and the onset of diversification in ecozone D3 (fig. 20.7).

Biogeographic Bioevents Support for the interpretation of eustatic changes associated with long-term ecozones is found in biogeographic data. The extent to which diversification or diversity reduction processes result from changes in the biogeographic ranges of taxa, either by immigration or failure to reestablish colonization in previously occupied habitat, reflect changes in migration corridors, or restriction of range resulting from regionally unfavorable conditions, such as oxygen deficiency. Biogeographic events are discernible through comparison of detailed range data within and outside the basin under study. Faunal studies of Cenomanian/Turonian strata in the Texas Gulf Coast (e.g., Stephenson and Reeside 1938; Moreman 1942; Stephenson 1952; Sohl 1967, 1971; Kauffman 1984a; Elder 1987) as well as in England, Germany, and Japan (e.g., Jefferies 1961, 1963; Dhondt 1971; Wright and Kennedy 1981; Kennedy and Cobban 1991; Matsumoto et al. 1991; Nishida et al. 1992) provide a preliminary comparative data base with which biogeographic events can be assessed for selected Western Interior taxa.

Within the study interval, eurytopic taxa whose stratigraphic ranges in the Western Interior are distinct from those of other areas provide examples of immigration and biogeographic reduction bioevents. For example, *Inoceramus ginterensis* Pergament has been documented in Middle Cenomanian strata in Japan (Matsumoto et al. 1991) but does not occur in the Western Interior until the Late Cenomanian Hartland Shale Member, suggesting an immigration event. Similarly, the ammonite *Calycoceras naviculare* (Mantell) has been documented in beds underlying the *Metoicoceras geslinianum* Biozone in Europe but does not occur until the base of the *Sciponoceras gracile* Biozone in the Western Interior, which is regarded as time equivalent with the *M. geslinianum* Biozone (Kennedy and Cobban 1991). In the case of the pectinid bivalve *Entolium membranaceum* (Nilsson), the opposite pattern is observed. The range of *E. membranaceum* terminates at the top of the *S. gracile* Biozone in the Western Interior Basin but is known to extend into the Maastrichtian in Europe (Dhondt 1971). Similarly, the gastropod *"Cerithiopsis sohli"* Kauffman (name in manuscript), which is present only in the Turonian interval of the Texas Gulf Coast, disappears in the Late Cenomanian *N. juddii* Biozone in the Western Interior Basin.

In each of these cases changes in the biogeographic range of taxa coincide generally with long-term ecozone boundaries. Although the preceding events constitute only a few examples for which accurate data were available, they suggest biogeographic bioevents at the horizons indicated in figure 20.7. Four immigration events (D_{bg}: fig. 20.7) coincide with the onset of long-term diversification ecozones (especially D2 and D3) and probably resulted from a combination of improved benthic conditions due to relative shallowing, and faunal influx associated with subsequent deepening and mixing of northern and southern water masses. Reduction in biogeographic range (R_{bg}: fig. 20.7) is correlated with onset of long-term reduction ecozones (fig. 20.7) and is consistent with the interpretation of intensified regional oxygen deficiency at these times. Thus the dominant controlling factors for biogeographic bioevents were probably changes in basin morphology (size, depth, and water mass inflow/outflow patterns) due to relative sea-level fluctuations, and concomitant changes in the salinity and temperature of surface waters and the oxygen content of deeper water masses in the Western Interior sea.

Evolution/Extinction Bioevents The Cenomanian/Turonian boundary interval contains one of the Phanerozoic mass extinctions described by Raup and Sepkoski (1982), and its macrofaunal extinction patterns have been studied extensively (Kauffman 1984b, 1986, 1988; Elder 1985, 1987, 1989). Biostratigraphic range data composited from 18 Western Interior localities by Elder (1989) indicate that 51% of the molluscan taxa documented within the Late Cenomanian (of a total of 84 species) disappear just below or at the C/T boundary. Ammonites and bivalves account for 85% of the species-level extinctions, and most of these are cosmopolitan taxa (Elder 1989). The mass extinction is expressed as a series of discrete horizons at which taxonomic ranges terminate (reduction bioevents) beginning in the *Sciponoceras gracile* Biozone and extending into the *Watinoceras* Biozone. Elder (1989) numbered the most pervasive extinction horizons 1–6 (D_{ex} bioevents: fig. 20.7). Each extinction bioevent is characterized by loss of diverse molluscan taxa, with progressive loss of shallower-dwelling species and more eurytopic species upsection. The extinctions reach a peak in the core of the OAE II interval (greatest species loss at extinction bioevent 3: fig. 20.7), which also coincides with the onset of reduction ecozone R3. Species loss declines in the overlying interval, with the exception of Early Turonian extinction bioevent 6, which is associated with rapid turnover of surviving and newly radiating biotas following the major extinction episode (Harries and Kauffman 1990).

The study interval is also characterized by a number of origination bioevents in which new taxa appear within narrow stratigraphic horizons (D_{ev} bioevents: fig. 20.7). These events occur dominantly at the base of or within diversification intervals. Speciation events occur at the base of interval D2, with the appearance of the diverse *S. gracile* fauna, and in the Lower Turonian *Watinoceras* Biozone and interval D3, with the first occurrences of *Mytiloides* species (Inoceramidae) and various new ammonites representing the recovery following the C/T mass extinction episode (Elder 1989, 1991; Harries and Kauffman 1990; Harries 1991). Trends in origination and extinction thus generally appear to follow the pattern of other bioevents: Diversification is greatest during the shallowest phases of inferred short-term relative sea-level fluctuations, when benthic oxygen levels and conditions for habitation were enhanced, and diversity reduction is highest during short-term deeper phases, when oxygen deficiency was presumably most extreme.

Discussion: Comparison of Local, Regional, and Global Events

This study defines a hierarchy of ecostratigraphic units based on the recognition of events ranging from local and/or geologically instantaneous to global and/or temporally long-ranging. The interpretation of events depends on the integration of detailed sedimentological, geochemical, and paleobiological data and the development of a basin-wide, high-resolution event stratigraphic framework. Species richness and faunal abundance data form the basis for ecostratigraphic analysis and are augmented by data on taphonomy, trophic complexity, community equitability, degree of bioturbation, and other factors incorporated into interpretations of resident and event biofacies. These data are correlated across a representative transect of the Western Interior Basin for an interval of Late Cenomanian/Early Turonian strata and are compared with similar but less detailed data from time-equivalent sequences of the Gulf Coast, Europe, and Asia, making possible recognition of local, regional, and global events.

The results of this study suggest that

1. Paleobiological data collected within a high-resolution chronstratigraphic framework from C/T strata of the Western Interior Basin can be used to define a gradient of relative "ecosystem well-being" with definitive endpoints (no living community vs. taxonomically and trophically diverse communities).
2. Paleontological event horizons, or thicker portions of a stratigraphic sequence, can be evaluated in terms of such a gradient and interpreted to represent relative diversification or relative diversity reduction. This study recognized horizons and very thin intervals (bioevents) and two scales of thicker stratigraphic intervals (short- and long-term ecozones).
3. Bioevents and ecozones are the products of colonization, immigration and speciation events (producing diversification), and mass mortality, emigration, and extinction events (producing diversity reduction) that occurred on local, regional, and global scales.
4. Western Interior diversification and diversity reduction events are similar or identical to the ecostratigraphic features (i.e., epiboles and outages) discussed by Brett and Baird (paper 10 in this volume).
5. Based on consideration of modern analogs as well as supporting evidence from sedimentologic and geochemical data, patterns of diversification and diversity reduction can be interpreted in terms of primary physical, chemical, and biological forcing mechanisms, resulting in an integrated reconstruction of paleoenvironments, paleocommunity dynamics, and basin history.

From the Cenomanian/Turonian interval described herein, we have illustrated local and regional colonization and mass mortality bioevents representing from months to years of geologic time, and longer-term sequences of colonization and diversity reduction that are correlative throughout the cross-basin transect, defining ecostratigraphic intervals or ecozones. These ecozones reflect repetitive environmental fluctuations corresponding to inferred Milankovitch climatic cycles (short-term ecozones) as well as longer-term changes in benthic environments that correspond to a record of relative sea-level fluctuations (long-term ecozones). Diversification zones characterize inferred dry-climate phases and the shallowest parts of relative sea-level cycles (including late regression and early transgression), and diversity reduction intervals characterize inferred wet-climate phases and the deepest parts of eustatic cycles (late transgression and early regression). The correspondence between positions and magnitudes of regional biogeographic and worldwide evolution/extinction bioevents and long-term ecozones suggests a relationship between species-level and community-level evolution and/or larval dispersal mechanisms, and changes in regional to global oceanographic factors related primarily to relative sea level.

At all levels of the ecostratigraphic hierarchy the data suggest that benthic communities of the Western Interior Basin were predominantly, if not entirely, controlled by changes in bottom-water oxygen levels and, to a lesser extent, substrate character; these were the parameters most strongly affected by changes in climate and sea level within the basin. From short-term events representing colonization following storm-induced mixing of a stratified water column in the Hartland Shale Member to widespread diversification reflecting an increase in origination and immigration rates at the base of the Bridge Creek Member (related to a transgressive pulse, increase in carbonate production, and full development of an open marine, well-mixed water column), to the progressive extinction of taxa during a period of global oxygen depletion and organic matter burial, and to the ultimate incursion of oceanic oxygen minimum zone waters in the basin near the C/T boundary, events at all scales of the hierarchy have similar characteristics.

We believe these paleoecological relationships are a direct consequence of the prevailing environmental regime of the mid-Cretaceous world. This was a time of peak Mesozoic global warming and "Greenhouse" conditions (Fischer 1982; Fischer and Arthur 1977), the causes of which have been linked tentatively to increased tectonic and volcanic activity (Fischer and Arthur 1977; Kauffman 1977, 1984a, 1985) and to elevated levels of atmospheric CO_2 (Arthur et al. 1985a, 1985b; Berner 1990; Budyko et al. 1987). The period was characterized by significant reduction in thermal gradients from pole to equator and from the surface to the bottom of world oceans, by a greatly expanded tropics, by apparent lack of permanent polar ice caps, by a global peak highstand of sea level for the Mesozoic, and by extensive development of epicontinental seas marginal to and within cratonic areas (Barron 1983; Barron and Washington 1982; Frakes 1979; Hancock and Kauffman 1979; Haq et al. 1987; Hay 1988; Kauffman 1984a). Oxygen deficiency may have been a prevalent condition in many such basins during the Mesozoic, for a variety of reasons, including basin size, depth, morphology, and connection to oceans. But oxygen deficiency became globally pervasive during the Late Cenomanian sea-level rise event, and appears to have been especially intense in conjunction with the burial of large volumes of organic matter near the C/T boundary (Arthur et al. 1987). Arthur et al. (1987) hypothesized that sea-level change was the driving force for intensified oxygen depletion during the C/T event. Because of flooding of arid, low-latitude regions and the formation of warm, saline bottom waters, large-scale displacement of deep, nutrient-rich oceanic water masses would have greatly stimulated productivity, causing intensification and expansion of mid-water oxygen minimum zones.

Under such conditions the most successful communities consisted of organisms that were highly tolerant, specially adapted opportunistic taxa with rapid dispersal mechanisms and extremely wide ranges. These taxa (dominantly molluscs) colonized whenever and wherever possible in a largely physically and chemically controlled marine realm and form the bulk of preserved C/T faunas. The prevalence of ecological bioevents in the study interval (D_{ec} and R_{ec}: fig. 20.7) indicates that C/T faunal history was extremely dynamic in the short term, with a near-constant succession of colonization and mortality events in the benthic realm. As a result, even subtle allocyclic changes in climate and sea level that affected physical/chemical conditions over large areas were recorded consistently by a highly sensitive ecosystem, resulting in a series of chronostratigraphically useful paleontological event horizons and intervals (bioevents and ecozones). The delicately balanced nature of this mid-Cretaceous ecosystem resulted in consistent local, regional, and global paleobiological responses to environmental fluctuations on short and long time scales.

Shallow-water marine environments in the postglacial Holocene world are characterized by high latitudinal temperature gradients, abundant oxygen and nutrients, and highly diverse, patchily distributed communities that are largely biologically accommodated. Under such conditions the relationship between widely distributed communities and changing environmental parameters may be far too complex to result in a consistent biological response, and a modern test of the ecostratigraphic method is difficult. But other intervals of Phanerozoic history (e.g., Silurian, Devonian) were characterized by nonglacially influenced "Greenhouse" conditions similar to those of the mid-Cretaceous. The application of high-resolution ecostratigraphy and a test of environmental sensitivity in the short- and long-term response of Silurian/Devonian communities might provide an important test of the consistency of paleobiological response in physically and chemically controlled ecosystems.

REFERENCES

Alvarez, W., E. G. Kauffman, F. Surlyk, L. W. Alvarez, F. Asaro, and H. V. Michel. 1984. The impact theory of mass extinctions and the marine invertebrate record across the Cretaceous-Tertiary boundary. *Science* 223:1135–41.

Arthur, M. A., H.-J. Bumsack, H. C. Jenkyns, and S. O. Schlanger. 1990. Stratigraphy, geochemistry, and paleoceanography of organic carbon-rich Cretaceous sequences. In R. N. Ginsburg and B. Beaudoin, eds., *Cretaceous Resources, Events and Rhythms: Background and Plans for Research,* pp. 75–119. Dordrecht, The Netherlands: Kluwer.

Arthur, M. A. and W. E. Dean. 1991. A holistic geochemical approach to cyclomania: Examples from Cretaceous pelagic sequences. In: G. Einsele, W. Ricken, and A. Seilacher, eds., *Cycles and Events in Stratigraphy,* pp. 126–67. Berlin: Springer-Verlag.

Arthur, M. A., W. E. Dean, R. Pollastro, P. A. Scholle, and G. E. Claypool. 1985a. A comparative geochemical study of two transgressive pelagic limestone units, Cretaceous Western Interior basin, U.S. In L. M. Pratt, E. G. Kauffman, and F. B. Zelts, eds., *Fine-Grained Deposits and Biofacies of the Cretaceous Western Interior Seaway: Evidence of Cyclic Sedimentary Processes,* pp. 16–27. Tulsa: Society for Economic Paleontology and Mineralogy, *Fieldtrip Guidebook* 4.

Arthur, M. A., W. E. Dean, and S. O. Schlanger. 1985b. Variations in the global carbon cycle during the cretaceous related to climate, volcanism, and changes in atmospheric CO_2. In E. T. Sundquist and W. S. Broecker, eds., *The Carbon Cycle and Atmospheric CO_2: Natural Variations Archean to Present,* pp. 504–29. *American Geophysical Union Monographs.*

Arthur, M. A. and I. Premoli-Silva. 1982. Development of widespread organic carbon-rich strata in the Mediterranean Tethys. In S. O. Schlanger and M. B. Cita, eds., *Nature and Origin of Cretaceous Carbon-Rich Facies,* pp. 7–54. London: Academic.

Arthur, M. A., S. O. Schlanger, and H. C. Jenkyns. 1987. The Cenomanian-Turonian oceanic anoxic event, II. Paleoceanographic controls on organic matter production and preservation. In J. Brooks and A. Fleet, eds., *Marine Petroleum Source Rocks*, pp. 401–20. Geological Society of London, *Special Publication* 26.

Barron, E. J. 1983. A warm, equable Cretaceous: The nature of the problem. *Earth Science Review* 19:305–38.

Barron, E. J., M. A. Arthur, and E. G. Kauffman. 1985. Cretaceous rhythmic bedding sequences: A plausible link between orbital variations and climate. *Earth and Planetary Science Letters* 72:327–40.

Barron, E. J. and W. M. Washington. 1982. Cretaceous climate: A comparison of atmospheric simulations with the geologic record. *Palaeogeography, Palaeoclimatology, Palaeoecology* 40:103–33.

Bates, R. L. and J. A. Jackson. 1987. *The Glossary of Geology*. Alexandria, VA: American Geological Institute.

Berner, R. A. 1990. Atmospheric carbon dioxide levels over Phanerozoic time. *Science* 249:1382–86.

Boucot, A. J. 1986. Ecostratigraphic criteria for evaluating the magnitude, character and duration of bioevents. In O. H. Walliser, ed., *Global Bioevents, Lecture Notes in Earth History*, vol. 8, pp. 25–45. Berlin: Springer-Verlag.

Bromley, R. G. and A. A. Ekdale. 1984. *Chondrites:* A trace fossil indicator of anoxia in sediments. *Science* 224:872–74.

Budyko, M. J., A. B. Ronov, and A. L. Yanshin. 1987. *History of the Earth's Atmosphere*. Berlin: Springer-Verlag/State Hydrol. Institute, Leningrad, U.S.S.R.

Cobban, W. A. 1984. Mid-Cretaceous ammonite zones, Western Interior, United States. *Bulletin of the Geological Society of Denmark* 33:71–89.

———. 1985. Ammonite record from Bridge Creek Member of Greenhorn Limestone at Pueblo Reservoir State Recreation Area, Colorado. In L. M. Pratt, E. G. Kauffman, and F. B. Zelt, eds., *Fine-Grained Deposits and Biofacies of the Cretaceous Western Interior Seaway: Evidence of Cyclic Sedimentary Processes*, pp. 135–38. Tulsa: Society for Economic Paleontology and Mineralogy, *Fieldtrip Guidebook* 4.

Cobban, W. A. and R. W. Scott. 1972. Stratigraphy and ammonite fauna of the Graneros Shale and Greenhorn Limestone near Pueblo, Colorado. *U.S. Geological Survey Professional Paper* 645:1–108.

Dhondt, A. 1971. Systematic revision of *Entolium, Propeamussium* (Amusiidae), and *Syncyclonema* (Pectinidae, Bivalvia, Mollusca) of the European Boreal Cretaceous. *Bulletin Institut Recherche Sciences Naturelle de Belgique* 47:95.

Eicher, D. L. and R. Diner. 1985. Foraminifera as indicators of water mass in the Cretaceous Greenhorn Sea, Western Interior. In L. M. Pratt, E. G. Kauffman, and F. B. Zelt, eds., *Fine-Grained Deposits and Biofacies of the Cretaceous Western Interior Seaway: Evidence of Cyclic Sedimentary Processes*, pp. 60–71. Tulsa: Society for Economic Paleontology and Mineralogy, *Fieldtrip Guidebook* 4.

Eicher, D. L. and P. Worstell. 1970. Cenomanian and Turonian foraminifera from the Great Plains, United States. *Micropaleontology* 16:269–324.

Elder, W. P. 1985. Biotic patterns across the Cenomanian-Turonian extinction boundary near Pueblo, Colorado. In L. M. Pratt, E. G. Kauffman, and F. B. Zelt, eds., *Fine-Grained Deposits and Biofacies of the Cretaceous Western Interior Seaway: Evidence of Cyclic Sedimentary Processes*, pp. 157–69. Tulsa: Society for Economic Paleontology and Mineralogy, *Fieldtrip Guidebook* 4.

———. 1987. The Cenomanian-Turonian (Cretaceous) Stage boundary extinctions in the Western Interior of the United States. Ph.D. dissertation, University of Colorado, Boulder.

———. 1988. Geometry of Upper Cretaceous bentonite beds: Implications about volcanic source areas and paleowind patterns, western interior, United States. *Geology* 16:835–38.

———. 1989. Molluscan extinction patterns across the Cenomanian-Turonian stage boundary in the Western Interior of the United States. *Paleobiology* 15:299–320.

———. 1991. Molluscan paleoecology and sedimentation patterns of the Cenomanian-Turonian extinction interval in the southern Colorado Plateau. In J. D. Nations and J. G. Eaton, eds., *Stratigraphy, Depositional Environments, and Sedimentary Tectonics of the Western Margin, Cretaceous Western Interior Seaway*, pp. 113–37. Geological Society of America, *Special Paper* No. 260.

Elder, W. P., E. R. Gustason, and B. B. Sageman. 1994. Correlation of basinal carbonate cycles to nearshore parasequences in the Late Cretaceous Greenhorn seaway, Western Interiors, U.S.A. *Geological Society of America Bulletin* 106:892–902.

Elder, W. P. and J. I. Kirkland. 1985. Stratigraphy and depositional environments of the Bridge Creek Limestone Member of the Greenhorn Limestone at Rock Canyon Anticline near Pueblo, Colorado. In L. M. Pratt, E. G. Kauffman, and F. B. Zelt, eds., *Fine-Grained Deposits and Biofacies of the Cretaceous Western Interior Seaway: Evidence of Cyclic Sedimentary Processes*, pp. 122–34. Tulsa: Society for Economic Paleontology and Mineralogy, *Fieldtrip Guidebook* 4.

Eldredge, N. and S. J. Gould. 1972. Punctuated equilibria: An alternative to phyletic gradualism. In T. J. M. Schopf, ed., *Models in Paleobiology*, pp. 82–115. San Francisco: Freeman, Cooper.

Fischer, A. G. 1982. Long-term climatic oscillations recorded in stratigraphy. In W. Berger and R. Crowell, eds., *Climate in Earth History*, pp. 97–104. Washington, D.C.: National Academy Press.

Fischer, A. G. and M. A. Arthur. 1977. Secular variations in the pelagic realm. In H. E. Cook and P. Enos, eds., *Deep Water Carbonate Environments*, pp. 19–50. Tulsa: Society for Economic Paleontology and Mineralogy, *Special Publication* 25.

Fischer, A. G., T. Herbert, and I. Premoli-Silva. 1985. Carbonate bedding cycles in Cretaceous pelagic and hemipelagic sediments. In L. M. Pratt, E. G. Kauffman, and F. B. Zelt, eds., *Fine-Grained Deposits and Biofacies of the Cretaceous Western Interior Seaway: Evidence of Cyclic Sedimentary Processes*, Fieldtrip Guidebook 4, pp. 1–10. Tulsa: Society for Economic Paleontology and Mineralogy.

Frakes, L. A. 1979. *Climates Through Geologic Time*. Amsterdam: Elsevier.

Frush, M. P. and D. L. Eicher. 1975. Cenomanian and Turonian foraminifera and paleoenvironments in the Big Bend region of Texas and Mexico. In W. G. E. Caldwell, ed., *The Cretaceous System in North America*, pp. 277–301. Geological Association of Canada, *Special Paper* No. 13.

Fürsich, F. and J. I. Kirkland. 1986. Biostratinomy and paleoecology of a Cretaceous brackish lagoon. *Palaios* 1(6):543–60.

Gilbert, G. K. 1895. Sedimentary measurement of geologic time. *Journal of Geology* 3:121–27.

Glancy Jr., T. J., M. A. Arthur, E. J. Barron, and E. G. Kaffman. 1993. A paleoclimate model for the North American Cretaceous (Cenomanian-Turonian) epicontinental sea. In W. G. E. Caldwell and E. G. Kauffman, eds., *The Evolution of the Western Interior Basin*, pp. 219–42. Geological Association of Canada, *Special Publication* No. 39.

Gustason, E. R. 1989. Stratigraphy and Sedimentology of the Middle Cretaceous (Albian-Cenomanian) Dakota Formation, Southwestern Utah. Ph.D. disertation, University of Colorado, Boulder.

Hallam, A. 1984. The causes of mass extinction. *Nature* 308:686–87.

Hancock, J. M. and E. G. Kauffman. 1979. The great transgressions of the Late Cretaceous. *Journal of the Geological Society of London* 136:175–86.

Haq, B. V., J. Hardenbol, and P. R. Vail. 1987. Chronology of fluctuating sea levels since the Triassic (250 million years ago to present). *Science* 235:1159–67.

Harries, P. J. 1993. Patterns of repopulation following the Cenomanian-Turonian (Upper Cretaceous) mass extinction. Unpublished Ph.D. dissertation, University of Colorado, Boulder.

Harries, P. J. and E. G. Kauffman. 1990. Patterns of survival and recovery following the Cenomanian-Turonian (Late Cretaceous) mass extinction in the Western Interior Basin, United States. In E. G. Kauffman and O. H. Walliser, eds., *Extinction Events in Earth History, Lecture Notes in Earth History,* vol. 30, pp. 277–98. Berlin: Springer-Verlag.

Hattin, D. E. 1971. Widespread, synchronously deposited, burrow-mottled limestone beds in the Greenhorn Limestone (Upper Cretaceous) of Kansas and southeastern Colorado. *American Association of Petroleum Geologists Bulletin* 55:412–31.

———. 1975. Stratigraphy and depositional environment of Greenhorn Limestone (Upper Cretaceous) of Kansas. *Kansas Geological Survey Bulletin* 209:1–128.

Hay, W. W. 1988. Paleoceanography: A review for the GSA Centennial. *Geological Society of America* 100:1934–56.

Hay, W. W., D. L. Eicher, and R. Diner. 1993. Paleoceanography of the Cretaceous Western Interior Seaway. In W. G. E. Caldwell and E. G. Kauffman, eds., *The Evolution of the Western Interior Basin,* pp. 297–318. Geological Association of Canada, *Special Publication* 39.

I.S.S.C. (International Subcommission on Stratigraphic Classification). 1976. *International Stratigraphic Guide.* New York: Wiley.

Jefferies, R. P. S. 1961. The paleoecology of the *Actinocamax plenus* Subzone (Lowest Turonian) in the Anglo-Paris Basin. *Palaeontology* 4(4):609–47.

———. 1963. The stratigraphy of the *Actinocamax plenus* Subzone (Turonian) in the Anglo-Paris Basin. *Proceedings of the Geological Association* 74(1):1–33.

Jordan, T. E. 1981. Thrust loads and foreland basin evolution, Cretaceous, western United States. *American Association of Petroleum Geologists Bulletin* 65:2506–20.

Kauffman, E. G. 1977. Geological and biological overview: Western Interior Cretaceous Basin. *Mountain Geologist* 13:75–99.

———. 1981. Ecological reappraisal of the German Posidonienschiefer (Toarcian) and the stagnant basin model. In J. Gray, A. J. Boucot, and W. B. N. Berry, eds., *Communities of the Past,* pp. 311–81. Stroudsburg, Pa.: Hutchinson Ross.

———. 1984a. Paleobiogeography and evolutionary response dynamic in the Cretaceous Western Interior Seaway of North America. In G. E. G. Westermann, ed., *Jurassic-Cretaceous Biochronology and Paleogeography of North America,* pp. 273–306. Geological Association of Canada, *Special Paper* 27.

———. 1984b. The fabric of Cretaceous marine extinctions. In W. A. Berggren and J. A. v. Couvering, eds., *Catastrophes in Earth History,* pp. 151–246. Princeton: Princeton University Press.

———. 1985. Cretaceous evolution of the Western Interior Basin of the United States. In L. M. Pratt, E. G. Kauffman, and F. B. Zelt, eds., *Fine-Grained Deposits and Biofacies of the Cretaceous Western Interior Seaway: Evidence of Cyclic Sedimentary Processes,* pp. iv–xiii. Tulsa: Society for Economic Paleontology and Mineralogy, *Fieldtrip Guidebook* 4.

———. 1986. High-resolution event stratigraphy: Regional and global bioevents. In O. H. Walliser, ed., *Global Bioevents,* pp. 279–335. Berlin: Springer-Verlag.

———. 1988a. Concepts and methods of high-resolution event stratigraphy. *Annual Reviews of Earth and Planetary Science* 16:605–54.

———. 1988b. The case of the missing community: Low-oxygen adapted Paleozoic and Mesozoic bivalves ("flat clams") and bacterial symbioses in typical Phanerozoic seas. Geological Society of America, Centennial Meeting 1988, Denver, Colo., *Abstracts with Programs* 20(7):48.

Kauffman, E. G., W. A. Cobban, and D. L. Eicher. 1978. Albian through lower Coniacian strata, biostratigraphy and principal events, Western United States, pp. xxiii1–xxiii24. Evènements de la Partie Moyenne du Crétecea (Mid-Cretaceous Events), Uppsala-Nice Symposia, 1975–1976. Nice, France: Musée d'Histoire Naturelle de Nice.

Kauffman, E. G., W. P. Elder, and B. B. Sageman. 1991. High-resolution correlation: A new tool in chronostratigraphy. In G. Einsele, W. Ricken, and A. Seilacher, eds., *Cycles and Events in Stratigraphy*, pp. 795–819. Berlin: Springer-Verlag.

Kauffman, E. G. and B. B. Sageman. 1990. Biological sensing of benthic environments in dark shales and related oxygen-restricted facies. In R. N. Ginsberg and B. Beaudoin, eds., *Cretaceous Resources, Events and Rhythms: Background and Plans for Research*, NATO–ASI Series C, Mathematical and Physical Sciences, vol. 304, pp. 121–38. Dordrecht, The Netherlands: Kluwer.

Kauffman, E. G., B. B. Sageman, W. P. Elder, and E. R. Gustason. 1987. *High-Resolution Event Stratigraphy, Greenhorn Cyclothem (Cretaceous: Cenomanian-Turonian), Western Interior Basin of Colorado and Utah:* pp. 1–198. GSA Rocky Mountain Section Regional Meeting, Boulder, Colorado: Geological Society of America.

Kauffman, E. G., B. B. Sageman, W. P. Elder, P. J. Harries, J. I. Kirkland, and T. Villamil, 1993. Cretaceous molluscan biostratigraphy of the Western Interior Basin, North America. In W. G. E. Caldwell and E. G. Kauffman, eds., *The Evolution of the Western Interior Basin*, pp. 397–434. Geological Association of Canada, *Special Publication* 39.

Kauffman, E. G. and R. W. Scott. 1976. Basic concepts of community ecology and paleoecology. In R. W. Scott and R. R. West, eds., *Structure and Classification of Paleocommunities*, pp. 1–28. Stroudsburg, Pa.: Dowden, Hutchinson, and Ross.

Kay, M. 1947. Analysis of stratigraphy. *American Association of Petroleum Geologists Bulletin* 31:162–68.

Kennedy, W. J. and W. A. Cobban. 1991. Stratigraphy and interregional correlation of the Cenomanian-Turonian transition in the Western Interior of the United States near Pueblo, Colorado: A potential boundary stratotype for the base of the Turonian stage. *Newsletters in Stratigraphy* 24(1/2):1–33.

Koch, C. 1977. Evolutionary and ecological patterns of Upper Cenomanian (Cretaceous) mollusc distribution in the Western Interior of North America. Ph.D. thesis, George Washington University.

Kowallis, B. J., E. H. Christiansen, and A. Deino. 1989. Multi-characteristic correlation of upper Cretaceous volcanic ash beds from southwestern Utah to central Colorado. *Utah Geological and Mineralogical Survey Miscellaneous Publication* 89-5:1–21.

Leckie, R. M. 1985. Foraminifera of the Cenomanian-Turonian boundary interval, Greenhorn Formation, Rock Canyon Anticline, Pueblo, Colorado. In L. M. Pratt, E. G. Kauffman, and F. B. Zelt, eds., *Fine-Grained Deposits and Biofacies of the Cretaceous Western Interior Seaway: Evidence of Cyclic Sedimentary Processes*, pp. 139–50. Tulsa: Society for Economic Paleontology and Mineralogy, *Fieldtrip Guidebook* 4.

Leckie, R. M., M. G. Schmidt, D. Finkelstein, and R. Yuretich. 1991. Paleoceanographic and paleoclimatic interpretations of the Mancos Shale (Upper Cretaceous), Black Mesa, Arizona. In J. D. Nations and J. G. Eaton, eds., *Stratigraphy, Depositional Environments, and Sedimentary Tectonics of the western Margin, Cretaceous Western Interior Seaway*, pp. 153–66. Geological Society of America, *Special Paper* No. 260.

Lowman, S. W. 1945. Research committee program. *American Association of Petroleum Geologists Bulletin* 29:1512–15.

Matsumoto, T., M. Noda, and S. Maiya. 1991. Towards an integrated ammonoid-, inoceramid-, and foraminiferal biostratigraphy of the Cenomanian and Turonian (Cretaceous) in Hokkaido. *Journal of the Geographical Society of Japan* 100(3):378–98.

Mayr, E. 1965. *Animal Species and Evolution*. Cambridge: Belknap/Harvard University Press.

McLeod, K. G. and K. A. Hoppe. 1992. Evidence that inoceramid bivalves were benthic and harbored chemosynthetic symbionts. *Geology* 20:117–20.

Moore, R. C. 1949. Meaning of facies. Geological Society of America, *Memoir* 39:1–34.

Moreman, W. L. 1942. Paleontology of the Eagle Ford Group of North and Central Texas. *Journal of Paleontology* 16(2):192–220.

N.A.C.S.N. (North American Commission on Stratigraphic Nomenclature). 1983. North American Stratigraphic Code. *American Association of Petroleum Geologists Bulletin* 67(5):841–75.

Nishida, T., T. Matsumoto, Y. Kyuma, and S. Maiya. 1992. Integrated inoceramid-foraminiferal biostratigraphy of the Cenomanian and Turonian (Cretaceous) in the Kotanbetsu Valley, Hokkaido. *Journal of the Faculty of Education, Saga University,* 39(2, pt. II):21–59.

Obradovich, J. D. 1993. A Cretaceous time scale. In W. G. E. Caldwell and E. G. Kauffman, eds., *The Evolution of the Western Interior Basin,* pp. 379–96. Geological Association of Canada, Special Publication No. 39.

Orth, C. J., M. Attrep, X. Mao, E. G. Kauffman, R. Diner, and W. P. Elder. 1988. Iridium abundance maxima in the Upper Cenomanian extinction interval. *Geophysical Research Letters* 15:346–49.

Orth, C. J., M. Attrep, Jr., L. R. Quitana, W. P. Elder, E. G. Kauffman, R. Diner, T. Villamil. 1993. Elemental abundance anomalies in the Late Cenomanian extinction interval: A search for the source(s). *Earth and Planetary Science Letters* 117(1–2):189–204.

Peterson, F. and A. R. Kirk. 1977. *Correlation of the Cretaceous rocks in the San Juan, Black Mesa, Kaiparowits and Henry Basins, southern Colorado Plateau,* pp. 167–78. San Juan Basin III: New Mexico Geological Society, *28th Annual Field Conference Guidebook.*

Pratt, L. M. 1985. Isotopic studies of organic matter and carbonate in rocks of the Greenhorn marine cycles. In L. M. Pratt, E. G. Kauffman, and F. B. Zelt, eds., *Fine-Grained Deposits and Biofacies of the Cretaceous Western Interior Seaway: Evidence of Cyclic Sedimentary Processes,* pp. 38–48. Tulsa: Society for Economic Paleontology and Mineralogy, *Fieldtrip Guidebook* 4.

Pratt, L. M., M. A. Arthur, W. E. Dean, and P. A. Scholle. 1993. Paleoceanographic cycles and events during the Late Cretaceous in the Western Interior Seaway of North America. In W. G. E. Caldwell and E. G. Kauffman, eds., *The Evolution of the Western Interior Basin,* pp. 333–54. Geological Association of Canada, *Special Publication* No. 39.

Pratt, L. M., E. R. Force, and B. Pomerol. 1991. Coupled manganese and carbon-isotopic events in marine carbonates at the Cenomanian-Turonian boundary. *Journal of Sedimentary Petrology* 61(3):370–83.

Pratt, L. M., E. G. Kauffman, and F. B. Zelt, eds. 1985. *Fine-Grained Deposits and Biofacies of the Cretaceous Western Interior Seaway: Evidence of Cyclic Sedimentary Processes.* 1985 Midyear Meeting, pp. 1–249. Tulsa: Society for Economic Paleontology and Mineralogy, *Fieldtrip Guidebook* 4.

Pratt, L. M. and C. N. Threlkeld. 1984. Stratigraphic significance of $^{13}C/^{12}C$ ratios in mid-Cretaceous rocks. In D. F. Stott and D. J. Glass, eds., *The Mesozoic of Middle North America,* pp. 305–12. Canadian Society of Petroleum Geology, *Memoir* 9.

Price, R. A. 1973. Large-scale gravitational flow of supracrustal rocks, southern Canadian Rockies. In K. A. DeLong and R. Scholten, eds., *Gravity and Tectonics,* pp. 491–502. New York: Wiley.

Raup, D. M. and J. J. Sepkoski. 1982. Mass extinctions in the marine fossil record. *Science* 215:1501–1503.

Sageman, B. B. 1985. High-resolution stratigraphy and paleobiology of the Hartland Shale Member: Analysis of an oxygen-deficient epi-continental sea. In L. M. Pratt, E. G. Kauffman, and F. B. Zelt, eds., *Fine-Grained Deposits and Biofacies of the Cretaceous*

Western Interior Seaway: Evidence of Cyclic Sedimentary Processes, pp. 112–21. Tulsa: Society for Economic Paleontology and Mineralogy, *Fieldtrip Guidebook* 4.

———. 1989. The benthic boundary biofacies model: Hartland Shale Member, Greenhorn Formation (Cenomanian) Western Interior, North America. *Palaeogeography, Palaeoecology, Paleoclimatology* 74 (1/2):87–110.

———. 1991. High-Resolution Event Stratigraphy, Carbon Geochemistry, and Paleobiology of the Upper Cenomanian Hartland Shale Member (Cretaceous), Greenhorn Formation, Western Interior, U.S. Ph.D. dissertation, University of Colorado, Boulder.

Sageman, B. B. and C. J. Johnson. 1985. Stratigraphy and paleobiology of the Lincoln Limestone Member, greenhorn Limestone, Rock Canyon Anticline, Colorado. In L. M. Pratt, E. G. Kauffman, and F. B. Zelt, eds., *Fine-Grained Deposits and Biofacies of the Cretaceous Western Interior Seaway: Evidence of Cyclic Sedimentary Processes*, pp. 100–109. Tulsa: Society for Economic Paleontology and Mineralogy, *Fieldtrip Guidebook* 4.

Sageman, B. B., P. B. Wignall, and E. G. Kauffman. 1991. Biofacies models for organic-rich facies in epicontinental seas: Tool for paleoenvironmental analysis. In G. Einsele, W. Ricken, and A. Seilacher, eds., *Cycles and Events in Stratigraphy*, pp. 542–64. Berlin: Springer-Verlag.

Savrda, C. E. and D. J. Bottjer. 1986. Trace-fossil model for reconstruction of paleooxygenation in bottom waters. *Geology* 14:3–6.

Schlanger, S. O. and H. C. Jenkyns. 1976. Cretaceous oceanic anoxic events: Causes and consequences. *Geologie MijnbouwKundige* 55:179–84.

Scholle, P. and M. A. Arthur. 1980. Carbon isotope fluctuations in Cretaceous pelagic limestones: Potential stratigraphic and petroleum exploration tool. *American Association of Petroleum Geologists Bulletin* 64:67–87.

Sloss, L. L., W. C. Krumbein, and E. C. Dapples. 1949. Integrated facies analysis, pp. 91–123. Geological Society of America, *Memoir* No. 39.

Sohl, N. F. 1967. Upper Cretaceous gastropod assemblages of the Western Interior of the United States. In E. G. Kauffmann and H. C. Kent, co-organizers, *Paleoenvironments of the Cretaceous Seaway*, pp. 1–37. Colorado School of Mines Special Publication.

———. 1971. North American Cretaceous biotic provinces delineated by gastropods. *Proceedings of the North American Paleontological Convention* 2(1):1610–38. Lawrence, Kansas: Allen Press.

Stanley, S. M. 1979. *Macroevolution: Pattern and Process*. San Francisco: W. H. Freeman.

Stephenson, L. W. 1952. Larger Invertebrate Fossils of the Woodbine Formation (Cenomanian) of Texas. U.S. Geological Survey Professional Paper No. 242, pp. 1–226.

Stephenson, L. W. and J. B. Reeside. 1938. Comparison of Upper Cretaceous deposits of Gulf region and Western Interior region. *American Association of Petroleum Geologists Bulletin* 22(12):1629–1638.

Van Wagoner, J. C., R. M. Mitchum, K. M. Campion, and V. D. Rahmanian. 1990. Siliciclastic sequence stratigraphy in well logs, cores, and outcrops. *AAPG Methods in Exploration*, series no. 7. Tulsa: American Association of Petroleum Geologists.

Vella, P. 1964. Biostratigraphic units. *New Zealand Journal of Geology and Geophysics* 7(3):615–25.

Weller, J. M. 1958. Stratigraphic facies differentiation and nomenclature. *American Association of Petroleum Geologists Bulletin* 42:609–39.

Williams, G. D. and C. R. Stelck. 1975. Speculation on the Cretaceous paleogeography of North America. In W. G. E. Caldwell, ed., *The Cretaceous System in the Western Interior of North America*, pp. 1–20. Geological Association of Canada, *Special Paper* No. 13.

Wright, C. W. and W. J. Kennedy. 1981. The Ammonoidea of the Plenus Marls and the Middle Chalk. *Monographs of the Palaeontological Society* 134:1–148.

Contributors

WILLIAM I. AUSICH
Department of Geological Sciences
Ohio State University, Columbus

GORDON C. BAIRD
Department of Geosciences
SUNY College at Fredonia

CARLTON E. BRETT
Department of Earth and Environmental Sciences
University of Rochester

ROGER J. CUFFEY
Department of Geosciences
Pennsylvania State University
University Park

KEITH E. EAGAN
Department of Geology
Utah State University, Logan

WILLIAM P. ELDER
U.S. Geological Survey
Denver, Colorado

HOWARD R. FELDMAN
Christopher G. Maples
Kansas Geological Survey
Lawrence, Kansas

ROBERT C. FREY
Ohio Department of Health
Bureau of Epidemiology and Toxicology

PETER J. HARRIES
Department of Geology
University of South Florida
Tampa

WILLIAM J. HICKERSON
Department of Geology
University of Iowa, Iowa City

STEVEN M. HOLLAND
Department of Geology
University of Georgia

LENNART JEPPSSON
Department of Historical Geology and Paleontolgy
Lunds Universitet
Stockholm, Sweden

MARKES E. JOHNSON
Department of Geology
Williams College
Willliamstown, Mass.

ERLE G. KAUFFMAN
Department of Geological Sciences
University of Colorado

W. DAVID LIDDELL
Department of Geology
Utah State University, Logan

STEVEN T. LODUCA
Department of Geology
Eastern Michigan University
Ypailanti

JAMES A. MACEACHERN
Earth Sciences Program
Simon Fraser University
Burnaby, B.C.

GEORGE R. MCGHEE JR.
Department of Geological Sciences
and Institute of Marine and Coastal Sciences
Rutgers University
New Brunswick, New Jersey

ARNOLD I. MILLER
Department of Geology
University of Cincinnati

S. GEORGE PEMBERTON
Department of Geology
University of Alberta

HAROLD B. ROLLINS
Department of Geology and Planetary Science
University of Pittsburgh

BRADLEY B. SAGEMAN
Department of Geological Sciences
Northwestern University
Evanston, Illinois

GREGORY A. SCHUMACHER
Division of Geological Survey
Ohio Department of Natural Resources

DOUGLAS L. SHRAKE
Division of Geological Survey
Ohio Department of Natural Resources

STEPHEN E. SPEYER
Encon Associates
Phoenix, Arizona

WENDY L. TAYLOR
Department of Earth and Experimental Sciences
University of Rochester

RONALD R. WEST
Department of Geology
Kansas State University
Manhattan, Kansas

THOMAS H. WOLOSZ
Center for Earth and Environmental Science
SUNY at Plattsburgh

Author Index

Subject Index

Page numbers referring to entries in tables or figures are set in **boldface**.